# carpentry

## Third Edition

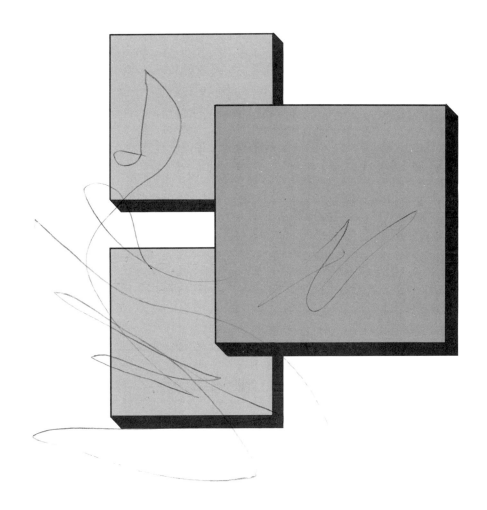

 AMERICAN TECHNICAL PUBLISHERS, INC.
HOMEWOOD, ILLINOIS 60430

**Leonard Koel**

3 4 5 6 7 8 9 – 97 – 9 8 7 6 5 4 3 2 1

Printed in the United States of America

**Library of Congress Cataloging-in-Publication Data**

Koel, Leonard.
    Carpentry / Leonard Koel. -- 3rd ed.
        p.    cm.
    Includes index.
    ISBN 0-8269-0735-0 (hard)
    1. Carpentry.    2. Building.    I. Title.
TH5604.K58    1997
694--dc21
                                                          97-16181
                                                             CIP

# ACKNOWLEDGMENTS

Many photographs and tables were provided courtesy of the following companies:

Acorn Structures, Inc.©
Air Master Corporation
The Aluminum Association, Inc.
American Institute of Timber Construction
American Plywood Association
AMPCOR
Andersen Corporation
Armstrong World Industries
Benchmark—General Products Company, Inc.
Bethlehem Steel Corporation
Binks Manufacturing Company
Blandex—Blandin Wood Products Company
Bostitch Division of Textron Inc.
Bruce Hardwood Floors, a division of Triangle Pacific
  Corp.
The Burke Company
Calweld, Inc.
California Redwood Association
Caterpillar Tractor Company
CertainTeed Corporation
Champion Paneling—Champion International
  Corporation
Conco Cement, Inc.
Continental Aluminum Products Company
Cronkhite Industries, Inc.
Cunningham & Walsh
David White, Inc.
Deck House, Inc.
Deere & Company
Dow Chemical U.S.A.
Duo-Fast Corporation
Economy Forms Corporation
Educational Lumber Company, Inc.
Emhart Hardware Group, Russwin Division
Forest Industries—Miller Freeman Publications
Forest Products Laboratory, Forest Service, USDA
Formica Corporation
Gang-Nail Systems, Inc.
The Garlinghouse Company
Georgia-Pacific Corporation
Glendale Optical Company
Greenlee Tool Company, a Division of Ex-Cell-O
Gypsum Association
HCB Contractors
Homelite Division of Textron Inc.
Hurd Millwork Company
Ingersoll-Rand
INRYCO, Inc.
Institute of Financial Education
Iron Age Shoe Company
Kaiser Cement Corporation
Keuffel and Esser Company
Klein Tools, Inc.
Koppers Company, Inc.
Laser Alignment, Inc.
The Lietz Company
The Lincoln Electric Company
Manville Building Materials Corporation

Masonite Corporation
May Wood Industries, Inc.
Millers Falls Tool Company
Milwaukee Electric Tool Corporation
Mobay Chemical Corporation
Morrow Crane Company, Inc.
National Gypsum Company
National Woodwork Manufacturers Association
Orem Research, Inc.
Overhead Door Corporation
Owens-Corning Fiberglas Corporation
The Panel-Clip Company
Paslode Company: Division of Signode Corporation
Patent Scaffolding Co.—A Division of Harsco
  Corporation
Pella—Rolscreen Company
Porter-Cable Corporation
Portland Cement Association
Precast Concrete Institute
Prestressed Concrete Institute
Ramset Fastening Systems
Red Cedar Shingle & Handsplit Shake Bureau
Research Products Corporation
Riviera Kitchens, an Evans Products Company
Rockwell International, Power Tool Division
Roto Frank of America, Inc.
Schlage Lock Company
Senco Products Fastening Systems
Shakertown Corporation
Simplex Products Division
Simpson Timber Company, Columbia Door Division
SKIL Corporation
Southern Forest Products Association
Southern Pine Council
Spencer, White & Prentis, Inc.
Sperry New Holland
Standard Structures, Inc.
Stanley Door Systems, division of The Stanley Works
Stanley Hardware, division of The Stanley Works
Stanley Tools, division of The Stanley Works
Star Expansion Company
The L. S. Starrett Company
Symons Corporation
Terminix International, Inc.
Timber Company, Columbia Door Division
Trus Joist Corporation, Micro = Lam Division
Truswal Systems
United Development
United States Gypsum
United States Steel Corporation
CPR Division, The Upjohn Company
The Upson Company
Ventarama Skylight Corporation
Von Duprin Exit Devices
Weather Shield Mfg., Inc.
Western Forms, Inc.
Western Wood Products Association
Weyerhaeuser Company

# table of contents

## SECTION 1 CARPENTRY AND CONSTRUCTION

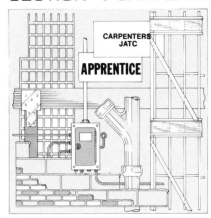

## SECTION 2 CONSTRUCTION MATERIALS

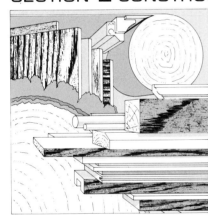

## SECTION 3 HAND TOOLS

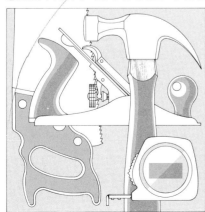

Contents

# SECTION **4** POWER TOOLS

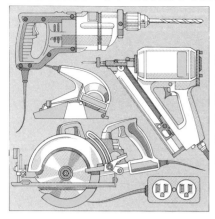

# SECTION **5** CONSTRUCTION EQUIPMENT, SITE CONDITIONS, AND SAFETY ON THE JOB

# SECTION **6** BUILDING DESIGN AND BLUEPRINT READING

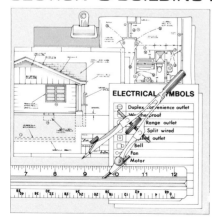

# SECTION **7** LEVELING INSTRUMENTS AND OPERATIONS

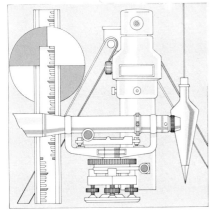

# SECTION **8** FOUNDATION AND OUTDOOR SLAB CONSTRUCTION

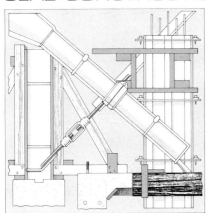

# SECTION **9** FLOOR, WALL, AND CEILING FRAME CONSTRUCTION

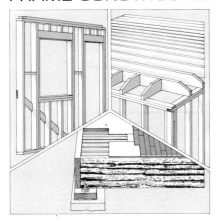

Contents

# SECTION **10** ROOF FRAME CONSTRUCTION

# SECTION **11** ENERGY CONSERVATION: INSULATION AND CONSTRUCTION METHODS

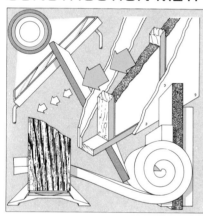

# SECTION **12** EXTERIOR FINISH

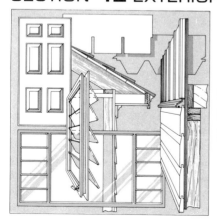

# SECTION **13** INTERIOR FINISH

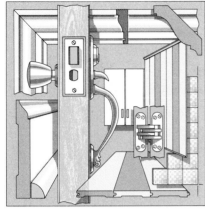

# SECTION **14** STAIRWAY CONSTRUCTION

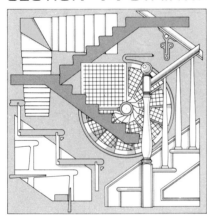

# SECTION **15** POST-AND-BEAM CONSTRUCTION

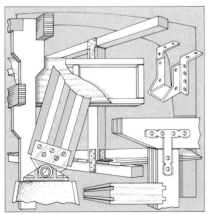

Contents

# SECTION **16** CONCRETE HEAVY CONSTRUCTION

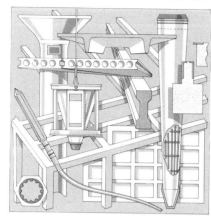

# SECTION 1

# Carpentry and Construction

Carpentry is a trade of the construction industry. This industry directly employs millions of people with different trades and skills. In addition, many other workers are employed by industries that manufacture building materials and appliances used in constructing and furnishing buildings. Plainly, the construction industry is a vital part of the nation's economy.

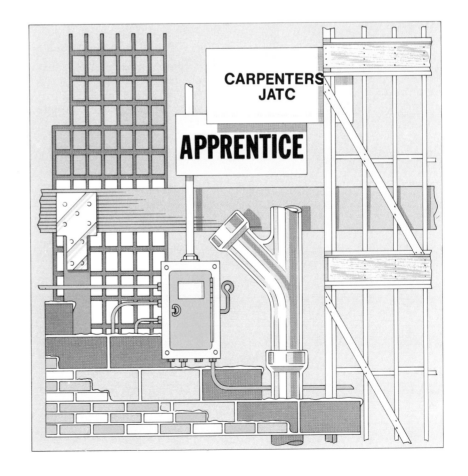

The bulk of the work force in the construction industry consists of about 25 different building trades. About one-third of all building trades workers are classified as carpenters. They are the largest single group on most construction projects.

Carpentry is one of the oldest trades known. When civilization first arose in the Mediterranean basin, methods of post-and-beam timber construction were used. The word *carpenter* comes from the French word *charpentier,* which in turn derives from the Latin word *carpentarius,* meaning chariot maker. A large group of woodworkers always accompanied the Roman armies in order to repair wagons and chariots and to build houses for fortifications. They also helped to build battering rams, catapults, and other instruments of war.

By the end of the Middle Ages (1453) carpentry had been established as an important and highly respected craft in the more advanced societies of western Europe. In 1333 a group of carpenters founded their own labor organization in London, England. It was known as the Carpenters' Guild of London. Most

carpenters in those days were called *master builders.* In addition to their own work, they directed the work of many other trades.

Carpenters participated in the building of the United States from the earliest colonization of the New World. There were more carpenters on the *Mayflower* than members of any other skilled craft. As the country began to develop westward, carpenters were a major work force in building its towns and cities, as well as the railroads and bridges that linked the towns together.

Carpenters continue to play a vital role in building America. They are not only an important part of the construction crews that build houses and apartments, but they also build the factories, mills, and offices that provide work. Carpenters help to erect schools, hospitals, and houses of worship. They are involved in the building of streets, highways, and freeways. They work on wharves and docks and help to construct dams and irrigation systems.

Many changes have taken place in the carpentry trade since

the old days of the woodworker who built and shaped everything with hand tools. Improved methods of production have been developed. Much of the work is done with power tools, which operate faster and more efficiently than hand tools.

Some people believe that the mechanization of the trade has done away with the old hand skills that gave carpenters pride in their work. It is true that certain types of skills are no longer needed by everyone who works in the trade. Yet, the industry will always need some carpenters who can do the fine work required for the interior finish of a building. Interior finish work includes such tasks as fitting molding and panels, placing cabinets, and hanging doors.

Skill, however, is not measured just by the ability to do finish work. Skill is required in all the different operations that are a part of the carpentry trade. Skill also includes keeping abreast of the new materials and methods that bring about constant change in the construction industry. It is this constant change that makes the carpentry trade so exciting and interesting today.

# UNIT 1

# Types of Construction

Carpenters work on two major types of building projects: *residential and other light construction* and *heavy construction.* In each of these categories many different methods are used to erect the buildings.

## RESIDENTIAL AND OTHER LIGHT CONSTRUCTION

Residential construction is one type of light construction. It in-cludes houses, condominiums, and small apartment buildings. See Figure 1–1. This branch of the industry employs the greatest number of carpenters. Other types of light construction include certain buildings of small to me-dium size that are used for com-mercial purposes, such as stores, restaurants, and ware-houses.

The wood-framing method used for most light construction is *platform framing.* It consists of wood studs, plates, joists, brac-ing, and other wood members. See Figure 1–2.

*Post-and-beam* construction, also known as *plank-and-beam* construction, is another type of wood-framing method. As its name implies, post-and-beam construction relies on posts and beams for its basic structure. See Figure 1–3. In most cases the beams are exposed, giving an attractive and open appear-ance to the interior of the building.

In *masonry* construction, the outside walls of the building are built of bricks, hollow concrete blocks, stone, or a combination

*Standard Structures, Inc.*

Figure 1–1. This four-story, wood-framed condominium comes under the category of light construction.

*Standard Structures, Inc.*

Figure 1–2. Example of platform fram-ing. Each floor unit provides a work-ing platform for the walls above.

3

Figure1–3. Post-and beam construction is characterized by exposed members that also serve as the basic structure of the building.

of masonry products. The interior walls and floors of the building are usually constructed of wood. Carpenters work very closely with bricklayers and stonemasons when building this type of structure, which is popular in the eastern and midwestern states.

In *brick-veneer* or *stone-ve-*

*neer* construction, brick or stone is used as an outside skin (veneer) over a conventionally framed wood stud wall. See Figure 1–4. Such buildings give the appearance of having masonry exterior walls.

Another type of construction is *alteration work*, or *remodeling*, in which a change or addition is made to a structure already built. Examples of alteration work are: adding a new room to an older house, modernizing a kitchen or bathroom, changing walls in an office, and redoing a store front. In large cities with limited space for new construction, alteration work is a large part of the carpentry trade.

## HEAVY CONSTRUCTION

Most heavy construction uses *reinforced concrete*. Reinforced concrete is concrete that contains steel bars (rebars) to strengthen it. Large office and apartment buildings, factories, bridges, freeways, and dams are included in the category. See Figures 1–5 through 1–7.

Figure 1–5. Carpenters are a major work force in the construction of highrise concrete buildings.

Figure 1–6. Carpenters build the forms required for the construction of a concrete dam.

## Monolithic Concrete Structures

The term *monolithic* describes the traditional method of concrete construction in which each major element of a building (such as a wall) is cast as a single piece. See Figures 1–8 through 1–10. Wood forms are built to the shapes of the walls, beams, columns, and floors of the build-

Figure 1–4. Brick-veneer walls create an attractive effect for the first story of this house. Redwood siding finishes off the second story.

4

ing. Steel reinforcing bars are placed inside the forms. Then the concrete is placed (poured) in the forms. After the concrete has hardened sufficiently, the forms are stripped away.

At one time all concrete buildings were constructed this way (monolithically poured). Today this method is used more often for buildings of small to medium size than for large buildings.

## Concrete Construction Using Precast Units

Large concrete buildings are usually constructed with precast (prefabricated) concrete units. These units of reinforced concrete are made at a factory (precast plant) and then transported to the job site. Many concrete buildings of small to medium size are also constructed with precast units.

**Steel-framed Concrete Structures.** High-rise buildings (skyscrapers) are erected with a steel framework. See Figure 1–11. Smaller buildings are also sometimes erected in this manner. At one time, carpenters built wood forms around the steel framework and then placed concrete in the forms. Today precast units are lifted into place by crane and fastened to the steel frame. Using precast units is more efficient and less costly than building forms and pouring concrete on the job site. Carpenters are often involved in the procedures used to attach the precast units to the frame.

**Concrete Structures without Steel Framework.** Many concrete buildings of small to medium size are constructed of precast units that are structurally tied together without requiring a steel framework. See Figure 1–12. Among the various methods used for this type of construction are the *tilt-up* and *lift-slab* methods.

In tilt-up construction the wall units are cast on the floor. They are raised into place by crane after the concrete has hardened. See Figure 1–13. In lift-slab construction the floor slabs are stack-cast around columns at the first floor and raised into place by hydraulic jacks. See Figure 1–14. Carpenters perform key operations in both tilt-up and lift-slab construction.

## Heavy Timber Construction

Heavy timber construction is a wood system considered to be part of heavy construction. This is one of the oldest construction methods used in North America to erect buildings, bridges, railroad trestles, and waterfront piers and docks. Today heavy timber systems are most often seen in commercial construction.

## INDUSTRIALIZED CONSTRUCTION

An important development in construction has been the in-

Kaiser Cement Corporation
Figure 1–7. Modern highway systems require the construction of many on-ramps, off-ramps, and overpasses. This is another area of concrete construction that employs many carpenters.

creasing use of *prefabricated* building units (industrialized construction). Any type of building unit that is manufactured in a plant and delivered to the job site is a prefabricated unit. Precast concrete units, roof trusses, floor trusses, glued and laminated beams, box beams, and stressed-skin panels are all prefabricated *structural* units. Packaged window assemblies, pre-

Portland Cement Association
Figure 1–8. A modern, monolithically constructed concrete building.

Figure 1–11. Most high-rise buildings have a steel framework.

Figure 1–9. Concrete office building under construction. Carpenters build the forms into which the concrete is placed. Carpenters will also do much of the finish work in the interior of the building.

hung doors, and exterior soffit systems are *nonstructural* units. Obviously, all site-constructed buildings today include some prefabricated units.

Completely prefabricated buildings usually consist of *panel systems* or *modular units*. These methods are particularly widespread in residential construction. Over one-third of new residential structures today are built

Figure 1–12. Raising the final wall panels of the building. This entire structure is made of precast concrete units.

Figure 1–10. After the concrete hardens, the forms are removed, as shown toward the front.

*The Burke Company*

Figure 1-13. In tilt-up concrete construction, the wall units are cast on the floor. They are raised into place by crane after the concrete has hardened.

*Portland Cement Association*

Figure 1-14. In lift-slab concrete construction, floor slabs are stack-cast around columns at the first floor and raised into place by hydraulic jacks.

this way. They are categorized as *manufactured housing*.

## Panel Systems

In addition to residential structures, light commercial structures such as office buildings, schools, and buildings in shopping centers are often constructed with panel systems.

The basic units of a panel system are the wall sections. They are constructed on an automated assembly line framing station. See Figure 1-15. When delivered by truck to the job site, they are installed by carpenters. See Figure 1-16. In addition to wall sections, a panel system may also include a roof system made up of roof trusses. Floor sections may also be provided, although usually the joists and subfloor are constructed on the job site.

**Open Panel System.** In an *open* panel system, the outside surface of the exterior wall panel is covered with sheathing or insulation board, but the inside surface of the wall is left exposed (open). Often the finish siding and completed window and door units are also installed at the factory. See Figures 1-17 and 1-18. However, plumbing parts and electrical wiring are installed on the job rather than at the factory.

**Closed Panel System.** In a *closed* panel system, the electrical wiring, plumbing, insulation, and interior wallboard are installed at the factory. As a result, the only work required on the job is usually some finish work where the panels are joined and hookup of plumbing and electrical fixtures.

Some closed panel systems come with a modular unit called a *mechanical core*. It contains the kitchen and bathroom equipment.

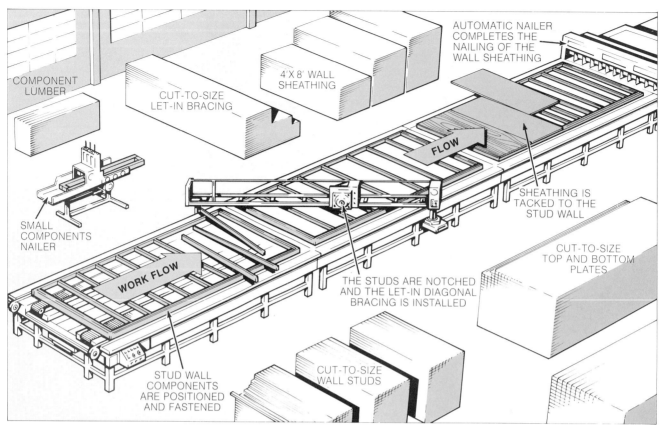

COMPONENT LUMBER

CUT-TO-SIZE LET-IN BRACING

4'X 8' WALL SHEATHING

AUTOMATIC NAILER COMPLETES THE NAILING OF THE WALL SHEATHING

FLOW

SMALL COMPONENTS NAILER

WORK FLOW

SHEATHING IS TACKED TO THE STUD WALL

CUT-TO-SIZE TOP AND BOTTOM PLATES

THE STUDS ARE NOTCHED AND THE LET-IN DIAGONAL BRACING IS INSTALLED

STUD WALL COMPONENTS ARE POSITIONED AND FASTENED

CUT-TO-SIZE WALL STUDS

Figure 1–15. A fully automated assembly line for constructing panel wall sections.

*American Plywood Association*
Figure 1–16. Carpenters install prefabricated panel wall sections.

*American Plywood Association*
Figure 1–17. View of the exterior of a panel wall section being lifted from the assembly table after completion. It is covered with grooved plywood siding.

*American Plywood Association*

Figure 1–18. View of the interior of a wall section being lifted from the assembly table after completion. A complete window unit is in place.

rectangular block. When all the blocks are fastened together, the house is completed. (All modular housing packages include a mechanical core containing the kitchen and bathroom equipment and almost all the mechanical equipment needed in a house.)

The installation procedure is called *setting* the house. Modular sections are placed on top of a foundation that has already been constructed on the job site. See Figure 1–19. The modules may be set in place by crane or slid into place on greased rails. As the modules are placed, carpenters on the job bolt them together and also fasten them to the foundation walls. The entire procedure of setting the house usually takes no more than a few hours.

A small amount of finish work is required after the modules are set. Carpenters must add exterior siding to the gable ends of the building to cover the areas

**Fold-out Panel System.** New types of panel systems are constantly being developed, such as the *fold-out* panel system, which features units of panelized floors, walls, and roof sections hinged together. The units fold flat during shipping and are quickly unfolded and set in place on the job.

## Modular Units

Modular construction is the most sophisticated and highly developed factory system. A modular house (also known as a *sectional* house) is 95% complete when it is delivered from the factory to the job site. The house is prefabricated in three-dimensional units, each consisting of a floor, walls, and a ceiling or roof. The walls are finished on both sides and have electrical wiring and plumbing installed. A modular unit can be compared to a

*American Plywood Association*

Figure 1–19. A modular unit of a house being placed on top of the basement foundation walls.

where the sections are joined. Trim boards may be necessary where the house rests on the foundation. Inside the house, wall joints must be finished and floor covering installed where the house sections have been joined. Electrical and plumbing utilities must also be hooked up.

## Advantages and Disadvantages of Manufactured Housing

Many people consider manufactured housing to be one of the answers to the nation's need for mass-produced, low-cost homes. The system is most profitable, however, when houses with a limited number of floor plans are mass-produced. Also, the job site should not be a great distance from the prefabricating plant. The financial savings resulting from lower labor costs can be cancelled by the cost of transportation problems.

Since manufactured housing is practical in some situations but not in others, it appears that both on-site and factory-produced houses will be necessary to meet the nation's growing housing needs for the foreseeable future.

## SPECIALIZATION IN THE TRADE

Improved production methods in construction have led to increased specialization of work. On housing tracts, for example, a large number of homes are constructed at the same time by crews of carpenters specializing in various divisions of carpentry work. See Figure 1–20. One crew of carpenters installs all the foundations of the houses. Other crews separately place the floor units, walls, roofs, and ceilings.

In certain areas of the country some carpenters (or carpentry

*Gang Nail Systems, Inc.*

Figure 1–20. Mass production methods are used on housing tracts. One crew installs all the foundations of the houses. Other crews separately install other elements of the houses.

firms) specialize in placing drywall finish on walls and ceilings. Others work exclusively on installing acoustical tile ceilings. Still others install only certain types of patented partitions (interior walls) in office buildings. The number of such single-specialty operations is growing.

Increasing specialization is probably necessary and inevitable in carpentry. As a result, though, the carpenter with "all-around" skills is becoming the exception rather than the rule. Still, it is possible for a young person entering the trade to become that exceptional carpenter if he or she gains experience in as many areas of construction work as possible. This means learning such skills as building the forms for concrete structures, erecting the framework for wood-framed buildings, and performing the trim work required in all types of construction. A variety of skills will give a person greater job security and earn the respect of employers and co-workers alike.

# UNIT 2

# The Building Trades

## CARPENTERS AND GENERAL CONTRACTORS

Carpentry is the single largest craft in the construction industry. Most carpenters, unless they are self-employed, work for a *general contractor,* also referred to as a *building contractor.* A general contractor is a licensed individual or firm that can enter into legal contracts to do construction work. Sometimes the general contractor is a single carpenter who either works alone to complete a project or hires extra carpenters to help. The general contractor may also be a giant corporation involved in massive construction projects all over the world. Whether the job is large or small, the general contractor is responsible for the overall organization and supervision of the construction project.

## OTHER BUILDING TRADES WORKERS

In addition to carpenters, workers from many other crafts work on construction projects. Most of them are employed by *subcontractors* who are licensed to perform work in their particular area.

For example, the electrical work in a building under construction is done by electricians employed by an electrical subcontractor.

Following are some of the major trades in the construction industry. Some examples of the type of work they do are also given:

*Construction laborers* are often directly employed by the general contractor. They carry materials to the carpenters, help place the concrete on form jobs, work on excavation, and so forth. Although considered an "unskilled" trade, this is an essential trade to any construction project.

*Bricklayers* do all the brick construction such as walls, fireplaces, and chimneys. They also work with other masonry materials such as concrete blocks and structural tile.

*Stonemasons* work on buildings that have solid stone or stone-veneer walls.

*Hod carriers and laborers* assist workers of other trades such as bricklayers, stonemasons and plasterers. Hod carriers mix the mortar and carry the masonry materials and mortar to the other workers. Laborers do excavations, compacting, and cleanup. They perform general assistance tasks for carpenters.

*Cabinetmakers* layout, build, and install kitchen cabinets, bathroom vanities, and bookcases. They also fabricate and install countertops.

*Cement masons,* or *cement finishers,* produce the finish on freshly poured concrete floor slabs and sidewalks. They can also set forms that are no more than one board high.

*Structural ironworkers* erect the steel framework for steel-framed buildings.

*Reinforcing ironworkers* place the reinforcing steel bars and wire mesh used in reinforced concrete construction.

*Electricians* install the conduit, wiring, and all other items that make up an electrical system.

*Plumbers* place all the pipes for the water, gas, sewage, and drainage systems. They install fixtures such as sinks, bathtubs, and toilets.

*Pile drivers* work on highways, buildings, dams, and bridges. They include divers who do underwater work.

*Pipefitters* usually work on industrial and commercial buildings. They install the pipe systems that carry hot water, steam, and various types of liquids and gases.

*Sheet-metal workers* build and install sheet-metal products such as the ducts used for heating and air conditioning. They also make and install gutters and flashing.

*Elevator constructors* install elevators, escalators, and dumbwaiters. NOTE: In some areas, escalators are installed by millwrights.

*Operating engineers* run and maintain heavy construction machinery such as bulldozers, tractors, cranes, and derricks.

*Millwrights* set up pumps, turbines, generators, conveyors, and other mechanical systems.

*Lathers* install the base wall materials such as wire or perforated gypsum boards to which plaster or stucco is applied. NOTE: Lathers are now members of the United Brotherhood of Carpenters and Joiners of America.

*Plasterers* apply the plaster and stucco finishes to buildings. These finishes may be on interior or exterior walls.

*Painters and paperhangers* apply the painted finishes to walls and ceilings. They also apply paper and other types of finish materials to walls. NOTE: In some areas they tape, float, and texture drywall.

*Glaziers* cut, fit, and install glass in windows, doors, and other types of glass units.

*Floor-covering installers* place materials such as carpeting linoleum, and vinyl and rubber tile. NOTE: In some areas, carpenters install resilient flooring.

*Tilesetters* place ceramic and other types of tiles on floors and walls. Some tilesetters specialize in placing marble or terrazzo tiles.

*Roofers* put the finish roof covering on buildings. They work with materials such as wood shingles, asphalt, slate, and tile.

## WORKING TOGETHER: CARPENTERS AND OTHER BUILDING TRADES WORKERS

A smoothly functioning construction job depends on cooperation among the workers of the different trades involved in the job. This is particularly true of carpenters in relation to workers in other crafts. The carpenters are the backbone of most types of construction jobs, as all the other crafts depend on the work done by the carpenters. Before the electrician can wire a house or the plumber place pipes, the carpenters must first frame the floors, walls, and ceilings.

During all stages of construction, the carpenters should keep in mind the work of the crafts that will follow. A good understanding of the work done by the other crafts will help the carpenters perform their work in a way that facilitates the work of the other crafts.

## INDUSTRY ORGANIZATIONS

The majority of building trades workers employed on larger jobs belong to labor unions. Other associations exist to represent the unified interests of the general contractors. All these organizations strongly influence the construction industry.

### The Carpenters' Union

The United Brotherhood of Carpenters and Joiners of America is the carpenters' union. Founded in 1881, it now has approximately 800,000 members. Local unions number more than 2,000. Every state of the United States and every province of Canada has at least one local union.

Most union members are construction carpenters. However,

the union also includes mill-cabinetworkers, pile drivers, millwrights, resilient floor-covering installers, and a growing number of industrial workers. The industrial workers are employed in lumber, plywood, and sawmill production, as well as in factories that produce prefabricated housing.

The carpenters' union includes members from many different racial, ethnic, and religious backgrounds. More than 27,000 women members are employed in industrial plants or work as carpenters on construction sites. The number of women working in carpentry as well as the other construction trades increases each year.

The purpose of the carpenters' union is to advance the interests of its membership. The union negotiates agreements with contractors' associations concerning wages, fringe benefits, working conditions, and provisions for apprenticeship training, hourly wage scale, and the number of hours in the work week. In addition, the agreement usually includes provisions for vacation pay and for a health and welfare plan that covers medical, dental, hospitalization, and prescription costs. Most agreements between the carpenters' union and the employers also provide for pension plans.

### Contractors' Associations

Many general contractors belong to industry associations that represent their interests as employers. The United States has numerous such groups. Two influential ones are the Associated General Contractors (AGC) and the National Association of Home Builders (NAHB). These organizations are national in scope and have state and local chapters as well. The AGC mainly represents contractors who do heavy construction work.

The NAHB mainly represents those in residential and light construction. It is with these two organizations that the carpenters' union negotiates its working agreement.

## ENTERING THE CARPENTRY TRADE

The carpentry trade offers many opportunities for a young person. The subject of carpentry is a standard offering of many post-secondary programs at vocational-technical institutes and community colleges. These programs are strongly recommended. However, it should be stressed that they are only preparatory courses. A person must actually work at a trade to learn it fully.

## Apprenticeship Programs

Apprenticeship is on-the-job training combined with other related instruction. The apprentice carpenter works with experienced journeymen while learning the trade.

**Origins of Apprenticeship.** Apprenticeship began in ancient times when the *master* of a skilled craft taught the craft to his sons or other young men. The young workers were completely under the direction of the master for the period of the apprenticeship. This method of learning a craft evolved into the *indenture* system practiced in Europe during the Middle Ages. An indenture was a written contract between the master craftsman and the apprentice. The standard period of apprenticeship was often as long as eight years. The in-

denture system was brought over to the New World and was common practice in this country until the development of craft labor organizations in the construction trades.

Carpentry apprenticeship in the United States was mainly a union-sponsored effort to preserve the craft from 1881 to 1939. During this period the United Brotherhood of Carpenters and Joiners of America developed a comprehensive program of on-the-job training and other related instruction.

**Labor-Management Apprenticeship Programs.** The passage of new labor laws in 1935 (the National Labor Relations Act) brought about improved collective bargaining between trade unions and employers. As a result, apprenticeship became the joint responsibility of both labor and management. Programs were set up wherever negotiated agreements existed between the carpenters' union and the employers.

Locally, these programs are supervised by a Joint Apprenticeship and Training Committee (JATC) made up of labor and employer representatives. An apprenticeship program in a given area works closely with the local school district as well as with the Department of Education for that state. On a national level, all apprenticeship programs must conform to the standards established by the Bureau of Apprenticeship and Training, which is a division of the United States Department of Labor. Information on apprenticeship programs can be obtained from the following address:

Bureau of Apprenticeship and Training
Employment and Training Administration
U.S. Department of Labor
601 D Street, NW
Washington, D.C. 20213

All persons enrolled in an apprenticeship program must attend classes a certain number of hours per year. A minimum of 144 hours for each year is recommended. These classes are held at local school facilities or special regional training centers. In addition to classroom hours, a minimum number of hours of work experience is also required. This number varies according to the trade (carpentry, electricity, plumbing, etc.).

**Merit Shop Apprenticeship Programs.** In recent years apprentice programs for a number of crafts including carpentry have been established by an employers' organization known as the Associated Builders and Contractors, Inc. These programs have also been approved by the Bureau of Apprenticeship and Training.

**Labor-Management Journeymen Programs.** In addition to supervising apprenticeship programs, many labor-management Joint Apprenticeship and Training Committees (JATCs) throughout the United States also conduct journeyman-level classes. These classes provide journeyman carpenters an opportunity to improve their skills in blueprint reading, estimating, use of leveling and transit instruments, welding, and other important areas.

# Construction Materials

Carpenters today work with many different types of building products, including the materials that compose a structure and the fasteners that hold it together. Wood, in the form of structural lumber or fabricated lumber products, is still the dominant material in the field of construction. Most homes and small commercial structures are framed and finished in wood. Wood is also used to build forms for concrete buildings.

# UNIT 3

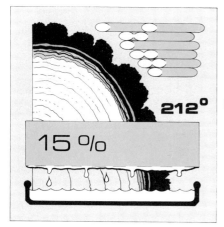

# The Nature of Wood

Wood used in carpentry (and other trades) is obtained mostly from the trunk of the tree. An understanding of the different parts of a treetrunk and of how a tree grows is useful to anyone who works with wood.

## STRUCTURE OF WOOD

Wood is composed of tiny cells (fibers). These cells are tubular in shape and are about as thick as a human hair. Cells in softwood trees are about ⅛″ long. Cells in hardwood trees are about ¹⁄₂₄″ long. See Figure 3–1.

The walls of each cell are composed of *cellulose* matter. The cells are held together with a natural cement called *lignin*. A tree grows by forming new wood cells. It reaches full maturity and stops growing when new cells stop forming.

*Annual rings* begin at the center of the trunk and continue outward to the bark. See Figure 3–2. Each ring represents a year of cellular growth. Therefore, the age of a tree is very close to the number of its annual rings. In drier seasons there is less growth, so some rings are narrower than others. A close look at each annual ring shows that it is made up of an inner, light-colored section and an outer, darker section. The light-colored section is the earlier growth of that year and is known as *springwood*. The darker part develops later in the growing season and is known as *summerwood*. Spring-

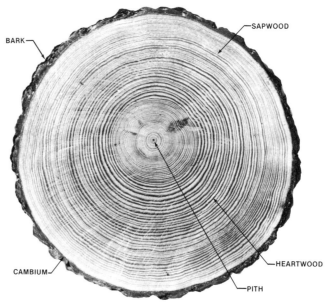

*Forest Products Laboratory—Forest Service USDA*

Figure 3–2. Cross section of treetrunk shows annual rings. Each ring represents a year of cellular growth.

Figure 3–1. Wood is composed of tiny cells. The cells are drawn here many times larger than their actual size.

wood is usually weaker and less dense than summerwood.

The outside covering of the tree consists of *bark*. A tree has two layers of bark. The outer layer is dry, dead tissue. Its purpose is to protect the tree from exterior damage. The inner bark is moist and soft. Its function is to help transport food from the leaves to all the growing areas of the tree.

Directly underneath the bark is a very thin layer called the *cambium*. The light-colored section under the cambium is *sapwood*. The darker layer that goes from the sapwood to the *pith* (center) of the trunk is *heartwood*. The *medullary rays*, also called *wood rays*, extend radially from the pith to the outer bark.

## Cambium

New cells are formed in the cambium. The inner part of the cambium develops the wood cells that become sapwood. The outer part of the cambium produces the new cells that form the bark.

## Sapwood

Sapwood is the growing portion of a tree. Food is stored and absorbed here. Sap, the watery fluid that circulates through a tree, travels from the roots, up through the sapwood, and to the leaves. A young tree consists entirely of sapwood.

## Heartwood

As a tree grows, the number of annual rings increases. Also, the layers of wood nearest the center of the trunk undergo certain changes. The wood cells become inactive and no longer conduct sap and food. When this happens, the sapwood in this central part of the tree changes into heartwood and usually begins to darken in color.

Lumber cut from sapwood and lumber cut from heartwood are about equal in strength. However, heartwood is more resistant to decay. As a result, it is more durable than sapwood when exposed to weather and is therefore a better outside finish material.

## Pith and Medullary Rays

The pith is the small central core of the tree. It is a soft, spongy material and does not produce a good structural grade of lumber.

The medullary rays start from the pith area and move toward the outside of the trunk. Their purpose is to store and transport food.

## MOISTURE CONTENT OF WOOD

Water accounts for a large percentage of the weight of a living tree. It is present in the cell cavities of the wood as well as in the walls of each cell. Recently cut lumber (*green lumber*) consequently has a very high amount of moisture.

Lumber begins to dry out as the water contained in the wood cells evaporates. The water evaporates first from the cell cavities, then from the cell walls. When the cell cavities are empty of water but the cell walls still contain water, the *fiber saturation point* has been reached. Wood does not begin to shrink until after it reaches the fiber saturation point. Only when the water begins to leave the cell walls do the cells begin to decrease in size, which causes the wood to shrink.

Lumber gives off moisture until the amount of moisture in the wood is the same as the amount of moisture in the surrounding air. When this occurs, the lumber has reached a state of *equilibrium moisture content*. At this point the lumber stops shrinking.

Moisture content can be tested in a laboratory by the oven-drying method shown in Figure 3–3. It can also be tested in the

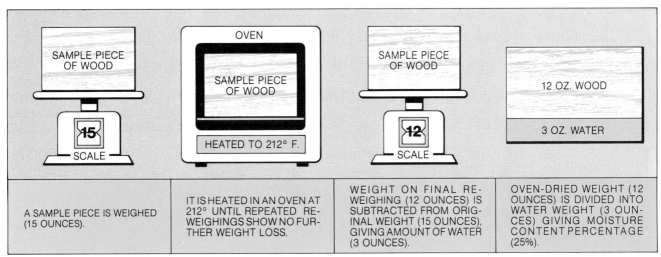

| A SAMPLE PIECE IS WEIGHED (15 OUNCES). | IT IS HEATED IN AN OVEN AT 212° UNTIL REPEATED RE-WEIGHINGS SHOW NO FURTHER WEIGHT LOSS. | WEIGHT ON FINAL RE-WEIGHING (12 OUNCES) IS SUBTRACTED FROM ORIGINAL WEIGHT (15 OUNCES), GIVING AMOUNT OF WATER (3 OUNCES). | OVEN-DRIED WEIGHT (12 OUNCES) IS DIVIDED INTO WATER WEIGHT (3 OUNCES) GIVING MOISTURE CONTENT PERCENTAGE (25%). |

Figure 3–3. The moisture content of wood is the percentage of its weight that is water. The oven-drying method gives an accurate measure of moisture content.

*Educational Lumber Company, Inc*

Figure 3–4. A moisture meter is a less accurate but more convenient way to check the moisture content of wood than the oven-drying method.

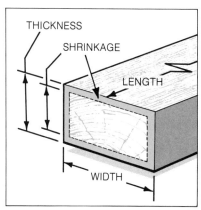

Figure 3–5. Most lumber shrinkage occurs across the grain. For example, a 2" x 4" x 8' piece of lumber will shrink more across its 2" thickness and its 4" width than along its 8' length.

field with a moisture meter. See Figure 3–4. A moisture meter gives an instant reading of moisture content by measuring the resistance to current flow between two points driven into the wood. It is not as accurate as the oven-drying method, but it is accurate enough for most construction purposes.

Lumber should have moisture content compatible with the air that will surround it after it is installed. In drier parts of the country the moisture content of lumber should be no more than 15%. In damper areas as much as 19% is acceptable. Interior finish materials in most areas should have a moisture content between 6% and 12%.

## Effects of Moisture Content

If lumber has too high a moisture content when used on a job, problems may develop from additional shrinkage. Framing members inside the walls may shrink as the wood dries out, causing plaster to crack, or nails to pop out if drywall is used. Most wood shrinkage occurs across the grain. A piece of material 2" thick, 4" wide, and 8' long will shrink very little along its 8' length. It will, however, noticeably shrink across its 2" thickness and 4" width. See Figure 3–5.

The strength of wood increases as moisture content decreases, because the cell fibers of the wood stiffen and become more compact as they dry out. Wood will not decay (rot) if the moisture content is below 20%. Wood installed under conditions where the moisture content will remain higher than 20% should be treated with chemicals that prevent decay.

Many wood products today (such as plywood) consist of layers glued together. The glue bond of these wood products improves as the moisture content of the wood decreases.

# UNIT 4

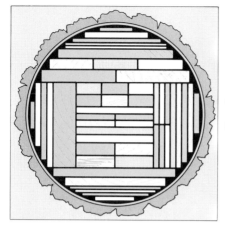

# Manufacture of Lumber

Trees are one of our greatest natural resources. As the population of our country continues to grow, however, more and more trees are needed to supply the lumber required for housing and other kinds of construction. Part of the answer to this problem is being provided by improved forest management, and another part by the elimination of waste in sawmill operations.

About half of the bulk of a round-shaped log fed into a modern sawmill comes out as usable construction lumber. The other half is *residue*, consisting of chips, bark, trimmings, shavings, and sawdust. At one time the residue was considered of no value and was burned as waste. Today almost all of the residue is converted to useful products. The wood chips are a major source of wood fiber used by paper mills. Planer shavings and chips are important ingredients in the manufacturing of panels used for sheathing and insulation. (These panel products are discussed in a later unit.)

The manufacture of lumber begins in the forest where the trees are cut. Limbs and branches are removed, and the tree is cut into sections (logs) small enough to be transported by truck to the

sawmill. See Figures 4–1 and 4–2. The logs are manufactured into the different sizes of lumber used in construction work. See Figure 4–3.

Initial sawmill operations include *debarking,* a process of stripping the bark from the log. See Figure 4–4. The debarked logs are cut into smaller sections, which are cut into boards by a bandsaw. See Figure 4–5. The boards are fed by conveyor belts into a *trimmer,* which cuts the boards to standard lengths and also cuts off pieces with defects. See Figure 4–6.

*Forest Industries—Miller Freeman Publications*
Figure 4–1. Log loader placing freshly cut logs on the bed of a lumber truck for transport to the sawmill. This loading equipment features a hydraulic power boom and grapple.

*Western Wood Products Association*
Figure 4–2. Mechanical log stacker at rear unloads trucks that have brought logs to the storage yard.

Figure 4–3. A log can be cut into many shapes and sizes of lumber.

*grained* softwood lumber. The other produces *quartersawn* hardwood lumber and *edge-grained* softwood lumber. See Figure 4–7.

## Plainsawn and Flat-grained lumber

In plainsawing, the annual growth rings of the log are at an angle of 45° or less to the wide surface of the boards being cut. This is sometimes called a *tangential* cut. Most lumber is produced in this manner. It provides the widest boards and results in the least amount of waste. The term *plainsawn* refers to hardwood lumber. *Flat-grained* refers to results of the same method on softwood.

## Quartersawn and Edge-grained Lumber

For quartersawn lumber, the log is first quartered lengthwise. Then boards are cut out of each quartered section. This method is much more expensive than plain-

## SAWING METHODS

There are two methods of cutting logs into pieces of lumber. One method produces *plainsawn* hardwood lumber and *flat-*

Figure 4–4. Logs being conveyed to the sawmill from a log merchandising deck. The logs first pass through a de-barker (upper right). They are then cut into smaller sections and moved to the conveyor belt shown in front of the merchandising deck. The conveyor belt carries the logs to their proper station in the sawmill.

Figure 4–5. Cutting a board out of a log. A carriage system moves the log through a bandsaw, which is shown completing the cut at the right end of the log.

*Forest Industries—Miller Freeman Publications*

Figure 4–6. The operator controls a trimmer (upper left), which consists of a series of saw blades that remove defects and cut the boards to standard lengths. The boards are fed into the trimmer by conveyor belts.

*Southern Forest Products Association*

Figure 4–8. One method of seasoning lumber is air drying. Note the stickers placed between rows of lumber to allow for complete circulation of air.

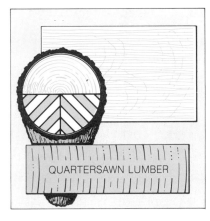

PLAINSAWN LUMBER

QUARTERSAWN LUMBER

Figure 4–7. One sawing method produces plainsawn lumber on hardwood and flat-grained lumber on softwood. The other sawing method produces quartersawn lumber on hardwood and edge-grained lumber on softwood.

sawing, as it produces narrower boards and creates more waste. However, it produces a more attractive grain pattern in some hardwoods. Also, there is less warpage in quartersawn wood. The term *quartersawn* refers to hardwood lumber. *Edge-grained* refers to results of the same method on softwood.

## SEASONING METHODS

Lumber must be *seasoned* (dried) before it is placed on the market. The two methods of sea-soning are *air drying* and *kiln drying*.

### Air Drying

For air drying, the newly produced lumber is stacked out in the open air. Wood strips are placed between the layers of wood so that air can circulate around each piece. See Figure 4–8. It takes several months for the lumber to season adequately. Most softwood used for rough construction is seasoned by this system.

### Kiln Drying

For kiln drying, the lumber is placed in a temperature-controlled building called a *kiln*, which acts like a large oven. First, steam is used to keep the humidity (amount of moisture in the air) high in the kiln. As the

21

temperature in the kiln is gradually increased, the humidity level is brought down. Lumber seasoned in this manner is stamped *kiln-dried* and is more expensive than air-dried material. This method of seasoning is usually used for the higher grades of hardwood lumber that are used for finish work.

## PLANING AND GRADING

When it is cut from the log, lumber has a rough surface. After seasoning, it must be finished off in a planing mill.

At the mill, it passes through *planers*, which are machines with rotating knives that smooth off (*surface*) the sides and edges of the lumber. As it moves along a conveyor belt, highly skilled workers called *graders* examine the boards for defects and mark each piece according to grade. See Figure 4–9. The graded pieces are later sorted according to thickness, width, and length. See Figure 4–10.

## LUMBER DEFECTS

Defects sometimes, but not always, affect the strength, stiffness, or appearance of lumber. Most individual pieces of lumber have some defects. The number and type of defects determine the grade of the lumber. Lumber with serious defects cannot be used for structural purposes.

## Natural Defects

Most defects occur from natural causes during the growth of the tree. *Knots* are one of the better known examples. As the tree grows, its upper limbs, as well as the limbs of surrounding trees, cast shadows upon its lower limbs. The lack of light causes some lower limbs to die, decay, and fall away. However, a small piece of the dead limb may remain attached to the tree. As the tree continues to grow and expand, new sapwood is added to the trunk. The pieces of dead limb are covered over and become knots.

Knots are found in most lumber. See Figure 4–11. If they are *sound* knots (those that remain firmly in place), a small number of them will not significantly affect the strength of the lumber. Knots are identified by their diameter as follows:

1. Pin knot: ½″ or less.
2. Small knot: More than ½″ but less than ¾″.
3. Medium knot: More than ¾″ but less than 1½″.
4. Large knot: More than 1½″.

Other defects brought about by natural causes are (see Figure 4–12):

Figure 4–10. The graded pieces of lumber are being removed from the conveyor belt and sorted according to thickness, width, and length.

Figure 4–9. Grader stamping lumber grades on the pieces emerging from the trimmer.

Figure 4–11. Knots are found in most grades of lumber. A small number of sound knots will not significantly affect the strength of the lumber.

WANES

SPLIT

SHAKE
*Western Wood Products Association*

Figure 4-12. Wanes, splits, and shakes are among the lumber defects that occur from natural causes.

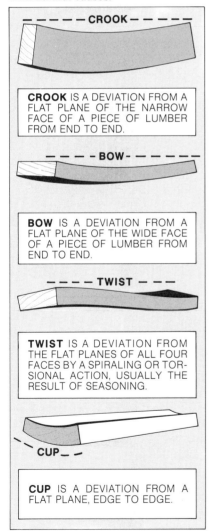

- - - - CROOK - - - -

**CROOK** IS A DEVIATION FROM A FLAT PLANE OF THE NARROW FACE OF A PIECE OF LUMBER FROM END TO END.

- - - - BOW - - - - - -

**BOW** IS A DEVIATION FROM A FLAT PLANE OF THE WIDE FACE OF A PIECE OF LUMBER FROM END TO END.

- - - - TWIST - - -

**TWIST** IS A DEVIATION FROM THE FLAT PLANES OF ALL FOUR FACES BY A SPIRALING OR TORSIONAL ACTION, USUALLY THE RESULT OF SEASONING.

CUP - -

**CUP** IS A DEVIATION FROM A FLAT PLANE, EDGE TO EDGE.

*Information from Western Wood Products Association*
Figure 4-13. Different kinds of warpage can occur during the evaporation of water from the cells of the wood.

*Wane:* An absence of wood or the presence of bark on the edge or corner of a piece of lumber.

*Shake:* A lengthwise separation of wood fibers between or through the annual growth rings.

*Check:* A separation of wood fibers across the annual growth rings.

*Split:* Same as a check but extending all the way through a piece of lumber.

*Pitch pocket:* An opening in the wood that contains solid or liquid pitch.

*Pitch streak:* A section of wood fibers saturated with enough pitch to be visible.

Defects caused by fungi are discussed later in this unit.

## Warping

Warping is a distortion that occurs during the evaporation (drying out) of water from the cells of the wood. Uneven shrinkage in the wood produces warping. This results in twisted and uneven shapes of lumber. Common warpage shapes are the bow, crook, twist, and cup. See Figure 4-13.

## Damage from Manufacturing Processes

Defects that mar the appearance of lumber may be caused during sawmill operations. A list of such defects follows:
1. Chipped, torn, raised, or loosened grain.
2. Skip marks that occur during surfacing.
3. Machine-caused burns.
4. Bite or knife marks.

## WOOD PROTECTIVE TREATMENT

Wood can be treated to make it resistant to fire, decay, and attacks from insects. Most wood treatments are designed to protect against fungi and termites, as these are the main causes of serious damage to lumber. The damage usually begins in the wood members near ground level. (Protection against termites is discussed in Section 8.)

## Damage Caused by Fungi

*Dry rot* causes wood tissue to break down, reducing the strength of a wood member. See Figure 4-14. It is the most common type of damage caused by a fungus. The fungus that causes dry rot lives in the wood and can be seen only with a microscope. Since this fungus must have water to live, it can survive only in wood with a moisture content of at least 20%. The term *dry rot* is misleading, since the decay begins under damp conditions. However, it is often not detected until after the wood has dried out.

Other types of fungi than the one that causes dry rot cause specks, molds, and stains. These fungi are fairly harmless, since they damage only the surface of the wood, affecting its appearance but not its structural quality. *White speck* and *honeycomb* are examples of this type of damage. Wood with white speck has small white spots or pits. See Figure 4–15. Honeycomb is similar, but the spots are larger or the pits are deeper. Another type of fungus causes *blue stain,* which is a blue-gray discoloration.

*Western Wood Products Association*
Figure 4–14. Decay (dry rot) is the breakdown of wood caused by a wood-destroying fungus.

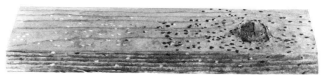

*Western Wood Products Association*
Figure 4–15. White speck affects only the surface of the lumber.

## Wood Preservatives

Wood preservatives come in liquid form. They contain chemicals that protect the wood against fungi decay and insect attack. Preservatives can be divided into three major types: water-borne, oil-borne, and creosote.

### Water-borne Preservatives.
Water-borne salt preservatives are used to treat lumber and plywood for residential construction. Two highly recommended mixtures are ammoniacal copper arsenate and chromated copper arsenate. After application of a water-borne preservative, the surface of the wood will be clear, odorless, and easy to paint.

### Oil-borne Preservative.
Best known as pentachlorophenol, an oil-borne preservative is highly toxic to fungi and insects. However, this type of preservative may affect the surface color of the treated material.

### Creosote.
Creosote is one of the oldest preservatives and is still one of the most widely used. It comes in a number of mixtures and leaves a slight odor after it has been applied. Surfaces coated with creosote cannot be painted.

**Methods of Application.** Preservatives may be applied by a *pressure* or a *non-pressure* process. The pressure process is considered to be the most effective. Lumber is loaded onto tram cars and rolled into long steel cylinders, which are then sealed off. The cylinders are filled with the preservative liquid. Intense pressure builds up inside the tank and causes the preservative to penetrate deep into the wood.

The non-pressure process gives less protection, but it is simpler and less expensive to apply. The wood is submerged in an open tank filled with preservative for a period of at least three minutes.

## Fire-retardant Treatment

Attention is being focused today on the fire hazards connected with building materials. Wood is highly combustible (it ignites and burns easily). Methods are continually being developed to give greater fire protection to lumber used in construction. Lumber that has been treated to make it fire-retardant is available for roof systems, beams, posts, studs, doors, hardwood paneling, and other interior trim products.

Fire retardants are applied the same way preservatives are applied. In a non-pressure process the wood receives a fire-retardant coating. In a pressure process it is impregnated with the fire-retardant chemicals.

The chemicals used in fire-retardant treatment react to heat slightly below the temperature required to ignite the wood. They release a vapor that surrounds the wood fibers and sets off a reaction in the wood. This reaction is the formation of a protective insulating *char* on the surface of the wood. The char prevents the wood from igniting and reduces the amount of smoke and toxic fumes caused by the fire. Fire-retardant wood products are also resistant to termites and decay.

# UNIT 5

# Softwood and Hardwood

All trees are divided into the two main classes of *softwood* and *hardwood*. Therefore, all lumber is referred to as either softwood or hardwood lumber. In addition, lumber is called by the same name as the tree it comes from. For example, Douglas fir lumber comes from a Douglas fir tree. Walnut lumber comes from a walnut tree.

The terms *softwood* and *hardwood* can be confusing, since some softwood lumber is harder than some hardwood lumber. Generally, however, hardwoods are more dense and harder than softwoods.

Softwood trees are called *conifers*. They have thin, needle-shaped leaves and they bear cones in which seeds germinate and grow. See Figure 5–1. These trees are called *evergreens* because they usually bear leaves all year long. Over 75% of the wood used for construction is softwood.

Hardwood trees are broad-leaved, *deciduous* trees, meaning they lose their leaves in the autumn. See Figure 5–2.

## SOFTWOOD LUMBER

All lumber used for rough construction is softwood. Examples

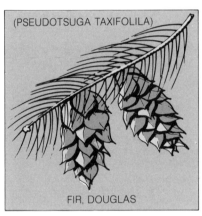

(PSEUDOTSUGA TAXIFOLILA)

FIR, DOUGLAS

Figure 5–1. Softwood lumber comes from evergreen trees, which bear cones and have needle-shaped leaves.

of rough construction are the framing and sheathing of wood-framed houses (Figure 5–3) and the building of forms for concrete structures.

Softwood species are also used for finish products such as moldings, doors, and cabinets. As a rule, softwood finish material is painted rather than stained. Some of the more frequently used softwoods are:

> Douglas fir
> White fir
> White pine
> Ponderosa pine
> Sugar pine
> Southern pine
> Hemlock
> Spruce
> Cypress

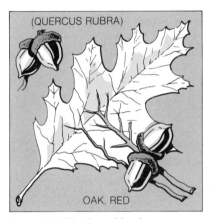

(QUERCUS RUBRA)

OAK, RED

Figure 5–2. Hardwood lumber comes from broad-leaved, deciduous trees. These trees lose their leaves in the fall.

> Redwood
> Western red cedar

Many species of softwood are about equal in quality. However, redwood and western red cedar are particularly recommended for exterior trim, siding, outside decks, and fences as they have a much stronger resistance to decay than most other kinds of wood.

In the western United States Douglas fir is used for most rough construction, since Douglas fir trees are abundant along the Pacific Coast. In the southeastern states various species of southern pine (longleaf, slash, shortleaf, loblolly) are widely used for rough construction.

Southern Forest Products Association
Figure 5-3. The framework of this building is constructed of softwood lumber.

These species grow in forests from Virginia to Texas.

## Softwood Grading Systems

The grade of a piece of lumber is based on its strength, stiffness, and appearance. A high grade of lumber has very few knots or other kinds of defects. A low grade of lumber may have knotholes along with many loose knots. The lowest grades are apt to have splits, checks, honeycombs, and some kind of warpage. The grade of lumber to be used on any construction job is usually stated in the specifications for a set of blueprints.

Grade systems are established by lumber-producing associations in different parts of the country. These associations must comply with all the provisions of the American Softwood Lumber Standard (PS 20-70) established by the U.S. Department of Commerce. One regional group is the Western Wood Products Association, which defines the grades of lumber used in the western states. See Appendix A. Another association is the Southern Forest Products Association, which serves the same function for the southeastern states. See Appendix B.

## Grade Marks

Grade marks are stamped on lumber to provide grading information. A typical grade mark includes the official trademark of the association, such as the Western Wood Products Association and the Southern Pine Inspection Bureau. Also included is the lumber grade, mill identification number, wood species, and surfacing and moisture designations. See Figure 5-4.

## Softwood Grading Categories

Three general grading categories, as defined in the Western Lumber Grading Rules, are Appearance, Framing, and Industrial Lumber.

**Appearance Lumber.** This "board lumber" consists of several nonstructural grades intended where strength is not the main consideration, such as High Quality Appearance, General Purpose, and Radius Edged Decking grades. High Quality Appearance grades (Selects, Finish, and Special Western Red Cedar) are used for finish applications. See Figures 5-5, 5-6, and 5-7. General Purpose Boards include several common board grades. The #1, #2, and #3 common grades are frequently used for shelving and paneling. The #4 common grade is used for sheathing, concrete forms, and low-cost fencing. A *board* is a piece of lumber less than 2″ thick and between 4″ and 12″ wide. Radius Edged Decking grades are used for patio decking. As this type of decking is considered load bearing and placed flat-wise, the joists below are spaced a maximum of 16″ on center.

**Framing Lumber.** This category includes Timbers, Dimension Lumber, and Special Dimension grades. *Timbers* are used for posts, beams, and girders. Timbers are lumber that is 5″ × 5″ or larger in nominal size with the width not more than 2″ greater than the thickness. See Figure 5-8. Dimension Lumber is used primarily for structural purposes. These include studs, joists, planks, roof rafters, trusses, and components that make up the framework of a

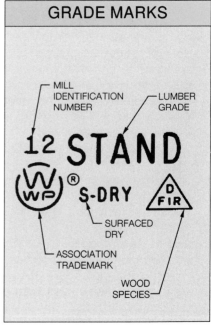

Western Wood Products Association
Southern Pine Inspection Bureau

Figure 5-4. Grade marks stamped on lumber provide grading and other relevant information.

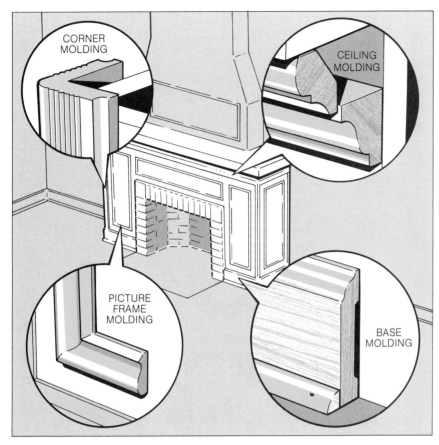

Figure 5–5. Appearance grade boards are used to make interior molding.

**Industrial Lumber.** This category includes Structural, Factory, and Nonstructural grades. The Structural grades are used for mining timbers, scaffold planks, and foundation lumber. Factory and Shop grades are intended for cut-up and manufacture. Nonstructural grades are used for fence pickets, lath, batten, and gutters.

## HARDWOOD LUMBER

Hardwood accounts for approximately 25% of total lumber production. Most of the hardwood

building. See Figures 5–9 and 5–10. Special Dimension Lumber includes structural decking. It is also designed for machine stress-rated lumber that must have a designed value such as light trusses, belt rails, box beams, and factory built homes.

*California Redwood Association*

Figure 5–7. Finish appearance grades of lumber are used for interior paneling.

*California Redwood Association*

Figure 5–6. Appearance grade boards are used for siding.

*Southern Forest Products Association*

Figure 5–8. These timber pieces will be used for heavy structural members such as posts, girders, and stringers.

Figure 5–9. Light framing grades of lumber are used for the construction of wood-framed walls.

*American Plywood Association*

Figure 5–10. Prefabricated roof trusses are often constructed of structural light framing grades of lumber.

tree species that are suitable for lumber manufacture grow in the eastern sections of the United States. Hardwood is much more expensive than softwood. It is used to make such products as molding, stair treads, the outside veneers of doors and wall paneling, and flooring. Better quality furniture and cabinets are often constructed from hardwood.

Since many of the hardwoods have a much more attractive grain pattern than softwoods, hardwood trim is normally used when a natural or stained finish is desired.

The hardwood species used most often are the oaks and walnuts. They account for over 50% of total hardwood production. Other hardwoods used frequently are:

Birch
White ash
Beech
Elm
Maple
Mahogany
Basswood
Butternut
Chestnut
Yellow poplar
Gum

## Hardwood Grading System

The standard grades of hardwood lumber established by the National Hardwood Lumber Association are:

First and Second (FAS)
Selects
    No. 1 Common
    No. 2 Common
    No. 3 Common

Appearance is of prime importance to the grade of hardwood. Another factor in the grade is the number of *cuttings* that can be obtained from the board. Cuttings are the smaller sections that are cut from the board. An FAS grade is usually required for hardwood trim materials that will have a natural or stained finish.

# UNIT 6

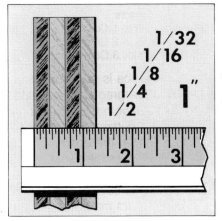

# Measurement of Lumber

The metric system is used in most countries of the world today. In the United States, however, we still use the *customary,* or *English,* system of measurement. In the customary system the basic units of measurements are the yard, foot, inch, and inch-fraction.

## ORDERING LUMBER

The specifications for a set of blueprints usually give the *species* and *grade* of lumber to be used on a construction job. The specifications may also include the stress rating and moisture content of the lumber. Unless otherwise mentioned, it is assumed that the lumber will be S4S (surfaced on all four sides).   When ordering materials from the lumber yard, it is necessary to give all the preceding information, plus the quantity of lumber required and the length of the pieces. The quantity (except for hardwood molding) is stated in *board feet*. (The board foot is discussed later in this unit.) A typical lumber order is:

> Douglas fir
> Standard grade
> S4S, 2 x 4 x 16
> 4,500 board feet

## Lumber Size

A piece of lumber is usually referred to by its *nominal* size, which differs from its *actual* size. See Figure 6–1. A 2 × 4, for example, is 2″ thick and 4″ wide when it is cut out of the log at the sawmill. However, it shrinks after being air-dried or kiln-dried. Then its measurements are further reduced by surfacing at the planing mill. When it is placed on the market, its *actual* size is 1½″ × 3½″. Nevertheless, it is called a 2 × 4 (its *nominal* size) because its original dimensions were 2″ × 4″.

Lumber measurements are stated in the following order: thickness, width, and length. For example, a piece of lumber 2″ (nominal size) thick, 4″ (nominal size) wide, and 16′ long is referred to as a 2 × 4 × 16. See Figure 6–2.

Softwood lumber is usually sold in even lengths ranging from 6′ to 24′. An extra (premium) charge is made for lumber over 20′ in length.

## The Board Foot

Lumber is usually ordered by the *board foot*, which is the *unit of*

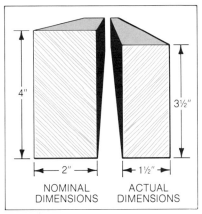

Figure 6–1. Nominal thickness and width compared to actual thickness and width.

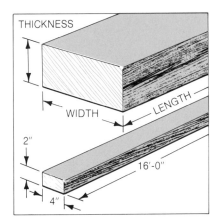

Figure 6–2. The abbreviated way of referring to a piece of lumber 2″ thick by 4″ wide by 16′ long is: 2 x 4 x 16.

*cost* of the material. The number of board feet in a piece of lumber is not the same as the number of lineal feet. For example, a 2 × 4 × 16 piece of lumber is 16 lineal feet long. However, its price is based on the number of board feet that it is equal to.

A board foot is 1″ (nominal size) thick, 12″ wide (nominal size) and 12″ long (1″ × 12″ × 12″) *or the equivalent.* For example, a piece of lumber 1″ thick, 6″ wide, and 24″ long equals one board foot. The amount of wood in a board foot is always 144 cubic inches. See Figure 6–3.

To find out how many board feet there are in a piece of lumber, use the following formula (in the formula, T = thickness, W = width, and L = length):

$$\frac{T \times W \times L}{12} = \text{board feet}$$

The example that follows shows how to find the number of board feet in a board that is 2″ thick, 4″ wide, and 16′ long.

$$\frac{2\,(T) \times 4\,(W) \times 16\,(L)}{12} =$$

$$\frac{2 \times \overset{1}{\cancel{4}} \times 16}{\underset{3}{\cancel{12}}} = \frac{32}{3} =$$

$$10\tfrac{2}{3} \text{ board feet}$$

There are 10⅔ board feet in a 2 × 4 × 16.

On any construction job many pieces of the same size lumber are ordered. The following example shows how to find the total number of board feet in 35 pieces of 2 × 4 × 16 lumber.

$$\frac{\text{number of pieces} \times T \times W \times L}{12} =$$

$$\frac{35 \times 2 \times 4 \times 16}{12} =$$

$$\frac{35 \times 2 \times \overset{1}{\cancel{4}} \times 16}{\underset{3}{\cancel{12}}} = \frac{1120}{3} =$$

373⅓ board feet

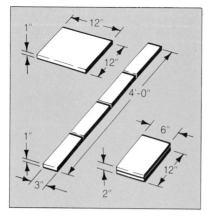

Figure 6–3. A board foot is equal to a piece of lumber 1″ x 12″ x 12″, or any other measurement that contains 144 cubic inches.

There are 373⅓ board feet in 35 pieces of 2 × 4 × 16 lumber.

To determine the cost of lumber, multiply the cost per board foot times the number of board feet. For the cost of the lumber in the preceding example, round off the total board feet to the next whole number (374) and multiply by the board foot price ($1.79).

```
    374     board feet
×  1.79     board foot price
  33 66
 261 8
  374
$669.46     total cost for
            374 board feet
```

## TYPES OF SURFACING

Almost all softwood lumber used in construction is surfaced on both sides and both edges. The *edge* of a piece of lumber is its narrowest dimension. The *side* is its widest dimension. (A 2 × 4 has a 2″ edge and a 4″ side.) Lumber can also be ordered with all rough surfaces, or with a combination of smooth and rough surfaces. Mill-cabinet shops often order materials that have rough edges and smooth sides. A board with unsurfaced edges is wider, so more pieces can be cut from it.

A special type of rough surfacing is applied to resawn lumber. See Figure 6–4. The pieces are

*California Redwood Association*

Figure 6–4. Resawn lumber has an attractive textured surface.

run through a special bandsaw that produces a coarse, textured pattern on the surface of the wood. Resawn lumber is most often used for exterior trim, siding, or paneling.

Abbreviations are used to indicate the type of surfacing required. For example, lumber surfaced on all four sides is ordered as S4S lumber. Some

common abbreviations for lumber surfacing are:

surfaced on:
S1S   one side
S2S   two sides
S4S   four sides
S1E   one edge
S2E   two edges
S1S1E   one side and one edge
S1S2E   one side and two edges
S2S1E   two sides and one edge
S/S   saw sized (resawn)

## STANDARD SIZES

Boards, dimension lumber, and timbers are available in standard sizes. The actual size is always smaller than the nominal size. For example, the actual size of a 2 × 4 is 1½″ thick and 3½″ wide. See Figure 6–5.

| STANDARD SIZES OF LUMBER | | | | |
|---|---|---|---|---|
| TYPE | NOMINAL SIZE | | ACTUAL SIZE | |
| | THICKNESS* | WIDTH* | THICKNESS* | WIDTH* |
| Common Boards 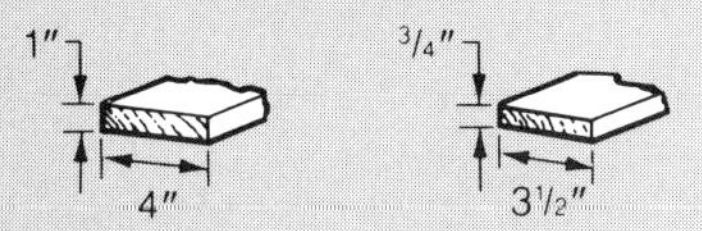 | 1 | 2 | ¾ | 1½ |
| | 1 | 4 | ¾ | 3½ |
| | 1 | 6 | ¾ | 5½ |
| | 1 | 8 | ¾ | 7¼ |
| | 1 | 10 | ¾ | 9¼ |
| | 1 | 12 | ¾ | 11¼ |
| Dimension 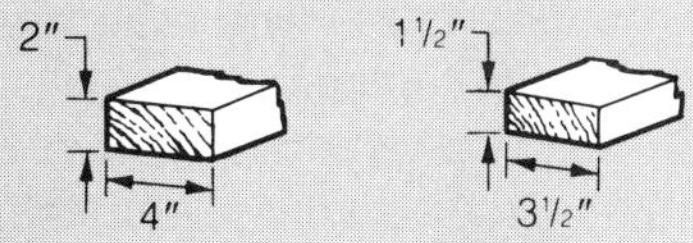 | 2 | 2 | 1½ | 1½ |
| | 2 | 4 | 1½ | 3½ |
| | 2 | 6 | 1½ | 5½ |
| | 2 | 8 | 1½ | 7¼ |
| | 2 | 10 | 1½ | 9¼ |
| | 2 | 12 | 1½ | 11¼ |
| Timbers 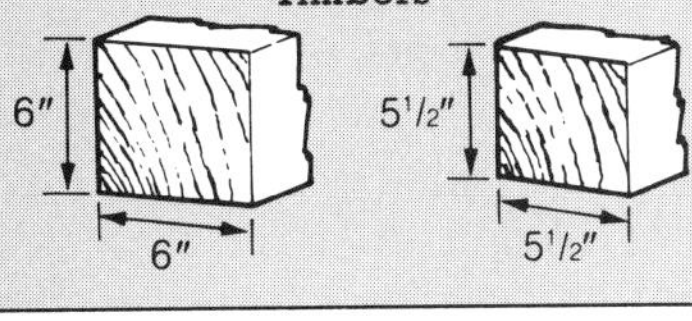 | 5 | 5 | 4½ | 4½ |
| | 6 | 6 | 5½ | 5½ |
| | 6 | 8 | 5½ | 7½ |
| | 6 | 10 | 5½ | 9½ |
| | 8 | 8 | 7½ | 7½ |
| | 8 | 10 | 7½ | 9½ |

* in inches

Figure 6–5. The actual size of lumber is always smaller than the nominal size.

# UNIT 7

# Wood Panel Products

Wood panel products are among the most important building materials in the construction industry. They are widely used for both structural and finish purposes. Structural wood panels have largely replaced board lumber as a sheathing material for walls and roofs (Figure 7–1) and as subflooring for wood-framed floors (Figure 7–2). They are also used for forms for concrete. Non-structural wood panels are used as a base for finish materials for floors, walls, and ceilings. They also help to insulate and strengthen a building.

## STRUCTURAL WOOD PANELS

The oldest and still the most widely used material for structural wood panels is *plywood*. Various *reconstituted* wood panel products have also been developed.

### Plywood

Plywood as we know it was first manufactured in the United States around 1905. At first it was used primarily for the manufacture of furniture. By 1940, however, plywood had been introduced as a valuable material in the construction of new homes. Today it is a major material in all types of construction.

Plywood allows for much more effective nailing than board lumber. Nails can be driven within ¼″ of the edge of a panel without causing splits in the wood.

Figure 7–1. Roofs are often sheathed with structural wood panels.

Figure 7–2. Structural wood panels are often used for subflooring. The plywood panels shown here have tongue-and-groove edges.

If stored properly, plywood also has less of a tendency to warp than board lumber. When delivered to the job, the panels should be laid flat until they are ready for use.

Structural plywood panels are manufactured from softwood lumber obtained from forests in the western and southern parts of the United States. At one time almost all structural plywood was manufactured from Douglas fir, a tree that grows abundantly on the West Coast. Today some of the pine species growing in the southeastern states are also widely used. In addition to Douglas fir and pine, more than 70 other species are sometimes used in the manufacture of plywood.

**Manufacture of Plywood.** A plywood panel is made up of thin layers of wood *(veneers)* that have been peeled off specially prepared logs *(peeler blocks)*. Peeler blocks are logs that have been cut from longer logs to a length that will fit into a veneer-

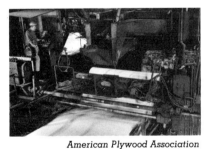

Figure 7–3. As the lathe chuck revolves the peeler block, the lathe knife peels a continuous, thin ribbon of veneer (up to 600 lineal feet per minute).

cutting lathe. See Figure 7–3. They are usually soaked in a hot-water vat or steamed before being peeled. The thicknesses of the veneers peeled from the blocks range from 1/10″ to 1/4″.

The veneer sheets are cut by clippers to usable widths of up to

American Plywood Association
Figure 7–4. Veneers are placed on a revolving table to be sorted by grade.

54″. Any sections with defects that cannot be repaired are cut out. The veneer sheets are then dried in ovens to bring their moisture content down to about 5%. Next, they are sorted by grade. See Figure 7–4.

Die-cutters are used to remove small defects such as knotholes. See Figure 7-5. In appearance-grade panels the holes are filled with wood patches. In panels of lower grades they are filled with synthetic (plastic) patches. After being patched, the sheets are glued and sandwiched together. See Figure 7–6. Then the veneer sandwiches are placed on racks in a hot press. See Figure 7–7. The sheets in each sandwich are bonded together by high heat and pressure in the hot press. The heat ranges from 230°F to 315°F. The pressure ranges from 175 to 200 pounds per square inch.

Finally, the panels are trimmed and stamped with a grade. If re-

American Plywood Association
Figure 7–5. Die cutters remove small defects such as knotholes.

quired, they are also sanded. Figure 7–8 shows panels ready for shipment.

Plywood panels are always made up of an odd number of layers, such as three, five, seven, and so forth. Each layer in turn consists of one or more *plies*. A ply is a single veneer

American Plywood Association
Figure 7–6. Veneer sheets are glued and sandwiched together in one continuous operation.

American Plywood Association
Figure 7–7. Veneer sandwiches are subjected to high heat and pressure in a hot press.

Deere & Company
Figure 7–8. Plywood panels stored in the warehouse and ready for shipment to building supply wholesalers and distributors. A forklift is being used to move strapped bundles of panels.

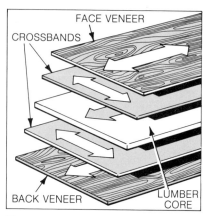

Figure 7–9. Cross-lamination of a plywood panel. The grain of each layer runs at a right angle to its adjacent layer. There are five layers shown in this example.

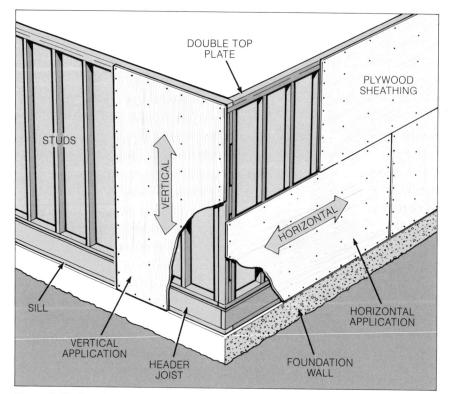

Figure 7–10. A wall sheathed with plywood is much stronger than a wall sheathed with board lumber.

sheet. The layers of a plywood panel are *cross-laminated*, meaning that each layer is placed with its grain running at a right angle to the adjacent layer. The outside layers of the panel are referred to as the *face veneer* and *back veneer*. Underneath the outside layers are the *crossbands*. The center layer is known as the *core*. See Figure 7–9. It is the cross-lamination of the veneer layers that makes plywood much stronger for wall racking than board lumber of the same thickness. For example, 5/16″ thick plywood wall sheathing provides twice the stiffness and more than twice the strength of 1″ thick diagonal wall sheathing. See Figure 7–10.

Plywood veneers are graded according to their appearance, natural growth characteristics (such as knots and splits), and the size and number of repairs made during manufacture. Veneers of higher grades are used on panels that will be exposed. If both sides of the panel will be exposed, high-grade veneers are used on both sides, and the grade is stamped on the edge. If only one side will be exposed, a veneer of a lower grade is used on the unexposed side, and the grade is stamped on that side. The table in Figure 7–11 gives a description of each veneer grade.

The species of wood used for plywood has an effect on the stiffness of the panel. The table in Figure 7–12 classifies various species according to strength, which determines stiffness.

## Table 1   Veneer Grades

| A | Smooth, paintable. Not more than 18 neatly made repairs, boat, sled, or router type, and parallel to grain, permitted. May be used for natural finish in less demanding applications. |
|---|---|
| B | Solid surface. Shims, circular repair plugs and tight knots to 1 inch across grain permitted. Some minor splits permitted. |
| C Plugged | Improved C veneer with splits limited to 1/8-inch width and knotholes and borer holes limited to 1/4 x 1/2 inch. Admits some broken grain. Synthetic repairs permitted. |
| C | Tight knots to 1-1/2 inch. Knotholes to 1 inch across grain and some to 1-1/2 inch if total width of knots and knotholes is within specified limits. Synthetic or wood repairs. Discoloration and sanding defects that do not impair strength permitted. Limited splits allowed. Stitching permitted. |
| D | Knots and knotholes to 2-1/2 inch width across grain and 1/2 inch larger within specified limits. Limited splits are permitted. Stitching permitted. Limited to Interior (Exposure 1 or 2) panels. |

*American Plywood Association*

Figure 7–11. Plywood veneer grades. Veneers of higher grades are used on panels that will be exposed.

## Table 2  Classification of Species

| Group 1 | Group 2 | Group 3 | Group 4 | Group 5 |
|---|---|---|---|---|
| Apitong | Cedar, Port | Alder, Red | Aspen | Basswood |
| Beech, | Orford | Birch, Paper | Bigtooth | Poplar, |
| American | Cypress | Cedar, Alaska | Quaking | Balsam |
| Birch | Douglas | Fir, | Cativo | |
| Sweet | Fir 2[a] | Subalpine | Cedar | |
| Yellow | Fir | Hemlock, | Incense | |
| Douglas | Balsam | Eastern | Western | |
| Fir 1[a] | California | Maple, | Red | |
| Kapur | Red | Bigleaf | Cottonwood | |
| Keruing | Grand | Pine | Eastern | |
| Larch, | Noble | Jack | Black | |
| Western | Pacific | Lodgepole | (Western | |
| Maple, Sugar | Silver | Ponderosa | Poplar) | |
| Pine | White | Spruce | Pine | |
| Caribbean | Hemlock, | Redwood | Eastern | |
| Ocote | Western | Spruce | White | |
| Pine, South. | Lauan | Engelmann | Sugar | |
| Loblolly | Almon | White | | |
| Longleaf | Bagtikan | | | |
| Shortleaf | Mayapis | | | |
| Slash | Red | | | |
| Tanoak | Tangile | | | |
| | White | | | |
| | Maple, Black | | | |
| | Mengkulang | | | |
| | Meranti, | | | |
| | Red[b] | | | |
| | Mersawa | | | |
| | Pine | | | |
| | Pond | | | |
| | Red | | | |
| | Virginia | | | |
| | Western | | | |
| | White | | | |
| | Spruce | | | |
| | Black | | | |
| | Red | | | |
| | Sitka | | | |
| | Sweetgum | | | |
| | Tamarack | | | |
| | Yellow- | | | |
| | Poplar | | | |

(a) Douglas Fir from trees grown in the states of Washington, Oregon, California, Idaho, Montana, Wyoming, and the Canadian Provinces of Alberta and British Columbia shall be classed as Douglas Fir No. 1. Douglas Fir from trees grown in the states of Nevada, Utah, Colorado, Arizona and New Mexico shall be classed as Douglas Fir No. 2.

(b) Red Meranti shall be limited to species having a specific gravity of 0.41 or more based on green volume and oven dry weight.

*American Plywood Association*

Figure 7–12. Wood species used to manufacture plywood are classified in five groups according to strength. Groups 1 and 2 both provide considerable strength.

**Panel Sizes.** The most popular size of plywood panel is 4′ wide and 8′ long. The next most popular size is 4′ wide and 10′ long. Some mills also produce panels up to 5′ wide and 12′ long. Panels are available with tongue-and-groove or square edges. Common thicknesses for flooring and sheathing purposes are ⁵⁄₁₆″, ³⁄₈″, ½″, ⅝″ and ¾″. A special thickness of 1⅛″ is used with post-and-beam floors.

## Reconstituted Wood Panels

Reconstituted wood panels, often called *non-veneered* panels, are a product of new and constantly improving technology applied to the manufacture of lumber. The word *reconstituted* refers to the use of wood particles, flakes, or strands that are bonded together and formed into full-size sheets. In many cases these particles or flakes are obtained from the waste (residue) from logs that have been sawed into lumber at the sawmill. For some types of reconstituted panels, flakes or strands are sliced to size out of short wood logs.

At one time reconstituted panels were not strong or durable enough to qualify as a structural material for subflooring or wall and roof sheathing. They were recommended only for such purposes as shelving, cabinets, interior wall paneling, and floor underlayment. Improved technology, however, has produced reconstituted panels strong enough for many of the same structural purposes as plywood. The most prominent of these products are *structural particleboard, waferboard, oriented strand board,* and *composite board.* See Figure 7–13.

**Structural Particleboard.** Also known as *flakeboard* or *chipboard*, structural particleboard is produced by combining wood particles (chips or flakes) with a resin binder and hot-pressing them into panels. Usually the wood chips are obtained from sawmill residue. The most commonly used resin is urea formaldehyde. Phenol-formaldehyde is used for more durable panels. Because particleboard has no grain, it presents a very smooth surface.

Particleboard was one of the first reconstituted wood products to be developed. Until recently its use was limited to non-structural applications such as shelving, countertops, kitchen cabinets, and floor underlayment. Its widest use is still for floor underlayment. (Underlayment is the material placed over the subfloor to provide a smooth, even surface for tile or rugs.)

Stronger, *structural* grades of particleboard have now been ap-

American Plywood Association

Figure 7–13. Non-veneered (reconstituted) structural wood panels are strong enough for many of the same structural purposes as plywood.

used for non-structural applications such as roof soffits, shelving, and built-in cabinets.

**Oriented Strand Board.** A recently developed structural panel used for sheathing purposes is oriented strand board. It is manufactured with *strands* of wood that are layered and are perpendicular to each other, much like plywood veneers. Usually three to five layers are used. See Figure 7–17. The layers are bonded together with a phenolic resin. The panels are stiff and present a smooth and uniform surface. Various panel sizes are available in thicknesses of ¼", ⅜", ⁷⁄₁₆", ½", ⅝", and ¾".

proved by many building codes as an acceptable material for subflooring and roof decking. Structural particleboard differs from the non-structural type by having layers of higher density on the surface. These layers contain extra resin and wax to provide greater strength and water repellency.

Particleboard comes in thicknesses ranging from ¼" to 1½".

Common panel sizes are 4' × 8', 4' × 12', 8' × 8', and 8' × 12'. The most frequently used panel size is 4' × 8'.

**Waferboard.** The manufacturing process for waferboard is similar to that for particleboard. The important difference between the two products is the source and size of the wood chips used and the type of resin used for a binder. The wood flakes used for waferboard are called *wafers*, and they are sliced from short logs to a specific size. Typical wafers are .030" thick and 1½" long, in random widths. An exterior grade of phenolic resin is used as the bonding agent for the wafers, which are hot-pressed into panels. See Figure 7–14.

Waferboard panels come in standard sizes of 4' × 8' or larger. They range in thicknesses from ¼" to ¾" and can be painted or stained. Waferboard panels are widely used for subflooring and for sheathing walls (Figure 7–15) and roofs (Figure 7–16). They are also

**Composite Board.** Composite board consists of a reconstituted wood center, such as particleboard, with a face and back veneer of softwood. See Figure 7–18. It is used for underlayment or for roof or wall sheathing. Various panel sizes are available in thicknesses of ½", ⅝" and ¾".

## PRODUCT STANDARDS FOR PANELS

Since 1966 product standards have been established in the United States for most grades of plywood. These product standards (published as *PSI-74, Construction and Commercial Plywood*) were cooperatively developed by the plywood industry and the U.S. Department of Commerce. The largest plywood industry organization in this country is the American Plywood Association (APA). Its membership represents about 85% of the mills that produce plywood and other panel products. Consequently, the APA was the most influential organization in establishing product performance

Forest Industries—Miller Freeman Publications

Figure 7–14. Pressed waferboard panels being trimmed to their proper size.

standards for plywood. Recently the APA expanded its role by establishing product and performance standards for composite panels and for non-veneered panels such as waferboard, oriented strand board, and certain classes of structural particleboard.

The APA has developed a series of charts for the construction industry that provide detailed information about the grades and performance ratings of different panel products. These charts enable builders to order materials that meet the structural requirements of a job. (See Appendices C through E at the back of this textbook.)

## APA Performance-rated Panels

Performance-rated panels are those generally recommended for sheathing and subflooring. They are divided into two categories: *exterior* and *interior*. However, the distinction between exterior and interior applies only to plywood panels and not to composite or reconstituted panels.

An exterior panel is recommended where there is continuous exposure to weather or moisture. The wood plies are of high-grade veneer and they are bonded together with a waterproof glue.

The plies of an interior panel may be bonded together with a moisture-resistant, rather than waterproof, glue. Most interior type panels, however, are manufactured with an exterior type of waterproof glue. In this case, the difference between an exterior and interior panel lies in the grade of veneer used. Lower grades of veneer are permitted for the backs and inner plies of interior panels. Since these lower-grade veneers may reduce

the bonding strength of the glue, interior panels should not be used where they would be continuously exposed to weather or moisture.

Interior panels are acceptable for wall and roof sheathing placed on the outside of the building because they will be covered (protected) by finish materials.

## Grade Trademarks

Grade trademarks are usually stamped on the backs of panels at the time of their manufacture. Examples of trademarks that appear on performance-rated panels manufactured by APA mills are shown in Figure 7–19. These trademarks give the following information:

1. *Panel Grade.* Panel grades are generally identified by veneer grades used on face and back (A-B, B-C, etc.) or by a name suggesting intended end-use (Rated Sheathing, Rated Sturd-I-Floor, etc.).

2. *Span Rating.* Span ratings for performance-rated structural panels give center-to-center spacing, in inches, of supports (studs, joists, roof rafters) over which the panel can be placed. Span ratings on APA-rated sheathing appear as two numbers separated by a slash mark as in 32/16. The left-handed number gives the maximum recommended spacing of supports when the panel is to be used for roof sheathing. The right-handed number gives the maximum recommended spacing over supports (floor joists) when the panel is to be used for subflooring.

The span ratings on APA-rated Sturd-I-Floor panels are shown as a single number, such as 16, 20, 24, and 48, indicating inches.

3. *Tongue-and-Groove.* The T&G width, if applicable, is given.

Figure 7–15. Waferboard is commonly used for sheathing exterior walls.

*Blandex—Blandin Wood Products Company*
Figure 7–16. Carpenters sheathing a roof with waferboard panels.

4. *Exposure Durability Classification.* Exposure 1 panels are recommended where the panels must be able to resist moisture during long construction delays and other severe conditions. These conditions

Figure 7–17. The layers of an oriented strand board panel are separated here to show how the strands of each layer are oriented at a right angle to the strands of the adjacent layer.

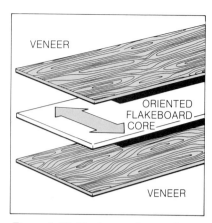

Figure 7-18. The layers of a panel are separated here to show how the reconstructed wood core layer is place between the two surface veneer layers.

usually apply for wall and roof sheathing. Exposure 2 panels are considered adequate for construction conditions where the panels will be exposed to moisture for shorter periods.

5. *Product Standard.* This is a voluntary commodity standard developed cooperatively by the U.S. Department of Commerce and the construction and plywood industry.

6. *Thickness.* The thickness measurement given is the actual, not the nominal, thickness of the panel.

7. *Mill Number.* The mill number identifies the mill where the panel was manufactured.

8. *APA's Performance Rated Panel Standard.* This indicates APA's satisfaction that the panel meets all construction requirements.

9. *Siding Face Grade.* Face grade, if applicable, is given.

10. Species Grade Number. Grade of interior or exterior plywood.

11. *HUD/FHA Recognition.* Panel is recognized for use by these agencies.

12. *Panel Grade, Canadian Standard.* Canadian grade.

13. *Panel Mark, Canadian Standard.* Rating and end-use designation.

14. *Canadian Performance-Rated Panel Standard.* Canadian performance rating.

15. *Panel Face Orientation Indicator.* Indicates strength axis direction.

## NON-STRUCTURAL WOOD PANELS

Non-structural wood panels are those used for purposes other than sheathing and subflooring. Some of the more important products in this group are the veneered *appearance* and *specialty* grades of plywood, hardboard, and non-structural particleboard. Some types of board insulation are also made of wood fibers, and these are discussed in Section 11.

## Non-structural Softwood Plywood

Most non-structural plywood panels have softwood face and back

veneers. The interior types are sanded on one or more sides. They are used for cabinets, built-ins, partitions, shelving, and other purposes where one or both sides of the panel will be exposed.

Exterior types of non-structural plywood panels have a high degree of strength and stiffness, even though they are not acceptable for wall or roof sheathing or for subflooring. Some exterior types of softwood plywood panels are sanded on the sides that are exposed, and others are not. *Specialty* exterior panels include *marine* panels, used for boat hulls, and *plyform*, used for concrete forms. Another specialty panel is the *siding* panel used to finish the outside surface of an exterior wall. This type of panel is discussed in Section 12.

## Hardwood Plywood

Hardwood plywood panels are used most often for interior finish. They are considerably more expensive than softwood panels. The face veneers of hardwood panels are made of lumber from the broad-leaved (deciduous) trees. This lumber gives the panels a very attractive grain pattern. As a rule, hardwood panels receive a natural or stained finish. Some of the more popular species are:

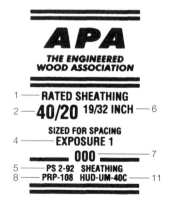

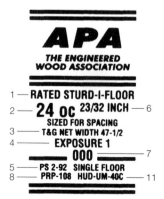

| 1. Panel Grade | 9. Siding Face Grade |
|---|---|
| 2. Span Rating | 10. Species Group Number |
| 3. Tongue-and-Groove | 11. HUD/FHA Recognition |
| 4. Exposure Durability Classification | 12. Panel Grade, Canadian Standard |
| 5. Product Standard | 13. Panel Mark — Rating and End-Use Designation, Canadian Standard |
| 6. Thickness | 14. Canadian Performance-Rated Panel Standard |
| 7. Mill Number | 15. Panel Face Orientation Indicator |
| 8. APA's Performance Rated Panel Standard | |

Figure 7-19. Trademarks are stamped on performance-rated panels.

*American Plywood Association*

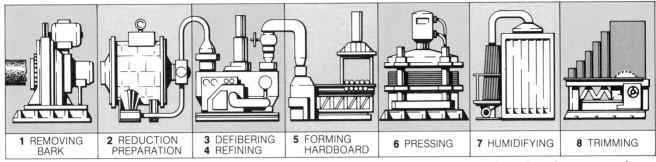

| 1 REMOVING BARK | 2 REDUCTION PREPARATION | 3 DEFIBERING 4 REFINING | 5 FORMING HARDBOARD | 6 PRESSING | 7 HUMIDIFYING | 8 TRIMMING |
|---|---|---|---|---|---|---|

Figure 7–20. Manufacture of hardboard. Wood chips for hardboard are obtained from sawmill residue or logs that are ground up in chopper machines.

Walnut
Luan
Oak
Maple
Ash
Mahogany
Birch
Gum

Construction carpenters are most familiar with hardwood plywood as a wall paneling material. Hardwood plywood is also used for cabinets, door skins (veneers), and face veneer for block flooring.

**Panel Sizes.** Hardwood panels can be ordered in different combinations of length, width, and thickness. The more common panel sizes are 48″ × 96″, 48″ × 48″, and 48″ × 120″, in thicknesses of 1/4″, 3/8″, 1/2″, and 3/4″.

**Grades.** A grading system established by the Hardwood Plywood Manufacturers Association is:

Premium Grade  (A)
Good Grade      (B)
Sound Grade    (2)
Industrial Grade (3)
Backing Grade  (4)

## Hardboard

Hardboard is one of the oldest reconstituted wood products used in construction today. Hardboard panels are manufactured of wood chips obtained from sawmill residue or logs that are ground up in large chopper ma-

chines. The wood chips must be broken down into wood fibers, and the fibers are held together by *lignin*, which is a natural wood-bonding agent. Figure 7–20 shows the procedure for producing hardboard. After bark is removed from the log (Step 1 in Figure 7–18), the debarked log goes to a chipper (Step 2), where whirling knives reduce it to chips about 5/8″ wide by 3/4″ long. The wood chips are reduced to a fibrous pulp (Step 3). The pulp is screened (Step 4) to ensure that individual fibers are within the proper size range. Undesirable waste materials (residue) are also eliminated. Chemicals may be added at this stage.

The pulp is poured onto a flat surface and formed into a mat (Step 5). The mat is pressed (Step 6) under heat (380°F to 550°F). (Pressure is 500 to 1,500 pounds per square inch.) To prevent later warping, the boards are raised to the moisture content of the surrounding atmosphere (Step 7). The boards are trimmed to standard sizes (Step 8).

Hardboard has many construction uses, including paneling, door skins, cabinets, underlayment, and exterior siding. Because it is a wood-based product, it can be sawed, routed, shaped, and drilled by standard

woodworking tools. The panels can be glued or fastened with screws, staples, or nails.

There are three grades of hardboard: tempered, standard, and service. The highest quality grade is tempered. Tempered panels receive an additional chemical and heat-treating process that increases their stiffness, water resistance, and durability. Wall paneling and exterior siding are often made of tempered hardboard.

The next highest quality hardboard is the standard grade. Standard panels are basically the same as tempered panels, except that they are not given the additional chemical and heat treatment. However, they still have good water resistance and a very smooth finish. They are often used for cabinets.

The lowest quality hardboard is the service grade. These panels have less strength than the other two grades, and their surface is not as smooth. They are economical to use where appearance and strength are not important.

Hardboard panels are manufactured with one side smooth (S1S) or both sides smooth (S2S). The panels are usually 4′ wide. Standard lengths are 8′, 10′, 12′, and 16′. Thicknesses vary from 1/12″ to 1 1/8″. The thicknesses used most often are ¼", ⅝", and ⅜".

## Non-structural Particleboard

Non-structural particleboard is manufactured the same way as structural particleboard. The manufacturing process is similar to the process for hardboard, except that the wood particles for particleboard are not broken down into fibers as is done for hardboard.

Particleboard is used for many of the same functions as hardboard, such as countertops, shelving, cabinets, and sliding doors. It is often used as the inner core (center layer) of plywood panels. It also can be shaped into molding. Exterior grades can be used for roof soffits.

All common fastening methods can be employed with particleboard, including nails, staples, rivets, screws, bolts, glue, or a combination of these methods. Sheet-metal screws have better holding power in particleboard than conventional wood screws. (Pilot holes should be drilled before driving the screws.) If nails are used, spiral-shank or ring-shank nails are recommended.

# UNIT 8

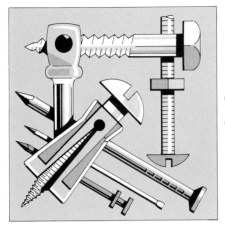

# Fastening Systems

Many kinds of metal fastening devices are used for construction purposes. Nails are still the most commonly used fastener; however, the use of staples to attach wood structural members is growing. For certain operations, screws and bolts are required. In addition, various metal devices exist for anchoring materials to concrete, masonry, and steel.

An important development in the building industry is the increasing use of adhesives (glues and mastics) in combination with, or in place of, nails and screws. Different types of adhesives are suitable for different types of jobs.

## NAILS

Nails come in many shapes and sizes, with a variety of heads, shanks, and points. See Figure 8–1. Some nails have greater holding power than others. Some have other special properties. Aluminum, stainless steel, and galvanized steel nails,

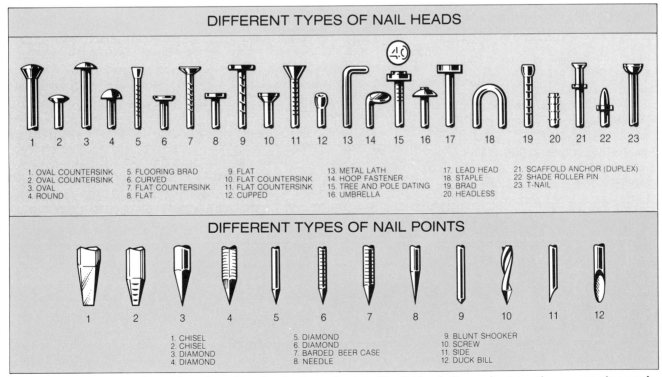

Figure 8–1. Nails come with a wide variety of heads, shanks, and points. The flat-head, diamond-point nails are most often used by carpenters.

for example, are used to fasten finish materials on the outside of a building because they are rust-resistant. These nails will not cause rust streaks on the surface of wood materials.

This unit covers the types of nails used most often for rough and finish carpentry work. Other, more specialized types of nails are discussed in later units.

## Penny System of Nail Size

Nail sizes are designated by a number and the letter *d*. Typical sizes are 6d, 8d, and 16d. A 6d nail is 2″ long, an 8d nail is 2½″ long, and a 16d nail is 3½″ long. See Figure 8–2. The letter *d* stands for *denarius*, an ancient Roman word for coin (or penny). At one time nail sizes were designated by the word *penny*, probably because when the penny system began in England, hundreds of years ago, nails were priced by how many pennies they cost per hundred. Smaller sizes of nails cost less per hundred than larger sizes.

## Nails for Rough Work

The nail used most often in wood frame and form construction is the *common wire nail*. See Figure 8–3. This type of nail is cut from wire and given a head and a point. It is available in sizes from 2d (1″ long) to 60d (6″ long).

The *box nail* is similar in appearance to the common nail, but its head and shank are thinner, so it is less likely to cause splits in wood. This type of nail is often used to fasten exterior insulation board and siding. It is available in sizes from 2d (1″ long) to 40d (5″ long). A disadvantage of the box nail is that it bends more easily when hammered.

The *double-head nail* (also known as the *duplex-head nail*

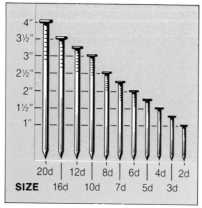
Figure 8–2. Examples of frequently used nail sizes.

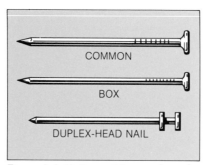
Figure 8–3. Nails used for rough work. The common wire nail is used most often.

or *staging nail*) is used for temporary construction such as formwork or scaffolding. The double head on this nail makes it easy to pull out when forms or scaffolding are torn down.

## Nails for Finish Work

Nails used for finish work are thinner than common nails. See Figure 8–4. They are used where appearance is important. The *finish nail* has a small, tulip-shaped head, which is easily driven below the surface of the lumber with a nail set. The finish nail is available in sizes from 2d (1″ long) to 20d (4″ long).

The *casing nail* is a thick version of the finish nail. Its head is slightly larger than that of the finish nail and is tapered toward the bottom. Casing nails are used to

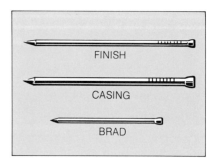
Figure 8–4. Nails used for finish work are thinner than nails used for rough work.

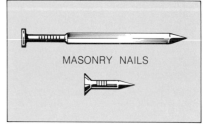
Figure 8–5. Masonry nails can be driven into concrete or masonry.

fasten heavier pieces of trim material.

Wire *brads* are identified by their length in inches rather than by the penny system. Their sizes range from ³⁄₁₆″ to 3″ long. They are thinner than finish or casing nails and are used with very light trim materials.

The *masonry nail* is made with a special hardened steel. See Figure 8–5. It is used to fasten wood to masonry (solid concrete, hollow concrete blocks, bricks, or stones). Masonry nails must be driven in perfectly straight or they may chip the masonry.

## Holding Power of Nails

When a nail is driven into wood, it compresses and pushes aside the wood fibers. After the nail is in place, the wood fibers spring back toward their original position. The pressure of the wood fibers against the surface of the nail gives the nail its holding

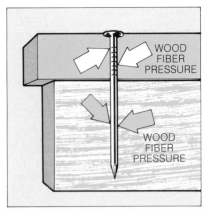

Figure 8–6. The arrows show how the wood fibers press against the shank of the nail. This pressure gives the nail its holding power.

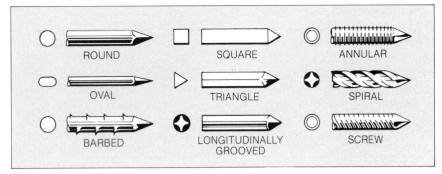

Figure 8–7. The first two shanks (round and oval) shown here are considered smooth shanks. The other types of shanks give greater holding power than smooth shanks.

power. See Figure 8–6.

Smooth-shank nails have sufficient holding power for most construction purposes. If greater holding power is needed, nails with different types of shanks are available. See Figure 8–7. Holding power is also greater with nails that are coated with cement, resin, or zinc.

## STAPLES

Today staples are often used where nails were used in the past. Staples are available in a variety of shapes and sizes. See Figure 8–8. They are usually used to fasten subflooring, sheathing, and paneling. Heavy-duty staples are driven in by electrically or pneumatically operated tools. Smaller staples are sometimes driven in by hand-operated tools.

## SCREWS

Screws provide greater holding power than nails. However, their use is limited, since they are too costly to be used as commonly as nails.

Carpenters use wood screws, sheet-metal screws, and machine screws. These screws are used for fastening hardware to

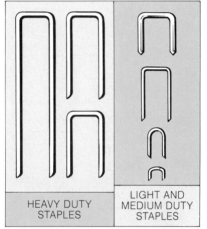

HEAVY DUTY STAPLES

LIGHT AND MEDIUM DUTY STAPLES

Figure 8–8. Heavy-duty staples are used to fasten plywood sheathing and subflooring. Light-duty and medium-duty staples are used for attaching molding and other interior trim.

wood or metal, attaching cabinets to walls, and fastening trim to metal surfaces. Additional types of screws used for special

purposes are discussed in later units.

Most screws are made of soft steel. Brass, bronze, and copper screws are also available. For decorative purposes and for matching different hardware finishes, steel screws come in many finishes, including nickel, chromium, silver plate, and gold plate.

### Wood Screws

Wood screws have flat, round, or oval heads. The screw head may have a single slot or a recessed cross slot (Phillips).See Figure 8–9. The Phillips screwhead is easier to grip than a single-slot screw head when driven by an electric screwdriver. It also has a more attractive appearance when in place.

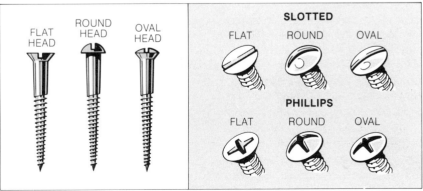

Figure 8–9. The three basic screwheads are flat, round, and oval. They have either a single slot or a recessed cross slot.

Wood screws range in size from ¼″ to 5″ long. The diameter (thickness) of the screw shank is identified by a gauge number. A higher gauge number indicates a thicker screw. Figure 8–10 shows screws with gauges ranging from 1 to 14.

## Self-tapping and Self-driving Metal Screws

In commercial construction, carpenters often work with materials that must be fastened to metal surfaces. Self-tapping screws are used for this purpose in metals from ¼″ to ½″ thick. See Figure 8–11. A hole smaller than the gauge of the screw must be drilled first. As the self-tapping screw is driven into the hole, it cuts threads in the metal.

Self-driving screws are a recent improvement over self-tapping screws. Mounted in an electric screwdriver, the self-driving screw drills a hole, cuts the threads, and fastens, all in one operation.

## Machine Screws

Machine screws are available with heads of various shapes. See Figure 8–12. They screw into threaded holes in the metal and have greater holding power than other types of screws that fasten to metal. Some examples of their uses are fastening door hinges, push plates, locks, and door closers to metal jambs and doors.

## BOLTS

Various bolts are used to fasten together heavy wood and metal materials. See Figure 8–13. All bolts require nuts. Whenever a nut bears against wood, a washer should also be used. The washer distributes the pressure over a wider area, and this prevents the nut from digging into the wood. A description of some types of bolts used in construction work follows:

*Carriage bolts* are used only in wood. The square section below the oval head of a carriage bolt is embedded in the wood as the nut is drawn up. This prevents the bolt from turning as the nut is tightened. See Figure 8–14.

*Machine bolts* have square or hexagonal heads. They are used to fasten together wood or metal pieces.

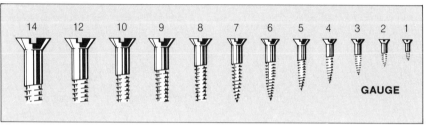

Figure 8–10. Wood screws of various gauges. A higher gauge number indicates a thicker screw.

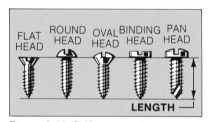

Figure 8–11. Self-tapping sheet-metal screws are used to fasten materials to metal surfaces.

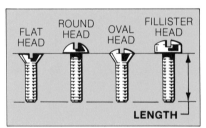

Figure 8–12. Machine screws have greater holding power than other types of screws that fasten to metal.

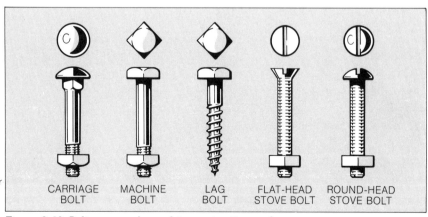

Figure 8–13. Bolts commonly used in construction work.

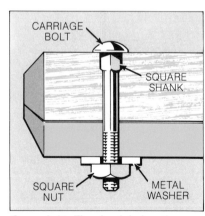

Figure 8–14. The shank below the head of a carriage bolt is embedded in the wood.

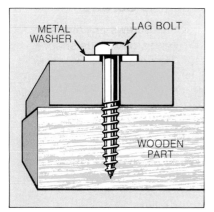

Figure 8–15. Lag bolts are often used when it is inconvenient or impossible to use a nut-and-bolt arrangement. Washers should be used under the head.

*Stove bolts* are used for lighter work. They have a smaller size range than the other types of bolts. They are available in lengths from ⅜″ to 6″ and in thicknesses from ⅛″ to ⅜″. Unlike other bolts, they have slotted flat or round heads. Stove bolts of shorter lengths, up to 2″, are threaded up to the head.

*Lag bolts* are not true bolts. They are actually heavy screws with square or hexagonal heads. See Figure 8–15. They are used to fasten heavy pieces of material into wood when a regular bolt-and-nut system will not work. A pilot hole must be drilled for the lag bolt, which is screwed in with a wrench.

## FASTENERS FOR HOLLOW WALLS

Several anchoring devices exist for fastening light materials to hollow walls. Examples of hollow walls are wood or metal partitions covered by plaster or plasterboard, and hollow-block ma-

sonry walls. One frequently used anchoring device is the *toggle bolt*. Another is the *screw anchor*, also known as the *expansion anchor*.

## Toggle Bolts

A toggle bolt consists of a machine screw with a wing-head nut that folds back as the entire assembly is pushed through a prepared hole in the wall. The wing head springs back inside the wall cavity. As the screw is tightened, the wing head is drawn against the inside surface of the finish wall material. See Figure 8–16. Spring-action wing-head toggle

*Star Expansion Company*

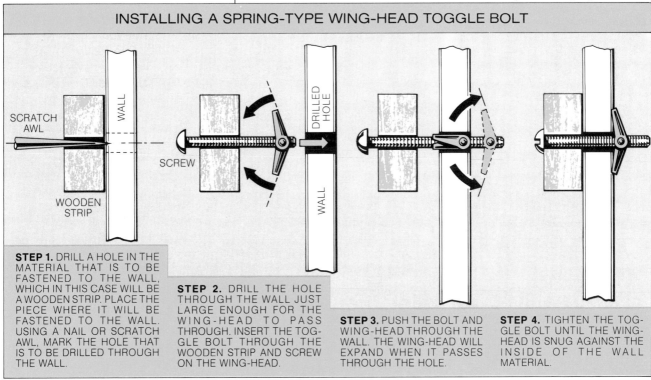

INSTALLING A SPRING-TYPE WING-HEAD TOGGLE BOLT

**STEP 1.** DRILL A HOLE IN THE MATERIAL THAT IS TO BE FASTENED TO THE WALL, WHICH IN THIS CASE WILL BE A WOODEN STRIP. PLACE THE PIECE WHERE IT WILL BE FASTENED TO THE WALL. USING A NAIL OR SCRATCH AWL, MARK THE HOLE THAT IS TO BE DRILLED THROUGH THE WALL.

**STEP 2.** DRILL THE HOLE THROUGH THE WALL JUST LARGE ENOUGH FOR THE WING-HEAD TO PASS THROUGH. INSERT THE TOGGLE BOLT THROUGH THE WOODEN STRIP AND SCREW ON THE WING-HEAD.

**STEP 3.** PUSH THE BOLT AND WING-HEAD THROUGH THE WALL. THE WING-HEAD WILL EXPAND WHEN IT PASSES THROUGH THE HOLE.

**STEP 4.** TIGHTEN THE TOGGLE BOLT UNTIL THE WING-HEAD IS SNUG AGAINST THE INSIDE OF THE WALL MATERIAL.

Figure 8–16. Installing a spring-action wing-head toggle bolt. The wing-head assembly folds back as it is pushed through the hole, then springs back inside the wall cavity.

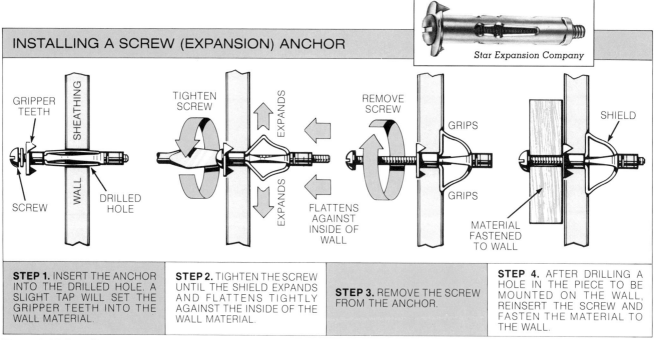

INSTALLING A SCREW (EXPANSION) ANCHOR

*Star Expansion Company*

GRIPPER TEETH

SHEATHING

WALL

SCREW

DRILLED HOLE

TIGHTEN SCREW

EXPANDS

EXPANDS

FLATTENS AGAINST INSIDE OF WALL

REMOVE SCREW

GRIPS

GRIPS

SHIELD

MATERIAL FASTENED TO WALL

**STEP 1.** INSERT THE ANCHOR INTO THE DRILLED HOLE. A SLIGHT TAP WILL SET THE GRIPPER TEETH INTO THE WALL MATERIAL.

**STEP 2.** TIGHTEN THE SCREW UNTIL THE SHIELD EXPANDS AND FLATTENS TIGHTLY AGAINST THE INSIDE OF THE WALL MATERIAL.

**STEP 3.** REMOVE THE SCREW FROM THE ANCHOR.

**STEP 4.** AFTER DRILLING A HOLE IN THE PIECE TO BE MOUNTED ON THE WALL, REINSERT THE SCREW AND FASTEN THE MATERIAL TO THE WALL.

Figure 8–17. Installing an expansion anchor in a hollow wall. As the screw is tightened, the shield spreads and flattens against the interior of the wall.

bolts are available with a variety of machine screw combinations. Common sizes range from ⅛″ to ⅜″ in diameter and 2″ to 6″ in length.

## Screw Anchors

The screw anchor (or expansion anchor) is used to fasten small cabinets, towel bars, drapery hangers, mirrors, electrical fixtures, and other lightweight items to hollow walls. It is inserted in a prepared hole. Prongs on the outside of the shield grip the wall surfaces to prevent the shield from turning as the anchor screw is being driven. As the screw is tightened, the shield spreads and flattens against the interior of the wall. See Figure 8–17. Various sizes of screw anchors can be used in hollow walls ⅛″ to 1¾″ thick.

## FASTENERS FOR SOLID CONCRETE (OR OTHER SOLID MASONRY) WALLS

An early method of fastening heavy materials to solid concrete

*Star Expansion Company*

Figure 8–18. Hand drill for drilling holes in masonry. The tapered shank fits into the socket of a drill holder.

*Star Expansion Company*

Figure 8–19. Two types of carbide-tipped, spiral-fluted masonry drills designed for use with a rotary drill motor.

(or other solid masonry) was to drive a wood plug into a hole drilled in the concrete, then fasten the material to the wall by means of a nail or screw driven into the plug. Another early method was to pour molten sulphur around a bolt set in a pre-drilled hole. These procedures were lengthy and undependable.

Today expansion anchors of various types make it possible to fasten heavy materials and

equipment securely and quickly to solid concrete (or other solid masonry) walls, floors, and ceilings. These devices range from light-duty lead, plastic, and fiber plugs to heavy-duty expansion shields.

Most expansion anchoring devices require that a hole equal to the outside diameter of the plug or shield be drilled in the concrete. The hole should be slightly deeper than the length of the plug or shield.

Holes in concrete can be drilled by hand with a drill such as the one shown in Figure 8–18. This drill is available in diameter sizes of ³⁄₁₆″ to ¾″. Its tapered shank fits into the socket of a drill holder. Drilling action is accomplished by striking the back end of the holder with a hammer, while rotating the holder about 30° to 40° in a continuous clockwise direction between each blow of the hammer.

A much easier way to drill holes in concrete is to use an electric drill. Figure 8–19 shows two types of carbide-tipped, spiral-fluted masonry drills designed for use with a rotary-type drill motor. The drills are available in lengths of 2½″ to 18″ and in diameter sizes of ⅛″ to 1½″. Figure 8–20 shows a core-type masonry drill that will cut through the hardest concrete and even through steel reinforcing rods inside the concrete. Holes (ports) on the sides permit the masonry debris to escape from the interior and be pushed up and carried to the surface. This drill is available in lengths of 6″ to 9½″.

Holes in masonry can also be drilled with an electrically driven rotary impact hammer. See Figure 8–21.

After a hole has been drilled, the anchoring device is placed inside the hole. As the screw, bolt, or nail is driven into the an-

chor, it *expands* and *presses* against the concrete. The pressure holds the anchor in place.

## Light-duty Anchors

Light-duty anchors are usually used to secure light items such as drapery brackets, telephones, electric fixtures, towel brackets, hooks, and other household accessories to solid masonry walls. They are often made of plastic or lead. The plastic anchors in Figure 8–22 are used with wood or sheet-metal screws. The molded anchor has a tapered cavity. It can be set in plasterboard as well as masonry. The tubular anchor has longitudinal slots that aid expansion and prevent turning or twisting of the plug.

The lead-alloy anchor in Figure 8–23 can be used with lag, wood, or sheet-metal screws. The longitudinal ribs of the anchor bite into the masonry as the screw progresses into the anchor.

**Hammer-driven Anchors.** A hammer-driven anchor does not

receive a screw or bolt. A special type of nail called an *expander pin* expands the anchor as it is hammered in. Figure 8–24 shows a medium-duty steel anchor. Its two-piece tubular shell is bound together with a lead-alloy band. The anchor expands as the nail-like expander is driven in.

To install this anchor, first drill a hole equal to the outside diameter of the anchor. The object to be fastened may be used as the template. Place the anchor, lead band first, through the object being fastened and into the hole in the masonry. The large flange should rest flush against the surface of the object being fastened. Place the nail into the anchor and drive it in.

Figure 8–25 shows a nylon

ANCHOR            EXPANDER PIN

*Star Expansion Company*

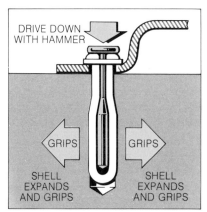

Figure 8–24. Installing a hammer-driven steel anchor. The anchor expands as the nail-like expander pin is driven in.

*Star Expansion Company*
Figure 8–20. Core-type masonry drill cuts through the hardest concrete and even through steel reinforcing rods. Holes on the sides permit the escape of masonry debris.

*Rockwell International, Power Tool Division*
Figure 8–21. Rotary impact hammers are widely used for drilling holes in concrete.

MOLDED HOLLOW ANCHOR

TUBULAR ANCHOR

*Star Expansion Company*
Figure 8–22. Light-duty plastic anchors are used with wood or sheet-metal screws.

*Star Expansion Company*
Figure 8–23. Light-duty lead-alloy anchor is used with lag, wood, or sheet-metal screws.

ANCHOR            EXPANDER PIN

*Star Expansion Company*
Figure 8–25. Hammer-driven nylon anchor. The slotted head on this anchor allows its removal with a screwdriver.

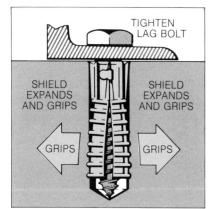

EXPANSION SHIELD        LAG BOLT

*Star Expansion Company*

MEDIUM-DUTY

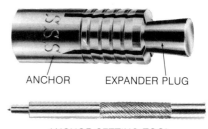

ANCHOR        EXPANDER PLUG

ANCHOR SETTING TOOL

*Star Expansion Company*

Figure 8–28. Steel anchor with expansion plug. This type of anchor is placed in a pre-bored hole and driven over the expander plug with a special setting tool.

TIGHTEN
LAG BOLT

SHIELD
EXPANDS
AND GRIPS

SHIELD
EXPANDS
AND GRIPS

GRIPS        GRIPS

Figure 8–26. Installing a lag-bolt expansion shield. As the bolt is tightened, the shield expands.

HEAVY-DUTY

*Star Expansion Company*

Figure 8–27. Machine-bolt expansion shields. The heavy-duty shield has expansion action at both ends to distribute the anchor load.

hammer-driven anchor. Its installation is similar to that of the steel anchor. The steel pin expander is like a nail except for the specially designed thread in the shank and the screwdriver slot in the head. The screwdriver head makes it possible to remove this fastener with a screwdriver.

## Expansion Shields

Various expansion shields are available for medium to heavy loads. They receive lag or machine bolts. As the bolts are tightened, the shields expand. Figure 8–26 shows a lag-bolt expansion shield. This medium-duty anchor is made of zinc alloy. It is available in sizes accommodating lag screws of ¼" to ¾" diameter.

To install this anchor, first drill a hole equal to the outside diameter of the shield. It should be as deep as the expansion shield is long, plus ½" or more. Place the expansion shield, ribbed end first, into the hole. A portion of the shield will protrude above the surface; therefore, hammer the

shield into the hole until it is flush with the surface. Position the object being fastened and begin to screw in the lag screw. If the lag screw torques up before the head of the lag screw is against the object being fastened, drive the screw by hammering against the head until it is flush with the object. After this, finish tightening the lag screw with a wrench.

Figure 8–27 shows two machine-bolt expansion shields. They require holes equal to their outside diameters and deep enough so that the shields will be flush or slightly below the concrete surface. The shields expand as the bolts are tightened. This type of fastener can be removed if desired, and the hole grouted to refinish the concrete surface.

The medium-duty anchor in Figure 8–27 is available in short-length sizes of 1½" to 2½" and in long-length sizes of 1¼" to 3½". Short-length sizes are for anchoring in good grades of concrete when the layer of concrete limits the depth of the hole. Long-length sizes are for anchoring in poorer grades of concrete (where extra anchoring strength is required) when the layer of concrete allows a deep enough hole for the longer bolt. Short-length anchors are available in

sizes accommodating bolts of 5⁄16" to 5⁄8" diameter. Long-length anchors are available in sizes accommodating bolts of ¼" to ¾" diameter.

The heavy-duty double expansion shield in Figure 8–27 is available in sizes accommodating bolts of ¼" to 1" diameter.

Figure 8–28 shows a steel anchor with an expansion plug and a setting tool. This type of anchor is placed in a pre-bored hole and driven over the expander plug with the special setting tool. Anchors are available in sizes accommodating bolts of ¼" to ¾" diameter.

To install this anchor, first drill a hole equal to the outside diameter of the anchor. Preassemble and place the anchor and plug into the hole. With the setting tool and a hammer, drive the anchor over the expander plug. Position the object to be fastened over the anchor and bolt it into place.

Another type of expansion shield often used today is the self-drilling anchor. The anchor is placed in a special chuck head that is adapted for a rotary impact hammer. The sharp teeth at the end of the anchor drill the hole in the concrete. The anchor is secured with an expander plug. Figure 8–29 shows the procedure for installing a self-drilling anchor. The chuck end of

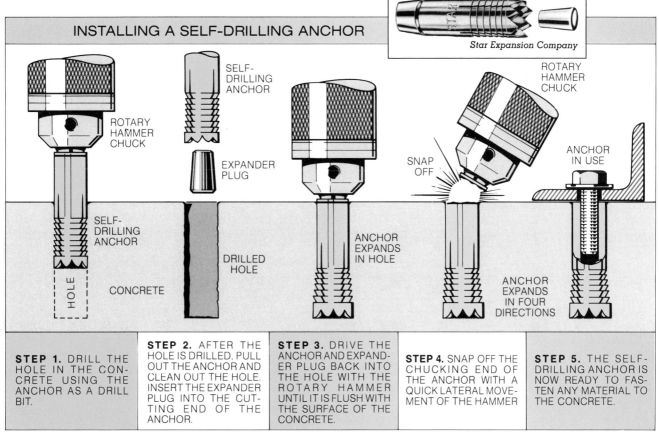

INSTALLING A SELF-DRILLING ANCHOR

*Star Expansion Company*

ROTARY HAMMER CHUCK

SELF-DRILLING ANCHOR

EXPANDER PLUG

SELF-DRILLING ANCHOR

HOLE

CONCRETE

DRILLED HOLE

ANCHOR EXPANDS IN HOLE

SNAP OFF

ROTARY HAMMER CHUCK

ANCHOR IN USE

ANCHOR EXPANDS IN FOUR DIRECTIONS

**STEP 1.** DRILL THE HOLE IN THE CONCRETE USING THE ANCHOR AS A DRILL BIT.

**STEP 2.** AFTER THE HOLE IS DRILLED, PULL OUT THE ANCHOR AND CLEAN OUT THE HOLE. INSERT THE EXPANDER PLUG INTO THE CUTTING END OF THE ANCHOR.

**STEP 3.** DRIVE THE ANCHOR AND EXPANDER PLUG BACK INTO THE HOLE WITH THE ROTARY HAMMER UNTIL IT IS FLUSH WITH THE SURFACE OF THE CONCRETE.

**STEP 4.** SNAP OFF THE CHUCKING END OF THE ANCHOR WITH A QUICK LATERAL MOVEMENT OF THE HAMMER

**STEP 5.** THE SELF-DRILLING ANCHOR IS NOW READY TO FASTEN ANY MATERIAL TO THE CONCRETE.

Figure 8–29. Installing a self-drilling anchor. This type of anchor is placed in a special chuck head that is adapted for a rotary impact hammer.

*Star Expansion Company*

Figure 8–30. Stud-bolt anchor with external thread. It is held in position with an expander plug.

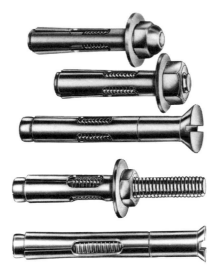

*Star Expansion Company*

Figure 8–31. Sleeve-type stud-bolt anchors. When the nut or screw head of the bolt is tightened, the expander sleeve wedges outward and locks the unit in place.

the anchor can be snapped off with a quick lateral movement of the rotary impact hammer. Another way to remove the chuck end is to use a snap-off tool. Self-drilling anchors are available in sizes accommodating bolts of ¼″ to ¾″ diameter.

## Stud-bolt Anchors

Stud-bolt anchors are considered the ideal fasteners for large equipment and machinery. The equipment to be fastened is set

in its final position, and the holes are drilled in the concrete through the mounting lugs of the

equipment. The stud bolts are inserted and secured.

Figure 8–30 shows a stud-bolt anchor with external threads. It is held in position with an expander plug. It is available in sizes accommodating stud bolts of ¼″ to ¾″.

To install this anchor, first drill holes through the mounting lugs of the equipment to be fastened. The holes should be the same diameter as the anchor to be used. Place the stud-bolt anchor, plug end first, into the hole. Using an anchor setting tool, drive the stud bolt over the expander plug. Fasten the equipment with a nut and washer.

Figure 8–31 shows several sleeve-type stud-bolt anchors. They come assembled in a number of different ready-to-use units. When the nut or screw

51

head of the bolt is tightened, the expander sleeve wedges outward and locks the unit into position.

Figure 8–32 shows a wedge-type stud-bolt anchor. This device is hammer-driven into the hole drilled in the wall. The anchor wedge is banded around the lower portion of the bolt. It is designed to allow the anchor to be driven into the hole yet offer resistance to it being pulled out. The wedge expands outwardly as the tapered portion of the stud bolt is drawn into it when the nut is tightened.

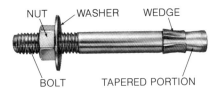

*Star Expansion Company*

Figure 8–32. Wedge-type stud-bolt anchor.

## Chemset Anchors

Chemset anchors consist of a sealed glass capsule, a specially designed threaded rod, and a nut and washer. The capsules are available in six different sizes to accommodate various rod diameters. Synthetic resin and quartz aggregate filler is contained in each capsule.

As the capsule is crushed, a chemical reaction occurs and the mixture fills the gap between the threaded rod and sides of the hole penetrating the pores of the concrete. Full holding is achieved after the resin hardens. See Figure 8–33.

## DRIVE PINS AND STUDS

*Drive pins* and *studs* are also widely used devices for fastening materials to concrete and other types of masonry walls. A drive pin is a nail made of hardened steel that is driven directly into

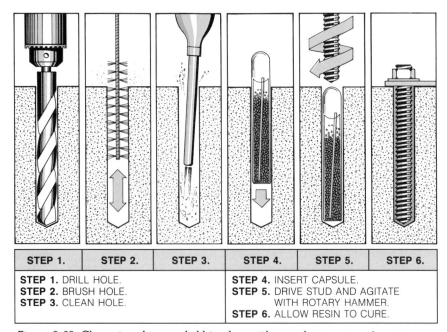

| STEP 1. | STEP 2. | STEP 3. | STEP 4. | STEP 5. | STEP 6. |
|---|---|---|---|---|---|

STEP 1. DRILL HOLE.
STEP 2. BRUSH HOLE.
STEP 3. CLEAN HOLE.
STEP 4. INSERT CAPSULE.
STEP 5. DRIVE STUD AND AGITATE WITH ROTARY HAMMER.
STEP 6. ALLOW RESIN TO CURE.

Figure 8–33. Chemset anchors are held in place with a synthetic resin and quartz aggregate filler.

the concrete. A stud consists of two parts. One end is the nail that is embedded in the concrete. The other end is a threaded bolt that will receive a nut. These fastening devices can be placed manually, although they are usually shot into the concrete with a *powder-actuated* fastening tool.

The hammer-driven tool shown in Figure 8–34 is used for manual placement. The drive pin or stud is placed in the recessed barrel of the tool. The steel plunger at the other end is struck with a hammer to drive the pin or stud into the wall.

ANVIL ROD IS PULLED BACK

*Ramset Fastening Systems*

Figure 8–34. Hammer-driven tool used to manually drive studs and pins into masonry walls.

The powder-actuated fastener with drive pins or studs is the quickest method for anchoring into concrete or other types of masonry. It can also be used to

drive pins and studs into steel beams and columns. The powder-actuated tool is a gun with a powder-filled shell that shoots the pin or stud into the wall. Many kinds of pins and studs can be driven in by a powder-actuated tool. See Figure 8–35.

## METAL CONNECTORS

Metal wood connectors strengthen the ties between members used in wood-framed construction. Strength is particularly needed in areas of the country where buildings may be subjected to earthquakes, hurricanes, or tornadoes. Evidence exists that when serious damage to a building occurs, it is usually a result of framing members being pulled loose at their *joints* rather than being broken in their *length.* Metal connectors used in addition to the usual nailing methods can help prevent or limit this type of structural damage.

Figure 8–36 shows some metal connectors used to help fasten structural members together. Many of these devices are discussed in detail in later units covering different stages of wood-framed construction.

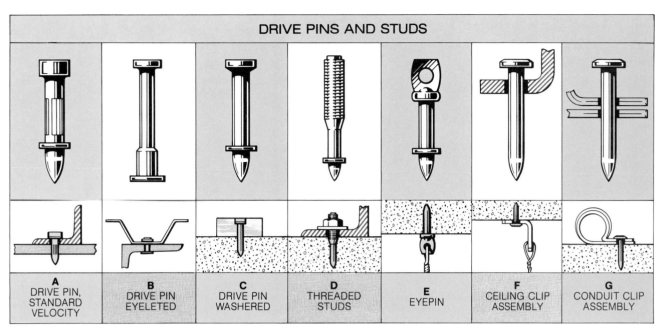

Figure 8-35. Many types of drive pins and studs can be driven into concrete or other masonry with a powder-actuated tool.

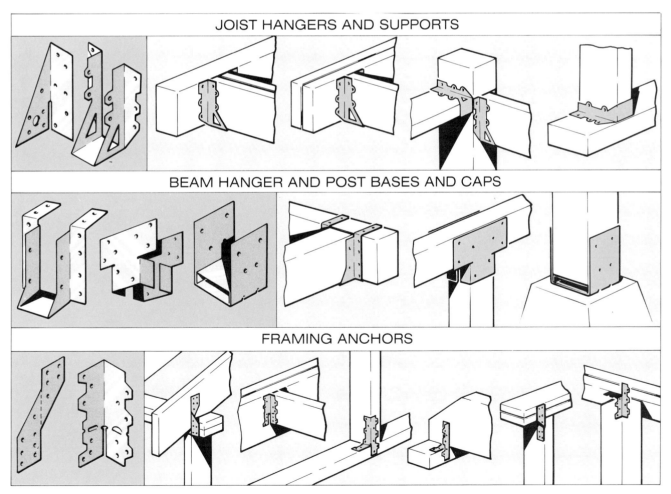

Figure 8-36. Some typical metal fasteners used to tie together members in wood-framed construction.

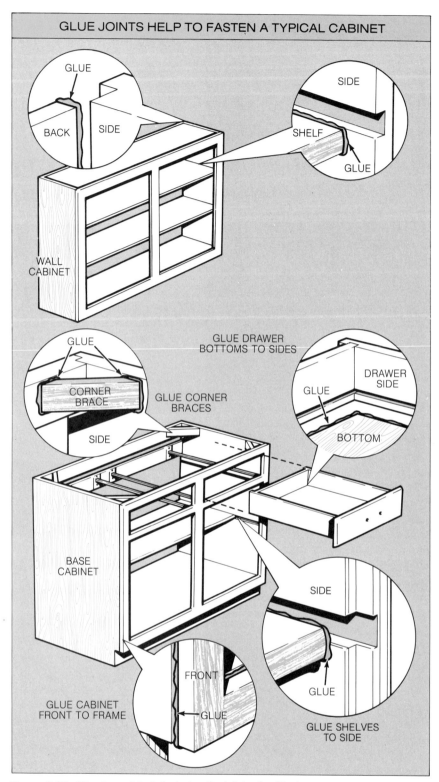

**GLUE JOINTS HELP TO FASTEN A TYPICAL CABINET**

GLUE

BACK    SIDE

SIDE

SHELF

GLUE

WALL CABINET

GLUE

CORNER BRACE

SIDE

GLUE CORNER BRACES

GLUE DRAWER BOTTOMS TO SIDES

GLUE

DRAWER SIDE

BOTTOM

BASE CABINET

SIDE

GLUE CABINET FRONT TO FRAME

FRONT

GLUE

GLUE

GLUE SHELVES TO SIDE

Figure 8–37. Glue is used to help fasten the joints of cabinets.

## ADHESIVES

Several types of adhesives are available for construction purposes. Some are *glues*, which have a plastic base, and others are *mastics*, which have an asphalt, rubber, or resin base. (The term *glue*, however, is often used for mastic systems.)

Method of application, drying time, and bonding characteristics vary among adhesives. Some are more resistant to moisture and to the extremes of hot and cold temperatures than others. Also, some are highly flammable, so the work area must be well ventilated. Others are highly irritating to the skin, so skin contact must be avoided. Manufacturer's instructions should always be followed in using adhesives.

### Glues

Glues are primarily used to hold together joints in mill and cabinet work. See Figure 8–37. Most glues have a plastic base. They are sold in a liquid form or as a powder to which water must be added. Many types are available under different brand names. A description of some of the more popular products follows:

*Polyvinyl resin*, better known as *white glue*, comes in different sizes of ready-to-use plastic squeeze bottles. It has a good rating for bonding wood together and sets up (dries) quickly after being applied. It is not waterproof. Do not use this glue on work that will be subject to constant moisture.

*Urea resin* is a plastic resin glue that comes in a powder form. The required amount is mixed with water at the time it is needed. It makes an excellent bond for wood and has fair water resistance.

*Phenolic resin* has excellent water resistance and temperature resistance. It is often employed for bonding the veneer layers of exterior grade plywood.

*Resorcinal resin* has excellent water resistance and temperature resistance and makes a very strong bond. It is frequently used for bonding the wood layers of glued, laminated timbers.

*Contact cement* is used to bond plastic laminates to wood surfaces. This adhesive has a neoprene rubber

Figure 8-38. Tubes that fit into caulking guns are the most convenient way of applying mastic adhesives.

board) directly to wall studs. They are also used to fasten gypsum board to furring strips or directly to concrete or masonry walls. Because nails are not used, there are no nail identations. See Figure 8-39.

**Wall Paneling.** Pre-finished wall panels, frequently used in both commercial and residential construction, present a neater appearance if nails are not driven through them. Mastic adhesives make it possible to apply paneling with very few or no nails at all. The panels can be bonded to studs, furring strips, or directly against concrete or masonry walls.

## Flooring Systems

In flooring systems commonly referred to as *glued flooring systems*, mastic adhesives (not glue) are used in addition to nails or staples to fasten plywood panels to the floor joists. The adhesive helps eliminate squeaks, bounce, and nail popping. It also increases the stiffness and strength of the floor unit.

base. Because it bonds rapidly, it is very useful for joining parts that cannot be clamped together.

## Mastics

The use of mastics has become widespread in construction. Mastics have a thicker consistency than glues and have an asphalt, rubber, or resin base. They are sold in cans, tubes, or canisters that fit into hand-operated or air-operated caulking guns. See Figure 8-38.

**Masonry and Concrete.** Mastic adhesives are used to bond materials directly to masonry or concrete walls. If furring strips are required on a wavy wall, they can be applied with mastic rather than by the more difficult procedure of driving in concrete nails. Also, insulation materials can be permanently fastened to masonry and concrete walls with a mastic adhesive.

**Drywall.** Mastic adhesives are used to bond drywall (gypsum

*National Gypsum Company*

Figure 8-39. Drywall panels may be applied with mastics to masonry walls.

## FLOOR SYSTEMS

Mastic adhesives can provide an important structural function for certain types of floor systems. They are generally called *glued floor systems,* although mastic adhesive (not glue) is used to fasten the floor panels to the joists. Field-glued floors can be laid down quickly and efficiently, even during cold weather conditions, using standard construction materials and procedures.

In glued floor systems, a bead of adhesive is applied with a caulking gun to the surface of the joists before each panel is placed. Glued floor systems frequently use panels with tongue-and-groove edges. In this case the adhesive should be spread in the grooved edges of a row

of floor panels before the tongues of the next row of panels are inserted in the grooves. The floor panels must then be nailed down before the adhesive sets. See Figure 8–40. The setting time varies, and manufacturers' recommendations should be followed. As a general rule, setting time accelerates during warm weather.

Using an adhesive in addition to nailing produces a bond strong enough to cause the floor and joists to behave like integral T-beam units. Floor stiffness is significantly increased, particularly when panels with tongue-and-groove edges are used. Glued tongue-and-groove edges also provide improved sound control by reducing the number of cracks that leak airborn noises. The

use of adhesives also helps to eliminate squeaks, bounce, and nail popping.

The adhesives used should conform to the performance specifications recommended for use with glued floor systems. A number of brands is available from building supply dealers. When using an adhesive, always follow the application recommendations of the adhesive manufacturer. (Additional information regarding glued floor systems is provided in Unit 42.)

Mastic adhesives are also used to fasten wood screeds to concrete floors. These screeds provide a nailing surface for wood flooring materials. This procedure is described and illustrated in Unit 62.

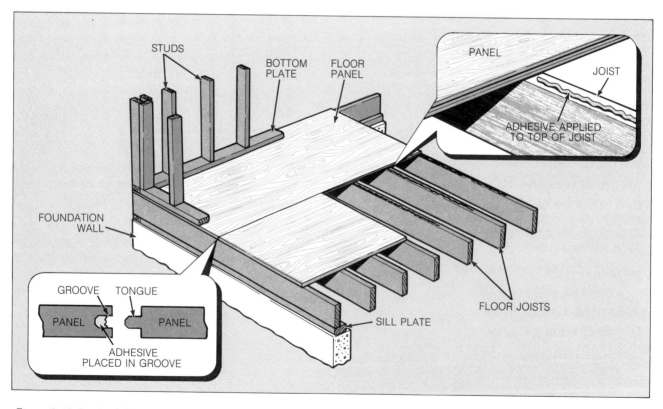

Figure 8–40. In glued floor systems, where panels with tongue-and-groove edges are used, mastic adhesive is placed in the grooves and on top of the joists.

# SECTION 3

# Hand Tools

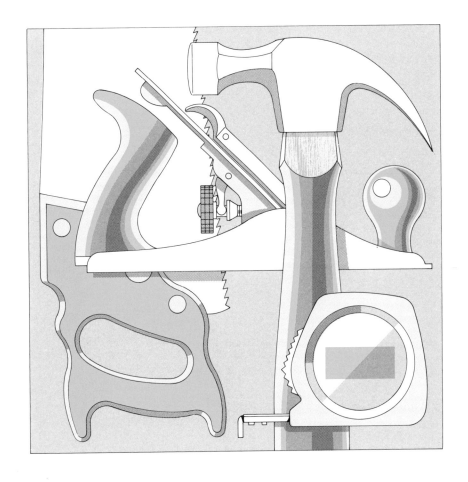

Good carpenters take pride in the quality and condition of their tools and are careful to use them properly. Tools in good condition, correctly selected for the job at hand, allow a carpenter to produce a greater amount of high quality work with less effort. Despite the increasing use of power tools, hand tools still play an important role in construction. They are used for many different operations. Tapes, levels, and squares are used for layout and measuring work. Hammers and screwdrivers are used to drive fasteners (nails, screws, and staples) that hold materials together. Other hand tools are used for cutting, boring, planing, and smoothing. Certain devices are also used to hold or clamp materials that are being worked upon. In addition, there are special hand tools for stripping and prying apart lumber.

This section does not attempt to describe all woodworking tools but instead concentrates on the basic ones needed to perform carpentry work. Other, more specialized tools used by carpenters are discussed where appropriate in units of other sections.

Most carpentry tools can be categorized as either *rough* or *finish* tools. The framing of a building and the construction of concrete forms require rough tools. Trim carpentry requires finish tools.

Working carpenters must have some kind of container in which to carry their tools. A large tool case is necessary to store a full set of tools. Since such a case is heavy to carry, it is usually kept in the carpenter's vehicle or in a safe storage area on the job site. Then an additional hand box is used to carry just the tools required for a particular job.

Many job-site injuries are related to the use of hand tools. Usually these injuries are a result of (1) failure to use the right tool for the job, (2) failure to use the tool properly, or (3) failure to keep the tool in a proper condition.

# UNIT 9

# Measuring and Layout Tools

Accurate measurement and layout are of vital importance in construction work. Various tapes, squares, levels, and marking devices are available for this purpose.

## TAPES AND RULES

Tapes and rules are used to measure the lengths and widths of materials to be cut. They are used to establish the locations of, and distances between, walls, floors, ceilings, and other structural parts of a building.

### Tape Rule

Also called a *pocket tape* or *push-pull rule,* the tape rule is probably used more often than any other measuring tool on the job. See Figure 9–1. It is available in many different lengths, although 12′, 16′, 20′, and 25′ seem to be the most popular sizes. The blade of the tape is usually ¾″ wide. Some 25′ types feature a 1″ blade. This wider blade offers the advantage of remaining rigid for an unsupported distance up to 7′. One edge of the blade in most tape rules is marked off in inches and the

other edge in feet and inches. A special identifying mark is placed every 16″ for the 16″ O.C. (on-center) layout of studs and floor or ceiling joists. Some tape rules give the English (customary) measurement (feet, inches, and inch-fractions) on one edge of the blade and the metric measurement (meters and millimeters) on the other edge.

## Steel Tape

The steel tape is used for longer distances, as shown in Figure 9–2. Most carpenters use a steel tape that has a blade measuring up to 50′ or 100′. The blade is normally marked off in feet, inches, and eighths of an inch.

Figure 9–1. A tape rule is used for many measuring jobs. Here it is being used to measure for a piece of drywall.

*Stanley Tools*

Figure 9–2. A steel tape is used to measure longer distances. Here it is being used to measure across a deck.

## Folding Wood Rule

The folding wood rule is an older measuring device, at one time considered a "must" in the carpenter's tool box. See Figure 9–3. Most construction workers now use tape rules instead of folding wood rules.

*Stanley Tools*

Figure 9–3. Traditional folding rule. Most construction workers now use tape rules instead.

## LEVELING AND PLUMBING TOOLS

The term *leveling* refers to horizontal planes. A perfectly level surface is a flat plane with no high or low points. The term *plumbing* refers to a vertical position. A plumb line will always be at a right angle (90°) to a level surface. See Figure 9–4.

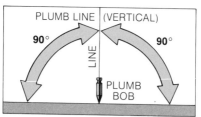

Figure 9–4. A plumb (vertical) line will always be at a 90° angle to a level (horizontal) line.

## Spirit Level

Also referred to as a *hand level,* the spirit level shown in Figure 9–5 is made of aluminum. Most carpenters today prefer these lightweight aluminum or magnesium types to the older wood levels.

Three slightly curved vials are set in the level and are protected by glass or plastic covers. The vials are partially filled with alcohol (spirit). The liquid contains an air bubble. The vials at each end of the level are used for plumbing purposes. When the level is held in a vertical position and the air bubbles are centered between the two lines marked on the end vials, the level is exactly plumb. The center vial of the level is always used for leveling purposes. The air bubble will be between the two lines when the tool is exactly level. See Figure 9–6.

Spirit levels are available in different lengths. The 24″ or 28″ level is used most often by carpenters. It fits in a standard size tool box. Levels in 48″, 72″, and 78″ sizes are also available. Figure 9–7 shows a level that has a

*Stanley Tools*

Figure 9–5. Aluminum spirit level. Most construction workers prefer these lightweight aluminum (or magnesium) spirit levels to older wood types.

magnetic strip on one side. This strip holds the tool to a steel surface.

A spirit level should be handled with care, since its vials and their protective covers are easily broken. At one time, it was necessary to return a level to the factory for repairs when a vial broke. Today levels have easily replaceable vials.

A new level should always be checked for exactness. Later it should be rechecked periodically, because it can become inaccurate when used over a period of

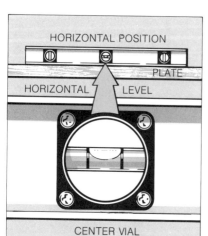

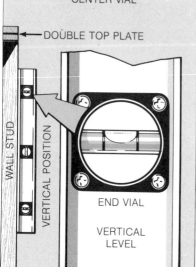

Figure 9–6. Plumbing and leveling with a spirit level. The vials at each end are used for plumbing. The vial in the center is used for leveling.

*Stanley Tools*

Figure 9–7. This magnetic level sticks to the side of the metal door frame, freeing both hands of the carpenter to set the frame in place.

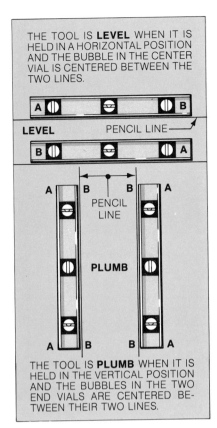

THE TOOL IS **LEVEL** WHEN IT IS HELD IN A HORIZONTAL POSITION AND THE BUBBLE IN THE CENTER VIAL IS CENTERED BETWEEN THE TWO LINES.

LEVEL       PENCIL LINE

PENCIL LINE

PLUMB

THE TOOL IS **PLUMB** WHEN IT IS HELD IN THE VERTICAL POSITION AND THE BUBBLES IN THE TWO END VIALS ARE CENTERED BETWEEN THEIR TWO LINES.

Figure 9-8. Checking a level for accuracy. This procedure should be repeated periodically because a level can become inaccurate over a period of time.

time. Figure 9-8 shows how to test a level for exactness.

To check the center vial, place the level against a flat, horizontal surface and adjust it so that the bubble is directly in the middle of the vial. Draw a line along the bottom of the level, marking the flat surface that it is resting upon. Reverse the level so that the end that was formerly on the left (labeled A in Figure 9-8) is now on the right. Align the bottom of the level with the line that was drawn on the surface the level is resting upon. If the bubble is directly in the middle of the vial, the level is accurate.

The top and bottom vials are checked the same way the center vial was checked, except that the level is held against a vertical rather than a horizontal surface.

**Using a Straightedge with a Level.** A hand level is accurate only up to its length. Use a straightedge along with the level

when leveling or plumbing over long distances. A straightedge is usually a piece of lumber that is perfectly straight and is the same width from one end to the other. A piece of ¾" plywood with a block at each end is often used for this purpose. See Figure 9-9.

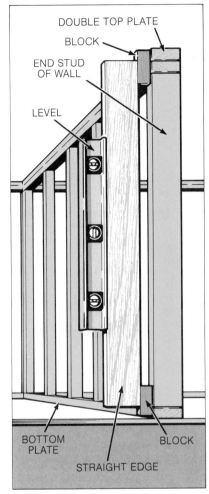

Figure 9-9. Plumbing a wall with a straightedge and level. A piece of ¾" plywood with a block at each end is often used for this purpose.

**Line Level.** Another way to level long distances is to use a line level. See Figure 9-10. It is hooked over a tightly stretched string. However, this is not always an accurate method, since there may be some sway in the line from which the level is hung.

## Plumb Bob and Line

A plumb bob and line level are used for plumbing greater

heights than can be handled by a level and straightedge. Figure 9-11 shows how a plumb bob is used. The wall is plumb when the distance between the line and the top of the wall (labeled A in Figure 9-11) is the same as the distance between the point of the plumb bob and the bottom of the wall (labeled B). Plumb bobs used by carpenters range in weight from 6 to 12 ounces.

*Stanley Tools*

Figure 9-10. A line level has hooks so that it can be attached to a tightly stretched string.

*Stanley Tools*

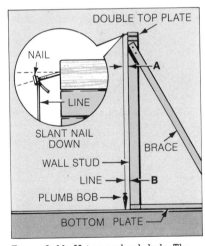

Figure 9-11. Using a plumb bob. The wall is plumb when the distance between the line and the top of the wall (A) is the same as the distance between the point of the plumb bob and the bottom of the wall (B).

## SQUARING TOOLS

A 90° (square) angle is the one used most often for cutting and

Figure 9–12. The pieces that are nailed together to make up these framed walls all have square cuts at their ends.

A

B

Figure 9–13. The pieces of casing being placed around this door opening have 45° miter cuts.

*Stanley Tools*

Figure 9–14. The combination square is the squaring tool used most often by carpenters.

fastening together construction materials. The pieces that are nailed together to make a framed wall all have square cuts at their ends. See Figure 9–12. A 45° angle is also used frequently, particularly when fitting finish pieces such as the trim around a window opening. See Figure 9–13.

## Combination Square

A combination square is used to mark both 90° and 45° angles. See Figures 9–14 and 9–15. It is small enough to fit into the back pocket of a pair of carpenter overalls, or it can be hung from a loop provided with leather carpenter aprons. The blades of most combination squares are 1′ long and are marked off in inches and inch-fractions. This tool can be used as a marking gauge as shown in Figure 9–15.

## Framing Square

Also called a *steel square* or *rafter square,* the framing square is a very valuable measuring and layout tool. The *blade* of the framing square (also called *body*) is 2″ wide and 24″ long. The *tongue* is

C

Figure 9–15. Using a combination square to (A) mark a 90°angle, (B) mark a 45° angle, and (C) draw a straight line.

1½″ wide and 16″ long. The outside corner of the square is the *heel.* The face of the square normally includes the manufacturer's name at the corner of the square and may include a rafter table on the blade and an octagon table on the tongue. The opposite side of the square is the *back* of the square and may include an essex table on the blade and a brace table on the tongue. See Figure 9–16.

The inches on the outside edges of the face side of the square are divided into standard ¹⁄₁₆″ graduations. The inches at the inside edges of the face side are divided into ⅛″ graduations. The inches at the outside edges of the back of the

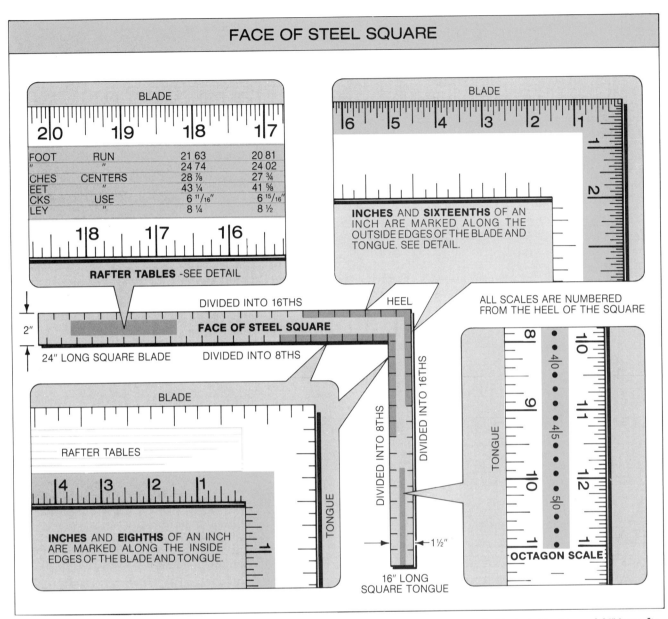

## FACE OF STEEL SQUARE

BLADE

| FOOT | RUN | 21 63 | 20 81 |
| " | " | 24 74 | 24 02 |
| CHES | CENTERS | 28 ⅞ | 27 ¾ |
| EET | " | 43 ¼ | 41 ⅝ |
| CKS | USE | 6 ¹¹⁄₁₆″ | 6 ¹⁵⁄₁₆″ |
| LEY | " | 8 ¼ | 8 ½ |

RAFTER TABLES -SEE DETAIL

BLADE

INCHES AND SIXTEENTHS OF AN INCH ARE MARKED ALONG THE OUTSIDE EDGES OF THE BLADE AND TONGUE. SEE DETAIL.

DIVIDED INTO 16THS

HEEL

ALL SCALES ARE NUMBERED FROM THE HEEL OF THE SQUARE

2″

FACE OF STEEL SQUARE

24″ LONG SQUARE BLADE

DIVIDED INTO 8THS

DIVIDED INTO 8THS

DIVIDED INTO 16THS

TONGUE

OCTAGON SCALE

BLADE

RAFTER TABLES

INCHES AND EIGHTHS OF AN INCH ARE MARKED ALONG THE INSIDE EDGES OF THE BLADE AND TONGUE.

TONGUE

1½″

16″ LONG SQUARE TONGUE

Figure 9–16. A carpenter's framing square (steel square) is used for many different purposes. Its blade is 2″ wide and 24″ long. Its tongue is 1½″ wide and 16″ long (continued next page).

square are divided into ¹⁄₁₂″ graduations. This is very useful when making scale layouts of 1″ = 1′-0″. In this case the inch represents a foot and each ¹⁄₁₂ graduation represents an inch. The inside edge of the blade of the back of the square is divided into ¹⁄₁₆″ graduations. However, the inside edge of the tongue is divided into ¹⁄₁₀″ graduations, also convenient for some types of layout.

Framing squares are used to check or square lines across wider boards or when greater

accuracy is required than is possible with a combination square. It is also used to mark 45° angles across wide boards and check inside corners for squareness. See Figure 9–17. To mark a 45° angle across a wide board, line up a number on the blade with the edge of the board, then line up the same number on the tongue with the edge of the board.

The framing square can also be used to lay out and mark the spacing of studs and joists 16″ and 24″ on centers (O.C.). To

lay out marks 16″ O.C., place the end of the tongue at the starting point of the layout and square a line at the heel. Move the square in the direction of the layout, place the end of the tongue on the first 16″ O.C. mark and square the next line at the heel. Continue this process until the layout is completed. To lay out marks 24″ O.C., follow the same procedure using the 24″ blade of the square instead of the tongue.

The framing square is used to lay out and mark the angle cuts

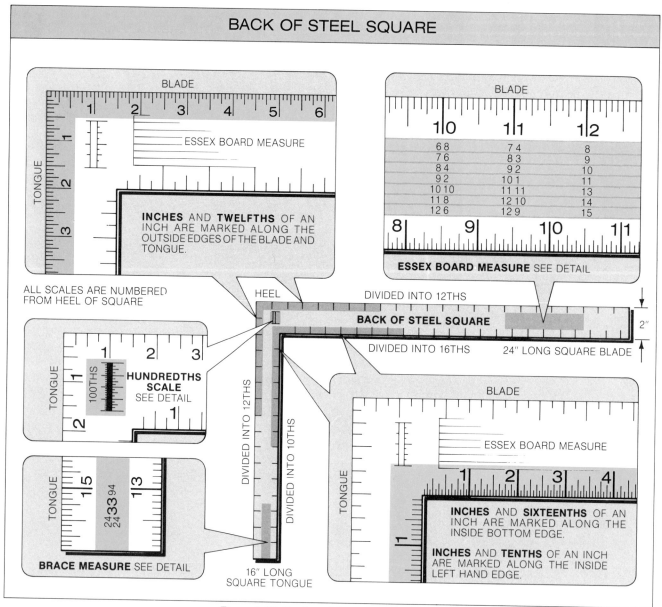

Figure 9–16. Continued from previous page.

for roof rafters. This is done by lining up figures corresponding to the unit rise and unit run of the roof on the tongue and blade of the square (see Section 10). Another important function of the framing square is for laying out and marking the risers and treads of stair stringers. This is done by marking the unit rise and run of the stairway by aligning the corresponding figures on the tongue and blade of the square (see Section 14).

The two surfaces of a traditional framing square contain a number of tables. (All these tables will not always appear on less expensive squares.) A complete set of tables consists of the *Rafter Table, Octagon Scale, Essex Board Measure Table,* and the *Brace Measure Table.* Except for the Rafter Table, the other tables are rarely used by construction carpenters today.

The Rafter Table, located on the face of the blade, enables the carpenter to find the lengths of the roof rafters by giving the *length per foot of run* for roof unit rises ranging from 2″ to 18″. The *difference in length of* *jacks* is also given. The procedures for using the Rafter Table are explained in Section 10.

The Octagon Scale is found at the center of the face of the tongue. Its main purpose is to lay out octagonal shapes. The Essex Board Measure Table is on the back of the blade. It is used to calculate and convert linear feet measurement to board feet measurement. The Brace Measure table, on the back of the tongue gives the diagonal lengths of braces based on some commonly used vertical and horizontal measurements.

Figure 9–17. Layout functions using a framing square include (A) squaring a line across a wide board, (B) marking a 45° angle across a wide board, and (C) checking an inside corner for squareness.

ferent rafter cuts. The details of this procedure are explained in Section 10.

## Sliding T-Bevel

The sliding T-bevel, also called a *bevel square,* is another basic layout tool considered essential by experienced carpenters. It can be set to any angle, and it is useful for transferring angles. See Figure 9–18. The wing nut tightens the blade in place.

## Other Squaring Tools

Another type of squaring tool used by carpenters is the *try square.* See Figure 9–19. It performs a similar function to the combination square. However, most carpenters find the combination square more convenient.

The *angle divider* is used to lay out cuts for joints that meet at angles other than 90° or 45°. See Figure 9–20. It is frequently used to make miter cuts on molding for walls that meet at angles other than 90° or 45°.

Two squaring tools that have recently appeared on the market are known by the brand names of *Speed Square* and *Super Square.*

The Speed Square can be used to mark 90° and 45° angles. It also enables the user to mark angle cuts for roof rafters. See Figure 9–21.

The Super Square combines

Figure 9–18. Marking a board with a sliding T-bevel. This tool can be set for any angle. The wing nut tightens the blade in place.

Figure 9–20. An angle divider is used to lay out miter cuts on molding for walls that meet at angles other than 90° or 45°. First, the angle divider is adjusted to fit the inside corner. Then it is used to mark the miter cut on the molding.

*Stanley Tools*

Figure 9–19. A try square, like a combination square, is used to mark 90° angles.

The two surfaces of the framing square also include a number of tables. The *Roof Rafter Framing Table* is the one used most often today. It enables the carpenter to find the lengths of the roof rafters and the angles of dif-

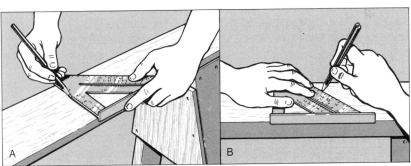

Figure 9–21. A Swanson Speed Square is used to mark 90° angles (A) or 45° angles (B).

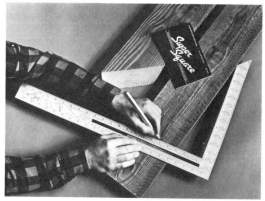

Orem Research, Inc.

Figure 9–22. A Super Square is used for roof and stair layout.

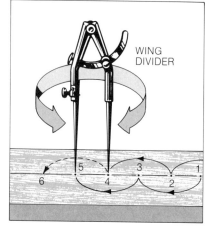

WING DIVIDER

Stanley Tools

Figure 9–24. Wing dividers can be used to draw arcs and circles and can be used as a scriber.

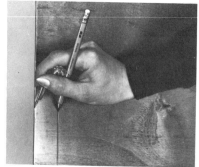

Figure 9–23. A scriber is used to mark a line when a close fit is required between the edge of a piece of material and an irregular surface. A finish saw should be used to cut the material along the scribed line.

Stanley Tools

Stanley Tools

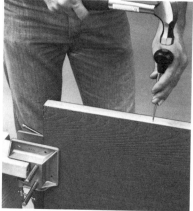

Figure 9–25. A scratch awl is usually used to start holes for wood screws.

Figure 9–26. A center punch is used to mark holes to be drilled in metal.

the basic design of the traditional framing square (refer to Figure 14–16) with new features that simplify the layout of angle cuts for roof rafters. See Figure 9–22.

## MARKING AND SCRIBING TOOLS

Many tools are available for marking materials for layout purposes. Some of the more commonly used tools are as follows:

A *scriber* (Figure 9–23) has two legs, one with a steel point and one holding a pencil. Materials that will not show a pencil line clearly are cut or scratched with the steel point. A scriber is used when a close fit is required between two pieces of material.

*Wing dividers* (Figure 9–24) help the carpenter draw arcs and circles or mark off even spaces. One metal leg is easily replaced by a pencil if desired.

A *scratch awl* (Figure 9–25) can be used to scratch lines on materials that will not show a pencil line. However, it is more often used, with a hammer, to start holes for wood screws.

A *center punch* (Figure 9–26) is struck with a hammer to mark holes to be drilled in metal.

*Trammel points* are used to lay

out circles of any size. See Figure 9–27. A pencil is clamped to one of the trammel points. Then both points are clamped to a piece of straight and narrow stock. The distance between the trammel points is equal to the radius of the circle. *Square gauges* (also called *stair gauges*) are used with a framing square to lay out roof rafters and stair stringers. See Figure 9–28. This

66

Figure 9–27. Trammel points can be used to lay out circles of any size. A pencil is clamped to one point, then both points are clamped to a piece of straight stock.

The L.S. Starrett Company

Figure 9–29. A chalk reel is used to snap lines on flat surfaces.

Figure 9–28. Square gauges (also called stair gauges) are attached to a steel square to lay out the angle cuts for roof rafters as well as the tread and riser cuts for stair stringers.

procedure is explained in Section 10.

A *chalk line reel* (also called a *chalk box*) is used to snap lines on flat surfaces. See Figure 9–29. The chalk box is filled with colored, powdered chalk, which coats the string (line) wound up in it. A ring at the top of the chalk line reel can be hooked on a nail while the line is unwound. The line should be stretched so that it is taut, then pulled straight up to snap a chalk line on the material beneath it.

# UNIT 10

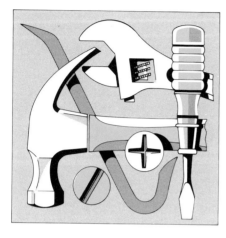

# Fastening and Prying Tools

## FASTENING TOOLS

Fastening tools are used to drive or otherwise apply the devices that hold together building materials. Nails, screws, staples, bolts, and adhesives all require different tools for application.

## Hammers

Hammers are striking tools used to drive nails. The main parts of a hammer are the *head, face, claw,* and *handle.* See Figure 10–1. The hammer head is forged of high quality steel and its face may be bell-shaped or plain-shaped. A bell-shaped face is curved slightly more than a plain-shaped face. Most carpenters prefer the bell-faced hammer because it can drive nails flush to the surface without leaving marks on the lumber.

The surface of the face may be smooth or serrated (cross-checked). Refer again to Figure 10–1. The serrated hammer face lessens the tendency of the hammer to slip off the nail head. However, it will scar the wood and should only be used when doing rough work.

Different materials are used for hammer handles. The older, traditional hammers have a hardwood handle that can be re-

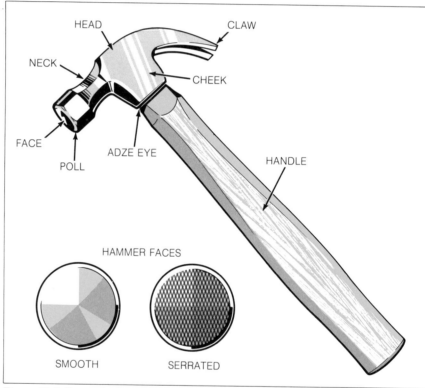

Figure 10–1. Parts of a hammer. The hammer face may be smooth or serrated.

placed if broken. Today most carpenters prefer hammers with steel or fiberglass handles that are cushioned with vinyl or neoprene grips. These handles are expected to last the lifetime of the tool.

Both *straight-claw* and *curved-claw* hammers are used by car-

penters. See Figure 10–2. The size of either kind is determined by the weight of the head.

**Straight-claw Hammer.** Also called a *ripping* hammer, the straight-claw hammer is well suited for rough work such as framing or the construction of

STRAIGHT-CLAW

CURVED-CLAW

*Stanley Tools*

Figure 10–2. Carpenters use both straight-claw and curved-claw hammers.

Figure 10–4. The curved-claw hammer is used to pull nails.

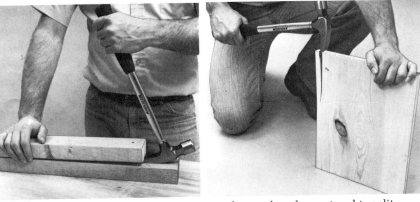

Figure 10–3. The straight-claw hammer is used to pry boards apart and to split boards.

forms for concrete. The straight claws can be used to pry boards apart and to split pieces of lumber. See Figure 10–3.

Although some straight-claw hammers are as light in weight as 16 ounces, 20-ounce or 22-ounce hammers are more often used in rough work. Some carpenters who specialize in framing work use hammers that weigh up to 28 ounces.

**Curved-claw Hammer.** The curved-claw hammer is well suited for finish work. The curved claws can be used to pull nails. See Figure 10–4. Some carpenters also feel that the curved-claw design makes a better balanced tool. Curved-claw hammers range in weight from 7 to 20 ounces. The 16-ounce hammer seems to be the most popular.

## Hammering Method

The beginning carpenter who learns good hammering technique will not have to break bad habits later on. A good technique results in faster work with less effort. Figure 10–5 shows the

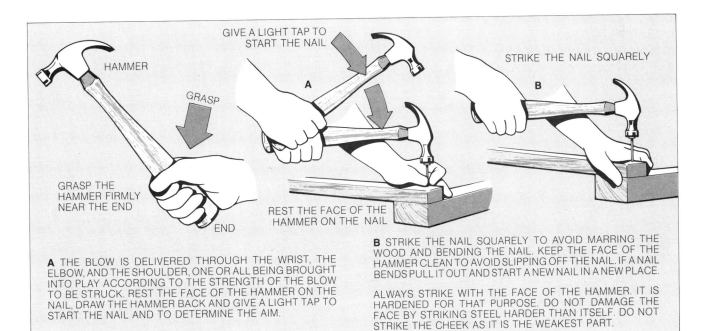

HAMMER

GRASP

GRASP THE HAMMER FIRMLY NEAR THE END

END

GIVE A LIGHT TAP TO START THE NAIL

A

REST THE FACE OF THE HAMMER ON THE NAIL

STRIKE THE NAIL SQUARELY

B

**A** THE BLOW IS DELIVERED THROUGH THE WRIST, THE ELBOW, AND THE SHOULDER, ONE OR ALL BEING BROUGHT INTO PLAY ACCORDING TO THE STRENGTH OF THE BLOW TO BE STRUCK. REST THE FACE OF THE HAMMER ON THE NAIL, DRAW THE HAMMER BACK AND GIVE A LIGHT TAP TO START THE NAIL AND TO DETERMINE THE AIM.

**B** STRIKE THE NAIL SQUARELY TO AVOID MARRING THE WOOD AND BENDING THE NAIL. KEEP THE FACE OF THE HAMMER CLEAN TO AVOID SLIPPING OFF THE NAIL. IF A NAIL BENDS PULL IT OUT AND START A NEW NAIL IN A NEW PLACE.

ALWAYS STRIKE WITH THE FACE OF THE HAMMER. IT IS HARDENED FOR THAT PURPOSE. DO NOT DAMAGE THE FACE BY STRIKING STEEL HARDER THAN ITSELF. DO NOT STRIKE THE CHEEK AS IT IS THE WEAKEST PART.

Figure 10–5. Proper use of the hammer results in more efficient driving of nails.

correct use of the hammer. Keep your wrist loose at all times while driving nails. The blow should be delivered by bringing into play the wrist, elbow, and shoulder. Start the nail with a light tap and then drive the nail as shown in Figure 10–5. Strike the nail squarely to avoid bending it and to avoid the hammer head bouncing off the nail and scarring the wood surface.

## Nailing Method

Improper nailing can result in weak ties between materials fastened together. For proper nailing, the following method should be used: (The different types of nails used in construction are discussed in Section 2.)

1. When possible, always nail from the thinner piece of material into the thicker piece. The nail should be long enough so that the upper third is in the thinner piece and the rest of the nail is driven into the thicker piece. See Figure 10–6.

2. When nailing near the end of a board, stagger the nails. Placing them in a straight line may split the board. See Figure 10–7.

3. To avoid splits in harder wood, either blunt the end of the nail with a hammer, cut off the point of the nail, or drill a pilot hole for the nail. See Figure 10–8.

4. Whenever possible, drive nails across the grain rather than into the end grain. Nails driven into end grain have less holding power. If it is necessary to nail into the end grain, drive the nails in at an angle to increase their holding power. See Figure 10–9.

5. When it is not possible to end-nail two pieces together, *toe-nail* them by driving the nail at such an angle that approximately half of the nail is in each piece of wood. See Figures 10–10 and 10–11.

6. For temporary nailing, *tack* the nails so that they stick out from the material and can be easily withdrawn. See Figure 10–12.

7. To increase the holding

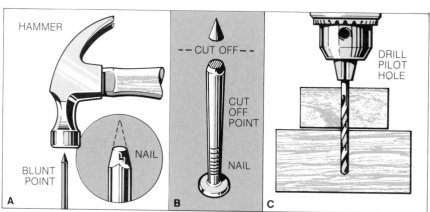

Figure 10–8. To avoid splits in harder wood, blunt the end of the nail with a hammer (A), cut off the point of the nail (B), or drill a pilot hole for the nail (C).

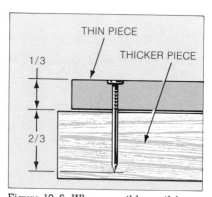

Figure 10–6. When possible, nail from the thinner piece into the thicker piece.

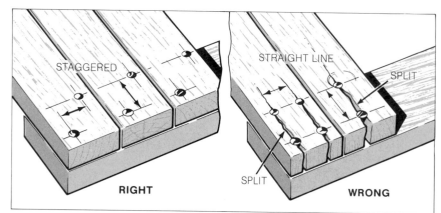

Figure 10–7. When nailing near the end of a board, stagger the nails. Placing them in a straight line may split the board.

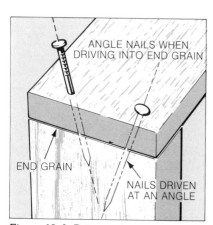

Figure 10–9. Drive nails in across the grain rather than into the end grain. If it is necessary to nail into end grain, drive the nails in at an angle to increase their holding power.

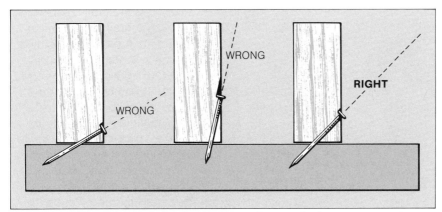

Figure 10–10. Toenails should be driven in at such an angle that approximately half of the nail is in each piece of wood.

Figure 10–13. Driving nails at an angle increases the holding power between two pieces fastened face to face.

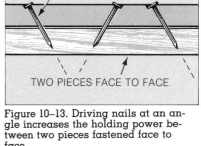

*Stanley Tools*

Figure 10–11. Toenailing a stud to a bottom plate. Approximately half of the nail is in each piece of wood.

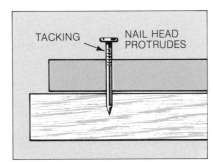

Figure 10–12. Tacking is a procedure for temporary nailing. The head of the nail should stick out so that it can be easily withdrawn.

Figure 10–14. A nail set is used to sink the head of a nail below the surface of the wood.

power between two pieces fastened face to face, drive in nails at an angle. See Figure 10–13.

**Nail Set.** A nail set is used to sink the head of a nail below the surface of the wood. See Figure 10–14. Nails are usually set at the time finish materials such as molding, paneling, or siding are applied. The holes left by the set nails are filled by the painter. Nail sets are usually 4″ long. Their tip sizes range from 1⁄32″ to 5⁄32″.

## Hatchets

Hatchets are striking tools that have a nailing face as well as a cutting edge. Three types of

hatchets used by carpenters are the *half-hatchet,* the *wallboard hatchet* (also called a *drywall hammer*), and the *shingle hatchet.* See Figure 10–15. All three types have nail-pulling slots at the side of the blade.

The half-hatchet is used in constructing wood forms for concrete. It is also used to sharpen stakes and trim boards. It has a beveled single or double blade that is usually 3½″ wide.

The wallboard hatchet has a nailing face that is curved enough to dimple the surface of gypsum wallboard without breaking through the paper covering.

The shingle hatchet is used to split and nail roof shingles. Its blade is usually 2½″ wide.

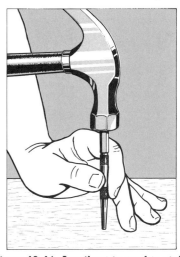

HALF-HATCHET

WALLBOARD HATCHET

SHINGLE HATCHET

*Millers Falls Tool Company*

Figure 10–15. Hatchets used in the carpentry trade. All three types shown here have a nail-pulling slot at the side of the blade.

## Use and Care of Striking Tools

Finger injuries are the most frequent accidents caused by striking tools. To avoid such injuries:

1. When using a hammer or hatchet with a wood handle, be sure the head of the tool fits tightly on the handle.

2. Always replace a cracked wood handle.

3. Use a flat-faced hammer for driving nails.

4. Hammer heads should be of proper hardness. Soft heads will mushroom and chips can break off.

5. Be wary of the nail claw of the hammer or the blade of a hatchet on the backswing.

6. Do not strike two hammer heads together in order to pry out nails.

7. Be sure mechanical staplers are pressed firmly against the work before releasing staples.

## Staplers

As a rule, hand-operated mechanical staplers (tackers) are not part of the carpenter's basic tool collection. They are usually provided by the building contractor. These heavy-duty stapling tools can perform many operations previously done by hammer

*Duo-Fast Corporation*

Figure 10–17. A hammer tacker is used to fasten insulation batts.

and nails.

The *strike tacker* is operated by striking the plunger with a rubber mallet. It is often used to fasten floor underlayment. The model shown in Figure 10–16 drives 18-ga. narrow-crown staples from ⅞″ to 1⅛″ long.

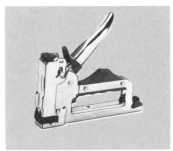

*Duo-Fast Corporation*

*Duo-Fast Corporation*

Figure 10–18. Heavy-duty staple guns are used to fasten vinyl flooring, insulation, roofing paper, carpet padding, tar paper, screening, carpeting, and ceiling tile.

The *hammer tacker* allows one-hand operation, since it releases a staple when it is struck against a surface. It is often used to fasten building paper, felt roof underlayment, and insulation. Insulation batts are shown being stapled in Figure 10–17.

The *gun tacker* is a heavy-duty stapler that also allows one-hand operation. It is used for a wide variety of fastening operations, including vinyl flooring, insulation, roofing paper, tar paper, screening, carpet padding, carpeting, and ceiling tile. The model shown in Figure 10–18 drives .050-ga. staples from ¼″ to ⁹⁄₁₆″ long.

## Screwdrivers

The parts of a screwdriver are the *head, handle, blade* (also called *shank*), and *tip.* See Figure 10–19. The size of a screwdriver is identified by the length of its blade. The more frequently used lengths are 3″, 4″, 6″, 8″, and 10″. Screwdrivers with longer blades allow the worker to apply greater force, which is required to drive larger screws. The various types of screws used for construction purposes are discussed in Section 2.

The two basic types of screwdrivers are the *standard* and the *Phillips* screwdrivers. See Figure 10–20. Both types are used by carpenters.

*Duo-Fast Corporation*

Figure 10–16. A strike tacker is used to fasten floor underlayment.

TIP    BLADE    HANDLE    HEAD

*Stanley Tools*

Figure 10–19. Parts of a screwdriver. The size of a screwdriver is identified by the length of its blade.

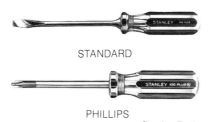

STANDARD

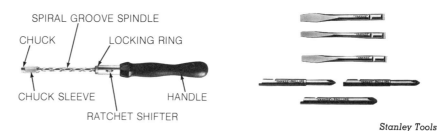

PHILLIPS

*Stanley Tools*

Figure 10–20. Both standard and Phillips screwdrivers are used by carpenters.

SPIRAL GROOVE SPINDLE

CHUCK          LOCKING RING

CHUCK SLEEVE          HANDLE

RATCHET SHIFTER

*Stanley Tools*

Figure 10–21. The spiral-ratchet screwdriver can be fitted with different types and sizes of bits.

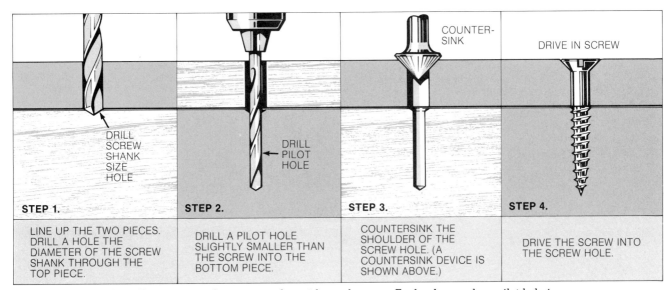

| STEP 1. | STEP 2. | STEP 3. | STEP 4. |
|---|---|---|---|
| DRILL SCREW SHANK SIZE HOLE | DRILL PILOT HOLE | COUNTERSINK | DRIVE IN SCREW |
| LINE UP THE TWO PIECES. DRILL A HOLE THE DIAMETER OF THE SCREW SHANK THROUGH THE TOP PIECE. | DRILL A PILOT HOLE SLIGHTLY SMALLER THAN THE SCREW INTO THE BOTTOM PIECE. | COUNTERSINK THE SHOULDER OF THE SCREW HOLE. (A COUNTERSINK DEVICE IS SHOWN ABOVE.) | DRIVE THE SCREW INTO THE SCREW HOLE. |

Figure 10–22. Fastening wood pieces together with wood screws. For harder woods, a pilot hole is necessary.

**Standard Screwdriver.** The tip of a standard screwdriver fits into a single slot in the head of the screw. Tips range in width from ⅛″ to ⅜″. Longer screwdrivers normally have wider tips, although there are exceptions to this rule. For best results, use a screwdriver with a tip the size of the screw slot.

**Phillips Screwdriver.** The tip of a Phillips screwdriver is shaped like a cross. It is used to drive the double-slotted head of the Phillips-type screw. The size of a Phillips screwdriver is determined by the length of the blade as well as by the tip size. Point numbers ranging from 0 to 24 specify the bit sizes.

**Spiral-ratchet Screwdriver.** The spiral-ratchet screwdriver is an effective hand tool for driving screws. See Figure 10–21. It is widely used to fasten door hinges and finish hardware. The spiral-grooved spindle turns as the screwdriver handle is pushed forward. This exerts a strong and rapid force on the screw. The *ratchet shifter* allows the spindle to either drive or withdraw screws. The ratchet unit can also be locked in place when desired. An important advantage of this tool is that different types and sizes of bits can be easily inserted in the chuck.

**Driving Wood Screws.** When driving screws into harder

woods, drill a pilot hole slightly smaller than the thickness of the screw. In the case of a flat-head screw, the shoulder of the hole must be countersunk with a countersink tool. This will allow the screw head to be even with or slightly below the surface. A pilot hole is not always required when driving screws into softer woods with a spiral-ratchet screwdriver. Also, flat-head screws driven into softer woods tend to countersink themselves. Rubbing wax or soap on the screw threads also makes it easier to drive a screw. Figure 10–22 describes the proper procedure for fastening materials together with wood screws.

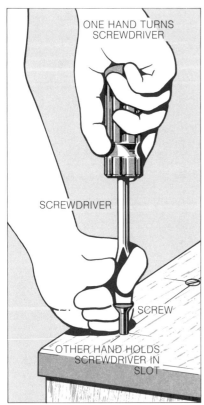

Figure 10–23. Proper way to hold a screwdriver. The right hand turns the screwdriver while the left hand holds it in position.

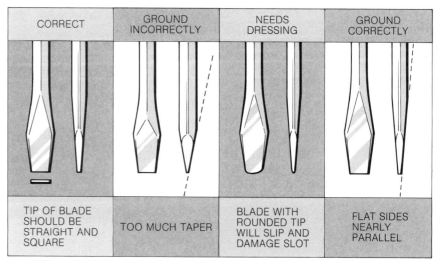

| CORRECT | GROUND INCORRECTLY | NEEDS DRESSING | GROUND CORRECTLY |
|---------|--------------------|----------------|-------------------|
| TIP OF BLADE SHOULD BE STRAIGHT AND SQUARE | TOO MUCH TAPER | BLADE WITH ROUNDED TIP WILL SLIP AND DAMAGE SLOT | FLAT SIDES NEARLY PARALLEL |

Figure 10–24. The screwdriver tip should be straight and square-cornered.

ADJUSTABLE WRENCH

C-JOINT PLIERS

SLIP-JOINT PLIERS

*Millers Falls Tool Company*
Figure 10–25. Various gripping tools (pliers and wrenches) are used for tightening bolts.

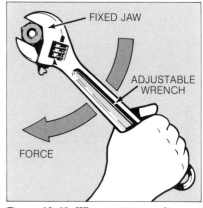

Figure 10–26. When using an adjustable wrench, be sure it is tightly adjusted to the nut. Pull the wrench so that the force is on the side of the fixed jaw.

## Use and Care of Screwdrivers.

Accidents with screwdrivers can cause puncture wounds in the hand. To avoid hand injuries, hold the screwdriver properly, as shown in Figure 10–23. Another important work and safety factor is the condition of the screwdriver tip. It should be straight and square-cornered, not rounded or excessively tapered. See Figure 10–24. Other safety rules are:

1. Do not use a hammer or wrench on a screwdriver.

2. Do not use a screwdriver as a punch, chisel, lever, or nail-puller.

3. Do not carry a screwdriver in your pants pocket.

## Pliers and Wrenches

Various types of bolts are used to fasten structural members to-gether. Gripping tools (pliers and wrenches) are needed to help install the bolts. See Figure 10–25. Two types of pliers used for this purpose are *slip-joint* and *C-joint* pliers.

## Use and Care of Wrenches.

Using the wrong type of wrench or using a wrench improperly can cause scraped knuckles, pulled back muscles, or a bad fall if the wrench slips. Rules to follow are:

1. Check for worn, cracked, or sprung jaws on the wrench.

2. Use the right size wrench for the job.

3. Whenever possible, always *pull,* rather than push, on a wrench. See Figure 10–26. There is a much greater danger of a wrench slipping and causing a hand injury when a pushing pressure is applied to the tool.

4. Never use a wrench as a hammer.

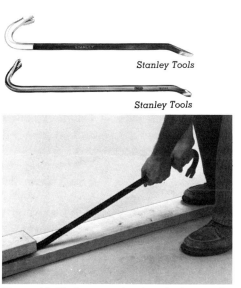

*Stanley Tools*

*Stanley Tools*

*Stanley Tools*

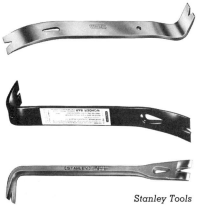

Figure 10-27. A ripping bar (pry bar) is used to pry boards apart and remove large nails or spikes.

Figure 10-29. A nail claw is used to pull nails above the surface of the lumber so that they can be pulled completely out with the claw of a hammer or ripping bar.

*Stanley Tools*

*Stanley Tools*

Figure 10-28. The nail slot at the end of a ripping chisel (flat bar) is used to pull nails in tightly enclosed areas.

## PRYING TOOLS

The opposite of a fastening tool is a prying tool, which is designed to pull apart materials that have been fastened. Carpentry work sometimes requires tearing apart structural members, especially in remodeling work. Small or large sections of a building may have to be torn down and removed. Also, in concrete form construction, wood forms must be stripped away after the concrete has hardened.

One type of prying tool is the *ripping bar* (or *pry bar.*) See Figure 10-27. It is available in lengths ranging from 12″ to 36″. The 30″ bar fits conveniently in the average size carpenter's tool box and is the most popular length.

The *ripping chisel* (or *flat bar*) has a nail slot at the end to pull nails out from tightly enclosed areas. It may also be used as a small pry bar. See Figure 10-28.

The *nail claw* is used solely for the removal of nails. See Figure 10-29. The sharpened nail slot is driven under a nail head to pull the nail head above the surface of the lumber. The nail is then pulled completely out with the claw of a hammer or with a ripping bar.

Most accidents related to the use of prying tools occur when a pry bar slips and the worker falls to the ground. To avoid this the worker should maintain a balanced footing and a firm grip on the tool. Proper use of prying tools not only reduces the chance of accidents but also reduces damage to materials that must be reused.

# UNIT 11

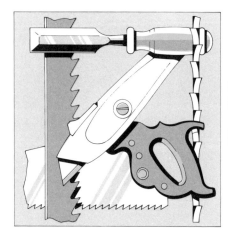

# Sawing and Cutting Tools

Sawing and cutting tools are used to trim construction materials to their proper dimensions. Saws and chisels are used most often. For metals and plastics, however, other implements such as tin snips are sometimes necessary.

## HANDSAWS

The main parts of a handsaw are the *blade* (including the *toe* and *heel* of the blade), *teeth, back,* and *handle.* See Figure 11–1. Although the basic construction of all handsaws is similar, there are many differences in the length and shape of the blade and the number and shape of the teeth. Also, although most handsaws have a straight back, the older type of curved-back saws (*skewback saws*) are still manufactured.

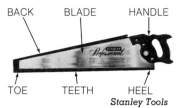

BACK    BLADE    HANDLE

TOE    TEETH    HEEL

*Stanley Tools*

Figure 11–1. Parts of a typical handsaw. Although basic construction is similar for all handsaws, there are many differences in the length and shape of the blade and the number and shape of the teeth.

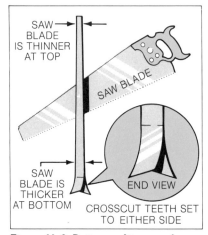

SAW BLADE IS THINNER AT TOP

SAW BLADE

SAW BLADE IS THICKER AT BOTTOM

END VIEW

CROSSCUT TEETH SET TO EITHER SIDE

Figure 11–2. Better quality saws have a taper-ground blade. The blade is thinner at its top than at its cutting edge.

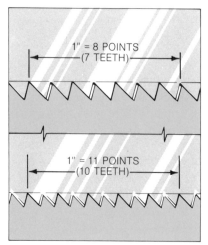

1" = 8 POINTS (7 TEETH)

1" = 11 POINTS (10 TEETH)

Figure 11–3. An 8-point saw has larger teeth, and less teeth per inch, than an 11-point saw.

The cut made by a saw is wider than the thickness of the saw blade. If this were not the case, the wood fibers pressing against the blade would cause the saw to *bind,* making the cutting action more difficult. To prevent this, the teeth of the saw are *set* so that they are alternately bent from side to side. Better quality saws have a *taper ground.* See Figure 11–2. The top side of the blade is thinner than it is at the cutting edge, requiring less set in the teeth.

A saw usually has a number printed on its blade giving the number of *teeth points* per inch. See Figure 11–3. The lower the number is, the larger the teeth are. For example, an 8-point saw has larger teeth than an 11-point saw.

The teeth of most handsaws are shaped to cut across the grain of the wood. These saws are called *crosscut saws.* Saws designed to cut with rather than across the grain are *ripsaws.*

## Crosscut Saws

The teeth on crosscut saws are shaped like knives, which are the most effective shape for cutting across the grain of wood. See

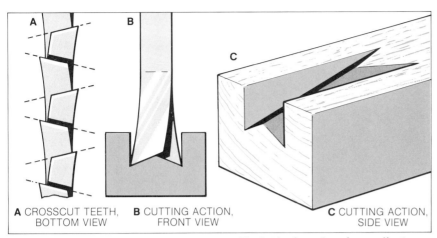

**A** CROSSCUT TEETH, BOTTOM VIEW   **B** CUTTING ACTION, FRONT VIEW   **C** CUTTING ACTION, SIDE VIEW

Figure 11-4. Cutting action of a crosscut saw. The knife-shaped teeth are effective for cutting across the grain.

Figure 11-4. Crosscut saws should be held at a 45° angle to the work.

The most popular type of crosscut saw for rough work has a 26″ blade with 8 points per inch. Crosscut saws for finish work usually have a shorter blade, such as 20″ or 22″, with 10 or 12 points per inch.

**Compass Saw.** The compass saw is used to cut curved lines and to saw holes. See Figure 11-5. It can start saw cuts in tight spaces where a regular saw will not fit. The compass saw's blade is 12″ or 14″ long, with 8 or 10 points per inch.

**Keyhole Saw.** The keyhole saw is similar to the compass saw, but it has a narrower and shorter blade and its teeth are finer. It is used to make curved cuts in areas too small for the compass saw to be used. In the past, one of the main functions of the key-hole saw was to cut out the key-holes for a type of mortise lock that is seldom used today.

**Backsaw.** The backsaw is used with a miter box to make very fine cuts in finish work. See Figure 11-6. The backsaw's blade is 10″ to 26″ long and 3¼″ to 6″

Figure 11-5. A compass saw is used to make curved cuts.

wide. Its teeth are 10 to 14 points per inch. A reinforcing strip at the top stiffens the blade.

The miter box that is used with the backsaw is not considered part of the carpenter's basic tool collection, but is normally placed on the job by the employer.

**Dovetail Saw.** The dovetail saw is similar to the backsaw, but it is smaller and has a round handle and a narrower blade. See Figure 11-7. It is used to make fine cuts in molding and other smaller types of trim materials. The type of dovetail saw used most often has a blade 10″ long and 2″ wide, with 15 points per inch.

**Coping Saw.** The coping saw is useful for cutting curves and irregular lines in thin material. See Figure 11-8. It is frequently used to cut

*Stanley Tools*

Figure 11-6. A backsaw is used with a steel miter box to make very fine cuts in finish work.

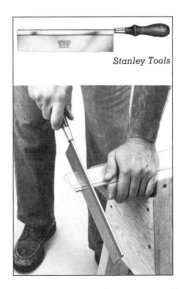

*Stanley Tools*

Figure 11-7. A dovetail saw is a smaller version of a backsaw.

77

Stanley Tools

Figure 11–8. A coping saw is used to make fine, irregular cuts in thin materials.

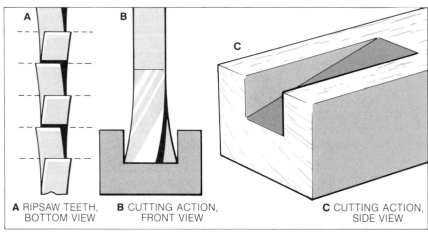

**A** RIPSAW TEETH, BOTTOM VIEW

**B** CUTTING ACTION, FRONT VIEW

**C** CUTTING ACTION, SIDE VIEW

Figure 11–9. Cutting action of a ripsaw. The chisel-shaped teeth are effective for cutting with the grain.

coped joints when fitting the inside corners of molding. The coping saw preferred by most carpenters has a blade 6⅜″ long and ⅛″ wide. The blade can be adjusted to make angle cuts easier.

## Ripsaws

The teeth on ripsaws are shaped like chisels, which are the most effective shape for cutting with the grain of wood. See Figure 11–9. Most ripsaws have a blade 26″ long, with 5½ points per inch. A ripsaw should be held at a 60° angle to the work.

Sawing with the grain takes much more time and effort than sawing across the grain. For this reason, power tools have generally replaced ripsaws for ripping lumber on the job. Even so, some carpenters still carry a ripsaw.

## Metal-cutting Saws

Two metal-cutting saws are the *hacksaw* and the *nail saw*. A hacksaw is used for cutting metal

Stanley Tools

Figure 11–10. A hacksaw is used to cut metals.

materials such as metal framing members, metal molding and exterior wall covering, and door thresholds. See Figure 11–10. A nail saw is often used in remodeling work where nails must be cut in order to tear apart framing members with a minimum of damage. One type of nail saw is shown in Figure 11–11. Its blade fits into a metal or plastic handle similar to that of a compass saw.

Figure 11–11. A nail saw is used to cut nails so that framing members can be more easily parted.

## Use and Care of Handsaws

Proper sawing methods enable accurate, fast cuts with less effort. Improper use of handsaws can cause the saw blade to jump out of the saw cut and bite into the non-sawing hand resting on the material. To avoid hand lacerations from a handsaw:

1. Press thumb lightly against the blade when starting the cut. See Figure 11–12.

2. Hold up on the handle so that the blade moves very lightly back and forth until a beginning saw kerf has been made.

3. Move the non-sawing hand a safe distance away from the blade. See Figure 11–13.

4. Do not *ride* (dig in with) the

blade. A sharp blade will cut quickly and accurately with very little pressure.

5. Always cut on the waste side of the cutting line.

Saws should be placed in the tool box so that the teeth of the saw are protected from contact with other metal objects. Many carpenters fasten a slotted, hardwood sawblock at the bottom and to one side of the tool box. The saws can then be placed in the slots of the sawblock.

Some carpenters wipe a thin film of oil on the saw blade at the end of the working day. This prevents the blade from rusting and prolongs the life of the tool.

Saws must be kept sharp. Dull saws are difficult to work with and they can cause injuries. In the past, some carpenters sharpened their saws using a saw set and appropriate files. Most carpenters today have their saws sharpened by professional saw filers who use special grinding machines.

## OTHER CUTTING TOOLS

In addition to saws, several other cutting tools are used by carpenters, including chisels, knives, and special implements for cutting tin and wire.

### Wood Chisels

Wood chisels are hand tools used for rapid removal of waste stock. Some are designed for very rough work, such as the all-metal *flooring chisel* shown in Figure 11–14. Others are designed for finish work, such as the *butt chisel,* which is used for mortising for door hinges, flush bolts, and other kinds of finish hardware. See Figure 11–15. (A detailed procedure for mortising for door hinges is discussed in Section 13.)

The butt chisels used today usually have a plastic handle that

Figure 11–12. Press thumb lightly against blade when starting a cut with a handsaw.

Figure 11–13. Move non-sawing hand a safe distance away from blade after making beginning saw kerf.

*Stanley Tools*

Figure 11–14. A flooring chisel is used for rough work.

holds the blade. Chisels are still available with the older type of wood handles, but they do not wear as well as plastic. The top of the handle is protected by a steel cap, which receives the direct hammer blow.

Chisels come in widths of 1/8″ to 2″ and blade lengths of 3″ to 6″. Carpenters who do finish work must carry chisels in an as-

*Stanley Tools*

*Stanley Tools*

Figure 11–15. A butt chisel is often used to mortise lumber for door hinges and other types of finish hardware.

sortment of sizes. The chisels are often stored in a plastic roll to protect the cutting edges.

**Use and Care of Wood Chisels.** Wood chisels must be kept sharp. A dull chisel requires greater effort to use and results in sloppy work. A dull chisel may also cause injury. The sharpening procedure for chisels is identical to the one described for plane irons in Unit 13.

To avoid injury from improper use of a wood chisel:

1. Do not carry a chisel with an exposed cutting edge in your pocket.

2. Do not use a wood chisel with a loose or cracked handle.

3. Do not use a chisel as a wedge or pry bar.

4. Always keep the hand holding the material in back of the cutting action of the chisel. Always cut away from your body. See Figure 11–16.

### Cold Chisels

Cold chisels are forged from special hardened and tempered

Figure 11-16. When using a wood chisel, keep free hand out of the way of the chisel, and cut away from body.

*Stanley Tools*

Figure 11-17. A cold chisel is used to cut metals and chip concrete.

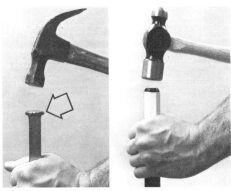

Figure 11-18. Chips can break off the mushroomed head (see arrow) of a cold chisel and cause a serious eye injury.

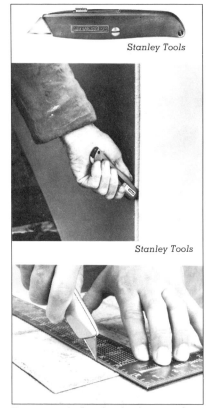
*Stanley Tools*

*Stanley Tools*

Figure 11-19. A utility knife is used to cut materials such as gypsum board, fiberboard, and insulation materials.

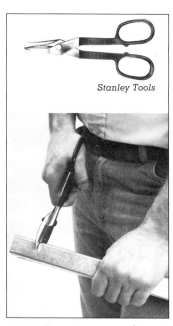
*Stanley Tools*

Figure 11-20. Tin snips are used to cut metal framing members.

*Millers Falls Tool Company*

Figure 11-21. End nippers are a handy tool for twisting or cutting wires and pulling or cutting nails.

alloy steel. See Figure 11-17. They are used to cut through nails and other metals, to chip concrete, and to cut through stucco or plaster. Cold chisels are available in widths of ¼" to 1" and blade lengths of 6" to 12".

Wear protective goggles when working with cold chisels. They develop mushroomed heads during use, and a mushroomed head creates the hazard of flying steel chips when it is struck. See Figure 11-18. As soon as a mushroomed head develops, it should be ground off.

## Knives

Carpenters use knives for cutting gypsum board, fiberboard, insulation materials, and many other items used in construction. The *utility knife* is frequently used. See Figure 11-19. It has a retractable blade, which is an important safety feature. The blade

is pulled back into the body of the knife when it is not being used. Additional blades are stored in the handle.

When using a knife, always cut away from your body. Keep your free hand out of the way of the knife blade.

## Tin Snips

Tin snips are available in several designs in addition to the type shown in Figure 11-20. A carpenter uses tin snips when working with metal framing members.

## End Nippers

Sometimes called *carpenters' pinchers,* end nippers are used for cutting and twisting wire. See Figure 11-21. They are also useful for cutting off or pulling out nails.

# UNIT 12

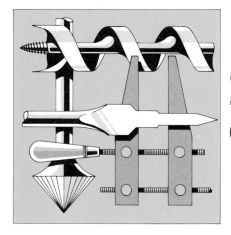

# Boring and Clamping Tools

## BORING TOOLS

Hand boring tools are often used in interior finish work. Although power-driven drills are more efficient for general boring purposes, they can be difficult to control. For this reason the hand tool is better suited for certain jobs.

## Ratchet Brace

The most frequently used hand boring tool is the ratchet brace. The main parts of the ratchet brace are the *jaws, shell, bow, handle, quill, head,* and *box ratchet.* See Figure 12–1. The *sweep* of a ratchet brace is the diameter of the circle made when the handle is turned. Most carpenters use a brace with a 10″ or 12″ sweep for general work.

The ratchet device can be adjusted to operate the brace in a clockwise or counterclockwise direction. This allows holes to be bored close to walls or corners where it is not possible to make a complete turn of the handle.

**Auger Bit.** The auger bit is the boring device that fits into the jaws of the ratchet brace. See

Figure 12–2. Its main parts are the *tang, shank, spur, feed screw,* and *twist.* A set of auger bits carried by the carpenter might include diameter sizes of ¼″ to 1″. The size of a bit is often marked as sixteenths of an inch on the tang. For example, a number 8 stamped on the tang identifies an ⁸⁄₁₆″, or ½″, bit. A common standard length for auger bits is 8″, although a type known as a *ship auger* is available in sizes from 18″ to 24″.

Auger bits have coarse-threaded or medium-threaded screws for rapid, rough boring operations. They are also available with fine-threaded screws for finish work.

**Other Bits.** In addition to auger bits, several other drilling devices are used with the ratchet brace as follows (Figure 12–3):

An *expansion bit* has two interchangeable cutters for drilling holes in two different size ranges.

A *bit extension* extends the length of a standard size bit by 18″ or 24″. It is sometimes necessary for drilling into a hard-to-reach surface or into stock thicker than the length of a standard size auger bit.

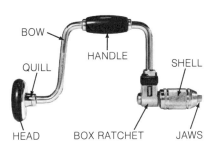

BOW

QUILL    HANDLE

SHELL

HEAD    BOX RATCHET    JAWS

*Stanley Tools*

*Stanley Tools*

Figure 12–1. Parts of a ratchet brace. This tool is the most frequently used hand boring tool.

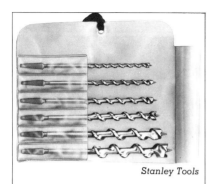

Stanley Tools

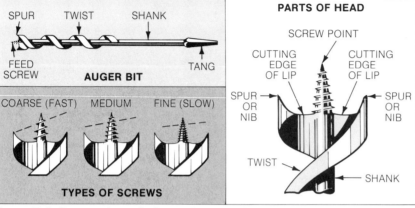

**Figure 12–2.** An auger bit is the boring device used most often with the ratchet brace. A set of auger bits usually includes diameter sizes of ¼″ to 1″.

A *countersink* is used to countersink holes for flat-head screws.

A *lockset bit* is specially designed to bore holes for cylindrical locks.

A *screwdriver bit* is used to drive large screws.

**Use and Care of Brace and Bits.** As with any other tool, proper handling of the ratchet brace is the key to producing the best results with the least amount of effort. The correct boring method is shown in Figure 12–4.

The moving parts of the brace should be periodically oiled. The bits should be kept sharp. They should be stored in a plastic bit roll or a box to protect the cutting edges when not in use. (Refer again to Figure 12–2.)

Figure 12–5 shows the procedure for sharpening auger bits. A specially designed auger bit file or a slim three-cornered file can be used.

## Hand-operated Drills

Hand-operated drills use the same type of twist drill that is used with power-driven drills. See Figure 12–6. Hand drills are normally used to drill holes that are ¼″ or less, although some have a ⅜″ chuck capacity.

The *automatic push drill* is a type of hand drill used to bore smaller holes, ranging from ¹⁄₁₆″

**Figure 12–3.** In addition to auger bits, several other drilling devices are used with a ratchet brace.

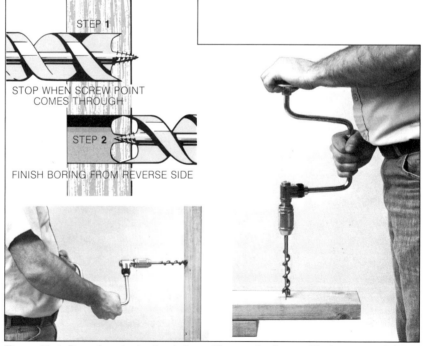

**Figure 12–4.** Procedure for boring holes with a ratchet brace. Moving parts of the brace should be periodically oiled, and the bits should be kept sharp.

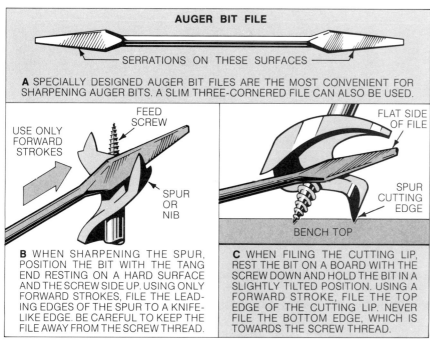

**AUGER BIT FILE**

SERRATIONS ON THESE SURFACES

**A** SPECIALLY DESIGNED AUGER BIT FILES ARE THE MOST CONVENIENT FOR SHARPENING AUGER BITS. A SLIM THREE-CORNERED FILE CAN ALSO BE USED.

FEED SCREW

USE ONLY FORWARD STROKES

SPUR OR NIB

FLAT SIDE OF FILE

SPUR CUTTING EDGE

BENCH TOP

**B** WHEN SHARPENING THE SPUR, POSITION THE BIT WITH THE TANG END RESTING ON A HARD SURFACE AND THE SCREW SIDE UP. USING ONLY FORWARD STROKES, FILE THE LEADING EDGES OF THE SPUR TO A KNIFE-LIKE EDGE. BE CAREFUL TO KEEP THE FILE AWAY FROM THE SCREW THREAD.

**C** WHEN FILING THE CUTTING LIP, REST THE BIT ON A BOARD WITH THE SCREW DOWN AND HOLD THE BIT IN A SLIGHTLY TILTED POSITION. USING A FORWARD STROKE, FILE THE TOP EDGE OF THE CUTTING LIP. NEVER FILE THE BOTTOM EDGE, WHICH IS TOWARDS THE SCREW THREAD.

Figure 12–5. Procedure for sharpening auger bits. Bits should be stored in a plastic bit roll or box to protect the cutting edges.

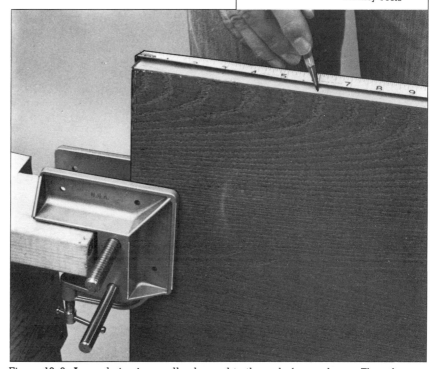

*Stanley Tools*

*Stanley Tools*

*Ingersoll-Rand*

Figure 12–6. Hand-operated drill. It uses the same type of twist drill that power-driven drills use.

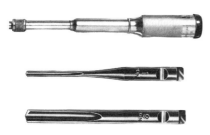

*Ingersoll-Rand*

Figure 12–7. Automatic push drill and fluted drill points. A common use for an automatic push drill is to bore pilot holes for wood screws.

Figure 12–8. A wood vise is usually clamped to the end of a sawhorse. Then the object to be worked on is inserted in the vise. Two styles of wood vises are shown here.

to $^{11}/_{64}$". See Figure 12–7. It operates with special types of fluted drill points that are often stored in the hollow handle of the tool. The primary use for the automatic push drill is boring pilot holes for wood screws.

## CLAMPING TOOLS

Clamping tools hold and support materials being worked upon or fastened together. A description

follows of some of these tools:

A *wood vise* is usually clamped to the end of a sawhorse. See Figure 12–8. Then the object to be worked on (such

*Stanley Tools*

*American Plywood Association*

Figure 12–9. C-clamps make many fitting operations easier to perform. Here they are shown holding together glued miter joints.

*Stanley Tools*

Figure 12–10. Bar clamps are used in cabinet construction to hold together glued pieces.

Figure 12–11. Wood screw clamps are also used to hold glued pieces together.

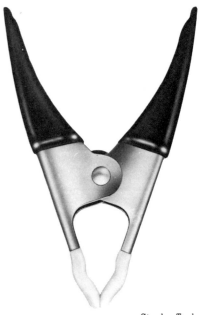

*Stanley Tools*

Figure 12–12. Spring clamps are used to hold narrow pieces together.

as a door) is inserted in the vise.

*C-clamps* (Figure 12–9), *bar clamps* (Figure 12–10), and *wood screw clamps* (Figure 12–11) are used to hold glued pieces together.

*Spring clamps* are used for very narrow pieces. See Figure 12–12.

*Locking clamps* with the trade name of *Vise Grips* are also sometimes used for clamping operations.

*Quick-Grip bar clamps* can be easily adjusted with one hand by squeezing the locking pistol-grip handle. A quick-release grip allows easy removal of the Quick-Grip bar clamps. Standard clamps with $3\frac{1}{4}''$ throat depths are available in 6″, 12″, 24″, 36″, and 50″ opening lengths. Micro-size clamps have a 1″ throat depth and $4\frac{1}{2}''$ opening length. Macro-size clamps have a $5\frac{1}{2}''$ throat depth and are available in 10″, 20″, and $33\frac{1}{2}''$ opening lengths. Non-mar pads protect the surface being clamped. See Figure 12–13.

Figure 12–13. Quick-Grip bar clamps are used to hold glued pieces together.

# UNIT 13

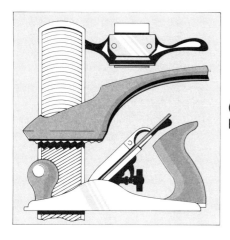

# Smoothing Tools

Smoothing tools include the traditional types of bench planes, scrapers, rasps, and serrated forming tools.

## BENCH PLANES

Bench planes are used for smoothing and jointing lumber. *Jointing* is an old term for the process of truing the edges of boards before they are fitted (joined) together.

The carpenter has a choice of different bench planes. The one chosen is often based on personal preference, since more than one may be equally effective for a certain operation. The main parts of a bench plane are the *knob, lever cap, cam, plane iron,* and *handle.* See Figure 13–1. Since the basic construction of most bench planes is the same, the major difference is size. Larger planes offer another difference—a choice of smooth or grooved bottoms.

## Jointer Plane and Fore Plane

The jointer plane and fore plane are longer and heavier than other planes. They are used for trimming long boards and fitting

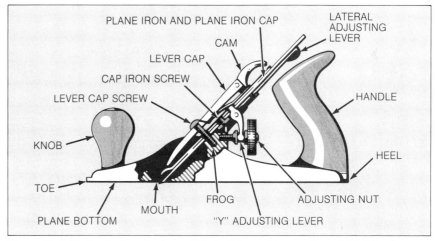

Figure 13–1. Parts of a bench plane. This tool is used to smooth and joint lumber.

doors. See Figure 13–2. The jointer plane ranges from 20″ to 24″ long, with a blade 2⅜″ to 2⅝″ wide. The fore plane is a shorter version of the jointer plane. It is 18″ long, with a blade normally 2⅜″ wide. Of the two planes, the fore plane is probably preferred by more carpenters.

## Jack Plane

The jack plane is an all-purpose tool for smoothing and fitting. It is the plane most carpenters use for fitting doors. The most common size is 14″ long with a blade 2″ wide.

*Stanley Tools*

Figure 13–2. A jointer plane is a large plane (20″ to 24″ long) used to trim long boards and fit doors. A fore plane is a shorter version (18″ long) of the jointer plane.

## Smooth Plane

Also known as a *smoothing plane*, the smooth plane is a

85

*Stanley Tools*

Figure 13–3. A smooth plane is an all-purpose plane that is 9¼", 9½", or 9¾" long.

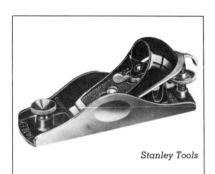

*Stanley Tools*

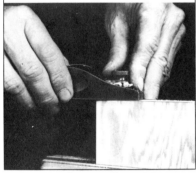

*American Plywood Association*

Figure 13–4. The block plane is 6" or 7" long and is used on small, narrow surfaces. Here it is shown smoothing the edge of a piece of plywood.

DUPLEX RABBET PLANE

*Stanley Tools*

BULLNOSE RABBET PLANE

*Stanley Tools*

Figure 13–5. A bullnose rabbet plane is used to reach into corners. A duplex rabbet plane can be adjusted for regular or bullnose work.

shorter all-purpose bench plane than the jack plane. See Figure 13–3. The smooth plane is available in lengths of 9¼", 9½", and 9¾", with a blade 1¾" or 2" wide. This tool is very useful for smoothing surfaces. It is not as effective as the larger planes for straightening surfaces.

## Block Plane

The block plane is used on small, narrow surfaces. It is more simply constructed than other types of planes. See Figure 13–4. The block plane is 6" or 7" long, with a blade 1⅜" or 1⅝" wide. The low angle of the blade produces clean cuts across the end grain of lumber and the edges of plywood.

## Rabbet Plane

The rabbet plane is used less frequently than other types of bench planes by carpenters. It is used for making rabbet joints on the ends of boards. Two types of rabbet plane are the *duplex* and the *bullnose* planes. See Figure 13–5.

The duplex rabbet plane is 8¼" long, with a blade 1½" wide.

The blade is placed in the rear seat for regular work or in the front seat for bullnose work. The fence on this plane can be adjusted for the desired width of the cut.

The bullnose rabbet plane is 4" long, with a blade 1³⁄₃₂" wide. It is used to reach into corners or other places that are hard to reach with other types of planes.

## Use and Care of Bench Planes

The correct procedure for using a bench plane is shown in Figure 13–6. A plane's effectiveness is determined by the condition and the sharpness of its blade. If the blade is in good condition, it may require only touching up on an oilstone. If the cutting edge is nicked or the bevel has worn down, a bench grinder can restore the blade to its proper condition. See Figure 13–7. (A bench grinder is also used to sharpen wood chisels.) The blade should be pressed lightly to the wheel and dipped frequently in water to prevent the steel from burning. Since burning softens steel, a burned blade will not retain a sharp edge when it is used.

A grinding attachment can be used to clamp the plane iron to the adjustable rest located in front of the grinding wheel. After the blade has been ground on a bench grinder, its fine edge is produced on an oilstone. See Figure 13–8.

## SCRAPERS

Scrapers are used to remove chip marks or to smooth torn grain that may have occurred during planing. See Figures

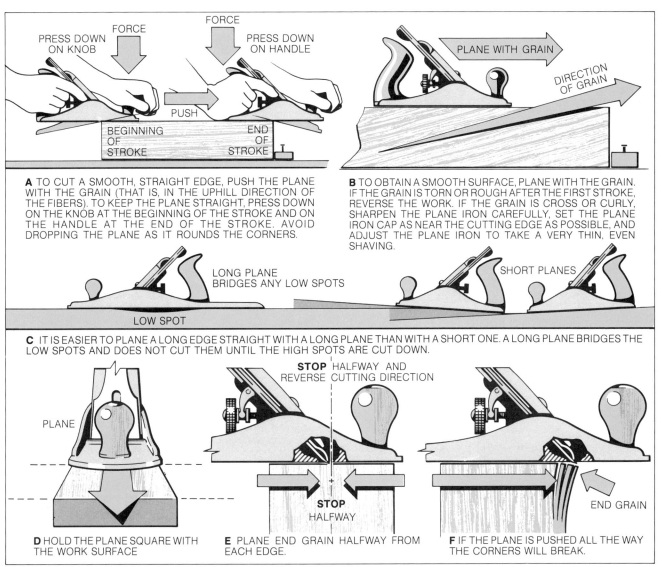

A TO CUT A SMOOTH, STRAIGHT EDGE, PUSH THE PLANE WITH THE GRAIN (THAT IS, IN THE UPHILL DIRECTION OF THE FIBERS). TO KEEP THE PLANE STRAIGHT, PRESS DOWN ON THE KNOB AT THE BEGINNING OF THE STROKE AND ON THE HANDLE AT THE END OF THE STROKE. AVOID DROPPING THE PLANE AS IT ROUNDS THE CORNERS.

B TO OBTAIN A SMOOTH SURFACE, PLANE WITH THE GRAIN. IF THE GRAIN IS TORN OR ROUGH AFTER THE FIRST STROKE, REVERSE THE WORK. IF THE GRAIN IS CROSS OR CURLY, SHARPEN THE PLANE IRON CAREFULLY, SET THE PLANE IRON CAP AS NEAR THE CUTTING EDGE AS POSSIBLE, AND ADJUST THE PLANE IRON TO TAKE A VERY THIN, EVEN SHAVING.

C IT IS EASIER TO PLANE A LONG EDGE STRAIGHT WITH A LONG PLANE THAN WITH A SHORT ONE. A LONG PLANE BRIDGES THE LOW SPOTS AND DOES NOT CUT THEM UNTIL THE HIGH SPOTS ARE CUT DOWN.

D HOLD THE PLANE SQUARE WITH THE WORK SURFACE

E PLANE END GRAIN HALFWAY FROM EACH EDGE.

F IF THE PLANE IS PUSHED ALL THE WAY THE CORNERS WILL BREAK.

*Information from Stanley Tools*

Figure 13–6. Proper procedure for using a bench plane. The blade should be kept sharp and free of nicks.

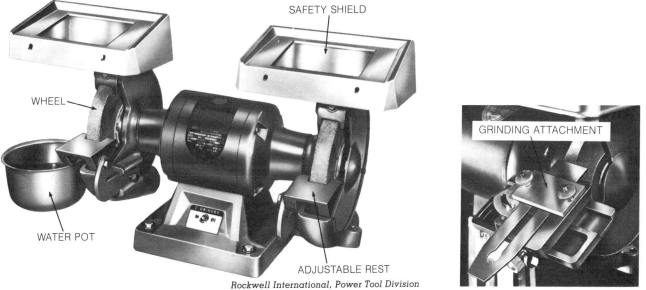

*Rockwell International, Power Tool Division*

Figure 13–7. A bench grinder is used to restore a plane blade or wood chisel to its proper condition (continued next page).

# WHEN TO GRIND A PLANE IRON OR CHISEL

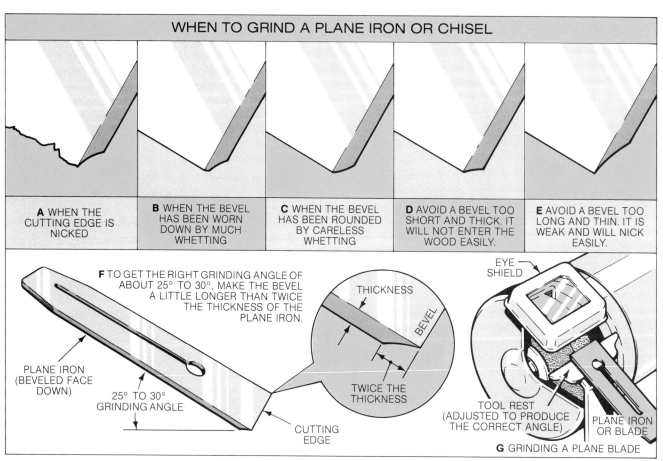

**A** WHEN THE CUTTING EDGE IS NICKED

**B** WHEN THE BEVEL HAS BEEN WORN DOWN BY MUCH WHETTING

**C** WHEN THE BEVEL HAS BEEN ROUNDED BY CARELESS WHETTING

**D** AVOID A BEVEL TOO SHORT AND THICK. IT WILL NOT ENTER THE WOOD EASILY.

**E** AVOID A BEVEL TOO LONG AND THIN. IT IS WEAK AND WILL NICK EASILY.

**F** TO GET THE RIGHT GRINDING ANGLE OF ABOUT 25° TO 30°, MAKE THE BEVEL A LITTLE LONGER THAN TWICE THE THICKNESS OF THE PLANE IRON.

PLANE IRON (BEVELED FACE DOWN)

25° TO 30° GRINDING ANGLE

CUTTING EDGE

THICKNESS

BEVEL

TWICE THE THICKNESS

EYE SHIELD

TOOL REST (ADJUSTED TO PRODUCE THE CORRECT ANGLE)

PLANE IRON OR BLADE

**G** GRINDING A PLANE BLADE

Figure 13–7. Continued from previous page.

*Information from Stanley Tools*

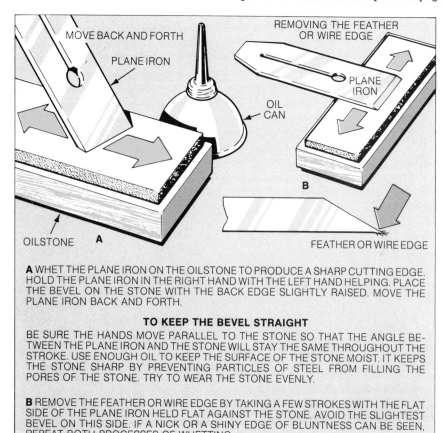

MOVE BACK AND FORTH

PLANE IRON

OIL CAN

OILSTONE **A**

REMOVING THE FEATHER OR WIRE EDGE

PLANE IRON

**B**

FEATHER OR WIRE EDGE

**A** WHET THE PLANE IRON ON THE OILSTONE TO PRODUCE A SHARP CUTTING EDGE. HOLD THE PLANE IRON IN THE RIGHT HAND WITH THE LEFT HAND HELPING. PLACE THE BEVEL ON THE STONE WITH THE BACK EDGE SLIGHTLY RAISED. MOVE THE PLANE IRON BACK AND FORTH.

### TO KEEP THE BEVEL STRAIGHT

BE SURE THE HANDS MOVE PARALLEL TO THE STONE SO THAT THE ANGLE BETWEEN THE PLANE IRON AND THE STONE WILL STAY THE SAME THROUGHOUT THE STROKE. USE ENOUGH OIL TO KEEP THE SURFACE OF THE STONE MOIST. IT KEEPS THE STONE SHARP BY PREVENTING PARTICLES OF STEEL FROM FILLING THE PORES OF THE STONE. TRY TO WEAR THE STONE EVENLY.

**B** REMOVE THE FEATHER OR WIRE EDGE BY TAKING A FEW STROKES WITH THE FLAT SIDE OF THE PLANE IRON HELD FLAT AGAINST THE STONE. AVOID THE SLIGHTEST BEVEL ON THIS SIDE. IF A NICK OR A SHINY EDGE OF BLUNTNESS CAN BE SEEN, REPEAT BOTH PROCESSES OF WHETTING.

*Information from Stanley Tools*

Figure 13–8. An oilstone is used to produce the final, sharp cutting edge on a plane blade or wood chisel.

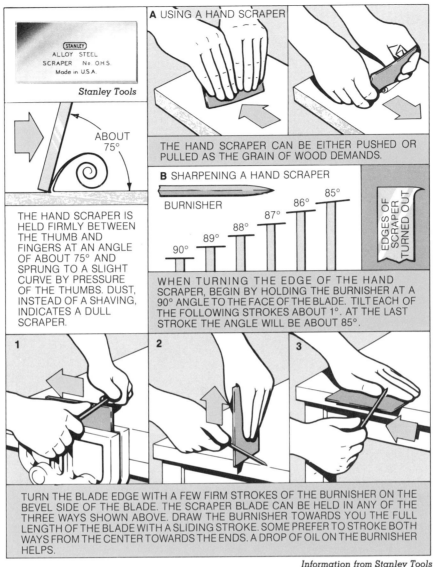

A USING A HAND SCRAPER

THE HAND SCRAPER CAN BE EITHER PUSHED OR PULLED AS THE GRAIN OF WOOD DEMANDS.

ABOUT 75°

THE HAND SCRAPER IS HELD FIRMLY BETWEEN THE THUMB AND FINGERS AT AN ANGLE OF ABOUT 75° AND SPRUNG TO A SLIGHT CURVE BY PRESSURE OF THE THUMBS. DUST, INSTEAD OF A SHAVING, INDICATES A DULL SCRAPER.

B SHARPENING A HAND SCRAPER

BURNISHER

90° 89° 88° 87° 86° 85°

EDGES OF SCRAPER TURNED OUT

WHEN TURNING THE EDGE OF THE HAND SCRAPER, BEGIN BY HOLDING THE BURNISHER AT A 90° ANGLE TO THE FACE OF THE BLADE. TILT EACH OF THE FOLLOWING STROKES ABOUT 1°. AT THE LAST STROKE THE ANGLE WILL BE ABOUT 85°.

1  2  3

TURN THE BLADE EDGE WITH A FEW FIRM STROKES OF THE BURNISHER ON THE BEVEL SIDE OF THE BLADE. THE SCRAPER BLADE CAN BE HELD IN ANY OF THE THREE WAYS SHOWN ABOVE. DRAW THE BURNISHER TOWARDS YOU THE FULL LENGTH OF THE BLADE WITH A SLIDING STROKE. SOME PREFER TO STROKE BOTH WAYS FROM THE CENTER TOWARDS THE ENDS. A DROP OF OIL ON THE BURNISHER HELPS.

*Information from Stanley Tools*

Figure 13–9. A hand scraper is used to produce a fine finish.

13–9 and 13–10. A very fine finish can be produced with scrapers. Scrapers are also used to remove light scratches in the surface veneers of doors and paneling.

The cutting action of the scraper blade is caused by a slight burr on the edge of the blade. This burr can be produced by using a special burnishing tool to turn the edges of the steel. The side of a nail set can also be used if a burnisher is not available.

## RASPS

Rasps are used for rapid removal of waste material. See Figure 13–11. They may also be used to dress curved edges and to enlarge or shape holes. The cutting action of the triangular-shaped teeth of a rasp produces a very rough surface, which must be finished by another tool. Rasps may be rectangular or half-round in cross section. *Combination* rasps have both coarse and fine teeth.

## SERRATED FORMING TOOLS

A recently developed line of forming tools features a serrated blade. The serrated blade has hundreds of pre-set, razor-sharp steel teeth that act as tiny chisels. The holes between the teeth permit the shavings to pass through, thus preventing clogging. Serrated blades cannot be sharpened, so they must be replaced when dull.

Serrated forming tools consist

**A** USING A CABINET SCRAPER

**Stanley Tools**

HANDLE

BLADE

ADJUSTING THUMB SCREW

CLAMP

BODY

CLAMP THUMB SCREWS

BOTTOM

HANDLE

**CABINET SCRAPER**

THE CABINET SCRAPER IS USED FOR THE FINAL SMOOTHING BEFORE SANDPAPERING. IT REMOVES THE SLIGHT RIDGES LEFT BY THE PLANE. IT IS ALSO USED TO SMOOTH SURFACES THAT ARE DIFFICULT TO PLANE BECAUSE OF CURLY OR IRREGULAR GRAIN.

TO ADJUST AND USE THE CABINET SCRAPER, LOOSEN THE ADJUSTING THUMB SCREW AND THE CLAMP THUMB SCREWS. INSERT THE BLADE FROM THE BOTTOM WITH THE BEVEL SIDE TOWARDS THE ADJUSTING THUMB SCREW.

TRY THE SCRAPER AND CHANGE THE ADJUSTMENT UNTIL IT TAKES A THIN, EVEN SHAVING. HOLD IT TURNED A LITTLE TO THE SIDE TO START A CUT. THE CABINET SCRAPER IS USUALLY PUSHED, BUT IT CAN BE PULLED. DUST, INSTEAD OF A SHAVING, INDICATES A DULL SCRAPER.

**B** SHARPENING A CABINET SCRAPER

55°    65°    75°

BURNISHER

45°   **A**    **B**    **C**    **D**

THE LEVEL OF THE BLADE SHOULD BE AT A 45° ANGLE WHEN TURNING THE EDGE. (A) BEGIN THE FIRST STROKE WITH THE BURNISHER HELD AT AN ANGLE A LITTLE GREATER THAN THE 45° LEVEL. (B) GRADUALLY INCREASE THE ANGLE WITH EACH CUT. (C) AT THE LAST STROKE THE ANGLE OF THE BURNISHER SHOULD BE AT ABOUT A 75° ANGLE TO THE FLAT FACE OF THE BLADE. (D) IF THE EDGE OF THE BLADE HAS BEEN TURNED TOO FAR, IT CAN BE RAISED BY DRAWING THE POINT OF THE BURNISHER ALONG THE EDGE UNDER THE BURR.

TURN THE BLADE EDGE WITH A FEW FIRM STROKES OF THE BURNISHER ON THE BEVEL SIDE OF THE BLADE. THE SCRAPER BLADE CAN BE HELD IN ANY OF THE THREE WAYS SHOWN ABOVE.

*Information from Stanley Tools*

Figure 13-10. Procedure for using a cabinet scraper. Cutting action is produced by a slight burr on the edge of the blade.

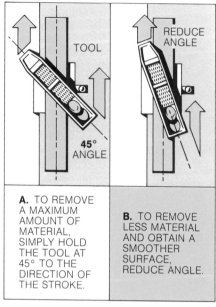

TOOL

REDUCE ANGLE

45° ANGLE

**A.** TO REMOVE A MAXIMUM AMOUNT OF MATERIAL, SIMPLY HOLD THE TOOL AT 45° TO THE DIRECTION OF THE STROKE.

**B.** TO REMOVE LESS MATERIAL AND OBTAIN A SMOOTHER SURFACE, REDUCE ANGLE.

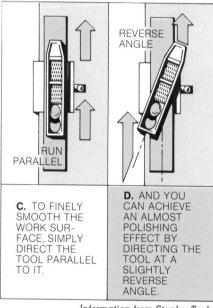

REVERSE ANGLE

RUN PARALLEL

**C.** TO FINELY SMOOTH THE WORK SURFACE, SIMPLY DIRECT THE TOOL PARALLEL TO IT.

**D.** AND YOU CAN ACHIEVE AN ALMOST POLISHING EFFECT BY DIRECTING THE TOOL AT A SLIGHTLY REVERSE ANGLE.

*Information from Stanley Tools*

Figure 13-13. Different cutting actions can be produced with a serrated forming tool.

Figure 13-11. Rasps are used to remove waste material quickly.

Figure 13-12. This serrated forming tool (trade name is Surform) is specially designed for curved surfaces.

of a body, a handle, and the serrated blade. See Figure 13-12. They can be adapted for rough cutting or for final smoothing by changing the angle at which they are held. See Figure 13-13. They are used on wood, plastic, vinyl, rubber, fiberglass, composition board, and soft metals such as aluminum.

# SECTION 4

# Power Tools

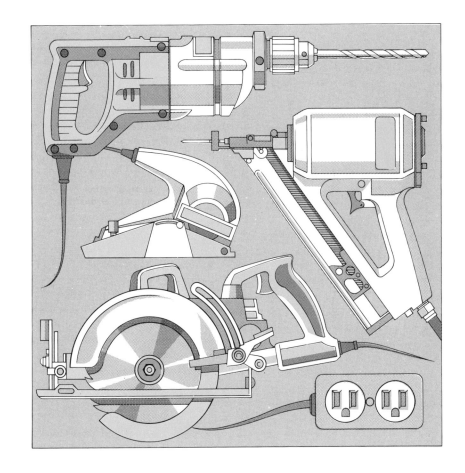

Power tools have greatly changed today's construction industry. Many operations formerly performed with hand tools are now performed with power tools. As a result, the individual carpenter can produce more work with greater efficiency than in the past.

On jobs under the jurisdiction of the United Brotherhood of Carpenters and Joiners of America, all power tools are provided by the building contractor. The carpenter must provide hand tools. In non-union jobs, the power tools may or may not be provided by the contractor.

*Portable* power tools are used more often than *stationary* power tools. Light enough to be easily carried by the carpenter, portable power tools are used during all stages of construction. Stationary power tools, which are heavier pieces of equipment that cannot be easily moved about, are usu-ally used in a shop rather than on the job. However, a few stationary power tools, such as the radial-arm saw and the table saw, may be set up on the job site.

Sawing and boring power tools are usually operated by electricity. If no electricity is available, gas-driven models are used. Nailers and staplers are often pneumatically operated. They receive their power from electrically powered or gas-powered air compressors. Powder-actuated tools are used to fasten materials to concrete and steel.

The increased use of power tools has produced a rise in job accidents. Most of these accidents have been caused by faulty equipment, careless handling, or lack of knowledge about the tool. One of the most important safety rules for using power tools is to read the *manufacturer's instructions* that are either printed on the tool or are provided in a booklet accompanying the tool.

# UNIT 14

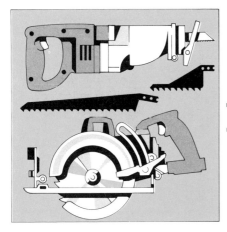

# Portable Power Saws

## SAFETY IN USING PORTABLE POWER SAWS AND OTHER POWER TOOLS

Many of the safety rules for using electrically powered saws apply to other electric tools as well. The greatest hazard related to electric tools is *electric shock,* which can cause serious injury or even death. Shock occurs when a defect in the electric system causes an electric current to pass through the outside housing of the tool. (A loose or exposed wire can cause this problem.) An electric current seeks the easiest path to the ground, and the human body is an excellent conductor of electricity.

Electric shock cannot occur if the tool is properly *grounded.* In order to be grounded, a power tool must have a three-wire conductor cord with a three-prong plug that fits into a grounded outlet. See Figure 14–1. A grounded outlet has a ground wire that is connected to a water pipe or a ground rod driven into the earth. If a fault occurs in the tool, the electric current will travel through the ground wire in the cord to the ground wire connected to the outlet. Figure 14–2

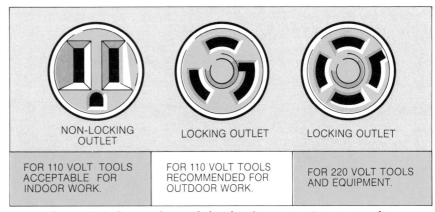

| NON-LOCKING OUTLET | LOCKING OUTLET | LOCKING OUTLET |
|---|---|---|
| FOR 110 VOLT TOOLS ACCEPTABLE FOR INDOOR WORK. | FOR 110 VOLT TOOLS RECOMMENDED FOR OUTDOOR WORK. | FOR 220 VOLT TOOLS AND EQUIPMENT. |

Figure 14–1. Approved grounded outlets for construction power tools.

shows a typical grounding system.

*Double-insulated* tools do not require a grounding system to be safe. The electric parts in the motors of these tools are covered by extra insulation that prevents electric current from reaching the surface of the tool.

Even a properly grounded tool or a double-insulated tool can be dangerous under wet conditions, since water is a conductor of electricity. Extra precautions should be taken when operating power tools near water or dampness.

General safety rules for using power tools of all types are:

1. Periodically inspect electric cords for cuts, kinks, worn insulation, and exposed wire.

2. Place electric cords so they do not present a tripping hazard. Do not expose cords to damage from mobile equipment, welding, or burning operations.

3. Wear safety goggles when using power tools.

4. Do not carry a tool by its cord. Do not yank the cord from a receptacle.

5. Avoid operating electrically powered tools in damp locations. When this is unavoidable, use insulating platforms, rubber mats, and rubber gloves.

6. Do not carry plugged-in

portable tools with a finger on the trigger switch.

7. Do not wear loose clothing, ties, rings, or other items that could be caught by moving parts.

8. Wear hair nets to keep long hair from becoming entangled in the moving parts of power-driven equipment.

9. Always disconnect electric tools when they are not in use.

10. Store portable power tools to protect them from damage, dampness, and dirt.

11. Keep the cord of a power tool safely away from the blade.

## CIRCULAR ELECTRIC HANDSAW

The circular electric handsaw is the power tool used most often by carpenters. Its main purpose is to cut lumber. Special blades can also be used to cut different types of non-wood materials.

The blade of the electric handsaw turns in an upward direction. It cuts from the underside of the board up through the top. This cutting action is opposite from that of a non-electric handsaw, which cuts down from the top of the board. Electric handsaws are equally efficient for crosscutting and ripping. They can be adjusted to cut angles ranging from 90° to 45°.

The two types of electric handsaws normally used in construction work are the *side-drive* and *worm-drive* saws. See Figure 14–3. The size of an electric handsaw is determined by the largest-diameter blade it can use. The chart in Figure 14–4 lists the different sizes available and the depths to which they can cut at a 90° or 45° angle. The saws used most often by construction carpenters have blades 7¼", 7½", or 8¼" in diameter.

## Blades for the Circular Electric Handsaw

Many kinds of blades are used with the electric handsaw. See Figure 14–5. *Rip* blades have teeth shaped for cutting in the direction of the grain. *Crosscut* blades are designed to cut across the grain. *Combination* blades are used for both crosscut and ripping operations. Combination blades are used more frequently than rip or crosscut blades, since most jobs require both kinds of sawing.

*Hollow-ground planer* blades provide a smoother (though slower) crosscut or ripping cut

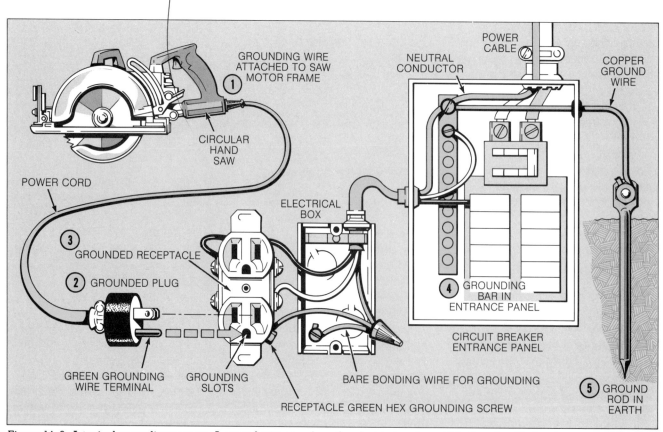

Figure 14–2. A typical grounding system. A ground wire runs from the electric handsaw (1) to the plug (2). Another ground wire runs from the grounded receptacle (3) to a grounding bar in the service panel (4). A copper ground goes from the electrical service box to the ground rod (5) in the earth.

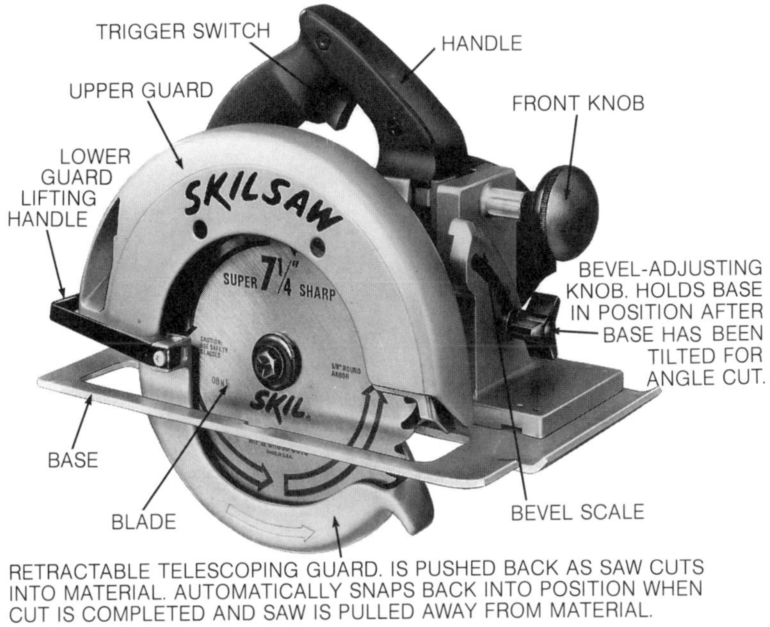
TRIGGER SWITCH
HANDLE
UPPER GUARD
FRONT KNOB
LOWER GUARD LIFTING HANDLE
SKILSAW
SUPER 7¼ SHARP
SKIL
BEVEL-ADJUSTING KNOB. HOLDS BASE IN POSITION AFTER BASE HAS BEEN TILTED FOR ANGLE CUT.
BASE
BLADE
BEVEL SCALE
RETRACTABLE TELESCOPING GUARD. IS PUSHED BACK AS SAW CUTS INTO MATERIAL. AUTOMATICALLY SNAPS BACK INTO POSITION WHEN CUT IS COMPLETED AND SAW IS PULLED AWAY FROM MATERIAL.

**SIDE-DRIVE**

*Skil Corporation*

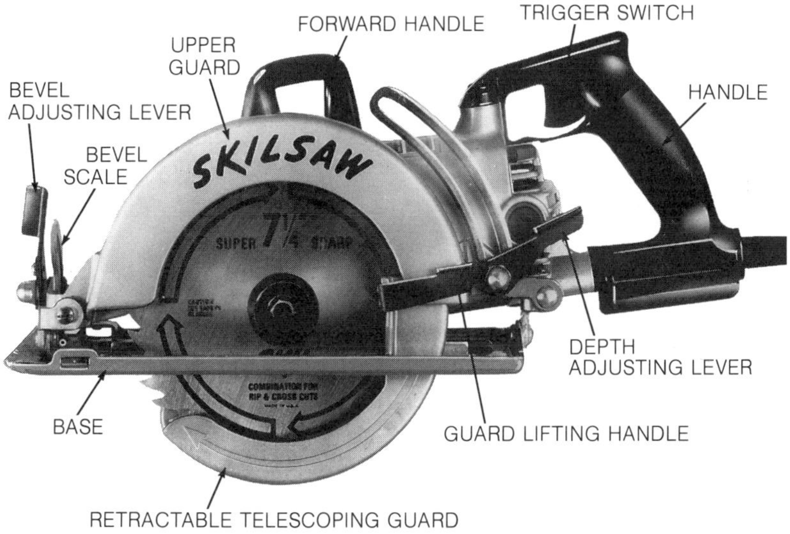
FORWARD HANDLE
TRIGGER SWITCH
UPPER GUARD
BEVEL ADJUSTING LEVER
HANDLE
BEVEL SCALE
SKILSAW
SUPER 7¼
DEPTH ADJUSTING LEVER
BASE
GUARD LIFTING HANDLE
RETRACTABLE TELESCOPING GUARD

**WORM-DRIVE**

*Skil Corporation*

Figure 14-3. Side-drive and worm-drive portable electric handsaws are used in construction work.

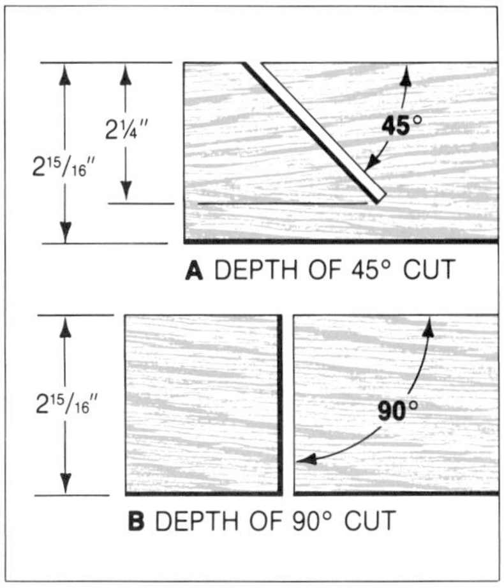
2¼"
2 15/16"
45°
A DEPTH OF 45° CUT
2 15/16"
90°
B DEPTH OF 90° CUT

| BLADE DIAMETER | CAPACITY 45° | CAPACITY 90° |
|---|---|---|
| $4\frac{1}{2}''$ | $1\frac{1}{16}''$ | $1\frac{5}{16}''$ |
| $6\frac{1}{2}''$ | $1\frac{5}{8}''$ | $2\frac{1}{16}''$ |
| $6\frac{3}{4}''$ | $1\frac{3}{4}''$ | $2\frac{7}{32}''$ |
| $7\frac{1}{4}''$ | $1\frac{7}{8}''$ | $2\frac{3}{8}''$ |
| $7\frac{1}{2}''$ | $2\frac{1}{16}''$ | $2\frac{17}{32}''$ |
| $8\frac{1}{4}''$ | $2\frac{1}{4}''$ | $2\frac{15}{16}''$ |
| $10\frac{1}{4}''$ | $2\frac{3}{4}''$ | $3\frac{5}{8}''$ |
| $12''$ | $3\frac{5}{16}''$ | $4\frac{3}{8}''$ |

Figure 14-4. This table gives the depths of 45° and 90° cuts by electric handsaws of different sizes. For example, an $8\frac{1}{4}''$ saw will cut to a 45° depth of $2\frac{1}{4}''$ and a 90° depth of $2\frac{15}{16}''$ as shown in drawings A and B.

than regular crosscut or rip blades. *Flat-ground plywood* blades cut plywood and fiberboard without tearing or splintering them. *Crosscut flooring* blades make smooth cross-grain cuts and can be used as a rip or cutoff blade on extremely hard woods. *Chisel-tooth combination* blades are especially good for tempered laminates, exterior plywood, and other materials that dull blades rapidly.

*Carbide-tipped* blades have tungsten-carbide tips braised to the teeth. They are effective for cutting through hard materials, and they remain sharp much longer than conventional blades. A major disadvantage of the carbide-tipped blade, however, is the expense of repairing it if the carbide tips are broken off by nails or other metals imbedded in wood being reused. Before cutting with a carbide-tipped blade, always check the material being cut for any material that could damage the tips.

*Abrasive* blades cut a variety

of masonry materials and metals. *Silicon carbide* abrasive blades cut concrete, marble, granite, glazed and ceramic tile, slate, terrazzo, acid-proof brick, silica, hard chrome brick, and magnesite. *Aluminum oxide* abrasive blades cut metals such as stainless steel, aluminum, bronze, and brass.

Figure 14–5. Different blades are used with a circular electric handsaw for different operations.

Figure 14-6. When cutting a compound angle with a circular electric handsaw, set the depth of the blade slightly more than the thickness of the material.

Figure 14-8. Use sawhorses when cutting across plywood panels.

*Milwaukee Electric Tool Corporation*
Figure 14-9. An adjustable rip fence attachment aids in making a rip cut with a circular electric handsaw.

Figure 14-7. When crosscutting long boards supported by sawhorses, do not cut between the sawhorses. Instead, make the cut past the end of a sawhorse.

Other blades are available for making rabbet and dado cuts, for cutting fabrics (such as carpet) and rubber, and for cutting glass, asbestos, and cement. Blades with abrasive grit make it possible to saw and sand in one operation.

## Cutting Methods with the Circular Electric Handsaw

When cutting a piece of material with a circular electric handsaw,

set the depth of the blade slightly more than the thickness of the material. The depth of the cut is adjusted by raising or lowering the blade in relation to the *saw base.* The blade is locked into place with a lever or knob. When adjusting to cut at an angle, loosen the *tilt knob (bevel-adjusting knob).* This allows the saw base to be tilted to the desired angle of the cut. The tilt knob is then retightened. See Figure 14-6.

When beginning any cut, hold the blade slightly back from the material. Start the saw and let it attain full speed before pushing it ahead. The *telescoping guard* will be pushed back as the saw advances into the material. When material is being cut at an angle, the telescoping guard may stick. If so, pull the guard up with the *guard-lifting handle.*

When cutting freehand along a straight line, follow the guide slot on the tool or watch the blade. Many carpenters prefer to watch the blade instead of following the guide slot, because guide slots are sometimes not accurate on older saws.

When crosscutting long pieces, support the material by sawhorses. See Figures 14-7 and 14-8. Do not make cuts between the sawhorses. This will cause the saw to bind and the lumber to split toward the end of the cut.

Before cutting across the width of plywood panels, some carpenters place 2" boards beneath the panels to prevent the cut pieces from dropping down and binding the saw.

Most electric handsaws can accommodate a fence attachment to make an accurate narrow rip. See Figure 14-9. To make an accurate cut for a wider piece, tack down a straightedge as shown in Figure 14-10.

## Safety in Using the Circular Electric Handsaw

More serious job injuries occur from the use of electric handsaws than from any other portable power tool. Most of these accidents are caused by carelessness or by a malfunction of the tool. More to blame than any other single factor is the retractable guard failing to snap back into position after the cut has been completed. The guard may break as a result of wear, or it may jam as a result of a wood chip lodging between the blade and the guard.

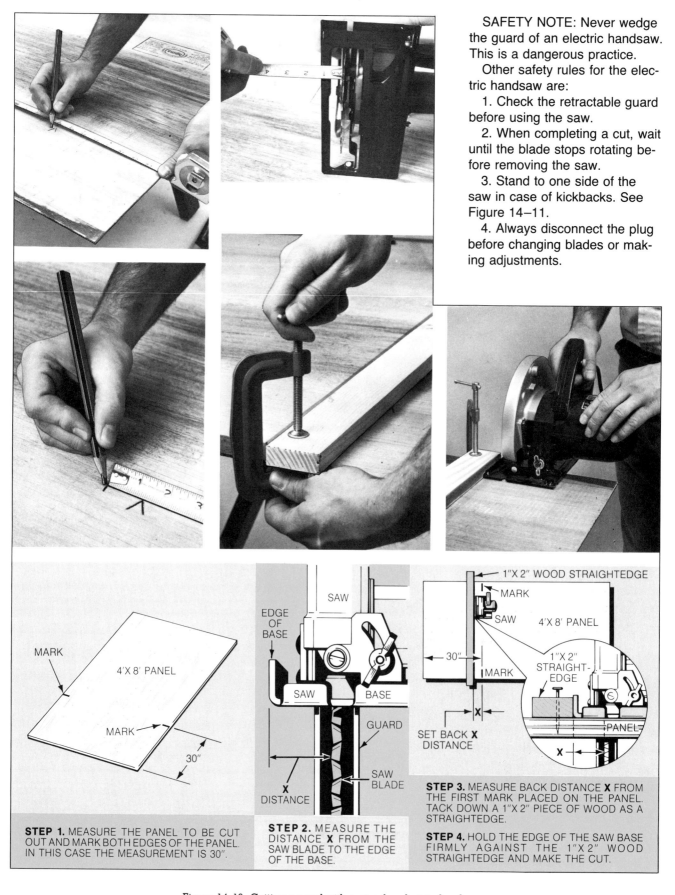

SAFETY NOTE: Never wedge the guard of an electric handsaw. This is a dangerous practice.

Other safety rules for the electric handsaw are:

1. Check the retractable guard before using the saw.

2. When completing a cut, wait until the blade stops rotating before removing the saw.

3. Stand to one side of the saw in case of kickbacks. See Figure 14–11.

4. Always disconnect the plug before changing blades or making adjustments.

MARK

4'X 8' PANEL

MARK

30"

**STEP 1.** MEASURE THE PANEL TO BE CUT OUT AND MARK BOTH EDGES OF THE PANEL. IN THIS CASE THE MEASUREMENT IS 30".

EDGE OF BASE

SAW

SAW    BASE

GUARD

SAW BLADE

X DISTANCE

**STEP 2.** MEASURE THE DISTANCE **X** FROM THE SAW BLADE TO THE EDGE OF THE BASE.

1"X 2" WOOD STRAIGHTEDGE

MARK

SAW

4'X 8' PANEL

30"

MARK

1"X 2" STRAIGHT-EDGE

X

SET BACK **X** DISTANCE

PANEL

X

**STEP 3.** MEASURE BACK DISTANCE **X** FROM THE FIRST MARK PLACED ON THE PANEL. TACK DOWN A 1"X 2" PIECE OF WOOD AS A STRAIGHTEDGE.

**STEP 4.** HOLD THE EDGE OF THE SAW BASE FIRMLY AGAINST THE 1"X 2" WOOD STRAIGHTEDGE AND MAKE THE CUT.

Figure 14–10. Cutting a panel with a circular electric handsaw.

Figure 14–11. Stand to one side of a circular electric handsaw in case of kickbacks.

## OTHER PORTABLE POWER SAWS

Although the circular electric handsaw is the "workhorse" of portable power cutting tools, it is usually not the only saw used on a construction job. The reciprocating saw, the saber saw, and the chain saw are better suited for certain tasks than the circular saw.

### Reciprocating Saw

A reciprocating saw (Figure 14–12) can be used under conditions that would be dangerous or impractical for a circular saw.

It is particularly useful for remodeling work where sections of framing members, sheathing, or inside wall covering must be cut out.

The reciprocating saw is named for its cutting action, which is a *reciprocating* (up-and-down) motion. Many different blades are available for cutting different materials and for various types of cutting operations. See Figure 14–13.

### Saber Saw

The saber saw is also called a *bayonet saw* or *jig saw.* See Figure 14–14. The narrow blade

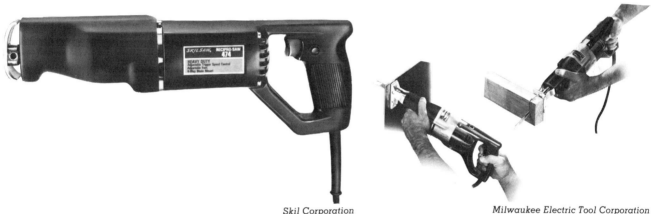

Skil Corporation                                   Milwaukee Electric Tool Corporation

Figure 14–12. A reciprocating saw is used where a circular saw would be dangerous or impractical to use.

Milwaukee Electric Tool Corporation

Figure 14–13. Different blades are used with a reciprocating saw for different operations.

of this saw extends from the base and cuts with an *orbital* (circular) movement. The blade cuts on the upstroke and moves slightly away from the material on the downstroke. See Figure 14–15.

The saber saw is used to cut a variety of thinner materials. It is well suited for sawing along curved lines and for cutting out circular and rectangular openings. Assorted blades are available for cutting different materials.

### Chain Saw

The chain saw is used for cutting heavy timbers and pilings. See

Figure 14-14. A saber (jig or bayonet) saw is used to saw along curved lines and to cut circular and rectangular openings.

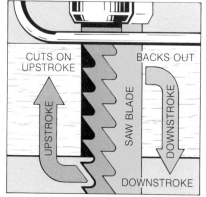

Figure 14-15. The saber saw cuts with an orbital movement. The blade cuts on the upstroke and moves slightly away from the material on the down-stroke.

*Homelite Division of Textron Inc.*

Figure 14-16. Parts of a gasoline-powered chain saw. This saw is used to cut heavy timbers and pilings.

Figure 14-16. It is also very useful on demolition projects. Some chain saws are electrically operated. However, gasoline-operated models are more powerful and are more frequently used.

## Safety in Using Reciprocating and Saber Saws

The main danger from reciprocating and saber saws is the exposed, unguarded blade. Observe these two major safety rules when using these saws:

1. Do not reach under the material being cut.

2. After completing a cut, wait until the motor stops before removing the saw from the material.

## Safety in Using Chain Saws

Safety rules for chain saws are:

1. Grip the handle bar when transporting the saw.

2. Carry the saw with the guide bar and chain pointing behind you.

3. Do not transport the saw with the engine running.

4. Do not fill a gasoline-operated saw when the engine is hot. After pouring gasoline, wipe off any spilled fuel before starting the engine. Do not start the saw in the area where it was refueled.

5. Do not smoke while refueling a gasoline-operated saw.

6. When checking the saw chain for tension, or when oiling the chain, always use gloves. Do not touch the cutters with your fingers or bare hands.

# UNIT 15

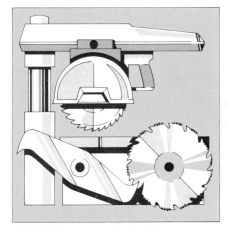

# Stationary Power Saws

Several types of stationary power saws may be used on the building site. These large saws are normally furnished by the contractor.

## RADIAL-ARM SAW

Many carpenters and building contractors consider the radial-arm saw essential for good production on any construction project of medium to large size. A radial-arm saw is also known as a *cutoff* saw. See Figure 15–1. It is placed in a convenient location on the job, and a platform is extended on both sides to support

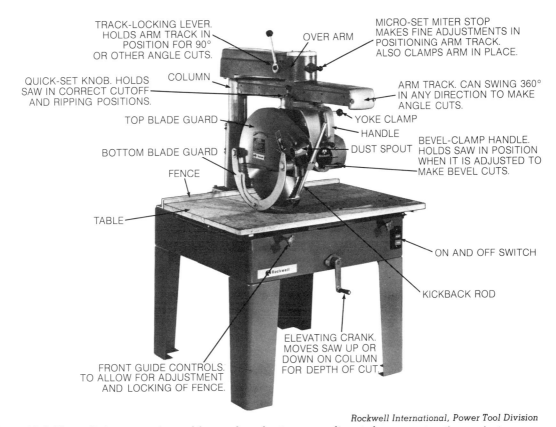

TRACK-LOCKING LEVER. HOLDS ARM TRACK IN POSITION FOR 90° OR OTHER ANGLE CUTS.

OVER ARM

MICRO-SET MITER STOP MAKES FINE ADJUSTMENTS IN POSITIONING ARM TRACK. ALSO CLAMPS ARM IN PLACE.

QUICK-SET KNOB. HOLDS SAW IN CORRECT CUTOFF AND RIPPING POSITIONS.

COLUMN

ARM TRACK. CAN SWING 360° IN ANY DIRECTION TO MAKE ANGLE CUTS.

TOP BLADE GUARD

YOKE CLAMP

HANDLE

BEVEL-CLAMP HANDLE. HOLDS SAW IN POSITION WHEN IT IS ADJUSTED TO MAKE BEVEL CUTS.

BOTTOM BLADE GUARD

DUST SPOUT

FENCE

TABLE

ON AND OFF SWITCH

KICKBACK ROD

FRONT GUIDE CONTROLS. TO ALLOW FOR ADJUSTMENT AND LOCKING OF FENCE.

ELEVATING CRANK. MOVES SAW UP OR DOWN ON COLUMN FOR DEPTH OF CUT.

*Rockwell International, Power Tool Division*

Figure 15–1. The radial-arm saw is used for good production on medium to large construction projects.

Figure 15–2. Cutting a long board on the job site with a radial-arm saw.

long pieces of material to be cut. See Figure 15–2.

The radial-arm saw can be used for most of the heavy-duty sawing required on most jobs. It can be set up for crosscutting, ripping, and many types of angle cuts. See Figure 15–3.

The size of a radial-arm saw is determined by the largest blade it will accommodate. Blades range from 8″ to 20″ in diameter.

The 14″ and 16″ sizes are recommended for heavy-duty construction work. Many types of blades are available for different cutting operations. See Figure 15–4.

## Safety in Using the Radial-arm Saw

All radial-arm saws today are provided with lower and upper blade guards. These guards are a vital safeguard against potential accidents. They should be checked regularly for breakage or other defects.

Danger can result from the forward crawl of the radial-arm saw caused by the rotation of the blade. The saw can usually be prevented from crawling forward if the saw table is slightly tilted back when the unit is set up. Pull-back devices are available and are strongly recommended. For example, a spring-loaded device that can be added to most radial-arm saw models features a cable that attaches to the head of the cutter. It ensures that the entire motor unit will return to a position behind the fence after the cut is made.

Safety rules for the radial-arm saw are:

1. Keep a firm grip on the yoke handle when pulling the saw blade through the material. When making crosscuts, keep the hand holding the material at least 8″ away from the cut.

2. Clear away scraps or other litter in the work area in front of the saw table.

3. Keep the saw table clear.

Figure 15–3. A radial-arm saw is used for crosscutting, ripping, and making angle cuts.

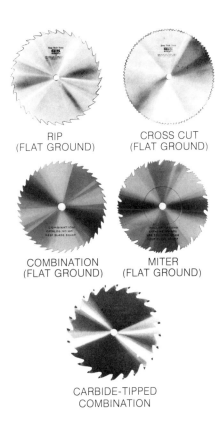

RIP
(FLAT GROUND)

CROSS CUT
(FLAT GROUND)

COMBINATION
(FLAT GROUND)

MITER
(FLAT GROUND)

CARBIDE-TIPPED
COMBINATION

*Rockwell International, Power Tool Division*
Figure 15–4. Different blades are used with a radial-arm saw for different operations.

Brush off wood chips and scraps.

4. When ripping, push the material through the blade as it rotates *toward* the operator.

5. Use a push stick when ripping.

## TABLE SAW

The table saw (Figure 15–5) is of great value in interior finish work. It is used to cut paneling and other trim materials such as molding. See Figures 15–6 and 15–7. With a dado setup of blades and inside cutters, it can also be used to make rabbet and dado cuts. See Figure 15–8. Rabbet and dado cuts of different sizes are made by placing a cutter or a combination of cutters between the blades. Paper inserts are used to make fine adjustments.

### Safety in Using the Table Saw

Most accidents occur with the table saw when it is used without the guard in place and without a push stick being used. Safety rules for the table saw are:

1. Use the rip fence for ripping. Use the miter gauge for crosscutting. Freehand crosscutting is very dangerous.

2. Keep the work area around the saw free of wood scraps and litter.

3. Set the saw blade no more than ⅛″ to ¼″ above the stock being cut.

4. Allow the blade to reach full speed before starting the cut.

5. Do not reach over the saw when it is running.

6. When ripping longer pieces, if there is no extension bench on the opposite side of the saw table, have another worker assist you to support the end of the material.

## MOTORIZED MITER BOX SAW

The motorized miter box saw is used mainly for finish work. It

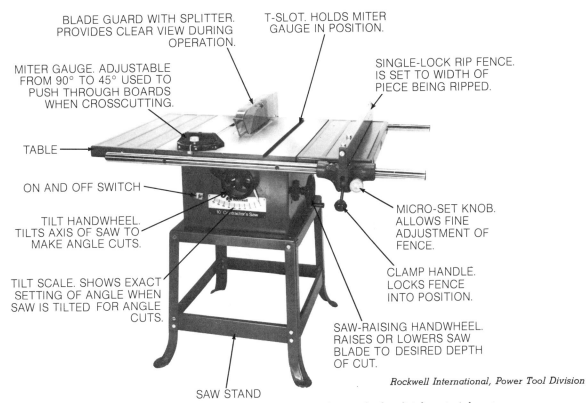

BLADE GUARD WITH SPLITTER. PROVIDES CLEAR VIEW DURING OPERATION.

T-SLOT. HOLDS MITER GAUGE IN POSITION.

MITER GAUGE. ADJUSTABLE FROM 90° TO 45° USED TO PUSH THROUGH BOARDS WHEN CROSSCUTTING.

SINGLE-LOCK RIP FENCE. IS SET TO WIDTH OF PIECE BEING RIPPED.

TABLE

ON AND OFF SWITCH

TILT HANDWHEEL. TILTS AXIS OF SAW TO MAKE ANGLE CUTS.

TILT SCALE. SHOWS EXACT SETTING OF ANGLE WHEN SAW IS TILTED FOR ANGLE CUTS.

MICRO-SET KNOB. ALLOWS FINE ADJUSTMENT OF FENCE.

CLAMP HANDLE. LOCKS FENCE INTO POSITION.

SAW-RAISING HANDWHEEL. RAISES OR LOWERS SAW BLADE TO DESIRED DEPTH OF CUT.

SAW STAND

*Rockwell International, Power Tool Division*

Figure 15–5 A tilting-arbor table saw is used to cut paneling and other finish materials.

Rockwell International, Power Tool Division
Figure 15–6. Ripping a board with a table saw.

saw, or it may be used as a portable power tool. As a stationary saw, it can be mounted on a stand and remain in place during the course of the job. Many models, however, weigh less than 50 pounds and can easily be moved about. See Figure 15–9.

## FRAME-AND-TRIM SAW

The manufacturers of power equipment are continually introducing new types of tools for the construction industry. One of the newest types of power saws is the all-aluminum frame-and-trim saw. See Figure 15–10.

The frame-and-trim saw com-

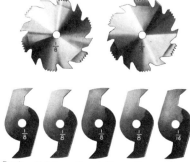

Rockwell International, Power Tool Division

makes accurate 90° to 45° angle cuts, which are used when fitting molding. It is suited for cutting wood, composition materials, plastic, and lightweight aluminum.

The motorized miter box saw may be set up as a stationary

Rockwell International, Power Tool Division
Figure 15–7. Crosscutting a miter with a table saw.

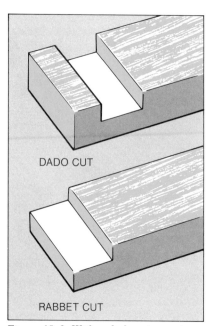

DADO CUT

RABBET CUT

Figure 15–8. With a dado setup, a table saw can be used to make rabbet and dado cuts. The set shown here has two hollow-ground blades and five inside cutters.

Figure 15–9. A motorized miter box saw is used to cut trim materials.

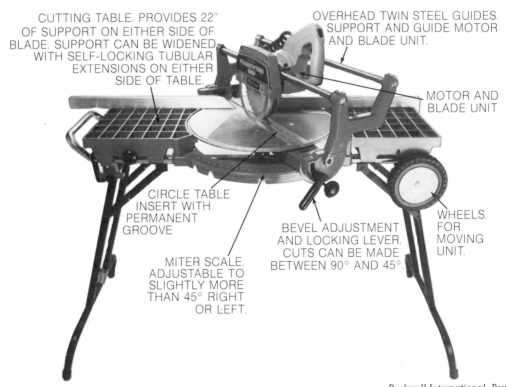

CUTTING TABLE. PROVIDES 22" OF SUPPORT ON EITHER SIDE OF BLADE. SUPPORT CAN BE WIDENED WITH SELF-LOCKING TUBULAR EXTENSIONS ON EITHER SIDE OF TABLE.

OVERHEAD TWIN STEEL GUIDES. SUPPORT AND GUIDE MOTOR AND BLADE UNIT.

MOTOR AND BLADE UNIT

CIRCLE TABLE INSERT WITH PERMANENT GROOVE

MITER SCALE. ADJUSTABLE TO SLIGHTLY MORE THAN 45° RIGHT OR LEFT.

BEVEL ADJUSTMENT AND LOCKING LEVER. CUTS CAN BE MADE BETWEEN 90° AND 45°.

WHEELS. FOR MOVING UNIT.

*Rockwell International, Power Tool Division*

Figure 15–10. Parts of a frame-and-trim saw. This saw combines many of the features of a radial-arm saw and a motorized miter box saw.

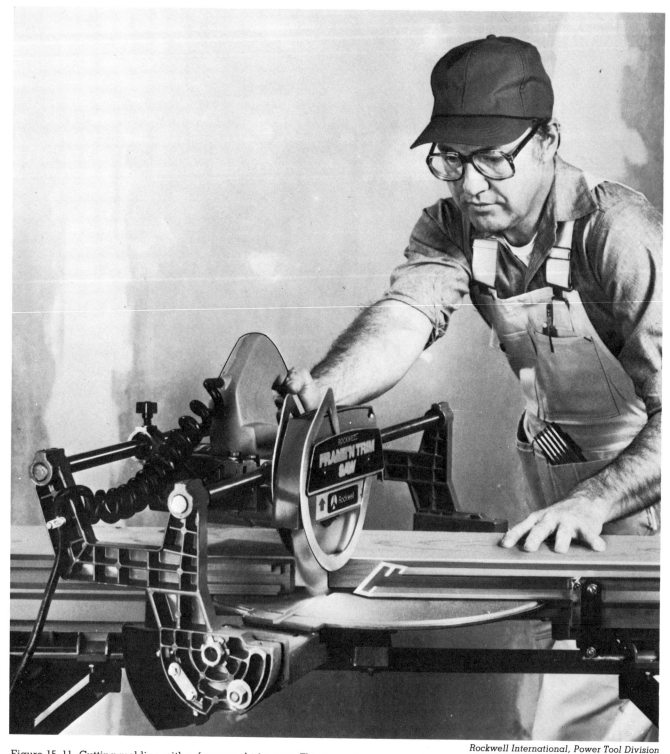

*Rockwell International, Power Tool Division*

Figure 15–11. Cutting molding with a frame-and-trim saw. This type of saw can perform crosscuts, miters, and bevel cuts on stock up to 2″ thick and 12″ wide.

bines many of the features of a radial-arm saw and a motorized miter box saw. It can perform crosscuts, miters, and bevel cuts on stock up to 2″ thick and 12″ wide. See Figure 15–11. It can-

not be used for ripping.

The frame-and-trim saw is a valuable tool on remodeling jobs and limited framing operations. An advantage is the ease with

which it can be moved and set up. The saw folds to a compact traveling storage size of approximately 22″ deep, 32″ wide, and 50″ long.

# UNIT 16

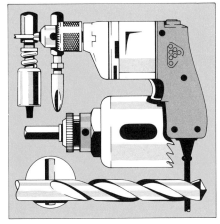

# Portable Power Drills and Power Screwdrivers

Portable power drills have generally replaced hand tools for drilling holes, because they are faster and more accurate. With variable-speed controls and special clutch-drive chucks, they can also be used as electric screwdrivers. More specialized power-driven screwdrivers are also available that have greatly increased the efficiency of many fastening operations in construction work.

## PORTABLE POWER DRILLS

Two basic designs for portable electric drills are the *spade* design for heavy-duty construction and the *pistol-grip* design for lighter work. See Figure 16–1. Sizes of power drills are based on the diameter of the largest drill shank that will fit into the chuck of the drill. Power drills are available in sizes from ¼″ to 1¼″. The sizes most often used by construction carpenters are the ¼″, ⅜″, or ½″ models. Spade-handle models usually are available in ⅜″ or ½″ chuck sizes. Pistol-grip models usually are available in ¼″ or ⅜″ chuck sizes.

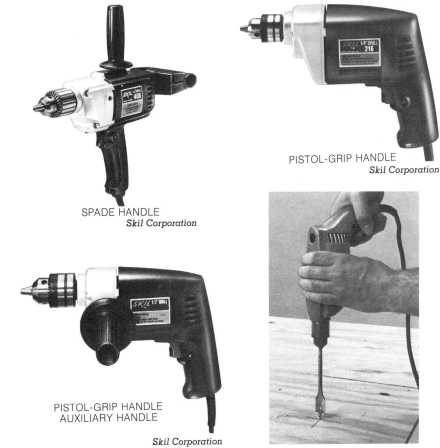

SPADE HANDLE
*Skil Corporation*

PISTOL-GRIP HANDLE
*Skil Corporation*

PISTOL-GRIP HANDLE
AUXILIARY HANDLE

*Skil Corporation*

Figure 16–1. The spade-handle drill is used for heavy-duty work, and the pistol-grip design is used for lighter work.

### Variable-speed Drill

Speed can be controlled by pressure on the trigger of a variable-speed drill. As a result, this kind of drill can be used on different materials. Slower drill speeds are more effective with harder mate-

Figure 16–2. Battery-powered cordless drill and charge unit.

REDUCED SHANK

STRAIGHT SHANK

Figure 16–3. A twist drill is the most frequently used device for boring holes in wood, metal, and other materials.

rials such as steel. Faster drill speeds work better with softer materials such as wood. Some variable-speed drills (such as the models shown in Figure 16–1) also have a *reversing switch* that is useful for withdrawing screws when a screwdriver attachment is used with the drill.

## Battery-powered Drill

The battery-powered drill receives its power from a battery located in the handle of the drill. See Figure 16–2. In some models the battery is carried in a pouch hooked to the worker's belt. A cord goes from the battery to the drill. Special battery chargers that plug into a conventional electric source are necessary, because the batteries lose their charge fairly quickly.

## Drilling Devices

The cutting devices that fit into the chuck of the drill are referred to as *drills* or *bits*. Note that the word *drill* can be used to refer to the cutting device that fits into a power drill or to the drill motor itself. The most frequently used type of device for drilling smaller holes is the *twist drill*. See Figure 16–3.

Twist drills may have a *straight shank* or a *reduced shank*. Straight-shank drills are available in sizes of 1/64" to 1/2". Reduced-shank drills are available in sizes up to 1 1/2". The reduced shank allows the drill to fit a chuck that is smaller in diameter than the drill. For example, a drill of 1/2" diameter will fit a chuck of 1/4" diameter if the drill has a reduced shank of 1/4" diameter. As another example, a drill of 1 1/2" diameter will fit a chuck of 1/2" diameter if the drill has a reduced shank of 1/2" diameter.

Although the twist drill is the most frequently used type of drill in construction work, various other types of drills are also used for boring into wood as well as metal, plastics, masonry, and other kinds of material. See Figure 16–4.

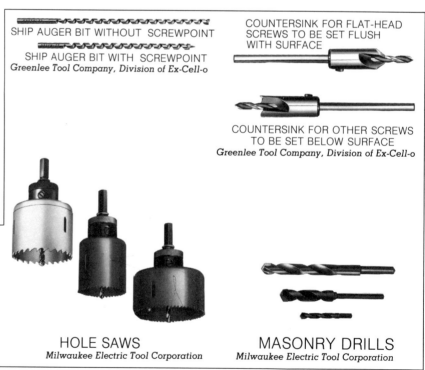

SHIP AUGER BIT WITHOUT SCREWPOINT

SHIP AUGER BIT WITH SCREWPOINT
*Greenlee Tool Company, Division of Ex-Cell-o*

COUNTERSINK FOR FLAT-HEAD SCREWS TO BE SET FLUSH WITH SURFACE

COUNTERSINK FOR OTHER SCREWS TO BE SET BELOW SURFACE
*Greenlee Tool Company, Division of Ex-Cell-o*

FEELER BIT
*Greenlee Tool Company, Division of Ex-Cell-o*

SPADE BIT
*Greenlee Tool Company, Division of Ex-Cell-o*

HOLE SAWS
*Milwaukee Electric Tool Corporation*

MASONRY DRILLS
*Milwaukee Electric Tool Corporation*

Figure 16–4. In addition to the twist drill, various other devices are used to bore into wood, masonry, plastic, and other materials.

The *feeler bit* is used to bore deep holes in wood. It is available in diameters of 3/16" to 3/8" and lengths of 12", 18", and 24".

The *ship auger bit* is also used to bore deep holes in wood. The type of ship auger used most often in construction has a screwpoint. Ship augers without screwpoints are designed especially for end-grain boring. Because there is no screwpoint and the heel is backed off, this type of auger has less tendency to drift in direction, following the grain of the wood. If a ship auger without a screwpoint is used, however, it is necessary to start the hole with a regular auger that has a screwpoint. Ship auger bits are available in diameters of 9/16" to 1" and lengths of 12", 18", and 24".

The carbide-tipped *masonry drill* is used to bore holes in plaster, slate, stone, brick, and other types of masonry. It is available in diameters of 1/8" to 1 1/4".

The *spade bit* (also known as a *flat bit*) is used to bore holes in wood, plastic, and various composition materials. It is available in diameters of 1/4" to 1 1/2".

The *hole saw* is used to cut larger holes than can be cut with drills. A pilot twist drill at the center of the hole saw acts as a guide for the saw blade around the rim. (The blade may have carbide tips.) The hole saw is used on wood, fiberglass, asbestos board, fiberboard, ceramic, tile, and soft metals such as copper piping. It is available in diameters of 9/16" to 6".

The adjustable *countersink cutter* is slipped over a twist drill and held in place with a socket setscrew. One type has cutting edges beveled 82° to match the taper of a flat-head screw so that it will be set flush with the surface of the wood. It is available in sizes that fit drills of 1/8" to 1/4" diameter. Another type cuts flat-

CARBIDE-TIPPED FLUTED PERCUSSION BIT

*Skil Corporation*                    *Skil Corporation*

Figure 16–5. A rotary hammer drills holes with a carbide-tipped fluted percussion or core bit.

bottomed holes for screws that must be set below the surface of the wood. It is available in sizes that fit drills of 3/16" to 1/2" diameter.

## Rotary Hammer and Hammer Drill

The rotary hammer and hammer drill are used by carpenters and other building trades workers to drill holes in concrete and other masonry materials. See Figure 16–5. Various types of anchors are then placed in these holes to receive bolts or screws used to fasten materials to the walls, floors, or ceilings. (Refer to Section 2 to review different types of expansion anchors).

Both the rotary hammer and the hammer drill operate with a rotation-hammering action. The rotary hammer is a heavy-duty tool used to drill large holes. The hammer drill is used to drill small holes.

With a carbide-tipped fluted percussion bit, the rotary hammer can drill holes of 3/16" to 1 1/4" diameter. With a carbide-tipped core bit, it can drill holes of 1 1/2" to 6" diameter. Some models of rotary hammers also accommodate a cold chisel bit for chipping

and edging concrete as shown in Figure 16–6.

The hammer drill can drill holes of up to 1/2" diameter in concrete and 3/8" diameter in steel. Many models of hammer drill have a mechanism that enables disengagement from the rotating-hammering motion, so that the tool can be used as a conventional drilling tool.

## POWER-DRIVEN SCREWDRIVERS

Electric screwdrivers (Figure 16–7) are used for rapid and efficient driving or removal of screws. These tools have become more important to carpenters in recent years because of the increased use of metal framing methods, which involve the increased use of screws. (Materials such as plasterboard and plywood must be attached to metal walls, floors, or ceilings with screws.)

Electric screwdrivers usually have an adjustable depth control to prevent overdriving the screws. Many models feature a clutch mechanism that automatically disengages when the screw has been driven in to the pre-set depth.

The *drywall screwdriver* is

*Skil Corporation*

Figure 16–6. A rotary hammer can be used to chip concrete with a cold chisel bit.

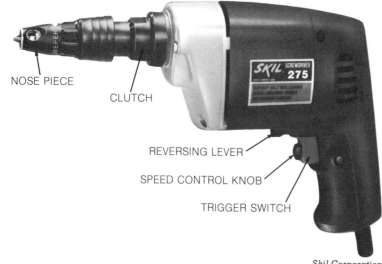

NOSE PIECE

CLUTCH

REVERSING LEVER

SPEED CONTROL KNOB

TRIGGER SWITCH

*Skil Corporation*

Figure 16–7. Parts of a variable-speed electric screwdriver. The drill direction can be reversed for withdrawing screws.

specifically designed for fastening drywall (gypsum board) to walls and ceilings. It can be adjusted so that the screw head will be driven just below the surface of the drywall material without cutting through the outside layer of paper.

For fastening materials into metal framing members, a *self-driving, self-tapping screw* is normally used. This type of screw, which drills and taps its own hole, is described in Section 2.

## SAFETY IN USING POWER DRILLS AND POWER SCREWDRIVERS

The main source of danger from electric drills and screwdrivers is the moving drill bits or screwdriver bits. Before the bit is withdrawn from a bored hole, the drill or screwdriver should be turned off and the chuck allowed to come to a complete stop.

The larger and more powerful the drill, the greater is the caution required from the operator. A large bit can bind in the hole, causing the drill handle to kick back and spin. This can cause hand injuries or knock the operator off balance. It is therefore important to maintain a firm, well-balanced position and a tight grip on the handle of the drill while operating it. Push steadily into the work and alternately pull back on the drill to clean the bit of wood chips that can cause binding. Keep the drill bit straight when reaming or drilling. Additional safety rules for electric drills and screwdrivers are:

1. Clamp small objects before drilling.

2. Keep drill bits sharp and properly shaped.

3. Tighten the chuck securely before using a drill.

4. Disconnect the drill when inserting or removing bits.

5. Do not lock the drill in an ON position while drilling.

6. Be sure the switch is in an OFF position before plugging in the drill.

# UNIT 17

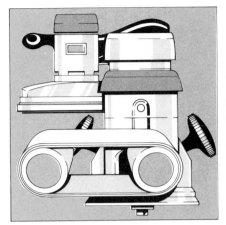

# Portable Power Planes, Routers, and Sanders

Portable power tools are particularly advantageous for finish work. Tools used most frequently are planes, routers, and sanders.

## PORTABLE POWER PLANES

Portable power planes reduce the amount of labor required for planing operations. A spiral cutter driven by an electric motor produces the cutting action of the plane. The depth of the cut is adjusted by a lever.

Cutters on different models of planes range from $2^{13}/_{32}''$ to $3''$ wide. Depth of cut can be adjusted from $1/_{32}''$ to $3/_{16}''$. An adjustable fence allows planing of bevels up to 45°.

A standard size electric plane (Figure 17–1) is used for fitting doors and window sashes and for smoothing rough surfaces on the sides and edges of boards. A smaller type, called a *block plane* (Figure 17–2), is used for planing molding, cabinet doors, and other thin-edged materials. It is also used for planing end grain, cleaning out rough rabbets, and cutting bevels.

## Safety in Using Power Planes

The main danger with power planes is from the exposed cutter. When a cut is completed, always turn off the motor. Do not set the plane down until the motor has stopped.

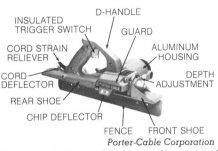
*Porter-Cable Corporation*

*Skil Corporation*

Figure 17–1. A portable power plane is used to fit doors and window sashes and smooth rough surfaces on the sides and edges of boards.

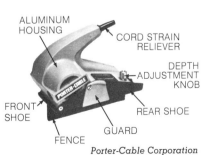
*Porter-Cable Corporation*

*Rockwell International, Power Tool Division*

Figure 17–2. A block plane is a small portable power plane used to fit molding, cabinet doors, and other thin-edged materials. It is also used to plane end grain, clean out rough rabbets, and cut bevels.

111

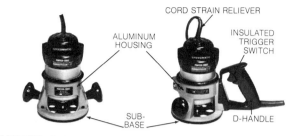

MICROMETER DEPTH ADJUSTMENT. RAISES OR
LOWERS ROUTER BIT TO THE DESIRED DEPTH OF CUT.

CORD STRAIN RELIEVER

ALUMINUM
HOUSING

INSULATED
TRIGGER
SWITCH

SUB-
BASE

D-HANDLE

ROUTER WITH TWO GUIDE KNOBS
AND RECESSED DOUBLE-
POLE SWITCH

ROUTER WITH D-HANDLE
AND TRIGGER SWITCH

COLLET-TYPE CHUCK HOLDS ROUTER BIT IN PLACE
AFTER COLLET HAS BEEN TIGHTENED.

LOCKING HANDLE HOLDS ROUTER BIT IN POSITION
AFTER IT HAS BEEN ADJUSTED TO THE DEPTH OF CUT.

*Porter-Cable Corporation*

Figure 17–3. Portable electric routers are used for mortising and shaping operations.

## PORTABLE POWER ROUTERS

On-the-job mortising and shaping operations are simplified by using portable power routers. See Figure 17–3. Routers are also used for mortising for door hinges.

Different kinds of bits can be inserted into the chuck of a router to make rabbet and dado cuts, as well as a variety of other shapes along the edge of material. Router bits and the types of cuts they produce are shown in Figure 17–4. For additional information on the use of this tool for finish work, see Section 13.

A specialized type of router is the *laminate trimmer*. See Figure 17–5. The sole purpose of this tool is to trim and shape the edges of plastic laminate material used for countertops.

### Safety in Using Power Routers

The main danger with power routers is from the exposed router bit. When a cut is completed, release the switch. Wait until the bit stops rotating before lifting the tool. Always disconnect the plug when inserting router bits and making depth adjustments.

## PORTABLE POWER SANDERS

As a rule, finish materials delivered to the job require little sanding. Occasionally, however, the surfaces of these materials are scratched or damaged while being transported. *Slash grain,* a condition in which the grain changes direction, may occur on the edge of a door, where it prevents a perfectly smooth finish after the door has been planed. For these reasons, portable electric sanders are used during the final trim stages of construction.

### Belt Sander

The belt sander (Figure 17–6) should be used with care. Careless handling can easily gouge the wood surface. The size of a belt sander is usually identified by the width of its sanding belt. Belt widths on heavier duty models are usually 3″ or 4″. Depending on the make and model, belt lengths vary from 21″ to 27″. Different grades of abrasives are available.

### Finish Sander

The finish sander is used for lighter and finer sanding than the belt sander. See Figure 17–7. Two kinds of finish sanders are

available. One operates with an *orbital* (circular) motion and the other has an *oscillating* (back-and-forth) movement. Finish sanders use regular abrasive paper (sandpaper) cut to size from full sheets.

### Sanding Abrasives

The sanding sheets and belts used with portable power sanders are made of four different types of material: flint, garnet, silicon carbide, and aluminum oxide. Flint and garnet are natural minerals. Silicon carbide and aluminum oxide are known as abrasives. They are recommended for most sanding operations done with electric sanders.

Abrasive materials are divided into fine, medium, and coarse grades. These grades are further divided into varying *grit numbers,* which are stamped on the back of the sheet or belt. A chart of grit numbers is shown in Figure 17–8.

### Safety in Using Power Sanders

Hand and arm abrasions can result from careless use of power sanders. Safety rules for these tools are:

1. Wear eye protection and a

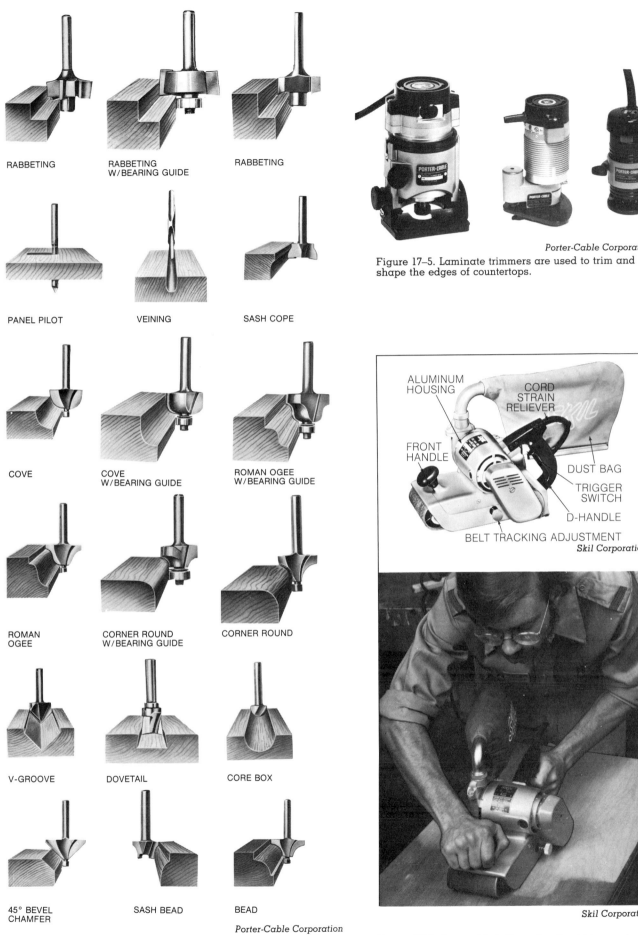

RABBETING

RABBETING
W/BEARING GUIDE

RABBETING

PANEL PILOT

VEINING

SASH COPE

COVE

COVE
W/BEARING GUIDE

ROMAN OGEE
W/BEARING GUIDE

ROMAN
OGEE

CORNER ROUND
W/BEARING GUIDE

CORNER ROUND

V-GROOVE

DOVETAIL

CORE BOX

45° BEVEL
CHAMFER

SASH BEAD

BEAD

*Porter-Cable Corporation*

Figure 17–4. Different router bits can be inserted into the chuck of a router to make different cuts, including rabbet and dado cuts.

*Porter-Cable Corporation*

Figure 17–5. Laminate trimmers are used to trim and shape the edges of countertops.

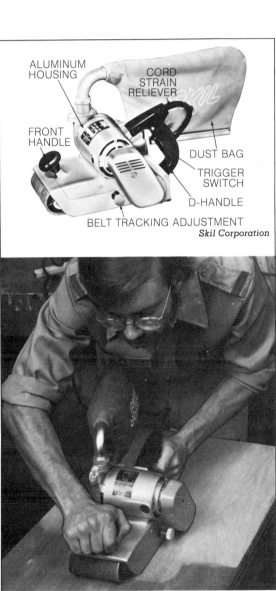

ALUMINUM
HOUSING

CORD
STRAIN
RELIEVER

FRONT
HANDLE

DUST BAG

TRIGGER
SWITCH

D-HANDLE

BELT TRACKING ADJUSTMENT

*Skil Corporation*

*Skil Corporation*

Figure 17–6. A portable belt sander should be used with care to avoid gouging the wood.

113

| ARTIFICIAL* | GARNET | FLINT | GRADE |
|---|---|---|---|
| 400–10/0 | — | — | |
| 360 | — | — | |
| 320–9/0 | — | 7/0 | |
| 280–8/0 | 8/0 | 6/0 | VERY FINE |
| 240–7/0 | 7/0 | 5/0 | |
| 220–6/0 | 6/0 | 4/0 | |
| — | — | 3/0 | |
| 180–5/0 | 5/0 | — | |
| 150–4/0 | 4/0 | — | |
| — | — | 2/0 | FINE |
| 120–3/0 | 3/0 | — | |
| — | — | 0 | |
| 100–2/0 | 2/0 | — | |
| — | — | ½ | |
| 80–0 | 0 | — | |
| — | — | 1 | MEDIUM |
| 60–½ | ½ | — | |
| 50–1 | 1 | 1½ | |
| — | — | 2 | |
| 40–1½ | 1½ | — | |
| — | — | 2½ | COARSE |
| 36–2 | 2 | — | |
| 30–2½ | 2½ | 3 | |
| 24–3 | 3 | — | |
| 20–3½ | 3½ | — | VERY COARSE |
| 16–4 | — | — | |
| 12–4½ | — | — | |

*Includes *silicon carbide* and *aluminum oxide*.

Figure 17–7. Finish sanders are used for lighter and finer sanding than belt sanders.

Figure 17–8. Grit numbers for the four materials used as sanding abrasives. These grit numbers are often stamped on the back of the sheets or belts.

face mask or respirator.

2. Do not light matches when sanding in confined areas. Dust from sanding can be explosive.

3. Lift the sander away from the work before turning off the motor. Wait until the movement has completely stopped before setting down the tool.

4. Always disconnect the plug when changing sanding belts.

5. Keep the electrical cord away from the area being sanded.

# UNIT 18

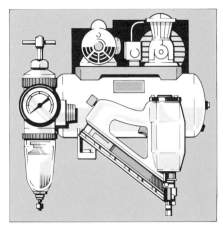

# Pneumatic and Power-Actuated Tools

## PNEUMATIC NAILERS AND STAPLERS

The use of pneumatically powered nailers and staplers on construction jobs continues to grow. These tools are driven by compressed air traveling through air lines connected to an air compressor. The amount of air pressure required to run a pneumatic tool depends on its size and the type of operation being performed. The air pressure is adjusted by a *regulator.* Figure 18–1 shows two types of air systems.

Air compressors are available in different sizes. Larger compressors can supply power for a number of tools at one time. Portable, lightweight air compressors, however, are more practical for most construction jobs. See Figure 18–2.

Many of the portable tools that could be driven only by electricity in the past can now be driven by compressed air. Pneumatic (airpowered) drills, sanders, impact wrenches, hacksaws, circular saws, routers, and laminate trimmers are available. However, these types of pneumatic tools are more practical for shop use than for on-site construction work. Nailers and staplers are the only pneumatic equipment widely used in the construction of buildings.

Pneumatic nailers and staplers are designed to fire in a number of different ways. With some types the trigger is squeezed after the tool is pressed against the material being fastened. With another model the trigger is squeezed first, and then the tool fires each time it is pressed against the material. Some pneumatic tools automatically fire up to a dozen fasteners per second with one pull of the trigger. An important safety feature of all pneumatic nailers and staplers is that they will not fire unless pressed against the material.

### Nailers

Most manufacturers of pneumatic tools produce an assortment of nailers. Heavy-duty models are used for such operations as wall and floor framing and fastening subflooring and sheathing. See Figure 18–3. Various nail sizes and types are available for use in a nailer. For example, a special hardened steel nail is used to fasten wood plates to a concrete floor slab. See Figure 18–4.

*Finish* and *brad* nailers are used to fasten interior trim such as baseboard and wall moldings, cabinet assembly, paneling, and door and window trim. See Figure 18–5. Some models can be adjusted to countersink nails below the surface of the material or flush with the surface. The finish nailer accommodates a variety of finish nail sizes, and the brad nailer accommodates a variety of brad nail sizes.

Nails for pneumatic nailers may be packaged in strips or coils. The coil-fed nailer in Figure 18–6 accommodates a coil of 300 nails and drives them at a rate of up to 10 nails per second. This model of nailer is popular for fastening panel subflooring.

### Staplers

Pneumatic staplers are used for many of the same operations as pneumatic nailers. Under certain conditions a staple has some advantages over a nail as a fastener. A staple will not split the wood as easily as a nail when driven in near the end of a

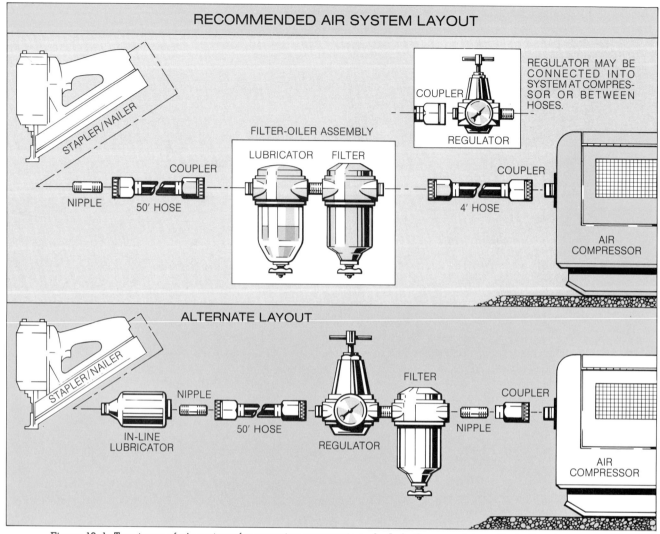

## RECOMMENDED AIR SYSTEM LAYOUT

STAPLER/NAILER

NIPPLE

50' HOSE

COUPLER

FILTER-OILER ASSEMBLY

LUBRICATOR    FILTER

COUPLER

REGULATOR

REGULATOR MAY BE CONNECTED INTO SYSTEM AT COMPRESSOR OR BETWEEN HOSES.

COUPLER

4' HOSE

AIR COMPRESSOR

## ALTERNATE LAYOUT

STAPLER/NAILER

IN-LINE LUBRICATOR

NIPPLE

50' HOSE

FILTER

REGULATOR

NIPPLE

COUPLER

AIR COMPRESSOR

Figure 18–1. Two types of air systems for powering pneumatic tools. In both, a regulator adjusts the air pressure.

Figure 18–2. A portable air compressor is used on the construction site to power pneumatic tools.

Figure 18–3. Heavy-duty pneumatic nailers are used for wall and floor framing, subflooring, and sheathing. They can accommodate a variety of nail types and sizes.

10¼-GA. PLAIN SHANK

11-GA. PLAIN SHANK

11½-GA. PLAIN SHANK

13-GA. PLAIN SHANK

12-GA. SCREW SHANK

11½-GA. RING SHANK

1    2    3

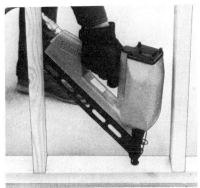

*Senco Products Fastening Systems*

Figure 18–4. A special hardened steel nail is used to fasten wood plates to a concrete floor slab with a pneumatic nailer.

HEAVY-DUTY FINISH NAILER

*Hilti, Inc.*

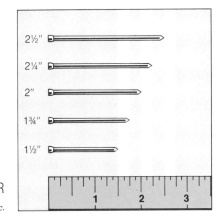

*Hilti, Inc.*

Figure 18–5. Finish and brad pneumatic nailers are used to fasten interior trim such as baseboard and wall moldings, cabinetwork, paneling, and door and window trim. They can accommodate a variety of nail sizes.

board. See Figure 18–7. Because it is a two-legged device covering a greater surface area, a staple makes an excellent fastener for sheathing, shingles, building paper, and other construction materials.

Like nailers, staplers are available in a range of models. A heavy-duty stapler is used for such operations as wall framing and fastening subflooring and sheathing. See Figure 18–8. A medium-duty nail stapler is used for interior trim work and cabinet assembly. See Figure 18–9. All models of staplers accommodate more than one size of staple.

## Safety in Using Pneumatic Nailers and Staplers

For all pneumatic tools, it is necessary to know how to interpret the gauges of the air compressor in order to ensure the proper air pressure. Be certain the unit has the required pressure controls. Safety rules for the air compressor are:

1. Use the correct air pressure for the particular tool. Do not exceed the recommended air pressure for any operation.

2. Check to make sure that the air hoses and all connections are in good condition.

3. Check to make sure that ex-

*Hilti, Inc.*

Figure 18–6. A coil-fed pneumatic nailer is popular for fastening panel subflooring and framing walls.

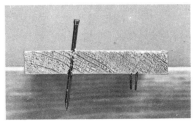

Figure 18–7. A staple is less likely than a nail to split a board when driven in near the end of the board.

posed belts on a portable compressor have guards on both sides to reduce the possibility of injuries to fingers and hands.

When pneumatic nailers and staplers are used on a construction job, the safety of not only the operator, but also the other people in the immediate area must be considered. For example, before panel materials are fastened to backing such as studs or joists, it is necessary to make sure that no one is on the other side of or below the backing. If the nail or staple misses the backing, it could go through the paneling and injure someone. Other safety rules for nailers and staplers are:

1. Follow the manufacturer's recommendations for the particular tool.

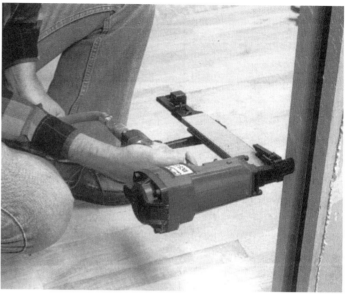

Hilti, Inc.

Figure 18–8. Heavy-duty pneumatic staplers, like heavy-duty pneumatic nailers, are used for wall framing and fastening subflooring and sheathing.

2. Use the right nailer or stapler for the job. Use the correct size and type of nail or staple.

3. Keep the nose of the tool pointed away from your body. Never point it in the direction of other workers.

4. Before starting work with a nailer or stapler, check to see that the mechanism and all other safety features are working properly. It is good practice to test-fire the tool into a block of wood.

5. Disconnect the stapler or nailer from the air supply when it is not in use.

## POWDER-ACTUATED TOOLS

Powder-actuated tools are used for fastening various materials to concrete and structural steel. Powder-actuated tools are available as *direct-acting* and *indirect-acting* types. See Figure 18–10. Both types have a firing mechanism similar to a gun. However, safety features on these tools, which are not found on guns, prevent accidental firing.

Powder-actuated tools will not fire unless fully depressed against the material to be fastened. When these tools are fired, the firing pin is released, striking a powder load specifically designed for use in the tool. The expanding gases generated act directly on the fastener, or on a special piston which in turn drives the fastener into the base material.

### Drive Pins and Studs

The fastener fired from a powder-actuated tool is either a drive pin or a threaded stud. Drive pins are used for permanent fastening. Studs are used for mounting materials that may require removal at a later time. Typical applications of drive pins and studs are shown in Figure 18–11. For additional information on this subject, see Section 2.

As a general rule, the penetration of the fastener should equal eight times the shank diameter to provide good holding power. For example, a drive pin with a $9/64''$ shank should penetrate the concrete $1^1/8''$ ($8 \times 9/64'' = 1^1/8''$). When

Figure 18–9. Medium-duty pneumatic staplers, like a medium-duty pneumatic nailers, are used for interior trim work and cabinet assembly.

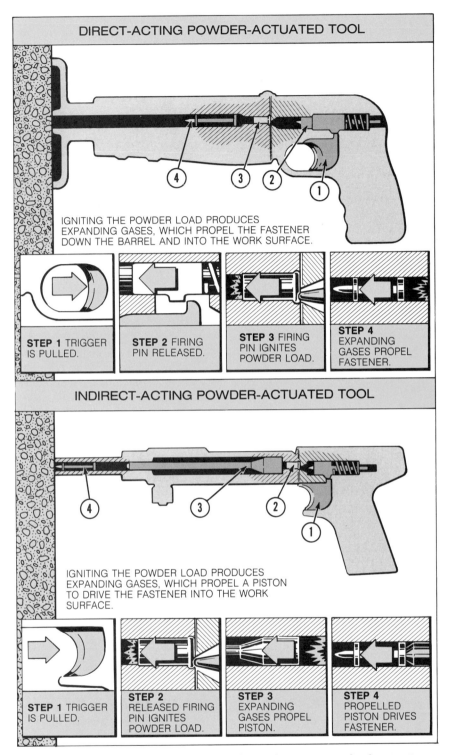

DIRECT-ACTING POWDER-ACTUATED TOOL

IGNITING THE POWDER LOAD PRODUCES EXPANDING GASES, WHICH PROPEL THE FASTENER DOWN THE BARREL AND INTO THE WORK SURFACE.

**STEP 1** TRIGGER IS PULLED.

**STEP 2** FIRING PIN RELEASED.

**STEP 3** FIRING PIN IGNITES POWDER LOAD.

**STEP 4** EXPANDING GASES PROPEL FASTENER.

INDIRECT-ACTING POWDER-ACTUATED TOOL

IGNITING THE POWDER LOAD PRODUCES EXPANDING GASES, WHICH PROPEL A PISTON TO DRIVE THE FASTENER INTO THE WORK SURFACE.

**STEP 1** TRIGGER IS PULLED.

**STEP 2** RELEASED FIRING PIN IGNITES POWDER LOAD.

**STEP 3** EXPANDING GASES PROPEL PISTON.

**STEP 4** PROPELLED PISTON DRIVES FASTENER.

Figure 18-10. Powder-actuated tools are available as direct-acting and indirect-acting types.

fastening into steel, the nose of the fastener should penetrate $1/4''$ through the steel. Consult the manufacturer for recommended specifications and procedures.

Generally, the depth of penetration obtained is controlled by the size of powder load. See Figure 18-12. The depth of penetration is also affected by the material driven into. For example,

the harder the concrete is, the higher the powder load strength required will be. Load strengths are identified by a standard color code system adopted by manufacturers and listed in the operator's manual. Additional information may be obtained from Powder-Actuated Tool Manufacturers Institute (PATMI).

Although some powder-actuated tools use .38 caliber powder loads, the most commonly used powder loads are .22, .25, and .27 caliber. Some powder-actuated tools have an adjustment feature that allows more efficient control of penetration. See Figure 18-13. This feature permits the adjustment of load strength for different applications and conditions.

Loading of any powder-actuated tool, whether direct- or indirect-acting, is always done by placing the fastener in the tool before placing the powder load in the tool chamber.

In a conventional single-shot powder-actuated tool, the stud or pin is inserted by hand into the barrel. The powder load is then inserted. Recent models of powder-actuated tools have semi-automatic powder load feeds that allow faster and more convenient fastening operations than conventional single-shot tools.

## Safety in Using Powder-actuated Tools

**Use extreme caution when using powder-actuated tools.** OSHA and many state and local jurisdictions require training and licensing of individuals before they can operate a powder-actuated tool.

Powder-actuated tools shoot a fastener designed to penetrate concrete and structural steel.

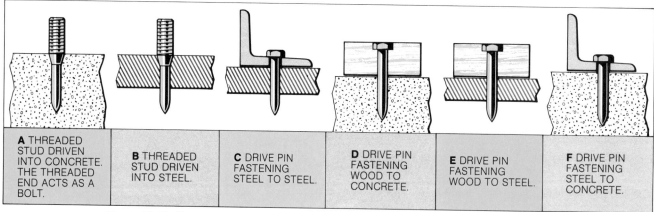

Figure 18-11. Typical applications of drive pins and studs fired from a powder-actuated tool.

| A THREADED STUD DRIVEN INTO CONCRETE. THE THREADED END ACTS AS A BOLT. | B THREADED STUD DRIVEN INTO STEEL. | C DRIVE PIN FASTENING STEEL TO STEEL. | D DRIVE PIN FASTENING WOOD TO CONCRETE. | E DRIVE PIN FASTENING WOOD TO STEEL. | F DRIVE PIN FASTENING STEEL TO CONCRETE. |

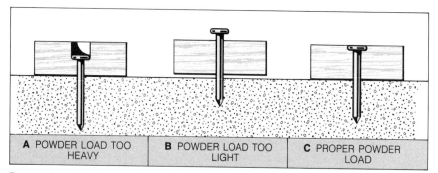

| A POWDER LOAD TOO HEAVY | B POWDER LOAD TOO LIGHT | C PROPER POWDER LOAD |

Figure 18-12. Proper penetration of a drive pin driven by a powder-actuated tool requires the proper powder load.

There have been cases where operators have mistakenly fired into plaster walls or where voids have been covered with plywood or drywall and seriously injured an individual on the other side of the wall. The American National Standards Institute (ANSI) publication, A10.3-1985 Safety Requirements for Powder-Actuated Systems, details safe procedures to be followed when using powder-actuated tools. General safety rules for powder-actuated tools are

1. Never attempt to use a powder-actuated tool until you have been properly trained and licensed.

2. Follow *all* manufacturer's instructions.

3. If the surface to which you are fastening material has a plaster or other type of finish, double-check to make sure that the wall itself is made of con-crete. This can be done by driving a nail through the surface covering. Do not fire into a concrete wall that is less than 2″ thick.

4. Wear safety goggles to protect your eyes from flying particles.

5. Wear ear protectors. Prolonged use of a powder-actuated tool can cause hearing damage.

6. Do not stand next to, or work in the immediate area of, a person operating the tool. The drive pin or stud shot into the concrete may glance off a piece of reinforcing steel in the concrete and fly to one side.

7. Do not leave a loaded tool lying around. Do not load the tool unless you intend to fire immediately.

8. Post warning signs where powder-actuated tools are in use.

ADJUSTMENT KNOB

*Ramset Fastening Systems*

Figure 18-13. On some powder-actuated tools, an adjustment knob provides penetration control.

# UNIT 19

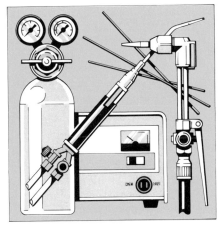

# Welding and Metal-cutting Equipment

Carpentry work often involves the use of metal materials that must be welded or cut. In welding, high heat is applied to adjacent metal surfaces to join them. In cutting metal, a hot jet of flame from a cutting torch burns through the metal.

Carpenters today perform many of the welding and metal-cutting operations required on a construction job. For example, carpenters weld anchor bolts, straps, J-bolts, and hangers in place. They may also weld reinforcing bars, make repairs on steel forms, and weld the structural members of light-gauge steel walls. See Figure 19–1.

In order to perform welding and metal-cutting operations, a carpenter must be *certified* by passing an examination conducted by the proper authorities. Many apprenticeship programs today include welding courses as part of their required training. Evening classes are often available for journeyman carpenters who want to add welding to their other skills.

The two basic types of welding processes are *electric arc welding* and *oxyacetylene welding.* Each method requires specialized skills and equipment. In construction work, electric arc welding is the method normally used for welding. Oxyacetylene equipment, which can be used for both welding and cutting, is usually used primarily for cutting rather than for welding in construction work.

## ELECTRIC ARC WELDING

In electric arc welding, metal surfaces are fused (melted together) by the heat of an electric arc. See Figure 19–2. The arc jumps between an electrode and the materials being welded. Arc welding machines (Figure 19–3)

*United States Steel Corporation*
Figure 19–1. Carpenters perform many welding and metal-cutting operations.

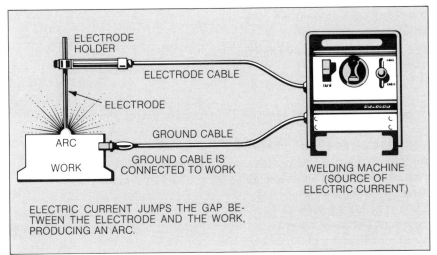

ELECTRODE HOLDER
ELECTRODE CABLE
ELECTRODE
GROUND CABLE
ARC
WORK
GROUND CABLE IS CONNECTED TO WORK
WELDING MACHINE (SOURCE OF ELECTRIC CURRENT)
ELECTRIC CURRENT JUMPS THE GAP BETWEEN THE ELECTRODE AND THE WORK, PRODUCING AN ARC.

Figure 19–2. In electric arc welding, metal surfaces are fused by the heat of an electric arc.

121

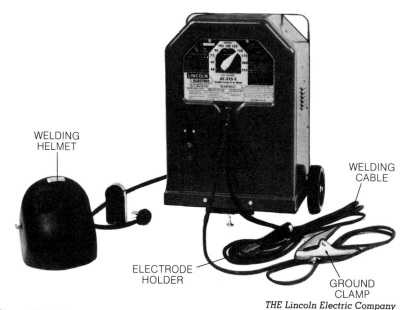

WELDING
HELMET

WELDING
CABLE

ELECTRODE
HOLDER

GROUND
CLAMP

*THE Lincoln Electric Company*

Figure 19–3. An arc welding machine and equipment used with it. Also available are DC (direct-current) models and transformer-rectifier types that can produce AC and DC power.

provide electric current to produce the welding arc. These machines may be alternating-current (AC) or direct-current (DC) models. Some transformer-rectifier types can produce both AC and DC power.

*Electrodes* are available in different types of metals. As a rule, the electrode should be the same type of metal as the work being welded.

*Electrode holders* are insulated clamps that hold the electrodes. *Welding cables* conduct the electrical current to the work through the electrode holder. A *ground clamp* is attached to the work.

## OXYACETYLENE WELDING AND CUTTING

In oxyacetylene welding, metal surfaces are fused (melted together) by the heat of a welding flame. The temperature of this flame may be adjusted as high as 6,300° Fahrenheit. When the edges of the metal being welded reach the melting point, the two edges flow together to form a

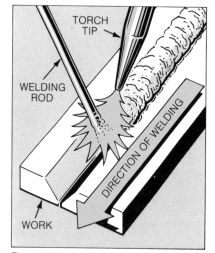

TORCH
TIP

WELDING
ROD

DIRECTION OF WELDING

WORK

Figure 19–4. Filler metal is being added in this oxyacetylene welding procedure.

solid piece. Some types of joints require that a filler metal be added. This is done with a filler rod. The filler metal is added to the molten puddle of the base metal. See Figure 19–4.

Oxyacetylene equipment in construction work is used primarily for cutting metal, not for welding. The setup of the equipment for cutting is similar to the setup for welding. Acetylene and oxy-

gen cylinders and hoses lead from the regulators to the torch. See Figure 19–5. Pressure regulators are at the tops of the cylinders, and adjusting valves are located where the hoses connect to the torch. The pressures at the cylinders and adjustment of the valves determine the type of flame produced.

The major difference between the welding setup and the cutting setup is a special attachment that connects to the torch for metal cutting. See Figure 19–6. The cutting process begins by preheating the metal with a flame fed by the oxygen and acetylene mixture. When the required temperature is reached, the oxygen lever on the cutting attachment is depressed to release pure oxygen. The pure oxygen causes the hot metal to burn. Oxyacetylene cutting is actually a burning-through process.

## SAFETY IN USING WELDING AND METAL-CUTTING EQUIPMENT

Potential hazards from electric arc welding and oxyacetylene equipment are great. Ultraviolet and infrared rays given off in electric arc welding can cause severe eye damage. Droplets of molten metal generated by the arc can cause burns and set fire to flammable materials. Flying sparks from both electric arc welding and oxyacetylene equipment are also a danger. Rules for operating both types of equipment safely are:

1. Wear eye protection and fireproof clothing and headgear.

2. Follow manufacturer's instructions for all welding and metal-cutting equipment.

3. Weld only in well-ventilated areas. When necessary, use an exhaust system to keep toxic

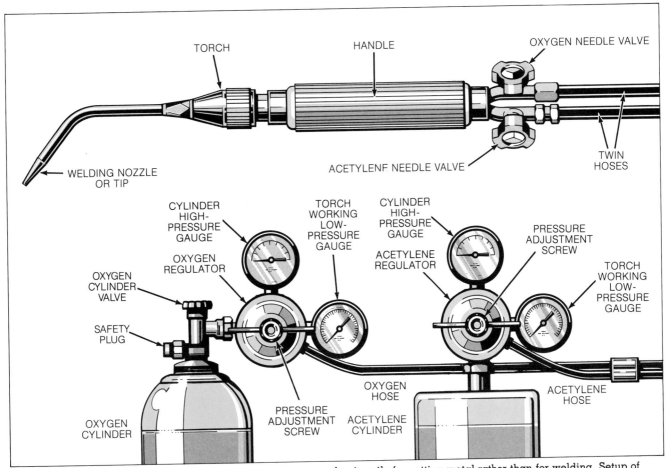

Figure 19–5. Oxyacetylene equipment is used in construction work primarily for cutting metal rather than for welding. Setup of equipment for cutting is similar to that for welding.

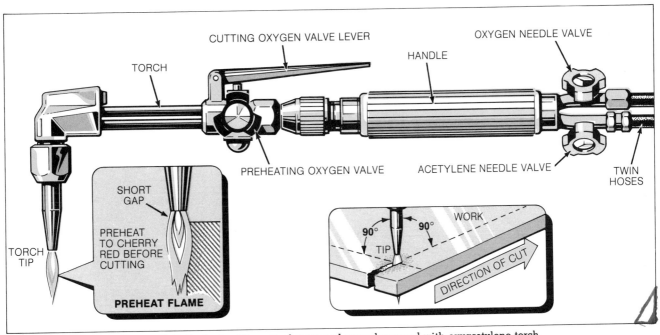

Figure 19–6. Typical cutting attachment and procedure used with oxyacetylene torch.

gases below the prescribed health limits. Also use a respirator when welding metals that produce toxic fumes.

4. Do not weld or cut used drums, barrels, tanks, or other containers unless they have been thoroughly cleaned of all combustible substances.

5. Do not weld or cut near flammable materials. If welding or cutting must be done near flammable materials, use fire-resisting guards, partitions, or screens.

6. Keep a fire extinguisher nearby any welding or cutting operation.

7. Keep flame and sparks away from oxygen cylinders and hoses.

8. Do not weld or cut near ventilators.

9. Do not use oxygen to dust off clothing or work.

10. Do not use oxygen as a substitute for compressed air.

11. Do not allow acetylene gas to come in contact with unalloyed copper except in a torch.

12. Open the cylinder valves on oxyacetylene equipment slowly.

13. Purge oxygen and acetylene lines before lighting the torch.

14. Install the equipment so that it is properly grounded, with a power disconnect switch nearby.

15. Do not make repairs to the equipment unless the power to the machine is shut OFF.

16. Periodically check cable connections to make sure they are tight.

17. Do not overload cables or allow them to come in contact with hot metal, water, oil, or grease. Avoid dragging the cables over or around sharp corners.

18. Do not stand in water or on a wet floor or use wet gloves when welding.

19. Dispose of used electrodes properly, in a stub can.

20. Do not pick up pieces of metal that have just been welded or heated.

21. Do not weld hollow (cored) castings unless they have been properly vented; otherwise, an explosion may occur.

22. Always turn equipment OFF before leaving the job site.

# SECTION 5

# Construction Equipment, Site Conditions, and Safety on the Job

Construction aids such as scaffolding, ladders, hoists, and jacks are used extensively by carpenters. Also, carpenters must work closely with the operating engineers who handle the heavy machinery used for excavating and grading, hoisting and lifting, placing concrete, and other moving operations. Carpenters should be aware of the safety factors involved in these operations.

# UNIT 20

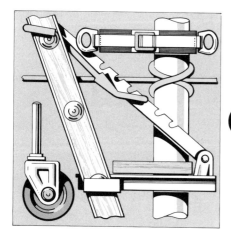

## Construction Aids

Much construction requires working at heights. These heights may range from a few feet to hundreds of feet above floor or ground level. Carpenters must be able to construct and safely use different kinds of scaffolding or ladders for performing high work.

## SCAFFOLDS

Scaffolds, also known as *staging,* are temporary platforms set up around the building in order to complete work that is out of reach from the ground or floor level. Such work includes exterior sheathing, exterior finish covering, and roof cornices. A scaffold must be strong enough to hold the weight of the workers, their tools, and building materials. A general rule is that a scaffold must have a safety margin of 4 to 1, meaning four times the weight to which it will be subjected.

In the past scaffolds were always constructed of wood. Today many builders prefer metal scaffolding, particularly on large construction projects. However, wood scaffolds are still used for smaller jobs, and a carpenter must know how to build them.

### Double-pole Wood Scaffold

*Poles* are the vertical upright members of the scaffold. In a double-pole wood scaffold one of the uprights, or poles, is close to the building and the other is directly opposite. One advantage of this type of scaffold is that it can stand free of the building. Figure 20–1 shows a light-trade,

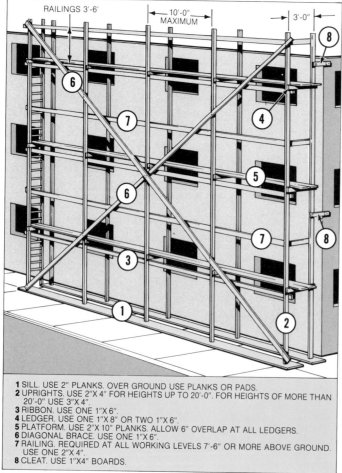

RAILINGS 3'-6'
10'-0" MAXIMUM
3'-0"

1 SILL. USE 2" PLANKS. OVER GROUND USE PLANKS OR PADS.
2 UPRIGHTS. USE 2"X 4" FOR HEIGHTS UP TO 20'-0". FOR HEIGHTS OF MORE THAN 20'-0" USE 3"X 4".
3 RIBBON. USE ONE 1"X 6".
4 LEDGER. USE ONE 1"X 8" OR TWO 1"X 6".
5 PLATFORM. USE 2"X 10" PLANKS. ALLOW 6" OVERLAP AT ALL LEDGERS.
6 DIAGONAL BRACE. USE ONE 1"X 6".
7 RAILING. REQUIRED AT ALL WORKING LEVELS 7'-6" OR MORE ABOVE GROUND. USE ONE 2"X 4".
8 CLEAT. USE 1"X4" BOARDS.

Figure 20–1. Typical light-trade double-pole wood scaffold. It stands free of the building.

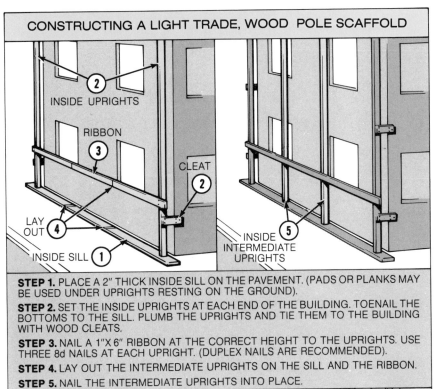

## CONSTRUCTING A LIGHT TRADE, WOOD POLE SCAFFOLD

INSIDE UPRIGHTS (2)

RIBBON (3)

CLEAT (2)

LAY OUT (4)

INSIDE SILL (1)

INSIDE INTERMEDIATE UPRIGHTS (5)

**STEP 1.** PLACE A 2″ THICK INSIDE SILL ON THE PAVEMENT. (PADS OR PLANKS MAY BE USED UNDER UPRIGHTS RESTING ON THE GROUND).

**STEP 2.** SET THE INSIDE UPRIGHTS AT EACH END OF THE BUILDING. TOENAIL THE BOTTOMS TO THE SILL. PLUMB THE UPRIGHTS AND TIE THEM TO THE BUILDING WITH WOOD CLEATS.

**STEP 3.** NAIL A 1″X 6″ RIBBON AT THE CORRECT HEIGHT TO THE UPRIGHTS. USE THREE 8d NAILS AT EACH UPRIGHT. (DUPLEX NAILS ARE RECOMMENDED).

**STEP 4.** LAY OUT THE INTERMEDIATE UPRIGHTS ON THE SILL AND THE RIBBON.

**STEP 5.** NAIL THE INTERMEDIATE UPRIGHTS INTO PLACE.

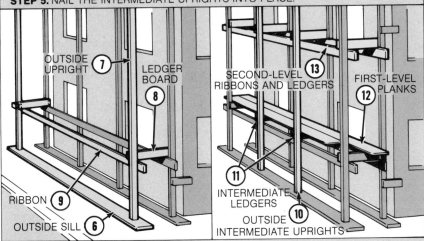

OUTSIDE UPRIGHT (7)

LEDGER BOARD (8)

SECOND-LEVEL RIBBONS AND LEDGERS (13)

FIRST-LEVEL PLANKS (12)

INTERMEDIATE LEDGERS (11)

OUTSIDE INTERMEDIATE UPRIGHTS (10)

RIBBON (9)

OUTSIDE SILL (6)

**STEP 6.** PLACE THE OUTSIDE SILL FOR THE OUTSIDE UPRIGHTS.

**STEP 7.** SET UP THE TWO END OUTSIDE UPRIGHTS AND NAIL IN PLACE.

**STEP 8.** LEVEL AND NAIL THE LEDGER BOARDS TO THE OUTSIDE UPRIGHTS. FASTEN THE LEDGER BOARDS WITH FIVE 8d NAILS WHEN USING 1″X 8″ MATERIAL.

**STEP 9.** PLACE THE RIBBON BELOW THE LEDGERS OF THE OUTSIDE UPRIGHTS.

**STEP 10.** NAIL THE INTERMEDIATE UPRIGHTS IN PLACE.

**STEP 11.** PLACE THE INTERMEDIATE LEDGER BOARDS.

**STEP 12.** PLACE THE PLANKS FOR THE PLATFORM.

**STEP 13.** NAIL THE SECOND-LEVEL RIBBONS AND LEDGERS IN PLACE.

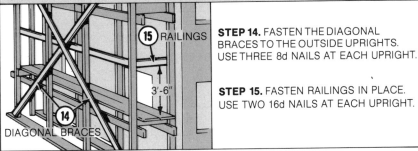

RAILINGS (15)

3′-6″

DIAGONAL BRACES (14)

**STEP 14.** FASTEN THE DIAGONAL BRACES TO THE OUTSIDE UPRIGHTS. USE THREE 8d NAILS AT EACH UPRIGHT.

**STEP 15.** FASTEN RAILINGS IN PLACE. USE TWO 16d NAILS AT EACH UPRIGHT.

Figure 20–2. Constructing a light-trade, double-pole wood scaffold. This type of scaffold can be set up for high buildings.

double-pole wood scaffold that can be set up for high buildings. It should be securely tied to the building with doubled 12-ga. iron wire or 1″ × 4″ boards with at least two nails at each connection. Maximum height between ledger boards is 7′-6″. Figure 20–2 shows the procedure for constructing a smaller scaffold of this type. Figure 20–3 shows another design used for double-pole scaffolds. This design uses

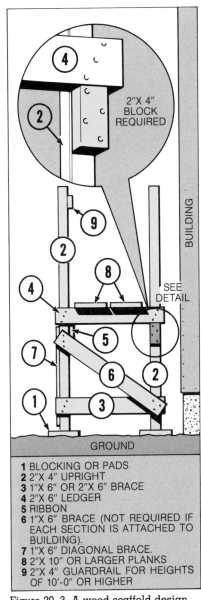

2″X 4″ BLOCK REQUIRED

BUILDING

SEE DETAIL

GROUND

**1** BLOCKING OR PADS
**2** 2″X 4″ UPRIGHT
**3** 1″X 6″ OR 2″X 6″ BRACE
**4** 2″X 6″ LEDGER
**5** RIBBON
**6** 1″X 6″ BRACE (NOT REQUIRED IF EACH SECTION IS ATTACHED TO BUILDING).
**7** 1″X 6″ DIAGONAL BRACE.
**8** 2″X 10″ OR LARGER PLANKS
**9** 2″X 4″ GUARDRAIL FOR HEIGHTS OF 10′-0″ OR HIGHER

Figure 20–3. A wood scaffold design acceptable in some areas uses 2″ × 6″ ledgers. A 2″ × 4″ block is placed under each ledger where it is fastened to the inside upright.

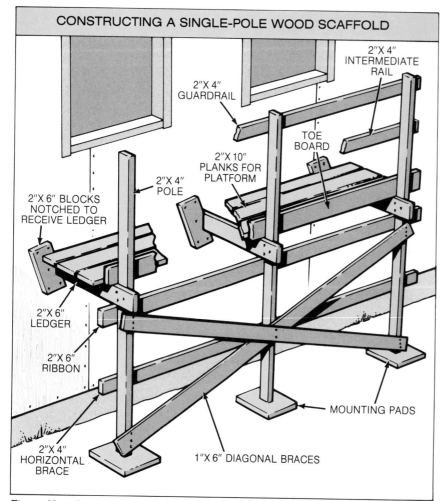

CONSTRUCTING A SINGLE-POLE WOOD SCAFFOLD

2"X 4" GUARDRAIL

2"X 4" INTERMEDIATE RAIL

TOE BOARD

2"X 10" PLANKS FOR PLATFORM

2"X 4" POLE

2"X 6" BLOCKS NOTCHED TO RECEIVE LEDGER

2"X 6" LEDGER

2"X 6" RIBBON

2"X 4" HORIZONTAL BRACE

1"X 6" DIAGONAL BRACES

MOUNTING PADS

Figure 20–4. A single-pole wood scaffold can be constructed more quickly than the double-pole type. However, it is only practical for one-story buildings.

Figure 20–5. Serious injury or death could result from working on this scaffold. Poor lumber has been used in its construction. The ledger boards are not supported by ribbons. The diagonal braces do not extend to the bottoms of the uprights. Only a single plank is being used as a platform. **Carpenters should never work on a scaffold such as this!**

*Patent Scaffolding Company—A Division of Harsco Corporation*

Figure 20–6. Sectional metal scaffolding placed around form work for concrete buildings.

2" × 6" ledgers. A 2" × 4" block is placed under each ledger where it is fastened to the inside upright.

## Single-pole Wood Scaffold

In a single-pole wood scaffold, the inside end of the ledger is nailed to a notched block that is fastened directly to the building. See Figure 20–4. This type of scaffold can be constructed more quickly than the double-pole type, but it can be used for only one-story buildings. After the higher work has been completed, the scaffold must be torn down to allow completion of the work at the lower level.

## Safety in Using Wood Scaffolds

Faultily constructed scaffolding may collapse, causing injuries and deaths. Figure 20–5 shows a highly dangerous scaffold. Safety rules for scaffold construction and use are:

1. Use a good grade of lumber that is free of loose knots.

2. Be sure that there are pads or a continuous sill under the poles.

3. Install guardrails and mid rails where required.

4. Use planks that are 2" thick and 10" wide or wider. Place them so that they overlap each other at least 12" and extend past the ledger boards exactly 6", but no more.

5. Be wary of power lines near the scaffold.

6. Provide overhead protection if work is being done over the scaffold.

## Metal Scaffold

A factory-produced steel or aluminum scaffold (Figure 20–6) offers the advantage of being easily assembled and disassembled.

128

Figure 20–7. Smaller rolling scaffolds are used for interior work.

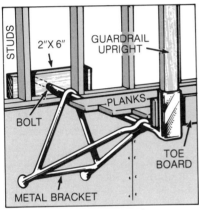

Figure 20–8. This metal bracket is secured with a bolt fastened to a piece of 2″ × 6″ set behind two studs in the wall. A 2″ × 4″ upright can be attached to the back of the bracket for fastening a guardrail.

Many builders find it more economical to rent metal scaffolding than to build it when a large amount is required. Often metal scaffolds are installed by the workers from the scaffolding firm. A small *rolling* type of metal scaffold is often used for interior work. See 20–7.

## BRACKETS AND JACKS

Other support systems employed for light scaffolding consist of metal devices that can be used

**A** OUTSIDE LADDER JACKS PLACE THE PLATFORM ON THE OUTSIDE OF THE LADDERS.

**B** INSIDE LADDER JACKS PLACE THE PLATFORM AT THE INSIDE OF THE LADDERS.

Figure 20–9. Inside and outside ladder jacks can be used to support a single worker doing light repair work.

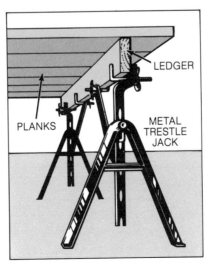

Figure 20–10. Metal trestle jacks are used when a low platform is needed. Ledger boards 2″ thick are clamped to the tops of the jacks and support the platform planks.

many times and moved from job to job.

Brackets for residential construction can be made of wood, although metal brackets are considered superior and are more often used. See Figure 20–8. Different metal types are available that are either bolted or nailed to the wall. One type of bracket is designed for use on forms for concrete. Bracket scaffolds can only support a limited amount of weight and can only be used for light operations.

*Ladder jacks* should be used only to support a single worker doing light repair work. See Figure 20–9. *Outside* ladder jacks place the platform outside the ladder, and *inside* ladder jacks place it inside the ladder.

*Trestle jacks* are used when a low working platform is required for interior or exterior work. See Figure 20–10. Ledger boards (2″ thick) are clamped to the tops of the trestle jacks to support the platform.

*Pump jacks* are adjustable scaffolding devices that can be used for heights up to 30′. See Figure 20–11. Uprights (4″ × 4″ material) are attached to the building with metal braces. Adjustable brackets that can be raised or lowered are attached to the uprights. The uprights provide a 24″ working platform. An attachment can be used with the pump jacks to provide a workbench area for tools and materials.

## LADDERS

Carpenters often work from ladders for light operations and repairs. Ladders are made of wood, aluminum, or fiberglass. A fiberglass ladder does not conduct electricity and costs less than a comparable wood ladder. Types of ladders used by carpenters include the rolling ladder, the stepladder, the adjustable

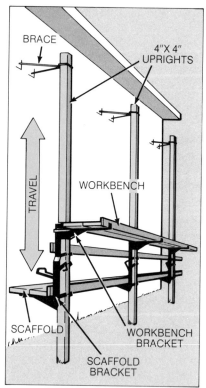

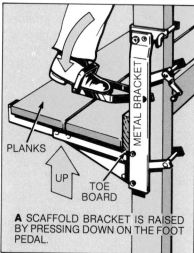

**A** SCAFFOLD BRACKET IS RAISED BY PRESSING DOWN ON THE FOOT PEDAL.

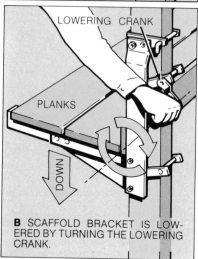

**B** SCAFFOLD BRACKET IS LOWERED BY TURNING THE LOWERING CRANK.

Figure 20–11. An adjustable pump jack scaffold can be used for heights up to 30'.

extension ladder, and the single ladder. See Figure 20–12. Some types are better suited for interior work and other types for exterior work.

## Job-built Ladders

Ladders are sometimes constructed on the job. They should be made of clear and straight lumber. The procedure for constructing a notched job-built ladder is shown in Figure 20–13. A similar ladder can be built with cleats nailed between the rungs instead of the rails being notched.

## Safety in Using Ladders

Careless use of ladders is a major cause of injuries in construction work. A ladder must be positioned correctly to be safe. The distance between the base of the ladder and the structure it leans

against should be one-fourth the distance from the floor or ground to the top support for the ladder. See Figure 20–14. Also, the ladder should extend at least 3' above the top support. Other safety rules for using ladders are:

1. Check to make sure the ladder has no broken, cracked, or otherwise defective parts.

2. Do not paint a ladder. The paint may hide defects that occur later.

3. Keep ladder rungs and side rails free of oil and grease to prevent slipping.

4. After the ladder is set in place, make sure the base of the ladder is level and firm. Most extension ladders have a swivel safety shoe at the base.

5. Do not place a ladder against a movable object.

6. Do not place a ladder against a window. Instead, tack a board across the window frame

LADDERS USED FOR INTERIOR AND EXTERIOR WORK

**A** ROLLING LADDER CONVENIENT FOR INTERIOR FINISH WORK.

**B** STEPLADDER WITH PLATFORM USEFUL FOR INTERIOR WORK.

**C** ADJUSTABLE EXTENSION LADDER USED FOR HIGH WORK.

**D** SINGLE LADDER

Figure 20–12. Types of ladders used by carpenters. Some types are better suited for interior work and other types for exterior work.

and place the ladder against the board.

7. Do not use a metal ladder where there is danger of contact with power lines or other electrical conductors.

8. Secure the ladder so that it will not slip while in use.

9. Always face the ladder when climbing.

10. Do not carry tools or materials in your hands when climbing up or down a ladder. Keep both hands free to hold the side rails. Carry tools in a work apron, or use a rope and bucket to hoist them up.

11. Place the ladder so that it is clear of doors or passageways.

12. Always stay below the top three ladder rungs.

13. Do not reposition an occupied ladder.

14. Do not lean or overreach from a ladder.

15. Do not use a ladder in a high wind.

## SAWHORSES

Wood sawhorses may be constructed on the job or in a shop. They serve as a portable workbench for carpenters and for other workers as well. Also, planks can be laid across sawhorses to form a low scaffold. The traditional carpenter's sawhorse is 2′ high, with a 2″ × 4″ or 2″ × 6″ top piece. The legs are made of 1″ × 6″ lumber, and plywood is recommended for the end pieces. A procedure for

building this type of sawhorse is shown in Figure 20–15.

## SAFETY BELTS AND NETS

Under certain conditions carpenters wear safety belts as an added safety precaution. See Figure 20–16. The safety belt may be a leather or nylon web belt equipped with a D-ring to which a lifeline (*lanyard*) can be attached. The other end of the lanyard is secured to the building. An improved method is to snap the end of the lanyard to a *rope grab*, which slides on a rope fastened to the building. As this device can be moved up and down, it makes it more convenient to move around the work area. Nylon rope capable of supporting a minimum dead weight of 5,400 pounds is recommended. The OSHA safety code states:

> "Safety belts will customarily be used when exposed to the hazards of falling from buildings, bridges, structures, or construction mem-

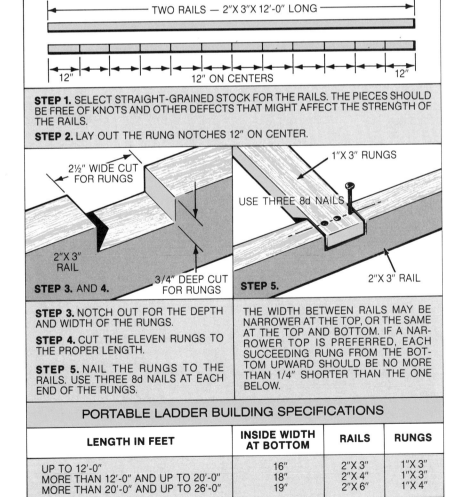

TWO RAILS — 2″X 3″X 12′-0″ LONG

12″ | 12″ ON CENTERS | 12″

**STEP 1.** SELECT STRAIGHT-GRAINED STOCK FOR THE RAILS. THE PIECES SHOULD BE FREE OF KNOTS AND OTHER DEFECTS THAT MIGHT AFFECT THE STRENGTH OF THE RAILS.

**STEP 2.** LAY OUT THE RUNG NOTCHES 12″ ON CENTER.

2½″ WIDE CUT FOR RUNGS

2″X 3″ RAIL

3/4″ DEEP CUT FOR RUNGS

**STEP 3.** AND **4.**

1″X 3″ RUNGS

USE THREE 8d NAILS

2″X 3″ RAIL

**STEP 5.**

**STEP 3.** NOTCH OUT FOR THE DEPTH AND WIDTH OF THE RUNGS.

**STEP 4.** CUT THE ELEVEN RUNGS TO THE PROPER LENGTH.

**STEP 5.** NAIL THE RUNGS TO THE RAILS. USE THREE 8d NAILS AT EACH END OF THE RUNGS.

THE WIDTH BETWEEN RAILS MAY BE NARROWER AT THE TOP, OR THE SAME AT THE TOP AND BOTTOM. IF A NARROWER TOP IS PREFERRED, EACH SUCCEEDING RUNG FROM THE BOTTOM UPWARD SHOULD BE NO MORE THAN 1/4″ SHORTER THAN THE ONE BELOW.

### PORTABLE LADDER BUILDING SPECIFICATIONS

| LENGTH IN FEET | INSIDE WIDTH AT BOTTOM | RAILS | RUNGS |
|---|---|---|---|
| UP TO 12′-0″ | 16″ | 2″X 3″ | 1″X 3″ |
| MORE THAN 12′-0″ AND UP TO 20′-0″ | 18″ | 2″X 4″ | 1″X 3″ |
| MORE THAN 20′-0″ AND UP TO 26′-0″ | 19″ | 2″X 6″ | 1″X 4″ |

Figure 20–13. Constructing a notched job-built ladder. A similar ladder can be built with cleats nailed between the rungs instead of the rails being notched.

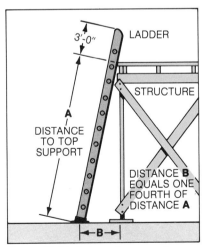

3′-0″

LADDER

A

DISTANCE TO TOP SUPPORT

STRUCTURE

DISTANCE **B** EQUALS ONE FOURTH OF DISTANCE **A**

B

Figure 20–14. Safe and proper placement of a ladder. The distance between the base of the ladder and the structure it leans against should be one-fourth the distance from the floor or ground to the top support for the ladder. Also, the ladder should extend at least 3′ above the top support.

# CONSTRUCTING A TYPICAL CARPENTER'S SAWHORSE

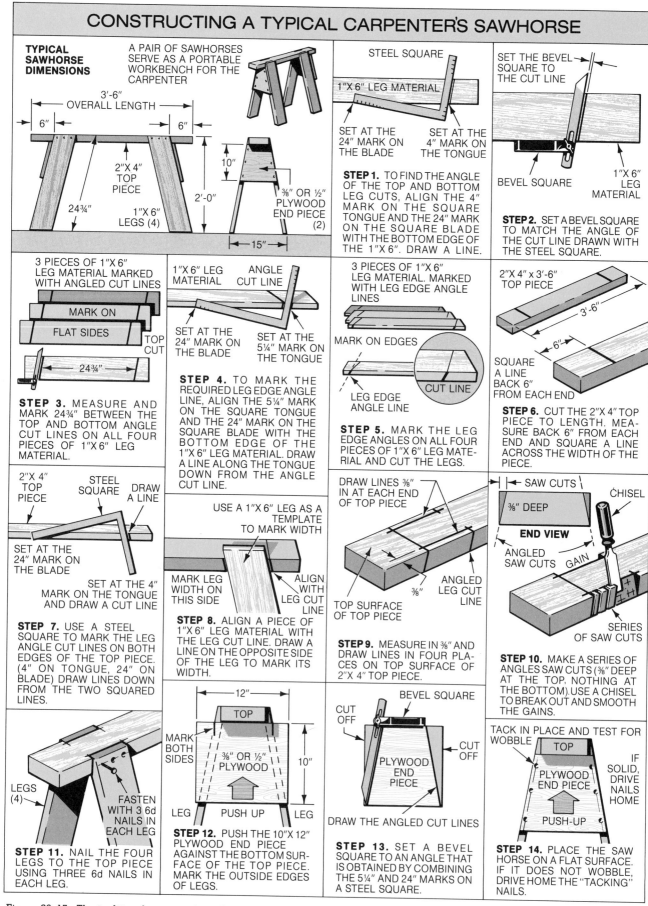

**TYPICAL SAWHORSE DIMENSIONS**

A PAIR OF SAWHORSES SERVE AS A PORTABLE WORKBENCH FOR THE CARPENTER

3'-6" OVERALL LENGTH

6"    6"

2"X 4" TOP PIECE

24¾"

1"X 6" LEGS (4)

2'-0"

10"

⅜" OR ½" PLYWOOD END PIECE (2)

15"

STEEL SQUARE

1"X 6" LEG MATERIAL

SET AT THE 24" MARK ON THE BLADE

SET AT THE 4" MARK ON THE TONGUE

**STEP 1.** TO FIND THE ANGLE OF THE TOP AND BOTTOM LEG CUTS, ALIGN THE 4" MARK ON THE SQUARE TONGUE AND THE 24" MARK ON THE SQUARE BLADE WITH THE BOTTOM EDGE OF THE 1"X 6". DRAW A LINE.

SET THE BEVEL SQUARE TO THE CUT LINE

BEVEL SQUARE

1"X 6" LEG MATERIAL

**STEP 2.** SET A BEVEL SQUARE TO MATCH THE ANGLE OF THE CUT LINE DRAWN WITH THE STEEL SQUARE.

3 PIECES OF 1"X 6" LEG MATERIAL MARKED WITH ANGLED CUT LINES

MARK ON FLAT SIDES

TOP CUT

24¾"

**STEP 3.** MEASURE AND MARK 24¾" BETWEEN THE TOP AND BOTTOM ANGLE CUT LINES ON ALL FOUR PIECES OF 1"X 6" LEG MATERIAL.

1"X 6" LEG MATERIAL

ANGLE CUT LINE

SET AT THE 24" MARK ON THE BLADE

SET AT THE 5¼" MARK ON THE TONGUE

**STEP 4.** TO MARK THE REQUIRED LEG EDGE ANGLE LINE, ALIGN THE 5¼" MARK ON THE SQUARE TONGUE AND THE 24" MARK ON THE SQUARE BLADE WITH THE BOTTOM EDGE OF THE 1"X 6" LEG MATERIAL. DRAW A LINE ALONG THE TONGUE DOWN FROM THE ANGLE CUT LINE.

3 PIECES OF 1"X 6" LEG MATERIAL. MARKED WITH LEG EDGE ANGLE LINES

MARK ON EDGES

LEG EDGE ANGLE LINE

CUT LINE

**STEP 5.** MARK THE LEG EDGE ANGLES ON ALL FOUR PIECES OF 1"X 6" LEG MATERIAL AND CUT THE LEGS.

2"X 4" x 3'-6" TOP PIECE

3'-6"

6"

SQUARE A LINE BACK 6" FROM EACH END

**STEP 6.** CUT THE 2"X 4" TOP PIECE TO LENGTH. MEASURE BACK 6" FROM EACH END AND SQUARE A LINE ACROSS THE WIDTH OF THE PIECE.

2"X 4" TOP PIECE

STEEL SQUARE

DRAW A LINE

SET AT THE 24" MARK ON THE BLADE

SET AT THE 4" MARK ON THE TONGUE AND DRAW A CUT LINE

**STEP 7.** USE A STEEL SQUARE TO MARK THE LEG ANGLE CUT LINES ON BOTH EDGES OF THE TOP PIECE. (4" ON TONGUE, 24" ON BLADE) DRAW LINES DOWN FROM THE TWO SQUARED LINES.

USE A 1"X 6" LEG AS A TEMPLATE TO MARK WIDTH

MARK LEG WIDTH ON THIS SIDE

ALIGN WITH LEG CUT LINE

**STEP 8.** ALIGN A PIECE OF 1"X 6" LEG MATERIAL WITH THE LEG CUT LINE. DRAW A LINE ON THE OPPOSITE SIDE OF THE LEG TO MARK ITS WIDTH.

DRAW LINES ⅜" IN AT EACH END OF TOP PIECE

TOP SURFACE OF TOP PIECE

⅜"

ANGLED LEG CUT LINE

**STEP 9.** MEASURE IN ⅜" AND DRAW LINES IN FOUR PLACES ON TOP SURFACE OF 2"X 4" TOP PIECE.

SAW CUTS

⅜" DEEP

END VIEW

CHISEL

ANGLED SAW CUTS

GAIN

SERIES OF SAW CUTS

**STEP 10.** MAKE A SERIES OF ANGLES SAW CUTS (⅜" DEEP AT THE TOP. NOTHING AT THE BOTTOM). USE A CHISEL TO BREAK OUT AND SMOOTH THE GAINS.

LEGS (4)

FASTEN WITH 3 6d NAILS IN EACH LEG

**STEP 11.** NAIL THE FOUR LEGS TO THE TOP PIECE USING THREE 6d NAILS IN EACH LEG.

12"

TOP

MARK BOTH SIDES

⅜" OR ½" PLYWOOD

10"

LEG    PUSH UP    LEG

**STEP 12.** PUSH THE 10"X 12" PLYWOOD END PIECE AGAINST THE BOTTOM SURFACE OF THE TOP PIECE. MARK THE OUTSIDE EDGES OF LEGS.

BEVEL SQUARE

CUT OFF

PLYWOOD END PIECE

CUT OFF

DRAW THE ANGLED CUT LINES

**STEP 13.** SET A BEVEL SQUARE TO AN ANGLE THAT IS OBTAINED BY COMBINING THE 5¼" AND 24" MARKS ON A STEEL SQUARE.

TACK IN PLACE AND TEST FOR WOBBLE

TOP

PLYWOOD END PIECE

IF SOLID, DRIVE NAILS HOME

PUSH-UP

**STEP 14.** PLACE THE SAW HORSE ON A FLAT SURFACE. IF IT DOES NOT WOBBLE, DRIVE HOME THE "TACKING" NAILS.

Figure 20-15. The traditional carpenter's sawhorse is 2' high, with a 2" × 4" or 2" × 6" top piece.

Klein Tools, Inc.
Figure 20–16. Safety belts are required when there is a danger of falling while working in high places.

bers such as trusses, beams, purlins, or plates of 4 inch nominal width or greater, at elevations exceeding 15 feet or 1 story above the ground, water surface, or a continuous floor level below.''

*Safety nets* are sometimes employed when safety belts are not practical. They may be used beneath work taking place 25' or more above the ground, water surface, or floor.

## HITCHES AND KNOTS

Every carpenter should be familiar with some of the approved hitches and knots used in construction work. These may be necessary for securing lines as well as for safely lifting timbers

and other types of loads. Some widely used hitches and knots are shown in Figure 20-17. They can be easily learned with a small amount of practice. A description follows of the uses of each:

The *square* knot is used to fasten together the ends of two ropes that are the same thickness.

The *single-sheet bend* knot is used if the ropes are of different thicknesses.

The *bowline* forms a loop that will not slip (loosen or tighten).

The *clove hitch* is used to fasten a rope to a stake or pole.

The *timber hitch* is used when timbers or other members need to be lifted.

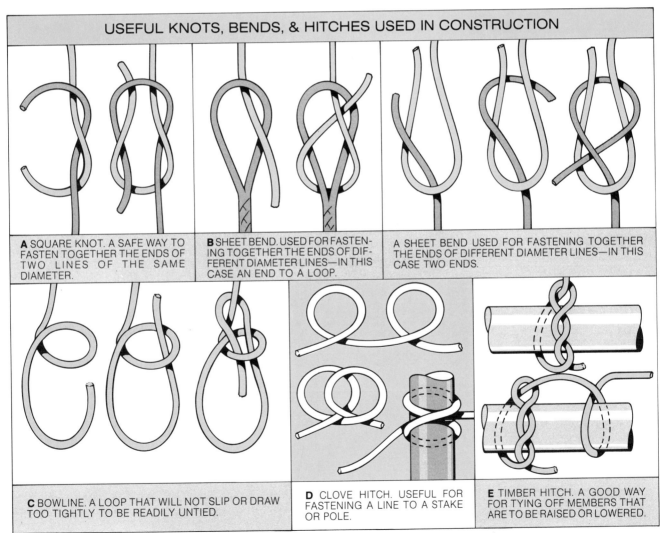

USEFUL KNOTS, BENDS, & HITCHES USED IN CONSTRUCTION

**A** SQUARE KNOT. A SAFE WAY TO FASTEN TOGETHER THE ENDS OF TWO LINES OF THE SAME DIAMETER.

**B** SHEET BEND. USED FOR FASTENING TOGETHER THE ENDS OF DIFFERENT DIAMETER LINES—IN THIS CASE AN END TO A LOOP.

A SHEET BEND USED FOR FASTENING TOGETHER THE ENDS OF DIFFERENT DIAMETER LINES—IN THIS CASE TWO ENDS.

**C** BOWLINE. A LOOP THAT WILL NOT SLIP OR DRAW TOO TIGHTLY TO BE READILY UNTIED.

**D** CLOVE HITCH. USEFUL FOR FASTENING A LINE TO A STAKE OR POLE.

**E** TIMBER HITCH. A GOOD WAY FOR TYING OFF MEMBERS THAT ARE TO BE RAISED OR LOWERED.

Figure 20–17. Typical knots used in construction work. Each has a special purpose.

# UNIT 21

# Construction Machinery

The amount of machinery and moving equipment used on a project depends on the type of construction taking place. The erection of a large concrete building, for example, involves the use of much more equipment than the building of a wood-framed house. Earth-moving machinery and hoisting and lifting machinery are discussed in this unit. Machinery used to place concrete into concrete forms is discussed in Section 16.

## EARTH-MOVING MACHINERY

Any new construction project usually requires some excavation or grading. The amount of such work depends on the size of the structure being erected. A lot for a small building may require only a small amount of grading and trenching for the foundation footings and perhaps holes drilled for piers. On a large project, tons of earth may have to be removed for deep footings and foundations, and pilings driven into the ground.

Carpenters are often required to set up lines and lay out the areas for excavation and trenching. Carpenters may also be called upon to check for the correct depths during digging operations.

One type of equipment used in excavation work is the *bulldozer,* which is a tractor with a blade mounted perpendicular to the line of travel. This machine is used to start excavations and to strip the rocks and topsoil from the excavation site. It is also used to move earth excavated from one part of the site to another part of the site where additional fill is required. See Figure 21–1.

*Deere & Company*

Figure 21–1. A bulldozer is used to start excavations and to strip rocks and topsoil from the excavation site. It is also used to move earth from one part of the site to another part.

134

*Deere & Company*

Figure 21-2. One of the chief functions of a loader is to pick up loose soil and rocks and deposit them into trucks.

*Deere & Company*

Figure 21-3. A motor grader is used for final grading operations. Its blade can be adjusted to various angles and positions.

*Kaiser Cement Corporation*

Figure 21-4. A power shovel is used to excavate earth and deposit it in trucks for removal or for transfer to elsewhere on the job site.

*Deere & Company*

Figure 21-5. A diesel, hydraulically powered excavator has many of the same functions as a power shovel. Different attachments allow it to be used for general excavation, trenching, and loading operations.

A *loader* is basically a wheeled tractor with hydraulic arms that control a bucket mounted in front. Its chief function is to pick up loose soil and rocks and deposit them into trucks for removal. It also may be used for stock-piling aggregate materials, and backfilling trenches. See Figure 21-2.

A *motor grader* is vital to the final grading operations required at large construction sites. The blade of this machine can be adjusted to various angles and positions. See Figure 21-3.

A *power shovel* is used to excavate earth and deposit it in trucks for removal or deposit it elsewhere on the job site. See Figure 21-4.

A diesel, hydraulically powered *excavator* is a recently developed machine used for many of the same functions as a power shovel. Different attachments make it useful for general excavation, trenching, and loading operations. See Figure 21-5.

A *backhoe loader* can perform two kinds of operations. The backhoe bucket is used to dig out the trenches for foundation footings. The loader bucket is used for smaller loading jobs.

An *earth auger* is a power-driven drill used to bore out holes for deep concrete piers or piles. See Figure 21-6.

## HOISTING AND LIFTING MACHINERY

Important to any successful building project is a well-organized flow of material on the job site. As the structure goes up, whether it is a one-story building or a skyscraper, materials or prefabricated sections

135

Figure 21–6. An earth auger is a power-driven drill used to bore out holes for deep concrete piers or piles.

*Caterpillar*

Figure 21–7. A fork lift is designed for rough-terrain construction work. Some models can lift and deposit materials up to one or two stories above the ground.

must be raised and placed. Various machines are used for hoisting and lifting.

The *forklift* is a heavy-duty lift truck designed for rough-terrain construction work. It is used to move and deposit materials where they are required on the job site. Some models are designed to lift and deposit materials up to one or two stories above the ground. See Figure 21–7. (A *skid steer loader* with a high-rise pallet fork attachment can be used to place roof trusses on low buildings. See Figure 21–8.)

Forklifts are most often used on light construction projects that are one or two stories high. For high-rise construction and for lifting very heavy loads, *cranes* are used instead of forklifts. They are either mobile or stationary in design.

Mobile cranes consist of a cab housing the power unit that can be moved about on crawler

tracks or wheels. See Figure 21–9. A crane operator in the cab works the boom and the control cables that raise and lower the loads.

Some of the more widely used cranes on heavy construction projects are the *tower cranes.* The two main types of tower cranes are *climbing* and *free-standing.* See Figure 21–10. Both types are transported to the job site by truck in sections. The

*Sperry New Holland*

Figure 21–8. A skid steer loader with a high-rise pallet fork attachment can be used to place roof trusses on low buildings.

sections are then assembled on the job site. A free-standing tower crane is secured in a stationary position next to the building. The base of the crane is bolted to a concrete pad. A climbing tower crane is secured to the floor structure of a high-rise building being erected. The crane is periodically raised as new floor levels are added to the structure.

Probably the most frequently

*Bethlehem Steel Corporation*

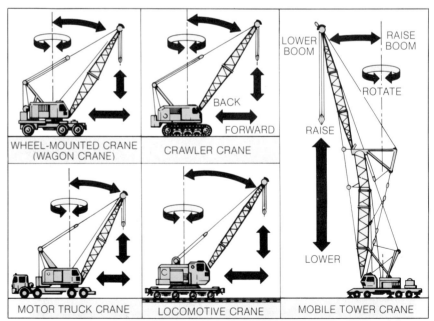

Figure 21–9. Types of mobile cranes used in heavy construction. A crane operator in the cab works the boom and the control cables.

used crane employed in high-rise structures is the inside-climbing tower crane. As the building gains in height, the crane moves up with the building. In order for this crane to be used, openings must be provided in the floor slabs for the tower sections of the crane. The entire crane is raised with hydraulic jacks. Steel collars that rest on the floor slab are secured to the tower section and support the weight of the crane. Additional shoring may be placed beneath the floor slab around the opening. The floor openings are filled in when they are no longer required for the lower sections. In some cases the building's elevator shaft can be used, thus eliminating the need for additional floor openings. Climbing tower cranes are normally raised after the completion of six or eight stories of the building.

Toward the end of the job, when the climbing crane is no longer required, the crane is dismantled into sections and lowered to the ground. For dismantling to be accomplished, a *derrick* must be set up on the roof. A derrick is a lifting and lowering unit that basically consists of a vertical mast and swinging boom. The type used to lower a dismantled crane is made up of 6′ sections. When the derrick in turn is taken apart, the sections can be hand-carried by stairs or elevator to the ground.

Two other types of tower cranes are the *traveling* and the *tie-in* cranes. The base of the traveling crane is mounted on wheels that move over tracks set next to the project. The base of the tie-in crane is bolted to a concrete pad. Unlimited height can be attained by tying sections to the building.

On small to medium jobs, small *self-erecting* tower cranes are used. See Figure 21–11. No

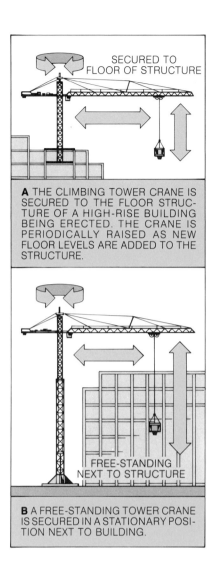

**A** THE CLIMBING TOWER CRANE IS SECURED TO THE FLOOR STRUCTURE OF A HIGH-RISE BUILDING BEING ERECTED. THE CRANE IS PERIODICALLY RAISED AS NEW FLOOR LEVELS ARE ADDED TO THE STRUCTURE.

**B** A FREE-STANDING TOWER CRANE IS SECURED IN A STATIONARY POSITION NEXT TO BUILDING.

HCB Contractors, Inc.

Morrow Crane Company, Inc.

Figure 21–10. Tower cranes may be either the (A) climbing or (B) free-standing type.

Morrow Crane Company, Inc.

Figure 21–11. Small self-erecting cranes require no other lifting equipment for assembly.

other lifting equipment is necessary to help assemble this type of crane.

## Hand Signals for Directing Crane Operators

Hand signals are used to direct boom operators of mobile and stationary cranes. See Figure 21–12. They should be given only by an experienced individual. A wrong signal can result in damage of materials or serious injury to people working near the crane. On large construction projects there is often a person classified to work with the operating engineer on the crane. On some jobs, however, the signal-

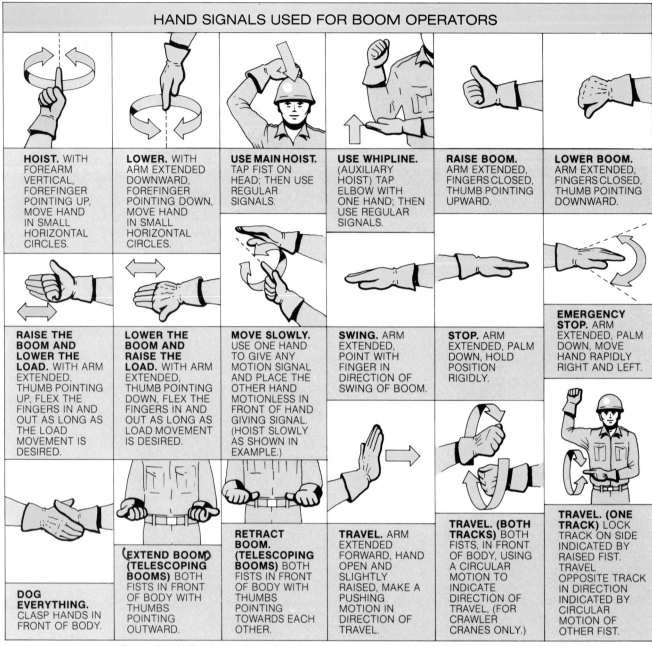

## HAND SIGNALS USED FOR BOOM OPERATORS

**HOIST.** WITH FOREARM VERTICAL, FOREFINGER POINTING UP, MOVE HAND IN SMALL HORIZONTAL CIRCLES.

**LOWER.** WITH ARM EXTENDED DOWNWARD, FOREFINGER POINTING DOWN, MOVE HAND IN SMALL HORIZONTAL CIRCLES.

**USE MAIN HOIST.** TAP FIST ON HEAD; THEN USE REGULAR SIGNALS.

**USE WHIPLINE.** (AUXILIARY HOIST) TAP ELBOW WITH ONE HAND; THEN USE REGULAR SIGNALS.

**RAISE BOOM.** ARM EXTENDED, FINGERS CLOSED, THUMB POINTING UPWARD.

**LOWER BOOM.** ARM EXTENDED, FINGERS CLOSED, THUMB POINTING DOWNWARD.

**RAISE THE BOOM AND LOWER THE LOAD.** WITH ARM EXTENDED, THUMB POINTING UP, FLEX THE FINGERS IN AND OUT AS LONG AS THE LOAD MOVEMENT IS DESIRED.

**LOWER THE BOOM AND RAISE THE LOAD.** WITH ARM EXTENDED, THUMB POINTING DOWN, FLEX THE FINGERS IN AND OUT AS LONG AS LOAD MOVEMENT IS DESIRED.

**MOVE SLOWLY.** USE ONE HAND TO GIVE ANY MOTION SIGNAL AND PLACE THE OTHER HAND MOTIONLESS IN FRONT OF HAND GIVING SIGNAL. (HOIST SLOWLY AS SHOWN IN EXAMPLE.)

**SWING.** ARM EXTENDED, POINT WITH FINGER IN DIRECTION OF SWING OF BOOM.

**STOP.** ARM EXTENDED, PALM DOWN, HOLD POSITION RIGIDLY.

**EMERGENCY STOP.** ARM EXTENDED, PALM DOWN, MOVE HAND RAPIDLY RIGHT AND LEFT.

**DOG EVERYTHING.** CLASP HANDS IN FRONT OF BODY.

**(EXTEND BOOM) (TELESCOPING BOOMS)** BOTH FISTS IN FRONT OF BODY WITH THUMBS POINTING OUTWARD.

**RETRACT BOOM. (TELESCOPING BOOMS)** BOTH FISTS IN FRONT OF BODY WITH THUMBS POINTING TOWARDS EACH OTHER.

**TRAVEL.** ARM EXTENDED FORWARD, HAND OPEN AND SLIGHTLY RAISED, MAKE A PUSHING MOTION IN DIRECTION OF TRAVEL.

**TRAVEL. (BOTH TRACKS)** BOTH FISTS, IN FRONT OF BODY, USING A CIRCULAR MOTION TO INDICATE DIRECTION OF TRAVEL, (FOR CRAWLER CRANES ONLY.)

**TRAVEL. (ONE TRACK)** LOCK TRACK ON SIDE INDICATED BY RAISED FIST. TRAVEL OPPOSITE TRACK IN DIRECTION INDICATED BY CIRCULAR MOTION OF OTHER FIST.

Figure 21–12. Hand signals used for boom operators of mobile and stationary cranes.

ing may be done by the supervisor of a trade such as carpentry or ironworking.

## SAFETY IN WORKING NEAR CONSTRUCTION MACHINERY

The following rules should be observed when working in the general area of earth-moving machinery and operating cranes:

1. Do not stand in a confined area when loads are being raised or lowered.

2. Do not ride on a load or sling attached to the cable of a crane.

3. Do not work or walk under skips or buckets.

4. Watch out for material that may fall while being raised or lowered by the crane.

5. Do not walk under a forklift.

6. Do not walk in the line of travel of a crane.

7. Watch out for swinging booms.

8. Always stay clear of an operating crane.

# UNIT 22

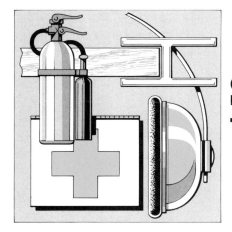

# Safety and Job-site Working Conditions

Accidents on the job can cause pain, disablement, and loss of pay to the worker injured. Sometimes they result in death. Accidents are also costly to the employer in terms of high insurance rates and a slowdown of production.

Experience has proven that most accidents can be prevented by proper safety practices. Various studies by government agencies stress the following facts:

1. Strain or over-exertion is the most common injury suffered by construction workers.

2. Slips or falls from work surfaces, high work areas, and ladders account for nearly one-third of all construction injuries.

3. Worker injuries resulting from use of machines and tools, or from the worker being struck by a tool or machine, account for one-fourth of injuries reported.

4. The most common cause of death from job-site accidents is an accident involving a moving motor vehicle.

In addition to following the safety rules for various tools and equipment, carpenters (and other workers) can take certain other precautions to reduce the likelihood of accidents. For example,

accidents can be caused or prevented by a worker's methods and clothing. Personal protective devices, good housekeeping, protective railings, proper storage of construction materials, and alertness on the part of every worker are all necessary factors in making the job site as safe as possible.

## PERSONAL PROTECTIVE EQUIPMENT

In the past personal protective equipment was mainly for eye, hand, and foot protection. Because of increasing concern about job conditions that are dangerous to the future health of the worker, protective equipment has increased. For example, the noise level of certain job operations can cause ear damage and hearing loss over a period of time, so noise-suppressing ear protectors are now available. Other health problems are caused by exposure to harmful dusts, fumes, mists, vapors, or gases, so respiratory protectors are also available. Figure 22–1 shows some of the personal protective equipment worn by carpenters.

## WORKING HABITS

Many accidents can be prevented by proper working habits such as:

1. Do not place tools where they may fall and injure someone below. See Figure 22–2.

2. Always watch where you step.

3. Look out for the safety of your fellow workers as well as your own safety.

4. Do not engage in horseplay on the job.

Many back injuries and hernias are caused by improper lifting or carrying of heavy objects. If an object is large or heavy, seek help in lifting or carrying it. Wear gloves if the object has splinters, slivers, or sharp edges. The proper procedure for lifting and carrying is shown in Figure 22–3.

## GOOD HOUSEKEEPING

Job hazards on a construction site are sometimes caused by general sloppiness, poor organization, and careless storage of materials. These conditions also reduce job efficiency. Rules for good housekeeping are:

1. Keep scrap lumber cleared

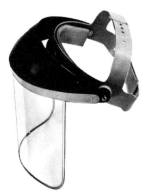

FACE SHIELD
*Glendale Optical Company*

SAFETY GOGGLES
*Glendale Optical Company*

HARD HAT AND
DISPOSABLE
RESPIRATORY
PROTECTOR
*Glendale Optical Company*

NOISE-SUPPRESSING EAR
PROTECTORS
*Glendale Optical Company*

SAFETY SHOES
*Iron Age Shoe Company*
Figure 22–1. Personal protective
equipment worn by carpenters.

DANGER
TOOL MAY FALL
AND CAUSE INJURY

Figure 22–2. Do not place tools where they may fall and injure someone.

from work areas, passageways, and stairs. Bend or pull out protruding nails.

2. Make sure ground areas within 6' of a building under construction are reasonably level. Bridge open ditches at convenient places to provide walkways for workers.

3. Keep material storage areas free of obstruction and debris.

4. Stack materials in such a way that they will not fall, slip, or collapse. A lumber pile should not exceed 16' in height if it is to be handled manually, or 20' if it is to be handled with equipment.

5. Maintain well-defined passageways and walkways on the construction site. Keep them well lit and free of tripping hazards.

6. Use clean-up crews to periodically remove all waste materials from the job site.

7. Store tools and equipment not being used in chests or tool sheds.

8. Remove slush or snow from work areas or walkways before it turns into ice. Slipping can be reduced by spreading sand, gravel, cinders, or other gritty materials over the work areas.

## SLOPING AND SHORING

The erection of a large building often requires deep excavations. Even when smaller jobs require only narrow trenches, the trenches may be very deep. Carpenters and other workers stand in these excavations while constructing the forms for footings and concrete walls. An ever-present danger to them is the possibility that the earth banks may collapse. Therefore, carpenters may either slope the banks of the excavation or construct a shoring system against the banks. The

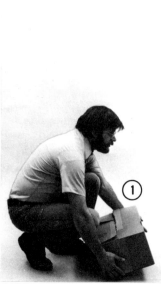

BEND KNEES AND GRASP
THE OBJECT FIRMLY.

LIFT OBJECT, STRAIGHTENING
LEGS, KEEPING BACK AS
STRAIGHT AS POSSIBLE.

DO NOT MOVE
FORWARD UNTIL
WHOLE BODY IS IN
A VERTICAL POSITION.

Figure 22–3. To avoid back injury, use proper lifting procedure.

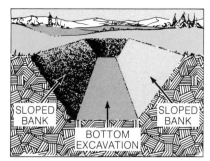

Figure 22-4. Whenever possible, the banks of a deep excavation should be sloped outward.

method used to protect against collapsing earth banks depends on the type of soil in the area, the depth of the excavation, the level of the water table, the type of foundations to be built, and the space around the excavation.

Federal OSHA regulations specify "Banks more than 5 feet high shall be shored, laid back to a stable slope, or some other equivalent means of protection shall be provided where employees may be exposed to moving ground or cave-ins."

## Sloping

The ground can be sloped to eliminate the possibility of the earth banks collapsing when there is room around the construction area. The sloping system cannot be used if other buildings

or streets are right next to the excavation. Federal OSHA regulations on sloping specify a 45° angle for excavations taking place under average soil conditions. Less slope may be permitted for solid rock, shale, or cemented sand and gravels. More slope may be required for compacted sharp sand or well-rounded loose sand. See Figure 22-5.

## Shoring Methods

In large and deep excavations, several methods of shoring are effective. *Interlocking sheet piling* consists of steel sections that can be reused many times. See Figure 22-6. This system is fairly watertight. The sheet piling is lowered by crane into templates that hold it in position. Then it is driven in with a pile-driving hammer. Braces may be installed to support the sheet piling. See Figure 22-7.

*Soldier piles* are driven into the ground with a pile-driving rig. Lagging, consisting of 3″ thick wood planks, is placed between the flanges of the soldier piles. A horizontal steel waler runs across the fronts of the soldier piles. It is held in position by *tie-backs*, which are steel strands placed into grouted holes drilled in the bank of the

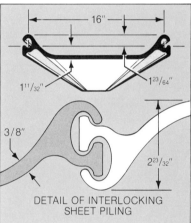

Figure 22-6. One shoring method is to place steel interlocking sheet piling around the area to be excavated. At top a section of sheet piling is being lowered by crane.

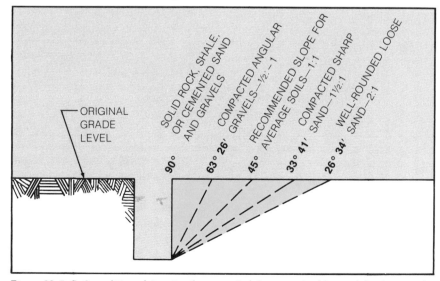

Figure 22-5. Soil conditions determine the amount of slope required for earth banks around excavations.

*Spencer, White & Prentis, Inc.*

Figure 22-7. Many braces have been installed to give lateral support to the sheet piling shoring up the banks of this excavation for a high-rise building.

141

excavation. The holes may extend as far back as 50' into the bank. See Figure 22–8.

Various shoring methods are recommended for ordinary trenching operations. Where hard, compact soil exists, trenches 5' or more in depth can be shored by placing vertical timbers on opposite sides of the trench. These timbers are held in place by cross braces or screw jacks. The uprights should be no more than 5' apart. See Figure 22–9.

In trenching operations in loose or running soil, carpenters may install wood sheet piling backed by stringers and bracing. See Figure 22–10.

## BARRICADES AND GUARDRAILS

On construction jobs where there is much public traffic in the adjoining streets, carpenters erect high fences (barricades) around the site to prevent unauthorized persons from entering the construction area. Often these barricades have overhead protection to guard the public from falling objects.

Guardrails help protect the safety of workers on the job. See Figure 22–11. A typical safety

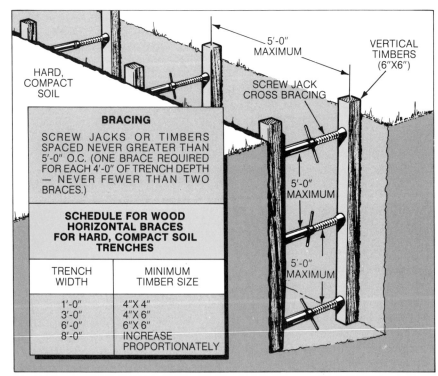

**BRACING**

SCREW JACKS OR TIMBERS SPACED NEVER GREATER THAN 5'-0" O.C. (ONE BRACE REQUIRED FOR EACH 4'-0" OF TRENCH DEPTH — NEVER FEWER THAN TWO BRACES.)

**SCHEDULE FOR WOOD HORIZONTAL BRACES FOR HARD, COMPACT SOIL TRENCHES**

| TRENCH WIDTH | MINIMUM TIMBER SIZE |
|---|---|
| 1'-0" | 4"X 4" |
| 3'-0" | 4"X 6" |
| 6'-0" | 6"X 6" |
| 8'-0" | INCREASE PROPORTIONATELY |

Figure 22–9. Trenches in hard, compact soil may require no more than spaced vertical timbers held in place by cross braces or screw jacks.

code requirement is that all work surfaces 7½' or higher must be guarded by railings. This requirement also applies to scaffolds and platforms. In addition, temporary guardrails should be placed around floor openings

being framed for stairwells. They should also be placed across openings for doors in exterior walls if there is a drop of more than 4' and if the bottom of a window opening is less then 3' above the working surface.

Top rails are made of smooth 2" × 4" material and placed 42" to 45" from the working surface. Mid rails are made of 1" × 6" material and placed midway between the top rails and the working surface.

Carpenters may also be required to install temporary guardrails on ramps and runways, elevator shafts, balconies, and other parts of the building under construction.

## RAMPS, RUNWAYS, AND TEMPORARY STAIRS

Heavy construction jobs usually require ramps, runways, and

*Spencer, White & Prentis, Inc.*

Figure 22–8. Steel soldier piles and wood lagging are often used for shoring excavations. Here the soldier piles have been driven into the ground and the lagging has been placed between the flanges of the soldier piles. The horizontal steel waler running across the fronts of the soldier piles is held in position by tie-backs.

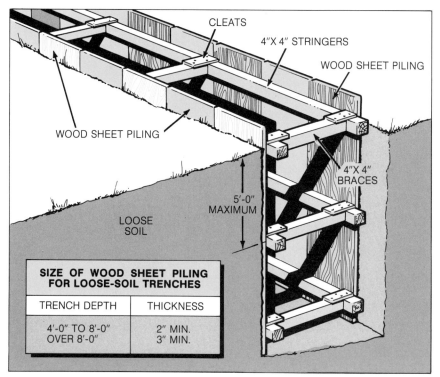

Figure 22–10. Trenches dug in loose soil may be shored with wood sheet piling backed by stringers and braces.

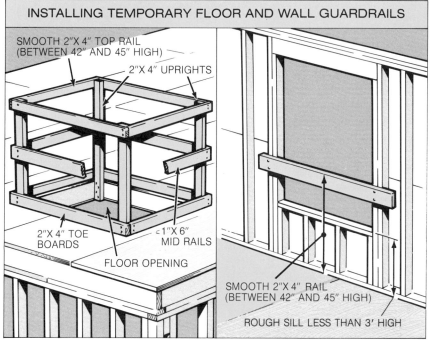

Figure 22–11. During construction, temporary guardrails should be placed around floor openings. They should also be placed across wall openings for exterior doors if there is a drop of more than 4' and across window openings less than 3' above the working surface.

sometimes temporary stairs. See Figure 22–12. Their purpose is to provide a convenient means for workers to move about on the job, and for materials to be transported. Ramps and runways are also constructed for the movement of wheelbarrows and power-driven buggies.

## FIRE PREVENTION

A serious concern for all construction workers is the ever-present danger of fire on the job site. An understanding of what creates a fire hazard and how this danger can be reduced is important.

### Types of Fires

There are four types of fires:

*Class A* fires occur with wood, paper, textiles, and similar materials. They can be extinguished with water and other water-based agents.

*Class B* fires occur with flammable liquids. They can be smothered with agents such as carbon dioxide and other chemical foams.

*Class C* fires occur with live electrical equipment. They can be extinguished by nonconductive dry chemical agents.

*Class D* fires occur with combustible metal materials such as magnesium, sodium, potassium, and others. They can be extinguished by a coarse powder agent that seals the burning surface and smothers the fire.

### Preventive Measures

The threat of fire can be decreased by the following preventive measures:

1. Do not allow rubbish and combustible material to accumulate on the job.

2. Use proper containers to burn rubbish.

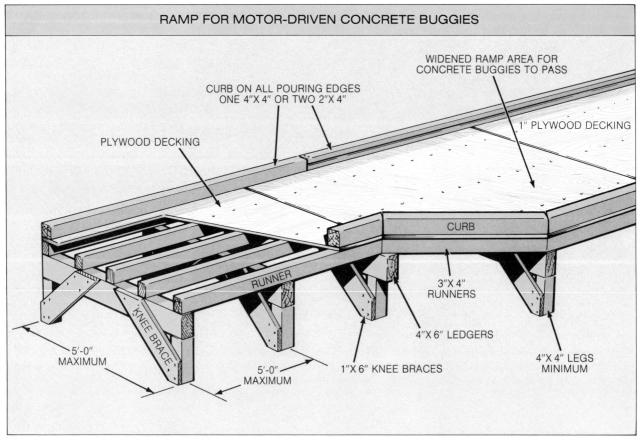

**RAMP FOR MOTOR-DRIVEN CONCRETE BUGGIES**

WIDENED RAMP AREA FOR
CONCRETE BUGGIES TO PASS

CURB ON ALL POURING EDGES
ONE 4"X 4" OR TWO 2"X 4"

1" PLYWOOD DECKING

PLYWOOD DECKING

CURB

RUNNER

3"X 4"
RUNNERS

KNEE BRACE

4"X 6" LEDGERS

5'-0"
MAXIMUM

5'-0"
MAXIMUM

1"X 6" KNEE BRACES

4"X 4" LEGS
MINIMUM

Figure 22–12. Construction specifications for a runway built to support motor-driven concrete buggies.

3. Keep volatile and flammable materials stored away from the immediate job site.

4. Do not smoke near volatile materials. These materials readily evaporate at normal temperatures and pressures, thus creating flammable vapors.

5. Keep all flammable liquids such as gas, paint thinner, oil, grease, and paint in tightly plugged or capped containers.

## Extinguishing Fires

If a fire occurs on the job site, an alarm should be sounded and the fire department called. However, small fires can be quickly extinguished if the proper equipment is present on the job. All safety codes include a requirement for portable fire extinguishers on the job. Fire extinguishers should be located in clear view

and should not be obstructed by building materials.

All approved types of fire extinguishers are coded for particular classes of fires. Using the wrong extinguisher for a particular class of fire can be dangerous. Never take a chance with a fire. **Always call the fire department if there is any chance of the fire spreading.**

# SECTION 6

# Building Design and Blueprint Reading

A set of *working drawings* called blueprints is required to construct a building. The blueprints in this section are drawn for instructional purposes and contain elements not found on actual blueprints. Dimensions are for instructional purposes and should not be scaled. Reduced blueprints, provided by Garlinghouse Company, are included in the *Carpentry Workbook*.

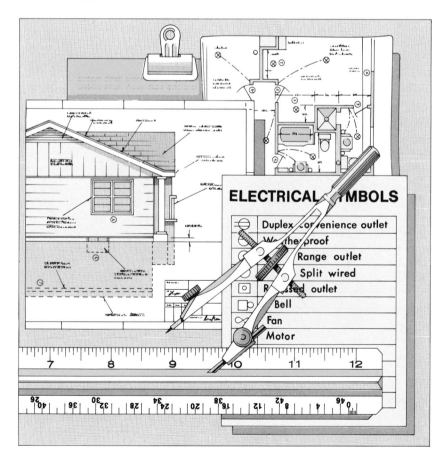

# UNIT 23

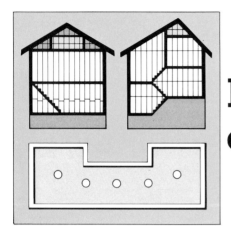

# Building Design, Plans, and Specifications

The design of a building determines its appearance and dictates the methods to be used in its construction. Most buildings are either *traditional* or *contemporary* in appearance. See Figures 23–1 and 23–2. Some traditional styles date back to colonial times in the United States yet are still popular in new construction.

*Architects* are people qualified to design buildings. Construction plans are drawn under their direction. A competent architect has a good understanding of how the building is built as well as how the building should look.

Blueprints for larger buildings often require drawings by *structural engineers.* These drawings include the parts of the building (beams, columns, floors, walls, and so forth) that support and hold together the entire structure. Architects also frequently employ the services of *soil engineers* to analyze the ground conditions at the building site.

## FACTORS THAT INFLUENCE BUILDING DESIGN

Many factors must be considered in designing a building. One fac-

*California Redwood Association*

Figure 23–1. The design of this newly built house is based upon traditional styles of architecture.

*California Redwood Association*

Figure 23–2. Residential building of contemporary design.

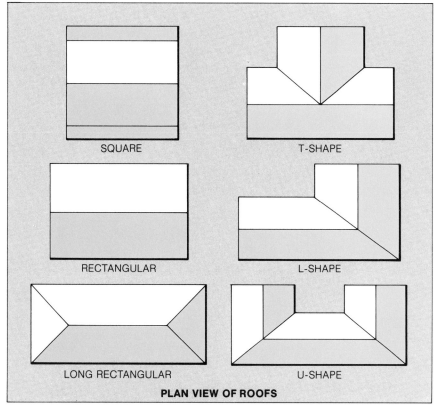

Figure 23–3. Common house shapes for one-family homes.

*The Garlinghouse Company*

Figure 23–4. A typical one-story house. Three types of foundations are possible.

tor is the size and shape of the lot. Another factor is the soil condition of the lot, which determines the type of foundation that can be used. Still another factor is the environment of the lot. If possible, a newly constructed building should conform to the sizes and styles of buildings already in the neighborhood. A newly constructed building must also conform to the building code and zoning regulations of that area. (Building code and zoning regulations are discussed in Unit 32.)

*Orientation* is the position of the building on the lot and the direction in which different walls will face. Wind conditions and the different positions of the sun during the day help determine the ideal orientation for a building. Privacy is another important concern in deciding on orientation.

The location of doors and windows, and even the type of windows, are largely determined by the orientation of a building and by the sun and wind factors of its environment.

## BASIC SHAPES AND TYPES OF BUILDING DESIGN

Whether traditional or contemporary, the design of a one-family home usually derives from one of several basic shapes: rectangular, T-shaped, L-shaped, or U-shaped. See Figure 23–3. The basic house types are: one-story, one-and-one-half-story, two-story, and split-level. Most house designs can be built over a full-basement, crawl-space, or slab-floor foundation system. See Figure 23–4.

### One-story House

In a one-story house all the *habitable* rooms are on the same

147

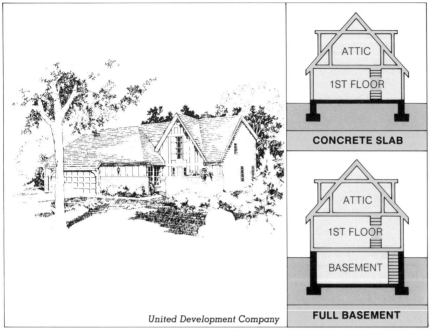

Figure 23–5. A one-and-one-half-story house. Note dormer windows in the roof. The basement is optional.

level. A habitable room is one that is used for living purposes, such as a bedroom, living room, dining room, family room, or bathroom. The attic of a one-story house is usually too small for living purposes but can sometimes be used for storage.

## One-and-one-half-story House

A one-and-one-half-story house has a high-pitched roof, which allows some of the attic space to be used for living purposes. See Figure 23–5. Dormers or roof windows provide light and ventilation. The rooms in the attic area are usually additional bedrooms and a second-floor bathroom.

## Two-story House

A two-story house consists of two full stories under a flat or low-pitched roof. See Figure 23–6. The second story usually consists of bedrooms and a second bathroom.

## Split-level House

A split-level house is more complicated in design. It is actually a one-story house with a basement under one section of the house

that is partly below and partly above the ground level. There are a number of variations on the split-level design, as shown in Figure 23–7. Split-level houses are practical for sloping lots.

## BUILDING PLANS

A set of building plans usually begins with a series of preliminary sketches that progress into the final *working drawings.* The original working drawings made under the direction of an architect are usually rendered in dark lines on white paper. At one time copies of the original drawings were made by a process that resulted in white lines against a blue background. The copies were therefore called *blueprints.* Today copies are usually produced (by a diazo printing machine) with dark lines against a white background. See Figure 23–8. However, the term *blueprints* is still generally used for any copies of the original working drawings.

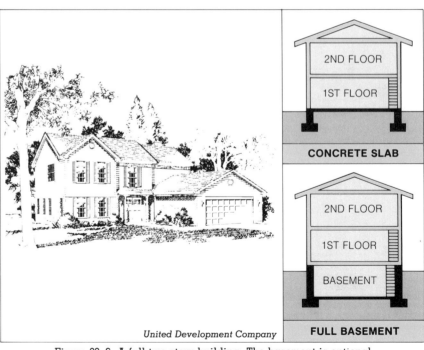

Figure 23–6. A full two-story building. The basement is optional.

Some buildings are *custom-built.* A person wanting to construct a building contacts an architect, who designs the building according to the requirements of the future owner and the amount of money the owner wants to spend.

Plans may also be prepared for general contractors who construct individual homes or apartments on *speculation.* These structures are financed and built by the contractors or real estate developers. They are sold during or after completion of the work. Housing tracts and condominiums are examples of this type of operation.

*Stock plans* are working drawings developed by architects and then purchased from concerns that produce a wide variety of working drawings for home construction. The Three-bedroom House Plan in this section is a stock plan from the Garlinghouse Plan Service of Topeka, Kansas.

## SPECIFICATIONS

A complete set of building plans usually includes a legal document called the *specifications* (or *specs*). The specifications contain data that help clarify the working drawings.

The form for writing specifications varies with different architects. A common practice is to divide the specifications into *divisions* that cover legal questions and pertain to different work areas on the construction project. A list of these divisions follows, along with examples of some of the information that would be included in a set of specifications for the Three-bedroom House Plan.

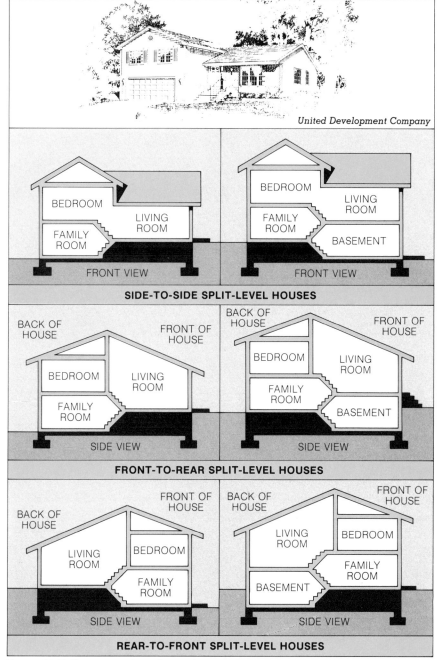

*United Development Company*

**SIDE-TO-SIDE SPLIT-LEVEL HOUSES**

**FRONT-TO-REAR SPLIT-LEVEL HOUSES**

**REAR-TO-FRONT SPLIT-LEVEL HOUSES**

Figure 23–7. Variations of a split-level house. In a split-level house, the basement is partly below and partly above the ground level.

**Division 1: General Conditions**
Contractor's and owner's legal responsibilities
Completion date of job
Contractor's guarantee of materials and workmanship

**Division 2: Site Work**
Excavation
Grading of lot
Backfill and fill

**Division 3: Concrete and Masonry**
Concrete mix
Placement of concrete in forms
Type of masonry work
Grout mixture for mortar work

**Division 4: Carpentry and Millwork**
Rough and finish carpentry work
Types and grades of materials to be used

**Division 5: Sheet Metal and Roofing**
Flashing used over exterior doors, windows, and on roof

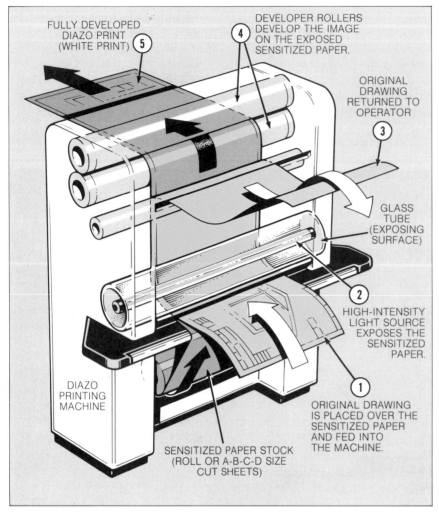

Figure 23-8. The diazo process of reproducing original working drawings produces copies (called blueprints) with dark lines on a white background.

FULLY DEVELOPED DIAZO PRINT (WHITE PRINT) ⑤

DEVELOPER ROLLERS DEVELOP THE IMAGE ON THE EXPOSED SENSITIZED PAPER. ④

ORIGINAL DRAWING RETURNED TO OPERATOR ③

GLASS TUBE (EXPOSING SURFACE)

HIGH-INTENSITY LIGHT SOURCE EXPOSES THE SENSITIZED PAPER. ②

DIAZO PRINTING MACHINE

ORIGINAL DRAWING IS PLACED OVER THE SENSITIZED PAPER AND FED INTO THE MACHINE. ①

SENSITIZED PAPER STOCK (ROLL OR A-B-C-D SIZE CUT SHEETS)

Information regarding installation of pipes and fixtures

The specific information given in each of the *divisions* of a set of specifications is further broken down into *sections.* An example of the sections of one of these divisions follows. Again, the example is taken from the Three-bedroom House Plan.

### Division 4: Carpentry and Millwork

Sec. 1. Scope. This division includes the furnishing and installation of all rough and finish carpentry work and millwork. It includes all related items necessary to complete the work as shown in the working drawings.

Sec. 2. Materials

a. Rough lumber shall be Standard grade Douglas fir for light framing purposes. No. 1 Select Structural is to be used for joists and rafters. Subflooring and sheathing shall be ¾″ APA rated sheathing, Exposure 1 or 2.

b. All exterior trim and siding shall be clear heart redwood.

c. Interior trim shall be C or better ponderosa pine. Patterns are shown in the trim details.

d. Interior and exterior doors are identified in the door schedule.

e. Exterior door frames shall be 1¼″ thick ponderosa pine.

f. Interior door frames shall be ¾″ thick ponderosa pine.

g. Kitchen cabinets shall be birch. Countertops shall be surfaced with Formica plastic laminate or equal.

h. Medicine cabinets shall have four adjustable glass shelves.

i. Closets and wardrobes shall have 1″ × 12″ shelving and 1⅜″ closet poles.

j. Bedrooms, living room, and hallway floors shall be finished with 1″ × 3″ clear oak. All other rooms shall receive linoleum.

Materials used for finish roof covering

**Division 6: Insulation**
Materials used for insulation
Application of materials for wall, floor, and ceiling insulation

**Division 7: Drywall**
Thickness, brand, and type of drywall
Installation information

**Division 8: Painting**
Preparation of wall surfaces
Brand name, type, and color of paints
Number of coats required

**Division 9: Finish Hardware**
Type of hardware used in building
Brand names of hardware
Installation information

**Division 10: Heating and Air Conditioning**
Equipment used, including make, style, and manufacturer's name
Information related to installation of heating ducts
Location of heating and air conditioning units

**Division 11: Electrical**
Sizes of wires used and number of circuits
Plugs, switches, and lighting outlets
Types, brand names, and catalog numbers of fixtures to be used

**Division 12: Plumbing**
Types, brand names, and catalog numbers of fixtures to be used
Types and sizes of pipes for gas, water, and waste lines

# UNIT 24

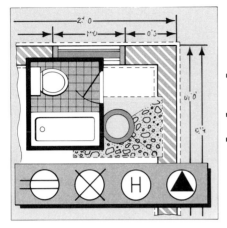

# Understanding the Language of Blueprints

A set of blueprints acts as a step-by-step guide to the construction of a building. Some blueprints have more detail than others, but all include certain basic types of information. The Three-bedroom House Plan contained in this section consists of the following:

Specifications
Plot plan
Foundation plan
Exterior elevations
Thru-house section view
Framing plans
Section views
Details
Door and window schedules
Finish schedule

In its original state, the Three-bedroom House Plan fills five 22″ × 36″ pages. This is a typical size plan for this type of building. The drawings reproduced in this text have been reduced in size, and in some cases rearranged, in order to fit the page.

Blueprint drawings give an *orthographic,* rather than a *pictorial,* view of each part of the building from above and from the sides. A comparison of orthographic and pictorial drawings is shown in Figure 24–1.

Although a complete set of blueprints consists of many different plans, all the plans relate to each other. During any stage of construction, carpenters may have to study several drawings in order to get a complete understanding of the work to be done.

## LINES, DIMENSIONS, AND SCALE

Different kinds of *lines* have different meanings in a blueprint drawing. Figure 24–2 shows a section of a foundation plan and identifies the lines used. A solid line indicates the visible outline of an object, while a broken line indicates an outline hidden from view. A center line establishes the center point of an area. A cutting plane line indicates where an object is "cut" so that interior features may be seen. A break line indicates a shortened view of a part that has a uniform shape. A leader line points from a note or measurement to a part of the building.

*Dimensions* are the measurements that give the distances between different points such as walls, columns, beams, and other structural parts. Dimensions also show the heights of

different sections of the building such as walls, window openings, and door openings. Figure 24–2 includes examples of how measurements are used with dimension lines.

Obviously, a set of plans cannot have drawings that are as large as the actual size of a building. The drawings must be made to *scale.* Inches or fractions of an inch are used to represent feet of the actual measurement of the building. For example, in a plan drawn to ¼″ scale, ¼″ on the drawing represents 1′ of the building. The scale for a drawing is usually explained directly below the drawing. The same scale may not be used for all the different drawings that make up a complete set of plans. Following are some of the scales that may be found in a set of plans:

$$\frac{1}{16}″ = 1′\text{-}0″$$
$$\frac{3}{32}″ = 1′\text{-}0″$$
$$\frac{1}{8}″ = 1′\text{-}0″$$
$$\frac{3}{16}″ = 1′\text{-}0″$$
$$\frac{1}{4}″ = 1′\text{-}0″$$
$$\frac{3}{8}″ = 1′\text{-}0″$$
$$\frac{1}{2}″ = 1′\text{-}0″$$
$$\frac{3}{4}″ = 1′\text{-}0″$$
$$1″ = 1′\text{-}0″$$
$$1\frac{1}{2}″ = 1′\text{-}0″$$
$$3″ = 1′\text{-}0$$

The scale used for a particular plan should be large enough to

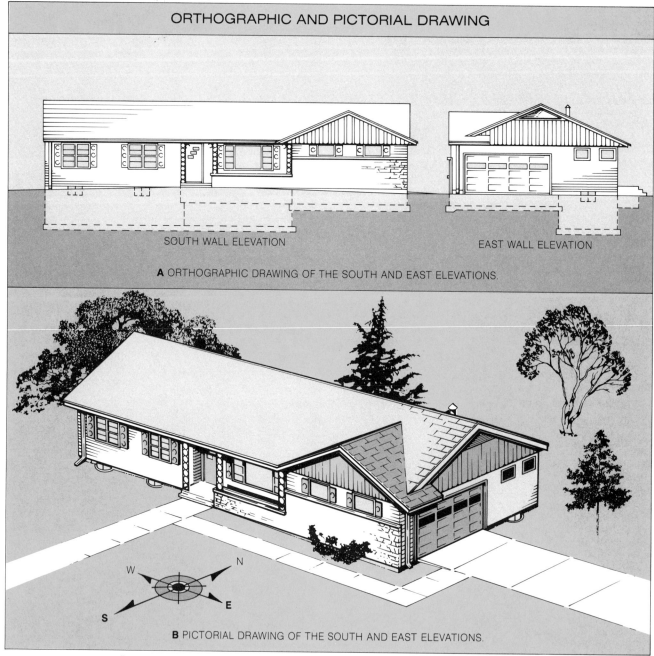

## ORTHOGRAPHIC AND PICTORIAL DRAWING

SOUTH WALL ELEVATION

EAST WALL ELEVATION

**A** ORTHOGRAPHIC DRAWING OF THE SOUTH AND EAST ELEVATIONS.

**B** PICTORIAL DRAWING OF THE SOUTH AND EAST ELEVATIONS.

Figure 24–1. Blueprints are orthographic drawings that give flat views of a building.

present a clear drawing. However, the scale used is also determined by the size or area being presented in the drawing. A ¼″ scale is used most often, although ³⁄₃₂″, ½″, and 3″ scales are also fairly typical for residential blueprints.

An *architect's scale* is used for scaling the dimensions on blueprint drawings. It contains several different scales. Three examples of how to use the ar-

chitect's scale are shown in Figure 24–3.

## SYMBOLS AND ABBREVIATIONS

Working drawings must graphically explain the different materials being used in a building and provide special details regarding structural parts. They must also show the locations of all electri-

cal switches, plugs, and outlets. All plumbing fixtures such as sinks (lavatories), bathtubs, shower enclosures, and water closets (toilets) must be identified. Symbols and abbreviations are used to show these items.

*Symbols* in architectural plans are figures, marks, or letters that represent a material, fixture, or structural part. Figures 24–4 through 24–8 show some common symbols.

# VARIOUS TYPES OF LINES USED ON ARCHITECTURAL DRAWINGS

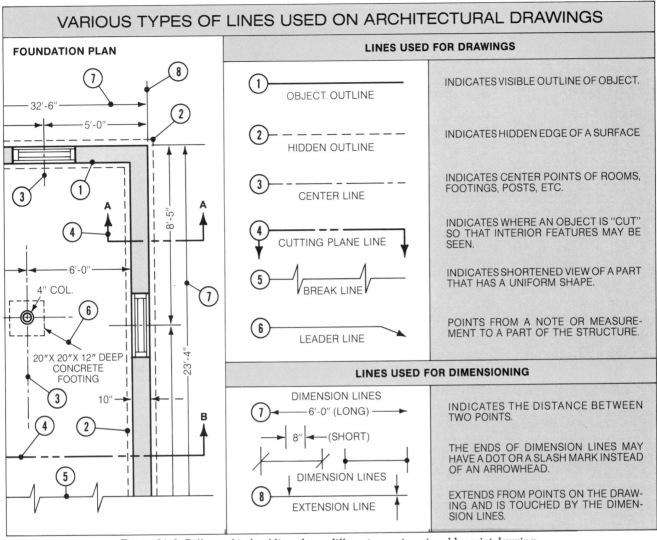

**FOUNDATION PLAN**

**LINES USED FOR DRAWINGS**

1. OBJECT. OUTLINE — INDICATES VISIBLE OUTLINE OF OBJECT.

2. HIDDEN OUTLINE — INDICATES HIDDEN EDGE OF A SURFACE

3. CENTER LINE — INDICATES CENTER POINTS OF ROOMS, FOOTINGS, POSTS, ETC.

4. CUTTING PLANE LINE — INDICATES WHERE AN OBJECT IS "CUT" SO THAT INTERIOR FEATURES MAY BE SEEN.

5. BREAK LINE — INDICATES SHORTENED VIEW OF A PART THAT HAS A UNIFORM SHAPE.

6. LEADER LINE — POINTS FROM A NOTE OR MEASUREMENT TO A PART OF THE STRUCTURE.

**LINES USED FOR DIMENSIONING**

7. DIMENSION LINES — 6'-0" (LONG) / 8" (SHORT) / DIMENSION LINES — INDICATES THE DISTANCE BETWEEN TWO POINTS. THE ENDS OF DIMENSION LINES MAY HAVE A DOT OR A SLASH MARK INSTEAD OF AN ARROWHEAD.

8. EXTENSION LINE — EXTENDS FROM POINTS ON THE DRAWING AND IS TOUCHED BY THE DIMENSION LINES.

Figure 24–2. Different kinds of lines have different meanings in a blueprint drawing.

# READING FEET, INCHES AND FRACTIONS OF AN INCH ON AN ARCHITECT'S TRIANGULAR SCALE

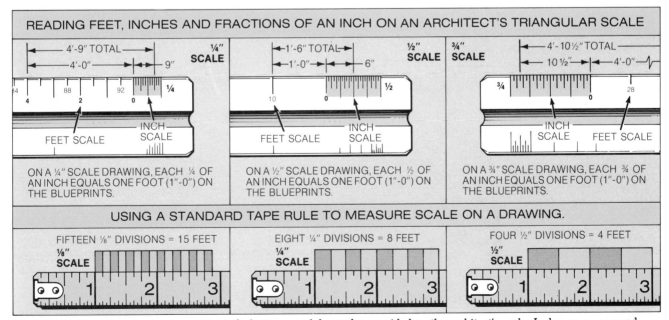

ON A ¼" SCALE DRAWING, EACH ¼ OF AN INCH EQUALS ONE FOOT (1"-0") ON THE BLUEPRINTS.

ON A ½" SCALE DRAWING, EACH ½ OF AN INCH EQUALS ONE FOOT (1"-0") ON THE BLUEPRINTS.

ON A ¾" SCALE DRAWING, EACH ¾ OF AN INCH EQUALS ONE FOOT (1"-0") ON THE BLUEPRINTS.

## USING A STANDARD TAPE RULE TO MEASURE SCALE ON A DRAWING.

FIFTEEN ⅛" DIVISIONS = 15 FEET

EIGHT ¼" DIVISIONS = 8 FEET

FOUR ½" DIVISIONS = 4 FEET

Figure 24–3. Examples of measurements made from some of the scales provided on the architect's scale. Inches are measured from one side of the zero point. Inch-fractions are measured from the opposite side of the zero.

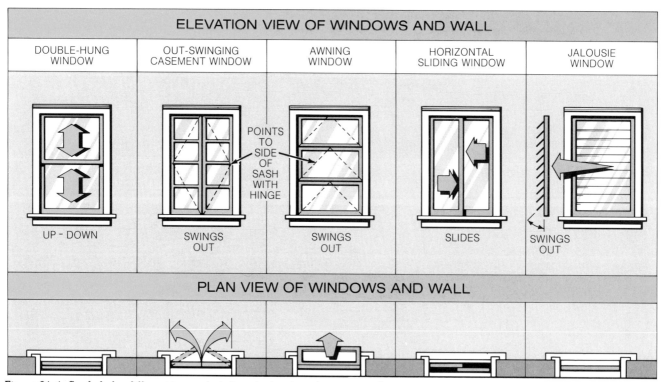

Figure 24-4. Symbols for different types of windows in drawings that give an elevation view (looking directly at the front of the window) and a plan view (looking down at the window from the top of the wall).

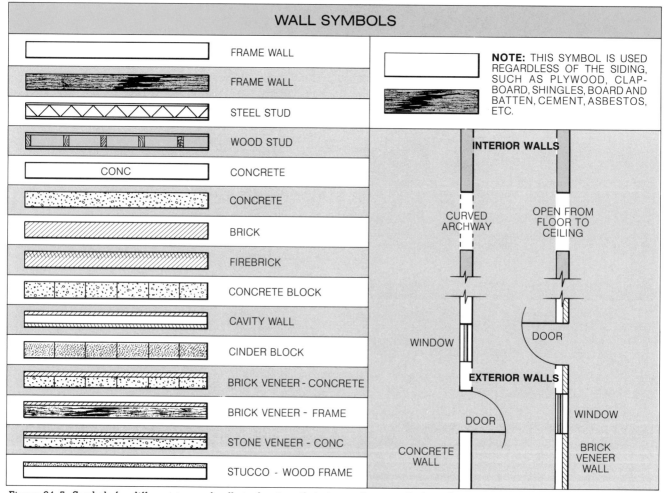

Figure 24-5. Symbols for different types of walls in drawings that give a plan view (looking-down view).

154

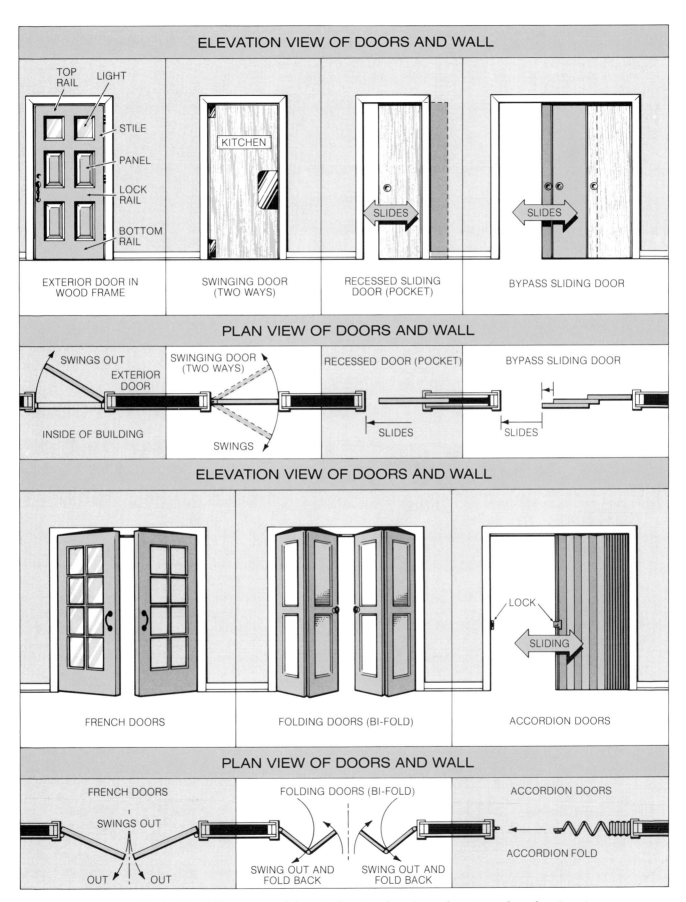

Figure 24–6. Symbols for different types of doors in drawings that give a plan view and an elevation view.

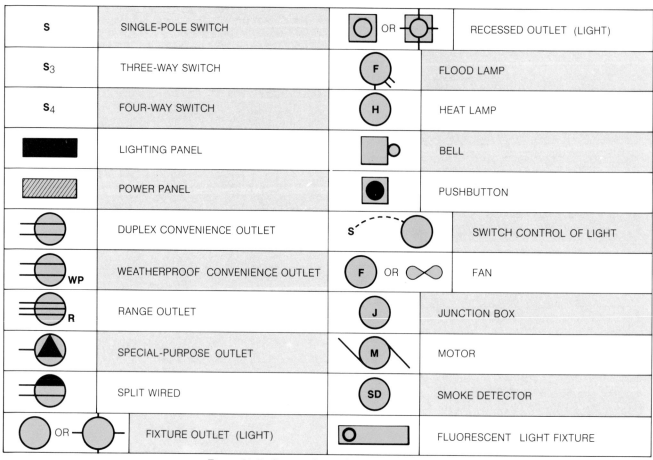

Figure 24–7. Commonly used electrical symbols.

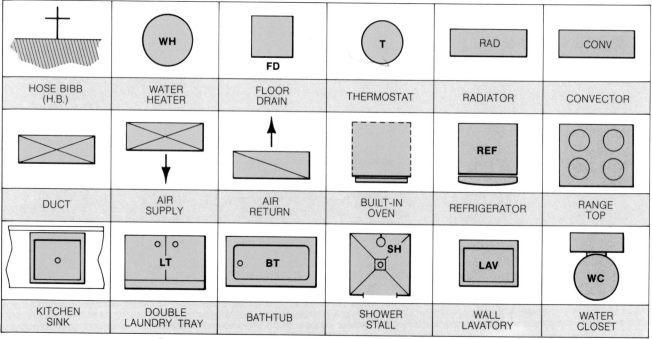

Figure 24–8. Symbols for plumbing fixtures and heating equipment.

| | | | | | |
|---|---|---|---|---|---|
| ALUMINUM | ALUM | FIXTURE | FIX | RIDGE | RDG |
| ANCHOR BOLT | AB | FLASHING | FL | RISER | R |
| ASPHALT TILE | AT | FLOOR | FL | ROOF | RF |
| BATHROOM | BATH | FOOTING | FTG | ROOFING | RFG |
| BATHTUB | BT | FOUNDATION | FDN | ROOF DRAIN | RD |
| BASEMENT | BSMT | FURNACE | FURN | ROOM | RM |
| BEAM | BM | GAUGE | GA | ROUGH OPENING | RO |
| BEDROOM | BR | GALVANIZED IRON | GI | SCREEN | SC |
| BENCH MARK | BM | GIRDER | GDR | SEWER | SEW |
| BLOCK | BLK | GLASS | GL | SHAKE | SHK |
| BRICK | BRK | GRADE | GR | SHEATHING | SHTH |
| BOARD | BD | GROUND | GRD | SHEET METAL | SM |
| BUILDING | BLDG | GYPSUM BOARD | GYP BD | SHINGLE | SHGL |
| BUILDING LINE | BL | HARDBOARD | HBD | SHOWER | SH |
| CABINET | CAB | HARDWOOD | HWD | SIDING | SDG |
| CASEMENT | CSMT | HEAD | HD | SILL | SL |
| CEDAR | CDR | HEAT | H | SINK | SK |
| CEILING | CLG | HOSE BIBB | HB | SKYLIGHT | SKL |
| CEMENT | CEM | HOT WATER HEATER | HWH | SLIDING DOOR | SL DR |
| CENTER | CTR | INSULATION | INS | SOIL PIPE | SP |
| CENTER LINE | CL | INTERIOR | INT | SOLAR PANEL | SLR PAN |
| CHIMNEY | CHIM | JAMB | JMB | SOFFIT | SOF |
| CLOSET | CLOS | JOIST | JST | SOUTH | S |
| COLUMN | COL | KITCHEN | KIT | STACK VENT | SV |
| CONCRETE | CONC | LAUNDRY | LAU | STAIRS | ST |
| CONCRETE BLOCK | CONC BLK | LAVATORY | LAV | STAIRWAY | STWY |
| CORNICE | COR | LIGHT | LT | STEEL | STL |
| CORRUGATED | CORR | LINEN CLOSET | LC | STONE | ST |
| DETAIL | DET | LINOLEUM | LINO | STREET | ST |
| DIAMETER | DIAM | LIVING ROOM | LR | TONGUE AND GROOVE | T & G |
| DINING ROOM | DR | LOUVER | LV | TOP HINGED | TH |
| DISHWASHER | DW | MEDICINE CABINET | MC | TREAD | TR |
| DOOR | DR | METAL | MET | UNEXCAVATED | UNEXC |
| DORMER | DRM | NORTH | N | UTILITY ROOM | UR |
| DOUBLE HUNG | DH | ON CENTER | OC | VENT | V |
| DOUGLAS FIR | DF | OPENING | OPNG | VENTILATION | VENT |
| DOWNSPOUT | DS | OVERHANG | OH | VENT STACK | VS |
| DRAIN | DR | OVERHEAD DOOR | OH DR | VINYL TILE | VA TILE |
| DRYWALL | DW | PANEL | PNL | WASHING MACHINE | WM |
| EAST | E | PARTITION | PTN | WATER | W |
| ELECTRIC | ELEC | PLATE | PL | WATERPROOF | WP |
| ELEVATION | EL | PLYWOOD | PLYWOOD | WATER CLOSET | WC |
| EXCAVATE | EXC | PORCH | P | WELDED WIRE MESH | WWM |
| EXTERIOR | EXT | PRESSURE TREATED | PT | WEST | W |
| FACE BRICK | FB | RAFTER | RFTR | WIDE FLANGE | WF |
| FILL | F | REDWOOD | RWD | WHITE PINE | WP |
| FINISH | FIN | REFRIGERATOR | REF | WINDOW | WDW |
| FINISH FLOOR | FIN FL | REINFORCED | REINF | WOOD | WD |
| FIREPLACE | FPL | REINFORCEMENT BAR | REBAR | YELLOW PINE | YP |
| FIREPROOF | FP | RETAINING WALL | RW | | |

Figure 24–9. Table of blueprint abbreviations.

Little space is allowed for writing on the sheets that make up a set of plans. Therefore, *abbreviations* are used whenever possible. The table in Figure 24–9 shows common abbreviations used in the construction trade. This table, however, is not an official listing. Abbreviations may vary.

# UNIT 25

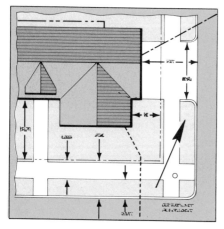

# Plot Plans

The *plot plan,* or *site plan,* is one of the first drawings to be considered in a building project. Before construction can begin, the exact location of the building on the property must be known. The high and low points must be determined so that the lot can be graded to provide proper water drainage away from the building. The plot plan gives this and other information.

The plot plan for the Three-bedroom House Plan is shown in Figure 25–1. The pictorial draw-

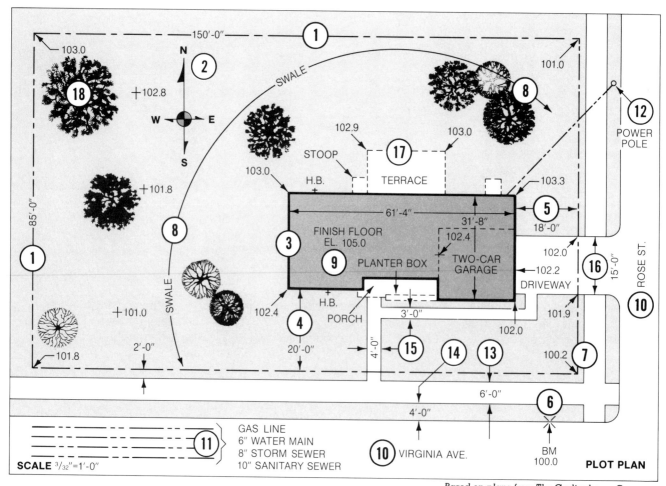

Figure 25–1. Plot plan for the Three-bedroom House Plan. Compare this plan with the pictorial drawings in Figures 25–2 and 25–3.

*Based on plans from The Garlinghouse Company*

ing in Figure 25–2 shows how the completed building appears on the property. The pictorial drawing in Figure 25–3 shows how the surface of the lot is shaped to allow proper water drainage.

## BASIC PLOT PLAN INFORMATION

An explanation of the different items shown on the plot plan in Figure 25–1 follows. The numbers in this explanation correspond to the numbers in the circles on the plot plan.

1. *Property lines:* Also known as *lot lines,* the property lines show the shape of the lot. This lot is rectangular. The width and length of the property are also given.

2. *Compass direction:* The plot plan usually includes an arrow pointing north. This is necessary to place the house in its correct position on the lot. The different sides of the building are referred to by compass directions (north, south, east, west) in other drawings of the blueprints.

3. *Building lines:* The outline of the building to be constructed on the lot is shown by building lines. Dimension lines show the width and length of the building.

4. *Front setback:* The distance (20′-0″) from the property line to the front (south wall) of the building is the front setback.

5. *Side yard:* The distance (18′-0″) from the side property line to the side (east wall) of the house is the side yard.

6. *Bench mark:* A point designated as 100.0′ has been established at the street curb as a bench mark. (Bench marks and how they relate to finish grades and elevations are discussed in greater detail later in this unit.)

7. *Finish grades:* A finish grade height of 100.2′ is noted at the southeast corner of the lot.

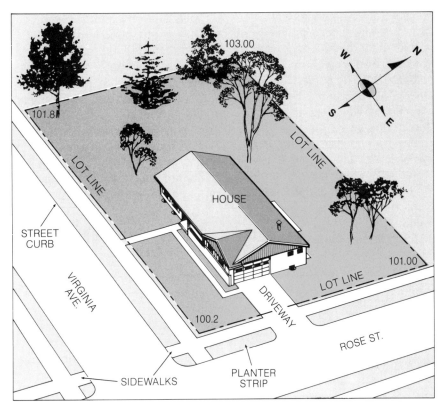

Figure 25–2. Pictorial drawings based on the plot plan in Figure 25–1.

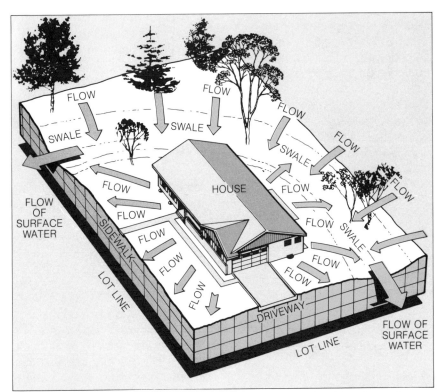

Figure 25–3. The arrows show how the flow of surface water (swale) is directed to flow away from the house.

Finish grade references are also noted at other points on the lot, including the other lot corners and the building corners.

8. *Swale:* The curved line with arrows at each end shows the direction of water drainage on the lot (the swale).

9. *Finish floor elevations:* A finish floor elevation of 105.0' is noted within the building outline. This is the height of the finished floor in relation to the bench mark.

10. *Roads:* South and east of the property are the roads.

11. *Plumbing utilities:* Public utilities such as water, gas, storm drainage, and sewer drainage are below the street surface.

12. *Electrical utilities:* A power pole is located near the northeast corner of the lot. Electric and telephone lines come from the power pole to the house. (Electrical utilities are sometimes placed underground.)

13. *Sidewalks:* The width of the sidewalk (6'-0") is given.

14. *Planter strip:* The distance (4'-0") from the street curb to the sidewalk is the planter strip.

15. *Walk:* The walk goes from the sidewalk to the front porch. It branches off toward the driveway.

16. *Driveway:* A concrete driveway goes from the garage to the street. Its width is given.

17. *Terrace:* The outline of a terrace is indicated. Measurements for the terrace are given on other drawings.

18. *Trees:* Before construction work can begin, some trees in the building area may have to be removed. Trees that will remain on the lot are shown on the plot plan.

## ADDITIONAL PLOT PLAN INFORMATION

In addition to the information shown in Figure 25–1, a plot plan may also indicate *retaining* *walls* and *easements*. Retaining walls made of concrete or concrete blocks keep earth from sliding. Easements are right-of-way provisions on the property. They may indicate, for example, the right of a neighbor to build a road or a public utility to install water and gas lines on the property. The property owner cannot build on an area where an easement has been provided.

## FINISH GRADES AND ELEVATIONS

The plot plan provides information about the surface shape of the lot. The lot may be flat, or it may have a steep or gradual slope. If it is sloped, it will be higher at some points and lower at others. In that case, it may have to be *graded.* A lot is graded by removing or adding soil so that surface water caused by rain or melting snow will be directed away from the building and into the street.

The finish grades and elevations shown on the plot plan are usually based on data provided by a professional surveyor or engineer. As a rule, these points are recorded on the plot plan in feet and *tenths of a foot* rather than in feet and inches.

### Bench Mark

A bench mark (also called the *job datum*) is a point established by a surveyor on or close to the property. It may also be placed at some point along the street curb next to the property. Often it is placed at one corner of the lot. It may be identified by a plugged pipe driven into the ground, a brass marker, or a wood stake.

The location of the job-site bench mark is shown on the plot plan with a grade figure next to it. The grade figure may be the number of feet above sea level at that point, or it may be the number 100.0'. In the plot plan in Figure 25–1, a 100.0' bench mark is shown at the street curb near the southeast corner of the lot.

### Finish Grades

The plot plan in Figure 25–1 shows finish grades at all four corners of the lot and at various other points on the lot, including at the corners of the building, in the garage, and at the driveway.

All these finish grades are based upon their relation to the 100.0' bench mark reference point. For example, the grade at the southeast corner of the lot is 100.2', meaning that the ground is 2/10' (100.2' minus 100.0' equals .2') higher at this point than at the bench mark point. The grade at the northwest corner of the lot is 103.0', meaning that the ground is 3' higher at this point than at the bench mark point.

The lot corner grades show that the lot slopes down from the northwest to the southwest corner. Notice that the grades closest to the building are higher than those farther out on the lot. The surface of the lot has been graded this way so that surface water will drain away from the house. See Figure 25–3.

Finish grades at the front and back of the garage (refer to Figure 25–1) indicate that the concrete slab in the garage will slope 2/10' (102.4' minus 102.2' equals .2') toward the door opening. The slope will allow any rainwater blown into the garage to drain out. In the driveway, finish grade points show a slope away from the garage and from north to south.

### Natural Grades and Contours

Some plot plans include the *existing* (natural) grades as well as the finish grades. The existing grade refers to the condition of the lot before grading. *Contour lines* are sometimes drawn on the plot plan to help show the existing as well as the finish surface shapes of the lot. (A plot plan that includes existing and finish grades along with contour lines is shown in Section 8).

## Elevations

The word *elevation* is often used to mean the same thing as *grade.* More precisely, however, *elevations* are the heights established for different levels of the building. A plot plan usually shows the finish floor elevation for a building. This is the level of the first floor of the building in relation to the job-site bench mark. The plot plan in Figure 25–1 shows the first floor elevation as 105.0′ (5′ higher than the bench mark grade). During the construction of the building, many important measurements are taken from the finish floor elevation.

## Changing Decimals of a Foot to Inches and Inch-Fractions

The grades on a plot plan are usually indicated in *feet* and *tenths of a foot.* Occasionally the grades are shown in feet and *hundredths of a foot.* Conversion charts are available that provide a quick and simple way to change these divisions of a foot to the inches and inch-fractions that appear on a carpenter's rule. Following is an example of how to use a conversion chart:

**Example 1.** Change .86 (86/100 of a foot) to inches, using the conversion chart in Figure 25–4.

Step 1. Find .86 in the chart.

Step 2. Record the number of inches (10″) shown at the top of the vertical column containing .86.

| 8TH | 0″ | 1″ | 2″ | 3″ | 4″ | 5″ | 6″ | 7″ | 8″ | 9″ | 10″ | 11″ |
|---|---|---|---|---|---|---|---|---|---|---|---|---|
| 0 | .00 | .08 | .17 | .25 | .33 | .42 | .50 | .58 | .67 | .75 | .83 | .92 |
| 1 | .01 | .09 | .18 | .26 | .34 | .43 | .51 | .59 | .68 | .76 | .84 | .93 |
| 2 | .02 | .10 | .19 | .27 | .35 | .44 | .52 | .60 | .69 | .77 | .85 | .94 |
| 3 | .03 | .11 | .20 | .28 | .36 | .45 | .53 | .61 | .70 | .78 | .86 | .95 |
| 4 | .04 | .13 | .21 | .29 | .38 | .46 | .54 | .63 | .71 | .79 | .88 | .96 |
| 5 | .05 | .14 | .22 | .30 | .39 | .47 | .55 | .64 | .72 | .80 | .89 | .97 |
| 6 | .06 | .15 | .23 | .31 | .40 | .48 | .56 | .65 | .73 | .81 | .90 | .98 |
| 7 | .07 | .16 | .24 | .32 | .41 | .49 | .57 | .66 | .74 | .82 | .91 | .99 |

Figure 25–4. Conversion chart for changing tenths and hundredths of a foot to inches. Equivalent inch-fractions are given to the nearest eighths of an inch.

Step 3. Record the number (3) shown at the far left of the horizontal column containing .86. (This number appears in the vertical column headed "8TH.")

Step 4. The number found in the vertical column headed "8TH" represents the number of *eighths of an inch.* Therefore, change the recorded number 3 to ⅜″.

Step 5. Combine the results of Steps 2 and 4: $^{86}/_{100}$ of a foot is equal to 10⅜″.

A carpenter should know how to make these conversions without a chart. Following are two mathematical examples of how this can be done:

**Example 2.** Change .7′ (⁷/₁₀ of a foot) to inches.

Step 1. Multiply the decimal .7 by 12. The answer will give inches and a decimal part of an inch.

$$\begin{array}{r} .7 \\ \times\ 12 \\ \hline 8.4\ \text{inches} \end{array}$$

Step 2. Multiply the decimal .4 by a denomination of an inch-fraction. The denomination can be expressed in sixteenths, eighths, or quarters. In this example the sixteenths denomination is used;

therefore, multiply .4 by 16.

$$\begin{array}{r} .4 \\ \times 16 \\ \hline 6.4\ \text{sixteenths} \\ \text{of an inch} \end{array}$$

Step 3. Round off the answer to the nearest sixteenth. This is ⁶/₁₆, or ⅜″.

Step 4. Combine the results of Steps 1, 2, and 3: ⁷/₁₀ of a foot is equal to 8⅜″.

**Example 3.** Change .84′ (⁸⁴/₁₀₀) to inches.

Step 1. Multiply the decimal .84 by 12. The answer will give inches and a decimal part of an inch.

$$\begin{array}{r} .84 \\ \times\ \ 12 \\ \hline 10.08\ \text{inches} \end{array}$$

Step 2. Multiply the decimal .08 by the sixteenth denomination of an inch-fraction.

$$\begin{array}{r} .08 \\ \times\ \ 16 \\ \hline 1.28\ \text{sixteenths of an inch} \end{array}$$

Step 3. Round off the answer to the nearest sixteenth. This is ¹/₁₆.

Step 4. Combine the results of Steps 1, 2, and 3: ⁸⁴/₁₀₀ of a foot is equal to 10¹/₁₆″.

# UNIT 26

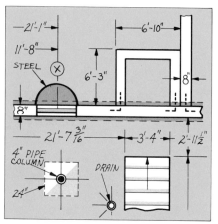

# Foundation Plans

The first stage in the construction of any building involves the foundation. The foundation must be designed to support its own weight as well as the rest of the building structure. Footings, walls, and piers are the basic features of a foundation. The *foundation plan* gives information

regarding these features and also the posts and beams that help support the floor unit above the foundation. If a joist system is used for the floor unit, the size and spacing of the joists are noted in the foundation plan.

Two types of foundation plans are described in this unit: the full-

basement foundation and the crawl-space foundation.

## FULL-BASEMENT FOUNDATION

The full-basement foundation that is part of the Three-bedroom House Plan is shown in Figure 26–1. The drawings show walls

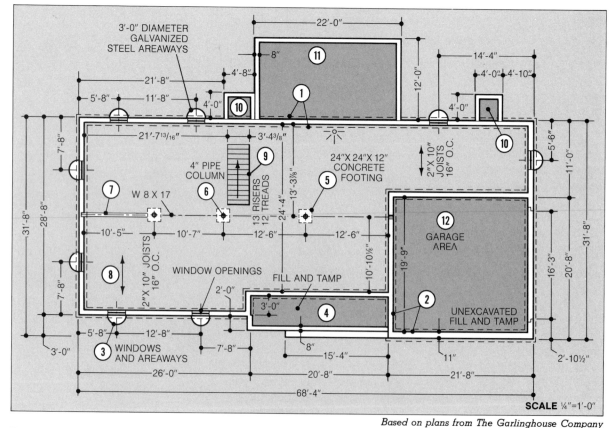

*Based on plans from The Garlinghouse Company*

Figure 26–1. Foundation plan for a full-basement foundation that could be used with the Three-bedroom House Plan. Compare this plan with the pictorial drawing in Figure 26–2.

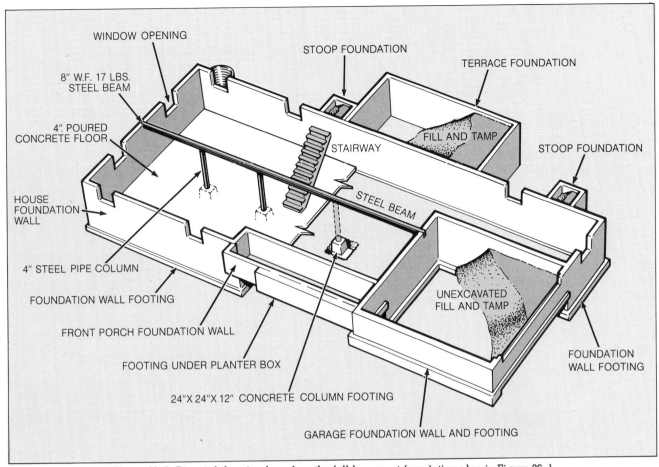

Figure 26–2. Pictorial drawing based on the full-basement foundation plan in Figure 26–1.

resting on footings. The footings extend from each side of the walls, making the shape of a T. This design, known as a T-foundation, is widely used. Carpenters construct the forms according to the foundation plan, and concrete is poured into the completed forms. Figure 26–2 is a pictorial view of the completed foundation.

An explanation of the different items shown on the foundation plan in Figure 26–1 follows. The numbers in this explanation correspond to the numbers in the circles on the foundation plan.

1. *Foundation footings:* Dashed (hidden) lines on each side of the foundation walls indicate the foundation footings. The hidden lines are used because one edge of each footing is covered by soil at the outside edge of the foundation walls. The other edge of the footing is covered by the basement concrete slab at the inside of the foundation walls. The footings are not visible in a plan view.

2. *Foundation walls:* These walls extend around the outside of the building and garage.

3. *Windows and areaway:* The basement windows extend below the grade level. The open space around each window is called an areaway. (Each areaway is 3'-0" diameter galvanized steel.)

4. *Front porch:* For the front porch, a concrete slab is supported at the front edge by a low

foundation wall with no footing.

5. *Column footing:* The column footings are placed below the pipe columns. The width and height of the footings are indicated.

6. *Pipe columns:* Hollow steel columns, 4" in diameter, rest on the footings. Dimensions from the outside walls to the centers of the columns, and from the center of one column to the next one, are given.

7. *Steel beam:* A steel beam is shown. It is an 8" wide-flange beam that weighs 17 pounds for each foot of length.

8. *Floor joists:* The direction in which the floor joists run is shown by an arrow. One end of the joists rests on the outside walls. The other end is notched

into the steel beam. The joists are spaced 16″ O.C. (on centers).

9. *Stairway:* The stairway leading from the basement to the main floor has 13 risers and 12 treads.

10. *Rear stoops:* The rear stoops are located at the two rear entrances to the house. For each stoop, a concrete slab is supported by a low concrete foundation wall around the outside edges. The stoops are filled and tamped.

11. *Terrace:* A low concrete foundation wall extends around the outside edges of the terrace. Its width is given. This area

should be filled and tamped.

12. *Garage area:* Note the "unexcavated, fill and tamp" instructions. Soil will not be removed in this area. If necessary, soil will be added and tamped (pressed down) before the concrete slab is poured.

## CRAWL-SPACE FOUNDATION

Although the Three-bedroom House Plan calls for a full-basement foundation, the same building could also be constructed

over a crawl-space foundation. This type of foundation does not have a basement area. *Crawl space* refers to the distance (usually 18″ or more) between the bottoms of the floor joists and the ground. The main difference between the full-basement and crawl-space foundation is the height of the walls. Also, no stairway or areaways are required in the crawl-space design.

An example of a crawl-space foundation plan that could be used as part of the Three-bedroom House Plan is shown in Figure 26–3. This plan shows that the girder and posts are made of wood. (6″ × 10″ girder

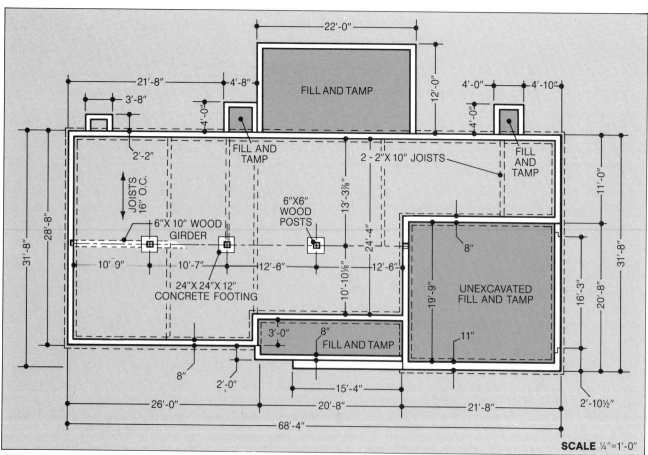

Figure 26–3. Foundation plan for a crawl-space foundation that could be used with the Three-bedroom House Plan. Note that the girder and posts are made of wood. Double joists run underneath the walls in the same direction as the walls. Compare this plan with the pictorial drawing in Figure 26–4.

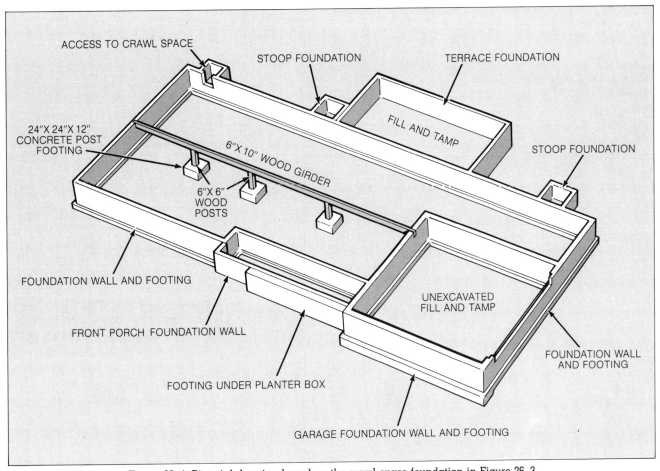

ACCESS TO CRAWL SPACE

STOOP FOUNDATION

TERRACE FOUNDATION

24"X 24"X 12"
CONCRETE POST
FOOTING

FILL AND TAMP

STOOP FOUNDATION

6"X 10" WOOD GIRDER

6"X 6"
WOOD
POSTS

FOUNDATION WALL AND FOOTING

FRONT PORCH FOUNDATION WALL

FOOTING UNDER PLANTER BOX

UNEXCAVATED
FILL AND TAMP

FOUNDATION WALL
AND FOOTING

GARAGE FOUNDATION WALL AND FOOTING

Figure 26–4. Pictorial drawing based on the crawl-space foundation in Figure 26–3.

and 6″ × 6″ posts). The wood posts rest on square concrete footings (24″ × 24″ × 12″).

Dashed lines indicate that the joists are doubled where the walls above run in the same direction as the joists. A pictorial drawing of the foundation is shown in Figure 26–4.

# UNIT 27

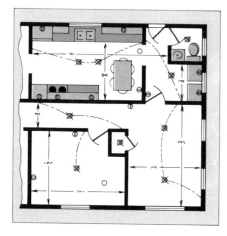

# Floor Plans

Floor and wall construction begins after the foundation has been completed. The section of the blueprints that provides most of the information for this stage of the work is the *floor plan.* See Figure 27–1. It gives a view looking down (a plan view) on the floor level above the foundation. A one-story house requires only one floor plan. Buildings with more than one story require

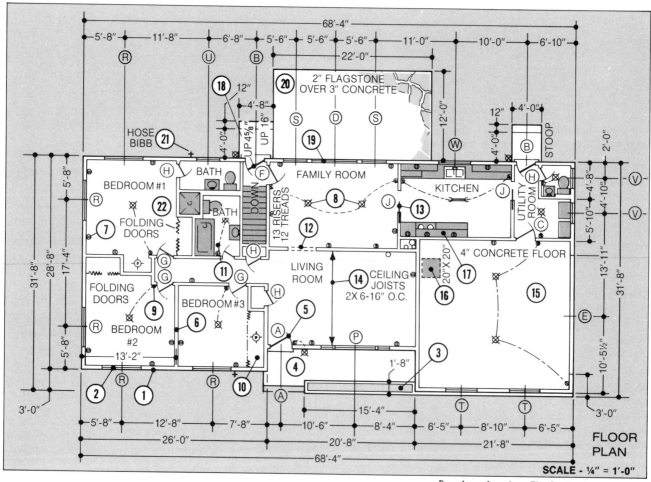

Figure 27–1. Floor plan for the Three-bedroom House Plan. Compare this plan with the pictorial drawing in Figure 27–2.

*Based on plans from The Garlinghouse Company*

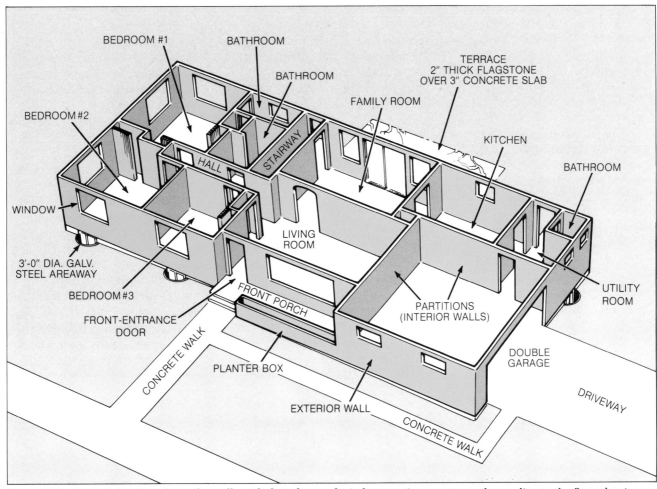

Figure 27–2. Pictorial drawing shows the walls with their door and window openings constructed according to the floor plan in Figure 27–1.

a separate floor plan for each level.

Floor plans give the positions of the exterior walls and the interior walls. The shape and arrangement of all the rooms can be seen by studying the floor plan. The floor plan also shows all the door and window openings. Electrical items such as plugs, light outlets, and switches are noted. The locations of plumbing fixtures such as sinks, water closets, bathtubs, shower stalls, and stoves are given.

Many floor plans also give information about the heating methods used in the building. Locations of wall heaters or the wall or floor openings (registers) that connect with a central heat-

ing system are shown.

Figure 27–1 is the floor plan for part of the Three-bedroom House Plan. A pictorial version is shown in Figure 27–2. An explanation of the main features of the floor plan in Figure 27–1 follows. The numbers in this explanation correspond to the numbers in the circles on the plan.

1. *Exterior walls:* The lengths of the exterior walls (the outside walls of the building) are shown by dimension lines. Note that the exterior walls are not in a straight line at the south side (front) of the house.

2. *Windows:* A pair of double-hung windows in the bedroom are shown here. The circled "R" below the window refers to a

window schedule. The window schedule gives the size of the window and other pertinent information. (Window and door schedules are discussed later in this unit.) The location of a window is established by a dimension line that measures to the center of the window unit. In this example, the "R" type window is laid out by measuring 5'-8" from the outside face of the west wall to the center of the window. The center of another window, in Bedroom #3 (also an "R" type window) is 12'-8" from the center of the window in Bedroom #2.

3. *Planter box:* The width and length of a planter box are given. Further details regarding the

structure and material used are given in section-view drawings that appear elsewhere in the plans.

4. *Front porch:* The porch area consists of a concrete slab 4″ thick. A step from the walk to the porch is indicated.

5. *Front-entrance door:* The front-entrance door swings into the living room. The circled "A" refers to a door schedule. The door schedule gives the size of the door and other information. There is a step from the porch to the living room level.

6. *Partition:* The interior wall (partition) pointed out here is between Bedroom #2 and Bedroom #3. Measurements to partitions are usually to the center of that wall. In this example, the partition is laid out by measuring 13′-2″ from the outside face of the west exterior wall to the center of the partition.

7. *Electrical receptacle:* The symbol indicates a duplex (double) wall plug. Symbols for these appear in all rooms of the house.

8. *Lighting outlet:* The symbol indicates an overhead (ceiling) light. This symbol appears in most of the rooms in the house, except for the living room. The symbol for a fluorescent ceiling light appears in the kitchen.

9. *Wall switch:* The symbol indicates a wall switch located next to the door opening. A line is drawn from the switch to the light it controls.

10. *Closet:* A bi-fold door opens to this closet. Inside the closet is an overhead light with a pull-chain switch. A closet is also shown in Bedroom #1.

11. *Bathroom:* This bathroom is entered from the hallway. Plumbing fixtures include a bathtub, a sink, and a water closet. The adjoining bathroom has a door leading from Bedroom #1. A third bathroom opens off the utility room.

12. *Wall opening:* The dashed lines indicate a wall opening with no door. The opening is between the living room and the family room.

13. *Pocket sliding door:* When opened, this door slides into a pocket inside the wall.

14. *Ceiling joists:* The arrow shows the direction in which the ceiling joists run. The size of the joists and their spacing are given.

15. *Garage area:* For the garage area, a 4″ thick concrete floor, 20′-8″ long, slopes 2″ down toward the door.

16. *Attic access:* Also known as a *scuttle,* the attic access is a ceiling opening covered by a removable panel. It provides access to the attic.

17. *Kitchen cabinets:* Cabinets are shown along the north and south walls of the kitchen. Cabinet details are provided in other drawings.

18. *Rear-entrance door:* The stoop pointed out here is 16″ above the ground level. It leads to a rear-entrance door. The floor level in the house is 4⅝″ above the stoop. Another stoop and rear-entrance door lead into the utility room.

19. *Sliding glass doors:* Two glass sliding doors open from the family room to the terrace. The terrace is 4⅝″ below the floor level.

20. *Terrace:* For the terrace, a 3″ thick concrete slab is covered by flagstone 2″ thick.

21. *Hose bibbs.* Hose bibbs are threaded water faucets to which hoses can be attached.

22. *Other doors:* Circled letters A through K identify these doors. See door schedule in Unit 31.

# UNIT 28

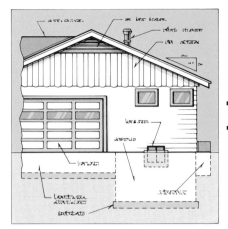

# Exterior Elevations

An *exterior elevation drawing* is a view from the side of an object. An elevation drawing of the side of a building includes the wall surface and the roof. Usually the outline of the foundation is

shown with dashed (hidden) lines. Elevation drawings make clearer much of the information on the floor plan. For example, the floor plan indicates where the doors and windows are located

in the outside walls. An elevation view of the same wall gives actual drawings of the doors and windows. See Figure 28–1.
Elevation drawings usually identify the materials used to fin-

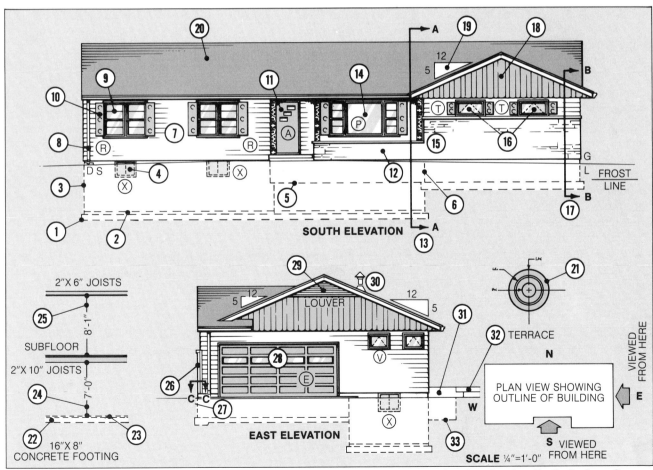

*Based on plans from The Garlinghouse Company*

Figure 28–1. South and east elevations of the Three-bedroom House Plan. Compare these drawings with the pictorial drawing in Figure 28–3.

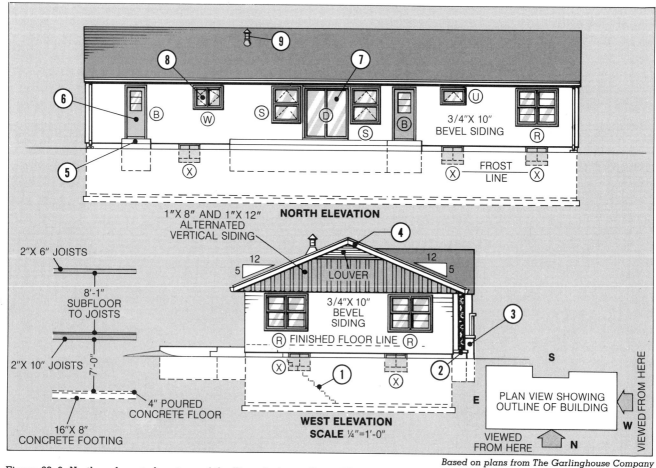

Figure 28–2. North and west elevations of the Three-bedroom House Plan. Compare these drawings with the pictorial drawing in Figure 28–4. Note that the plan view building outline has been reversed so that the north (back) wall is at the bottom of the drawing and the west (side) wall is at the right.

ish the outside surfaces of the walls and roof. The height from the finish floor to the finish ceiling, and the height from the floor to the top of the door and window openings are given.

The downspouts leading from the roof gutters and the roof vents also appear on elevation drawings. Flashing required over doors and windows and on the roof is described. The locations of diagonal bracing may be indicated by dashed lines.

The exterior elevation drawings for the Three-bedroom House Plan are shown in Figures 28–1 and 28–2. Pictorial drawings of these elevation views appear in Figures 28–3 and 28–4. Each elevation drawing is identified by its compass direction,

such as north, south, east, or west. The plot plan (Figure 25–1 of Unit 25) shows that the rear wall of the house is toward the north side of the lot. For this reason, the rear wall is referred to as the *north wall*. Another way to identify elevation drawings is to refer to them as the front, rear, left, and right elevations.

## SOUTH ELEVATION

An explanation follows of the main features of the south elevation drawing in Figure 28–1. The numbers in this explanation correspond to the numbers in the circles on the drawing. The exterior wall in this drawing runs across Bedroom #2 and Bedroom #3, the living room, and

one side of the garage.

1. *Foundation footings:* The tops and bottoms of the footings are identified by two dashed lines.

2. *Basement floor slab:* Dashed lines show the top of the concrete slab for the basement floor.

3. *Foundation walls:* The outlines for the foundation walls are shown by vertical dashed lines.

4. *Basement window and areaway:* Steel-framed basement windows extend below the surface of the ground.

5. *Front-porch foundation wall:* Dashed lines identify the low concrete foundation that supports the front edge of the concrete floor slab. No footing is required.

6. *Garage foundation wall:*

There is no basement under this area of the house. Less depth is required for the walls and footings.

7. *Bevel siding:* Redwood ¾″ × 10″ boards are used for the outside finish at this section of the wall.

8. *Downspout:* A metal pipe that carries rainwater from the roof gutters to the ground is the downspout. The *splash block* shown directly below the downspout directs the water away from the building.

9. *Double-hung window units:* The windows shown here are the same "R" type of windows shown in the south bedroom walls on the floor plan.

10. *Shutters:* Wood shutters are placed at each side of the window units.

11. *Front door:* A flush door with three glass lights (panes) is shown at the front porch.

12. *Planter box:* A front view of the planter box is shown.

13. *Cutting plane line A:* This cutting plane line refers to the Thru-House Section A-A drawing (Figure 29–1 in the next unit).

14. *Picture window:* Also known as a *fixed-sash window,* the picture window cannot be opened. Double-hung windows are shown at each side.

15. *Stone veneer:* Cut stone, 4″ thick, finishes off the outside wall in this area. The cut stone extends to the front of the planter box.

Figure 28–3. Pictorial drawing based on the south and east elevation drawings in Figure 28–1.

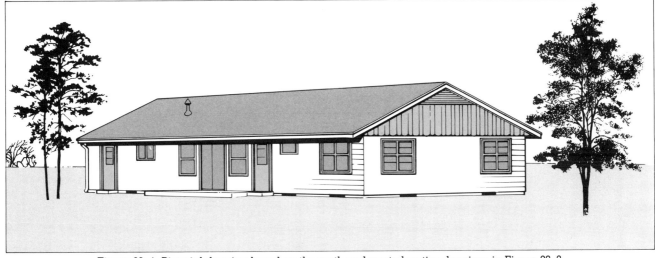

Figure 28–4. Pictorial drawing based on the north and west elevation drawings in Figure 28–2.

16. *Awning windows:* Opening off the south garage wall are awning windows. The dashed lines indicate that the windows are hinged at the top.

17. *Cutting plane line B:* This cutting plane line refers to the Garage Section B-B drawing (Figure 29–3 in the next unit).

18. *Vertical siding:* The gable end of the intersecting roof is finished with 1″ × 8″ and 1″ × 12″ vertical redwood or cedar boards.

19. *Unit rise:* The small triangle indicates a 5″ unit rise. The roof slope rises vertically 5″ for every 12″ of horizontal run. (Section 10 includes a full discussion of unit rise and run.)

20. *Asphalt shingles:* The material used as the finish roof covering is asphalt shingles.

21. *Plywood appliqué detail:* Decorations cut from ½″ plywood are fastened to the surfaces of the shutters.

## EAST ELEVATION

The east wall of the house runs along the front of the garage and one side of the utility room. It is shown on the east elevation drawing of the Three-bedroom House Plan (Figure 28–1). A pictorial drawing is shown in Figure 28–3.

An explanation follows of the main features of the east elevation drawing in Figure 28–1. Features similar to those described for the south elevation are not repeated. Numbers in this explanation correspond to the numbers in the circles on the drawing.

22. *Concrete footing:* The width and height of the footing are given.

23. *Concrete floor:* The thickness of the slab for the concrete floor is 4″.

24. *Basement ceiling height:* The dimension line gives the distance from the concrete floor to the bottom of the floor joists.

25. *Floor ceiling height:* The distance from the subfloor to the ceiling joists is given.

26. *Stone veneer:* The end of the stone veneer on the south wall can be seen.

27. *Cutting plane line C:* This cutting plane line refers to the Section C-C drawing (Figure 29–4 in the next unit).

28. *Garage door:* A paneled overhead garage door is shown. It has four lights.

29. *Louver:* A wood louver is located under the roof ridge to allow ventilation under the roof.

30. *Roof vent:* The exhaust from gas appliances (stoves and heaters) is carried away through the roof vent.

31. *Stoop:* A side view of one of the rear stoops is shown.

32. *Terrace:* A side view of the terrace is shown.

33. *Stoop foundation:* The dashed lines show a small foundation that provides support around the perimeter of the stoop.

## NORTH AND WEST ELEVATIONS

The north and west elevations in Figure 28–2 show the back and left sides of the house. (A pictorial drawing is shown in Figure 28–4.) The same type of bevel siding shown on the south and east elevation drawings is also shown here. Asphalt shingles are again identified on the roof.

An explanation follows of the main features of the north and west elevation drawings. Features similar to those already described for the south and east elevations are not repeated. Numbers in this explanation correspond to the numbers in the circles on the drawing.

1. *Stairway:* Dashed (hidden) lines show the outline of the stairway inside the basement. This stairway leads up to the main floor.

2. *Front porch:* From the west side of the building, a side view of the porch can be seen. Dashed lines indicate the porch foundation.

3. *Planter box:* A side view of the planter box is shown. Dashed lines indicate the foundation of the planter box.

4. *Barge rafters and trim:* The gable end of the roof is finished with rafters known as *fascia rafters* or *barge rafters.*

5. *Rear stoop:* A front view of the rear stoop is shown. Dashed lines indicate the foundation. Another rear stoop is located directly to the right of the terrace.

6. *Rear-entrance door:* A panel door with three lights (glass panes) leads into the utility room. An identical door over the other stoop leads into the hallway off the family room.

7. *Sliding glass doors:* Sliding doors lead from the family room to the terrace. One door is usually stationary, and the other door slides to open and close.

8. *Casement window:* The only casement windows used in the building are located in the wall area above the kitchen sink. Dashed lines indicate side hinges. Casement windows swing out like doors.

9. *Roof vent:* This is another view of the roof vent shown on the east elevation drawing.

# UNIT 29

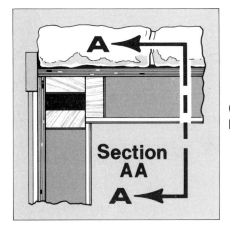

# Section Views

A *section view* is a drawing showing the part of a building that would be revealed if a vertical or horizontal cut were made through the building. It gives important information that cannot be obtained from other drawings in a set of blueprints. For exam-

ple, section views show the structural members and materials used inside the walls and on the outside surfaces. The height, thickness, and shape of the walls are shown. Often window and door heights are given. The Three-bedroom House Plan con-

tains several section-view drawings. See Figure 29–1.

To be useful, section-view drawings must be related to the other parts of the blueprints. The *cutting plane lines* (also called *section lines*) that are usually found on foundation plans, floor

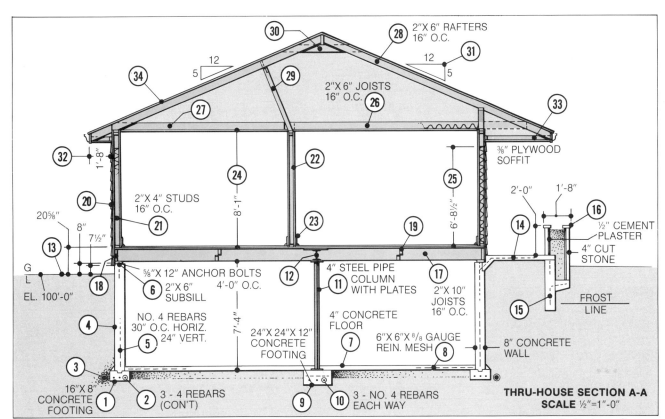

THRU-HOUSE SECTION A-A
SCALE ½"=1'-0"

*Based on plans from The Garlinghouse Company*

Figure 29–1. Section view (Thru-House section AA) representing a transverse cut across the width of the house from north to south. It shows the foundation, first floor, walls, ceiling, and roof. Compare this section view with the pictorial drawing in Figure 29–2.

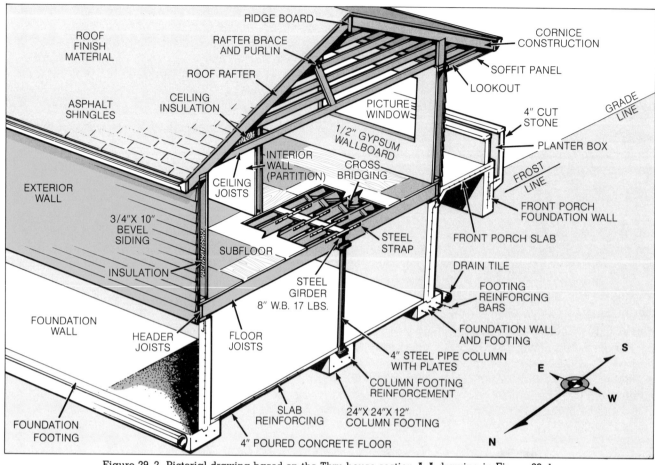

Figure 29–2. Pictorial drawing based on the Thru-house section A-A drawing in Figure 29–1.

plans, and exterior elevation plans serve this purpose.

## THRU-HOUSE SECTION A-A

Refer to the south elevation drawing in Figure 28–1 of Unit 28. To the right of the living room window, a cutting plane line extends through the roof, wall, and foundation. At each end of the line, an arrow points to the right with an "A" next to it. The arrows point in the direction of what would be seen if a *transverse-section* cut were made across the house. A transverse section presents a view across the width of the house. (A *longitudinal section* presents a view across the length of the house.) The "A" refers to the drawing called Thru-house Section A-A, shown in

Figure 29–1. A pictorial drawing based on this section view is shown in Figure 29–2.

An explanation of the main features of the Thru-house Section A-A drawing follows. Numbers in this explanation correspond to the numbers in the circles on the drawing in Figure 29–1).

1. *Foundation footings:* The heights and widths of the footings are given.

2. *Footing reinforcing bars:* The number of steel bars used and their gauge are shown. (A #4 bar is $\frac{4}{8}$" in diameter, or $\frac{1}{2}$".)

3. *Drain tiles:* To take water away from the foundation, drain tiles are placed in a layer of crushed rock (gravel) next to the footing.

4. *Foundation walls:* The height of each wall from the top

of the footing to the bottom of the first floor joist is shown. The thickness of each wall is also given.

5. *Foundation wall reinforcement:* The size of the steel bars and their horizontal and vertical spacing are given.

6. *Foundation sill plate:* The wood plate referred to as the *sub sill* in the drawing is the foundation sill plate. It is also called a *mudsill*. The diameter and length of the bolts used to attach the sill plate are given. The spacing of the bolts is also shown.

7. *Basement floor:* The thickness of the concrete slab for the basement floor is given. It will be placed over a layer of crushed stone.

8. *Slab reinforcement:* Wire mesh used for slab reinforcement is shown by the broken lines inside the slab. The gauge

of the mesh and the distance between its wires are given.

9. *Column footings:* The square concrete base for each column is the column footing. The length of the sides and height of the footing are given.

10. *Column footing reinforcement:* The spacing and number of reinforcing bars in the column footings are given.

11. *Columns:* Steel pipe columns are used to support the floor beam. The spacing of the columns is shown in the foundation plan. (Refer to Figure 26–1 of Unit 26.)

12. *Floor beam:* A wide-flange steel beam is used. The vertical dimension of the beam is 8″. It weighs 17 pounds per foot.

13. *Grade line:* The grade line is the level of the ground at the outside of the foundation walls. It is shown by a horizontal line marked "GL." From left to right, measurements are given from the grade line to the finish floor, to the top of the foundation wall, and to the bottom of the wood bevel siding.

14. *Front porch:* A concrete slab with steel reinforcement is shown for the front porch.

15. *Front porch foundation wall:* The thickness and depth of the foundation wall for the front porch are given. The bottom of the wall is below the frost line.

16. *Planter box:* The inside wall of the planter box rests on the front porch. The outside wall is supported by a footing. The height of the planter box is measured from the porch slab. A dimension line at the top of the planter box gives the width. Material used for its construction is also identified.

17. *Floor joists:* The size and spacing of the floor joists are given. One end of each joist rests on the outside wall. The other end notches into the beam.

18. *Header joists:* The pieces that nail into the ends of the regular joists are the header joists.

19. *Cross bridging:* Cross bridging is placed between the joists at the center of their spans.

20. *Exterior wall:* The size and spacing of the studs for the exterior wall are given.

21. *Insulation:* Blanket or batt insulation is placed between the studs.

22. *Interior wall:* One interior partition is shown.

23. *Base and shoe:* The finish pieces placed at the bottom of the wall are the base and shoe moldings.

24. *Floor-to-ceiling height:* The distance from the subfloor to the bottom of the ceiling joists is the floor-to-ceiling height.

25. *Window and door heights:* A dimension is given from the subfloor to the top of the doors and windows. Usually the tops of all doors and windows align.

26. *Ceiling joists:* The size and spacing of the ceiling joists are given.

27. *Ceiling insulation:* Blanket or batt insulation is placed between the ceiling joists.

28. *Roof rafters:* The size and spacing of the roof rafters are given.

29. *Rafter braces and purlin:* The rafter braces hold up a purlin, which is a horizontal member that provides additional support to the rafters. It is notched into the top ends of the braces.

30. *Collar beams:* Collar beams help tie opposite rafters together.

31. *Roof slope:* The unit rise of the roof is given.

32. *Roof overhang:* The horizontal distance is given from the side of the house to the end of the rafters.

33. *Cornice construction:* The area under the roof overhang is the cornice. When this area is closed in, as shown in the drawing, nailing blocks must be placed between the ends of the rafters and the wall. The material

that closes the cornice is shown in the drawing.

34. *Roof finish:* The wood sheathing is covered with building paper and then with asphalt shingles.

## GARAGE SECTION B-B

A section view of the garage wall is also included in the Three-bedroom House Plan. Refer to Figure 28–1 of Unit 28, which shows the south elevation. At the far right of that drawing is a cutting plane line with the arrows "B" pointing to the right. This refers to the Garage Section B-B drawing shown in Figure 29–3. A pictorial drawing based on the section view is included in Figure 29–3.

An explanation follows of the main features of the Garage Section B-B drawing. The numbers in this explanation correspond to the numbers in the arrows on the drawing.

1. *Foundation footings:* The width and height of the footings are given. Information about the steel reinforcement is also given.

2. *Frost line:* The tops of the footings must be below the frost line in the area.

3. *Foundation walls:* The thickness of the foundation walls and the minimum distance that they must extend above the ground are shown.

4. *Garage floor:* The thickness of the concrete slab for the garage floor is given. A layer of crushed stone is shown beneath the slab.

5. *Garage wall:* The size and spacing of the studs for the garage wall are given.

6. *Exterior finish:* The lower portion of the exterior walls has a stone veneer. The upper part of the walls has a bevel siding. Sheathing covered with building paper is placed against the stud wall before the stone veneer or bevel siding is applied. A 1″ air

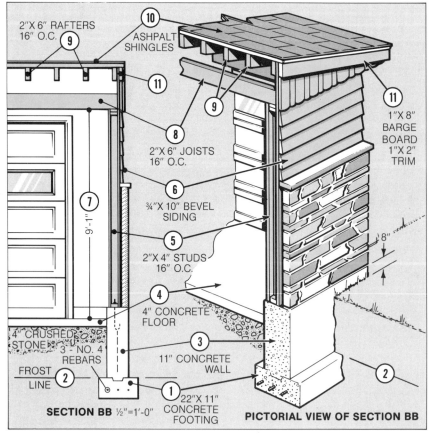

2"X 6" RAFTERS
16" O.C.

ASHPALT
SHINGLES

2"X 6" JOISTS
16" O.C.

¾"X 10" BEVEL
SIDING

2"X 4" STUDS
16" O.C.

4" CONCRETE
FLOOR

4" CRUSHED
STONE

3 - NO. 4
REBARS

FROST
LINE

11" CONCRETE
WALL

**SECTION BB** ½"=1'-0"

22"X 11"
CONCRETE
FOOTING

9'-1"

1"X 8"
BARGE
BOARD
1"X 2"
TRIM

8"

**PICTORIAL VIEW OF SECTION BB**

*Based on plans from The Garlinghouse Company*

Figure 29–3. Garage Section B-B section view is shown at left. Pictorial drawing based on that section view is shown at right.

space is behind the stone veneer.

7. *Ceiling height:* The distance from the top of the foundation wall to the bottom of the ceiling joists is given.

8. *Ceiling joists:* The size and spacing of the ceiling joists are given.

9. *Roof rafters:* The size of the material and the spacing of the rafters are given.

10. *Roof finish material:* Asphalt shingles are again identified for the roof.

11. *Cornice trim:* The size of the bargeboard and the trim for the cornice are given.

## CORNER SECTION C-C

The Three-bedroom House Plan includes a section view that helps clarify the corner construction at the southeast corner of the garage. Refer to Figure 28–1 of Unit 28, which shows the east elevation. The cutting plane line C-C at the left side of the drawing refers to the Corner Section C-C section view in Figure 29–4. Note that the arrows are pointing down, which indicates that Corner Section C-C will be a view looking down. A pictorial drawing based on this section view is included in Figure 29–4.

An explanation follows of the main features of the Corner Section C-C drawing. The numbers in this explanation correspond to the numbers in the circles on the drawings.

1. *Corner framing:* Three 2 × 4 studs are shown.

2. *Sheathing:* Panels or board sheathing are nailed against the stud wall.

3. *Masonry veneer:* The edge of the cut stone veneer is shown.

4. *Siding:* A top view of the bevel siding is shown.

5. *Corner trim:* A board fits against the siding to cover the gap between the sheathing and the cut stone.

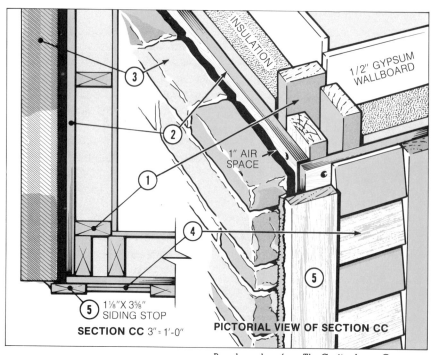

INSULATION

1/2" GYPSUM
WALLBOARD

1" AIR
SPACE

1⅛"X 3⅝"
SIDING STOP

**SECTION CC** 3"=1'-0"

**PICTORIAL VIEW OF SECTION CC**

*Based on plans from The Garlinghouse Company*

Figure 29–4. Corner Section C-C section view is shown at left. Pictorial drawing based on that section view is shown at right.

# UNIT 30

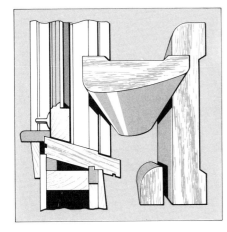

# Details and Framing Plans

A *detail* usually gives an enlarged picture of a part of the building that cannot be fully explained in other drawings of the blueprints. Most details appear on separate sheets. Examples of construction features that often require details are: door and window units, kitchen cabinets, stairways, fireplaces, roof cornices, and trim materials.

A set of blueprints for a wood-framed house may also include *framing plans.* These plans provide information on constructing the framework of the building.

## DETAILS

Details for doors and windows in the Three-bedroom House Plan are shown in Figures 30–1 through 30–6.

## Windows

The floor plans and elevation drawings show three types of windows. *Double-hung windows* are located in the bedrooms and on each side of the picture window in the living room (Figure 30–1). *Awning windows* are located in the garage, utility room, family room, and bathrooms (Figure 30–2). *Casement windows* are located over the sink in the kitchen; however, no detail drawing is provided for these casement windows in the plans.

## Doors

The floor plans and elevation drawings show three types of doors. *Hinged doors* are located at all the outside entrances (Figure 30–3) to the house and at many of the openings between rooms inside the house (Figure 30–4). These doors swing back and forth. (Doors opening to the outside are called *exterior doors,* and those inside the house are called *interior doors.*)

*Pocket (recessed) sliding doors* are located in two walls of the kitchen (Figure 30–5). These doors slide into a space (pocket) framed in the wall.

*Folding doors* are located at the entrance of all bedroom closets. The closets in Bedrooms #1 and #2 have single folding doors. The closet in Bedroom #3 has a pair of folding doors (Figure 30–6).

## Stairs

A good example of a structural detail drawing is the stairway detail in Figure 30–7. This stairway (shown in both the floor plan and foundation plan) goes from the hallway next to the family room, down into the basement. A plan view of the framing around the stairwell is included in the detail.

An explanation follows of the main features of the stairway detail drawing. The numbers in this explanation correspond to the numbers in the circles on the drawing.

1. *Carriage:* Also known as *stringers,* the carriages are the main support for the stairway. They are located at the center and on both sides of the stairway.

2. *Riser and tread:* The height of the riser and the width of the tread are given.

3. *Railing posts:* The thickness and width of the railing posts are given.

4. *Railing:* The thickness and width of the railing are given.

5. *Stairwell:* The floor opening for the stairway is the stairwell. Its length is given.

6. *Head room:* The minimum vertical distance from a stair tread to any part of the ceiling

# DOUBLE-HUNG WINDOW AND FRAME SECTION

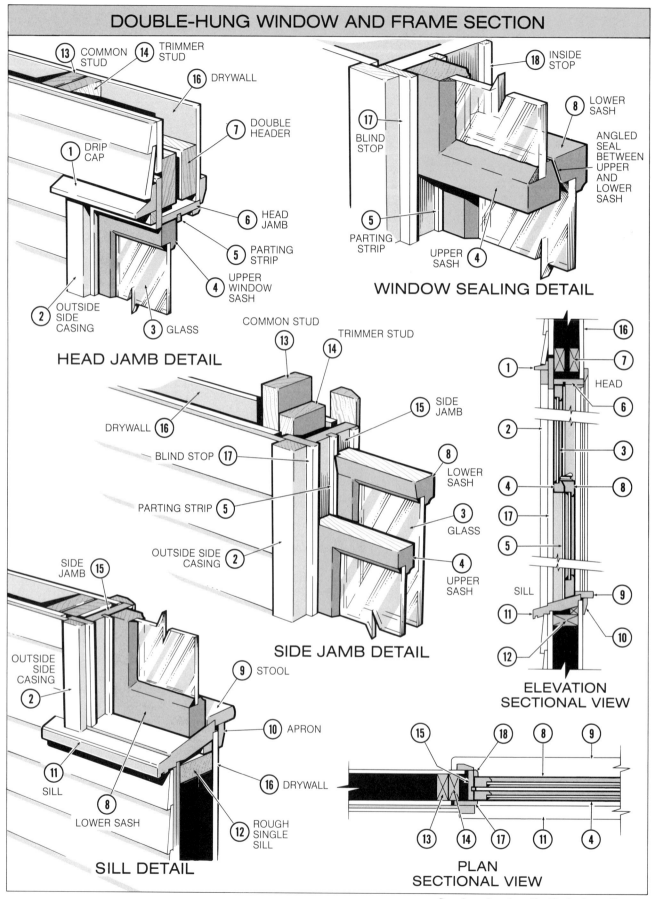

HEAD JAMB DETAIL

WINDOW SEALING DETAIL

SIDE JAMB DETAIL

SILL DETAIL

ELEVATION SECTIONAL VIEW

PLAN SECTIONAL VIEW

*Based on plans from The Garlinghouse Company*

Figure 30–1. Details and pictorial drawings of double-hung window and frame section.

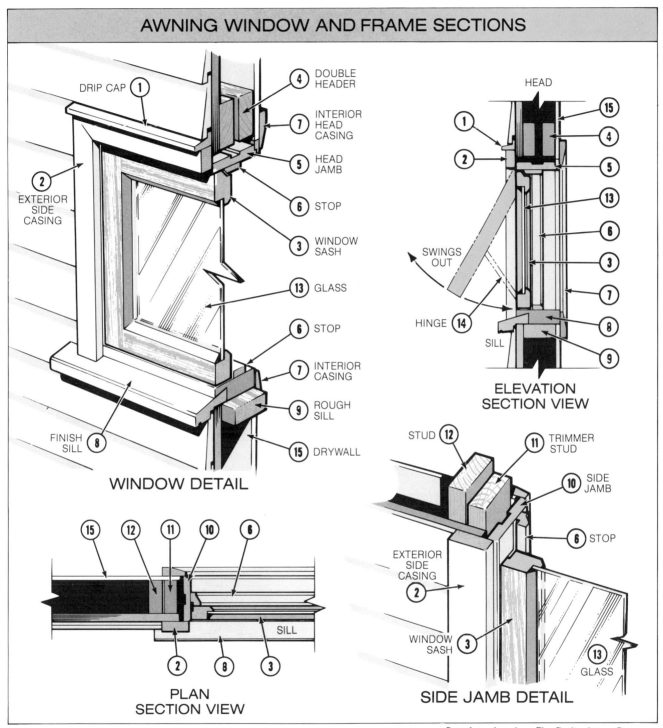

**AWNING WINDOW AND FRAME SECTIONS**

DRIP CAP ① 1

④ 4 DOUBLE HEADER

⑦ 7 INTERIOR HEAD CASING

⑤ 5 HEAD JAMB

② 2 EXTERIOR SIDE CASING

⑥ 6 STOP

③ 3 WINDOW SASH

⑬ 13 GLASS

⑥ 6 STOP

⑦ 7 INTERIOR CASING

⑨ 9 ROUGH SILL

FINISH SILL ⑧ 8

⑮ 15 DRYWALL

**WINDOW DETAIL**

HEAD

① 1  ⑮ 15
② 2  ④ 4
⑤ 5
⑬ 13
⑥ 6
③ 3
⑦ 7
SWINGS OUT
⑧ 8
HINGE ⑭ 14
SILL
⑨ 9

**ELEVATION SECTION VIEW**

STUD ⑫ 12  ⑪ 11 TRIMMER STUD

⑩ 10 SIDE JAMB

⑥ 6 STOP

EXTERIOR SIDE CASING ② 2

WINDOW SASH ③ 3

⑬ 13 GLASS

**SIDE JAMB DETAIL**

⑮ 15  ⑫ 12  ⑪ 11  ⑩ 10  ⑥ 6

SILL

② 2  ⑧ 8  ③ 3

**PLAN SECTION VIEW**

*Based on plans from The Garlinghouse Company*

Figure 30–2. Detail and pictorial drawings of an awning window section.

179

# EXTERIOR DOOR AND FRAME SECTIONS

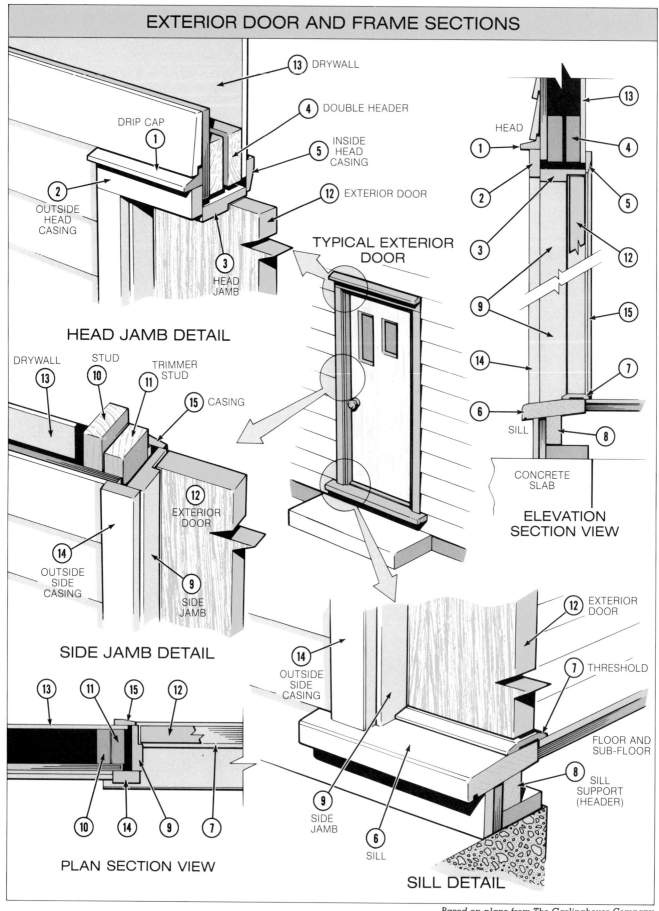

(13) DRYWALL

(4) DOUBLE HEADER

(5) INSIDE HEAD CASING

(12) EXTERIOR DOOR

DRIP CAP (1)

(2) OUTSIDE HEAD CASING

(3) HEAD JAMB

## HEAD JAMB DETAIL

## TYPICAL EXTERIOR DOOR

HEAD

(13)

(4)

(1)

(2)

(3)

(5)

(12)

(9)

(15)

(14)

(7)

(6)

SILL

(8)

CONCRETE SLAB

## ELEVATION SECTION VIEW

DRYWALL

STUD

TRIMMER STUD

(13)

(10)

(11)

(15) CASING

(12) EXTERIOR DOOR

(14)

OUTSIDE SIDE CASING

(9) SIDE JAMB

## SIDE JAMB DETAIL

(13)

(11)

(15)

(12)

(10)

(14)

(9)

(7)

## PLAN SECTION VIEW

(14)

OUTSIDE SIDE CASING

(9) SIDE JAMB

(6) SILL

(12) EXTERIOR DOOR

(7) THRESHOLD

FLOOR AND SUB-FLOOR

(8) SILL SUPPORT (HEADER)

## SILL DETAIL

Based on plans from The Garlinghouse Company

Figure 30–3. Detail and pictorial drawings of an exterior door and frame section.

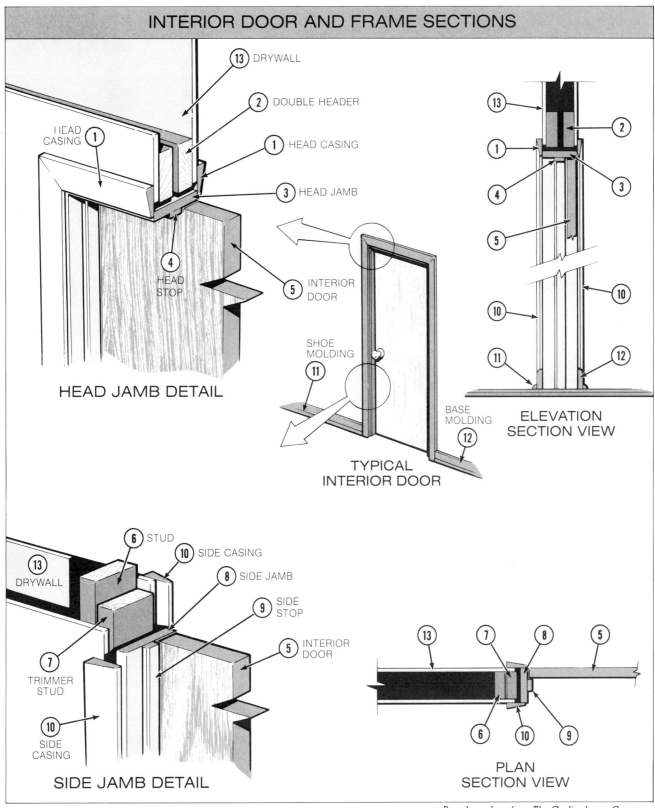

HEAD JAMB DETAIL

SIDE JAMB DETAIL

TYPICAL
INTERIOR DOOR

ELEVATION
SECTION VIEW

PLAN
SECTION VIEW

Based on plans from The Garlinghouse Company
Figure 30–4. Detail and pictorial drawings of an interior door and frame section.

# RECESSED (POCKET) SLIDING DOOR AND FRAME SECTIONS

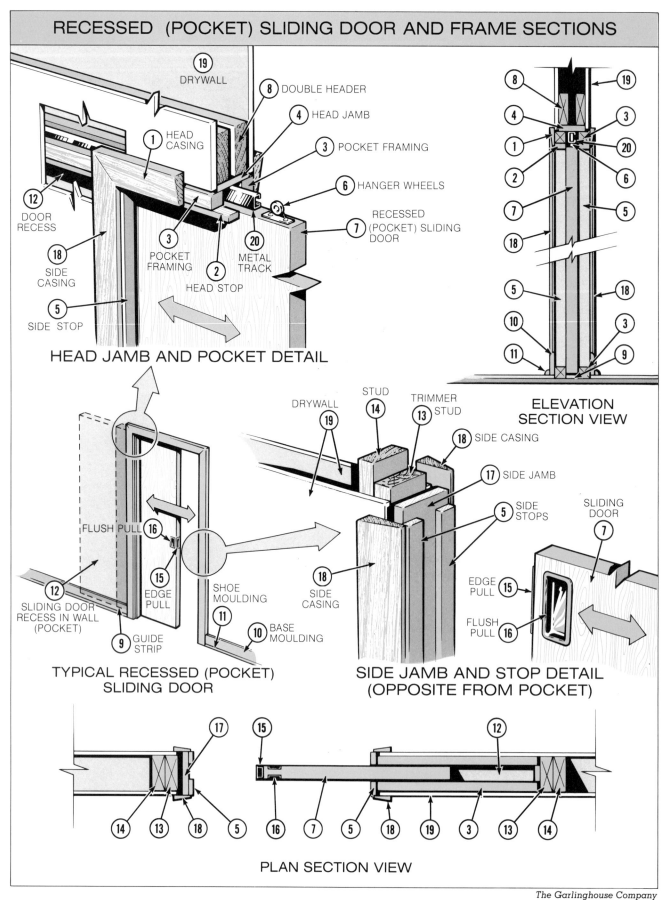

**HEAD JAMB AND POCKET DETAIL**

- 19 DRYWALL
- 8 DOUBLE HEADER
- 4 HEAD JAMB
- 3 POCKET FRAMING
- 1 HEAD CASING
- 6 HANGER WHEELS
- 7 RECESSED (POCKET) SLIDING DOOR
- 12 DOOR RECESS
- 18 SIDE CASING
- 5 SIDE STOP
- 3 POCKET FRAMING
- 2 HEAD STOP
- 20 METAL TRACK

**ELEVATION SECTION VIEW**

**TYPICAL RECESSED (POCKET) SLIDING DOOR**

- FLUSH PULL 16
- 15 EDGE PULL
- 12 SLIDING DOOR RECESS IN WALL (POCKET)
- 9 GUIDE STRIP
- 11 SHOE MOULDING
- 10 BASE MOULDING

**SIDE JAMB AND STOP DETAIL (OPPOSITE FROM POCKET)**

- 19 DRYWALL
- 14 STUD
- 13 TRIMMER STUD
- 18 SIDE CASING
- 17 SIDE JAMB
- 5 SIDE STOPS
- 18 SIDE CASING
- 7 SLIDING DOOR
- 15 EDGE PULL
- 16 FLUSH PULL

**PLAN SECTION VIEW**

The Garlinghouse Company

Figure 30–5. Detail and pictorial drawings of a pocket (recessed) sliding door and frame section.

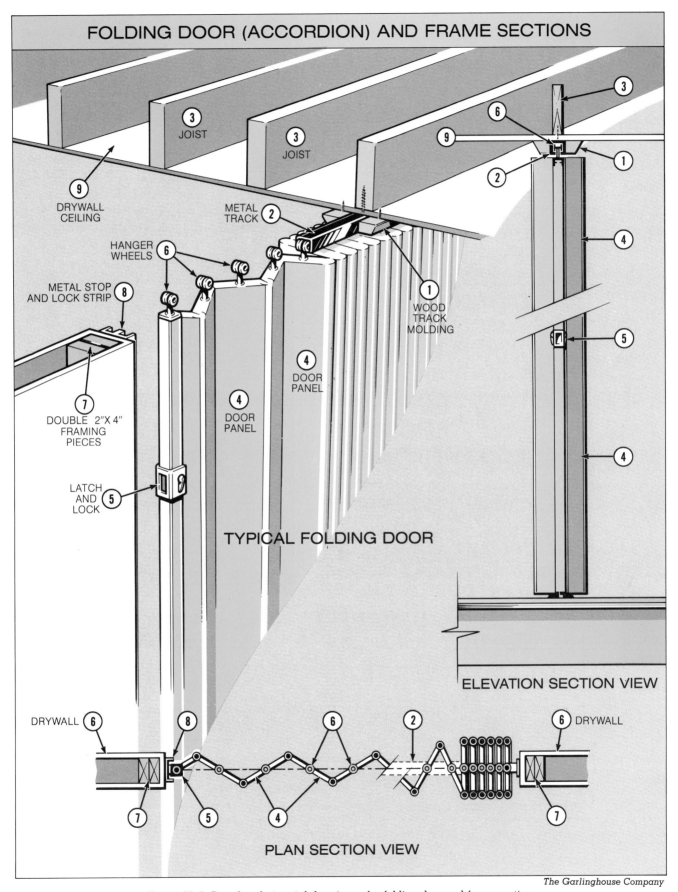

FOLDING DOOR (ACCORDION) AND FRAME SECTIONS

③ JOIST

③ JOIST

③

⑥

⑨

② ①

④

⑤

⑨ DRYWALL CEILING

METAL TRACK ②

HANGER WHEELS ⑥

① WOOD TRACK MOLDING

METAL STOP AND LOCK STRIP ⑧

④ DOOR PANEL

④ DOOR PANEL

④

⑦ DOUBLE 2" X 4" FRAMING PIECES

⑤

LATCH AND LOCK ⑤

TYPICAL FOLDING DOOR

ELEVATION SECTION VIEW

DRYWALL ⑥

⑧

⑥

②

⑥ DRYWALL

⑦

⑤

④

⑦

PLAN SECTION VIEW

The Garlinghouse Company

Figure 30–6. Detail and pictorial drawings of a folding door and frame section.

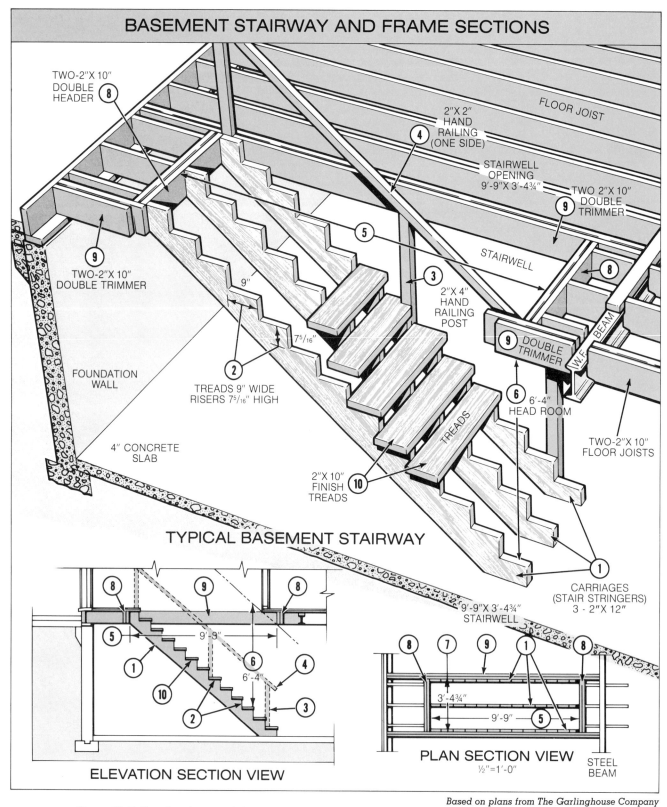

## BASEMENT STAIRWAY AND FRAME SECTIONS

TWO-2"X 10" DOUBLE HEADER ⑧

2"X 2" HAND RAILING (ONE SIDE) ④

FLOOR JOIST

STAIRWELL OPENING 9'-9"X 3'-4¾"

TWO 2"X 10" DOUBLE TRIMMER ⑨

⑤

⑨ TWO-2"X 10" DOUBLE TRIMMER

9"

STAIRWELL

③

⑧

2"X 4" HAND RAILING POST

⑨ DOUBLE TRIMMER

W.F. BEAM

7⁵/₁₆"

② TREADS 9" WIDE RISERS 7⁵/₁₆" HIGH

FOUNDATION WALL

⑥ 6'-4" HEAD ROOM

TREADS

TWO-2"X 10" FLOOR JOISTS

4" CONCRETE SLAB

2"X 10" FINISH TREADS ⑩

### TYPICAL BASEMENT STAIRWAY

①

CARRIAGES (STAIR STRINGERS) 3 - 2"X 12"

9'-9"X 3'-4¾" STAIRWELL

⑧   ⑨   ⑧

⑤   ⑨'-9"   ⑥   ④

① 10   ② 6'-4"   ③

### ELEVATION SECTION VIEW

⑧   ⑦   ⑨   ①   ⑧

3'-4¾"

9'-9"   ⑤

### PLAN SECTION VIEW
½"=1'-0"

STEEL BEAM

*Based on plans from The Garlinghouse Company*

Figure 30–7. Detail and pictorial drawings of a basement stairway. A stairwell framing plan is included.

over the stairway is the head room.

## Kitchen Cabinets

On the floor plan of the Three-bedroom House Plan, kitchen cabinets are indicated along both the north and south kit-

chen walls. A detail for these cabinets is shown in Figure 30–8. A pictorial drawing of the detail for the north wall cabinets is shown in Figure 30–9.

An explanation follows of the main features of the detail drawing for the kitchen cabinets. The numbers in this explanation correspond to the numbers in the circles on the drawing.

1. *Length of base cabinets:* The total length of the base cabinets is given. A space to the left of the north wall cabinets is provided for a refrigerator.

2. *Width of base sections:* The base cabinets consist of a number of sections that are fastened together. The width is given for each section.

3. *Toe space:* At the bottom of

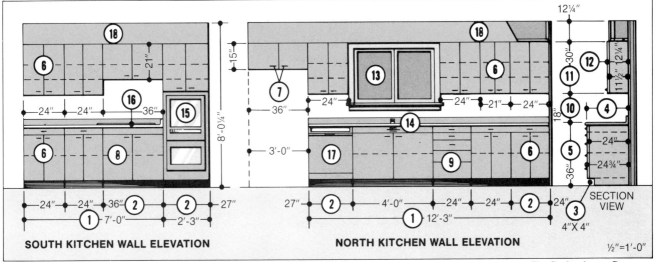

*The Garlinghouse Company*

Figure 30–8. Detail of cabinets on the north and south walls of the kitchen. A pictorial drawing based on the detail of the south wall cabinets is shown in Figure 30–9.

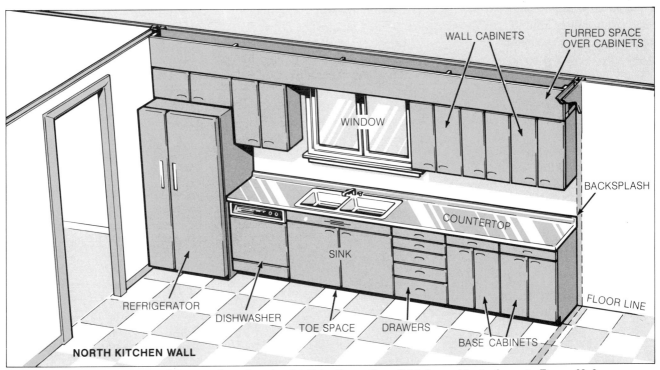

Figure 30–9. Pictorial drawing based on the detail for the north wall kitchen cabinets shown in Figure 30–8.

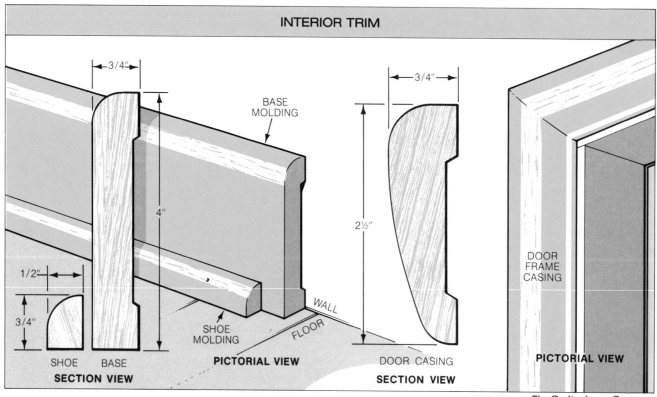

Figure 30–10. Detail and pictorial drawings of interior trim materials. The base and shoe molding is fitted to the floor at the bottom of the wall. The casing molding goes around the door and window openings.

the base cabinet, toe space is provided.

4. *Depth of base cabinet:* The distance from the front to the back of the base cabinet is given.

5. *Height of base cabinet:* The distance from the floor to the countertop is given.

6. *Cabinet doors:* All the cabinet sections have double doors.

7. *Door handles:* Short lines toward the top of the base cabinet doors indicate handles. (Handles are at the bottoms of the wall cabinets.)

8. *Base cabinet shelves:* Dashed lines indicate shelves in each section.

9. *Drawers:* The section to the right of the sink consists of five drawers. Drawers are at the top of other sections. There are no drawers in the sink area (north wall) and the range area (south wall).

10. *Distance between base and wall cabinets:* The distance from the countertop to the bottom of the wall cabinet is given.

11. *Height of wall cabinets:* The distance from the bottom to the top of the wall cabinets is given.

12. *Depth of wall cabinets:* The section view gives the depth of the wall cabinets.

13. *Window space:* The plan gives the location of the window. The schedule gives the width and height of the window.

14. *Sink area:* Part of the faucet fixture is visible above the counter.

15. *Oven:* The correct amount of space must be provided for the oven.

16. *Countertop range:* The range fits into the countertop. (Refer to the part of the floor plan that shows the south

kitchen wall.)

17. *Dishwasher:* An automatic dishwasher is located under the counter to the left of the sink.

18. *Furring:* The space between the top of the wall cabinets and the ceiling will be closed off flush to the face of the cabinets.

## Trim Materials

A set of blueprints often includes details showing the size and shape of the molding for the interior finish of the house (Figure 30–10).

## FRAMING PLANS

The Three-bedroom House Plan includes framing plans for the exterior walls, the roof, and the floor and ceiling units (Figures 30–11 through 30–15).

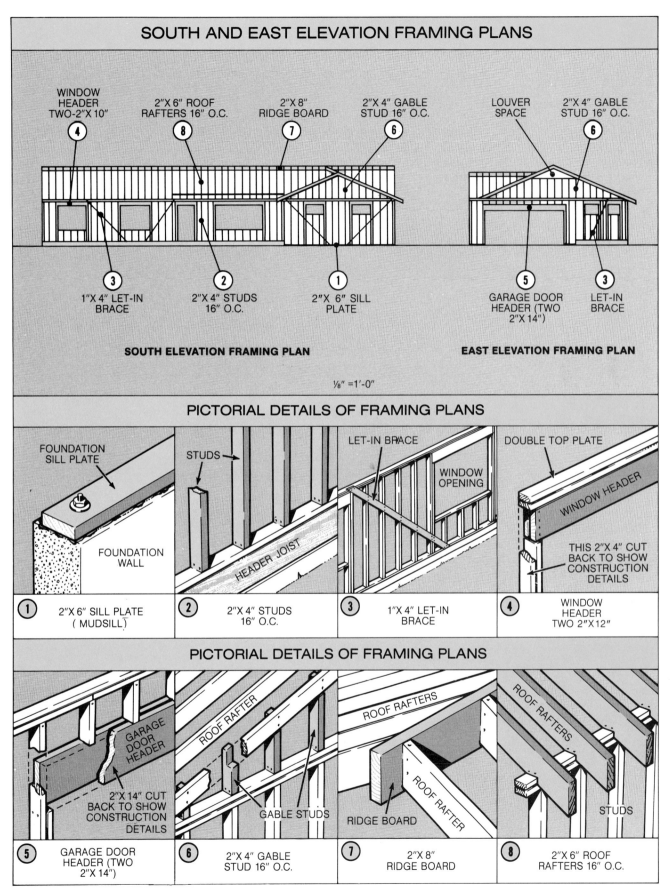

Figure 30–11. South and east elevation framing plans and pictorial drawings.

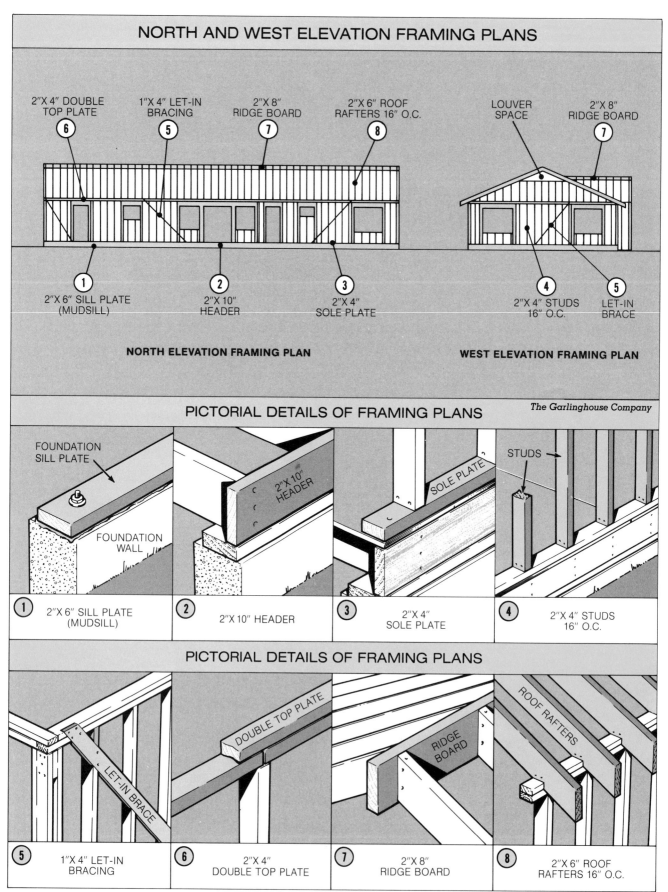

**NORTH AND WEST ELEVATION FRAMING PLANS**

2"X 4" DOUBLE TOP PLATE ⑥

1"X 4" LET-IN BRACING ⑤

2"X 8" RIDGE BOARD ⑦

2"X 6" ROOF RAFTERS 16" O.C. ⑧

LOUVER SPACE

2"X 8" RIDGE BOARD ⑦

① 2"X 6" SILL PLATE (MUDSILL)

② 2"X 10" HEADER

③ 2"X 4" SOLE PLATE

④ 2"X 4" STUDS 16" O.C.

⑤ LET-IN BRACE

**NORTH ELEVATION FRAMING PLAN**

**WEST ELEVATION FRAMING PLAN**

**PICTORIAL DETAILS OF FRAMING PLANS**    *The Garlinghouse Company*

FOUNDATION SILL PLATE

FOUNDATION WALL

2"X 10" HEADER

SOLE PLATE

STUDS

① 2"X 6" SILL PLATE (MUDSILL)

② 2"X 10" HEADER

③ 2"X 4" SOLE PLATE

④ 2"X 4" STUDS 16" O.C.

**PICTORIAL DETAILS OF FRAMING PLANS**

LET-IN BRACE

DOUBLE TOP PLATE

RIDGE BOARD

ROOF RAFTERS

⑤ 1"X 4" LET-IN BRACING

⑥ 2"X 4" DOUBLE TOP PLATE

⑦ 2"X 8" RIDGE BOARD

⑧ 2"X 6" ROOF RAFTERS 16" O.C.

Figure 30–12. North and west elevation framing plans and pictorial drawings.

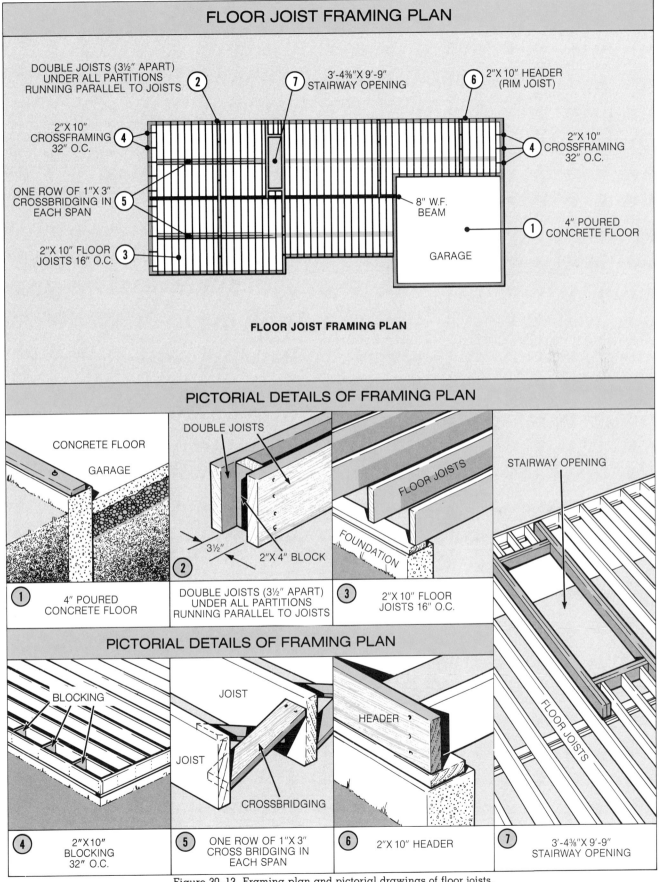

## FLOOR JOIST FRAMING PLAN

DOUBLE JOISTS (3½" APART) UNDER ALL PARTITIONS RUNNING PARALLEL TO JOISTS ②

⑦ 3'-4⅜"X 9'-9" STAIRWAY OPENING

⑥ 2"X 10" HEADER (RIM JOIST)

2"X 10" CROSSFRAMING 32" O.C. ④

④ 2"X 10" CROSSFRAMING 32" O.C.

ONE ROW OF 1"X 3" CROSSBRIDGING IN EACH SPAN ⑤

8" W.F. BEAM

① 4" POURED CONCRETE FLOOR

2"X 10" FLOOR JOISTS 16" O.C. ③

GARAGE

**FLOOR JOIST FRAMING PLAN**

## PICTORIAL DETAILS OF FRAMING PLAN

CONCRETE FLOOR

GARAGE

DOUBLE JOISTS

FLOOR JOISTS

FOUNDATION

3½"

2"X 4" BLOCK

STAIRWAY OPENING

① 4" POURED CONCRETE FLOOR

DOUBLE JOISTS (3½" APART) UNDER ALL PARTITIONS RUNNING PARALLEL TO JOISTS

③ 2"X 10" FLOOR JOISTS 16" O.C.

## PICTORIAL DETAILS OF FRAMING PLAN

BLOCKING

JOIST

JOIST

JOIST

CROSSBRIDGING

HEADER

FLOOR JOISTS

④ 2"X 10" BLOCKING 32" O.C.

⑤ ONE ROW OF 1"X 3" CROSS BRIDGING IN EACH SPAN

⑥ 2"X 10" HEADER

⑦ 3'-4⅜"X 9'-9" STAIRWAY OPENING

Figure 30–13. Framing plan and pictorial drawings of floor joists.

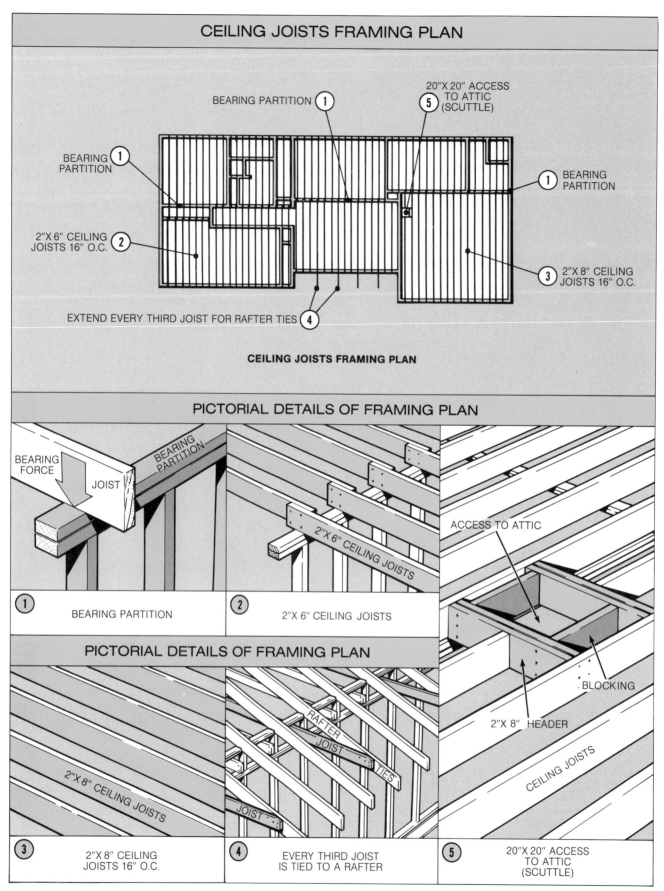

Figure 30–14. Framing plan and pictorial drawings of ceiling joists.

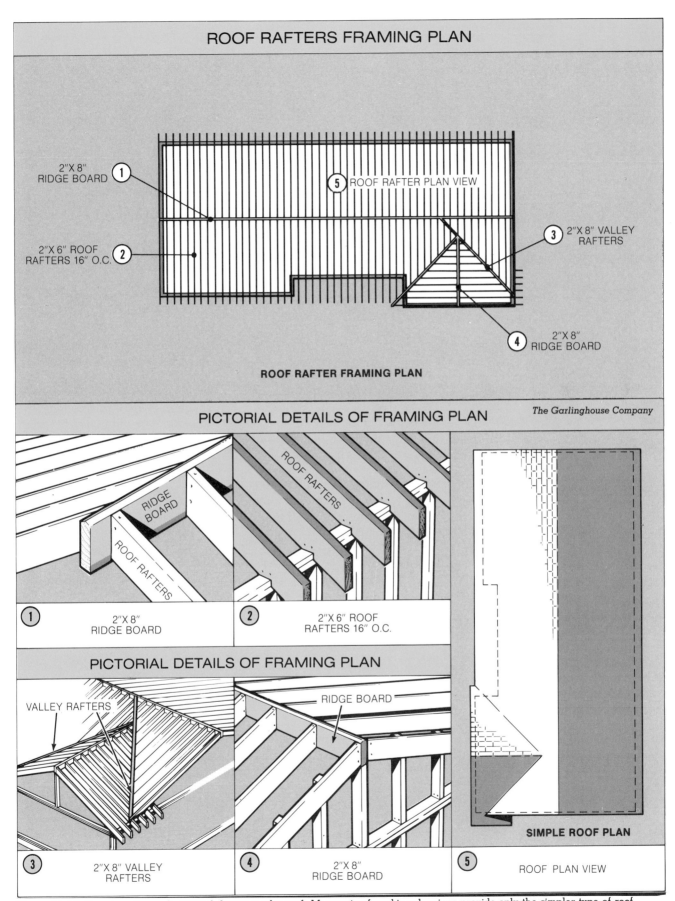

ROOF RAFTERS FRAMING PLAN

2"X 8" RIDGE BOARD ①

② 2"X 6" ROOF RAFTERS 16" O.C.

⑤ ROOF RAFTER PLAN VIEW

③ 2"X 8" VALLEY RAFTERS

④ 2"X 8" RIDGE BOARD

**ROOF RAFTER FRAMING PLAN**

PICTORIAL DETAILS OF FRAMING PLAN      *The Garlinghouse Company*

RIDGE BOARD
ROOF RAFTERS

ROOF RAFTERS

① 2"X 8" RIDGE BOARD

② 2"X 6" ROOF RAFTERS 16" O.C.

PICTORIAL DETAILS OF FRAMING PLAN

VALLEY RAFTERS

RIDGE BOARD

③ 2"X 8" VALLEY RAFTERS

④ 2"X 8" RIDGE BOARD

**SIMPLE ROOF PLAN**

⑤ ROOF PLAN VIEW

Figure 30-15. Framing plan and pictorial drawings of a roof. Many sets of working drawings provide only the simpler type of roof plan shown at lower right.

# UNIT 31

| Door Schedule | | | |
|:---:|:---:|:---:|:---:|
| **CODE** | **QUAN** | **SIZE** | **TH** |
| A | 1 | 3'-0" X 6'-8" | 1¾ |
| B | 1 | 2'-8" X 6'-8" | 1¾ |
| C | 2 | 6'-0" X 6'-10" | |
| D | 1 | 10'-0" X 7'-0" | 2" |
| E | 1 | 3'-0" X 6'-8" | 1½ |
| F | 2 | 4'-0" X 7'-11" | |
| G | 5 | 3'-0" X 6'-8" | 1¾ |

# Door, Window, and Finish Schedules

A full set of working drawings usually includes a *door schedule* and a *window schedule*. Often a *room finish schedule* is also included.

## DOOR AND WINDOW SCHEDULES

The locations of all door and window openings are shown on the floor plan. The only exceptions are the windows in a full-basement foundation. These appear on the foundation plan instead of the floor plan. In some cases, the sizes (widths and heights) of the doors and windows are noted on the floor plans next to the openings. Usually, however, these dimensions are not printed on the plans but are given in special door and window schedules. Schedules vary with different plans. Some give more information than others.

## Door Schedule

The door schedule for the Three-bedroom House Plan is divided into the following columns (Figure 31–1):

*Code:* The letter used to identify the door is given. (Some plans use a number.) For example, a circled letter "A" is shown by the entrance door in the floor plan and the south elevation drawings. Information for a door of this type is given on the first line of the door schedule (after the code letter "A").

*Quan:* The number (quantity) of doors of this type and size is given. For example, there is only one "A" door in this house.

*Size:* The width and height of the door are given. (The "A" door is 3'-0" wide and 6'-8" high.)

*Thk:* The thickness of the door is given. (The "A" door is 1¾" thick.)

*Rough Opening:* The width and height are given for the opening that must be provided in the wall to accommodate the door and the door frame (jamb). A ½" clearance is usually allowed at the sides and top of the jamb. Clearance must also be provided beneath the door. (The rough opening for the "A" door is 3'-3" wide and 6'-10¼" high.)

Some door and window schedules do not include rough open-

ings. In that case the carpenter must know how to calculate the rough opening from information found in different parts of the plans. (Methods for figuring rough openings are discussed in Section 13.)

*Jamb Size:* The thickness and width of the door frame are given. (The jamb for the "A" door is 1³⁄₁₆" thick and 4⅞" wide.)

*Type:* The door action is indicated by its type. (The "A" door is a hinged door. It swings open and shut.)

*Design:* The appearance and construction of the door are indicated by its design. (The "A" door is a flush, solid-core door with three lights.)

*Remarks:* Special information such as the location of the door, special operating instructions, the brand name of the door, and so forth, are given. (The "A" door is identified as a front-entrance door.)

## Window Schedule

The window schedule gives the same type of information about windows that the door schedule gives about doors. The window schedule for the Three-bedroom

| DOOR SCHEDULE | | | | | | | | | |
|---|---|---|---|---|---|---|---|---|---|
| CODE | QUAN | SIZE | THK | ROUGH OPENING | MASONRY OP'G | JAMB SIZE | TYPE | DESIGN | REMARKS |
| A | 1 | 3'-0" × 6'-8" | 1³/₄" | 3'-3" × 6'-10¹/₄" | | 1³/₁₆" × 4⁷/₈" | HINGED | 3 LITES FLUSH SOLID-CORE | FRONT ENTRANCE DOOR |
| B | 2 | 2'-8" × 6'-8" | 1³/₄" | 2'-11" × 6'-10¹/₄" | | 1⁵/₁₆" × 4⁷/₈" | '' | 3 LTS; 1 PANEL | REAR SERVICE DOORS |
| C | 1 | 2'-8" × 6'-8" | 1³/₈" | 2'-11" × 6'-10¹/₂" | | 1⁵/₁₆" × 4⁵/₈" | '' | FLUSH HOLLOW-CORE | DOOR BETWEEN UTILITY ROOM & GARAGE |
| D | 1 | 6'-0¹/₈" × 6'-10" | | 6'-0¹/₂" × 6'-10³/₈" | | | SLIDING | GLASS | SLIDING GLASS DOOR |
| E | 1 | 16'-0" × 7'-0" | 1³/₈" | 16'-3" × 7'-1¹/₂" | | ³/₄" × 6" | OVERHEAD | 4 LTS; 16 PANELS | GARAGE DOOR |
| F | 1 | 2'-8" × 6'-8" | 1³/₈" | 2'-10¹/₂" × 6'-10¹/₂" | | ³/₄" × 4⁵/₈" | HINGED | FLUSH HOLLOW-CORE | INTERIOR DOOR |
| G | 3 | 2'-6" × 6'-8" | 1³/₈" | 2'-8¹/₂" × 6'-10¹/₂" | | ³/₄" × 4⁵/₈" | '' | '' '' | '' '' |
| H | 5 | 2'-0" × 6'-8" | 1³/₈" | 2'-2¹/₂" × 6'-10¹/₂" | | ³/₄" × 4⁵/₈" | '' | '' '' | '' '' |
| J | 2 | 2'-6" × 6'-8" | 1³/₈" | 5'-2" × 7'-0" | | | SLIDING | '' '' | RECESSED DOOR |
| K | 1 | 3'-0" × 6'-8" | | 3'-2¹/₂" × 6'-10¹/₂" | | ³/₄" × 4⁵/₈" | | | CASED OPENING |
| L | 1 | 7'-7⁹/₁₆" × 7'-11" | | 7'-8⁹/₁₆" × 7'-11³/₄" | | WALLBOARD | FOLDING | WOOD SLATS | PELLA WOOD FOLDING DOORS (2)3'-11¹/₄" SIZE |
| M | 1 | 5'-3³/₈" × 7'-11" | | 5'-4³/₈" × 7'-11³/₄" | | '' | '' | '' '' | PELLA WOOD FOLDING DOORS (1)5'-6" SIZE |
| N | 1 | 4'-11⁹/₁₆" × 7'-11" | | 5'-0⁹/₁₆" × 7'-11³/₄" | | '' | '' | '' '' | PELLA WOOD FOLDING DOORS (1)5'-2¹/₄" SIZE |
| O | 1 | 3'-11³/₈" × 7'-11" | | 4'-0³/₈" × 7'-11³/₄" | | '' | '' | '' '' | PELLA WOOD FOLDING DOORS (1)4'-3" SIZE |

Figure 31–1. A door schedule provides the information necessary to lay out the door openings.

House Plan is divided into the following columns (Figure 31–2):

*Code:* The letter used to identify the window is given. For example, a circled letter "R" appears next to each bedroom window unit on the floor plan and elevation drawings. Information for a window of this type is given on the second line of the window schedule.

*Quan:* The number (quantity) of windows of this type and size is given. (There are five "R" windows in this house.)

*No. Lts.:* The number of lights (panes of glass) in each window is given. (The "R" window is a double-hung window. It has two lights in the top section and two in the bottom section. Refer to Figure 28–1 of Unit 28 for a view of these windows on the south elevation drawing.)

| WINDOW SCHEDULE | | | | | | | |
|---|---|---|---|---|---|---|---|
| CODE | QUAN | NO LTS | GLASS SIZE | SASH SIZE | ROUGH OPENING | MASONRY OP'G | REMARKS |
| P | 1 | 1 / 4 | 56" × 49" / 20" × 24" | (1)5'-0" × 4'-6" / (2)2'-0" × 4'-6" | 9'-8" × 4'-10" | | PICTURE WINDOW FLANKED EACH SIDE BY ONE (1) D.H. WINDOW |
| R | 5 | 4 | 28" × 24" | (2)2'-8" × 4'-6" | 5'-10" × 4'-10" | | DOUBLE HUNG WINDOW- DOUBLE UNIT 2" MULLION |
| S | 2 | 2 | 39" × 22" | 3'-8" × 4'-6" | 3'-9" × 4'-7³/₁₀" | | NO. A-12-84 CURTIS CONVERTIBLE AWNING WINDOW UNIT |
| T | 2 | 1 | 39" × 17" | 3'-8" × 1'-10" | 3'-9" × 1'-11³/₈" | | NO. A-11-83 '' '' '' |
| U | 1 | 1 | 33" × 17" | 3'-2" × 1'-10" | 3-3" × 1'-11³/₈" | | NO. A-11-73 '' '' '' |
| V | 2 | 1 | 27" × 17" | 2'-8" × 1'-10" | 2'-9" × 1'-11³/₈" | | NO. A-11-63 '' '' '' |
| W | 1 | 3 | 17" × 27" | 3'-8" × 2'-8" | 3'-9" × 2'-9³/₈" | | NO. C-21-56 CURTIS CONVERTIBLE CASEMENT WINDOW UNIT |
| X | 8 | 2 | 15" × 20" | 2'-8¹/₂" × 1'-10³/₄" | | 2'-0" × 1'-11" | STEEL BASEMENT WINDOWS |

Figure 31–2. A window schedule provides the information necessary to lay out the window openings.

| ROOM FINISH SCHEDULE | | | | | |
|---|---|---|---|---|---|
| ROOM | FLOOR | WALLS | CEILING | BASE | TRIM |
| LIVING ROOM | 1" × 3" OAK | $\frac{1}{4}$" PANELING OVER $\frac{1}{2}$" DRYWALL | $\frac{1}{2}$" DRYWALL | WOOD | WOOD |
| FAMILY ROOM | 1" × 3" OAK | $\frac{1}{2}$" DRYWALL-PAPERED | $\frac{1}{2}$" DRYWALL | WOOD | WOOD |
| BEDROOMS | 1" × 3" OAK | $\frac{1}{2}$" DRYWALL-PAPERED | $\frac{1}{2}$" DRYWALL | WOOD | WOOD |
| KITCHEN | LINOLEUM | $\frac{1}{2}$" DRYWALL-PAINT | $\frac{1}{2}$" DRYWALL | LINO. COVE | WOOD |
| UTILITY ROOM | LINOLEUM | $\frac{1}{2}$" DRYWALL-PAINT | $\frac{1}{2}$" DRYWALL | LINO. COVE | WOOD |
| HALL | 1" × 3" OAK | $\frac{1}{2}$" DRYWALL-PAINT | $\frac{1}{2}$" DRYWALL | WOOD | WOOD |
| BATHROOM | LINOLEUM | $\frac{1}{2}$" DRYWALL-PAINT | $\frac{1}{2}$" DRYWALL | LINO. COVE | WOOD |
| GARAGE | CONCRETE | $\frac{1}{2}$" DRYWALL-PAINT | $\frac{1}{2}$" DRYWALL | WOOD | WOOD |

Figure 31–3. A room finish schedule identifies the finish materials for the walls, floors, and ceilings.

*Glass Size:* The dimensions of the entire glass area in each window section are given. (In an "R" window the glass is 28" wide and 24" high in both the top and bottom sections.)

*Sash Size:* The dimensions of the entire window unit after the glass has been set in its frame give the window sash size. (The sash size of an "R" window is 2'-8" wide by 4'-6" high.)

*Rough Opening:* The width and height are given for the opening that must be provided in the wall to accommodate the window, the frame, and the clearance around the frame. (The "R" window, which is a double-hung window, requires a rough opening 5'-10" wide by 4'-10" high.)

*Remarks:* Special information such as the location of the window, special operating instructions, the brand name of the window, and so forth, are given. (The "R" windows are identified as double-hung windows installed as a double unit. The units are separated by a 2" vertical piece called a *mullion.*)

## ROOM FINISH SCHEDULE

Some blueprints also have a room finish schedule, which specifies the interior finish materials for each room in the house. The room finish schedule for the Three-bedroom House Plan provides this information for the floor, walls, ceiling, base, and trim in each room (Figure 31–3). The *base* is the molding nailed at the bottom of the walls. *Trim* refers to the material around the door and window openings.

# UNIT 32

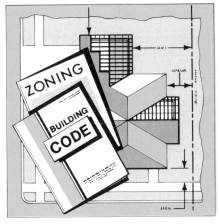

# Building Codes, Zoning, Permits, and Inspections

## BUILDING CODES

In most sections of the country, particularly in and near towns and cities, strict rules exist for the erection of new buildings. These rules ensure that high quality workmanship, proper materials, and sound construction methods are used. They are written in pamphlets or books known as *building codes.* No single factor has a more direct effect on building design than the building code. The architect must have a thorough knowledge of all local code practices before drawing a set of plans. The building contractor and the workers on the job must be familiar with the local code regulations as they apply to construction procedures.

Building codes establish the minimum standards required to protect the health, safety, and welfare of persons who will be living or working in the building. Fire-resistant materials, adequate lighting, ventilation, and insulation are among the areas covered in up-to-date code regulations. Often separate code books exist for electrical and plumbing work.

## State and Local Codes

Most states in the United States have a building code that can be applied in all areas of the state. In addition, larger cities in the state, such as San Francisco or Los Angeles in California, may have their own set of code regulations. City codes are usually more detailed and often have stricter requirements than statewide codes. Any officially adopted code is enforceable by law.

## Model codes

In addition to state and local codes, a number of national codes have been developed through conferences between building officials and industry representatives from all parts of the nation. These codes serve as *models* that can be adopted into law by states or local communities.

Model codes are constantly revised to keep up with changing conditions and the new materials being developed for the construction industry. A prominent model code is the *Uniform Build-*

*ing Code,* which strongly influences construction procedures in the western states. Another regional model code is the *Southern Standard Building Code,* which affects practices in southern states. National model codes are the *Basic Building Code* and the *National Building Code.*

An important guide to code requirements in all parts of the country is the *Minimum Property Standards,* published by the U.S. Department of Housing and Urban Development. This code is issued in three volumes, entitled *One and Two Family Dwellings, Multifamily Housing,* and *Care Type Housing.*

## ZONING REGULATIONS

Like the building code, zoning regulations strongly influence building design. Most cities and counties have specific laws concerning the type of buildings that can be constructed in different areas of the community. A city is usually divided into different zones. The boundaries of the zones are shown on special *zoning maps.* When developing a set of blueprints, the architect must be certain that the type of building designed is permitted in the area where it is to be built.

There are three major types of zones in larger communities: *residential, commercial,* and *manufacturing.* Residential zones are districts limited to buildings in which people live and buildings that serve the neighborhood, such as schools, churches, libraries, and playgrounds. Some residential zones may permit only single-family dwellings while others may permit apartment houses.

Commercial zones permit buildings such as stores, shopping centers, movie theatres, bars, bowling alleys, and hospitals.

Manufacturing zones are areas set aside for factories, warehouses, and other types of industry. It is not uncommon for older residential buildings to be found in manufacturing zones.

In addition to the type of occupancy and uses allowed for buildings in a particular zone, most cities specify the minimum lot area required for that zone. For example, a zoning regulation for a residential section of a city might state that any new building must be located on a lot of no less than 5,000 square feet. See Figure 32–1.

Another regulation might specify the maximum amount of the lot that the building is allowed to cover. For example, if the building is allowed to cover no more than 40% of the lot, and if the lot is 5,000 square feet, then the building must cover no more than 2,000 square feet (40% of 5,000 = 2,000). Consequently, this zoning regulation affects the size of the house designed for that lot.

Other regulations may establish the maximum height of a building in a particular zone. For example, a residential zone may

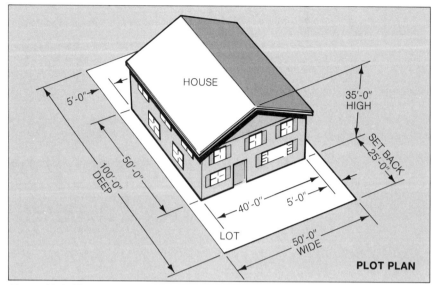

Figure 32–1. The house and lot in this drawing comply with the examples of zoning regulations described in this unit.

allow one-family buildings up to three stories that do not exceed 35' in height.

Another set of regulations usually contained in zoning ordinances concerns the minimum distance the building can be placed from the property lines. For example, the regulation may specify that the distance from the front of the house to the front property line (setback) can be no less than 20', and that the distance from the side of the house to the side property line (side yard) can be no less than 4'. An example of a house built according to regulations for all of these aspects is shown in Figure 32–1.

Zoning regulations are subject to change. Residential zones limited to one-family dwellings may be reclassified to allow apartment houses. They may even be changed to commercial zones. Many cities and counties have *planning commissions* that review requests for changes in zoning regulations and make recommendations to the local government. Since zoning changes may occur at any time, current information should be ob-

tained before the drawings for a building are finalized. Changing a building to conform to a zoning change once construction has begun will not only delay construction but can be very costly.

## PERMITS

Before construction begins, the owner or building contractor must apply to local building authorities for a *building permit.* An application must be filled out, stating the kind of construction proposed. See Figure 32–2. The application also requires a legal description of the land upon which the construction is to take place and the estimated cost of the project. In addition, other information may be requested by the local building officials.

One or two sets of the working drawings and specifications must be submitted with the application. The plans are examined by the proper authorities to make sure they conform to the local building code and zoning ordinances. If everything appears to be in order, the plans are approved. A permit is then granted to begin

NO. _____
**DO NOT WRITE IN THIS SPACE**
For Village Clerk's Use Only

DATE _____

## APPLICATION FOR BUILDING PERMIT
### VILLAGE OF LANSING, ILLINOIS

FOR USE OF ASSESSOR
VOLUME _____
PAGE _____ LINE _____
BLK. _____ PARCEL _____

SKETCH OF APPLICANT'S LOT SHOWING EXISTING AND PROPOSED IMPROVEMENTS, INCLUDING CONSTRUCTION DETAILS.

FRAME ☐   BRICK ☐   BASEMENT: YES ☐   NO ☐

WIDTH OF LOT _____

REAR YARD

SIDE YARD                    SIDE YARD

LENGTH OF LOT

SET BACK LINE

FRONT YARD

**DO NOT FOLD**

I HEREBY APPLY FOR A BUILDING PERMIT FOR THE CONSTRUCTION (REPAIR) OF A BUILDING SPECIFIED AS FOLLOWS:

OWNER _____

OWNER'S PRESENT ADDRESS _____

ADDRESS OF PROPOSED STRUCTURE _____

TYPE OF BUILDING _____

_____

_____

GENERAL CONTRACTOR _____

ADDRESS OF GENERAL CONTRACTOR _____

ELECTRICAL CONTRACTOR _____

ADDRESS OF ELECTRICAL CONTRACTOR _____

PLUMBING CONTRACTOR _____

ADDRESS OF PLUMBING CONTRACTOR _____

PLANS PREPARED BY _____

HEATING, AIR COND. CONTRACTOR _____

ADDRESS HEATING, AIR COND. CONT. _____

LEGAL DESCRIPTION: LOT _____ BLOCK _____

SUBDIVISION _____

SEC. _____ TOWN ____36____ RANGE ____14____

COLLECTOR'S TAX BILL VOL. NO. _____ ITEM _____

PERMANENT REAL ESTATE INDEX NO. _____

COST OF BUILDING COMPLETE $ _____
AS THE APPLICANT FOR THIS PERMIT I EXPRESSLY AGREE TO CONFORM TO ALL APPLICABLE ORDINANCES, RULES AND REGULATIONS OF THE VILLAGE OF LANSING.

_____
SIGNATURE OF APPLICANT

## BUILDING INSPECTOR'S ANALYSIS AND APPROVAL

**ANALYSIS**

| | APPLICATION SATISFACTORY | NOT SATISFACTORY |
|---|---|---|
| 1. PLOT PLAN RECEIVED | ☐ | ☐ |
| 2. DUPLICATE SET OF PLANS RECEIVED | ☐ | ☐ |
| 3. WATER CONNECTION LOCATION DESCRIBED IN PLAN | ☐ | ☐ |
| 4. SEWAGE DISPOSAL PLAN | ☐ | ☐ |
| 5. ZONING ORDINANCE COMPLIANCE | ☐ | ☐ |
| 6. ELECTRICAL PLANS, LICENSE AND BOND | ☐ | ☐ |
| 7. PLUMBING PLANS, LICENSE AND BOND | ☐ | ☐ |
| 8. HEATING, AIR COND. & REFRIG. LICENSE & BOND | ☐ | ☐ |
| 9. PUBLIC SIDEWALK PLAN | ☐ | ☐ |
| 10. IF PUBLIC OR APT. BLDG., FIRE CHIEF'S APPROVAL | ☐ | ☐ |
| 11. IF PAVING TO BE CUT, IS REQUIRED SURETY CO. BOND SUPPLIED AND IN ORDER? | ☐ | ☐ |
| 12. _____ | ☐ | ☐ |
| 13. _____ | ☐ | ☐ |
| 14. _____ | ☐ | ☐ |

**APPROVED:**

DATE _____
BUILDING INSPECTOR, LANSING, ILL.

CERTIFICATE OF OCCUPANCY ISSUED:

DATE _____
BUILDING INSPECTOR, LANSING, ILL.

**COMPUTATION OF FEES**

| | |
|---|---|
| BUILDING PERMIT NO. | $ _____ |
| ELECTRICAL PERMIT No. | _____ |
| SEWER TAP & PLUMBING PERMIT NO. | _____ |
| WATER TAP | _____ |
| HEATING, AIR COND. & REFRIGERATION NO. | _____ |
| WATER METER | _____ |
| WATER USE DURING CONSTRUCTION (THIS DOES NOT AUTHORIZE USE OF WATER FROM ANY FIRE HYDRANT. ANYONE TAMPERING WITH A FIRE HYDRANT WILL BE ARRESTED) | |
| SIDEWALK CONSTRUCTION PERMIT No. | _____ |
| EXCAVATION FOR WATER TAP | _____ |
| EXCAVATION FOR SEWER TAP | _____ |
| BUILDING INSPECTOR'S FEE: | |
| PLANS AND SPECIFICATIONS | _____ |
| OCCUPANCY & FIELD INSPECTIONS | _____ |
| STREET & ALLEY | _____ |
| TOTAL $ | _____ |

PENALTY - $5.00 FOR CHANGE OF SUB-CONTRACTOR

Figure 32–2. Example of an application for a building permit. Local building authorities may also request additional information.

construction. See Figure 32–3.

In addition to the building permit, separate *electrical* and *plumbing* permits are usually required. The owner or building contractor must pay fees for the permits. The amount is usually based on the total cost of the project.

## INSPECTION

A building project for which a permit has been granted will be inspected a number of times as the work progresses. The overall structural inspection is conducted by a local official called a building inspector. The plumbing and electrical work is usually checked by separate plumbing and electrical inspectors. If inspectors discover a violation of the building code, they have the legal authority to have the work torn down and reconstructed properly. This can be very costly to the owner or building contractor.

The Uniform Building Code recommends the following sequence of inspection:

1. *Foundation inspection:* To be made after the trenches have been excavated, the forms constructed, and all the required bolts and reinforcing steel are placed inside the form. The inspection must take place before the concrete is poured.

2. *Concrete slab or under-floor inspection:* To be made where a concrete slab floor is to be placed. This mainly concerns the inspection of electrical conduits, plumbing, duct work, and other equipment that will be covered by the slab.

3. *Frame inspection:* To be made after the floors, walls, ceilings, and roof have been framed and all blocking and bracing set in place. All the rough electrical, plumbing, and duct work must be completed and visible for inspection before the walls can be covered up.

4. *Lath and–or gypsum board inspection:* To be made after all exterior and interior lath or gypsum boards are in place, and before any plaster is applied or any of the gypsum board joints are taped and finished.

5. *Final inspection:* To be made after the building has been completed and is ready for occupancy.

### Inspection Record Card

Many local code authorities require that an inspection record card be posted in a conspicuous place on the job. This card lists all the inspections required for the job. Inspectors sign the card as the different stages of construction are completed.

Figure 32–3. Example of a building permit. Construction cannot begin until a permit is obtained.

# SECTION 7

# Leveling Instruments and Operations

Most building contractors provide several types of leveling instruments for on-site layout purposes. The older, more traditional instruments are the *builder's level* and the *transit-level* (sometimes called *level-transit*, or simply *transit*). They are still the most commonly used leveling instruments in construction work. New instruments are the self-leveling *automatic level* and the *laser transit-level*.

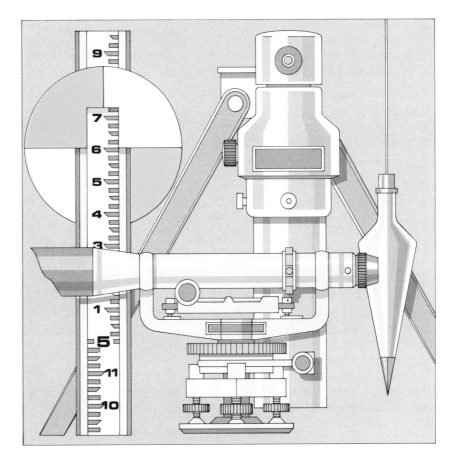

# UNIT 33

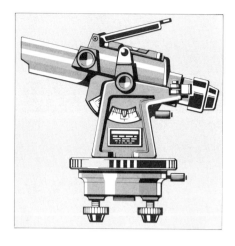

# Builder's Level, Automatic Level, and Transit-level

## BUILDER'S LEVEL

A *builder's level* is the tool used most often to check and establish grades and elevations and to set up level points over long distances. The main parts of a builder's level are the *telescope, spirit level,* and *leveling screws.* The whole assembly is mounted over a circular base. See Figure

33–1. The telescope includes a focusing knob. It has vertical and horizontal crosshairs within the barrel. The sensitive spirit level is parallel with the telescope and is located above or below the barrel. It is adjusted with four leveling screws. (Some models require only two or three leveling screws.)

A typical builder's level has a

*horizontal clamp screw* to hold the instrument in a fixed horizontal position. A *horizontal tangent screw* allows slight movement of the telescope in a left or right horizontal direction. A graduated *horizontal circle* and a *vernier scale* are situated over the circular base. (The *transit-level,* discussed later in this unit, also has a horizontal circle and vernier

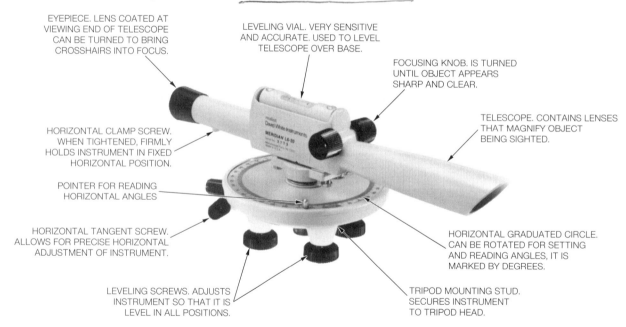

EYEPIECE. LENS COATED AT VIEWING END OF TELESCOPE CAN BE TURNED TO BRING CROSSHAIRS INTO FOCUS.

LEVELING VIAL. VERY SENSITIVE AND ACCURATE. USED TO LEVEL TELESCOPE OVER BASE.

FOCUSING KNOB. IS TURNED UNTIL OBJECT APPEARS SHARP AND CLEAR.

HORIZONTAL CLAMP SCREW. WHEN TIGHTENED, FIRMLY HOLDS INSTRUMENT IN FIXED HORIZONTAL POSITION.

TELESCOPE. CONTAINS LENSES THAT MAGNIFY OBJECT BEING SIGHTED.

POINTER FOR READING HORIZONTAL ANGLES

HORIZONTAL TANGENT SCREW. ALLOWS FOR PRECISE HORIZONTAL ADJUSTMENT OF INSTRUMENT.

HORIZONTAL GRADUATED CIRCLE. CAN BE ROTATED FOR SETTING AND READING ANGLES, IT IS MARKED BY DEGREES.

LEVELING SCREWS. ADJUSTS INSTRUMENT SO THAT IT IS LEVEL IN ALL POSITIONS.

TRIPOD MOUNTING STUD. SECURES INSTRUMENT TO TRIPOD HEAD.

*David White, Inc.*

Figure 33–1. Parts of a builder's level. The instrument is secured to the tripod head by hand tightening the cup assembly on the tripod head to the instrument's mounting stud.

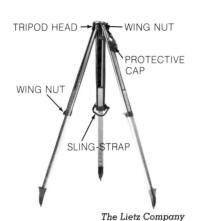

*The Lietz Company*

*David White, Inc.*

Figure 33–2. A builder's level is mounted on a tripod. When the tripod is not in use, its head should be covered with a protective cap.

scale. These features are explained in detail when the transit-level is discussed.)

A chain and hook are attached to the bottom of the leveling plate so that a plumb bob can hang beneath the builder's level. A plumb bob is used with the builder's level when the instrument must be set up over a specific point.

A dust cap to protect the *objective lens* of the telescope when it is not in use is usually included with the instrument. (The objective lens is at the front end of the telescope.) Sun shades are furnished with many models to be slipped over the end of the telescope when it is in use.

Various models of the builder's level are available. They range from the less expensive types with 12-power telescopes to the highly sensitive 32-power models. The *power* of a telescope determines how much closer an object will appear when viewed through the telescope. A target seen through an 18-power telescope will seem 18 times closer than when it is seen with the naked eye.

The builder's level is mounted

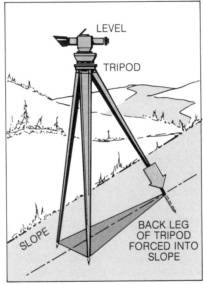

Figure 33–3. Set up a tripod on a slope so that one leg of the tripod faces into the slope.

Figure 33–4. On soft or marshy soil, stakes can be driven into the ground to support the tripod.

on a *tripod.* See Figure 33–2. The *tripod head* is supported by three legs. Wing nuts, located under the tripod head, tighten the legs into position. Some tripods have adjustable extension legs, which accommodate sloping or uneven ground. Tripods are available in hardwood or aluminum.

## Setting up the Builder's Level

Place the builder's level where it will provide an unobstructed view of the work area. Set the tripod in a stable position, then fasten the builder's level to the tripod. Adjust the instrument until it is exactly level. A poorly adjusted instrument is of no value in a layout operation.

**Placing the Tripod.** Spread the legs of the tripod about 3' apart. Keep the tripod head as level as possible. Push the legs firmly into the ground, and tighten the wing nuts. On sloping ground, place one leg of the tripod into the slope. See Figure 33–3. A tripod set on very soft or marshy soil may require a base of three stakes driven into the ground. See Figure 33–4. For placing a tripod over concrete or any other smooth surface, a triangular wood base is helpful. See Figure 33–5.

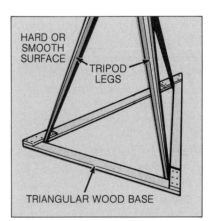

Figure 33–5. A triangular wood base helps to stabilize a tripod placed on a hard concrete surface.

**Fastening the Builder's Level to the Tripod.** All models of builder's levels have a carrying case. When lifting the instrument from the case, grasp it by the base plate or *standard*. The standard is the frame in which the telescope is mounted. Hold the instrument directly over the tripod head. If a plumb bob chain and hook are fastened to the bottom of the instrument, they must hang freely through the hole in the tripod head before the base plate is screwed down.

Two types of fastening arrangements are generally used, depending on the design of the tripod head. See Figure 33–6. If the tripod head is the threaded type, the base of the builder's level is screwed directly onto it. If the tripod head has a cup assembly, a threaded mounting stud at the base of the builder's level is screwed into the cup assembly.

**Adjusting the Builder's Level.** The telescope of the builder's level rotates on top of the circular base. To function properly, it must be level in all positions over the base. This is accomplished by adjusting the leveling screws. The procedure for this operation is shown in Figure 33–7.

Although there should be firm contact between the screws and the base, overtightening the screws can damage the instrument. The leveling procedure becomes quick and simple with practice. Key points to remember are that the both screws must be turned (1) equally, (2) at the same time, and (3) in opposite directions. The direction that your left thumb moves is the direction that the bubble will move. See Figure 33–8.

When the builder's level is used to lay out horizontal angles, it must be positioned directly over a specific point while it is

THREADED TRIPOD HEAD

TRIPOD HEAD WITH CUP ASSEMBLY

*David White, Inc.*

Figure 33–6. Two types of tripod heads are the threaded type and the cup assembly type.

being adjusted. However, the builder's level is not often used for this purpose, because it is much easier to lay out horizontal angles with a transit-level. The procedure for setting up over a point is discussed later in this unit when the transit-level is discussed.

**Focusing and Sighting.** To view an object through the builder's level, focus the telescope by turning the focusing knob until the object sighted is sharp and clear. The focusing knob adjusts the lenses inside the telescope barrel.

The builder's level is used to focus on a very small target (a leveling rod, discussed later in this unit), and the *field of vision* is very small. See Figure 33–9. The field of vision is the total magnified area seen through the telescope. Some builder's levels have an eyepiece focusing ring for bringing into sharp focus the *crosshairs* within the telescope. The crosshairs are the fine horizontal and vertical lines in the telescope that permit the object being sighted to be placed exactly in the center of the field of vision.

A good procedure for locating a target and focusing on it with the builder's level follows:

1. Aim the telescope at the

target by looking across the top of the barrel. Some instruments have devices similar to gun sights at the top of the barrel.

2. Focus the telescope until the target is clear.

3. When the target is sighted as closely as possible, tighten the horizontal clamp screw. Make final adjustments by using the horizontal tangent screw to move the telescope into position.

The dashed line in Figure 33–10 shows the principle of the *line of sight.* If this line could be seen, it would be perfectly level. It would extend from the horizontal crosshair at the center of the telescope barrel to the target.

## Leveling Rods and Targets

When a carpenter uses a builder's level, a second person must hold a vertical measuring device in the area where the grade or elevation is being checked or established. See Figure 33–11. The horizontal crosshair is then lined up with a measurement or mark on the rod. Refer again to Figure 33–9.

**Manufactured Leveling Rods.** Rods of wood, plastic, or aluminum have been specially designed for use with leveling instruments. Most manufactured leveling rods have adjustable

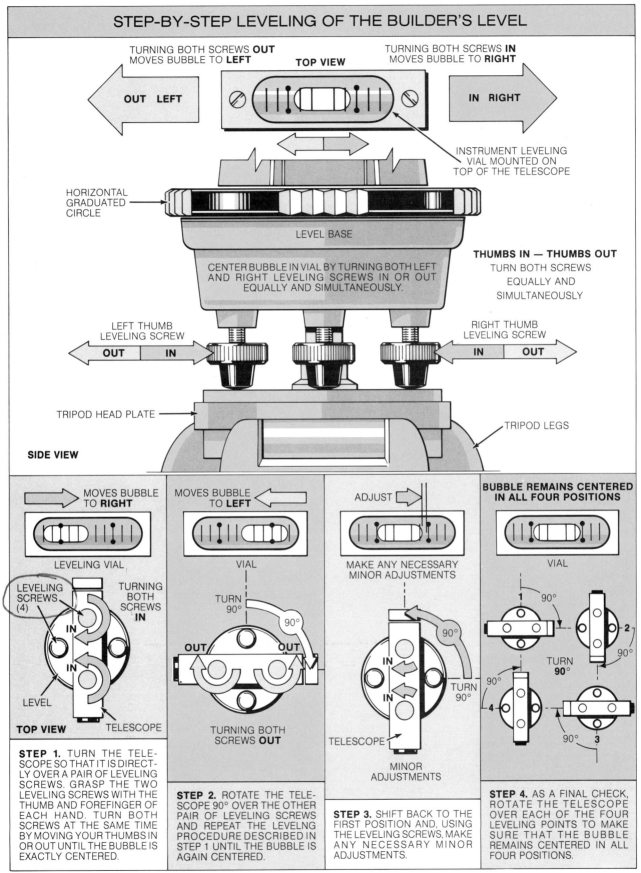

## STEP-BY-STEP LEVELING OF THE BUILDER'S LEVEL

TURNING BOTH SCREWS **OUT** MOVES BUBBLE TO **LEFT**

TOP VIEW

TURNING BOTH SCREWS **IN** MOVES BUBBLE TO **RIGHT**

OUT LEFT

IN RIGHT

INSTRUMENT LEVELING VIAL MOUNTED ON TOP OF THE TELESCOPE

HORIZONTAL GRADUATED CIRCLE

LEVEL BASE

**THUMBS IN — THUMBS OUT**
TURN BOTH SCREWS EQUALLY AND SIMULTANEOUSLY

CENTER BUBBLE IN VIAL BY TURNING BOTH LEFT AND RIGHT LEVELING SCREWS IN OR OUT EQUALLY AND SIMULTANEOUSLY.

LEFT THUMB LEVELING SCREW

OUT    IN

RIGHT THUMB LEVELING SCREW

IN    OUT

TRIPOD HEAD PLATE

TRIPOD LEGS

**SIDE VIEW**

---

MOVES BUBBLE TO **RIGHT**

LEVELING VIAL

LEVELING SCREWS (4)

TURNING BOTH SCREWS **IN**

IN

IN

LEVEL

**TOP VIEW**    TELESCOPE

**STEP 1.** TURN THE TELE-SCOPE SO THAT IT IS DIRECT-LY OVER A PAIR OF LEVELING SCREWS. GRASP THE TWO LEVELING SCREWS WITH THE THUMB AND FOREFINGER OF EACH HAND. TURN BOTH SCREWS AT THE SAME TIME BY MOVING YOUR THUMBS IN OR OUT UNTIL THE BUBBLE IS EXACTLY CENTERED.

---

MOVES BUBBLE TO **LEFT**

VIAL

TURN 90°

90°

OUT    OUT

TURNING BOTH SCREWS **OUT**

**STEP 2.** ROTATE THE TELE-SCOPE 90° OVER THE OTHER PAIR OF LEVELING SCREWS AND REPEAT THE LEVELNG PROCEDURE DESCRIBED IN STEP 1 UNTIL THE BUBBLE IS AGAIN CENTERED.

---

ADJUST

MAKE ANY NECESSARY MINOR ADJUSTMENTS

90°

IN

IN

TURN 90°

TELESCOPE

MINOR ADJUSTMENTS

**STEP 3.** SHIFT BACK TO THE FIRST POSITION AND, USING THE LEVELING SCREWS, MAKE ANY NECESSARY MINOR ADJUSTMENTS.

---

**BUBBLE REMAINS CENTERED IN ALL FOUR POSITIONS**

VIAL

1    90°

90°    2

TURN **90°**

90°    90°

4    3

90°

**STEP 4.** AS A FINAL CHECK, ROTATE THE TELESCOPE OVER EACH OF THE FOUR LEVELING POINTS TO MAKE SURE THAT THE BUBBLE REMAINS CENTERED IN ALL FOUR POSITIONS.

Figure 33–7. Leveling a builder's level. Be careful to avoid overtightening the adjusting screws.

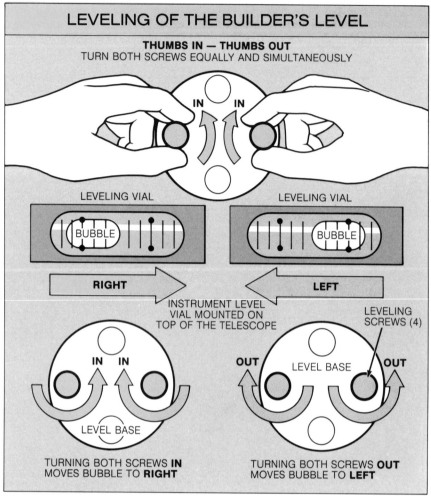

## LEVELING OF THE BUILDER'S LEVEL

**THUMBS IN — THUMBS OUT**
TURN BOTH SCREWS EQUALLY AND SIMULTANEOUSLY

IN          IN

LEVELING VIAL          LEVELING VIAL

BUBBLE          BUBBLE

RIGHT          LEFT

INSTRUMENT LEVEL
VIAL MOUNTED ON
TOP OF THE TELESCOPE

LEVELING
SCREWS (4)

IN   IN          OUT          OUT

LEVEL BASE

LEVEL BASE

TURNING BOTH SCREWS **IN**
MOVES BUBBLE TO **RIGHT**

TURNING BOTH SCREWS **OUT**
MOVES BUBBLE TO **LEFT**

Figure 33–8. In leveling a builder's level, the direction your left thumb moves is the direction that the bubble will move.

*David White, Inc.*

Figure 33–11. Holding a leveling rod in position for use with a builder's level.

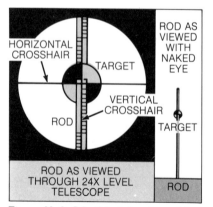

HORIZONTAL
CROSSHAIR

TARGET

ROD

ROD AS VIEWED
WITH
NAKED
EYE

VERTICAL
CROSSHAIR

TARGET

ROD AS VIEWED
THROUGH 24X LEVEL
TELESCOPE

ROD

Figure 33–9. A builder's level provides a very small field of vision.

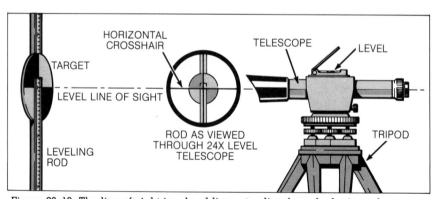

HORIZONTAL
CROSSHAIR

TARGET

LEVEL LINE OF SIGHT

LEVELING
ROD

ROD AS VIEWED
THROUGH 24X LEVEL
TELESCOPE

TELESCOPE          LEVEL

TRIPOD

Figure 33–10. The line of sight is a level line extending from the horizontal crosshair at the center of the telescope barrel to the target.

sections. Typical two-section rods extend from 8′ to 9½′. Three-section rods extend from 12′ to 14′. The numbers and graduation marks on a rod are large so that they can be easily read. The foot numbers are the largest and are usually printed in red. The figures and graduations between the foot numbers are usually printed in black. Movable metal targets are fitted to the rods to make sighting easier at longer distances. See Figure 33–12.

Leveling rods are available with three types of graduations. One type is metric and two types are customary (English). The two rods with customary measurements are the *architect's rod* and the *engineer's rod*. See Figure 33–13. The architect's rod is graduated in feet, inches, and eighths of an inch. It is used by carpenters and other construction workers. The engineer's rod is graduated in feet, tenths of a foot, and hundredths of a foot. It is used by surveyors and engineers.

**Stick-and-Rule Method.** When sighting short distances, carpenters frequently use an ordinary measuring tape or rule held against a wood ripping. See Figure 33–14.

**Plain-stick Method.** Often the most convenient way to perform layout operations with a builder's level is to use an ordinary unmarked wood rod. The line of sight is marked on the rod as it is held over an established point.

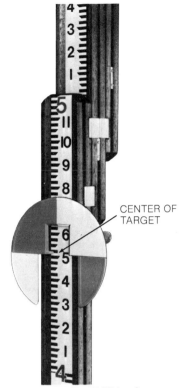

CENTER OF TARGET

*David White, Inc.*

Figure 33–12. An adjustable manufactured leveling rod with a movable metal target. The center of the target is lined up with the reading being taken on the rod (in this example, 4'–5⅛').

After this the rod can be moved about to establish grades at other locations.

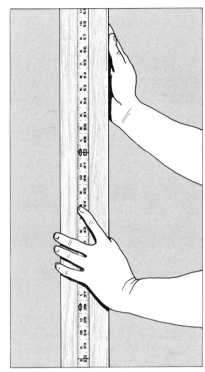

Figure 33–14. Carpenters often use a stick and measuring tape as a leveling rod.

**Arm Signals.** For accurate readings with the builder's level, the person holding the leveling rod must hold it in a plumb position. When the rod is a great distance away from the builder's level, the

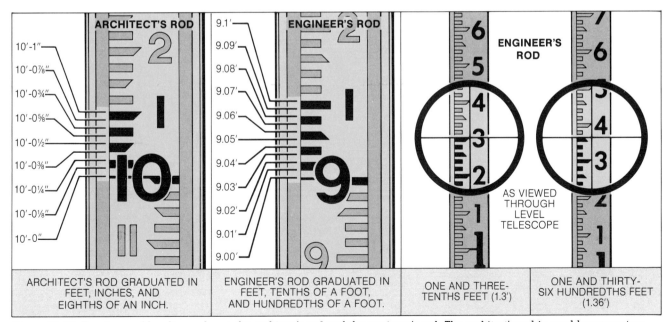

| ARCHITECT'S ROD | ENGINEER'S ROD | ENGINEER'S ROD | |
|---|---|---|---|
| 10'-1" | 9.1' | | 6 |
| 10'-0⅞" | 9.09' | 6 | |
| 10'-0¾" | 9.08' | 5 | 5 |
| 10'-0⅝" | 9.07' | 4 | |
| 10'-0½" | 9.06' | 3 | 4 |
| 10'-0⅜" | 9.05' | 2 | 3 |
| 10'-0¼" | 9.04' | | |
| 10'-0⅛" | 9.03' | 1 | 2 |
| 10'-0" | 9.02' | | |
| | 9.01' | AS VIEWED THROUGH LEVEL TELESCOPE | 1 |
| | 9.00' | | |
| ARCHITECT'S ROD GRADUATED IN FEET, INCHES, AND EIGHTHS OF AN INCH. | ENGINEER'S ROD GRADUATED IN FEET, TENTHS OF A FOOT, AND HUNDREDTHS OF A FOOT. | ONE AND THREE-TENTHS FEET (1.3') | ONE AND THIRTY-SIX HUNDREDTHS FEET (1.36') |

Figure 33–13. Two types of leveling rods are the architect's rod and the engineer's rod. The architect's rod is used by carpenters.

David White, Inc.

Figure 33-15. Instrument operator signaling to lower target.

operator of the level may use arm signals to instruct the person holding the rod to bring the rod into a plumb position or otherwise move it. See Figure 33-15.

Figure 33-16 shows standard arm signals for bringing the rod into plumb position. Figure 33-17 shows standard arm signals for bringing the target on the rod into an on-grade position.

## Operations with the Builder's Level

Some of the more common operations performed with a builder's level are checking grade differences, establishing elevations and level points, and calculating stud lengths for stepped foundations.

### Checking Grade Differences.

All lots have low and high grade points. These grade points should be known before the first corner height of the foundation wall is established. If *grading* (reshaping by removing or adding soil) is required on the lot, a builder's level and rod can be used to check the heights while the grading work is carried out. Figure 33-18 shows how the builder's level is used to check grade differences. When a lot has a steep slope, the procedure is more complicated. See Figure 33-19.

**Establishing Elevations.** Jobsite elevations determine the major structural levels of the building, such as the top of the foundation walls and the finish floor height of the first floor. All jobsite elevations relate to a *bench mark* or *datum* (sometimes called *point of beginning*) that has been established for the job. On many jobs the top of a stake driven at one corner of the lot, or a chiseled mark at the top of a concrete curb, identifies the bench marks. The location of this reference point is also marked on the plot plan drawing of the

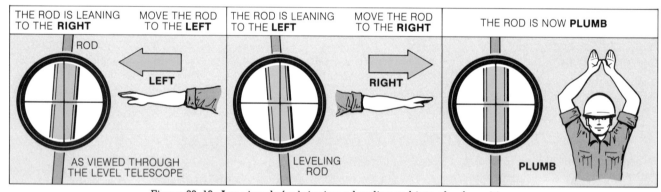

Figure 33-16. Arm signals for bringing a leveling rod into plumb position.

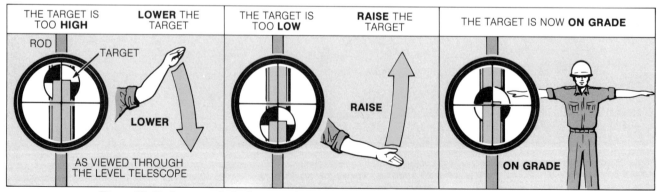

Figure 33-17. Arm signals for bringing the target on a leveling rod into an on-grade position.

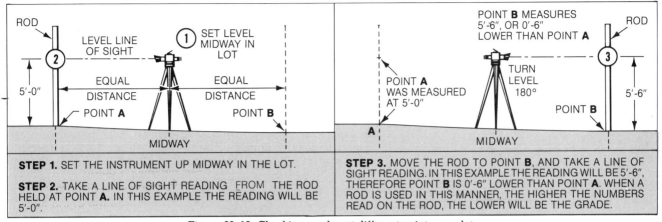

Figure 33–18. Checking grades at different points on a lot.

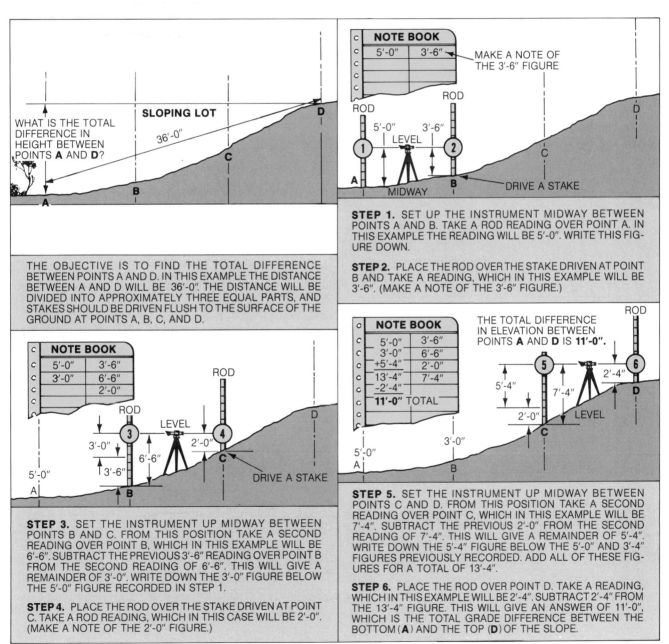

Figure 33–19. Finding the total grade difference between the bottom and top of a steep slope.

blueprints (discussed in Section 6).

Figures 33–20 and 33–21 show how the builder's level is used to establish an elevation point in relation to a bench mark. The plain-stick method is used in Figure 33–20, and the stick-and-rule method is used in Figure 33–21.

A method of checking the dif-ferences in elevation levels of a building under construction is shown in Figure 33–22.

**Establishing Level Points.** After an elevation point has been set in relation to the bench mark, it often must be leveled across to other points. For example, level points must be established at the corner stakes when forms are constructed for the concrete foundation of a building. (Form construction for foundations is discussed in Section 8.) Figure 33–23 shows how level points are established with a builder's level.

**Calculating Stud Lengths for a Stepped Foundation.** The foundation walls on a sloped lot are

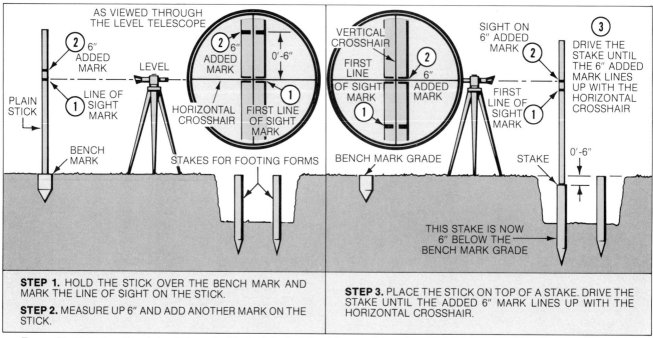

**STEP 1.** HOLD THE STICK OVER THE BENCH MARK AND MARK THE LINE OF SIGHT ON THE STICK.

**STEP 2.** MEASURE UP 6″ AND ADD ANOTHER MARK ON THE STICK.

**STEP 3.** PLACE THE STICK ON TOP OF A STAKE. DRIVE THE STAKE UNTIL THE ADDED 6″ MARK LINES UP WITH THE HORIZONTAL CROSSHAIR.

Figure 33–20. Using the plain-stick method to find the height of footing forms that are to be 6″ below the bench mark grade.

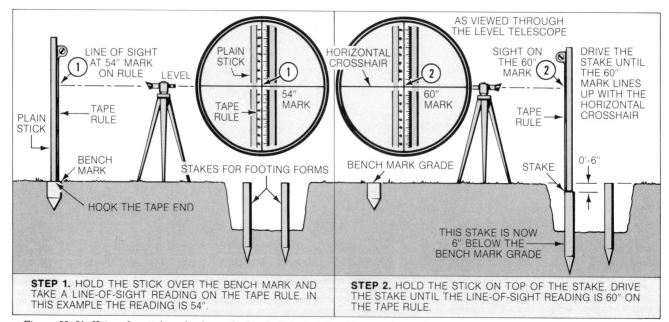

**STEP 1.** HOLD THE STICK OVER THE BENCH MARK AND TAKE A LINE-OF-SIGHT READING ON THE TAPE RULE. IN THIS EXAMPLE THE READING IS 54″.

**STEP 2.** HOLD THE STICK ON TOP OF THE STAKE. DRIVE THE STAKE UNTIL THE LINE-OF-SIGHT READING IS 60″ ON THE TAPE RULE.

Figure 33–21. Using the stick-and-rule method to find the height of footing forms that are to be 6″ below the bench mark grade.

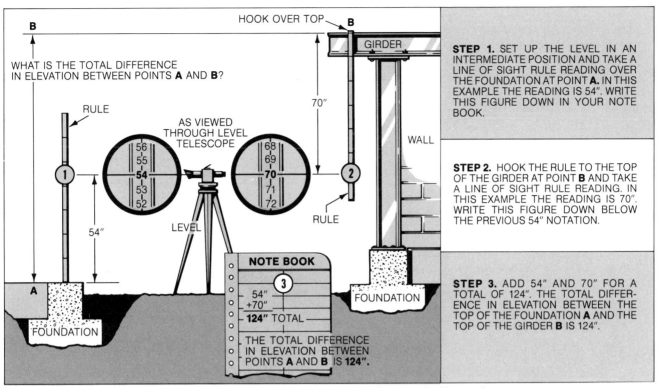

Figure 33-22. Measuring differences in elevations of existing building and new construction.

often *stepped* (shaped like steps). (Stepped foundations are discussed in Section 8.) The stepped design saves labor and material when the forms are being constructed. Less concrete is also required. In order for the floor above a stepped foundation to be level, a stud wall *(underpinning)* must be built. A system for using the builder's level to find the different lengths of the wall studs, which are then cut on a radial-arm saw, is shown in Figure 33-24.

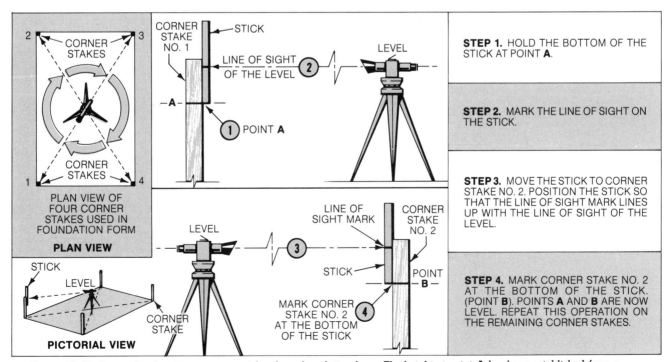

Figure 33-23. Setting level points on corner stakes for a foundation form. The height at point **A** has been established from a bench mark.

# USING A BUILDER'S LEVEL TO CALCULATE STUD LENGTHS

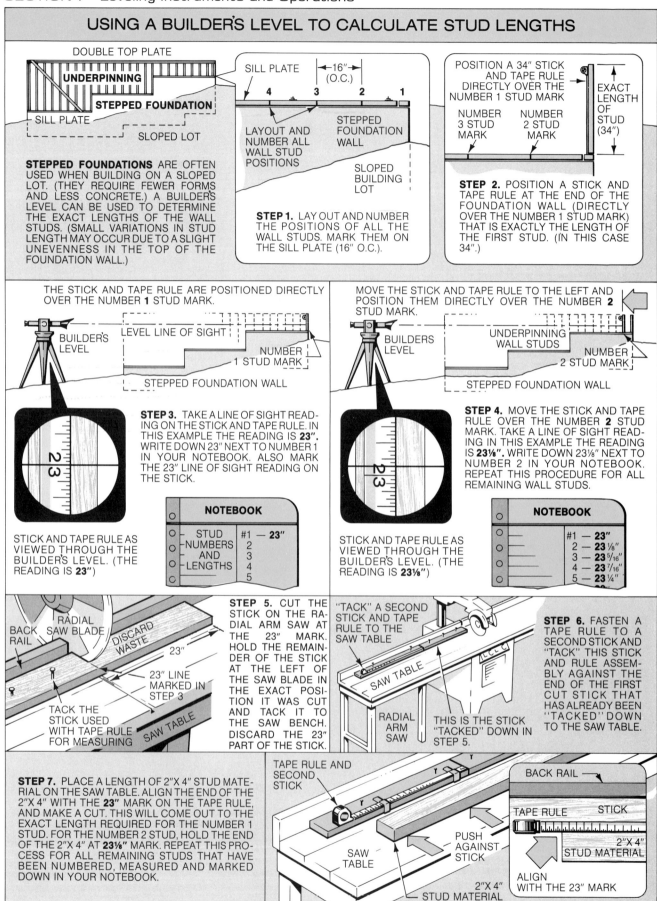

**STEPPED FOUNDATIONS** ARE OFTEN USED WHEN BUILDING ON A SLOPED LOT. (THEY REQUIRE FEWER FORMS AND LESS CONCRETE.) A BUILDER'S LEVEL CAN BE USED TO DETERMINE THE EXACT LENGTHS OF THE WALL STUDS. (SMALL VARIATIONS IN STUD LENGTH MAY OCCUR DUE TO A SLIGHT UNEVENNESS IN THE TOP OF THE FOUNDATION WALL.)

**STEP 1.** LAY OUT AND NUMBER THE POSITIONS OF ALL THE WALL STUDS. MARK THEM ON THE SILL PLATE (16" O.C.).

**STEP 2.** POSITION A STICK AND TAPE RULE AT THE END OF THE FOUNDATION WALL (DIRECTLY OVER THE NUMBER 1 STUD MARK) THAT IS EXACTLY THE LENGTH OF THE FIRST STUD. (IN THIS CASE 34".)

**STEP 3.** TAKE A LINE OF SIGHT READING ON THE STICK AND TAPE RULE. IN THIS EXAMPLE THE READING IS **23"**. WRITE DOWN 23" NEXT TO NUMBER 1 IN YOUR NOTEBOOK. ALSO MARK THE 23" LINE OF SIGHT READING ON THE STICK.

STICK AND TAPE RULE AS VIEWED THROUGH THE BUILDER'S LEVEL. (THE READING IS **23"**)

| NOTEBOOK | |
|---|---|
| STUD NUMBERS AND LENGTHS | #1 — **23"** |
| | 2 |
| | 3 |
| | 4 |
| | 5 |

**STEP 4.** MOVE THE STICK AND TAPE RULE OVER THE NUMBER **2** STUD MARK. TAKE A LINE OF SIGHT READING IN THIS EXAMPLE THE READING IS **23⅛"**. WRITE DOWN 23⅛" NEXT TO NUMBER 2 IN YOUR NOTEBOOK. REPEAT THIS PROCEDURE FOR ALL REMAINING WALL STUDS.

STICK AND TAPE RULE AS VIEWED THROUGH THE BUILDER'S LEVEL. (THE READING IS **23⅛"**)

| NOTEBOOK | |
|---|---|
| #1 — **23"** | |
| 2 — **23⅛"** | |
| 3 — **23 5/16"** | |
| 4 — **23 7/16"** | |
| 5 — **23¼"** | |

**STEP 5.** CUT THE STICK ON THE RADIAL ARM SAW AT THE 23" MARK. HOLD THE REMAINDER OF THE STICK AT THE LEFT OF THE SAW BLADE IN THE EXACT POSITION IT WAS CUT AND TACK IT TO THE SAW BENCH. DISCARD THE 23" PART OF THE STICK.

**STEP 6.** FASTEN A TAPE RULE TO A SECOND STICK AND "TACK" THIS STICK AND RULE ASSEMBLY AGAINST THE END OF THE FIRST CUT STICK THAT HAS ALREADY BEEN "TACKED" DOWN TO THE SAW TABLE.

**STEP 7.** PLACE A LENGTH OF 2"X 4" STUD MATERIAL ON THE SAW TABLE. ALIGN THE END OF THE 2"X 4" WITH THE **23"** MARK ON THE TAPE RULE, AND MAKE A CUT. THIS WILL COME OUT TO THE EXACT LENGTH REQUIRED FOR THE NUMBER 1 STUD. FOR THE NUMBER 2 STUD, HOLD THE END OF THE 2"X 4" AT **23⅛"** MARK. REPEAT THIS PROCESS FOR ALL REMAINING STUDS THAT HAVE BEEN NUMBERED, MEASURED AND MARKED DOWN IN YOUR NOTEBOOK.

Figure 33–24. Using a builder's level to calculate stud lengths on a stepped foundation.

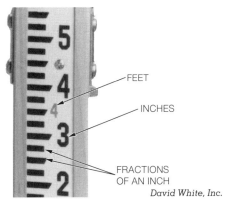

FEET

INCHES

FRACTIONS
OF AN INCH

*David White, Inc.*

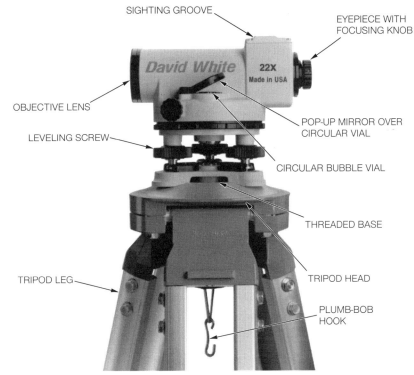

SIGHTING GROOVE

EYEPIECE WITH
FOCUSING KNOB

22X
Made in USA

OBJECTIVE LENS

POP-UP MIRROR OVER
CIRCULAR VIAL

LEVELING SCREW

CIRCULAR BUBBLE VIAL

THREADED BASE

TRIPOD HEAD

TRIPOD LEG

PLUMB-BOB
HOOK

*David White, Inc.*

Figure 33–25. An automatic level can be adjusted very quickly and maintains a high degree of accuracy.

## AUTOMATIC LEVEL

The self-leveling *automatic level* (Figure 33–25) was introduced in this country in 1948 by the Zeiss Company of West Germany. Since then other manufacturers have developed similar self-leveling models. These instruments are very popular on construction jobs.

To set up the automatic level, secure it to the tripod head with a tripod draw screw. Three (rather than four) leveling screws are used to make a *rough ad-justment*. The instrument is in rough adjustment when the bubble in the *circular level* is centered. After the instrument has been roughly adjusted, an *internal compensator* aligns the instrument into a level position.

The self-leveling automatic level offers some important advantages over traditional levels. It can be adjusted more quickly and maintains a high degree of accuracy. Recent models include an electronic safety system that alerts the user if the instrument has shifted out of its level position.

## TRANSIT-LEVEL

The transit-level (also known as a *level-transit*) is similar in appearance to the builder's level. See Figure 33–26. A major difference between the two is that the telescope of the transit-level can be tilted up and down (vertically) as well as moved from side to side (horizontally). This allows a number of operations with the transit-level that are not possible with the builder's level.

Transit-levels are available in a variety of models. All have a locking lever or clamp to hold the telescope in a fixed, level position. When the telescope is locked in a level position, the transit-level can perform all the functions of the builder's level.

When the locking device is released, allowing up-and-down movement of the telescope, a vertical clamp screw is tightened to hold the telescope in any vertical position desired. A vertical tangent screw is provided for fine adjustments.

## Horizontal Circle and Horizontal Vernier Scale

A transit-level has a *horizontal circle* and a *horizontal vernier scale.* See Figure 33–27. These are intersecting scales used to measure horizontal angles. The horizontal circle can be turned by hand. It does not move when the telescope rotates. The horizontal vernier scale is attached to the instrument frame. It moves around the inside of the horizontal setting circle as the telescope is turned to the right or left. The horizontal circle is divided into four sections (quadrants), each reading from zero to 90°. The vernier scale has 12 graduations of zero to 60 minutes at each side of the zero index.

## Vertical Arc and Vertical Vernier Scale

All transit-levels are also equipped with a *vertical arc,* which is similar to the horizontal circle. Refer again to Figure 33–26. It is used to measure vertical angles. The vertical arc is graduated from zero to 45° in two directions and will move with the up-and-down motion of the telescope. Some models have a *vertical vernier scale,* which is attached to the frame. It does not move. Refer again to Figure 33–26. Other models do not have a vertical vernier scale but instead have a pointer to indicate the vertical degrees to which the telescope has been set. See Figure 33–28.

## Setting up the Transit-Level

The procedure for setting up the transit-level is similar to that for the builder's level. The telescope must be locked into its horizontal postion before the instrument is adjusted with the leveling screws.

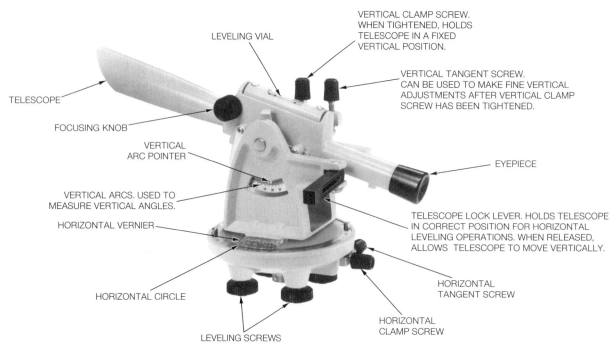

VERTICAL CLAMP SCREW. WHEN TIGHTENED, HOLDS TELESCOPE IN A FIXED VERTICAL POSITION.

LEVELING VIAL

VERTICAL TANGENT SCREW. CAN BE USED TO MAKE FINE VERTICAL ADJUSTMENTS AFTER VERTICAL CLAMP SCREW HAS BEEN TIGHTENED.

TELESCOPE

FOCUSING KNOB

VERTICAL ARC POINTER

EYEPIECE

VERTICAL ARCS. USED TO MEASURE VERTICAL ANGLES.

HORIZONTAL VERNIER

TELESCOPE LOCK LEVER. HOLDS TELESCOPE IN CORRECT POSITION FOR HORIZONTAL LEVELING OPERATIONS. WHEN RELEASED, ALLOWS TELESCOPE TO MOVE VERTICALLY.

HORIZONTAL CIRCLE

HORIZONTAL TANGENT SCREW

LEVELING SCREWS

HORIZONTAL CLAMP SCREW

*David White, Inc.*

Figure 33–26. A transit level resembles a builder's level. However, the telescope can be moved vertically and horizontally, making more operations possible than with a builder's level.

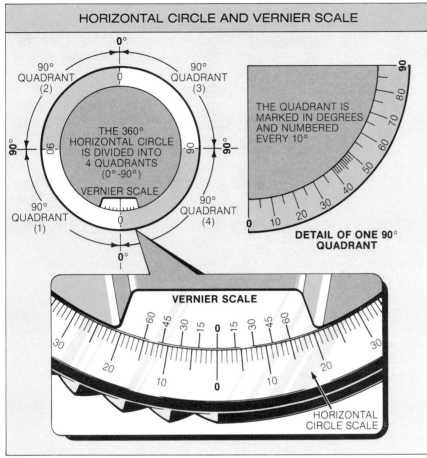

## HORIZONTAL CIRCLE AND VERNIER SCALE

90° QUADRANT (2)

90° QUADRANT (3)

THE 360° HORIZONTAL CIRCLE IS DIVIDED INTO 4 QUADRANTS (0°–90°)

90° QUADRANT (1)

90° QUADRANT (4)

VERNIER SCALE

THE QUADRANT IS MARKED IN DEGREES AND NUMBERED EVERY 10°

**DETAIL OF ONE 90° QUADRANT**

**VERNIER SCALE**

HORIZONTAL CIRCLE SCALE

Figure 33–27. A close-up view of the horizontal circle and vernier scale on a transit-level.

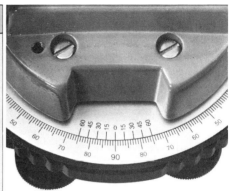

*David White, Inc.*

**Setting up over a Point.** For measuring and laying out horizontal angles, the transit-level must be set up over a specific point on the ground. A plumb bob is used in this procedure as follows (Figure 33–29):

1. Spread the legs of the tripod. Place the tripod head as closely as possible over the point.

2. Secure the transit-level to the tripod. Attach the plumb bob. A slip knot permits the plumb bob to be raised or lowered by sliding the knot along the cord. Adjust the line length so that the plumb bob is about ¼″ above the

*David White, Inc.*

Figure 33–28. This transit-level has a pointer instead of a vernier scale. It can measure vertical angles only to whole degrees, not to fractions of a degree.

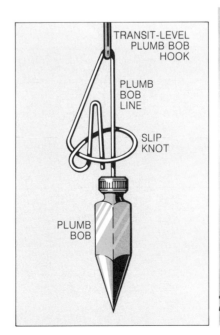

TRANSIT-LEVEL PLUMB BOB HOOK

PLUMB BOB LINE

SLIP KNOT

PLUMB BOB

*Keuffel & Esser Company*

Figure 33–29. A plumb bob is used to set up a transit over a point on the ground. A slip knot permits the plumb bob to be raised or lowered.

point.

3. If necessary, shift the tripod to bring the plumb bob closer over the point.

4. Roughly level the transit-level with the leveling screws. Do not tighten the screws.

5. Most transit-levels have a shifting center, which allows for some movement of the instrument on the leveling plate. When the plumb bob is close enough to the point, make the final adjustment by shifting the transit-level on the leveling plate.

6. Adjust and tighten the leveling screws. The transit-level should be perfectly level in all positions.

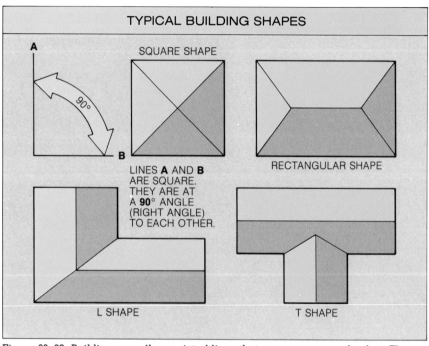

Figure 33–30. Buildings usually consist of lines that are square to each other. The shapes shown here are plan views.

## Operations with the Transit-Level

The right angle (90°) is the angle used most often in construction work. Most buildings are made up of one or more rectangular parts. Each line in a rectangular shape forms a 90° corner at the point where it meets another line. Walls constructed according to these lines are referred to as being *square* to each other. See Figure 33–30.

The first stage of a building project is the construction of the foundation. Before any foundation work can occur, building lines (usually of nylon string) must be set up giving the exact outline of the foundation. These lines are usually at a 90° angle to each other. They can be laid out by a number of methods, but the fastest way is to use the transit-level. The transit-level can also be used to lay out angles other than 90°, to plumb, and to establish points in a straight line.

**Measuring a 90° Horizontal Angle.** The points at both ends of a line must be established before a 90° horizontal angle can

be measured. The transit-level is set up over one of the points. A sight is taken on the second point. The transit-level is then rotated 90° to establish a second line that is square to the first line. This procedure is shown in Figure 33–31.

**Measuring Horizontal Angles Other Than 90°.** The circle on the transit-level is marked into four quadrants, each reading from zero to 90°. Any angle can be measured by rotating the telescope to the right or left and then performing addition or subtraction as necessary. Angles of 90° or less can be measured directly on the horizontal circle without any addition or subtraction. Examples of procedures for measuring various angles are shown in Figure 33–32. (Remember, the vernier scale moves with the swing of the instrument.)

**Measuring Horizontal Angles to a Fraction of a Degree.** *Min-*

*utes* and *seconds* are used to express fractions of a degree. The breakdown is as follows:

Circle = 360°
1° = 60 minutes
1 minute = 60 seconds

The transit-level (or the builder's level) is used to measure angles in degrees and minutes by correlating the degree graduations on the horizontal circle to the graduations that represent minutes on the vernier scale. The numbers on the vernier scale run from zero to 60 to the right and left of the zero index. Each mark on the vernier scale represents 5 minutes. The symbol for minutes is the same as the symbol for feet ('). In Figure 33–33 an explanation and comparison are given between a degree-only reading and a degree-and-minute reading. Additional examples of degree-and-minute readings are shown in Figure 33–34.

**Plumbing.** A column, building corner, or any other vertical

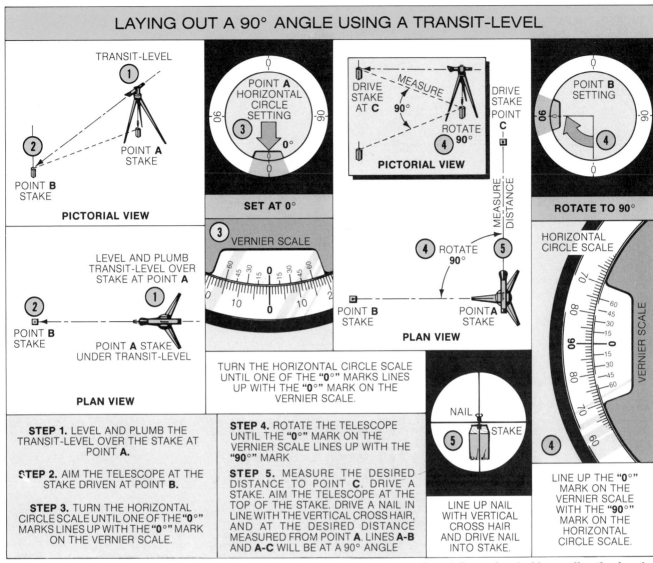

Figure 33–31. Laying out a 90° angle with a transit-level. In this example, points A, B, and C are identified by small nails placed at the top of wood stakes.

member can be plumbed with great accuracy with the transit-level. When using the instrument for this purpose, place it at a convenient distance from the object being plumbed. If possible, this distance should be greater than the height of the object. The procedure for plumbing a column is shown in Figure 33–35.

**Establishing Points in a Straight Line.** The transit-level simplifies the task of establishing points in a straight line. It can be applied to setting stakes for a form wall, establishing centers of piers or footings, lining up drain tiles, and so forth. The procedure for establishing points in a straight line with the transit-level is shown in Figure 33–36.

## CARE OF LEVELING INSTRUMENTS

Leveling instruments are expensive precision instruments. They can withstand many years of normal use on a construction job, but only if they are handled carefully. When workers are unfamiliar with a particular brand or model, they should refer to the instruction manual before using the instrument.

Most problems that occur with leveling instruments are caused by improper care or use. For example, continued exposure to rain and dust may cloud the telescope lenses. A slight blow or continual vibration may throw the instrument out of alignment and reduce its accuracy. A hard blow or a fall can ruin the instrument.

The following precautions should be taken with all leveling instruments:

1. Keep the instrument in the

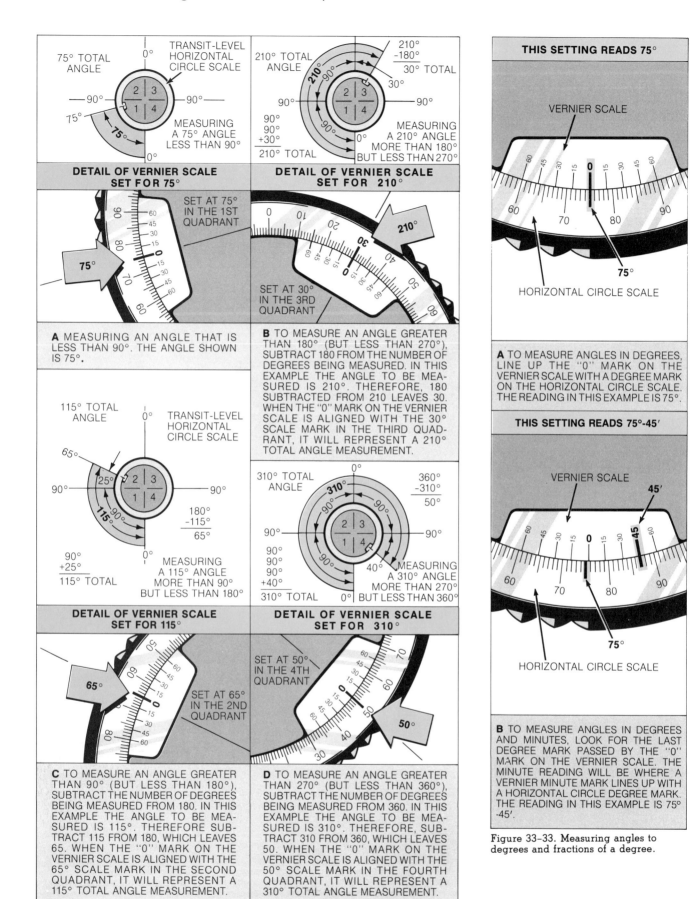

**THIS SETTING READS 75°**

VERNIER SCALE

HORIZONTAL CIRCLE SCALE

75°

**A** TO MEASURE ANGLES IN DEGREES, LINE UP THE "0" MARK ON THE VERNIER SCALE WITH A DEGREE MARK ON THE HORIZONTAL CIRCLE SCALE. THE READING IN THIS EXAMPLE IS 75°.

**THIS SETTING READS 75°-45'**

VERNIER SCALE

45'

75°

HORIZONTAL CIRCLE SCALE

**B** TO MEASURE ANGLES IN DEGREES AND MINUTES, LOOK FOR THE LAST DEGREE MARK PASSED BY THE "0" MARK ON THE VERNIER SCALE. THE MINUTE READING WILL BE WHERE A VERNIER MINUTE MARK LINES UP WITH A HORIZONTAL CIRCLE DEGREE MARK. THE READING IN THIS EXAMPLE IS 75° -45'.

Figure 33-33. Measuring angles to degrees and fractions of a degree.

---

**75° TOTAL ANGLE**

TRANSIT-LEVEL HORIZONTAL CIRCLE SCALE

MEASURING A 75° ANGLE LESS THAN 90°

**DETAIL OF VERNIER SCALE SET FOR 75°**

SET AT 75° IN THE 1ST QUADRANT

**A** MEASURING AN ANGLE THAT IS LESS THAN 90°. THE ANGLE SHOWN IS 75°.

**115° TOTAL ANGLE**

TRANSIT-LEVEL HORIZONTAL CIRCLE SCALE

$$\begin{array}{r} 180° \\ -115° \\ \hline 65° \end{array}$$

$$\begin{array}{r} 90° \\ +25° \\ \hline 115° \text{ TOTAL} \end{array}$$

MEASURING A 115° ANGLE MORE THAN 90° BUT LESS THAN 180°

**DETAIL OF VERNIER SCALE SET FOR 115°**

SET AT 65° IN THE 2ND QUADRANT

**C** TO MEASURE AN ANGLE GREATER THAN 90° (BUT LESS THAN 180°), SUBTRACT THE NUMBER OF DEGREES BEING MEASURED FROM 180. IN THIS EXAMPLE THE ANGLE TO BE MEASURED IS 115°. THEREFORE SUBTRACT 115 FROM 180, WHICH LEAVES 65. WHEN THE "0" MARK ON THE VERNIER SCALE IS ALIGNED WITH THE 65° SCALE MARK IN THE SECOND QUADRANT, IT WILL REPRESENT A 115° TOTAL ANGLE MEASUREMENT.

---

**210° TOTAL ANGLE**

$$\begin{array}{r} 210° \\ -180° \\ \hline 30° \text{ TOTAL} \end{array}$$

$$\begin{array}{r} 90° \\ 90° \\ +30° \\ \hline 210° \text{ TOTAL} \end{array}$$

MEASURING A 210° ANGLE MORE THAN 180° BUT LESS THAN 270°

**DETAIL OF VERNIER SCALE SET FOR 210°**

210°

SET AT 30° IN THE 3RD QUADRANT

**B** TO MEASURE AN ANGLE GREATER THAN 180° (BUT LESS THAN 270°), SUBTRACT 180 FROM THE NUMBER OF DEGREES BEING MEASURED. IN THIS EXAMPLE THE ANGLE TO BE MEASURED IS 210°. THEREFORE, 180 SUBTRACTED FROM 210 LEAVES 30. WHEN THE "0" MARK ON THE VERNIER SCALE IS ALIGNED WITH THE 30° SCALE MARK IN THE THIRD QUADRANT, IT WILL REPRESENT A 210° TOTAL ANGLE MEASUREMENT.

**310° TOTAL ANGLE**

$$\begin{array}{r} 360° \\ -310° \\ \hline 50° \end{array}$$

$$\begin{array}{r} 90° \\ 90° \\ 90° \\ +40° \\ \hline 310° \text{ TOTAL} \end{array}$$

MEASURING A 310° ANGLE MORE THAN 270° BUT LESS THAN 360°

**DETAIL OF VERNIER SCALE SET FOR 310°**

SET AT 50° IN THE 4TH QUADRANT

50°

**D** TO MEASURE AN ANGLE GREATER THAN 270° (BUT LESS THAN 360°), SUBTRACT THE NUMBER OF DEGREES BEING MEASURED FROM 360. IN THIS EXAMPLE THE ANGLE TO BE MEASURED IS 310°. THEREFORE, SUBTRACT 310 FROM 360, WHICH LEAVES 50. WHEN THE "0" MARK ON THE VERNIER SCALE IS ALIGNED WITH THE 50° SCALE MARK IN THE FOURTH QUADRANT, IT WILL REPRESENT A 310° TOTAL ANGLE MEASUREMENT.

Figure 33-32. Measuring angles other than 90°. Any angle can be measured by rotating the telescope to the right or left and then performing addition or subtraction as necessary.

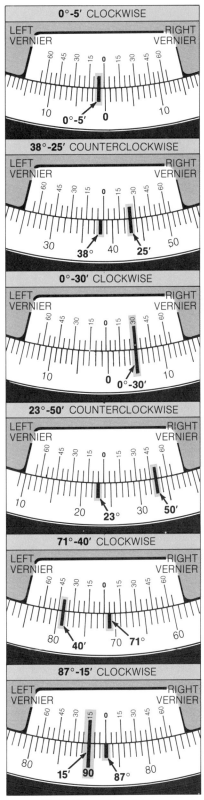

Figure 33–34. Examples of degree and minute readings. When the degree numbers increase to the left, clockwise angles are being read. When the degree numbers increase to the right, counterclockwise angles are being read.

carrying case when it is not in use.

2. Adjust the leveling and clamp screws so that when the instrument is in the case, its parts will not move about freely. However, do not tighten the screws enough to put the instru-ment in a rigid position.

3. When transporting the instrument in a vehicle, do not place the carrying case on a hard floor where it will be subjected to vibration.

4. Never set up the instrument

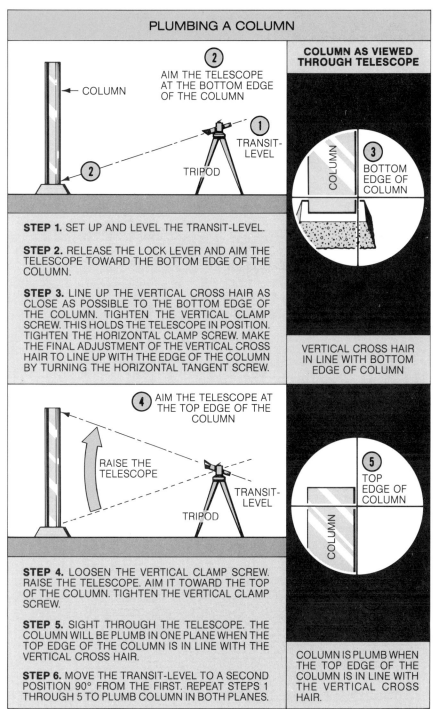

PLUMBING A COLUMN

COLUMN AS VIEWED THROUGH TELESCOPE

STEP 1. SET UP AND LEVEL THE TRANSIT-LEVEL.

STEP 2. RELEASE THE LOCK LEVER AND AIM THE TELESCOPE TOWARD THE BOTTOM EDGE OF THE COLUMN.

STEP 3. LINE UP THE VERTICAL CROSS HAIR AS CLOSE AS POSSIBLE TO THE BOTTOM EDGE OF THE COLUMN. TIGHTEN THE VERTICAL CLAMP SCREW. THIS HOLDS THE TELESCOPE IN POSITION. TIGHTEN THE HORIZONTAL CLAMP SCREW. MAKE THE FINAL ADJUSTMENT OF THE VERTICAL CROSS HAIR TO LINE UP WITH THE EDGE OF THE COLUMN BY TURNING THE HORIZONTAL TANGENT SCREW.

VERTICAL CROSS HAIR IN LINE WITH BOTTOM EDGE OF COLUMN

STEP 4. LOOSEN THE VERTICAL CLAMP SCREW. RAISE THE TELESCOPE. AIM IT TOWARD THE TOP OF THE COLUMN. TIGHTEN THE VERTICAL CLAMP SCREW.

STEP 5. SIGHT THROUGH THE TELESCOPE. THE COLUMN WILL BE PLUMB IN ONE PLANE WHEN THE TOP EDGE OF THE COLUMN IS IN LINE WITH THE VERTICAL CROSS HAIR.

STEP 6. MOVE THE TRANSIT-LEVEL TO A SECOND POSITION 90° FROM THE FIRST. REPEAT STEPS 1 THROUGH 5 TO PLUMB COLUMN IN BOTH PLANES.

COLUMN IS PLUMB WHEN THE TOP EDGE OF THE COLUMN IS IN LINE WITH THE VERTICAL CROSS HAIR.

Figure 33–35. Plumbing a column with a transit-level. If possible, the distance between the transit-level and the object being plumbed should be greater than the height of the object.

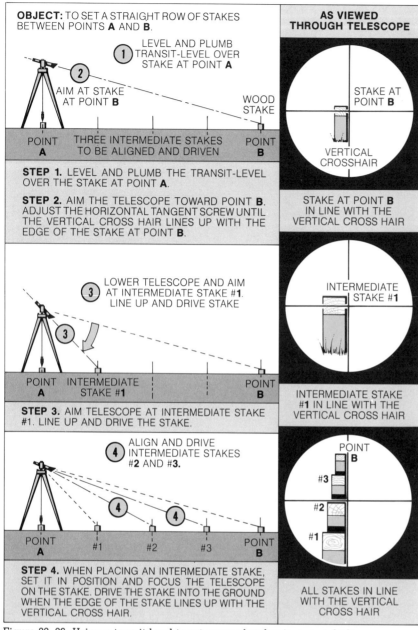

**OBJECT:** TO SET A STRAIGHT ROW OF STAKES BETWEEN POINTS **A** AND **B**.

① LEVEL AND PLUMB TRANSIT-LEVEL OVER STAKE AT POINT **A**

② AIM AT STAKE AT POINT **B**

WOOD STAKE

| POINT **A** | THREE INTERMEDIATE STAKES TO BE ALIGNED AND DRIVEN | POINT **B** |

**AS VIEWED THROUGH TELESCOPE**

STAKE AT POINT **B**

VERTICAL CROSSHAIR

**STEP 1.** LEVEL AND PLUMB THE TRANSIT-LEVEL OVER THE STAKE AT POINT **A**.

**STEP 2.** AIM THE TELESCOPE TOWARD POINT **B**. ADJUST THE HORIZONTAL TANGENT SCREW UNTIL THE VERTICAL CROSS HAIR LINES UP WITH THE EDGE OF THE STAKE AT POINT **B**.

STAKE AT POINT **B** IN LINE WITH THE VERTICAL CROSS HAIR

③ LOWER TELESCOPE AND AIM AT INTERMEDIATE STAKE #1. LINE UP AND DRIVE STAKE

③

| POINT **A** | INTERMEDIATE STAKE #1 | | POINT **B** |

INTERMEDIATE STAKE #1

INTERMEDIATE STAKE #1 IN LINE WITH THE VERTICAL CROSS HAIR

**STEP 3.** AIM TELESCOPE AT INTERMEDIATE STAKE #1. LINE UP AND DRIVE THE STAKE.

④ ALIGN AND DRIVE INTERMEDIATE STAKES #2 AND #3.

④  ④

| POINT **A** | #1 | #2 | #3 | POINT **B** |

POINT **B**

#3

#2

#1

ALL STAKES IN LINE WITH THE VERTICAL CROSS HAIR

**STEP 4.** WHEN PLACING AN INTERMEDIATE STAKE, SET IT IN POSITION AND FOCUS THE TELESCOPE ON THE STAKE. DRIVE THE STAKE INTO THE GROUND WHEN THE EDGE OF THE STAKE LINES UP WITH THE VERTICAL CROSS HAIR.

Figure 33–36. Using a transit-level to set a row of stakes in a straight line.

without spreading the tripod legs. Press the legs firmly into the ground.

5. Do not overtighten the leveling and clamp screws. Overtightening can cause the plate to become pitted and warped over a period of time. A firm adjustment of the screws is adequate.

6. If the instrument must be left standing in rain or heavy dust for an extended period of time, protect it with a waterproof cover.

7. Use the sun shade to help protect the objective lens from dust.

8. Never rub dust or dirt off a lens. Blow the dust or dirt off or use a camel's hair brush.

9. If the leveling screws or other movable parts require cleaning, wipe them with an oiled cloth or brush. Use a light instrument oil.

10. Do not allow unauthorized persons to use the instrument.

11. In order to move the instrument to another location on the job while it is still mounted on the tripod, slightly loosen the leveling screws to take the pressure off the tripod plate. Hold the instrument in front of you with the tripod under your arm.

# UNIT 34

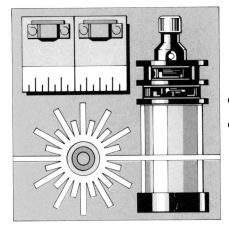

# Laser Transit-Level

An important new tool for construction work is the laser transit-level. It performs most of the functions of the conventional transit-level, but it requires only one person (instead of two) to carry out any layout operation.

## HOW THE LASER TRANSIT-LEVEL WORKS

The barrel of a typical laser model contains a helium-neon, sealed-in tube. When the instrument is turned on, it emits a highly concentrated red beam. The beam is about ⅜″ in diameter. It is directed toward a light-sensitive target (a sensor). See Figure 34–1. Under proper conditions the laser beam is highly accurate. At a range of 500′ the expected accuracy of the beam is within ⅛″. Close readings can be taken up to 1,500′. Strong air disturbances, however, can bend the laser light and lessen its accuracy.

The main parts of the laser transit-level are its rotating head and the leveling screws and vials required to properly adjust the whole unit. See Figure 34–2. A *gimbal* attachment permits mounting the instrument on a tripod or column.

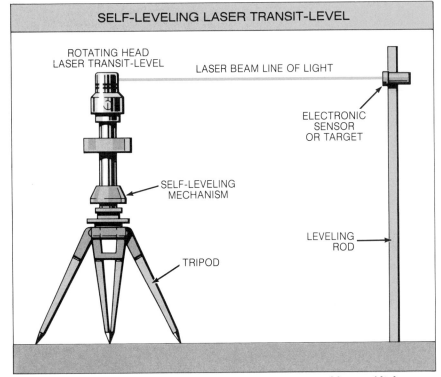

Figure 34–1. A laser transit-level sends out a highly concentrated beam of light that strikes a light-sensitive target.

A laser transit-level is adjusted with leveling screws the same way a conventional transit-level is adjusted. The more refined laser models have a self-leveling mechanism that keeps the instrument at its original setting despite changes in thermal conditions or minor jolts.

## Setting up the Laser Transit-Level

The laser unit must be mounted securely enough that it cannot be easily jarred or shifted. This is done by placing the instrument in a tripod designed for the laser barrel. See Figure 34–3. The in-

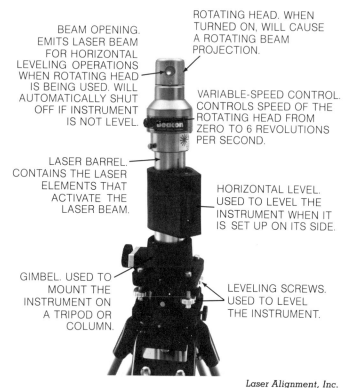

BEAM OPENING. EMITS LASER BEAM FOR HORIZONTAL LEVELING OPERATIONS WHEN ROTATING HEAD IS BEING USED. WILL AUTOMATICALLY SHUT OFF IF INSTRUMENT IS NOT LEVEL.

ROTATING HEAD. WHEN TURNED ON, WILL CAUSE A ROTATING BEAM PROJECTION.

VARIABLE-SPEED CONTROL. CONTROLS SPEED OF THE ROTATING HEAD FROM ZERO TO 6 REVOLUTIONS PER SECOND.

LASER BARREL. CONTAINS THE LASER ELEMENTS THAT ACTIVATE THE LASER BEAM.

HORIZONTAL LEVEL. USED TO LEVEL THE INSTRUMENT WHEN IT IS SET UP ON ITS SIDE.

GIMBEL. USED TO MOUNT THE INSTRUMENT ON A TRIPOD OR COLUMN.

LEVELING SCREWS. USED TO LEVEL THE INSTRUMENT.

*Laser Alignment, Inc.*

Figure 34–2. Main parts of a self-leveling laser transit-level.

*Laser Alignment, Inc.*

Figure 34–4. A laser transit-level can be secured to a stationary column such as a wood post with a column mount. The post has been set up on the construction site where the instrument will be used for grading and leveling operations. The electrical cord is connected to a 12-volt battery.

*Laser Alignment, Inc.*               *Laser Alignment, Inc.*

Figure 34–3. A laser transit-level is set up in a tripod for many operations. Note the electrical cord leading to a power source.

## Power Source and Beam Operation

Most laser transit-levels operate on 120-volt AC electrical current

strument can also be fastened to a stationary column with a column mount. See Figure 34–4.

or a 12-volt DC portable storage battery. After the laser instrument is connected to its power source and turned on, it can be operated in either a stationary or a sweeping position. If it is operated in a stationary position, the beam is aimed at one spot. If it is operated in a sweeping posi-

tion, the laser head can be adjusted to rotate at various speeds up to 360 revolutions per minute (RPM). This gives off a continuous flat plane of light that permits leveling operations at many different points on the job.

## TYPES OF TARGETS FOR THE LASER TRANSIT-LEVEL

The laser transit-level uses various types of targets for different operations. Some are called *sensors*. These devices are usually attached to leveling rods with brackets designed for this purpose. See Figure 34–5. As the sensor bracket is moved up or down, the sensor will light up when it is in a direct, level line with the laser beam. The sensor unit is powered with a rechargeable battery pack.

Various types of *magnetic targets* with graduated lines are also available. See Figure 34–6. Magnetic targets are mainly used for interior operations. The target is raised or lowered until the laser line of light hits the desired graduation on the target.

## OPERATIONS WITH THE LASER TRANSIT-LEVEL

The laser transit-level is highly efficient for establishing grades and leveling over long distances. After the instrument has been set up and leveled, its rotating head is set at its highest speed of 360 RPM. A rod with an attached sensor device is then placed at the point where a grade reading is required. The rotating laser light beam will hit the sensor when it is in line with the light beam. See Figures 34–7 and 34–8.

Highly accurate plumbing of walls and columns is also possible with the laser transit-level. When the rotating head is re-

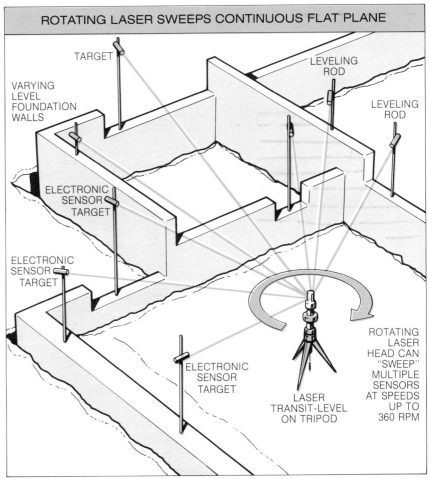

Figure 34–6. Types of graduated metal targets used in laser operations.

STANDARD

HORIZONTAL RANGE

NON-GRID

ALUMINUM GRID

ACCESS FLOOR

moved, and the instrument is set in a plumb vertical position, it will project a plumb vertical line to a target. When used in this way, the laser instrument can be secured in a short tripod (see Figure 34–9) or by a column-mount assembly.

The laser transit-level is used in many types of interior work. For example, it is often used in installing suspended grid ceilings and drywall track partitions. After proper adjustment, the laser beam can be seen by the naked eye and points can be marked where the beam hits the wall, floor, or ceiling. Depending on the operation, the laser unit is set in a horizontal or vertical position.

Figure 34–10 shows a laser transit-level being used to help

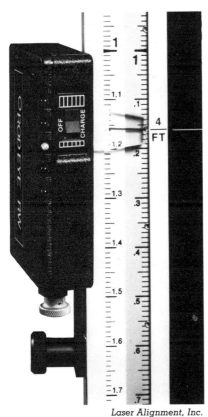

*Laser Alignment, Inc.*

Figure 34–5. Rod eye sensor attached to a leveling rod. This type of device may also be mounted on a wood stick or a metal pole. A light goes on when the sensor reads the exact center of the laser beam.

### ROTATING LASER SWEEPS CONTINUOUS FLAT PLANE

TARGET

LEVELING ROD

LEVELING ROD

VARYING LEVEL FOUNDATION WALLS

ELECTRONIC SENSOR TARGET

ELECTRONIC SENSOR TARGET

ELECTRONIC SENSOR TARGET

ROTATING LASER HEAD CAN "SWEEP" MULTIPLE SENSORS AT SPEEDS UP TO 360 RPM

LASER TRANSIT-LEVEL ON TRIPOD

Figure 34–7. When the head of the laser transit-level is rotating, the laser beam can hit many targets on the job site.

Laser Alignment, Inc.                Laser Alignment, Inc.

Figure 34–8. Establishing grade points on grade stakes (left) and on concrete form panels (right) with a laser transit-level.

Laser Alignment, Inc.

Figure 34–10. Using a laser transit-level (wall-mounted) to level ceiling grids for suspended ceilings.

install ceiling grids for a suspended ceiling. For this operation the instrument is wall-mounted (as shown in Figure 34–10) or mounted on a tall tripod. It is set with the laser beam revolving either at the finish ceiling height or at a reference point. A special target is snapped to the grid, and the grid is moved up or down until the line of light crosses the target's offset mark. The grid is then secured in place.

When leveled and adjusted in a horizontal position (Figure 34–11), a laser transit-level can

Laser Alignment, Inc.

Figure 34–11. When leveled and adjusted in a horizontal position, a laser transit-level can be used to lay out 90° cross lines for walls and partitions and to plumb floor lines to the ceiling.

be used to lay out 90° cross lines for walls and partitions. It can also be used to plumb floor lines to the ceiling.

## LASER HAND LEVEL

The laser hand level can be used as an ordinary carpenter's level or switched on to emit a laser beam. It can be mounted on a tripod or placed on a level surface. The laser hand level is far less expensive than other laser instruments, and accurate enough to establish level and plumb marks for closer indoor and outdoor operations. See Figure 34–12.

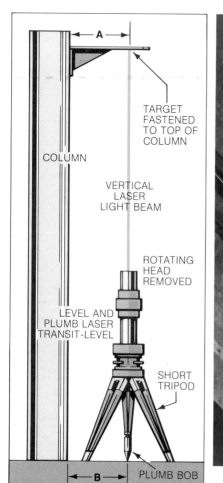

Laser Alignment, Inc.

Figure 34–9. Plumbing operation with a laser transit-level secured in a short tripod. In this example a column is being plumbed. When the rotating head is removed, the barrel shoots a straight beam upward. The instrument is plumbed over a point on the ground (note the plumb bob). The beam is then aimed at a target fastened to the top of the column. When distance A is the same as distance B, the column is plumb.

David White, Inc.

Figure 34–12. A laser hand level can be used for closer operation.

# SECTION 8

# Foundation and Outdoor Slab Construction

The foundation is the base for the entire building structure. It must be designed strong enough to support its own weight and the weight of the rest of the building. This section discusses foundation systems (and outdoor slabs such as sidewalks) used for houses and other types of light construction. Foundations for heavy construction are discussed in Section 16.

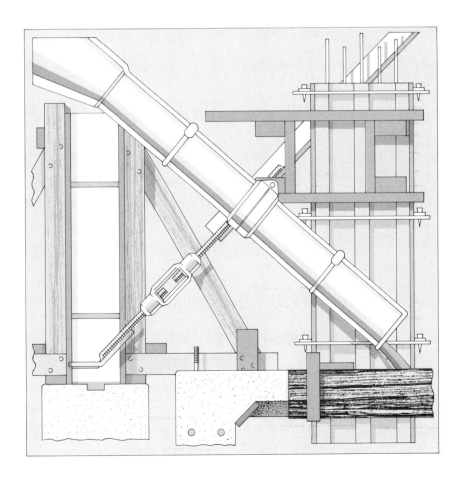

# UNIT 35

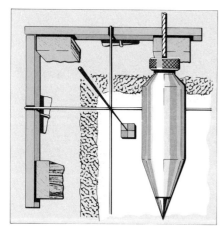

# Building Site and Foundation Layout

## BUILDING SITE

In residential areas where streets have been established, a piece of property is referred to as a *lot*. When a building is to be constructed on a lot the lot becomes a *building site.* Various features of the building site help determine the type of foundation best suited for the building. For example, the shape and size of the lot and the degree to which the ground is sloped must be considered in designing the foundation. Other factors to be considered are the weather conditions of the area and the soil conditions of the lot. The soil conditions of one lot may be different from conditions of others in the same general area.

## Soil Conditions

Since the full load of a building rests on the ground below it, the condition of the soil is important to the foundation of the building. Most buildings are constructed on soils classified as *sand, silt, or clay.* The main difference among the three is the size of the grains (particles) that make up the soil. Clay is composed of the smallest particles. Sand is composed of the largest. Silt particles are larger than clay but smaller than sand.

Because sand particles are larger, they will press together (compress) less than clay particles when subjected to heavy pressure. See Figure 35–1. Consequently, there will be less downward movement (settlement) of a building built on sand than of one built on clay.

A certain amount of settlement can be expected with any newly constructed building, unless it has been built on solid rock *(bedrock).* No problems arise if the settlement is minimal and takes place evenly. A large amount of uneven settlement, however, can cause cracks in the foundation and structural damage to the rest of the building.

In drawing the plans for a foundation, an architect must consider the soil conditions of the building site. Sometimes the architect consults with a soil engineer, who makes test bores and analyzes samples of soil taken from the lot.

## Earthquake and Weather Conditions

In some areas of the United States the danger of earthquakes is high *(seismic risk zones), so* a foundation must be designed to withstand greater stress. Reinforcing steel bars are normally required in all concrete or masonry foundation walls constructed in these zones.

A weather condition that affects foundations is the *frost line.* In colder regions, moisture from

Figure 35–1. Sand grains are larger than clay grains and less space exists between them. Therefore, sandy soil compresses less than clay soil.

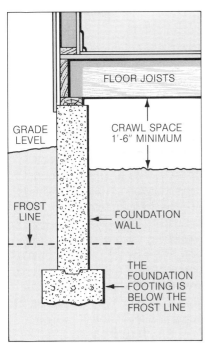

Figure 35–2. The bottom of a footing must be below the frost line.

rain and snow penetrates the ground and freezes during the winter season. The frost line is the depth to which the soil freezes.

The freezing and thawing actions of soil cause it to expand and recede. If the foundation footing is placed below the frost line, there will be no movement of the foundation when these actions take place. See Figure 35–2. The local building code usually specifies how deep the foundation footings should be placed below the prevailing frost line in the area.

## LAYING OUT FOUNDATION WALLS

Before construction of a foundation can begin, the precise boundaries of the building site must be verified by means of a *lot survey*. Next, lines must be set up to show the exact location of the building. The information required to lay out the foundation walls is in the plot plan of the blueprints. The plot plan not only shows the location of the building on the lot but also gives different grade elevations indicating the high and low points of the ground. See Figure 35–3.

## Establishing Lot Boundaries by the Lot Survey

Lots are usually mapped, and the maps are recorded by the local building or zoning authorities. The lot surveyor studies the zoning maps and records to deter-

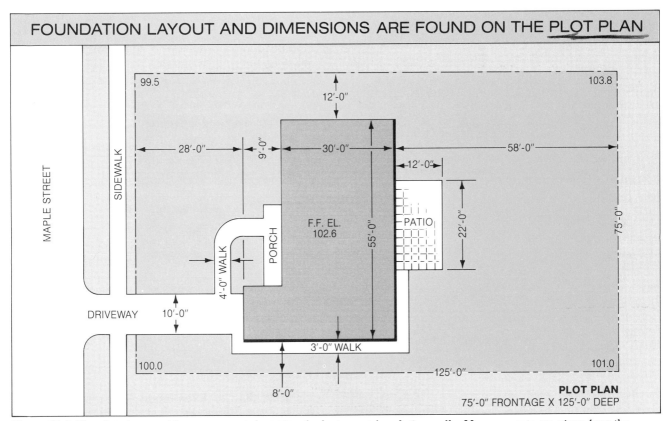

## FOUNDATION LAYOUT AND DIMENSIONS ARE FOUND ON THE PLOT PLAN

Figure 35–3. The plot plan provides necessary information for laying out foundation walls. Measurements are given from the property lines to the sides of the building. Also, different grade elevations are given to indicate the high and low points of the ground.

mine the precise boundaries of the lot. By measuring from the proper reference points, the surveyor can establish the two front corners of the lot. Often street curbs are used as reference points. In some areas of the United States however, the front corners are established by measuring from the center of the road or from special monuments placed in the sidewalks. A transit-level is usually used to establish the two rear corners of the lot.

**Corner Stakes and Property Points.** The surveyor must place some kind of marker at the lot corners. Often a 2″ × 2″ wood stake, sometimes called a *hub,* is used. It is driven in so that the top of the stake is flush with the surface of the ground. A pipe with a lead plug may also be used for this purpose. A small nail is driven into the top of the stake or pipe to identify the exact corner of the property. This is called a *property point* or *survey point.* The property lines drawn on the plot plan represent lines extending from the four corners of the lot.

## Locating the Building on the Lot

Local zoning regulations or construction codes give the minimum *setbacks* permitted for buildings. The setback is the distance between the building and the property lines. The setbacks allowed vary from one city or county to another. Often they vary among different sections of the same city.

A common setback for buildings located in a single family residential zone is: 20′ minimum from the front property line to the front of the house, and 4′ minimum from the side property lines to the side of the house. In contrast, buildings in commercial zones are sometimes built within inches of the property lines.

**Establishing Building Corners.** The most efficient way to establish building corners is to use the transit-level. Figure 35–4 shows this procedure for a particular building. An older procedure for establishing building corners, without a transit-level, is to use the 3-4-5 method of determining right corners. In this method one stake is driven in line 3′ from the corner stake and another stake is driven in line 4′ from the corner stake on the adjoining wall. The diagonal distance between these two stakes should be exactly 5′. A greater degree of accuracy can be maintained by doubling the distances that the auxiliary stakes are driven from the corner stake. For example, the first stake is driven 6′ from the corner stake, and the second stake is driven 8′ from the corner stake on the adjacent wall. The diagonal distance across these two stakes is 10′.

## Building Lines and Batter Boards

After the building corners have been established, building lines (usually nylon string) must be set up to mark the boundaries of the building. The forms for the foundation walls will be set to these building lines. Since it is not practical to attach the strings to the low stakes that mark the building corners, *batter boards* are constructed to hold the building lines while the form work is in progress. See Figure 35–5. These batter boards are level

ledger boards nailed to stakes driven into the ground. When forms are being constructed for foundations on fairly level lots, the batter boards at all corners of the building should be level to each other.

Batter boards are usually placed 4′ to 6′ behind each building corner. This distance is necessary to provide working room between the batter boards and the form construction.

Building lines are set up by stretching each one tightly from one batter board to the opposite batter board. The line is positioned directly over the building corner by means of a straightedge and level or with a plumb bob. See Figure 35–6.

After the lines are secured, carpenters should check to make sure the measurements between the lines are accurate. Also, the diagonals across the lines should be measured to make sure they are square. Any errors in layout can be easily corrected at this time. A foundation built out of square, or to a wrong dimension, will create many problems for the construction that follows.

## Ground Work

Before forms for the foundation can be built, preliminary ground work must be completed. Excavation may be needed, as well as finish grading. Trenches may have to be dug for the footings. Most of the ground work is done by earth-moving machinery run by operating engineers. See Figure 35–7. Carpenters often work with the operating engineer by setting up lines to guide the excavation and trenching operations.

**Excavation.** An excavation is the removal of a large quantity of earth at the construction site. It might involve leveling a slope or

# USING A TRANSIT-LEVEL TO LAY OUT BUILDING LINES

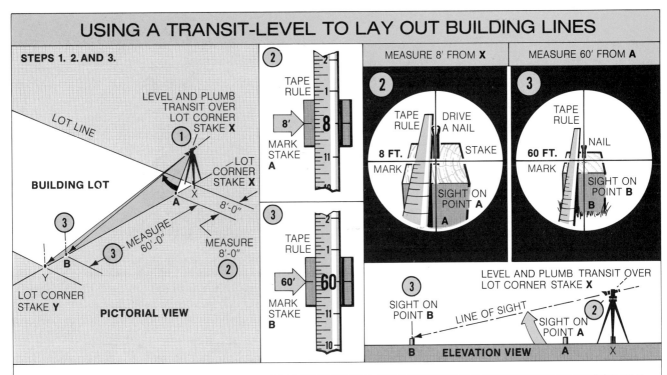

**STEP 1.** LEVEL AND PLUMB THE TRANSIT OVER LOT CORNER STAKE **X**. SIGHT DOWN TO THE OPPOSITE LOT CORNER STAKE **Y**.

**STEP 2.** THE DISTANCE FROM THE PROPERTY LINE TO ONE SIDE OF THE BUILDING IS 8'-0". DRIVE A STAKE 8'-0" FROM CORNER STAKE **X** THAT WILL BE IN LINE WITH LOT CORNER STAKES **X** AND **Y**. THIS CAN BE DONE BY LOWERING THE TELESCOPE UNTIL THE VERTICAL AND HORIZONTAL CROSSHAIRS ARE CLOSE TO THE CENTER OF THE STAKE TOP AT THE 8'-0" MEASUREMENT. USING THE FINE ADJUSTMENT TANGENT SCREWS OF THE TRANSIT, SIGHT THROUGH THE TRANSIT AND DRIVE A NAIL INTO THE STAKE TOP EXACTLY IN LINE WITH THE VERTICAL AND HORIZONTAL CROSSHAIRS AT THE 8'-0" READING OF THE TAPE.

**STEP 3.** MEASURE 60'-0" FROM POINT **A**, WHICH IS THE WIDTH OF THE BUILDING. DRIVE ANOTHER STAKE THAT IS IN LINE WITH LOT CORNER STAKES **X**, **Y** AND POINT **A**. HOLDING THE TAPE RULE IN POSITION OVER THIS STAKE, RAISE THE TELESCOPE UNTIL THE HORIZONTAL CROSSHAIR COINCIDES WITH THE 60'-0" MARK ON THE TAPE. NOW USING THE VERTICAL CROSSHAIR, ALIGN AND DRIVE A NAIL INTO THE TOP OF THE STAKE, ESTABLISHING POINT **B**.

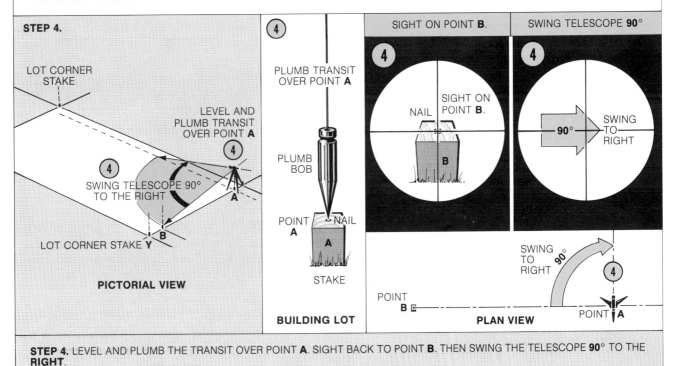

**STEP 4.** LEVEL AND PLUMB THE TRANSIT OVER POINT **A**. SIGHT BACK TO POINT **B**. THEN SWING THE TELESCOPE **90°** TO THE **RIGHT**.

Figure 35–4. Establishing building corners with a transit-level (continued next page).

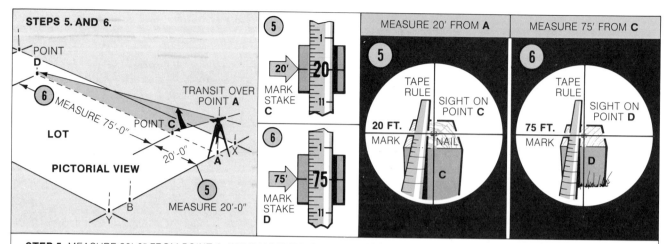

**STEPS 5. AND 6.**

**STEP 5.** MEASURE 20'-0" FROM POINT **A**, WHICH IS THE DISTANCE FROM THE FRONT PROPERTY LINE TO THE FRONT OF THE BUILDING. LOWER THE TELESCOPE UNTIL THE HORIZONTAL CROSSHAIR COINCIDES WITH THE 20'-0" MARK ON THE TAPE. DRIVE A STAKE AND NAIL, ESTABLISHING POINT **C**, WHICH WILL BE THE FIRST CORNER OF THE BUILDING.

**STEP 6.** MEASURE 75'-0" FROM POINT **C**, WHICH IS THE LENGTH OF THE BUILDING. RAISE THE TELESCOPE UNTIL THE HORIZONTAL CROSSHAIR COINCIDES WITH THE 75'-0" MARK ON THE TAPE. DRIVE A STAKE AND NAIL, ESTABLISHING POINT **D**, WHICH WILL BE THE SECOND CORNER OF THE BUILDING.

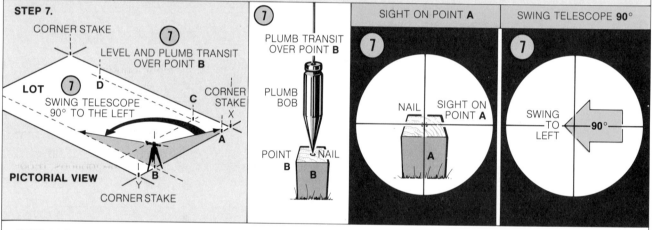

**STEP 7.**

**STEP 7.** LEVEL AND PLUMB THE TRANSIT OVER POINT **B**. SIGHT BACK TO POINT **A**, THEN SWING THE TELESCOPE **90°** TO THE LEFT.

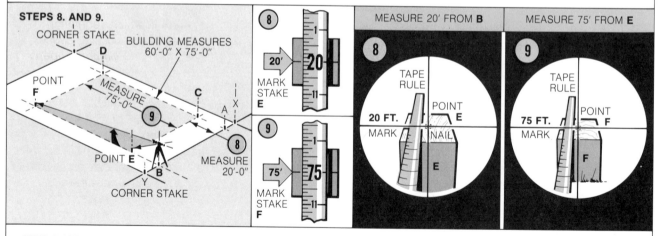

**STEPS 8. AND 9.**

**STEP 8.** MEASURE 20'-0" FROM POINT **B**. LOWER THE TELESCOPE UNTIL THE HORIZONTAL CROSSHAIR COINCIDES WITH THE 20'-0" MARK ON THE TAPE. DRIVE A STAKE AND NAIL, ESTABLISHING POINT **E**, WHICH WILL BE THE THIRD CORNER OF THE BUILDING.

**STEP 9.** MEASURE 75'-0" FROM POINT **E**. RAISE THE TELESCOPE UNTIL THE HORIZONTAL CROSSHAIR COINCIDES WITH THE 75'-0" MARK ON THE TAPE. DRIVE A STAKE AND NAIL, ESTABLISHING POINT **F**, WHICH WILL BE THE FOURTH CORNER OF THE BUILDING.

Figure 35-4. Continued from previous page.

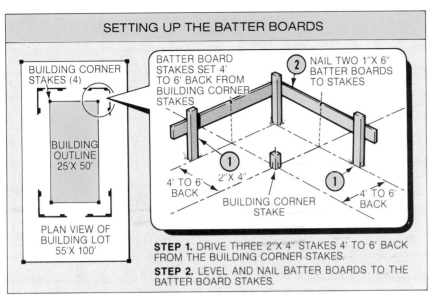

## SETTING UP THE BATTER BOARDS

BUILDING CORNER STAKES (4)

BUILDING OUTLINE 25'X 50'

PLAN VIEW OF BUILDING LOT 55'X 100'

BATTER BOARD STAKES SET 4' TO 6' BACK FROM BUILDING CORNER STAKES

② NAIL TWO 1"X 6" BATTER BOARDS TO STAKES

2"X 4"

4' TO 6' BACK

BUILDING CORNER STAKE

4' TO 6' BACK

**STEP 1.** DRIVE THREE 2"X 4" STAKES 4' TO 6' BACK FROM THE BUILDING CORNER STAKES.

**STEP 2.** LEVEL AND NAIL BATTER BOARDS TO THE BATTER BOARD STAKES.

Figure 35–5. Constructing batter boards to hold building lines in place to mark boundaries of foundation.

*Deere & Company*
Figure 35–7. Most ground work is done by earth-moving machinery such as this bulldozer run by an operating engineer. The boom seen in the background is part of a mobile crane also run by an operating engineer.

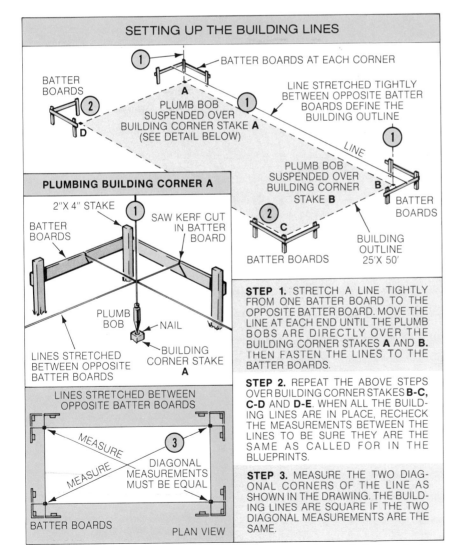

## SETTING UP THE BUILDING LINES

BATTER BOARDS AT EACH CORNER

BATTER BOARDS

A

PLUMB BOB SUSPENDED OVER BUILDING CORNER STAKE **A** (SEE DETAIL BELOW)

D

LINE STRETCHED TIGHTLY BETWEEN OPPOSITE BATTER BOARDS DEFINE THE BUILDING OUTLINE

LINE

PLUMB BOB SUSPENDED OVER BUILDING CORNER STAKE **B**

B

BATTER BOARDS

C

BUILDING OUTLINE 25'X 50'

BATTER BOARDS

### PLUMBING BUILDING CORNER A

2"X 4" STAKE

BATTER BOARDS

SAW KERF CUT IN BATTER BOARD

PLUMB BOB

NAIL

LINES STRETCHED BETWEEN OPPOSITE BATTER BOARDS

BUILDING CORNER STAKE **A**

LINES STRETCHED BETWEEN OPPOSITE BATTER BOARDS

MEASURE

MEASURE

③

DIAGONAL MEASUREMENTS MUST BE EQUAL

BATTER BOARDS

PLAN VIEW

**STEP 1.** STRETCH A LINE TIGHTLY FROM ONE BATTER BOARD TO THE OPPOSITE BATTER BOARD. MOVE THE LINE AT EACH END UNTIL THE PLUMB BOBS ARE DIRECTLY OVER THE BUILDING CORNER STAKES **A** AND **B**. THEN FASTEN THE LINES TO THE BATTER BOARDS.

**STEP 2.** REPEAT THE ABOVE STEPS OVER BUILDING CORNER STAKES **B-C**, **C-D** AND **D-E**. WHEN ALL THE BUILDING LINES ARE IN PLACE, RECHECK THE MEASUREMENTS BETWEEN THE LINES TO BE SURE THEY ARE THE SAME AS CALLED FOR IN THE BLUEPRINTS.

**STEP 3.** MEASURE THE TWO DIAGONAL CORNERS OF THE LINE AS SHOWN IN THE DRAWING. THE BUILDING LINES ARE SQUARE IF THE TWO DIAGONAL MEASUREMENTS ARE THE SAME.

Figure 35–6. Setting up building lines (usually nylon string) to mark the boundaries of a building.

digging the ground for a full-basement foundation.

**Trenching.** Fairly level ground beneath a crawl-space foundation generally requires little excavation but will probably require *trenching* for the footings. (Footings are discussed in the next unit.) Trenches must be dug to the depth called for by the local building code. Usually the depth required means removing all the *topsoil* (the softer, surface layer of earth in which vegetation grows). Some building codes, such as the *Minimum Property Standards,* in the volume entitled *One and Two Family Dwellings* (published by the U.S. Department of Housing and Urban Development), require that the trench extend an additional 6" deeper, into natural, undisturbed soil. In areas where the ground will freeze, the bottom of the trench must be below the frost line.

The depth of the trenches is usually given in the section views of the foundation plan. Trenches must be wide enough to allow room for the construction of the

Figure 35–8. A backhoe is used to dig trenches for footings.

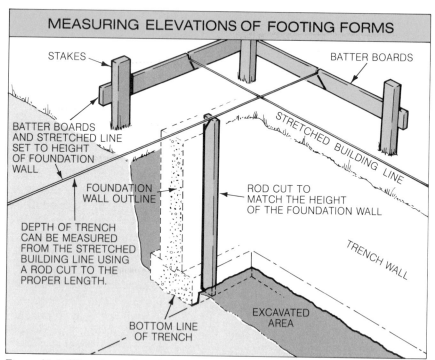

MEASURING ELEVATIONS OF FOOTING FORMS

STAKES

BATTER BOARDS

BATTER BOARDS AND STRETCHED LINE SET TO HEIGHT OF FOUNDATION WALL

FOUNDATION WALL OUTLINE

ROD CUT TO MATCH THE HEIGHT OF THE FOUNDATION WALL

STRETCHED BUILDING LINE

DEPTH OF TRENCH CAN BE MEASURED FROM THE STRETCHED BUILDING LINE USING A ROD CUT TO THE PROPER LENGTH.

TRENCH WALL

BOTTOM LINE OF TRENCH

EXCAVATED AREA

Figure 35–9. When batter boards have been set to the height of the foundation walls, the depth of the trench can be measured from the building line.

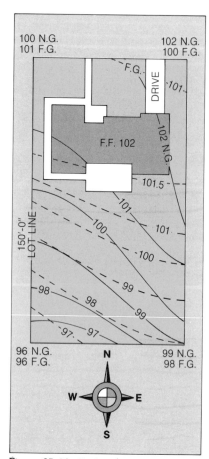

Figure 35–10. A plot plan shows natural and finish grade levels. In this plan, the lot is to be graded so that it slopes toward the southwest and northeast corners. Soil will be added at the northwest corner and removed from the southeast and northeast corners. The southwest corner will remain the same.

foundation forms. The bottom of the trenches must be level. Most lots have some slope; therefore, the starting point from which the trenches are leveled should be at the lowest point of the excavation.

A trench can be dug with a pick and shovel, but it is more convenient to use heavy equip-ment such as a backhoe. See Figure 35–8. As the digging proceeds, the depths are checked with a builder's level and rod. If the batter boards have been set close to the height of the foundation walls, and the trenches are being dug by hand, a line can be stretched and the depth measured from the line. A rod cut to

the required measurement is more convenient to use than a tape rule. See Figure 35–9. Trenches for stepped footings must also be level.

**Grading.** Many lots require *grading*, which is reshaping the surface of the lot. Soil is removed from some sections and added to others. Lots are graded so that surface water (rain or melting snow) will flow away from the building. The recommended minimum slope is 6" in 10', or 5%. However, if the area surrounding the foundation is to be paved, a 1% slope (1/8" in 1') is adequate.

The plot plan in Figure 35–10

Figure 35-11. Lot grades are checked with a builder's level and leveling rod.

Figure 35-12. Backfill is placed against completed foundation walls. Gravel is often used instead of soil, since it allows better water drainage away from the building.

shows *natural grades* (NG) at all four corners of the lot and *finish grades* (FG) at three corners. The natural grade is the level of the ground before the work begins. The finish grade is the level of the ground after soil has been added or taken away. Soil on the lot is removed or added to match the finish grades on the plot plan.

The plot plan also shows *contour lines.* A contour line shows the shape produced by the varying grades of the lot. The solid lines represent the natural contour (shape) of the ground. The dashed lines show the finished contour. Lot grades can be checked with a builder's level and leveling rod, as discussed in Section 7. See Figure 35-11.

**Backfilling.** Backfilling is done after the forms have been removed and work such as waterproofing and placing drain tiles has been completed. The soil used for backfilling should be free of wood scraps and any other type of waste material. It is placed carefully against the foundation walls and compacted (pressed down). See Figure 35-12. On many jobs, gravel is required for backfill since it allows better water drainage away from the building.

# UNIT 36

# Types of Foundations

## WALLS AND FOOTINGS

Most foundation designs consist of walls and footings. Footings rest directly on the soil and act as a base for the walls. In most foundation designs, they serve to spread the weight of the building over a wider soil area.

The most visible parts of a foundation are the walls around the perimeter (outside) of the building. Additional foundation walls are often located under interior sections of the building to help support loads from the framed units above. Concrete *piers* (square, round, or battered structures that support posts or girders) are also frequently used for this purpose.

Common designs are the *T-shaped, (technically, an inverted T), L-shaped, battered,* and *rectangular* foundations. See Figure 36–1. The T-shaped foundation has walls with spread footings. The L-shaped foundation also has walls with spread footings, but the footings extend from only one side rather than both sides of the wall. The battered foundation has walls that are wider at their base. The rectangular foundation, designed for light wall loads and firm soil conditions, has walls that are not enlarged at their base.

The dimensions of walls and footings are normally provided in the *section views* of the foundation plan. This information may also appear in the type of "Thru-house Section" drawing discussed in Section 6. A formula for determining the dimensions of foundation footings to be constructed for normal soil conditions is given in Figure 36–2.

Variously shaped concrete footings, often called *piers*, are used as pedestals to support and anchor the bottoms of steel columns and wood posts. These

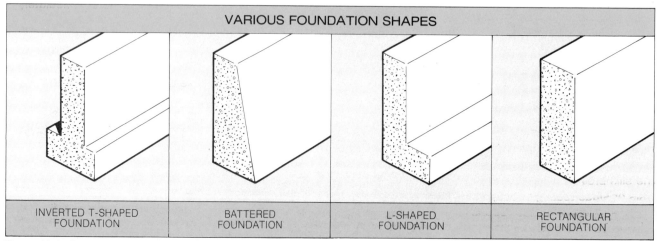

**VARIOUS FOUNDATION SHAPES**

| INVERTED T-SHAPED FOUNDATION | BATTERED FOUNDATION | L-SHAPED FOUNDATION | RECTANGULAR FOUNDATION |

Figure 36–1. Common foundation designs. The rectangular foundation can be used only for light wall loads and only in firm soil.

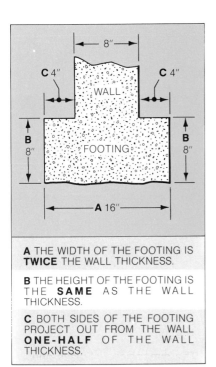

A THE WIDTH OF THE FOOTING IS **TWICE** THE WALL THICKNESS.

B THE HEIGHT OF THE FOOTING IS THE **SAME** AS THE WALL THICKNESS.

C BOTH SIDES OF THE FOOTING PROJECT OUT FROM THE WALL **ONE-HALF** OF THE WALL THICKNESS.

Figure 36–2. Calculating the dimensions of a footing (for normal soil conditions).

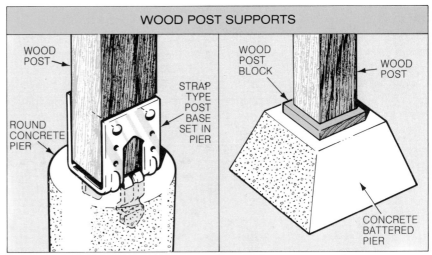

Figure 36–3. Round or battered piers are often used to support wood posts.

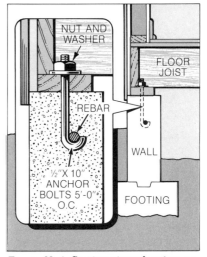

Figure 36–4. Section view showing anchor bolts (J-bolts) used to fasten sill plates to foundation walls. In this example, the floor joists are nailed to the sill plates.

columns and posts support beams or girders that are part of the framing system of a building. A square footing is usually placed beneath a steel column. (A "Thru-house Section" drawing in Section 6 shows this type of footing used with a steel column.) Footings of other shapes, such as round or battered concrete piers, are often placed beneath wood posts. See Figure 36–3.

Form construction for various foundation designs is discussed in Unit 39.

## FOUNDATION SILLS

Wood sill plates, often called *mudsills,* are usually fastened to the tops of a foundation wall. The sills provide nailing for the joists or studs resting directly over the foundation. Redwood is highly recommended for foundation sills because of its superior resistance to decay and insect attack. Other kinds of lumber

may be used if they have been treated with a preservative by a pressure process (see Section 2). In residential and other light construction, foundation sills are usually 2 × 4s or 2 × 6s. Heavier construction may require 4 × 6s or larger pieces.

## Fastening with Anchor Bolts

An old but very effective method of fastening sill plates is to use anchor bolts, also called *J-bolts* because of their shape. See Figure 36–4. Anchor bolts should have a diameter of ½" or more. Information regarding the bolts is usually found in the section views of the foundation plan.

The Uniform Building Code (UBC) states that the bent end of the anchor bolt should be embedded a minimum of 7" into the concrete. In unreinforced masonry it should penetrate 15". The UBC further states that the anchor bolts must be spaced no farther apart than 6' O.C. and that there must be a bolt within 12" of the ends of any piece. A short section of sill must have at least two anchor bolts, spaced no more than 6' apart.

Sills are sometimes secured to

the foundation walls immediately after concrete is placed in the wall forms. See Figure 36–5. The pieces are prepared and cut to length ahead of time, and the holes are drilled. The anchor bolts are placed in the holes, and the nuts are started on the threads. Washers should always be used under the nuts. When the concrete poured into the forms reaches the required height, the sill pieces are pressed down into the soft concrete. At the same time, the bolts are tapped down, using a wood block to prevent damage to the threads. The sills are then leveled and cleats are nailed across the form to hold them in place. When the concrete has hardened for several days, the nuts can be tightened.

Another way to fasten sills is to set the bolts in position before the concrete pour. An advantage of this method is that the bolts can be hooked beneath the horizontal reinforcing bars (*rebars*), giving greater holding power. (See Unit 37 for an explanation of reinforcing bars.) The bolts are placed in templates. This allows the bolts to remain in a plumb, vertical position while the concrete is being poured. It also keeps them extending the proper amount above the concrete. See Figure 36–6.

After the concrete has set up, holes are laid out and drilled in the sill pieces that will be slipped over the bolts. Often a layer of a mixture of sand, cement, and water, called *grout,* is put down to provide a level and even base for the sills. The nuts can be tightened on the anchor bolts after the grout has hardened.

### Fastening to Concrete Slabs.

When anchor bolts are used to fasten sills to concrete slabs, they are set in the concrete at the time of the pour. The sill is placed and bolted down after the concrete sets up. Sills for interior framed walls are often "pinned" down with a powder-actuated fastener.

## Fastening with Studs or Anchor Clips

Sometimes sill plates are attached with powder-driven studs. In this case, the word *stud* refers to a special kind of concrete nail driven by the powder-actuated tool described in Section 4.

Today several manufacturers offer a new, strap type of anchor clip. See Figure 36–7. The anchor clips are embedded in the concrete wall or in the mortar joints of a masonry wall.

## FOUNDATION SYSTEMS

The three main types of foundation systems are the *crawl-space* foundation, the *full-basement* foundation, and the *slab-at-grade* foundation. The major difference between the crawl-space and full-basement types is the height of the walls. The slab-at-

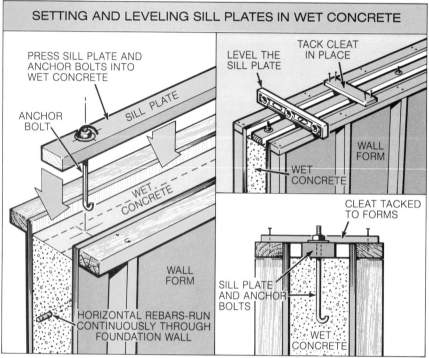

Figure 36–5. Sill plates are sometimes placed immediately after the concrete pour. Freshly poured concrete meets the required level in the wall form. In this example, the walls are made of panels reinforced by studs and a top plate.

*Portland Cement Association*

Figure 36–6. Anchor bolts extending from the top of a foundation wall.

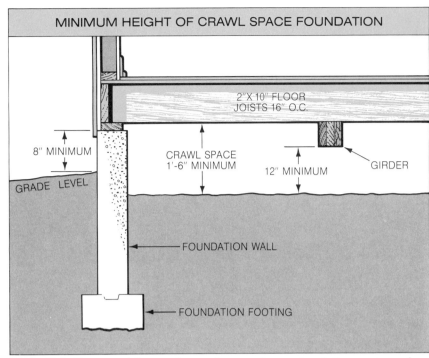

Figure 36–8. The area of the crawl space in a crawl-space foundation must comply with local building code requirements.

The Panel-Clip Company

Figure 36–7. An anchor clip is sometimes used to fasten sill plates to a foundation wall. The clip is embedded in solid concrete or in the mortar joints of a concrete-block wall.

grade foundation features a concrete slab floor instead of a framed floor unit.

## Crawl-space Foundation

A crawl space is a narrow space between the bottom of the floor unit and the ground. See Figure 36–8. Crawl-space foundations are also known as *basement-less* foundations. Local building codes usually give a minimum crawl-space distance for buildings in the area. A common minimum distance is 18″ from the bottom of the floor joists to the ground, and 12″ below the girders that support the joists. Crawl-space foundations are often constructed with 2′ clearance beneath the joists to provide easier access for plumbing, electrical, or furnace repairs.

## Full-basement Foundation

In a full-basement foundation, the foundation wall also serves as a basement wall. See Figure 36–9. The top of the wall should extend at least 8″ above the finish grade. The walls are usually 7′ to 8′ from the floor slab to the ceiling. The concrete walls must withstand the lateral pressure of the soil. The walls must also be fully waterproof. (Another type of full-basement design is described in Section 9.)

**Areaways.** When basement windows are located below the finish-grade level, areaways are required. Areaway walls are constructed of poured concrete, hollow concrete blocks, brick, or even sheet metal. Areaways must project above the finish grade and below the bottom of the window.

## Slab-at-Grade Foundation

In a slab-at-grade foundation, the foundation walls are combined with a concrete floor slab. The surface of the slab is at the same level as the tops of the walls. In this type of foundation,

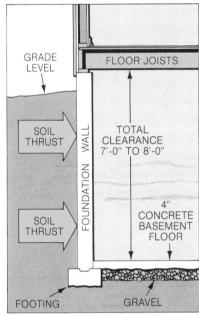

Figure 36–9. A full-basement foundation must have adequate room clearance between the basement floor and ceiling.

the floor slab receives its main support from the ground. A slab-at-grade foundation is not practical over steeply sloped lots or where the water table is near the

Figure 36–10. A stepped foundation is used on a steeply sloped lot.

side lots. See Figure 36–13. Grade beams are also used on level lots where unstable soil conditions exist.

## HOLLOW-BLOCK MASONRY FOUNDATION WALLS

In many areas of the country hollow concrete blocks are frequently used to build foundation walls. This work is normally performed by skilled workers called *masons.* See Figure 36–14. The walls rest on solid concrete footings. Forms for the concrete footings are laid out and placed by carpenters. Carpenters also plumb and brace any window or door frames in the hollow-block masonry wall. See Figure 36–15.

Many shapes and sizes are used for hollow concrete blocks. They are usually 7⅝″ high and 15⅝″ long. The thickness of the block varies according to the thickness required for the wall. See Figure 36–16. Since hollow blocks are lightweight and re-

ground surface.

Slab-at-grade floors can be poured at the same time as the foundation walls, or they can be poured separately and then used together with grade beams as well as with T-shaped, rectangular, or battered foundation walls.

another. See Figure 36–11.

**Grade Beams.**  Grade beams are foundation walls that receive their main support from piers extending deep into the ground. See Figure 36–12. They are often used with stepped or ramped foundations erected on steep hill-

### Foundations for Sloped Lots

A *stepped* (shaped like a series of steps) foundation is used on a steeply sloped lot. See Figure 36–10. A level foundation for such a lot would require more concrete and more excavation of the hillside than a stepped foundation. A stepped foundation may provide both a crawl space and a full basement.

Stepped foundations require vertical (plumb) and horizontal (level) footings. Many building codes require that the distance between one horizontal step and another be no less than 2′. The vertical footing must be at least 6″ thick and no higher than three-quarters of the distance between one horizontal step and

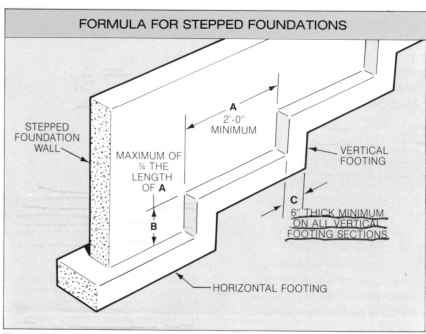

**FORMULA FOR STEPPED FOUNDATIONS**

STEPPED FOUNDATION WALL

A
2′-0″ MINIMUM

VERTICAL FOOTING

MAXIMUM OF ¾ THE LENGTH OF A

B

C
6″ THICK MINIMUM ON ALL VERTICAL FOOTING SECTIONS

HORIZONTAL FOOTING

Figure 36–11. Many building codes specify the following formula for stepped foundations. The distance between one horizontal step and another (A) must be no less than 2′. The vertical footing (B) must be no higher than three-quarters of the distance between one horizontal step and another. The vertical footing must be at least 6″ thick (C).

quire less labor than poured concrete, hollow-block construction is often less costly than poured concrete.

## WOOD FOUNDATION

In some sections of the United States and Canada the use of a wood foundation in residential construction is increasing. The *All-Weather Wood Foundation System* is responsible for this trend. This system was researched and developed through a joint effort by the National Forests Product Association and the American Wood Preservers Institute. It has since been approved by the U.S. Department of Housing and Urban Development (HUD) and a number of national and regional building code agencies.

A major advantage of the All-Weather Wood Foundation system is that, as its name implies, it can be installed in any kind of weather. In contrast, construction of poured concrete and hollow-block foundations may be delayed by certain weather conditions.

The wood foundation system can be used to construct a full-basement or a crawl-space foundation. It is basically a stud wall with a top and bottom plate. The studs are normally 2 × 4s placed 16″ O.C. (on centers). Under conditions where greater soil pressures exist, 2 × 6s are placed 12″ O.C. Plywood sheathing is nailed to the outside. The entire wall rests on a 2″ thick wood footing. See Figure 36–17.

All wood used for the wall must be pressure-preservative treated. The moisture content of the material must be no more than 18% for the plywood and 19% for the framing lumber. Nails and other types of fasteners used must be corrosion-resistant.

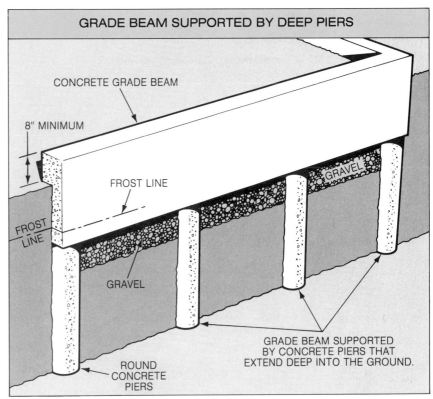

Figure 36–12. A grade beam is supported by piers that extend deep into the ground.

### Preparing the Site for a Wood Foundation

Footings in a wood foundation must rest on firm soil and be below the frost line, as required for concrete footings. For a crawl-space wood foundation, a bed of gravel is placed where the footings will rest. The thickness of

Figure 36–13. A ramped grade-beam foundation wall follows the slope of the hill.

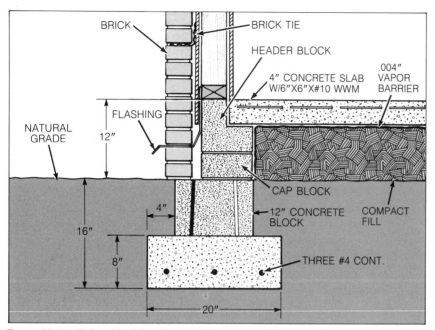

Figure 36-14. Hollow-block foundation walls rest on solid concrete footings.

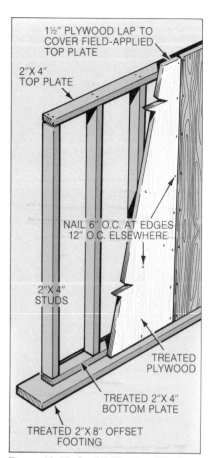

Figure 36-17. An All-Weather Wood Foundation wall is basically a stud wall with a top and bottom plate. Plywood sheathing is nailed to the outside, and the wall rests on 2" thick wood footing.

Figure 36-15. Carpenters plumb and brace door frames in masonry foundation walls.

*Kaiser Cement Corporation*

Figure 36-16. A typical modular hollow concrete block is 7⅝" high and 15⅝" long.

36-18. The gravel bed helps support the floor and provides proper drainage of any water that collects below the slab.

## Constructing a Wood Foundation

Wood foundation walls are built in sections. These sections may be constructed in a shop and delivered to the job, or they may be made up on the construction site. First the footing plates are placed on the gravel. Then the panels are set in their proper positions, aligned with a string, fas-

*American Plywood Association*

Figure 36-18. Carpenters preparing to place a section of a wood foundation wall. Gravel has been spread and leveled over the entire basement area. Note footing plates in position at left.

the gravel bed must be at least three-quarters the width of the footing plate. The gravel helps distribute the load from the foundation wall to the underlying soil.

For a full-basement wood foundation, a layer of gravel is also spread over the area where the concrete slab will be poured after the foundation walls have been completed. See Figure

tened to each other, and temporarily braced. See Figure 36-19. Where the wall sections are joined, caulking is applied to the

*American Plywood Association*

*American Plywood Association*

Figure 36–19. Setting up wood foundation walls, which are delivered to the construction site in sections.

butt edges of the plywood. A waterproof 6 mil polyethylene film is placed on the outside of the wall areas that will be below grade.

Moisture control is extremely important for a wood foundation.

All the procedures described in Unit 41 for a concrete foundation should also be followed for a wood foundation. The ground surrounding the foundation should be sloped to allow surface water to move away. Sometimes a sump may also be required. Water collecting in the sump can flow through a drain and away from the foundation.

239

# UNIT 37

# Concrete

Concrete is the strongest and most durable material for a foundation. It is *poured* (placed) in wood forms built to the shape of the foundation walls. See Figure 37–1. After it has *set up* (dried), the forms are stripped away. See Figure 37–2.

A large amount of carpentry work today concerns the construction forms for concrete. Therefore, carpenters should understand the make-up and placement of concrete. In addition to the information presented in this unit, further information about concrete is presented in Unit 69 of Section 16.

## COMPOSITION OF CONCRETE

The process employed today to manufacture concrete has been in use (with modifications) for more than 150 years. A more primitive but similar concrete mixture to that used today was used by the ancient Romans to construct buildings, roads, and aqueducts. Some of their structures are still standing.

Concrete is composed of *portland cement, sand,* and *gravel.* When water is added to these ingredients, a bonding chemical

*Georgia-Pacific Corporation*

Figure 37–1. Foundation forms constructed by carpenters. This patented panel system of forming requires very little bracing. At rear, a transit-mix concrete truck is placing ready-mixed concrete into the forms. Note the chute extending from the truck and positioned over the form. Inside the foundation walls, pier forms are also being filled with concrete.

action called *hydration* takes place that results in the hardening of the concrete.

## Portland Cement

Cement acts as the paste that bonds together the sand and gravel in a concrete mixture. Portland cement was so named by Joseph Aspden, an English-man, who developed the modern process of manufacturing cement. He used rocks quarried (dug out) from limestone deposits on the island of Portland, located off the coast of England. Limestone is obtained by digging into the earth's surface or from mining underneath the ground. It may also be dredged from deposits covered by water.

After limestone has been extracted from the ground, it is broken down and transported to a cement mill. The rock is then pulverized into a hard powder. The powder is mixed with other chemicals such as silica, iron oxide, and alumina. The combined ingredients are heated to a very high temperature (2,600° F to 3,000° F) in a rotary kiln until they form into small rock-like lumps called *clinkers.* The clinkers are then ground to a fine powder. Gypsum, which affects the setting time (hardening) of the cement, is added at this time.

## Aggregates—Sand and Gravel

The greatest part of the concrete mixture consists of *fine* and *coarse* aggregates. The fine aggregate is sand. The coarse aggregate is gravel or crushed stone. The particles of gravel or crushed stone are ¼″ to 1½″ in diameter. Good aggregate material comes from hard and durable rock. It should be free of harmful chemicals, soil, or vegetable matter.

## Water

Of great importance is the amount of water combined with the portland cement and aggregates. Too much water dilutes the cement and causes the aggregates to separate, producing a weakened concrete. Too little water results in poor mixing action of the cement, sand, and gravel, again producing a weakened concrete. The water should be clean and free of oil, alkali, or acid. Drinking water is recommended for mixing concrete.

## VARIATIONS IN THE CONCRETE MIX

The proportions of cement, aggregate, and water vary according to the type of concrete mixture (called the *mix*) needed for a particular job. The table in Figure 37–3 gives the formulas for several concrete mixtures. The pie chart illustrates the first formula in the table. Quantities in the table are measured by the cubic foot. Water is normally measured by gallons but is shown in the table as ½ cubic foot, which equals 3½ gallons of water. (One cubic foot contains 7 gallons of water.)

The table also shows the maximum diameter (in inches) of the particles of coarse aggregate for each concrete mix. The size of coarse aggregate chosen for a concrete mixture depends partly on the spacing between any steel reinforcing bars (*rebars*) required in a wall and on the thickness of the wall. Walls that are narrow or that have a greater amount of steel placed in them might require a concrete mixture with smaller size gravel.

Although engineers or concrete field specialists usually calculate the proportions for the concrete mix, blueprint specifications often stipulate the *minimum cement content* in relation to the *lowest water-cement ratio.* The

Figure 37–2. Section of a concrete foundation after the forms have been removed.

water-cement ratio largely determines the *compression strength* of concrete. See Unit 69 in Section 16 for a discussion of compression strength. In addition, the size and amount of coarse and fine aggregate and types of *admixtures* (added ingredients) are often given in the blueprint specifications. For smaller concrete structures, such as residential foundations, local building codes may also furnish this information. Another source of information is the local batch plants that produce *ready-mixed* concrete (discussed later in this unit).

| MAXIMUM-SIZE COARSE AGGREGATE | CONCRETE | | | |
|---|---|---|---|---|
| | CEMENT | SAND | COARSE AGGREGATE | WATER |
| ⅜″ | 1 | 2½ | 1½ | ½ |
| ½″ | 1 | 2½ | 2 | ½ |
| ¾″ | 1 | 2½ | 2½ | ½ |
| 1″ | 1 | 2½ | 2¾ | ½ |
| 1½″ | 1 | 2½ | 3 | ½ |

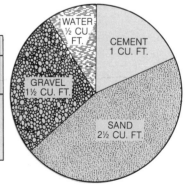

Figure 37–3. The table gives proportions for several concrete mixtures. The pie chart illustrates the cubic foot quantities of the first concrete mixture listed in the table. Water is normally measured by the gallon but in the table is shown as ½ cubic foot, which equals 3½ gallons. The table also gives the maximum diameter of the particles of coarse aggregate (gravel) for each mixture.

Figure 37-4. Concrete is usually delivered to the job site in a transit-mix truck. Here it is being chuted into forms directly from the truck.

## DELIVERING CONCRETE

Very small amounts of concrete can be mixed by hand or with small mechanical mixers. For the large amounts required by most construction projects, however, it is standard procedure to use *ready-mixed* concrete, which is prepared to specification at a *batch plant* and delivered to the job site by truck.

The estimated amount of concrete required is ordered from the batch plant by the cubic yard. (One cubic yard equals 27 cubic feet.) Automatic controls assure a proper mix of cement, sand, and gravel. If the job site is a short distance away, water is also added at the plant. If it is a long distance away, water is added enroute or upon arrival at the site. Under normal conditions, concrete must be discharged within 1½ hours after the water has been added.

Various kinds of trucks are used for delivering concrete, but usually a *transit-mix* truck is used. See Figure 37-4. It is equipped with a large, revolving drum operated by an auxiliary engine. The drum rotates to mix the concrete during its transportation. Drum capacities range from 1 to 12 cubic yards. The transit-mix truck also has a water tank so that water can be added enroute if necessary.

## POURING CONCRETE

Concrete is not a liquid; therefore, technically, it is placed rather than poured. However, most construction workers refer to this operation as *pouring concrete,* or the *concrete pour,* or simply *the pour.*

Concrete can be placed at or below ground level by chuting it directly from the truck (refer again to Figure 37-4) or by using a pumping system. The concrete flows from the pumping apparatus into hoses supported by a boom extending from the pump truck.

Concrete should always be deposited as close as possible to its final location. It cannot be discharged in one corner and then be expected to flow into the rest of the form. Also, it should not be placed in high piles and then leveled off. Rather, it should be discharged from different positions until an even *lift* (layer) 12″ to 20″ thick has been placed in the entire form. The procedure is repeated until enough lifts of concrete have been deposited to fill the form. See Unit 69 of Section 16 for further information on lifts.

As the concrete is placed, it should be *spaded* with long, narrow rods. Spading helps settle the concrete and eliminate *honeycombs.* Honeycombs are open spaces (voids) caused by trapped air pockets formed during the pour. On larger jobs, a mechanical vibrator is used to spade concrete. (See Unit 69 of Section 16 for further information on mechanical vibrators.)

## CURING CONCRETE

Concrete hardens because of a chemical action called *hydration.* This action occurs between the water and the cement in the concrete mix. It begins as soon as the water and cement are combined. If the water in the concrete mix evaporates too quickly, the water-cement hydration process will end before the concrete attains its full strength. Rapid water loss may also cause the concrete to shrink and crack.

*Curing* is the process of keeping the concrete moist long enough to allow proper hydration to occur. Under normal conditions this is accomplished for concrete walls by allowing the wall forms to remain in place for a sufficient period of time (usually three days to a week) after the concrete has been placed. During hot, dry weather, the forms should be dampened by sprinkling them with water. Floor slabs are more difficult to cure than walls. They may need to be misted or flooded with water, covered with waterproof paper or polyethylene film, or sprayed with chemical sealing compounds.

The first three days after discharge are the most critical to the quality of concrete. During this early period, it is most vulnerable to damage. At seven days after discharge, it reaches about 70% of its strength and at fourteen days about 85%. Under normal conditions maximum strength is reached at twenty-eight days.

Special considerations such as unusual weather conditions, the size of the structural members being poured, and the proportions of ingredients used for the mix may make it necessary to allow a longer curing period. See Unit 69 of Section 16 for further discussion of curing.

## REINFORCED CONCRETE

Two different kinds of pressure are exerted on a foundation wall: vertical and lateral. See Figure 37–5. A wall made of concrete has a great deal of *compression strength,* which is its ability to hold up under vertical pressure. (Compression strength is discussed further in Unit 69 of Section 16.) However, it has far less resistance to lateral forces, which push against the sides of the wall. Steel reinforcing bars *(rebars)* are used to help overcome this weakness. See Figure 37–6. Concrete with rebars is called *reinforced concrete.*

### Rebars

Rebars are usually *deformed,* meaning they have ridges on the

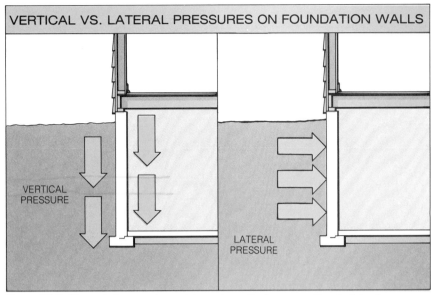

VERTICAL VS. LATERAL PRESSURES ON FOUNDATION WALLS

VERTICAL PRESSURE

LATERAL PRESSURE

Figure 37-5. Concrete has greater resistance to vertical pressure than to lateral pressure.

surface. These ridges give a better gripping action with the concrete. The rebars are positioned inside the form walls before the concrete is poured. They may be placed vertically, horizontally, or in a combination of both positions. They are identified by numbers, from #2 through #18. The diameter of the bar is found by multiplying the number designation by ⅛″. For example, a #4 bar is ⁴⁄₈″, or ½″. The size of the bars and their placement and spacing are shown in the foundation plan drawings of the blueprints.

Concrete-block walls may be reinforced with rebars laid in the mortar joints. In seismic risk zones, vertical rebars extend from the concrete footing and are tied into the horizontal bars. The block cavities are filled with concrete as the wall is being constructed.

### Wire Mesh

Wire mesh reinforcement is placed in concrete slabs resting directly on the ground. It helps prevent cracks from occurring later in the concrete. It is often used for slab-at-grade floors,

Figure 37-6. Rebars are placed in a low foundation form.

sidewalks, and driveways.

The mesh consists of wire welded together. It comes in rolls that are laid in position before the concrete is poured. Wire mesh is identified by the spacing and gauge diameter of its wires.

# UNIT 38

# Forming Methods and Materials

Various methods are used today for foundation form construction. All require some type of sheathing, studs or walers (or both), bracing, and a method of tying the form walls together. See Figure 38–1. (See Unit 69 of Section 16 for information on forms for concrete heavy construction.)

Since a form is a temporary structure, it should be constructed for easy dismantling. Duplex nails are used wherever practical as they can be quickly removed. Sheathing is fastened to the stakes or studs with just enough nails to hold it in place. There should be no skimping, however, on the number of braces and ties necessary to keep the walls aligned and in place.

## SHEATHING

Although different materials are available for sheathing form walls, plywood is probably used most often. Almost any exterior grade plywood of proper thickness can be used; however, a special product called *Plyform* is manufactured specifically for concrete form construction. It is

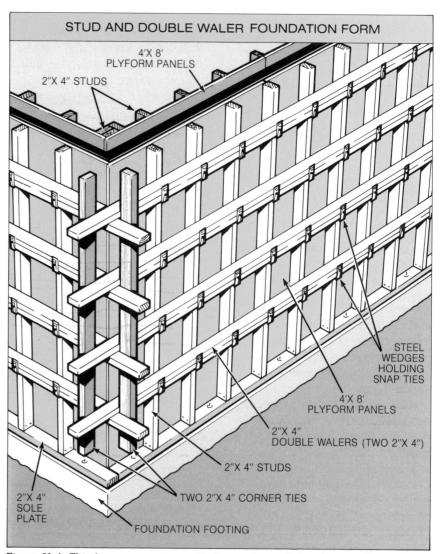

Figure 38–1. This form construction method uses plywood sheathing, studs, and double walers. Note the use of corner ties to lace the walers together.

Figure 38–2. This form construction method uses planks with wedge ties to form foundation walls.

available in 4' × 8' sheets, in ⁵/₈" or ¾" thicknesses, and in different grades. It can be reused many times.

Spread footings for T-shaped walls are usually formed with *planks* of 2" nominal thickness. Planks are also used to form foundation walls in some forming systems. See Figure 38–2.

## FRAMING AND BRACING MATERIALS

A wall form is subjected to great pressure when the concrete is placed. The pressure increases as the height or thickness of the wall increases. A faster rate of pour also places a greater strain on the forms.

*Walers* (also known as *wales*) reinforce and stiffen the wall. They run horizontally and are toe-nailed to the studs. They may also be fastened directly to the sheathing. The distance between walers depends on the thickness and height of the wall being poured. See Figure 39–9 in Unit 39 for a traditional double-waler system.

When ¾" plywood is used with

low walls, the studs or stakes are usually spaced 2' apart. Higher walls may require a spacing of 16" or 12".

Adequate *bracing* is essential to hold the wall in position during the concrete pour. One end of the brace is fastened to the wall. The other end is usually nailed to a stake driven into the ground.

Lumber used for walers, studs, and bracing is usually cut from 2 × 4s. A good grade of wood should be used. Often this lumber is reused. Metal stakes may be used in place of wooden

stakes to hold the forms in place. See Figure 38–3.

## TIES

Form walls must be tied together so they will not shift during the concrete pour. Small form walls may be tied together by braces and wood cleats. See Figure 38–4. Larger walls require some type of metal ties. At one time wire was used for this purpose. Today various patented ties are available that not only hold the form walls together, but also

Figure 38–3. Reusable round metal stakes hold the planks in this low foundation form. Duplex nails are driven through the holes provided in the stake and into the planks.

Figure 38–4. Braces and wood cleats are sufficient to tie these form walls together.

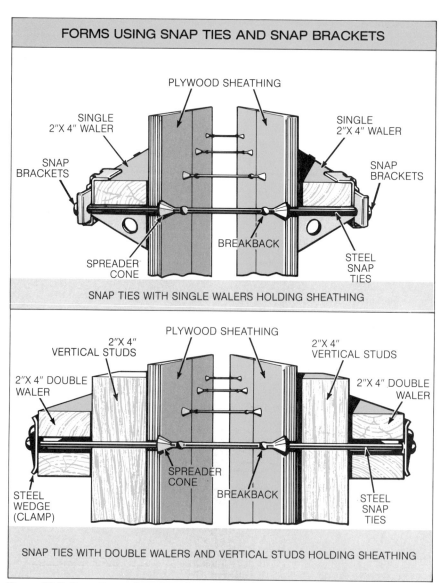

## FORMS USING SNAP TIES AND SNAP BRACKETS

PLYWOOD SHEATHING

SINGLE 2"X 4" WALER

SINGLE 2"X 4" WALER

SNAP BRACKETS

SNAP BRACKETS

SPREADER CONE

BREAKBACK

STEEL SNAP TIES

SNAP TIES WITH SINGLE WALERS HOLDING SHEATHING

PLYWOOD SHEATHING

2"X 4" VERTICAL STUDS

2"X 4" VERTICAL STUDS

2"X 4" DOUBLE WALER

2"X 4" DOUBLE WALER

STEEL WEDGE (CLAMP)

SPREADER CONE

BREAKBACK

STEEL SNAP TIES

SNAP TIES WITH DOUBLE WALERS AND VERTICAL STUDS HOLDING SHEATHING

Figure 38–6. Snap ties and steel wedges are widely used to hold form walls together. Two cones act as spreaders that set the walls to the correct width. The buttons at the ends of the tie hold steel wedges (clamps) that are driven behind the walers. Snap ties are designed for both single-waler and double-waler systems. Breakbacks are grooves behind the spreader cones. After the forms are stripped from the foundation wall, the section of the snap tie protruding from the wall can be snapped off at the breakback points when moved back and forth.

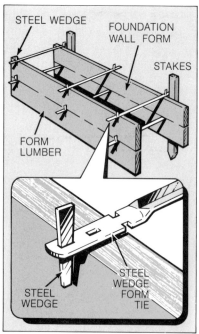

STEEL WEDGE

FOUNDATION WALL FORM

STAKES

FORM LUMBER

STEEL WEDGE FORM TIE

STEEL WEDGE

Figure 38–5. Wedge ties are used with this plank type of form construction method.

maintain the proper spacing between the walls.

Figure 38–5 shows a plank type of forming system with wedge ties. Figure 38–6 shows a system of snap ties and steel wedges used with single and double walers. *Spreader cones keep the walls the correct distance apart.* Buttons at the ends of the snap ties hold the steel wedges (clamps) that are driven behind the walers. Behind the spreader cones are grooves called *breakbacks.* After the forms are stripped from the foundation wall, the section of the snap tie that protrudes from the wall can be snapped off at the breakback points.

## BUILT-IN-PLACE FORMS

The oldest method of form construction is the built-in-place method. First, stakes or studs are positioned. Then the sheathing material is fastened in place.

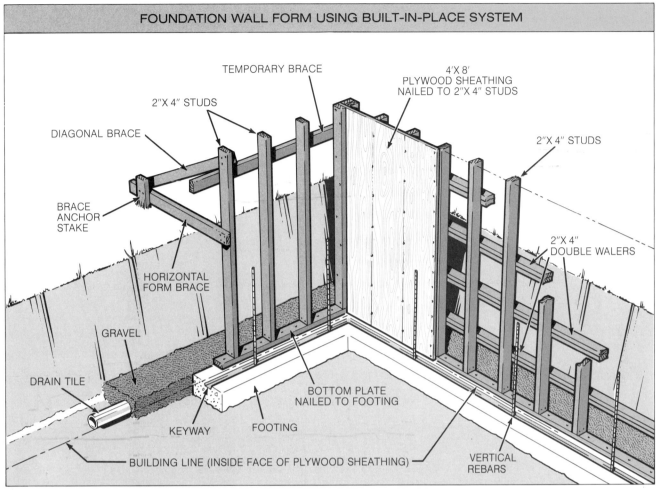

FOUNDATION WALL FORM USING BUILT-IN-PLACE SYSTEM

TEMPORARY BRACE

2"X 4" STUDS

DIAGONAL BRACE

4'X 8'
PLYWOOD SHEATHING
NAILED TO 2"X 4" STUDS

2"X 4" STUDS

BRACE
ANCHOR
STAKE

2"X 4"
DOUBLE WALERS

HORIZONTAL
FORM BRACE

GRAVEL

DRAIN TILE

BOTTOM PLATE
NAILED TO FOOTING

KEYWAY     FOOTING

VERTICAL
REBARS

BUILDING LINE (INSIDE FACE OF PLYWOOD SHEATHING)

Figure 38–7. Example of the outside wall construction of a built-in-place form construction method. The sole (bottom) plate is fastened to the concrete. Studs are set up, temporarily tied together, and braced. Sheathing is then applied to the inner face of the studs. The walers are placed last.

See Figure 38–7. A method recently introduced on the West Coast is a built-in-place system using 2" planks and a type of wedge tie. This system does not require any walers, and the stakes are spaced farther apart than in older built-in-place systems. The planks can be reused.

## PANEL FORMS

Many carpenters and builders consider panel forms a more efficient form construction method than built-in-place forms. The forms consist of sections made up of studs and plates nailed to a plywood panel. See Figure 38–8. When the sections are set in place, the end studs are fastened to each other with duplex nails.

Panel form sections can be constructed in the shop or on the job. They are convenient on housing tracts, where one foundation design is repeated. The panel sections can be reused. Patented panel systems of wood or metal can be rented or purchased.

Figure 38–9 shows lightweight aluminum footing forms. Round stakes driven through holes in the aluminum sections hold them in place. Wedge locks at the stakes make it possible to adjust the height of each section. Figure 38–10 shows a lightweight panel system for foundation walls. See Unit 39 for further information on panel systems.

## DOOR AND WINDOW OPENINGS

Full-basement concrete foundations often have windows. Sometimes they also have a door to provide access from the outside of the building. Preparations must be made for any door and window openings when the forms are built. A traditional method is to construct a well-braced frame called a *buck*. (See Unit 68 in Section 16 for more information on door and window bucks.) The buck is set in place and fastened to the outside wall with duplex nails. After the inside form walls are placed, duplex nails are also driven through the inside form walls into the frame. (A trade term for placing inside form walls

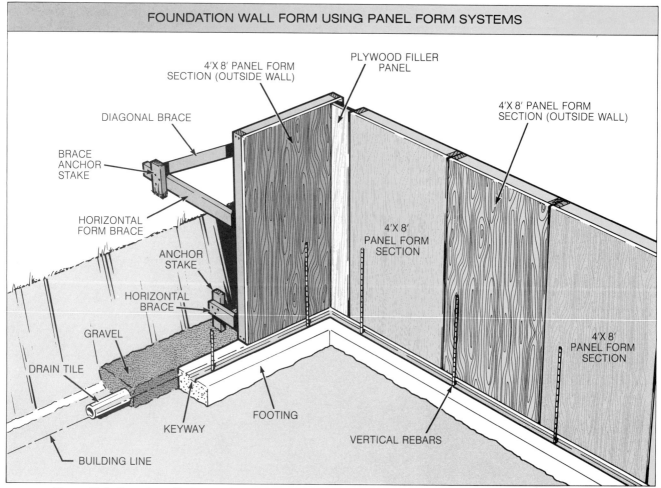

FOUNDATION WALL FORM USING PANEL FORM SYSTEMS

Figure 38–8. Outside wall of a panel form construction method. The panel sections can be reused many times.

*Western Forms, Inc.*

Figure 38–10. A lightweight aluminum panel system was used to form this basement foundation wall. Some of the panels are still in place on the wall at right.

*Western Forms, Inc.*

Figure 38–9. Reusable aluminum footing forms are used here to hold the concrete that will be placed for the footing. Round stakes are driven through holes in the aluminum sections to hold them in place. Wedge locks at the stakes make it possible to adjust the height of each form section.

is *doubling up the walls.*) Door and window bucks are removed after the concrete has set up and the outside and inside form walls have been stripped away.

Finish frames for doors and windows may be metal or wood, but metal jambs are used more often. There are several ways to fasten finish frames into the opening. An older method is to position nailing strips in the con-crete at the time the bucks are set in place. Today finish frames are usually attached with a powder-actuated fastener instead. An-other method is to drive bolts or screws into expansion devices embedded in the concrete.

Also available are metal frames that can be set in place at the time the wall is being formed. This eliminates the need for a door or window buck. After the concrete hardens, the frame is permanently fastened in posi-tion.

# UNIT 39

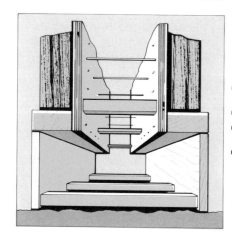

# Foundation Designs— Form Construction

## T-FOUNDATION FORMS

T-foundations (which are shaped like an inverted T) can be used for full basements or crawl spaces. The spread footing of the T-foundation provides good bearing on all types of soil. The forming procedure used depends on the climate, the soil condition, and the height of the foundation walls. For a low T-foundation, the concrete for footings and walls is poured at the same time (mono-lithically). For a high T-founda-tion, the concrete for footings and walls is poured separately.

## Footings

Where the soil condition is firm and stable, forms for the footings may not be needed. Instead, a trench may be dug to the width and depth of the footing and the concrete placed directly in the trench. Where the soil condition is unstable, a footing form must be built. See Figure 39–1. Either 1″ or 2″ (nominal thickness) boards can be used for this pur-pose. With the thicker boards, stakes can be placed farther apart, and less bracing is re-quired. A procedure for con-

structing a footing form is shown in Figure 39–2.

**Keyways.** Immediately after the concrete has been placed, pieces of chamfered 2 × 4s called *key strips* are often pressed into the concrete toward the center of the footing. When these pieces are removed after the concrete has hardened, a groove called a *keyway* has

been formed in the concrete. See Figure 39–3 and 39–4. The keyway helps to secure the bot-tom of the foundation wall to the footing.

**Reinforcing Steel.** In seismic risk areas, reinforcing steel is po-sitioned in the footing forms be-fore the concrete is placed. Verti-cal rebars project out of the footing forms and are later tied to

Figure 39–1. Footing forms filled with concrete. Forms are necessary (instead of just a trench) when the soil condition is unstable.

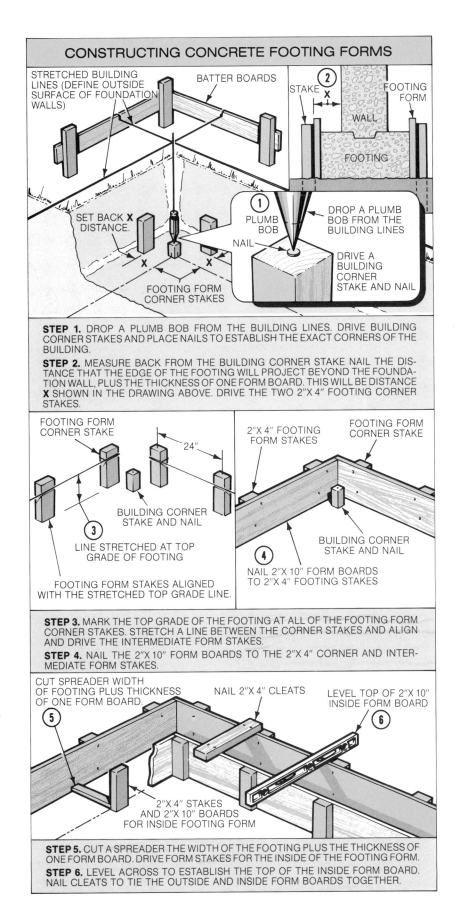

## CONSTRUCTING CONCRETE FOOTING FORMS

STRETCHED BUILDING LINES (DEFINE OUTSIDE SURFACE OF FOUNDATION WALLS)

BATTER BOARDS

STAKE

FOOTING FORM

WALL

FOOTING

SET BACK **X** DISTANCE.

PLUMB BOB

NAIL

① DROP A PLUMB BOB FROM THE BUILDING LINES

DRIVE A BUILDING CORNER STAKE AND NAIL

FOOTING FORM CORNER STAKES

**STEP 1.** DROP A PLUMB BOB FROM THE BUILDING LINES. DRIVE BUILDING CORNER STAKES AND PLACE NAILS TO ESTABLISH THE EXACT CORNERS OF THE BUILDING.

**STEP 2.** MEASURE BACK FROM THE BUILDING CORNER STAKE NAIL THE DISTANCE THAT THE EDGE OF THE FOOTING WILL PROJECT BEYOND THE FOUNDATION WALL, PLUS THE THICKNESS OF ONE FORM BOARD. THIS WILL BE DISTANCE **X** SHOWN IN THE DRAWING ABOVE. DRIVE THE TWO 2"X 4" FOOTING CORNER STAKES.

FOOTING FORM CORNER STAKE

24"

BUILDING CORNER STAKE AND NAIL

③ LINE STRETCHED AT TOP GRADE OF FOOTING

FOOTING FORM STAKES ALIGNED WITH THE STRETCHED TOP GRADE LINE.

2"X 4" FOOTING FORM STAKES

FOOTING FORM CORNER STAKE

BUILDING CORNER STAKE AND NAIL

④ NAIL 2"X 10" FORM BOARDS TO 2"X 4" FOOTING STAKES

**STEP 3.** MARK THE TOP GRADE OF THE FOOTING AT ALL OF THE FOOTING FORM CORNER STAKES. STRETCH A LINE BETWEEN THE CORNER STAKES AND ALIGN AND DRIVE THE INTERMEDIATE FORM STAKES.

**STEP 4.** NAIL THE 2"X 10" FORM BOARDS TO THE 2"X 4" CORNER AND INTERMEDIATE FORM STAKES.

CUT SPREADER WIDTH OF FOOTING PLUS THICKNESS OF ONE FORM BOARD.

⑤

NAIL 2"X 4" CLEATS

LEVEL TOP OF 2"X 10" INSIDE FORM BOARD

⑥

2"X 4" STAKES AND 2"X 10" BOARDS FOR INSIDE FOOTING FORM

**STEP 5.** CUT A SPREADER THE WIDTH OF THE FOOTING PLUS THE THICKNESS OF ONE FORM BOARD. DRIVE FORM STAKES FOR THE INSIDE OF THE FOOTING FORM.

**STEP 6.** LEVEL ACROSS TO ESTABLISH THE TOP OF THE INSIDE FORM BOARD. NAIL CLEATS TO TIE THE OUTSIDE AND INSIDE FORM BOARDS TOGETHER.

Figure 39–2. Constructing forms for a concrete footing. Either 1" or 2" boards may be used.

the rebars placed in the wall above. See Figures 39–5 and 39–6.

After the footing forms have been completed, the reinforcing steel positioned, and provisions made for the keyways, the concrete is placed. See Figure 39–7.

## Walls

Wall forms are built after the concrete has set up in the footings. Plates can easily be nailed into the freshly placed (green) concrete as a base for the outside walls of either a built-in-place forming system or a panel forming system. See Figure 39–8.

Except in the case of very low forms, reinforcing steel bars are usually installed after the outside form walls are set. When all the rebars have been placed, the inside walls of the form can be constructed. If a large amount of reinforcement is required, it is usually done by steelworkers specializing in this type of work.

Rebars should be clean and free of loose rust when installed in the form. They must be positioned and held in place so that they will be covered by an adequate protective layer of concrete.

Preparations must be made for any door and window openings when the wall forms are built. See Unit 38 for information on constructing a door or window buck and on installing finish frames.

A traditional double-waler wall system is shown in Figure 39–9. A procedure for building a stud-and-waler panel wall form for a T-foundation is shown in Figure 39–10.

Patented panel systems are also widely used in the construction of all types of foundation wall forms. They often consist of

Figure 39-5. Corner section of a footing form with reinforcing steel bars. The key strips on this job are wired to the vertical steel bars or tacked to cleats nailed to the top of the form boards.

*Portland Cement Association*

Figure 39-3. Footing form filled with concrete. The keyway at the center was formed by key strips that have been removed.

Figure 39-6. Footing forms with reinforcing steel bars in place. Note vertical bars extending from reinforcing steel bars.

Figure 39-4. A concrete footing after the forms have been stripped. Note the keyway close to the vertical rebars.

Figure 39-7. Concrete being pumped into footing forms.

Figure 39-8. Outside form panel sections being fastened to the top of the footing. Note the snap ties projecting from the top of the panels.

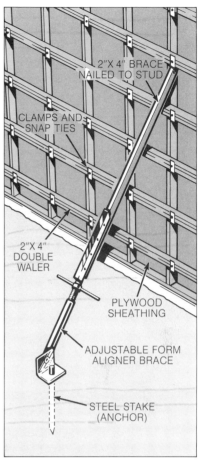

Figure 39-9. Double-waler system used with wedges and snap ties to hold walls together.

plywood panel sections set in metal frames. See Figure 39-11. These sections are secured to each other with wedge bolts. See Figure 39-12. One wedge bolt is inserted in slots provided in the side rails, and the other is inserted in a slot provided in the first wedge bolt.

The patented panel system described here is aligned and secured with wood braces that have metal turnbuckles attached to their upper ends. See Figure 39-13. The lower end of each brace is nailed to a wood or metal stake driven into the ground. The turnbuckle is secured with wedge bolts to a panel side rail. It is then adjusted to hold the wall in position.

After the walls have been set for one side of the form, the looped ends of wire ties are inserted into dadoed slots provided in the side rails. See Figure 39-14. These wire ties space and hold together opposite wall forms. See Figure 39-15.

In the Steel-Ply system, walers are required only at the upper section of the wall forms. See Figure 39-16. The walers are tightened and secured to the wall forms with wedge bolts driven through slots provided in waler clamps and through the looped ends of the wire ties.

## Monolithic T-Foundation

When low walls are required for a crawl-space foundation, the forms for the walls and footings may be built as one unit. Since the concrete for the walls and footings is poured at the same time, the foundation is called a *monolithic* foundation. An advantage of a monolithic foundation is that there is no possibility of moisture seeping through a *cold joint* afterwards. A cold joint occurs where the concrete for a wall section is placed over the already hardened concrete of a footing.

Figure 39-17 shows a monolithic T-foundation form built over an open trench. Forms are not required for the footings. Figure 39-18 shows the forming procedure used when soil conditions require footings to be formed along with the wall.

## Establishing Height of Pour

Concrete is placed in a form until it reaches the level required for the top of the foundation wall. Often the form is constructed to the actual height of the finished foundation wall. However, when the panel system is used, form walls are often higher than the level required for the foundation. In that case, a builder's level is used to establish elevation points at intervals and at all corners of the form. Lines are snapped, and a narrow strip of wood called a *pour strip* is tacked above the line. See Figure 39-19. When the concrete has been poured to the proper level, the pour strip is removed.

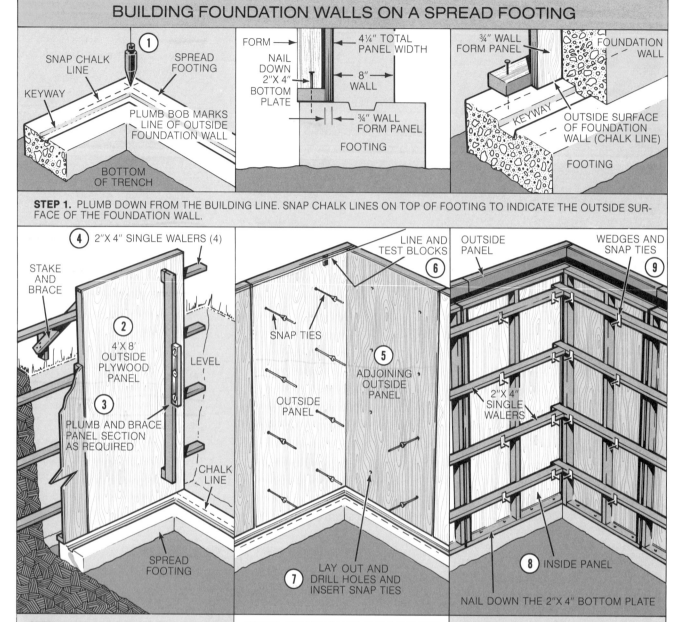

**BUILDING FOUNDATION WALLS ON A SPREAD FOOTING**

SNAP CHALK LINE

SPREAD FOOTING

KEYWAY

PLUMB BOB MARKS LINE OF OUTSIDE FOUNDATION WALL

BOTTOM OF TRENCH

FORM

NAIL DOWN 2"X 4" BOTTOM PLATE

4¼" TOTAL PANEL WIDTH

8" WALL

¾" WALL FORM PANEL

FOOTING

¾" WALL FORM PANEL

FOUNDATION WALL

KEYWAY

OUTSIDE SURFACE OF FOUNDATION WALL (CHALK LINE)

FOOTING

**STEP 1.** PLUMB DOWN FROM THE BUILDING LINE. SNAP CHALK LINES ON TOP OF FOOTING TO INDICATE THE OUTSIDE SURFACE OF THE FOUNDATION WALL.

2"X 4" SINGLE WALERS (4)

STAKE AND BRACE

4'X 8' OUTSIDE PLYWOOD PANEL

LEVEL

PLUMB AND BRACE PANEL SECTION AS REQUIRED

CHALK LINE

SPREAD FOOTING

LINE AND TEST BLOCKS

SNAP TIES

ADJOINING OUTSIDE PANEL

OUTSIDE PANEL

LAY OUT AND DRILL HOLES AND INSERT SNAP TIES

OUTSIDE PANEL

WEDGES AND SNAP TIES

2"X 4" SINGLE WALERS

INSIDE PANEL

NAIL DOWN THE 2"X 4" BOTTOM PLATE

**STEP 2.** SET THE FIRST OUTSIDE PANEL IN PLACE AND NAIL IT TO THE BOTTOM PLATE.

**STEP 3.** PLUMB AND BRACE THE PANEL SECTION. SET THE REST OF THE OUTSIDE PANELS FOR THE WALL, BRACING WHERE NECESSARY.

**STEP 4.** TOENAIL THE 2"X 4" SINGLE WALERS TO THE BACK OF THE OUTSIDE PANEL SECTIONS.

**STEP 5.** SET THE PANEL SECTIONS AND WALERS FOR THE ADJOINING OUTSIDE WALLS.

**STEP 6.** WHEN ALL THE OUTSIDE WALLS HAVE BEEN CONSTRUCTED, ALIGN THE WALLS WITH A LINE AND TEST BLOCKS. ADD BRACES WHERE NECESSARY.

**STEP 7.** LAY OUT AND DRILL HOLES FOR SNAP TIES. INSERT THE SNAP TIES THROUGH THE HOLES.

**STEP 8.** DOUBLE UP THE WALLS. HOLES FOR THE SNAP TIES ARE LAID OUT AND DRILLED FOR EACH INSIDE PANEL BEFORE IT IS SET IN PLACE. THE SNAP TIES ARE FED INTO THE HOLES AS EACH INSIDE PANEL IS TILTED INTO PLACE.

**STEP 9.** TOENAIL THE 2"X 4" SINGLE WALERS TO THE INSIDE WALLS AND DRIVE THE CLAMPS OVER THE BUTTON ENDS OF THE SNAP TIES.

Figure 39–10. Building a T-foundation wall where concrete has already been placed for a spread footing. (The panel system is used in this example).

## RECTANGULAR AND BATTERED FORMS

Under certain conditions rectangular or battered foundation walls provide adequate support for a building. They are often used as grade beams and for the wall sections of slab-at-grade foundation systems.

Either the built-in-place method or the panel method can be used to build rectangular or battered forms. The components that make up the form (stakes, sheathing, walers, bracing) are the same as those used for the T-foundation. As there is no

Figure 39-11. Patented panel system (Steel-Ply) consisting of plywood panel sections set in metal frames.

Figure 39-13. This patented panel system uses wood braces with metal turnbuckles that are secured to a side rail with wedge bolts.

Figure 39-12. Wedge bolts secure panel sections of a patented panel system such as Steel-Ply.

spread footing, the wood or metal stakes holding the sheathing are driven directly into the ground. See Figures 39–20 and 39–21. Where hard soil makes stake driving difficult, wood plates can be set on the ground. These are held in position with steel dowels driven through holes bored in the plates. Some stages in the construction of rectangular, built-in-place plank forms held together with wedge ties are shown in Figures 39–22 and 39–23.

## PIER FORMS

*Piers* are square, round, or battered structures (footings) that act as a base for wood posts or steel columns that are placed under beams or girders to help support the floor and wall units of a building. There are various designs of pier forms. Whatever type is used, the ground must be dug so that the bottoms of the

piers rest on firm soil. Also, the bottoms of the piers must be below the frost line. The lower end of a post or column supported by a pier must be fastened securely to the pier. Two methods of fastening are described in Unit 42 of Section 9.

A circular type of pier is popular in some areas. The form for the pier is cut from treated, waterproof cardboard or other fibrous material. See Figure 39–24. When the form is set in position, soil is placed around it to hold it in place while it is being filled with concrete.

Other piers are square in

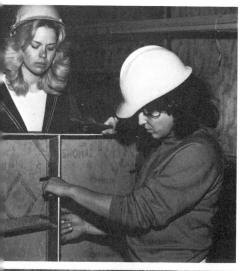

Figure 39–14. The carpenter on the opposite side of the wall is inserting wire ties into dado slots provided in the side rails of the panels. Her partner will drive wedges through the looped ends.

Figure 39–15. End view of doubled wall sections showing wire ties in place.

Figure 39–16. Walers are placed toward the top of a Steel-Ply patented panel wall form.

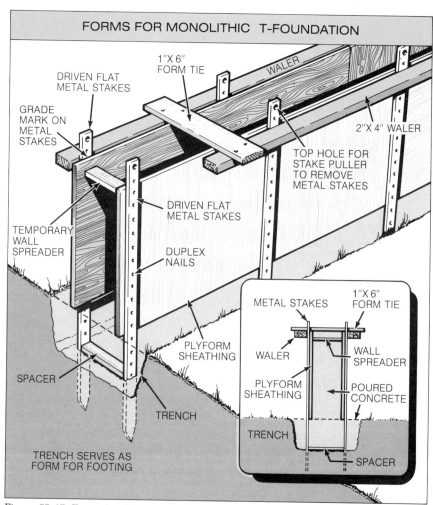

FORMS FOR MONOLITHIC T-FOUNDATION

DRIVEN FLAT METAL STAKES

1"X 6" FORM TIE

WALER

GRADE MARK ON METAL STAKES

2"X 4" WALER

TOP HOLE FOR STAKE PULLER TO REMOVE METAL STAKES

DRIVEN FLAT METAL STAKES

TEMPORARY WALL SPREADER

DUPLEX NAILS

SPACER

PLYFORM SHEATHING

TRENCH

METAL STAKES

1"X 6" FORM TIE

WALER

WALL SPREADER

PLYFORM SHEATHING

POURED CONCRETE

TRENCH

SPACER

TRENCH SERVES AS FORM FOR FOOTING

Figure 39–17. Example of one method used to construct walls for a monolithic T-foundation where footing forms are not required. The flat metal stakes are pried out after the concrete has set up.

## MONOLITHIC T-FOUNDATION REQUIRING FOOTING FORMS

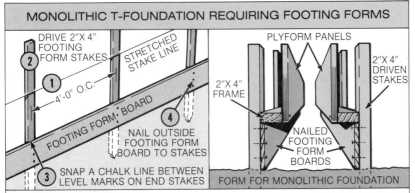

DRIVE 2"X 4" FOOTING FORM STAKES

STRETCHED STAKE LINE

4'-0" O.C.

FOOTING FORM BOARD

NAIL OUTSIDE FOOTING FORM BOARD TO STAKES

SNAP A CHALK LINE BETWEEN LEVEL MARKS ON END STAKES

PLYFORM PANELS

2"X 4" FRAME

2"X 4" DRIVEN STAKES

NAILED FOOTING FORM BOARDS

FORM FOR MONOLITHIC FOUNDATION

**STEP 1.** STRETCH A LINE FOR THE OUTSIDE ROW OF 2"X 4" FOOTING FORM STAKES. THE STAKES SHOULD BE SPACED 4'-0" O.C.

**STEP 2.** ALIGN THE STAKES ALONG THE STRETCHED LINE AND DRIVE THEM INTO THE GROUND.

**STEP 3.** ESTABLISH LEVEL MARKS ON THE TWO END STAKES. THIS CAN BE DONE WITH A BUILDER'S LEVEL OR TRANSIT. SNAP A LEVEL CHALK LINE ON THE OTHER FOOTING FORM STAKES.

**STEP 4.** SET THE TOP OF THE OUTSIDE FOOTING FORM BOARD EVEN WITH THE LEVEL CHALK LINE AND NAIL IT TO THE 2"X 4" STAKES. BE SURE THE BOARD IS LEVEL AND AT THE CORRECT HEIGHT.

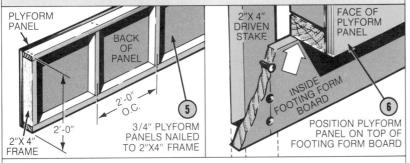

PLYFORM PANEL

BACK OF PANEL

2'-0" O.C.

2'-0"

2"X 4" FRAME

3/4" PLYFORM PANELS NAILED TO 2"X4" FRAME

2"X 4" DRIVEN STAKE

FACE OF PLYFORM PANEL

INSIDE FOOTING FORM BOARD

POSITION PLYFORM PANEL ON TOP OF FOOTING FORM BOARD

**STEP 5.** MAKE UP PANEL SECTIONS USING PLYFORM SHEATHING NAILED TO 2"X 4" FRAMES.

**STEP 6.** POSITION THE PANELS ON TOP OF THE FOOTING FORM BOARD. DRIVE NAILS THROUGH THE 2"X 4" STAKES INTO THE PANEL FRAME.

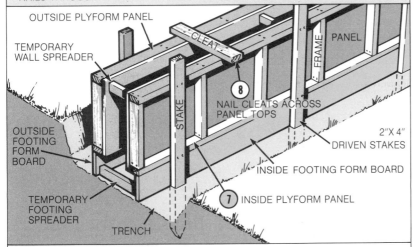

OUTSIDE PLYFORM PANEL

TEMPORARY WALL SPREADER

CLEAT

FRAME

PANEL

NAIL CLEATS ACROSS PANEL TOPS

OUTSIDE FOOTING FORM BOARD

2"X 4" DRIVEN STAKES

INSIDE FOOTING FORM BOARD

TEMPORARY FOOTING SPREADER

STAKE

TRENCH

INSIDE PLYFORM PANEL

**STEP 7.** AFTER THE OUTSIDE FORM WALLS HAVE BEEN COMPLETED, CONSTRUCT THE INTERIOR WALLS. USE TEMPORARY SPREADERS AT THE TOP AND BOTTOM TO MAINTAIN THE CORRECT WIDTHS AT THE FOOTING AND WALL. SPREADERS ARE REMOVED WHEN POURING CONCRETE.

**STEP 8.** TIE THE PLYFORM PANEL WALLS TOGETHER WITH 2"X 4" CLEATS NAILED ACROSS THE PANEL TOP FRAMES.

Figure 39-18. Constructing the forms for a monolithic T-foundation built over an open trench. Different wall thicknesses require other sizes of framing materials.

POUR STRIP TACKED TO SIDE OF PLYFORM PANEL

OUTSIDE PLYFORM PANEL

POUR STRIP

CONCRETE POURED FLUSH TO BOTTOM OF POUR STRIP

INSIDE PLYFORM PANEL

Figure 39-19. A pour strip tacked to one side of the form indicates the height to which the concrete will be poured.

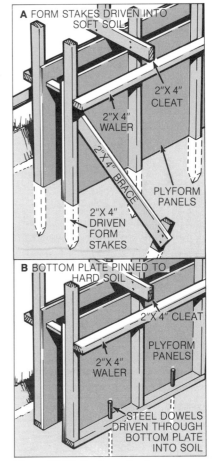

**A** FORM STAKES DRIVEN INTO SOFT SOIL

2"X 4" CLEAT

2"X 4" WALER

2"X 4" BRACE

2"X 4" DRIVEN FORM STAKES

PLYFORM PANELS

**B** BOTTOM PLATE PINNED TO HARD SOIL

2"X 4" CLEAT

PLYFORM PANELS

2"X 4" WALER

STEEL DOWELS DRIVEN THROUGH BOTTOM PLATE INTO SOIL

Figure 39-20. Rectangular forms can be constructed with stakes driven directly into the ground (A). In hard soil, plates can be pinned to the surface of the ground with steel dowels (B).

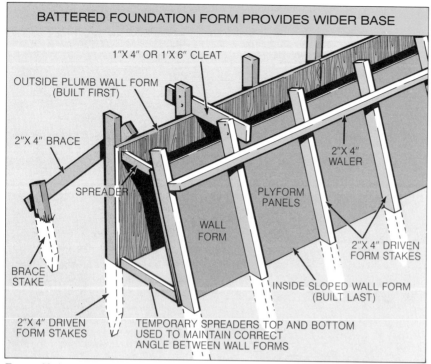

**BATTERED FOUNDATION FORM PROVIDES WIDER BASE**

1"X 4" OR 1"X 6" CLEAT

OUTSIDE PLUMB WALL FORM
(BUILT FIRST)

2"X 4" BRACE

SPREADER

2"X 4"
WALER

PLYFORM
PANELS

WALL
FORM

BRACE
STAKE

2"X 4" DRIVEN
FORM STAKES

2"X 4" DRIVEN
FORM STAKES

INSIDE SLOPED WALL FORM
(BUILT LAST)

TEMPORARY SPREADERS TOP AND BOTTOM
USED TO MAINTAIN CORRECT
ANGLE BETWEEN WALL FORMS

Figure 39–21. This battered foundation has a plumb outside wall. The inside wall slopes to provide a wider base against the ground. The plumb wall is built first. Spreaders are used to drive stakes at the correct angles for the battered wall.

Figure 39–23. The form walls have now been doubled using wedge ties that space the walls the correct distance apart and hold them in place. No walers are required with this system. The planks are nailed to flat, upright 2 × 4s spaced 4' to 6' feet apart. Metal or wood stakes are driven into the ground behind the vertical pieces. Braces are fastened at the top. Note the shut-off pieces at the vertical steps of the form.

Figure 39–22. Outside form wall for a stepped-up foundation constructed of 2" × 12" planks. The dark vertical pieces are redwood nailing strips that will remain attached to the concrete and provide nailing for exterior finish material. Note reinforcing bars in position.

Figure 39–24. Circular pier with a strap type of post base connector.

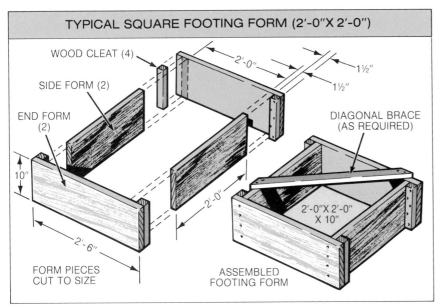

Figure 39–25. Form for a square pier to support a post or column. In this example, the dimensions of the pier will be 2'-0" × 2'-0" × 10".

shape, tapered, or stepped. Figure 39–25 shows the forming procedure for a square pier to support a post or column. Figure 39–26 shows a procedure for setting the bolts that secure the base of a steel (Lally) column to a square pier. The bolt holes are laid out on a template that is centered on the pier form.

Figure 39–27 shows the forming procedure for a tapered pier. A pier block is used here to provide a nailing base for the bottom of a wood post. Spikes are driven at angles into the pier block that is pressed into the wet concrete. The angled nails will hold the block securely after the concrete has set up.

Figure 39–28 shows the forming procedure for a stepped pier. Cleats hold the top section of the pier in place over the bottom section. A metal dowel is sometimes used to hold a wood post in position. This type of pier is also used to support a steel (Lally) column.

Piers must be level and positioned according to the dimensions provided in the foundation plan. A typical layout procedure is shown in Figure 39–29.

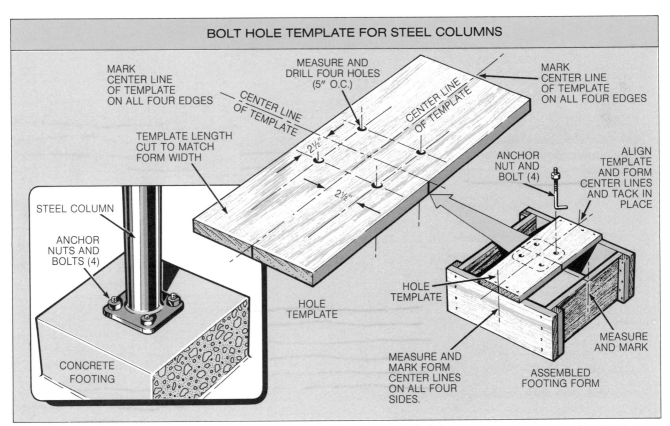

Figure 39–26. Setting the bolts that secure the base of a steel (Lally) column. The bolt holes are laid out on a template that is centered on the form.

## GRADE-BEAM FORMS

The basic design of a grade-beam wall is the same as that of a foundation wall resting on a spread footing. The top of the grade beam should extend at least 8″ above the finish grade of the lot. The bottom must be below the frost line. Many regional building codes require that the soil directly beneath grade beams be removed and replaced with a layer of coarse rock or gravel. The gravel provides drainage and reduces the chance of freezing action causing movement in the foundation walls.

Pier holes must be sunk before the forms for grade beams can be built. The holes are made with a mechanical drilling rig (Figure 39–30) that drives a large soil auger into the ground. Steel reinforcing bars extending above the ground are placed in the pier holes. In firm ground, concrete can be poured into the open hole. If the soil is soft and unstable, a round fiber form should be dropped into the hole before the concrete is placed. See Figure 39–31.

The wall form is built directly over the piers. The rebars extending from the piers are tied to the steel bars that are placed in the grade-beam forms.

### Precast Grade Beams

In some parts of the United States *precast* grade beams are used in residential foundation construction. Precast concrete members are formed at a precast plant and transported by truck to the job site, where they are lifted and set in place by cranes. Precast grade beams enable fast and efficient construction of crawl-space foundations that are placed on fairly level ground.

The precast system shown in Figures 39–32 through 39–36

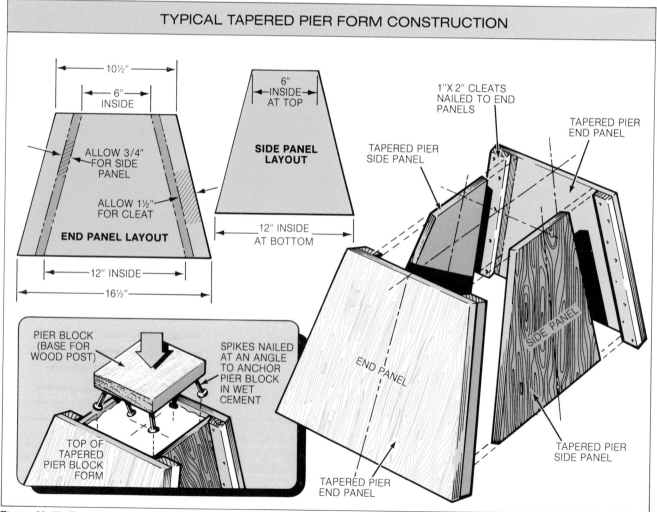

### TYPICAL TAPERED PIER FORM CONSTRUCTION

10½″
6″ INSIDE
ALLOW 3/4″ FOR SIDE PANEL
ALLOW 1½″ FOR CLEAT
**END PANEL LAYOUT**
12″ INSIDE
16½″

6″ INSIDE AT TOP
**SIDE PANEL LAYOUT**
12″ INSIDE AT BOTTOM

1″X 2″ CLEATS NAILED TO END PANELS
TAPERED PIER SIDE PANEL
TAPERED PIER END PANEL
SIDE PANEL
TAPERED PIER SIDE PANEL

PIER BLOCK (BASE FOR WOOD POST)
SPIKES NAILED AT AN ANGLE TO ANCHOR PIER BLOCK IN WET CEMENT
TOP OF TAPERED PIER BLOCK FORM
END PANEL
TAPERED PIER END PANEL

Figure 39–27. Form for a tapered pier. A pier block is used in this example to provide a nailing base for the bottom of a wood post. Spikes are driven at angles into the pier block that is pressed into the wet concrete. The angled nails will hold the block securely after the concrete has set up.

## TYPICAL FORM CONSTRUCTION FOR STEPPED POST FOOTING

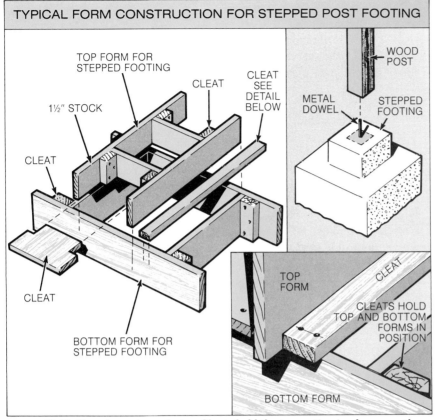

TOP FORM FOR STEPPED FOOTING

CLEAT

CLEAT SEE DETAIL BELOW

1½" STOCK

CLEAT

CLEAT

BOTTOM FORM FOR STEPPED FOOTING

WOOD POST

METAL DOWEL

STEPPED FOOTING

TOP FORM

CLEAT

CLEATS HOLD TOP AND BOTTOM FORMS IN POSITION

BOTTOM FORM

Figure 39–28. Form for a stepped pier. Cleats hold the top section in place over the bottom section. A metal dowel is sometimes used to hold a wood post in position as shown in the top drawing. This type of footing is also used as a base for a steel (Lally) column.

Conco Cement, Inc.

Figure 39–30. Pier holes are dug with a mechanical drilling drill rig that drives a large soil auger bit.

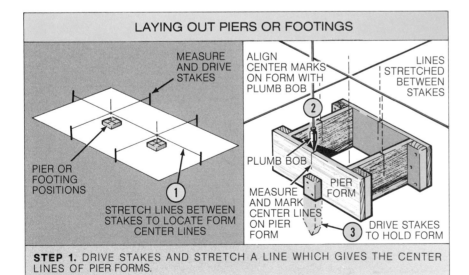

## LAYING OUT PIERS OR FOOTINGS

MEASURE AND DRIVE STAKES

ALIGN CENTER MARKS ON FORM WITH PLUMB BOB

LINES STRETCHED BETWEEN STAKES

PLUMB BOB

PIER FORM

PIER OR FOOTING POSITIONS

MEASURE AND MARK CENTER LINES ON PIER FORM

DRIVE STAKES TO HOLD FORM

STRETCH LINES BETWEEN STAKES TO LOCATE FORM CENTER LINES

**STEP 1.** DRIVE STAKES AND STRETCH A LINE WHICH GIVES THE CENTER LINES OF PIER FORMS.

**STEP 2.** MEASURE AND MARK THE CENTER LINES ON ALL FOUR SIDES OF THE PIER FORM. USING A PLUMB BOB OR HAND LEVEL, POSITION THE PIER FORM BOX SO THAT THE CENTER LINES OF THE BOX ARE PLUMB WITH THE STRETCHED CENTER LINES.

**STEP 3.** DRIVE STAKES AND PLACE SOIL AGAINST THE FORM BOX TO HOLD IT IN POSITION WHILE THE CONCRETE IS BEING POURED.

Figure 39–29. Laying out a pier to support a post or column.

Figure 39–31. Tubing extending from a pier hole. After the concrete has been placed, this pier will help support a grade-beam foundation. Note the reinforcing bars projecting from the tubing.

Figure 39–32. Precast grade-beam sections are delivered to the job site.

*Conco Cement, Inc.*

*Conco Cement, Inc.*

Figure 39–33. Grade-beam sections are placed by crane over the pier holes. They are temporarily supported by blocks and wedges.

Figure 39–34. Carpenter maneuvering the grade beam to its correct position.

uses concrete grade beams that are 6″ wide and 12″ high. The beams are temporarily supported by wood blocks when they are placed over the pier forms. A ½″ steel bar is inserted through a pre-drilled hole in the beam. It is tied to another steel bar that has been placed in the pier. After the grade beams have been straightened and aligned, concrete is pumped by hose into the pier forms up to the bottom of the beams. The forms must be held securely in position during the concrete pour.

## SLAB-AT-GRADE FORMS

In slab-at-grade construction the concrete floor slab rests directly on a bed of gravel that has been placed on the ground. Forms are constructed around the perimeter of the slab. See Figure 39–37.

Two methods are used in slab-at-grade foundation construction. In one method, the foundation walls are poured first. The slab is poured separately after the walls have set up. See Figure 39–38. This method is recommended for colder climates. Concrete floors

*Conco Cement, Inc.*

Figure 39–35. When all the grade-beam sections have been placed, they are adjusted to a line.

*Conco Cement, Inc.*

Figure 39–36. Section of the completed grade-beam foundation. Construction of the floor unit is in progress. Note that the plastic pier forms are still in place.

*Portland Cement Association*

Figure 39–37. Transit-mix trucks preparing to place concrete in slab-at-grade concrete forms. Note rebars placed over the ground and screeds running between the outer form boards.

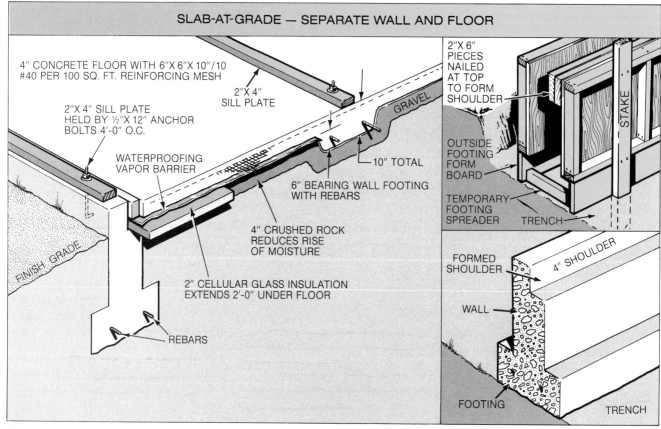

Figure 39–38. In this slab-at-grade system, the concrete for the foundation walls and floor slab is poured separately. This design provides a shoulder at the top of the wall, which helps support the edges of the floor slab.

tend to be cold because of heat loss occurring around the edge of the slab and exterior foundation walls. To help prevent such heat loss, a rigid type of insulation should be placed around the perimeter of the slab. The insulation should also extend 2′ beneath the concrete floor.

In warmer climates, the usual method is to place the concrete for the walls and floor at the same time (a monolithic pour). See Figure 39–39. The foundation walls and floor slab are one unit.

General rules to observe when constructing slab-at-grade foundations are:

1. Remove the topsoil in the slab area. Replace it with a 4″ to 6″ layer of gravel to help prevent ground water from collecting beneath the floor.

2. After the topsoil has been removed, but before the gravel is laid down, install all required

pipes and ducts.

3. Place a moisture-resistant *vapor barrier* over the gravel. A material often used for this purpose is 6 mil polyethylene plastic. An older method is to place 45-pound or heavier roofing paper. Whatever material is placed, joints should be lapped at least 4″. Care must be taken not to puncture the material while it is being spread over the gravel.

4. The slab should be at least 4″ thick and reinforced by 6″ × 6″ × 10/10 wire mesh.

## Post-Tensioned Slabs

Post-tensioned slabs are used in conjunction with slab-at-grade floor systems in residential as well as industrial construction. Post-tensioned slabs improve crack control with fewer joints and also reduce

slab deflection. Less excavation is required and installation of the slab is faster because less reinforcing material is needed.

In post-tensioned slabs, high strength tendons (cables) are used in place of wire mesh or conventional rebars. In a typical 4″ slab, the tendons are positioned at the center of the slab thickness and run in both directions at 2′ to 5′ intervals. See Figure 39–40. Slabs constructed over less stable soil conditions may require post-tensioned ground beams spaced 10′ to 20′ apart.

The concrete is placed after the tendons have been positioned. When the concrete has reached sufficient strength, the tendons are stressed by hydraulic jacks to an effective force of about 25,000 pounds. Anchoring devices at the ends of the tendons permanently transfer this force to the concrete slab. See Figure 39–41.

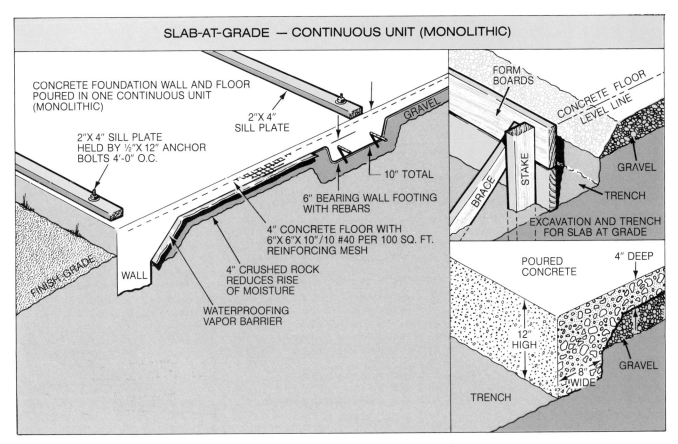

## SLAB-AT-GRADE — CONTINUOUS UNIT (MONOLITHIC)

CONCRETE FOUNDATION WALL AND FLOOR POURED IN ONE CONTINUOUS UNIT (MONOLITHIC)

2"X 4" SILL PLATE HELD BY ½"X 12" ANCHOR BOLTS 4'-0" O.C.

2"X 4" SILL PLATE

GRAVEL

10" TOTAL

6" BEARING WALL FOOTING WITH REBARS

4" CONCRETE FLOOR WITH 6"X 6"X 10"/10 #40 PER 100 SQ. FT. REINFORCING MESH

4" CRUSHED ROCK REDUCES RISE OF MOISTURE

WATERPROOFING VAPOR BARRIER

FINISH GRADE

WALL

FORM BOARDS

CONCRETE FLOOR LEVEL LINE

GRAVEL

BRACE

STAKE

TRENCH

EXCAVATION AND TRENCH FOR SLAB AT GRADE

POURED CONCRETE

4" DEEP

12" HIGH

8" WIDE

GRAVEL

TRENCH

Figure 39-39. The walls and floor slab form a one continuous unit in this monolithic slab-at-grade system.

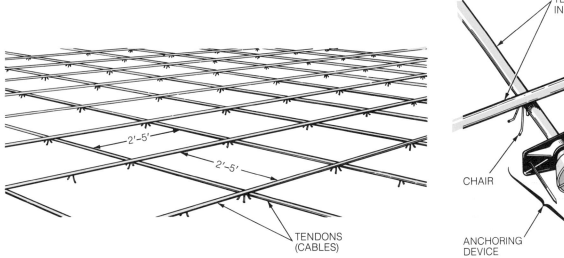

2'-5'

2'-5'

TENDONS (CABLES)

Figure 39-40. Tendons run in both directions at 2' to 5' intervals in a typical 4" thick post-tensioned slab.

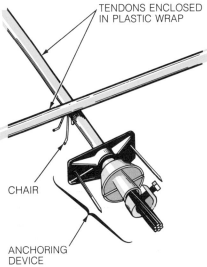

TENDONS ENCLOSED IN PLASTIC WRAP

CHAIR

ANCHORING DEVICE

Figure 39-41. Anchoring devices hold the ends of the stressed tendons.

# UNIT 40

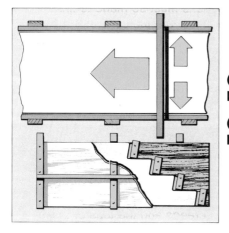

# Stairway and Outdoor Slab Forms

## STAIRWAY FORMS

The height of the first floor of a building may require construction of an entrance platform and stairway. Concrete is a very practical material for this purpose. It will hold up indefinitely and does not rot under damp conditions.

Figure 40–1 shows a procedure for building the forms for an entrance platform and stairway. The riser height is 7½″ and the tread width is 10″. (Additional information on stairways in concrete buildings is given in Section 16.)

## OUTDOOR SLAB FORMS

Outdoor concrete slabs include sidewalks, patios, and driveways. Information regarding the widths and locations of these areas is given in the plot plan of the blueprints. Forms for sidewalks, patios, and driveways are usually set up near the end of the construction work on a building. Final grading of the soil should be completed and the ground settled and compacted before these forms are set up.

Slabs for sidewalks, patios,

and driveways are usually placed directly on the soil. If the site has much surface water, or if problems could result from frost conditions, a layer of gravel should

be laid before the concrete is placed. In most areas of the United States either carpenters or cement masons may set the forms for the slabs. These forms

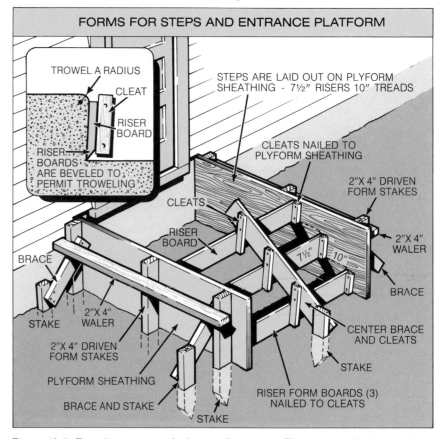

Figure 40–1. Form for entrance platform and stairway. The steps are laid out on the plywood sheathing. Cleats are nailed to the sheathing. The riser form boards are placed against the cleats. For wide steps a brace is placed at the center of the riser boards.

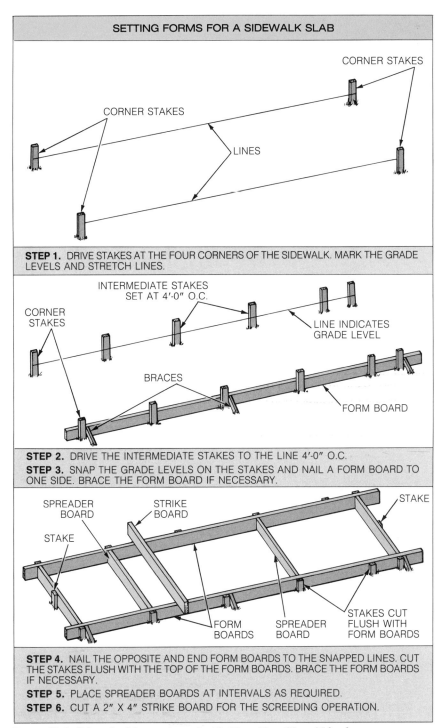

**SETTING FORMS FOR A SIDEWALK SLAB**

CORNER STAKES

CORNER STAKES

LINES

**STEP 1.** DRIVE STAKES AT THE FOUR CORNERS OF THE SIDEWALK. MARK THE GRADE LEVELS AND STRETCH LINES.

INTERMEDIATE STAKES
SET AT 4'-0" O.C.

CORNER
STAKES

LINE INDICATES
GRADE LEVEL

BRACES

FORM BOARD

**STEP 2.** DRIVE THE INTERMEDIATE STAKES TO THE LINE 4'-0" O.C.
**STEP 3.** SNAP THE GRADE LEVELS ON THE STAKES AND NAIL A FORM BOARD TO ONE SIDE. BRACE THE FORM BOARD IF NECESSARY.

SPREADER
BOARD

STRIKE
BOARD

STAKE

STAKE

STAKES CUT
FLUSH WITH
FORM BOARDS

FORM
BOARDS

SPREADER
BOARD

**STEP 4.** NAIL THE OPPOSITE AND END FORM BOARDS TO THE SNAPPED LINES. CUT THE STAKES FLUSH WITH THE TOP OF THE FORM BOARDS. BRACE THE FORM BOARDS IF NECESSARY.
**STEP 5.** PLACE SPREADER BOARDS AT INTERVALS AS REQUIRED.
**STEP 6.** CUT A 2" × 4" STRIKE BOARD FOR THE SCREEDING OPERATION.

Figure 40–2. Sidewalk slabs are formed with 2" × 4" or 2" × 6" lumber.

shown in Figure 40–2.

When concrete is placed for any type of outdoor slab, provisions must be made to finish off *(strike off)* the concrete to the required level. For a sidewalk form, a straightedge known as a *screed board* (also called a *strike board*) is laid across the top of the outer form boards. Refer again to Figure 40–2.

For a wider slab, such as for a patio or terrace, temporary wood or metal pieces must be placed at intervals in the slab area. These are also referred to as *screeds* and the tops of these pieces are set to the desired finish surface of the concrete. The actual screeding operation is accomplished by moving the screed board back and forth with a saw-like motion after the concrete had been placed. See Figure 40–3. The screeding operation can also be done with power-driven screeds that make the job much easier. See Figure 40–4. When all the concrete has been placed and struck off, cement masons work the concrete to the desired finish.

## Sidewalks and Patios

The width required for a walk depends on whether it is a sidewalk, front walk, or service walk. Often local building codes provide these specifications. Usually walks and patios are 4" thick, although walks should be thicker if trucks will frequently cross over them.

The slope across the width of the walk, known as the *cross slope*, should be ⅛" to ¼" per foot to allow water to drain. A slope of 1" in 12' is sufficient for patios.

## Driveways

Small-car driveways are usually 8' to 9' wide. Double-car drive-

usually are 2" × 4" or 2" × 6" pieces placed on edge and held in place by stakes. *Spreader boards* are placed at intervals to act as spacers between the outside form boards. They also tem-porarily hold back the concrete while it is being poured in different sections of the form. The spreader boards are removed as the concrete is poured. A procedure for setting sidewalk forms is

Cronkhite Industries, Inc.

Figure 40-3. Screeding concrete for an outdoor patio. The screed board is moved back and forth with a sawlike motion.

Portland Cement Association

Figure 40-5. Cement mason cutting a control joint with a straightedge and hand groover.

ways range from 15' to 18' wide. Driveways used for passenger cars should be at least 4" thick. Those used for trucks or other heavy vehicles should be 5" thick.

Driveways can be reinforced with wire mesh to prevent cracking. Sometimes reinforcing bars are used instead of wire mesh. There should be a minimum cross slope of ¼" to the foot for water drainage.

## Control and Isolation Joints

The purpose of a *control joint* is to control cracking in the concrete slab. Grooves are cut at intervals to a depth of one-fourth the slab thickness. This can be done with a special grooving tool at the time the concrete is being finished off. See Figure 40-5. Control joints may also be cut in after the concrete has set up with power saws equipped with special abrasive blades.

Control joints should be spaced not more than 40" apart for walks 2' wide and 60" apart for walks 3' or wider. In driveways, control joints running across the width should be no more than 10' apart. For driveways 12' wide or more, a control joint is also recommended along the length of the drive.

*Isolation* joints should be provided to separate dissimilar construction. An isolation joint runs through the entire slab thickness. For example, an isolation joint should be used where the side of a walk butts up against a foundation wall, steps, or driveway. An asphalt-impregnated strip or other type of material is normally placed at the juncture of the isolation joint.

Cronkhite Industries, Inc.

Figure 40-4. Screeding operations are more efficient with mechanical screeding equipment.

# UNIT 41

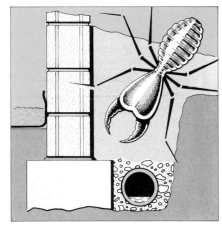

# Moisture Control and Termite Prevention for Foundations

## MOISTURE CONTROL

Water conditions on and below the surface of the ground must be considered in the construction of a foundation. Precautions must be taken to make sure that water does not leak into the living area of a full-basement foundation. Also, water collecting on the ground inside and surrounding a crawl-space foundation creates dampness that can cause wood decay, unpleasant odors, mold, and rust.

Water from rain and melting snow that stays on the surface of the ground is called *surface water.* The finished grades around the foundation should be sloped so that this water flows away from the building. (See discussion of *swale* in Section 6.)

Most problems related to water conditions of the ground concern the *water table,* which is located beneath the surface of the ground. The water table is the highest point below the surface of the ground that is normally saturated with water in a given area. Water tables tend to rise during wet seasons and subside during dry seasons.

| CAPILLARY RISE | SOIL TYPE | SATURATION ZONE |
|---|---|---|
| 11.5' | CLAY | 5.7' |
| 11.5' | SILT | 5.7' |
| 7.5' | FINE SAND | 4.5' |
| 2.6' | COARSE SAND | 2.2' |
| 0.0 | GRAVEL | 0.0' |

FLOOR SURFACE
CONCRETE SLAB
VAPOR ONLY
CAPILLARY RISE
VAPOR AND WATER
SATURATION ZONE
WATER TABLE

Figure 41–1. Water moves toward the surface of most soils as a result of capillary action. It rises the highest in fine-grained soils such as clay or silt.

## Capillary Action

A certain amount of water and water vapor rise from the water table through a process known as *capillary action.* This occurs in all types of soil, although water will rise higher in more porous soils such as silt and clay than in less porous soils. See Figure 41–1. Even when the water table is well below the finish grade, capillary action can cause damp conditions at the surface of the ground.

Capillary action does not occur in coarse types of soil such as gravel. Therefore, one method of water control is to dig out some of the porous soil around a foundation and replace it with gravel.

## Ground Vapor Barriers

One approach to solving the problem of water collecting under a crawl-space foundation is to cover the ground with a vapor barrier. The material used must be resistant to decay and insect

269

attack. The material often used for this purpose is 4 mil polyethelene film. After the film is placed, it should be covered with a layer of pea gravel or sand. See Figure 41–2.

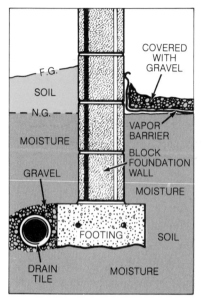

Figure 41-2. A vapor barrier helps prevent moisture from accumulating underneath a house with a crawl-space foundation.

## Waterproofing Foundation Walls

Basement foundation walls should be waterproofed from the edge of the footing up to the finish-grade line. A bituminous material such as asphalt is often applied to the outside surface of concrete and masonry walls for this purpose. See Figures 41–3 and 41–4. In some areas of the United States one or two layers of cement plaster are applied to a masonry wall before the asphalt is applied.

## Drain Tiles and Drain Pipes

One of the most effective ways to prevent problems from water below the surface of the ground is to collect the water and move it away from the foundation. This can be done by placing sections of pipe called *drain tile* or *drain pipe* beside the base of the

Figure 41-3. Asphalt is being applied to waterproof this masonry wall below the finish-grade line.

Figure 41-4. Celotex panels are placed over the asphalt to protect it from being punctured by the soil or gravel that will later be placed against the wall.

footing. All types of drain tiles or drain pipes must be surrounded by a layer of gravel.

One type of *drain tile* is positioned with a 1/4″ space between each section. Water enters the pipe through this open space and then flows along the pipe. A strip of asphalt building paper is placed over the top of the joint to prevent soil or gravel from falling into the pipe.

Another kind of perforated drain tile has holes at the bottom of the pipe sections. The water enters through the holes and then flows down the inside of the pipe.

Plastic drain pipes are available with corrugated and plain surfaces. Some types of plastic drain pipes have perforations, and others do not. See Figure 41–5.

A type of porous drain pipe has been introduced that is made of a mixture of portland cement, basaltic traprock, and sand. Water flows into the pipe through tiny channels too small to allow the passage of dirt, gravel, etc. The pipes are joined by self-sealing slip joints. Mortar or wrapping is

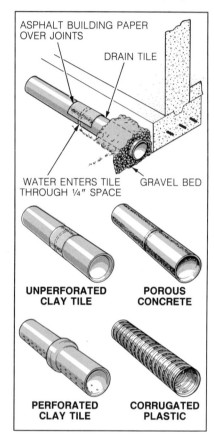

Figure 41-5. Various types of drain tiles and pipes are used to collect and move water away from the foundation.

not required. Porous drain pipes range in diameter from 4″ to 24″.

There should be a gradual slope to the drain tile or pipe of 1″ to every 20′. The collected water may simply drain toward a lower elevation and be absorbed by the ground some distance away from the building.

## Sump Well

Sometimes it is necessary to install a *sump well*. See Figure 41–6. The sump well is a pit that collects the water, which is then drained by gravity or by a pump to a storm sewer. Some local building codes do not allow sump water to be directed to a storm sewer, because the sewage system could become overloaded during periods of heavy rainfall. In this case the sump water is simply directed outside the building where it can drain away.

## TERMITE PREVENTION

Termites are small insects that can cause serious structural damage to wood buildings. Methods of preventing termite attack must be considered during the early stages of excavation and foundation construction.

There are two kinds of termites, *subterranean* and *non-subterranean*. The subterranean type accounts for 95% of all termite damage in the United States. These termites require warm and moist conditions to survive. In the past they were a serious problem only in the warmer climate areas of the South and West. However, in recent years subterranean termites have become more active in the colder, northern regions as well.

Improved methods of heating and insulating buildings have provided warm and favorable conditions for year-round termite activity.

Subterranean termites live in underground nests. They build tunnels to get to the surface of the earth. Termites attack the wood portion of a building in several ways. If the wood is in direct contact with the ground, it is a simple matter for the termite to penetrate the material. Termites also work their way up through small cracks in the foundation walls. They often build small mud tunnels up the outside of the foundation wall until they reach wood. The termites consume the wood but leave the outside shell to protect themselves. For this reason termite attack may go unnoticed until very serious damage has occurred. See Figures 41–7 and 41–8.

Non-subterranean termites are also called *drywood* termites because of their ability to live without moisture or contact with the ground. They, too, can cause damage, but they are responsible for much less damage than the subterranean type.

## Proper Building Construction for Termite Prevention

Termite prevention begins with the construction of the foundation. Any wood materials placed too close to the ground will increase the probability of termite attack. Local building codes usually specify the clearances required between wood members and the finish grades around the foundation.

## Moisture Control for Termite Prevention

Since subterranean termites need moist conditions to survive, moisture control is an effective preventive measure. This is particularly important for buildings with crawl-space foundations. The ground should be graded to prevent the accumulation of wa-

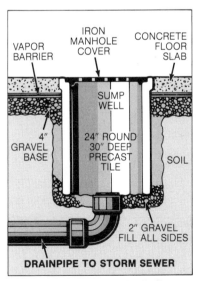

Figure 41-6. Sump well in the floor of a full-basement foundation system. Water seepage into the basement will flow toward the sump and collect in the well. It will then be drained by pipe to a storm sewer or to the outside of the building. Pumps are sometimes installed in the sump well.

*Terminix International, Inc.*

Figure 41-7. The arrows point to a mud termite tunnel coming from the ground in the basement area. The tunnel runs up a concrete column that supports one end of a wood girder. The termites have already caused serious damage to the girder.

*Terminix International, Inc.*

Figure 41-8. The termite damage in this floor went unnoticed until the floor began to crumble.

ter (ponding) within the foundation walls. The vapor barrier described earlier in this unit is helpful. An adequate number of vents should be provided to ensure good circulation of air beneath the building. (Methods of venting crawl-space areas are discussed in Section 9.)

## Treated Wood for Termite Prevention

An effective preventive measure against termites is to use lumber that has been treated with pre-servatives by a pressure process. The chemicals forced into the wood fibers increase the wood's resistance against termites.

## Termite Shields

Some local building codes call for metal termite shields. These are placed between the sill plate and the top of the foundation wall. The shield should be made of 24-ga. galvanized iron. It should extend 2″ on each side of the wall and be bent down at a 45° angle. Holes punched in the metal for foundation bolts must be properly sealed.

## Soil Treatment for Termite Prevention

The use of soil treatment to prevent termites has increased in recent years. Special soil poisons are applied to the ground during construction. These poisons form a chemical barrier that is toxic to termites and other harmful soil insects. However, because of environmental concerns, some local building codes prohibit this measure.

# Floor, Wall, and Ceiling Frame Construction

The *framework* of a building is its structural skeleton. It is covered with finish materials to complete the building. Because of the abundant supply of lumber in North America, framing with wood is the most common method of light construction in the United States and Canada. Recently there has been an increase in the use of steel as a framing material. The construction procedure for steel-framed walls is basically the same as for wood-framed walls.

# UNIT 42

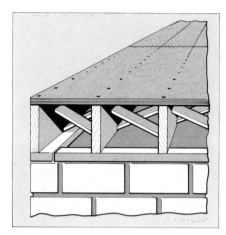

# Floor Framing

Floor framing begins after the foundation work has been completed. In platform construction the floor unit is framed directly over the foundation walls or short studded walls known as *underpinning*. Most floor units include *posts, girders, joists, bridging* or *blocking*, and the *subfloor*. See Figure 42–1. Posts and girders hold up the lapped or butted ends of the joists, or they may provide central support for longer joist spans. The subfloor is the wood deck that rests on top of the joists.

## FLOOR UNIT RESTING ON MUDSILLS

The floor unit may be framed directly on the mudsills of a building with a crawl-space foundation and low foundation walls. See Figure 42–2. In this case one end of the joists rests on the outside foundation walls. The lapped ends are on top of a foundation wall running down the center of the building. Posts and girders give central support to the long span of the girders. There should be at least 18″ clearance between the bottoms of the floor joists and the ground, and at least 12″ between the bottom of the girder and the ground.

The floor unit may also be framed directly on the mudsills of a building with a full-basement foundation and high foundation walls. In this case one end of the joists laps over a girder.

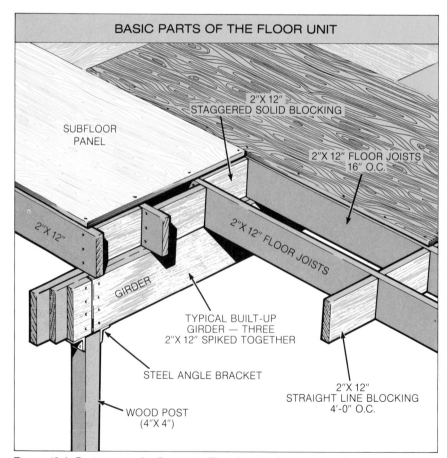

BASIC PARTS OF THE FLOOR UNIT

SUBFLOOR PANEL

2″X 12″ STAGGERED SOLID BLOCKING

2″X 12″ FLOOR JOISTS 16″ O.C.

2″X 12″

2″X 12″ FLOOR JOISTS

GIRDER

TYPICAL BUILT-UP GIRDER — THREE 2″X 12″ SPIKED TOGETHER

STEEL ANGLE BRACKET

WOOD POST (4″X 4″)

2″X 12″ STRAIGHT LINE BLOCKING 4′-0″ O.C.

Figure 42–1. Basic parts of a floor unit. Floor framing begins after the foundation work is completed.

274

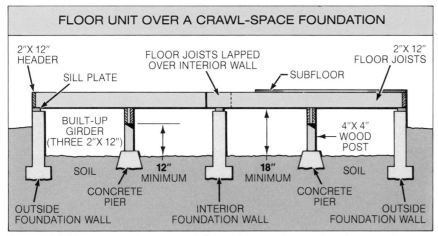

Figure 42-2. Floor unit resting on mudsills, with a crawl-space foundation underneath. There should be 18″ minimum clearance between the floor joists and the ground, and 12″ between the girder and the ground.

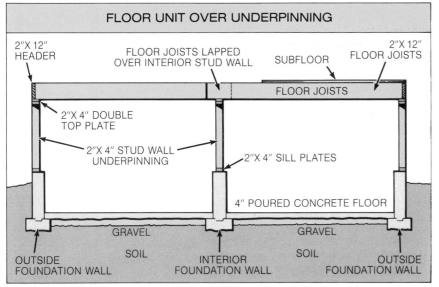

Figure 42-3. Floor unit resting on underpinning, with a full-basement foundation underneath.

## FLOOR UNIT RESTING ON UNDERPINNING

Underpinning is used to extend the height of low foundation walls, usually for a full-basement house. See Figure 42-3. Under some conditions this is less costly than the greater amount of concrete and formwork required for a higher foundation wall. The underpinning studs are toe-nailed to the mudsills and are commonly spaced 16″ O.C. Un-derpinning studs should be no smaller in cross section than wall studs above the floor. Floor joists placed on top of the wall should be directly above the underpinning stud.

A stepped foundation is an-other example of where under-pinning is used to reduce con-struction costs. This type of foundation is needed on hillside and steeply sloping lots. In a typical stepped foundation, the floor unit rests directly on the mudsill at the highest area of the foundation. Underpinning is placed over the lower walls, and in that area the floor unit rests on the underpinning.

## POSTS AND GIRDERS

Posts and girders help support the floor joists and subfloor material. Refer again to Figure 42-1. Posts and girders may be steel or wood, and their size depends upon the load they are required to carry. The dimensions and placement of the girders are shown on the foundation plan.

### Wood Posts

Wood posts are placed directly below wood girders. As a general rule, the width of the wood post should be equal to the width of the girder it supports. A 4″ wide girder requires 4″ × 4″ or 4″ × 6″ posts. An 8″ wide girder requires 6″ × 8″ or 8″ × 8″ posts.

The bottom of each post may be nailed to a pier block that is secured to the top of a concrete pier. Another method is to place a ½″ steel dowel in the concrete pier at the time of the pour. The dowel fits into a hole drilled at the bottom of the post and holds the post in position. See Figure 42-4. The dowel should extend at least 3″ into the concrete and 3″ into the post.

The top of each post is fastened to the bottom of the girder. An angle iron or some type of metal connector is usually used to make a stronger tie. Refer again to Figure 42-4.

Figure 42-5 shows two types of metal bases that are set into the concrete pier at the time of the pour. The wood post fits into the base, and no bolts or other fasteners need be attached to the concrete. These bases must be positioned in the concrete very accurately.

An adjustable metal base may also be used. It is secured to the concrete by means of a ½″ J-bolt set in the concrete at the time of the pour. A stand-off plate provides a flat bearing area for the post and keeps the post 1³⁄₁₆″ above the surface of the concrete. This serves to guard against wood rot and termite damage at the bottom of the post. A slotted adjustment plate permits movement for plumbing the post.

Figure 42–6 shows two types of metal post caps used to tie together posts and girders. These post caps can be nailed or bolted to the timbers. The twin-design post cap can be installed after the girder has been placed on top of the post since it comes in two pieces.

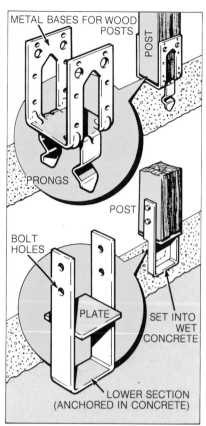

Figure 42–5. No bolts or other types of fasteners are necessary to secure these metal post bases to the concrete. The lower section of the post base is set in the concrete at the time of the pour.

## Wood Girders

Wood girders may consist of a solid piece of timber or they may be *built up* of more than one plank. A built-up girder may, for example, consist of three 2″ planks. See Figure 42–7. The joints between the planks are staggered. In framing, a built-up girder is placed so that the joints on the outsides of the girder fall directly over a post. Two 16d nails are driven at the ends of the planks, and other nails are staggered 32″ O.C.

The ends of a girder often rest in pockets prepared in the concrete wall. Refer again to Figure 42–7. When this method is used, the girder ends must bear at least 4″ on the wall, and the pocket should be large enough to provide ½″ air space around the sides and end of the girder. To protect against termites, it is advisable to treat the ends of the girder with a preservative. As a further precaution, the concrete

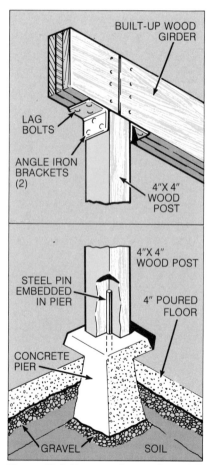

Figure 42–4. Posts are fastened at the top of the pier and at the bottom of the girder.

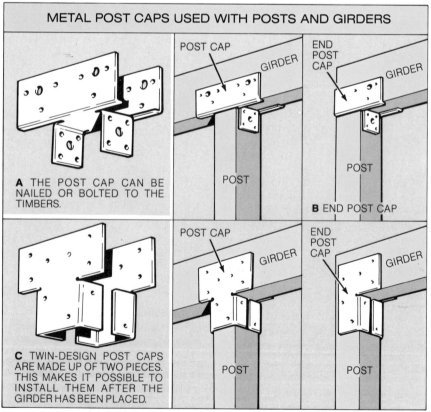

### METAL POST CAPS USED WITH POSTS AND GIRDERS

**A** THE POST CAP CAN BE NAILED OR BOLTED TO THE TIMBERS.

**C** TWIN-DESIGN POST CAPS ARE MADE UP OF TWO PIECES. THIS MAKES IT POSSIBLE TO INSTALL THEM AFTER THE GIRDER HAS BEEN PLACED.

POST CAP
GIRDER
POST

**B** END POST CAP

END POST CAP
GIRDER
POST

Figure 42–6. Examples of metal post caps used to tie together posts and girders.

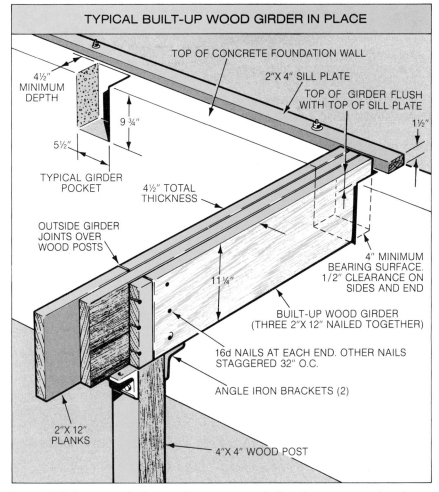

TYPICAL BUILT-UP WOOD GIRDER IN PLACE

- 4½" MINIMUM DEPTH
- 5½"
- 9¾"
- TYPICAL GIRDER POCKET
- OUTSIDE GIRDER JOINTS OVER WOOD POSTS
- 4½" TOTAL THICKNESS
- 11¼"
- 2"X 12" PLANKS
- TOP OF CONCRETE FOUNDATION WALL
- 2"X 4" SILL PLATE
- TOP OF GIRDER FLUSH WITH TOP OF SILL PLATE
- 1½"
- 4" MINIMUM BEARING SURFACE. 1/2" CLEARANCE ON SIDES AND END
- BUILT-UP WOOD GIRDER (THREE 2"X 12" NAILED TOGETHER)
- 16d NAILS AT EACH END. OTHER NAILS STAGGERED 32" O.C.
- ANGLE IRON BRACKETS (2)
- 4"X 4" WOOD POST

Figure 42-7. Joints on a built-up girder are staggered. Outside joints occur directly over a post.

wall pocket should be lined with metal.

Girders are classified as *bearing* and *non-bearing* according to the amount and type of load they support. Non-bearing girders must support the *dead load* and the *live load* of the floor system directly above. The dead load is the weight of the material used for the floor unit itself. The live load is the weight created by people, furniture, appliances, and so forth. Bearing girders must support a wall framed directly above, as well as the live load and dead load of the floor.

**Allowable Spans for Girders.** The *allowable span* is the distance between supporting posts or walls permitted for different size girders. Some examples of allowable spans are shown in the table in Figure 42-8.

## Steel Posts (Pipe Columns)

Steel pipe *(Lally)* columns are often used as posts in wood-framed buildings. They can be

| WIDTH OF STRUCTURE | GIRDER SIZE (SOLID OR BUILT-UP) | MAXIMUM SPAN | | | |
|---|---|---|---|---|---|
| | | SUPPORTING BEARING PARTITION | | SUPPORTING NON-BEARING PARTITION | INTERMEDIATE GIRDERS (OTHER THAN MAIN GIRDER) |
| | | 1 story | 1½ or 2 story | | |
| UP TO 26' WIDE....... | 4 × 6 .............. | ........ | ............. | 5' 6" | 7' 6" |
| | 4 × 8 .............. | ........ | ............. | 7' 6" | 9' 6" |
| | 6 × 8 .............. | 7' 0" | 6' 0" | 9' 0" | 12' 0" |
| | 6 × 10 ............ | 9' 0" | 7' 6" | 11' 6" | ................. |
| | 6 × 12 ............ | 10' 6" | 9' 0" | 12' 0" | ................. |
| 26' TO 32' WIDE ....... | 4 × 6 .............. | ........ | ............. | ........ | 6' 6" |
| | 4 × 8 .............. | ........ | ............. | 7' 0" | 8' 6" |
| | 6 × 8 .............. | 6' 6" | 5' 6" | 8' 6" | 10' 6" |
| | 6 × 10 ............ | 8' 0" | 7' 0" | 10' 6" | 13' 6" |
| | 6 × 12 ............ | 10' 0" | 8' 0" | 11' 6" | ................. |

NOTE: The above spans are based upon an allowable fiber stress of 1,500 psi. These allowable stresses are average values, taking into consideration upgrading for members in built-up beams. Where conditions vary from these assumptions, design girders in accordance with standard engineering practice.

*FHA Minimum Property Standards for One and Two Family Dwellings*

Figure 42-8. Allowable spans for girders. The first underlined example shows that a 6 × 10 bearing girder covering a distance up to 26' requires posts every 9' in a one-story house. The second underlined example shows that a 6 × 8 intermediate girder covering a distance of 26' to 32' requires posts every 10'-6".

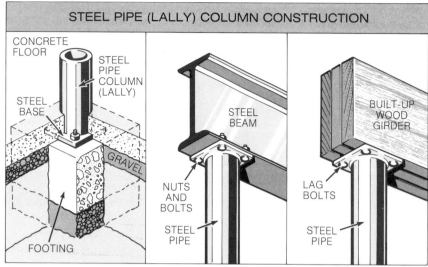

STEEL PIPE (LALLY) COLUMN CONSTRUCTION

Figure 42–9. Steel pipe (Lally) columns are frequently used to support wood girders or steel beams.

Figure 42–10. These steel pipe (Lally) columns will support the ends of steel beams. The columns will be inside a masonry foundation constructed over the concrete footings. Note the vertical rebars extending from the footings.

placed beneath either wood girders or steel beams. See Figure 42–9. The cap at the top of a steel pipe column is secured with lag bolts when it is attached to a wood girder. Machine bolts are used when it is attached to a steel beam. The base of the column is bolted to the top of the pier or floor slab. The bolts holding the base must be accurately set into the concrete at the time of the pour.

Figure 42–10 shows steel pipe columns positioned over concrete footings in readiness to support the ends of steel beams. The columns will be inside a masonry foundation constructed over the concrete footings. Figure 42–11 shows a steel pipe column supporting a steel beam.

shows wood plates attached to the top of the steel beams.

## Placing Wood Posts and Girders

Posts must be cut to length and set up before the girders can be installed. The upper surface of the girder may be in line with the foundation mudsill, or the girder

## Steel Beams

A standard steel beam called a *wide-flange beam* is used most often in connection with wood framing. The wood joists either rest on top of the beam or butt up against the sides of the beam. See Figure 42–11 and Figure 42–12. Figure 42–12

Figure 42–11. This wide-flange steel beam is supported by steel pipe columns. The far end of the beam is supported by a pipe column inside the masonry foundation wall.

ends may rest on top of the walls. Long girders must be placed in sections. Solid girders must be measured and cut so that the ends will fall over the center of a post. Built-up girders should be placed so that their outside joints fall over the posts. Refer again to Figure 42–7.

One procedure for placing posts and girders is shown in Figure 42–13. The top of the girder should be in line with the surface of the mudsill.

## FLOOR JOISTS

In platform framing, one end of the floor joist rests directly on the mudsill of the exterior foundation wall or on the top plate of a framed outside wall. The bearing should be at least 1½". The opposite end of the joist laps over or butts into an interior girder or wall. The size of joist material (2" × 6", 2" × 10", 2" × 12", and so forth) must be chosen with consideration for the span and the amount of load to be carried. The foundation plan usually specifies the joist size, the spacing between joists, and what direction the joists should travel.

The usual spacing of floor joists is 16" O.C. However, some newer systems have been developed that allow wider spacing (24" to 32") between floor joists.

### Allowable Spans for Joists

The table in Figure 42–14 gives the allowable joist spans for various species of framing lumber. These spans are based on a strength requirement of 10 pounds per square foot of dead load, plus 40 pounds per square foot of live load. This table can be used to discover, for example, that 2" × 8" joists of Douglas fir or larch, No. 2 grade, spaced 16" O.C., have an allowable span of 13'-1". As another example, 2" × 12" joists of Idaho

white pine, No. 3 grade, spaced 24" O.C., have an allowable span of 10'-9".

## Support of Joists over Exterior Walls

Floor joists are supported and held in position over exterior walls by *header joists* or by *solid blocking* between the joists. The header-joist system is used more often.

**Header Joists.** Header joists, also known as *rim* or *band* joists,

run along the outside walls. Three 16d nails are driven through the header joists into the ends of the regular joists. See

Figure 42–12. Wood plates may be attached to the top of steel beams with a powder-actuated tool.

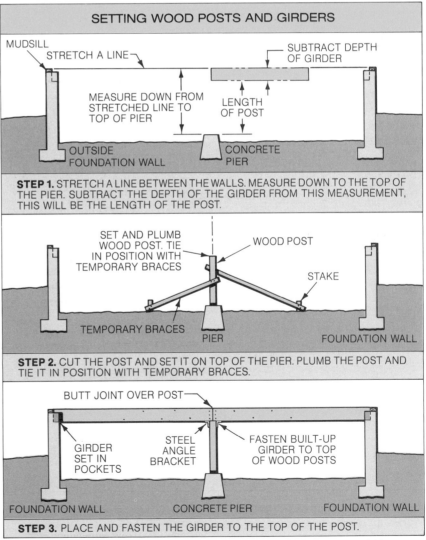
Figure 42–13. Placing wood posts and girders. The top of each girder should be in line with the surface of the mudsill.

| SPECIES | GRADE | SPAN (FEET AND INCHES) | | | | | |
|---|---|---|---|---|---|---|---|
| | | 2 × 8 | | 2 × 10 | | 2 × 12 | |
| | | 16" oc | 24" oc | 16" oc | 24" oc | 16" oc | 24" oc |
| DOUGLAS FIR-LARCH | 2 | 13–1 | 11–3 | 16–9 | 14–5 | 20–4 | 17–6 |
| | 3 | 10–7 | 8–8 | 13–6 | 11–0 | 16–5 | 13–5 |
| DOUGLAS FIR SOUTH | 2 | 12–0 | 10–6 | 15–3 | 13–4 | 18–7 | 16–3 |
| | 3 | 10–3 | 8–4 | 13–1 | 10–8 | 15–11 | 13–0 |
| HEM-FIR | 2 | 12–3 | 10–0 | 15–8 | 12–10 | 19–1 | 15–7 |
| | 3 | 9–5 | 7–8 | 12–0 | 9–10 | 14–7 | 11–11 |
| MOUNTAIN HEMLOCK | 2 | 11–4 | 9–11 | 14–6 | 12–8 | 17–7 | 15–4 |
| | 3 | 9–7 | 7–10 | 12–3 | 10–0 | 14–11 | 12–2 |
| MOUNTAIN HEMLOCK-HEM-FIR | 2 | 11–4 | 9–11 | 14–6 | 12–8 | 17–7 | 15–4 |
| | 3 | 9–5 | 7–8 | 12–0 | 9–10 | 14–7 | 11–11 |
| WESTERN HEMLOCK | 2 | 12–3 | 10–6 | 15–8 | 13–4 | 19–1 | 16–3 |
| | 3 | 9–11 | 8–1 | 12–8 | 10–4 | 15–5 | 12–7 |
| ENGELMANN SPRUCE ALPINE FIR (Engelmann Spruce-Lodgepole Pine) | 2 | 11–2 | 9–1 | 14–3 | 11–7 | 17–3 | 14–2 |
| | 3 | 8–6 | 6–11 | 10–10 | 8–10 | 13–2 | 10–9 |
| LODGEPOLE PINE | 2 | 11–8 | 9–7 | 14–11 | 12–3 | 18–1 | 14–11 |
| | 3 | 9–1 | 7–5 | 11–7 | 9–5 | 14–1 | 11–6 |
| PONDEROSA PINE-SUGAR PINE (Ponderosa Pine-Lodgepole Pine) | 2 | 11–4 | 9–3 | 14–5 | 11–9 | 17–7 | 14–4 |
| | 3 | 8–8 | 7–1 | 11–1 | 9–1 | 13–6 | 11–0 |
| WHITE WOODS (Western Woods) | 2 | 11–0 | 9–0 | 14–0 | 11–6 | 17–0 | 14–0 |
| | 3 | 8–6 | 6–11 | 10–10 | 8–10 | 13–2 | 10–9 |
| IDAHO WHITE PINE | 2 | 11–0 | 9–0 | 14–0 | 11–6 | 17–1 | 14–0 |
| | 3 | 8–6 | 6–11 | 10–10 | 8–10 | 13–2 | 10–9 |
| WESTERN CEDARS | 2 | 11–0 | 9–7 | 14–0 | 12–3 | 17–0 | 14–11 |
| | 3 | 9–1 | 7–5 | 11–6 | 9–5 | 14–0 | 11–6 |

Design Criteria:
  Strength—10 lbs. per sq. ft. dead load plus 40 lbs. per sq. ft. live load.
  Deflection—Limited to span in inches divided by 360 for live load only.

*Western Wood Products Association*

Figure 42–14. Allowable spans for floor joists. The allowable span is the distance permitted between supports.

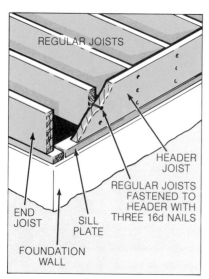

Figure 42–15. Regular joists are fastened to the header joist.

Figure 42–15. The header joists prevent the regular joists from rolling or tipping. They also help support the wall above and fill in the spaces between the regular joists.

**Blocking between Joists.** Another system of providing exterior support to joists is to place solid blocking between the outside ends of the joists. In this way the end of the joist has fuller bearing on the outside walls.

## Interior Support of Joists

Floor joists usually run across the full width of the building. However, extremely long joists are expensive and difficult to handle. Therefore, two or more shorter joists are used instead. The ends of these joists are supported by lapping or butting them over a girder, butting them against a girder, lapping them over a wall, attaching them over a steel beam, or butting them against a steel beam.

**Joists Lapped over a Girder or Wall.** Often joists are lapped over a girder running down the center of a building. The lapped ends of the joists may also be supported by an interior foundation or framed wall. It is standard procedure to lap joists the full width of the girder or wall. The minimum lap should be 4". Figure 42–16 shows lapped joists resting on a steel girder. Solid blocking is installed between the lapped ends after all the joists have been nailed down. See figure 42–17. Another system is to put in the blocks at the time the joists are placed.

**Joists Butted over a Girder.** The ends of the joists can also be

280

Figure 42–16. Carpenters setting the floor joists. The lapped ends of the joists rest on a steel girder supported by round steel pipe columns.

*Georgia-Pacific Corporation*

vide additional support by nailing a ledger board beneath the joists. See Figure 42–19.

**Joists Supported by Steel Beam.** Often wood joists are supported by a steel beam instead of a wood girder. The joists may rest on top of the steel beam (Figure 42–20), or they may be butted (and notched to fit) against the sides of the beam (Figure 42–21).

If the joists are to rest on top of the steel beam, a plate is fastened to the beam, and the joists are toenailed into the plate.

When joists are notched to fit against the sides of the beam,

butted (rather than lapped) over a girder. The joists should then be scabbed together with a wood or metal tie. See Figure 42–18. The ties can be left out if the line of panels from the plywood subfloor straddles the butt joints.

**Joists Butted against a Girder.** Butting joists against (rather than over) a girder allows for more headroom below the

girder. Metal joist hangers, as well as other types of metal framing anchors, are usually used to support and tie the ends of the joists to the girders. These are the same types of metal anchors used for floor openings and shown in Figure 42–29 on page 286. An older method that does not require metal anchors is to notch the joists over the girder and pro-

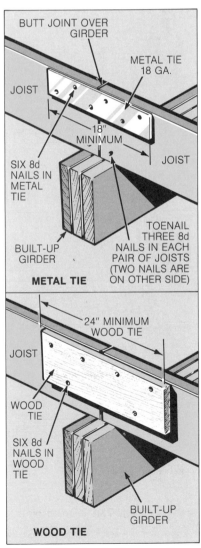

Figure 42–18. Joists butting over the girder. A wood or metal tie is used.

Figure 42–17. Blocking is placed between the lapped ends of joists. (These joints are lapped over wood girders.)

281

⅜" clearance should be allowed above the top flange. This permits the wood some shrinkage without splitting. Sometimes opposite pairs of joists are scabbed across the top with a metal strap. Also, a wide plate may be welded to the bottom of the beam to provide better support, and wood blocks may be placed at the bottom of the joists to help keep them in position.

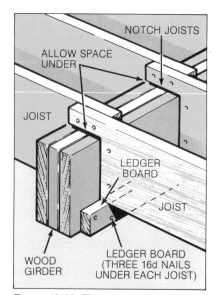

Figure 42–19. The joists are notched over the girder. A ledger board nailed to the girder provides support underneath the joists. Three 16d nails are driven into the ledger under each joist. The joists are toenailed into the girder with three 10d nails.

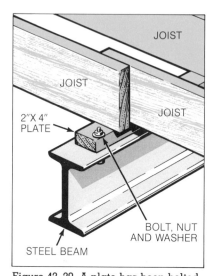

Figure 42–20. A plate has been bolted to the top of the steel beam. The joists are toenailed into the plate. (The plate may also be attached to the steel beam with a powder-actuated tool).

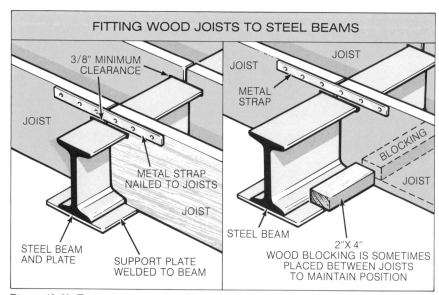

Figure 42–21. Fitting joists against the sides of a steel beam. Sometimes the opposite pairs are scabbed across the top with a metal strap. Also, a wide plate may be welded to the bottom of the beam to provide better support. Wood blocks are sometimes placed at the bottom of the joists to help keep them in position.

## Doubled Joists

Joists should be doubled under partitions that run in the same direction as the joists. Some walls have water pipes, vent stacks, or heating ducts coming up from the basement or the floor below. Blocks can be placed between the doubled joists to allow space for these purposes.

## Cantilevered Joists

Cantilevered joists are used when a floor or balcony of a building projects past the wall below it. See Figure 42–22. A header piece is nailed to the ends of the joists.

When regular floor joists run parallel to the intended overhang of the floor, the inside ends of the cantilevered joists are fastened to a pair of double joists. See Figure 42–23. Nailing should be through the first regular joist into the ends of the cantilevered joists. Framing anchors are strongly recommended and often required by the building code. A header piece is also nailed to the outside ends of the cantilevered joists.

## Bridging between Joists

Many local building codes call for bridging between the joists. The bridging holds the joists in line and helps distribute the load carried by the floor unit. It is usually required when the joist spans are more than 8'. Joists having a 15' span need one row of bridging down the center of the span. For longer spans, such as 18', two rows of bridging spaced 6' apart are required.

**Cross Bridging.** Also known as *herringbone* bridging, cross bridging usually consists of 1" × 3" or 2" × 3" wood. It is installed as shown in Figure 42–24.

Cross bridging is toenailed at each end with 6d or 8d nails. The pieces are usually pre-cut on a radial-arm saw. The nails are started at each end before the cross bridging is placed between the joists. The usual procedure is to fasten only the top end of the cross bridging. The nails at the bottom end are not driven in until the subfloor has been placed. Otherwise the joist could be pushed out of line when the bridging is nailed in. An effi-

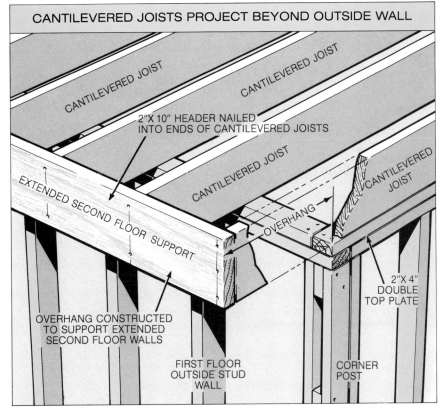

Figure 42–22. Cantilevered joists provide support for a wall that will project past the wall below.

CANTILEVERED JOISTS PROJECT BEYOND OUTSIDE WALL

CANTILEVERED JOIST

CANTILEVERED JOIST

CANTILEVERED JOIST

CANTILEVERED JOIST

CANTILEVERED JOIST

2"X 10" HEADER NAILED INTO ENDS OF CANTILEVERED JOISTS

EXTENDED SECOND FLOOR SUPPORT

OVERHANG

2"X 4" DOUBLE TOP PLATE

OVERHANG CONSTRUCTED TO SUPPORT EXTENDED SECOND FLOOR WALLS

FIRST FLOOR OUTSIDE STUD WALL

CORNER POST

CANTILEVERED JOIST FRAMING (PARALLEL JOISTS)

REGULAR JOISTS

DOUBLE JOISTS

PLYWOOD SUBFLOOR

FRAMING ANCHORS OR JOIST HANGERS

CANTILEVERED JOISTS

CANTILEVERED JOISTS

CANTILEVERED JOISTS

HEADER

OVERHANG

2"X 10" HEADER

2"X 4" DOUBLE TOP PLATE

HEADER NAILED INTO ENDS OF CANTILEVERED JOISTS

OUTSIDE STUD WALL

OUTSIDE STUD WALL

Figure 42–23. Framing for cantilevered joists where regular joists are parallel to the projecting wall.

cient method for placing wood cross bridging is shown in Figure 42–25.

Another approved system of cross bridging uses metal pieces instead of wood and requires no nails. The pieces are available for 12″, 16″, and 18″ joist spacing. The installation of this type of cross bridging is shown in Figure 42–26.

**Solid Bridging.** Also known as *solid blocking,* solid bridging serves the same purpose as cross bridging. This method is preferred by many carpenters and builders to cross bridging. The pieces are cut from lumber that is the same width as the joist material. They can be installed in a straight line or staggered. If staggered, the blocks can be spiked from both ends, resulting in a faster nailing operation. Straight lines of blocking may be required every 4′ O.C. to provide a nailing base for the plywood subfloor.

## Floor, Fireplace, and Chimney Openings in Joists

A floor opening must be framed where stairs rise to the floor. Fireplaces and chimneys also require special framed floor openings. When the joists are cut for such openings, there is a loss of

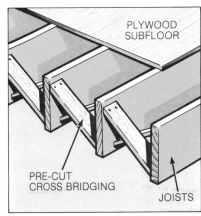

PLYWOOD SUBFLOOR

PRE-CUT CROSS BRIDGING

JOISTS

Figure 42–24. Cross bridging is pre-cut to fit between the joist spans.

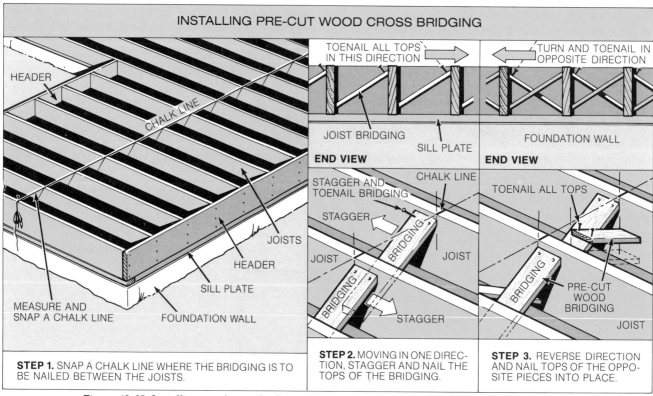

Figure 42–25. Installing wood cross bridging. Pieces are toenailed at each end with 6d or 8d nails.

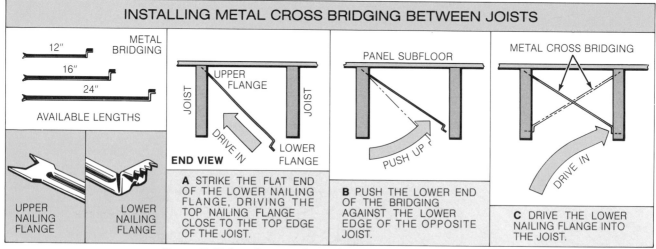

Figure 42–26. Installing metal cross bridging. No nails are required.

strength in the area of the opening. The opening must be framed in a way that restores this lost strength. Figure 42–27 shows the procedure for framing an opening.

A pair of joists called *trimmers* are placed at each side of the opening. These trimmers support the headers. The headers should be doubled if the span is more

than 4'. Nails supporting the ends of the headers are driven through the trimmer joist into the ends of the header pieces.

*Tail joists* run from the header to a supporting wall or girder. Nails are driven through the header into the ends of the tail joist. Figure 42–28 shows headers nailed to trimmers, and tail joists to headers.

Various metal anchors are also used to add strength to the framed floor opening. See Figure 42–29.

## Placing Floor Joists

Before the floor joists are placed, the sill plates and the girders must be marked to show where the joists are to be nailed. Floor

# TYPICAL FRAMING PROCEDURE FOR FLOOR OPENING

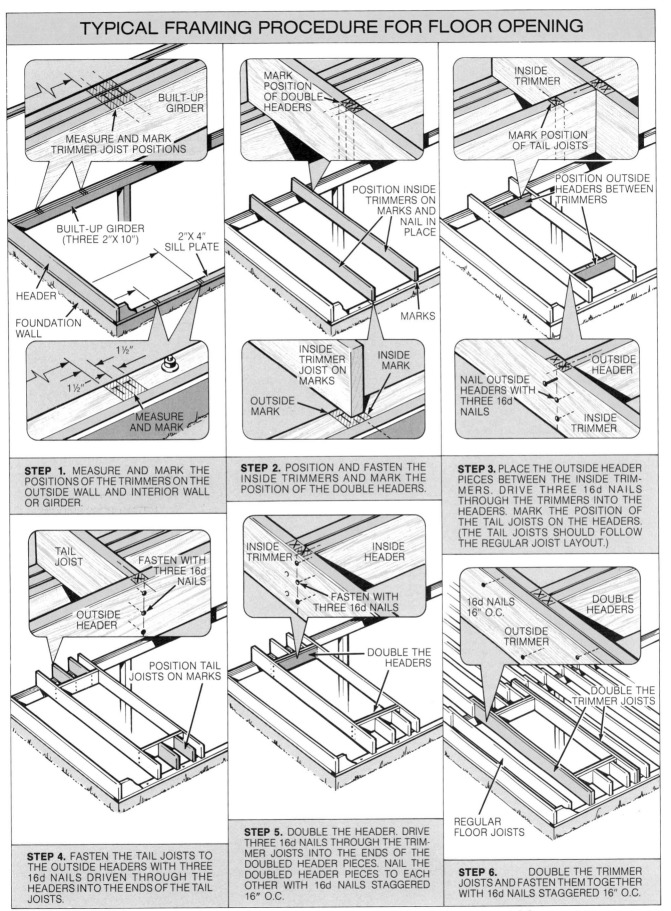

BUILT-UP GIRDER

MEASURE AND MARK TRIMMER JOIST POSITIONS

BUILT-UP GIRDER (THREE 2"X 10")

2"X 4" SILL PLATE

HEADER

FOUNDATION WALL

1½"

1½"

MEASURE AND MARK

MARK POSITION OF DOUBLE HEADERS

POSITION INSIDE TRIMMERS ON MARKS AND NAIL IN PLACE

MARKS

INSIDE TRIMMER JOIST ON MARKS

INSIDE MARK

OUTSIDE MARK

INSIDE TRIMMER

MARK POSITION OF TAIL JOISTS

POSITION OUTSIDE HEADERS BETWEEN TRIMMERS

OUTSIDE HEADER

NAIL OUTSIDE HEADERS WITH THREE 16d NAILS

INSIDE TRIMMER

**STEP 1.** MEASURE AND MARK THE POSITIONS OF THE TRIMMERS ON THE OUTSIDE WALL AND INTERIOR WALL OR GIRDER.

**STEP 2.** POSITION AND FASTEN THE INSIDE TRIMMERS AND MARK THE POSITION OF THE DOUBLE HEADERS.

**STEP 3.** PLACE THE OUTSIDE HEADER PIECES BETWEEN THE INSIDE TRIMMERS. DRIVE THREE 16d NAILS THROUGH THE TRIMMERS INTO THE HEADERS. MARK THE POSITION OF THE TAIL JOISTS ON THE HEADERS. (THE TAIL JOISTS SHOULD FOLLOW THE REGULAR JOIST LAYOUT.)

TAIL JOIST

FASTEN WITH THREE 16d NAILS

OUTSIDE HEADER

POSITION TAIL JOISTS ON MARKS

INSIDE TRIMMER

INSIDE HEADER

FASTEN WITH THREE 16d NAILS

DOUBLE THE HEADERS

16d NAILS 16" O.C.

DOUBLE HEADERS

OUTSIDE TRIMMER

DOUBLE THE TRIMMER JOISTS

REGULAR FLOOR JOISTS

**STEP 4.** FASTEN THE TAIL JOISTS TO THE OUTSIDE HEADERS WITH THREE 16d NAILS DRIVEN THROUGH THE HEADERS INTO THE ENDS OF THE TAIL JOISTS.

**STEP 5.** DOUBLE THE HEADER. DRIVE THREE 16d NAILS THROUGH THE TRIMMER JOISTS INTO THE ENDS OF THE DOUBLED HEADER PIECES. NAIL THE DOUBLED HEADER PIECES TO EACH OTHER WITH 16d NAILS STAGGERED 16" O.C.

**STEP 6.** DOUBLE THE TRIMMER JOISTS AND FASTEN THEM TOGETHER WITH 16d NAILS STAGGERED 16" O.C.

Figure 42–27. Framing a floor opening. Stairways, fireplaces, and chimneys require framed floor openings.

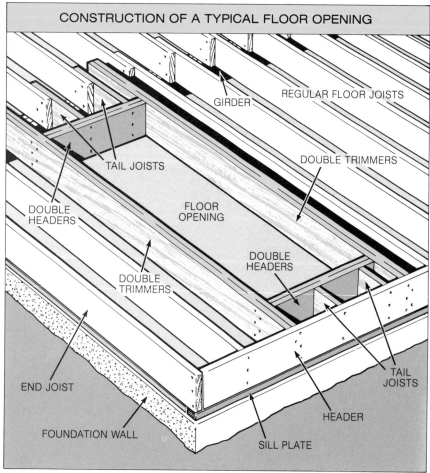

CONSTRUCTION OF A TYPICAL FLOOR OPENING

GIRDER

REGULAR FLOOR JOISTS

TAIL JOISTS

DOUBLE TRIMMERS

DOUBLE HEADERS

FLOOR OPENING

DOUBLE HEADERS

DOUBLE TRIMMERS

TAIL JOISTS

END JOIST

HEADER

FOUNDATION WALL

SILL PLATE

Figure 42–28. Framework of a typical floor opening. Headers are nailed to trimmers, and tail joists to headers.

joists are usually placed 16″ O.C. Most carpenters use a 12′ or longer steel tape rule for layout purposes, and these tape rules have special markings every 16″ to aid in layout. Framing systems have been introduced in some areas of the United States, however, that permit joist spacing of 24″ O.C. Figure 42–30 shows on-center layout.

For joists that rest directly on the foundation walls, layout marks may be placed on the sill plates or the header (rim) joists. Lines must also be marked on top of the central beams or walls over which the joists will lap. If framed walls are below the floor unit, the joists are laid out on top of the double plate. The floor layout should also show where any joists are to be doubled because of partitions resting on the floor that run in the same direction as the floor joists. Floor openings for stairwells must also be marked.

Joists should be laid out so that the edges of standard size subfloor panels break over the centers of the joists. This layout eliminates additional cutting of panels when they are being fitted and nailed into place. One method of laying out joists this way is to mark the first joist 15¼″ O.C. from the edge of the building. From then on the layout is 16″ O.C. See Figure 42–31. A procedure for laying out the entire floor is shown in Figure 42–32.

Most of the framing members should be pre-cut before construction begins. The joists

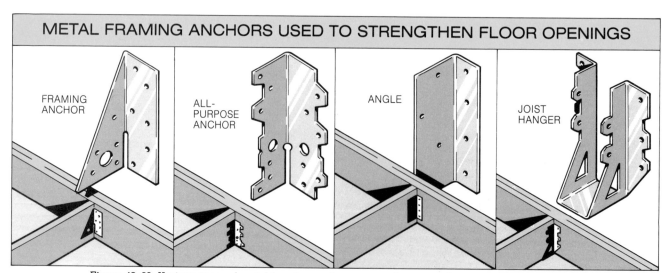

METAL FRAMING ANCHORS USED TO STRENGTHEN FLOOR OPENINGS

FRAMING ANCHOR

ALL-PURPOSE ANCHOR

ANGLE

JOIST HANGER

Figure 42–29. Various types of metal framing anchors are used to strengthen ties of floor openings.

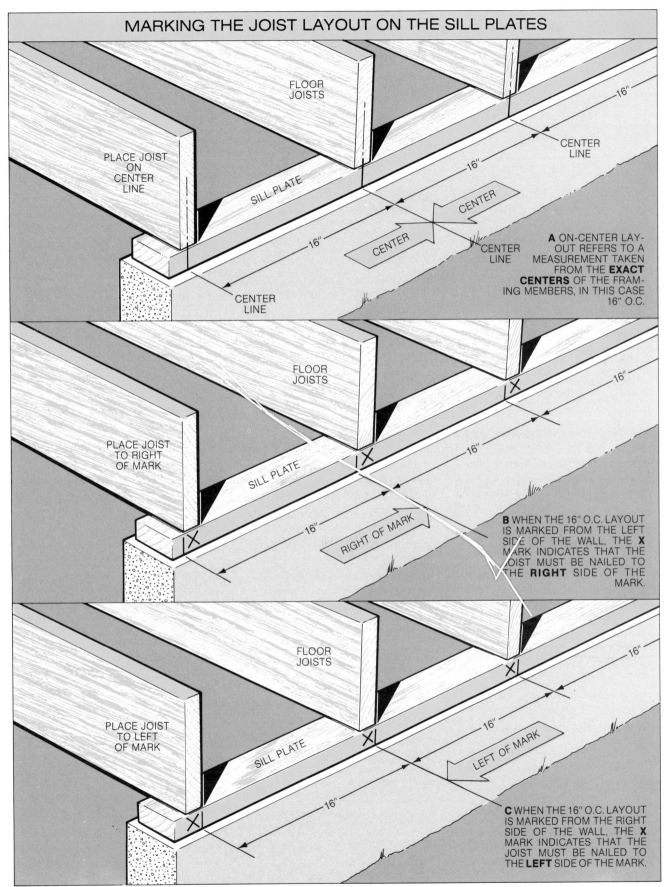

**MARKING THE JOIST LAYOUT ON THE SILL PLATES**

FLOOR JOISTS

PLACE JOIST ON CENTER LINE

SILL PLATE

16"

16"

16"

CENTER LINE

CENTER

CENTER

CENTER LINE

CENTER LINE

**A** ON-CENTER LAY-OUT REFERS TO A MEASUREMENT TAKEN FROM THE **EXACT CENTERS** OF THE FRAMING MEMBERS, IN THIS CASE 16" O.C.

FLOOR JOISTS

PLACE JOIST TO RIGHT OF MARK

SILL PLATE

16"

16"

16"

RIGHT OF MARK

**B** WHEN THE 16" O.C. LAYOUT IS MARKED FROM THE LEFT SIDE OF THE WALL, THE **X** MARK INDICATES THAT THE JOIST MUST BE NAILED TO THE **RIGHT** SIDE OF THE MARK.

FLOOR JOISTS

PLACE JOIST TO LEFT OF MARK

SILL PLATE

16"

16"

16"

LEFT OF MARK

**C** WHEN THE 16" O.C. LAYOUT IS MARKED FROM THE RIGHT SIDE OF THE WALL, THE **X** MARK INDICATES THAT THE JOIST MUST BE NAILED TO THE **LEFT** SIDE OF THE MARK.

Figure 42–30. Laying out joists. An "X" mark shows the side where the joists are to be nailed.

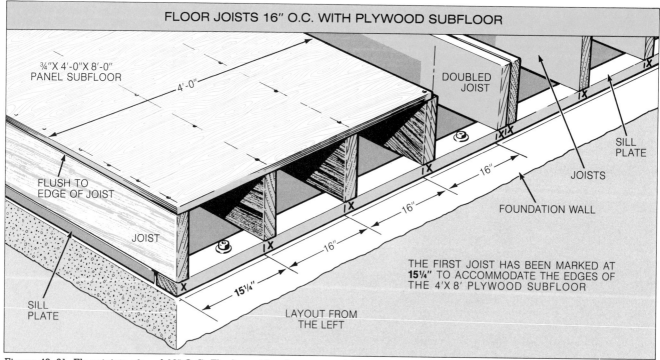

FLOOR JOISTS 16" O.C. WITH PLYWOOD SUBFLOOR

¾"X 4'-0"X 8'-0"
PANEL SUBFLOOR

4'-0"

DOUBLED
JOIST

FLUSH TO
EDGE OF JOIST

SILL
PLATE

JOIST

JOISTS

FOUNDATION WALL

16"

16"

16"

15¼"

SILL
PLATE

15¼"

LAYOUT FROM
THE LEFT

THE FIRST JOIST HAS BEEN MARKED AT
15¼" TO ACCOMMODATE THE EDGES OF
THE 4'X 8' PLYWOOD SUBFLOOR

Figure 42–31. Floor joists placed 16" O.C. The layout has been started from the left side. Note that the first joist has been marked at 15¼". The layout for the following joists is then 16". The first joist is marked at 15¼" to ensure that the 4' or 8' edge of a panel will fall on the center of a joist. Doubled joists are laid out at the far right.

should all be trimmed to their proper lengths. Cross bridging and solid blocks should be cut to fit between the joists that have a common spacing. The distance between joists is usually 14½" for joists spaced 16" O.C. Blocking for the odd spaces is cut afterwards. A typical procedure for framing is shown in Figure 42–33.

**Crowns.** Most joists have a *crown* (a bow shape) on one side. Each joist should be sighted before being nailed in place in order to make certain that the crown is turned up. The joist will later settle from the weight of the floor and straighten out.

## SUBFLOOR

The subfloor, also known as *rough flooring,* is nailed to the top of the floor frame. See Figure 42–34. It strengthens the entire floor unit and serves as a base for the finish floor material. The

walls of the building are laid out, framed, and raised into place on top of the subfloor.

At one time, board lumber (common or shiplap) was always used for subflooring. Today panel products are usually used instead. Less labor is required with panels than with board lumber.

Plywood is the oldest type of panel product. It is still the most widely used subfloor material in residential and other light framed construction. However, the use of non-veneered (reconstituted wood) panels such as structural particleboard, waferboard, oriented strand board, and composite board is growing. (The manufacture and composition of non-veneered panel products is discussed in Section 2.)

Panels used for the subfloor must be the proper grade and thickness for the floor system of the building. The APA performance-rated panels meet the code requirements for subfloor-

ing in all parts of the United States. (Refer to Section 2.) Subfloor panels may have square edges or tongue-and-groove (T & G) edges.

## Applying Subfloor Panels

Subfloor panels are applied with the longer edge at a right angle to the joists. Blocking should be nailed under the long edges of the panels if tongue-and-groove panels are not used. Be sure that there is a ¹⁄₁₆" space between the end joints, and a ⅛" space between the edge joints. See Figure 42–35.

The nailing schedule for most types of subfloor panels calls for 6d common nails for materials up to ⅞" thick, and for 8d nails for heavier panels up to 1⅛" thick. Deformed-shank nails are strongly recommended. They are usually spaced 6" O.C. along the edges of the panel and 10" O.C. over the intermediate joists.

Carpenters usually use just

## MARKING THE JOIST LAYOUT ON THE SILL PLATES

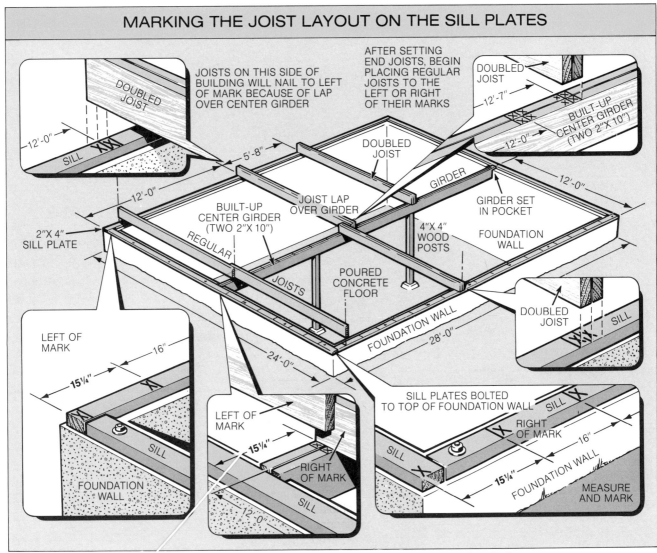

Figure 42-32. Laying out a floor unit. Joist placement is marked on the sill plates.

enough nails to hold the panels in position. When all the panels are placed, lines can be snapped to locate the centers of the joists below. The entire subfloor is then nailed in one operation. Pneumatic nailers or staplers are often used. See Figure 42–36. A procedure for placing subfloor panels is shown in Figure 42–37.

**Glue-nailing Panels.** In an improved system for placing floor panels, a construction adhesive is used in addition to nails for fastening the panels to the joists. Although this system is commonly referred to as *glue-nailing*, a mastic adhesive rather than a

glue is used. The mastic is applied to the joists, and the panel is then set in place and nailed. See Figure 42–38. The combination of an adhesive with nails helps to further stiffen the entire floor system. There is less squeak to the floor, and the nails are less likely to pop over a period of time. The nailing schedule for a glue-nailed panel requires nailing only 12″ O.C. along the panel edges and over the intermediate joists.

## Post-and-Beam Subfloor System

In a post-and-beam subfloor system the floor unit receives its

main support from floor beams rather than from floor joists. See Figure 42–39. The beams are usually spaced 4′ O.C. and are supported by posts resting on concrete piers. The subfloor panels are 1⅛″ or 1¼″ thick and have tongue-and-groove edges. Nails are 10d common, spaced 6″ apart. Post-and-beam systems are discussed in greater detail in Section 14.

## FLOOR UNDERLAYMENT

Floor underlayment consists of thin panels, usually plywood or particleboard, that are placed directly over the subfloor. See Fig-

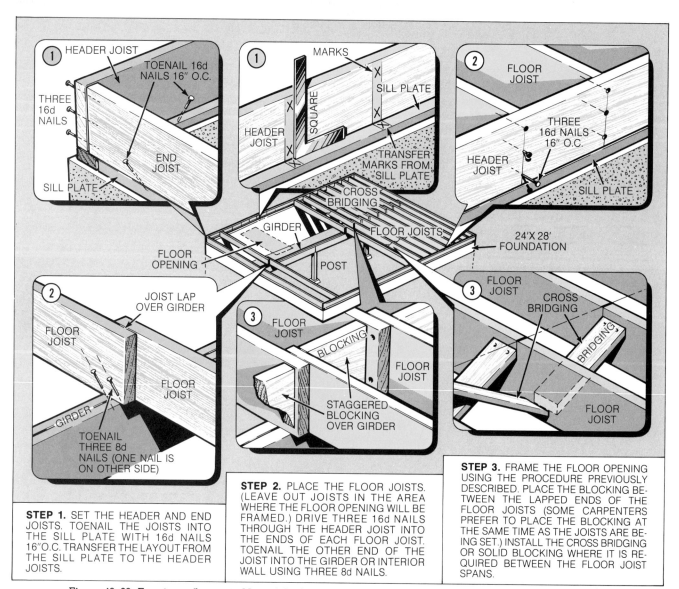

**STEP 1.** SET THE HEADER AND END JOISTS. TOENAIL THE JOISTS INTO THE SILL PLATE WITH 16d NAILS 16"O.C. TRANSFER THE LAYOUT FROM THE SILL PLATE TO THE HEADER JOISTS.

**STEP 2.** PLACE THE FLOOR JOISTS. (LEAVE OUT JOISTS IN THE AREA WHERE THE FLOOR OPENING WILL BE FRAMED.) DRIVE THREE 16d NAILS THROUGH THE HEADER JOIST INTO THE ENDS OF EACH FLOOR JOIST. TOENAIL THE OTHER END OF THE JOIST INTO THE GIRDER OR INTERIOR WALL USING THREE 8d NAILS.

**STEP 3.** FRAME THE FLOOR OPENING USING THE PROCEDURE PREVIOUSLY DESCRIBED. PLACE THE BLOCKING BETWEEN THE LAPPED ENDS OF THE FLOOR JOISTS (SOME CARPENTERS PREFER TO PLACE THE BLOCKING AT THE SAME TIME AS THE JOISTS ARE BEING SET.) INSTALL THE CROSS BRIDGING OR SOLID BLOCKING WHERE IT IS REQUIRED BETWEEN THE FLOOR JOIST SPANS.

Figure 42–33. Framing a floor unit. Most of the framing pieces should be pre-cut before construction begins.

Georgia-Pacific Corporation
Figure 42–34. Plywood subfloor panels are placed over the joist so that long sides of panels run at a right angle to the joists. Note that the joints are staggered.

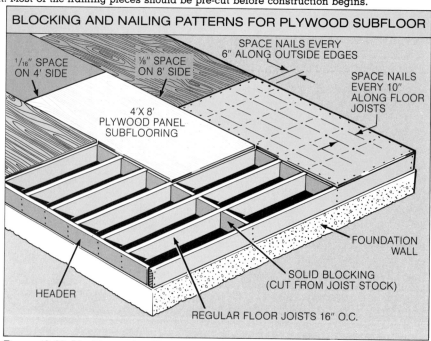

Figure 42–35. Blocking and nailing patterns for plywood subfloor panels. Deformed-shank nails are strongly recommended.

ure 42–40. The purpose of underlayment is to provide a smooth, even surface for finish floor materials such as tile, lino-

Figure 42–36. A heavy-duty pneumatic nailer is often used to nail a plywood subfloor.

Figure 42–38. In the glue-nailed panel method, a construction adhesive is applied to the top of the joists before the subfloor panel is placed.

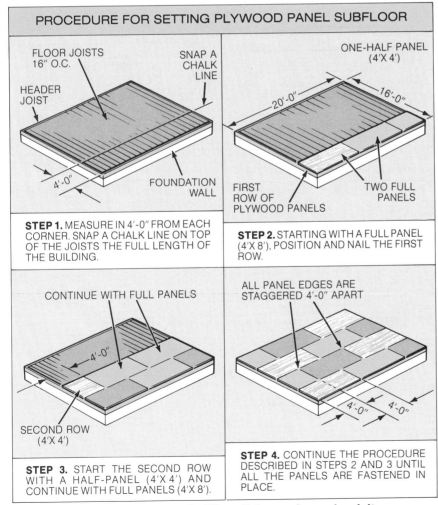

## PROCEDURE FOR SETTING PLYWOOD PANEL SUBFLOOR

FLOOR JOISTS 16" O.C.

SNAP A CHALK LINE

HEADER JOIST

4'-0"

FOUNDATION WALL

**STEP 1.** MEASURE IN 4'-0" FROM EACH CORNER. SNAP A CHALK LINE ON TOP OF THE JOISTS THE FULL LENGTH OF THE BUILDING.

ONE-HALF PANEL (4'X 4')

20'-0"    16'-0"

FIRST ROW OF PLYWOOD PANELS

TWO FULL PANELS

**STEP 2.** STARTING WITH A FULL PANEL (4'X 8'), POSITION AND NAIL THE FIRST ROW.

CONTINUE WITH FULL PANELS

4'-0"

SECOND ROW (4'X 4')

**STEP 3.** START THE SECOND ROW WITH A HALF-PANEL (4'X 4') AND CONTINUE WITH FULL PANELS (4'X 8').

ALL PANEL EDGES ARE STAGGERED 4'-0" APART

4'-0"    4'-0"

**STEP 4.** CONTINUE THE PROCEDURE DESCRIBED IN STEPS 2 AND 3 UNTIL ALL THE PANELS ARE FASTENED IN PLACE.

Figure 42–37. Placing subfloor panels. When all the panels are placed, lines are snapped to locate the centers of the joists below.

leum, and rugs. Underlayment panels must be a high grade of material that has good resistance to dents and punctures from concentrated loads.

Some types of subfloor panels have surfaces that are smooth enough for the direct application of the finish floor materials. An example of such a product is the APA performance-rated Sturd-I-Floor panel. However, the surface of any subfloor becomes scratched, dented, and roughened during construction. For this reason it may be necessary to later place underlayment panels over the Sturd-I-Floor panels in the kitchen and bathroom areas, or anywhere else that resilient materials such as tile or linoleum will be used to finish the floor. Underlayment is also placed under tile and linoleum to bring the finished floor up to the level of carpeted or hardwood floors in other parts of the building.

## Fastening Methods for Underlayment

The APA nailing schedule for ¼" thick plywood underlayment rec-

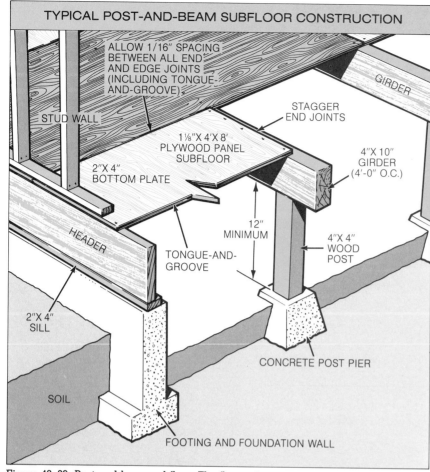

**TYPICAL POST-AND-BEAM SUBFLOOR CONSTRUCTION**

ALLOW 1/16" SPACING BETWEEN ALL END AND EDGE JOINTS (INCLUDING TONGUE-AND-GROOVE)

GIRDER

STAGGER END JOINTS

STUD WALL

1⅛" X 4' X 8' PLYWOOD PANEL SUBFLOOR

4" X 10" GIRDER (4'-0" O.C.)

2" X 4" BOTTOM PLATE

12" MINIMUM

4" X 4" WOOD POST

HEADER

TONGUE-AND-GROOVE

2" X 4" SILL

CONCRETE POST PIER

SOIL

FOOTING AND FOUNDATION WALL

Figure 42–39. Post-and-beam subfloor. The floor unit receives its main support from beams rather than from floor joists.

mastic adhesive is used in addition to nails, the spacing of the nails can be increased to 16" O.C. around the edges and throughout the body of the panel.

## FLOOR TRUSSES

The use of floor truss systems is increasing in residential and commercial wood-framed construction. The typical floor truss is made up of top and bottom chords tied together by web members. See Figure 42–41. The chord material is usually 2" × 4" lumber. The web members may also be 2" × 4" lumber and are joined to the top and bottom chords with metal connector plates.

Another type of truss design uses metal web material. See Figure 42–42. Although the top and bottom chords in this type of truss are 2" × 4" lumber, the web is tubular steel. A 2" × 8"

ommends 3d ring-shank nails or 16-ga. to 18-ga. staples. Either type of fastener should be spaced 3" apart along the edges of the panel, and 6" each way at the intermediate sections.

The National Particleboard Association suggests 4d nails for fastening particleboard panels thinner than ⅜", spaced 3" apart around the edges, and 6" O.C. each way throughout the body of the panel. For panels ⅜" to ⅝" thick, it recommends 6d nails spaced 6" O.C. around the edges, and 10" O.C. each way throughout the body of the panel. If staples are used with particleboard, they should be a minimum of ⅞" long, 18 ga., with a ³⁄₁₆" crown for ¼" thick panels. Staples should be 1⅛" long, 16 ga., with a ⅜" crown for ½" or ⅝" thick panels.

Glue-nailing is also recommended for underlayment. When

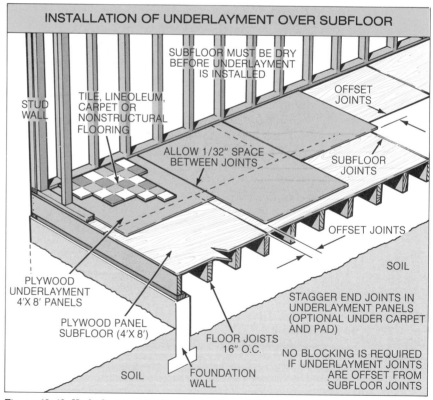

**INSTALLATION OF UNDERLAYMENT OVER SUBFLOOR**

SUBFLOOR MUST BE DRY BEFORE UNDERLAYMENT IS INSTALLED

OFFSET JOINTS

STUD WALL

TILE, LINOLEUM, CARPET OR NONSTRUCTURAL FLOORING

ALLOW 1/32" SPACE BETWEEN JOINTS

SUBFLOOR JOINTS

OFFSET JOINTS

SOIL

PLYWOOD UNDERLAYMENT 4' X 8' PANELS

PLYWOOD PANEL SUBFLOOR (4' X 8')

FLOOR JOISTS 16" O.C.

STAGGER END JOINTS IN UNDERLAYMENT PANELS (OPTIONAL UNDER CARPET AND PAD)

SOIL

FOUNDATION WALL

NO BLOCKING IS REQUIRED IF UNDERLAYMENT JOINTS ARE OFFSET FROM SUBFLOOR JOINTS

Figure 42–40. Underlayment is placed over the subfloor to provide a smooth and even surface for finish floor materials such as tile, linoleum, and rugs.

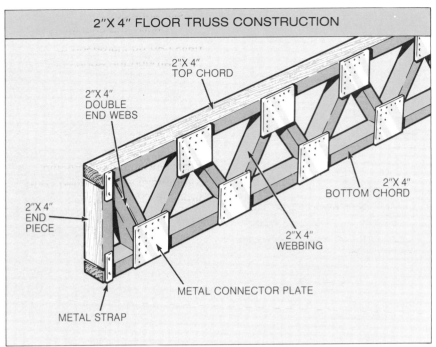

## 2"X 4" FLOOR TRUSS CONSTRUCTION

2"X 4"
TOP CHORD

2"X 4"
DOUBLE
END WEBS

2"X 4"
END
PIECE

2"X 4"
BOTTOM CHORD

2"X 4"
WEBBING

METAL CONNECTOR PLATE

METAL STRAP

Figure 42–41. A floor truss has top and bottom chords tied together by web members. Usually the chords and web members are 2 × 4s. They are joined together with metal connector plates.

piece of lumber runs under the top chords at intervals to provide lateral support to the truss.

Floor trusses are usually built in the shop by the truss manufacturer. The design and length of the trusses are determined by information on the blueprints of the building under construction. The finished trusses are delivered to the job site and installed by carpenters.

Floor trusses offer the advantage of covering long spans without requiring the intermediate support of a wall or girder. Also, the space between the top and bottom chords makes it convenient to run pipes, wires, and ducts through the flooring system. Floor trusses are usually spaced 24" O.C. They provide a wide nailing surface (3½") for the subfloor. See Figure 42–43.

*Trus Joist Corporation, Micro = Lam Division*

Figure 42–42. These floor trusses have top and bottom chords of 2" × 4" lumber and a tubular steel web. The ends of the top chords rest on the foundation wall. The 2" × 8" piece shown under the top chord at the center of the picture stabilizes and ties the trusses together.

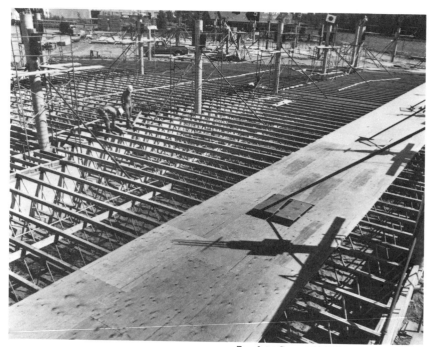

*Trus Joist Corporation, Micro=Lam Division*
Figure 42–45. A metal joist hanger is used to fasten the end of a veneered joist to a supporting beam. The beam shown here is, like the joist, made of laminated veneer pieces.

*Trus Joist Corporation, Micro=Lam Division*
Figure 42–43. Floor trusses provide a wide nailing surface for the plywood sub-flooring.

## VENEERED JOIST SYSTEMS

A recently developed veneered joist system features a plywood web pressed between a top and bottom flange. See Figures 42–44 and 42–45. Both the web and the flanges are made of laminated-veneer lumber, resulting in a very strong structural member.

Veneered joists are available in 9½" and 11" depths. The flanges are 1¾" wide. The joists can be as long as 60′, which eliminates lapping over central beams or walls. They are lightweight and easy to grip. A 26′ joist weighs approximately 50 pounds.

*Trus Joist Corporation, Micro=Lam Division*
Figure 42–44. Carpenters placing veneered floor joists over the foundation walls.

# UNIT 43

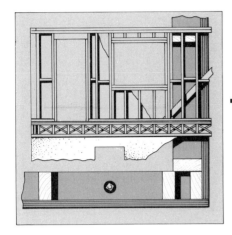

# Wall Framing

Wall construction begins after the subfloor has been nailed in place. The wall system of a wood-framed building consists of exterior (outside) and interior (inside) walls. The exterior walls have door and window openings. See Figure 43–1. The interior walls are usually referred to as *partitions.* They divide the living area of the house into separate rooms. Some have door openings or archways.

Partitions are either *bearing* or *non-bearing.* Bearing partitions support the ends of the floor joists or ceiling joists. Non-bearing partitions run in the same direction as the joists and therefore carry little weight from the floor or ceiling above.

Traditionally, 2″ × 4″ structural lumber is used for the framed walls of one-story buildings, although the use of 2″ × 6″ lumber is increasing. The 2″ × 6″ lumber allows for thicker insula-

## TYPICAL EXTERIOR 2″X 4″ STUD WALL SHOWING CONSTRUCTION DETAILS

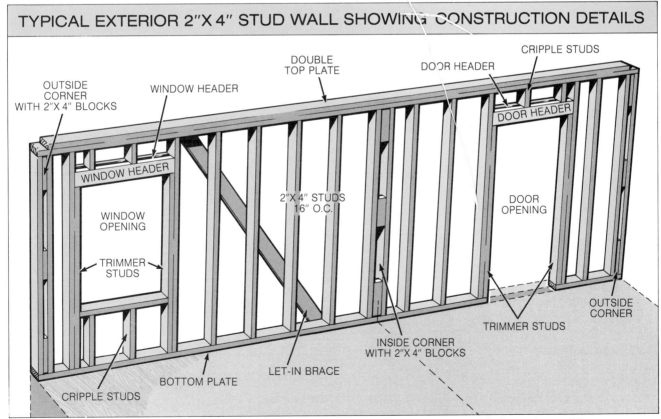

Figure 43–1. A typical exterior wall, showing corners, door and window openings, and a let-in diagonal brace.

tion materials inside the walls. Buildings of more than one floor, such as multi-story apartment buildings, require 2″ × 6″ or 3″ × 4″ lumber on the lower levels in order to support the weight of the floors above.

## COMPONENTS OF A WOOD-FRAMED WALL

A wood-framed wall consists of structural parts referred to as *wall components* or *framing members.* These components are: *studs, plates, headers, trimmers, sills, corner assemblies,* and *diagonal braces.* Refer again to Figure 43–1. Each component has a special function within the total wall structure.

### Studs, Plates, and Corners

Studs are upright (vertical) framing members that run between the wall plates. Studs are usually spaced 16″ O.C. (on centers). In

some areas of the country *engineered* framing systems are used that allow 24″ O.C. stud spacing in the framed walls of one-story buildings.

The plate at the bottom of a wall is the *sole plate,* or *bottom plate.* The plate at the top of the wall is the *top plate.* Usually a *double top plate* is used. The topmost plate of a double top plate is sometimes called a *doubler.* It strengthens the upper section of the wall and helps carry the weight of the joists and roof rafters. Since wall plates are nailed into all the vertical framing members of the wall, including trimmer studs and cripple studs, they serve to tie the entire wall together.

Corner assemblies, also called *corner posts,* are constructed wherever a wall ties into another wall. *Outside corners* are at the ends of a wall. *Inside corners* occur where a partition ties into a wall at some point between the ends of the wall.

Three typical designs for corner assemblies are shown in Figure 43–2. All corner assemblies should be constructed from straight stud material and should be well nailed.

## Door and Window Openings

A rough opening must be framed in a wall wherever a door or window is planned. The dimensions of the rough opening must allow for the finish frame into which the door or window will fit, and for the required clearance around the frame. Methods of calculating the dimensions are explained later in this unit.

The rough opening for a typical door is framed with a *header, trimmer, studs,* and, in some cases, top *cripple studs.* The rough opening for a typical window includes the same members as for a door, plus a rough windowsill and bottom cripples. See Figure 43–3.

The header is placed at the

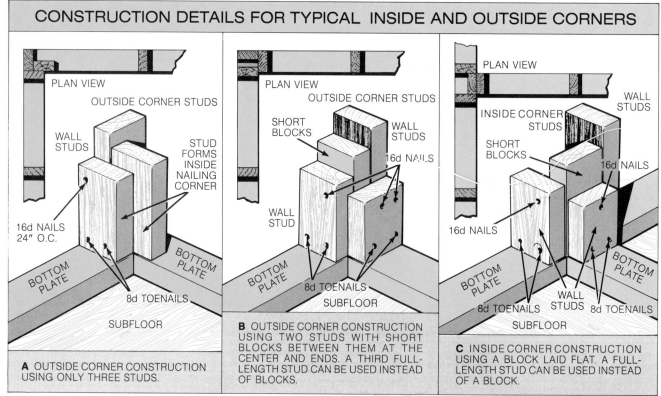

CONSTRUCTION DETAILS FOR TYPICAL INSIDE AND OUTSIDE CORNERS

PLAN VIEW
OUTSIDE CORNER STUDS
WALL STUDS
STUD FORMS INSIDE NAILING CORNER
16d NAILS 24″ O.C.
BOTTOM PLATE
BOTTOM PLATE
8d TOENAILS
SUBFLOOR

**A** OUTSIDE CORNER CONSTRUCTION USING ONLY THREE STUDS.

PLAN VIEW
OUTSIDE CORNER STUDS
SHORT BLOCKS
WALL STUDS
16d NAILS
WALL STUD
BOTTOM PLATE
8d TOENAILS
SUBFLOOR

**B** OUTSIDE CORNER CONSTRUCTION USING TWO STUDS WITH SHORT BLOCKS BETWEEN THEM AT THE CENTER AND ENDS. A THIRD FULL-LENGTH STUD CAN BE USED INSTEAD OF BLOCKS.

PLAN VIEW
WALL STUDS
INSIDE CORNER STUDS
SHORT BLOCKS
16d NAILS
16d NAILS
BOTTOM PLATE
BOTTOM PLATE
8d TOENAILS
WALL STUDS
8d TOENAILS
SUBFLOOR

**C** INSIDE CORNER CONSTRUCTION USING A BLOCK LAID FLAT. A FULL-LENGTH STUD CAN BE USED INSTEAD OF A BLOCK.

Figure 43–2. Methods of constructing typical corners. Straight stud material is used.

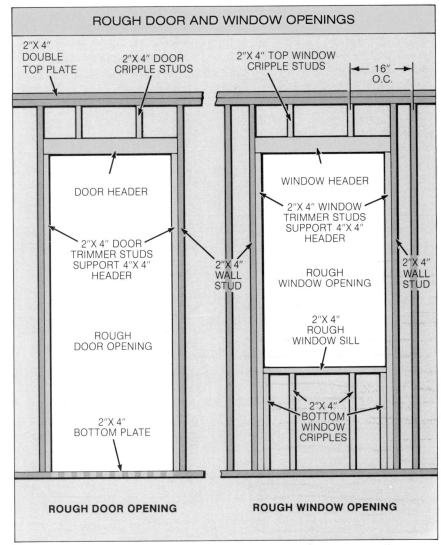

Figure 43–3. Rough openings for doors and windows in a wood-framed wall.

required is shown in the blueprints. Header size is determined by the width of the opening and by how much weight is bearing down from the floor above. See Figure 43–4.

The tops of all door and window openings in all walls are usually in line with each other. Therefore, all headers are usually the same height from the floor. The standard height of the walls in most residential wood-framed buildings is either 8'-0¾" or 8'-1" from the subfloor to the ceiling joists. The standard height of the doors is 6'-8".

Cripple studs are nailed between the header and the double top plate of a door opening in order to carry the weight from the top plate to the header. The cripple studs are generally spaced 16" O.C.

For door openings of standard height, cripples are not necessary if a 4" × 12" header is placed directly below the top plate. The distance between the bottom of a 4" × 12" header and the subfloor allows for the 6'-8" door height, the door jamb, the clearance above the jamb, and the required clearance below the door.

If a header less than 12" wide is used, cripples are necessary. In today's construction, 12" wide headers are usually used to save the labor cost of placing cripples. Header-cripple combinations are usually used only where walls are more than 8'-1" high. In walls this high, a 4" × 12"

top of a rough opening. It must be strong enough to carry the weight bearing down on that section of the wall. The header is supported by trimmer studs that fit between the sole plate and the bottom of the header. The trimmer studs are nailed into the regular studs at each side of the header. Nails are also driven through the regular studs into the ends of the header.

The header may be a solid piece or may be built up of two pieces with a ½" spacer block between them. The spacer blocks are needed to bring the width of the header to 3½", which is the actual width of a nominal 2" × 4" stud wall. A

built-up header is just as strong or stronger than a solid piece. However, it involves extra labor. Another type of built-up header is made of two planks, such as 2" × 10" planks, with a 2" × 4" piece nailed at the bottom.

The type and size of header

| Lumber size | Species | Minimum grade | Maximum allowable span |
|---|---|---|---|
| 4 × 4 | DF | No. 2 | 4' |
| 4 × 6 | DF | No. 2 | 6' |
| 4 × 8 | DF | No. 2 | 8' |
| 4 × 10 | DF | No. 2 | 10' |
| 4 × 12 | DF | No. 2 | 12' |
| 4 × 14 | DF | No. 1 | 16' (commonly used for garage door) |

Figure 43–4. Header sizes for door and window openings. For example, a 4" × 6" header can be used for an opening up to 6'-0" in width. A 4" × 12" header is required for a 12'-0" opening.

Figure 43–5. The door opening at top has a 4″ × 12″ header. The one at bottom has a smaller header with cripple studs.

header placed directly under the top plate would provide a rough opening that would be too high for standard size doors. Cripple studs must be used to provide a rough openings of the correct size. See Figures 43–5 and 43–6.

**Rough Windowsills and Bottom Cripples.** A rough windowsill is added to the bottom of a rough window opening. The sill provides support for the finished window and frame that will later be placed in the wall. The distance between the sill and the header is determined by the di-

mensions of the window, the window frame, and the necessary clearances at the top and bottom of the frame. Cripple studs, spaced 16″ O.C., are nailed between the sill and sole plate. Additional cripple studs may be placed under each end of the sill. Some local building codes require that sills be doubled if the width of the opening is more than 4′.

## Diagonal Bracing

Diagonal bracing is necessary for the lateral strength of a wall. In all exterior walls and main interior partitions, bracing should be placed at both ends (where possible) and at 25′ intervals. An exception to this requirement is an outside wall that is to be covered with structural sheathing nailed according to building code requirements. This type of wall does not need to be braced.

Diagonal bracing is most effective when installed at a 45° to 60° angle. Usually the braces are placed in the wall after the wall has been squared and is still lying on the subfloor. The most widely used bracing system is the 1″ × 4″ *let-in* type. The studs are notched so that the 1″ × 4″ piece will be flush with the surface of the studs. See Figure 43–7.

## Fire Blocks

Some local building codes require that fire blocks (also known as *fire stops*) be placed in walls that are over 8′-1″ high. Fire blocks slow down a fire traveling inside the walls. They are nailed between the studs, either before or after the wall is raised. They can be nailed in a straight line or staggered.

Fire blocks do not have to be nailed at the midpoint of the wall. They can be positioned so they

Figure 43–6. The window opening at top has a 4″ × 12″ header. The one at bottom has a smaller header with cripple studs.

provide additional backing for nailing the edges of plasterboard or plywood.

## CONSTRUCTING A WOOD-FRAMED WALL

All the major parts of a wall should be cut before any are assembled. By studying the blueprints, a carpenter can determine the number and lengths of studs, trimmer studs, cripple studs, headers, and rough windowsills required for the building. These

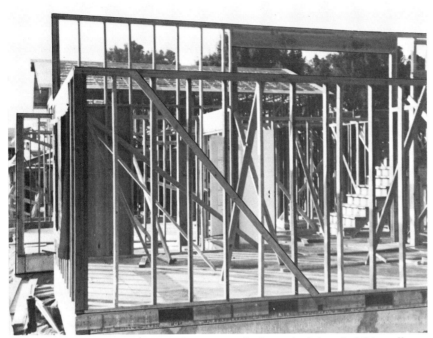

Figure 43-7. A let-in diagonal brace is notched into the left end of this wall.

Figure 43-8. Framing members are cut with a radial-arm saw on the job site.

Figure 43-9. A partially framed wall lies on the subfloor. A completed wall has been raised into place at the left.

pieces are cut with a radial-arm saw on the job site. See Figure 43-8. For buildings on large housing tracts, framing members are often cut to length at the mill or lumber yard and then delivered to the job.

The different parts of the wall are assembled on the subfloor. The pieces are then nailed together, and the completed wall is raised into place. Figure 43-9 shows a partially assembled wall lying on the subfloor of a building under construction. Figure 43-10 shows a building with all walls raised into place but with temporary wall bracing not yet removed.

## Laying out Walls

Carpenters must know where each wall is to be placed before construction can begin. Wall layout requires the ability to read blueprints and a thorough understanding of wall construction. It is usually done by one of the more competent carpenters on the job

or by the job supervisor. Two types of procedures are involved in wall layout: *horizontal* plate layout and *vertical* layout.

*Southern Forest Products Association*

Figure 43-10. Rough framing of this building has been completed. Note the temporary wall bracing still in place. Permanent let-in diagonal braces have been set in both ends of the wall at left. Panel sheathing is being placed on the roof rafters.

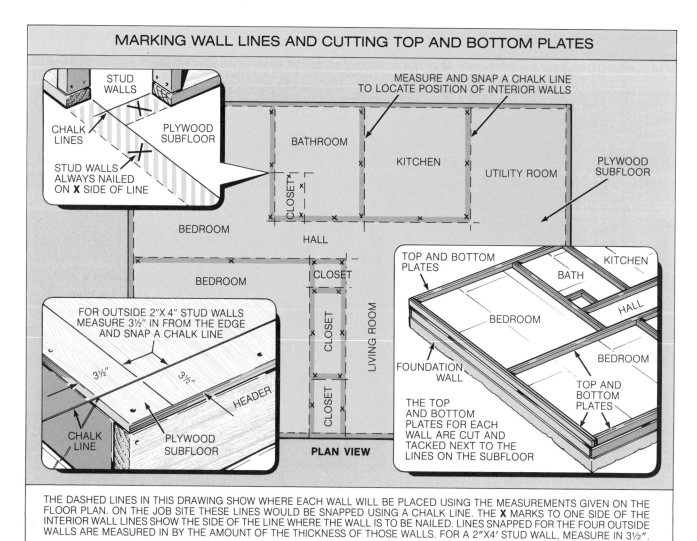

Figure 43–11. Marking the wall lines and cutting the wall plates.

Figure 43–12. Top and bottom plates cut and tacked next to the lines snapped on the subfloor.

**Horizontal Plate Layout.** The first step in wall layout is to snap lines on the floor to show the exact locations of the walls. These locations are determined from the measurements given in the floor plan of the blueprints.

After all the lines are snapped, the wall plates are cut and tacked next to the lines. See Figures 43–11 and 43–12. The plates can then be marked off for corner posts and regular studs, as well as for the studs, trimmers, and cripples for the rough openings. All framing members must be clearly marked on the plates for the carpenters to frame the wall efficiently and without

error. Figure 43–13 shows a wall with framing members nailed in place according to layout markings.

A procedure for marking outside and inside corners for *stud-and-block* corner post construction is shown in Figure 43–14. A procedure for laying out studs for the first exterior wall is shown in Figure 43–15. In this layout method, which is often used, the plates are marked for the first stud from a corner to be placed 15¼" from the end of the corner. Other studs, after the first stud, follow 16" O.C. layout. This layout method ensures that the edges of standard size panels used for sheathing or wallboard will fall on the centers of the studs. Cripples are laid out to follow the layout of the studs.

A procedure for laying out studs for the second exterior wall is shown in Figure 43–16. The plates are marked for the first stud to be placed 15¼″ from the outside edge of the panel thickness on the first wall. This layout allows the corner of the first panel on the second wall to line up with the edge of the first panel on the first wall. Also, the opposite edge of the panel on the second wall, like the panel on the first wall, will break on the center of a stud.

A procedure for laying out studs for interior walls (partitions) is shown in Figure 43–17. If panels are placed on the exterior wall first, and then on the interior wall, the wall plates for the interior wall are marked for the first stud to be placed 15¼″ from the edge of the panel thickness on the exterior wall. If panels are to be placed on the interior wall before they are placed on the exterior wall, then the wall plates of the interior wall are marked for the first stud to be placed 15¼″ from the unpaneled exterior wall.

Rough openings for doors and windows must also be marked on the wall plates. The rough opening for a wood door or window is calculated on the basis of the door or window width, the thickness of the finish frame, and ½″ clearance for shim shingles at the sides of the frame. See Figures 43–18 and 43–19.

Some blueprint door and window schedules give the rough opening dimensions. This simplifies layout for the carpenter. When rough openings are not given, the carpenter must know how to calculate the rough openings. This procedure is given in Figures 43–20 and 43–21.

A rough opening for a metal window often requires a ½″ clearance around the entire frame. If the measurements are not given in the window schedule, they can be obtained from

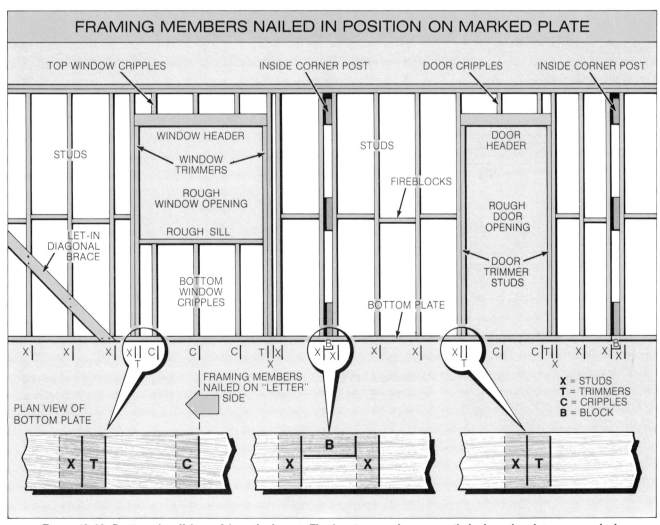

Figure 43–13. Section of wall framed from the layout. The framing members are nailed where the plates are marked.

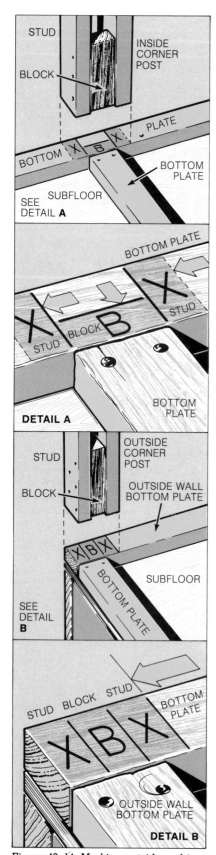

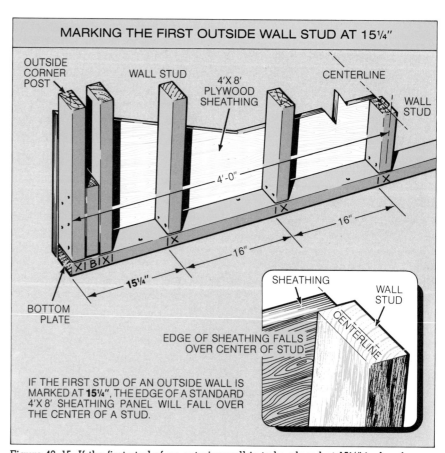

**MARKING THE FIRST OUTSIDE WALL STUD AT 15¼"**

EDGE OF SHEATHING FALLS OVER CENTER OF STUD

IF THE FIRST STUD OF AN OUTSIDE WALL IS MARKED AT **15¼"**, THE EDGE OF A STANDARD 4'X 8' SHEATHING PANEL WILL FALL OVER THE CENTER OF A STUD.

Figure 43–15. If the first stud of an exterior wall is to be placed at 15¼" inches from the end of the corner, and the other studs follow 16" O.C. layout, the edges of standard size panels will fall over the centers of the studs. In this example, a 4' wide panel is shown. If the panel were placed horizontally, an 8' or 12' length would also break over the center of a stud.

Figure 43–14. Marking outside and inside corners in stud-and-block construction.

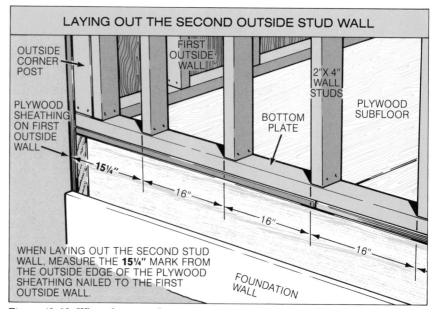

**LAYING OUT THE SECOND OUTSIDE STUD WALL**

WHEN LAYING OUT THE SECOND STUD WALL, MEASURE THE **15¼"** MARK FROM THE OUTSIDE EDGE OF THE PLYWOOD SHEATHING NAILED TO THE FIRST OUTSIDE WALL.

Figure 43–16. When the second exterior wall is laid out, the 15¼" mark should be measured from the outside edge of the panel thickness on the first wall. The corner of the starting panel of the second wall will then line up with the edge of the first wall panel. The opposite edge of the panel will break on the center of a stud.

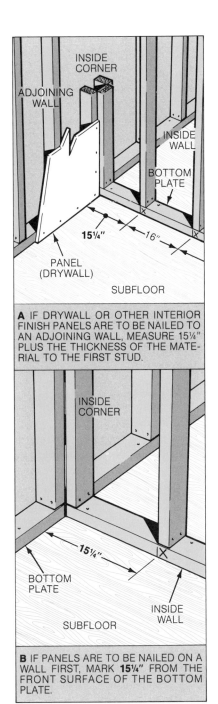

**A** IF DRYWALL OR OTHER INTERIOR FINISH PANELS ARE TO BE NAILED TO AN ADJOINING WALL, MEASURE 15¼" PLUS THE THICKNESS OF THE MATERIAL TO THE FIRST STUD.

**B** IF PANELS ARE TO BE NAILED ON A WALL FIRST, MARK **15¼"** FROM THE FRONT SURFACE OF THE BOTTOM PLATE.

Figure 43-17. On interior walls the 15¼" starting measurement ensures that standard size wallboard or interior finish panels will break over the center of a stud.

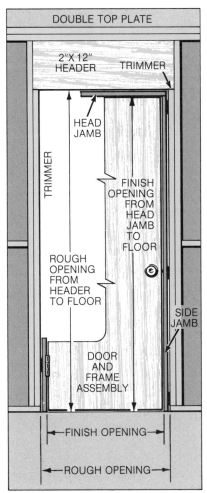

Figure 43-18. The finish door opening is the width of the door and the distance from the head jamb to the floor. The rough door opening is the distance between the trimmers and the height from the floor to the header.

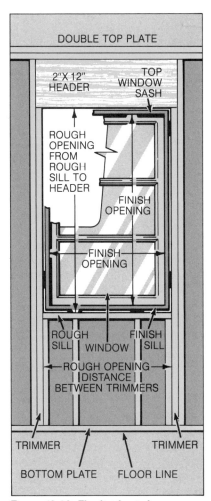

Figure 43-19. The finish window opening width and length is the frame to frame dimension. The rough window opening is the distance between the trimmers and the height from the rough window sill to the header.

the manufacturer's instructions supplied with the windows.

A completely laid out wall plate includes markings for corner posts, rough openings, studs, and cripples. A procedure for laying out walls according to a specific floor plan (Figure 43–22) is shown in Figures 43–23 through 43–26. The corner posts are laid out first. Next, the 16" marks for the studs and cripples are marked, and then the rough openings.

Some carpenters prefer to lay out the rough openings before the studs and cripples are marked. There is, however, an advantage to laying out the 16" O.C. marks first. Often the studs and trimmers framing a door and window fall very close to a 16" O.C. stud mark. Slightly shifting the position of the rough opening may eliminate an unnecessary stud from the wall frame.

**Vertical Layout.** Vertical layout is a procedure for calculating the lengths of the different vertical members of a wood-framed wall. This makes it possible to pre-cut all of the studs, trimmers, and cripples required for a building.

The most efficient vertical layout procedure involves a wall framing *story pole*. A story pole is a 1" × 2" or 1" × 4" ripping marked off to give the lengths of the studs, trimmers, and cripples. A story pole reduces the possibility of making mathematical errors. In addition, it allows measurements to be retained for future use. The information required to mark off the story pole is taken from the wall section views, the window schedule, the door schedule, the window details, and the door details of the blueprints.

Some blueprints contain section views that give the exact rough heights of walls. (The rough height is the distance from the subfloor to the bottom of the ceiling joists.) The rough height to the top of the door (the distance from the subfloor to the bottom of the door header) may

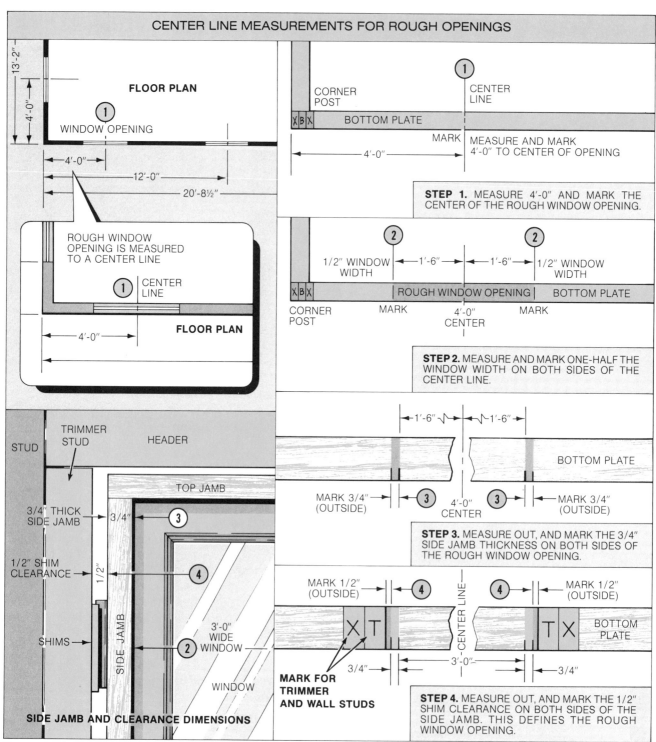

Figure 43–20. Laying out the width of a door or window rough opening when the blueprint measurement is to the center of the opening.

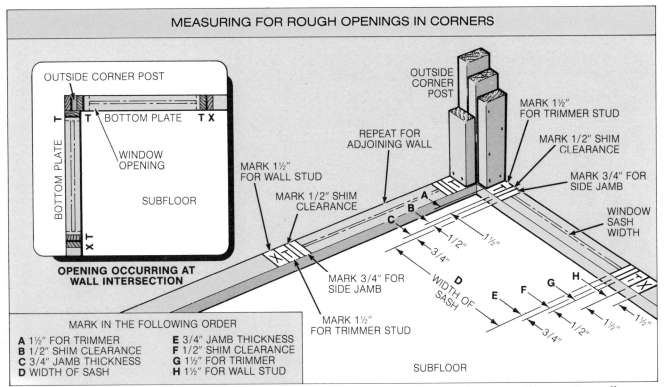

## MEASURING FOR ROUGH OPENINGS IN CORNERS

OUTSIDE CORNER POST

OUTSIDE CORNER POST

MARK 1½" FOR TRIMMER STUD

MARK 1½" SHIM CLEARANCE

MARK 3/4" FOR SIDE JAMB

WINDOW SASH WIDTH

REPEAT FOR ADJOINING WALL

MARK 1½" FOR WALL STUD

MARK 1/2" SHIM CLEARANCE

T  T  TX

BOTTOM PLATE

WINDOW OPENING

SUBFLOOR

BOTTOM PLATE

XT

**OPENING OCCURRING AT WALL INTERSECTION**

MARK 3/4" FOR SIDE JAMB

MARK 1½" FOR TRIMMER STUD

1/2"  1½"

3/4"

WIDTH OF SASH

E  F  G  H

3/4"  1/2"  1½"  1½"

SUBFLOOR

### MARK IN THE FOLLOWING ORDER

**A** 1½" FOR TRIMMER     **E** 3/4" JAMB THICKNESS
**B** 1/2" SHIM CLEARANCE    **F** 1/2" SHIM CLEARANCE
**C** 3/4" JAMB THICKNESS    **G** 1½" FOR TRIMMER
**D** WIDTH OF SASH       **H** 1½" FOR WALL STUD

Figure 43-21. Laying out the widths of door or window rough openings that occur at the corners of intersecting walls.

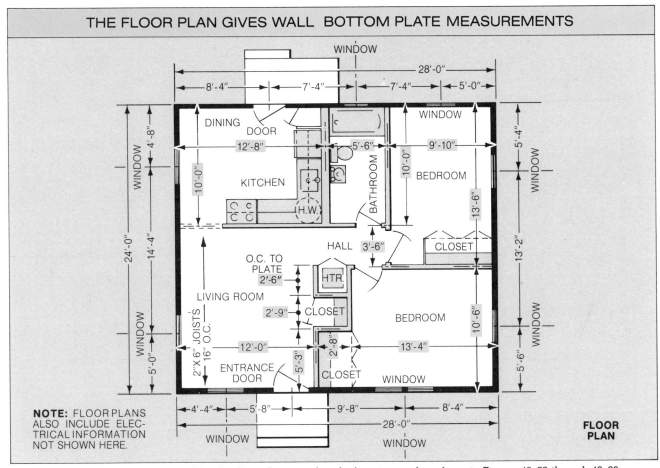

## THE FLOOR PLAN GIVES WALL BOTTOM PLATE MEASUREMENTS

WINDOW

28'-0"

8'-4"  7'-4"  7'-4"  5'-0"

WINDOW

DINING

DOOR

12'-8"

5'-6"

9'-10"

5'-4"

WINDOW

KITCHEN

10'-0"

H.W.

BATHROOM

BEDROOM

10'-0"

13'-6"

WINDOW

24'-0"

14'-4"

HALL

3'-6"

CLOSET

13'-2"

O.C. TO PLATE
2'-6"

HTR.

2'-9"

CLOSET

LIVING ROOM

2'-8"

BEDROOM

10'-6"

WINDOW

2"x6" JOISTS 16" O.C.

12'-0"

5'-3"

CLOSET

13'-4"

WINDOW

5'-0"

ENTRANCE DOOR

WINDOW

5'-6"

**NOTE:** FLOOR PLANS ALSO INCLUDE ELECTRICAL INFORMATION NOT SHOWN HERE.

4'-4"  5'-8"  9'-8"  8'-4"

28'-0"

WINDOW

WINDOW

**FLOOR PLAN**

Figure 43-22. Information provided by this floor plan is used in the layout procedure shown in Figures 43-23 through 43-26.

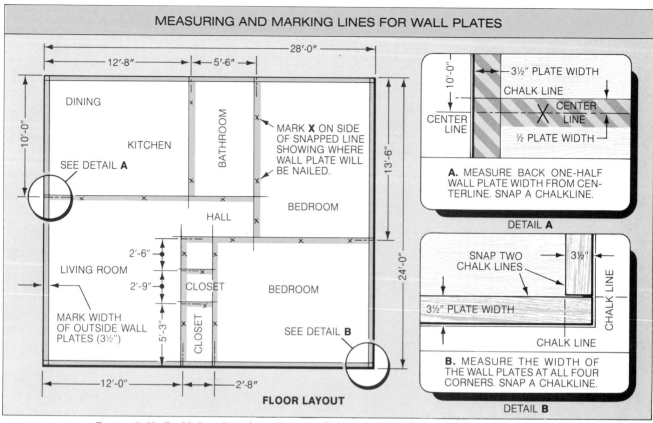

Figure 43-23. Establishing lines for wall plates. (Refer to Figure 43–22 for blueprint information.)

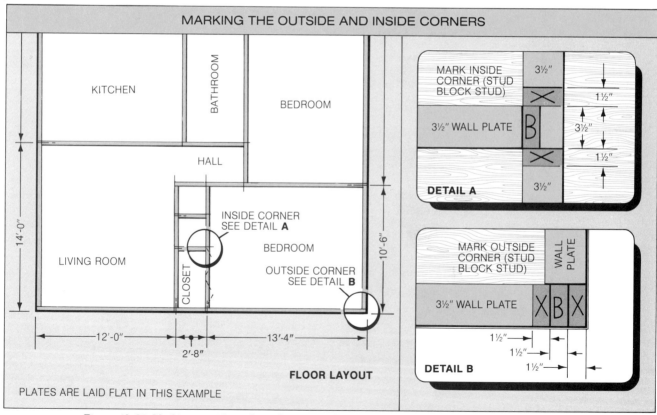

Figure 43–24. Marking outside and inside corners. (Refer to Figure 43–22 for blueprint information.)

also be noted on the section drawing. In addition, it may be given in the column for rough opening measurements on the door schedule. The rough height to the top of the door establishes the measurement for the rough height to the top of the window, as window headers are usually in line with door headers.

The distance from the bottom of a window rough opening to the top (the rough length of the window) can be found by measuring down from the bottom of the window header, using dimensions provided in the rough opening column of the window schedule. A building usually contains windows of more than one length, but only the dimensions of the type of window used most frequently in the building are marked on the story pole.

Figure 43–27 shows a typical wall section view and door and window schedules. Figure 43–28 shows a procedure for laying out a story pole based on information obtained from the section

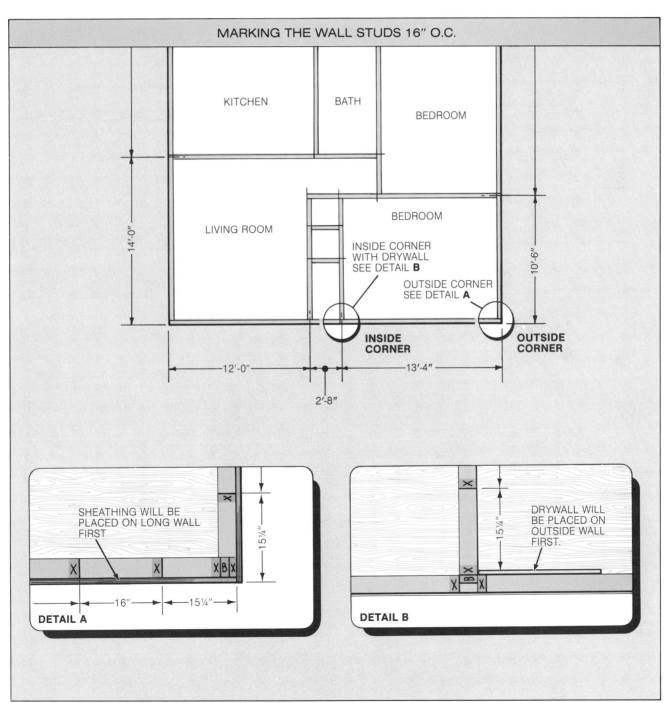

Figure 43–25. Laying out 16″ O.C. marks. (Refer to Figure 43–22 for blueprint information.)

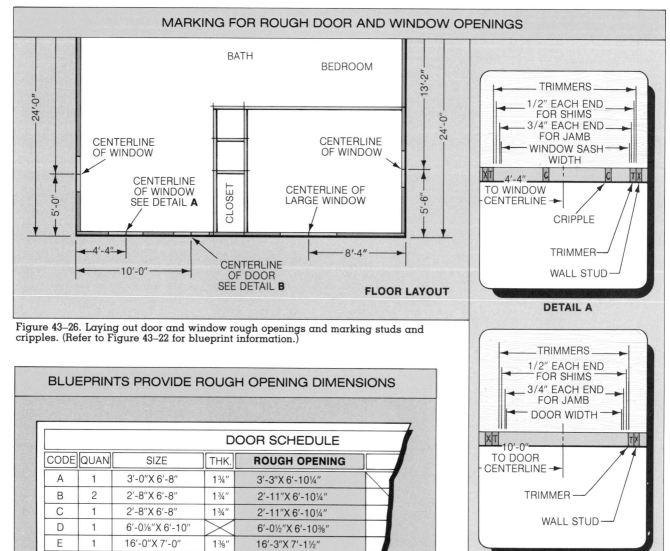

Figure 43–26. Laying out door and window rough openings and marking studs and cripples. (Refer to Figure 43–22 for blueprint information.)

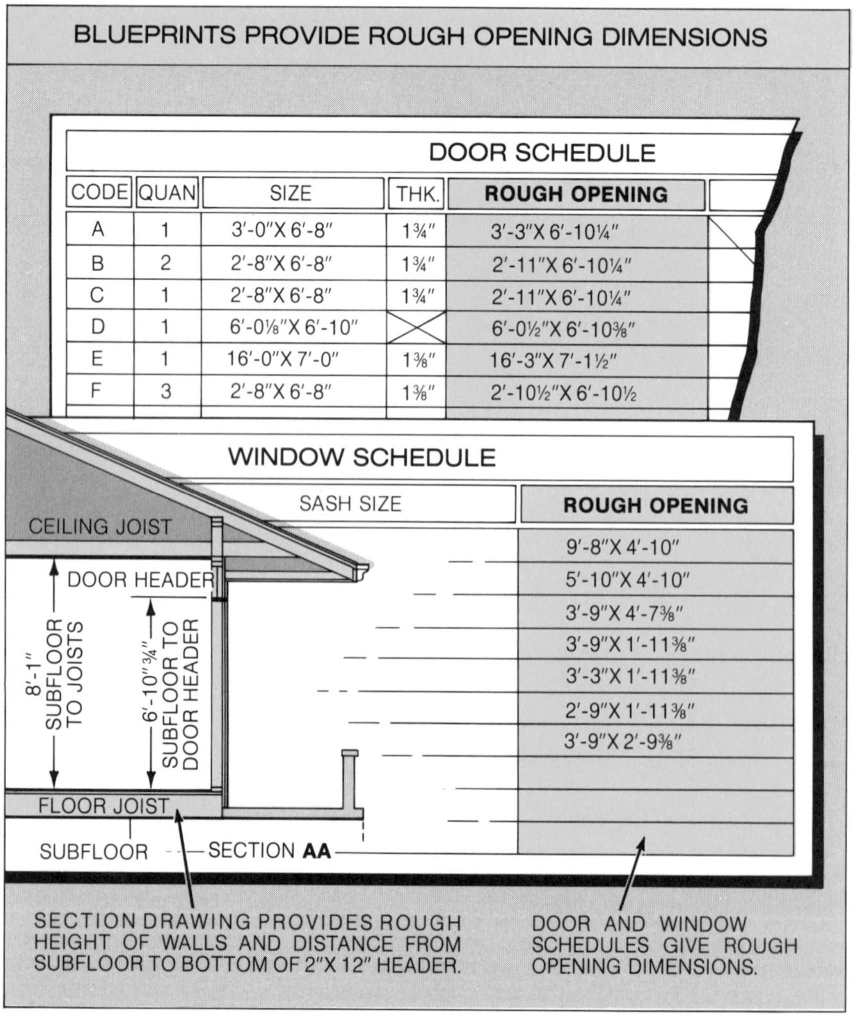

## BLUEPRINTS PROVIDE ROUGH OPENING DIMENSIONS

### DOOR SCHEDULE

| CODE | QUAN | SIZE | THK. | ROUGH OPENING | |
|------|------|------|------|---------------|---|
| A | 1 | 3'-0"X 6'-8" | 1¾" | 3'-3"X 6'-10¼" | |
| B | 2 | 2'-8"X 6'-8" | 1¾" | 2'-11"X 6'-10¼" | |
| C | 1 | 2'-8"X 6'-8" | 1¾" | 2'-11"X 6'-10¼" | |
| D | 1 | 6'-0⅛"X 6'-10" | ✕ | 6'-0½"X 6'-10⅜" | |
| E | 1 | 16'-0"X 7'-0" | 1⅜" | 16'-3"X 7'-1½" | |
| F | 3 | 2'-8"X 6'-8" | 1⅜" | 2'-10½"X 6'-10½ | |

### WINDOW SCHEDULE

| SASH SIZE | ROUGH OPENING |
|-----------|---------------|
| | 9'-8"X 4'-10" |
| | 5'-10"X 4'-10" |
| | 3'-9"X 4'-7⅜" |
| | 3'-9"X 1'-11⅜" |
| | 3'-3"X 1'-11⅜" |
| | 2'-9"X 1'-11⅜" |
| | 3'-9"X 2'-9⅜" |

SECTION AA

SECTION DRAWING PROVIDES ROUGH HEIGHT OF WALLS AND DISTANCE FROM SUBFLOOR TO BOTTOM OF 2"X 12" HEADER.

DOOR AND WINDOW SCHEDULES GIVE ROUGH OPENING DIMENSIONS.

Figure 43–27. Rough door and window opening dimensions are obtained from the door and window schedules on the blueprints.

view and door and window schedules in Figure 43–27.

Some blueprints provide only finish (not rough) opening measurements in the wall section views and door and window schedules. In this case, additional information must be obtained from detail drawings and from the finish schedule to lay out the story pole. Figure 43–29 shows a wall section view that provides only finish dimensions. Parts of door and window schedules, as well as detail drawings, are also shown in the figure. Figure 43–30 shows a procedure for marking a story pole based

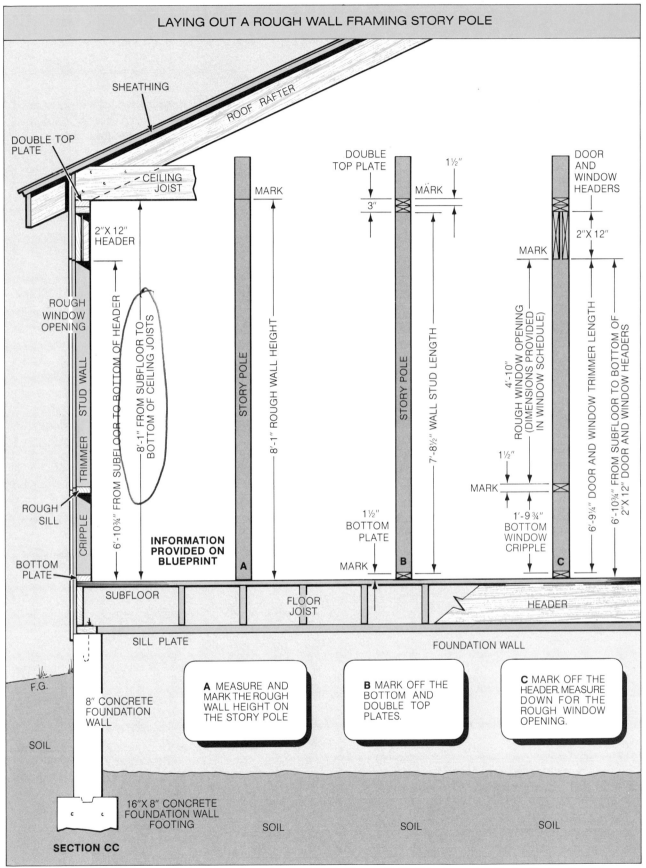

LAYING OUT A ROUGH WALL FRAMING STORY POLE

SHEATHING

ROOF RAFTER

DOUBLE TOP PLATE

CEILING JOIST

2" X 12" HEADER

ROUGH WINDOW OPENING

STUD WALL

TRIMMER

CRIPPLE

ROUGH SILL

BOTTOM PLATE

6'-10¾" FROM SUBFLOOR TO BOTTOM OF HEADER

8'-1" FROM SUBFLOOR TO BOTTOM OF CEILING JOISTS

INFORMATION PROVIDED ON BLUEPRINT

STORY POLE

MARK

8'-1" ROUGH WALL HEIGHT

A

SUBFLOOR

FLOOR JOIST

DOUBLE TOP PLATE

3"

MARK

1½"

STORY POLE

7'-8½" WALL STUD LENGTH

1½" BOTTOM PLATE

MARK

B

DOOR AND WINDOW HEADERS

2" X 12"

MARK

4'-10" ROUGH WINDOW OPENING (DIMENSIONS PROVIDED IN WINDOW SCHEDULE)

1½"

MARK

1'-9¾" BOTTOM WINDOW CRIPPLE

6'-9¼" DOOR AND WINDOW TRIMMER LENGTH

6'-10¾" FROM SUBFLOOR TO BOTTOM OF 2" X 12" DOOR AND WINDOW HEADERS

C

HEADER

SILL PLATE

FOUNDATION WALL

F.G.

8" CONCRETE FOUNDATION WALL

SOIL

16" X 8" CONCRETE FOUNDATION WALL FOOTING

SECTION CC

A MEASURE AND MARK THE ROUGH WALL HEIGHT ON THE STORY POLE

B MARK OFF THE BOTTOM AND DOUBLE TOP PLATES.

C MARK OFF THE HEADER. MEASURE DOWN FOR THE ROUGH WINDOW OPENING.

SOIL          SOIL          SOIL          SOIL

Figure 43-28. Laying out a story pole from the information provided in Figure 43-27. This story pole will give the lengths of the studs, trimmers, and bottom window cripples.

on the information provided in Figure 43–29.

## Constructing Corners

All the corner assemblies (posts) for a building can be constructed at one time. An efficient way to do this is to set up a bench for this purpose as shown in Figure 43–31. Figure 43–32 shows a procedure for assembling inside and outside corners. Blocks should be held slightly back from the ends of the studs. Before nailing, be sure that the ends of the studs line up with each other.

## Framing Door and Window Openings

Many carpenters prefer to frame the door and window openings before assembling the rest of the wall. Figure 43–33 shows a typical procedure for framing a door opening. In this example cripple studs are required over the header. To frame a door opening with a 4″ × 12″ header in a wall no more than 8′-1″ high, the procedure is the same except that cripple studs are not needed.

A rough windowsill is added to the bottom of a window opening.

Cripple studs that follow the stud layout (usually 16″ O.C.) are nailed between the sill and the sole plate. Good framing practice requires that cripples also be placed under each end of the sill. The steps for framing a window opening follow the same order as for a door opening. The rough windowsill and bottom cripple studs are added as shown in Figure 43–34.

## Assembling the Wall

After the corners and door and window openings have been

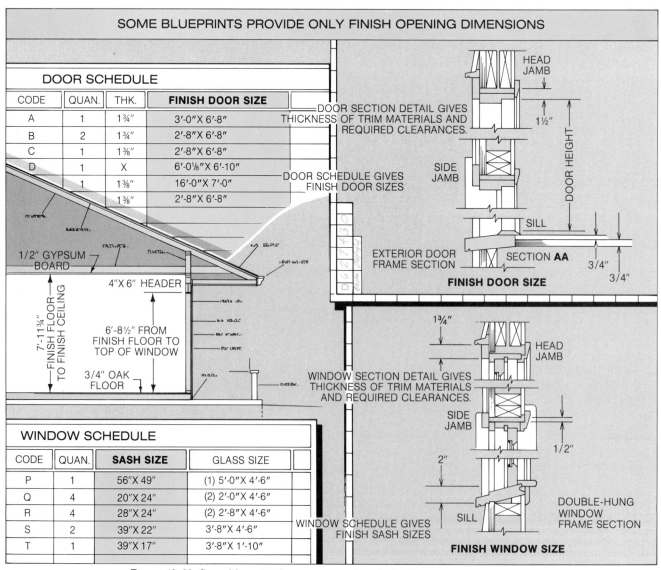

Figure 43–29. Some blueprint drawings provide only finish opening dimensions.

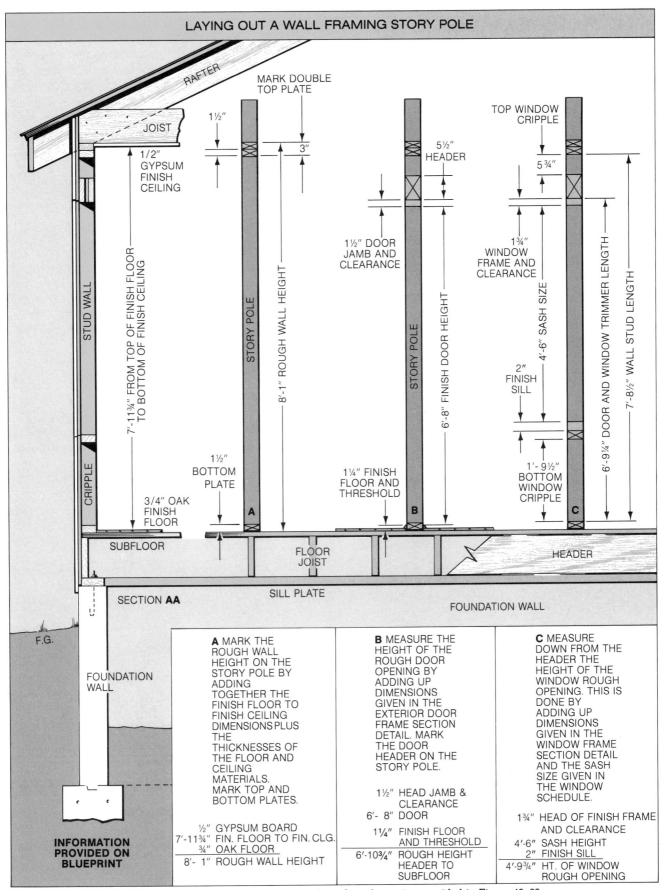

**LAYING OUT A WALL FRAMING STORY POLE**

RAFTER

JOIST

MARK DOUBLE TOP PLATE

1½"

3"

1/2" GYPSUM FINISH CEILING

5½" HEADER

TOP WINDOW CRIPPLE

5¾"

1½" DOOR JAMB AND CLEARANCE

1¾" WINDOW FRAME AND CLEARANCE

STUD WALL

7'-11¾" FROM TOP OF FINISH FLOOR TO BOTTOM OF FINISH CEILING

STORY POLE

8'-1" ROUGH WALL HEIGHT

STORY POLE

6'-8" FINISH DOOR HEIGHT

4'-6" SASH SIZE

6'-9¼" DOOR AND WINDOW TRIMMER LENGTH

7'-8½" WALL STUD LENGTH

2" FINISH SILL

CRIPPLE

1½" BOTTOM PLATE

3/4" OAK FINISH FLOOR

A

1¼" FINISH FLOOR AND THRESHOLD

B

1'-9½" BOTTOM WINDOW CRIPPLE

C

SUBFLOOR

FLOOR JOIST

HEADER

SECTION **AA**

SILL PLATE

FOUNDATION WALL

F.G.

FOUNDATION WALL

**INFORMATION PROVIDED ON BLUEPRINT**

**A** MARK THE ROUGH WALL HEIGHT ON THE STORY POLE BY ADDING TOGETHER THE FINISH FLOOR TO FINISH CEILING DIMENSIONS PLUS THE THICKNESSES OF THE FLOOR AND CEILING MATERIALS. MARK TOP AND BOTTOM PLATES.

½" GYPSUM BOARD
7'-11¾" FIN. FLOOR TO FIN. CLG.
¾" OAK FLOOR
8'-1" ROUGH WALL HEIGHT

**B** MEASURE THE HEIGHT OF THE ROUGH DOOR OPENING BY ADDING UP DIMENSIONS GIVEN IN THE EXTERIOR DOOR FRAME SECTION DETAIL. MARK THE DOOR HEADER ON THE STORY POLE.

1½" HEAD JAMB & CLEARANCE
6'-8" DOOR
1¼" FINISH FLOOR AND THRESHOLD
6'-10¾" ROUGH HEIGHT HEADER TO SUBFLOOR

**C** MEASURE DOWN FROM THE HEADER THE HEIGHT OF THE WINDOW ROUGH OPENING. THIS IS DONE BY ADDING UP DIMENSIONS GIVEN IN THE WINDOW FRAME SECTION DETAIL AND THE SASH SIZE GIVEN IN THE WINDOW SCHEDULE.

1¾" HEAD OF FINISH FRAME AND CLEARANCE
4'-6" SASH HEIGHT
2" FINISH SILL
4'-9¾" HT. OF WINDOW ROUGH OPENING

Figure 43–30. Laying out a story pole from the information provided in Figure 43–29.

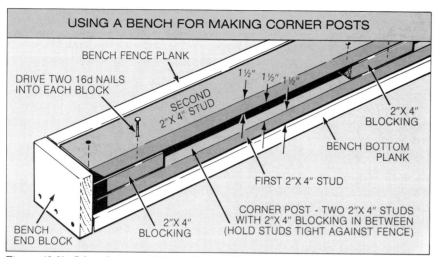

## USING A BENCH FOR MAKING CORNER POSTS

BENCH FENCE PLANK

DRIVE TWO 16d NAILS INTO EACH BLOCK

SECOND 2"X 4" STUD

2"X 4" BLOCKING

1½" 1½" 1½"

BENCH BOTTOM PLANK

BENCH END BLOCK

2"X 4" BLOCKING

FIRST 2"X 4" STUD

CORNER POST - TWO 2"X 4" STUDS WITH 2"X 4" BLOCKING IN BETWEEN (HOLD STUDS TIGHT AGAINST FENCE)

Figure 43-31. A bench is useful for making up large quantities of corner posts. The material should be held tightly against the fence and end block when the pieces are nailed together.

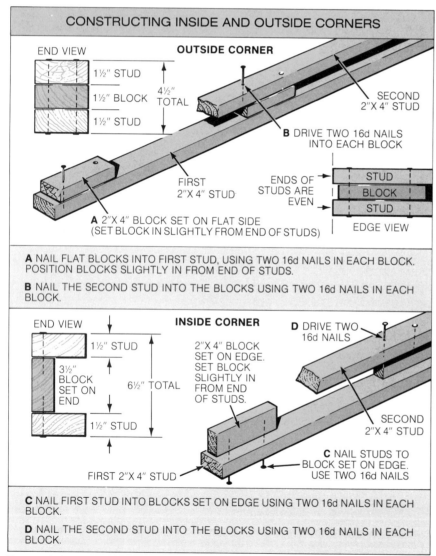

## CONSTRUCTING INSIDE AND OUTSIDE CORNERS

**OUTSIDE CORNER**

END VIEW

1½" STUD
1½" BLOCK
1½" STUD

4½" TOTAL

SECOND 2"X 4" STUD

**B** DRIVE TWO 16d NAILS INTO EACH BLOCK

FIRST 2"X 4" STUD

ENDS OF STUDS ARE EVEN

STUD
BLOCK
STUD

EDGE VIEW

**A** 2"X 4" BLOCK SET ON FLAT SIDE (SET BLOCK IN SLIGHTLY FROM END OF STUDS)

**A** NAIL FLAT BLOCKS INTO FIRST STUD, USING TWO 16d NAILS IN EACH BLOCK. POSITION BLOCKS SLIGHTLY IN FROM END OF STUDS.

**B** NAIL THE SECOND STUD INTO THE BLOCKS USING TWO 16d NAILS IN EACH BLOCK.

**INSIDE CORNER**

END VIEW

1½" STUD

3½" BLOCK SET ON END

1½" STUD

6½" TOTAL

2"X 4" BLOCK SET ON EDGE. SET BLOCK SLIGHTLY IN FROM END OF STUDS.

**D** DRIVE TWO 16d NAILS

SECOND 2"X 4" STUD

FIRST 2"X 4" STUD

**C** NAIL STUDS TO BLOCK SET ON EDGE. USE TWO 16d NAILS

**C** NAIL FIRST STUD INTO BLOCKS SET ON EDGE USING TWO 16d NAILS IN EACH BLOCK.

**D** NAIL THE SECOND STUD INTO THE BLOCKS USING TWO 16d NAILS IN EACH BLOCK.

Figure 43-32. Assembling outside and inside corners. Before nailing, align the ends of the studs.

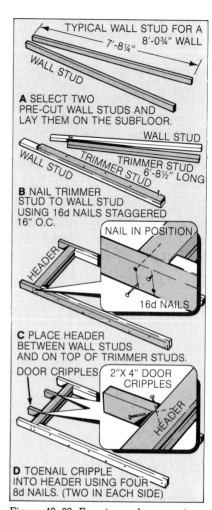

TYPICAL WALL STUD FOR A 8'-0¾" WALL
7'-8¼"

WALL STUD

**A** SELECT TWO PRE-CUT WALL STUDS AND LAY THEM ON THE SUBFLOOR.

WALL STUD

WALL STUD

TRIMMER STUD 6'-8½" LONG

TRIMMER STUD

**B** NAIL TRIMMER STUD TO WALL STUD USING 16d NAILS STAGGERED 16" O.C.

NAIL IN POSITION

HEADER

16d NAILS

**C** PLACE HEADER BETWEEN WALL STUDS AND ON TOP OF TRIMMER STUDS.

DOOR CRIPPLES

2"X 4" DOOR CRIPPLES

HEADER

**D** TOENAIL CRIPPLE INTO HEADER USING FOUR 8d NAILS. (TWO IN EACH SIDE)

Figure 43-33. Framing a door opening. In this example, cripple studs are required over the header.

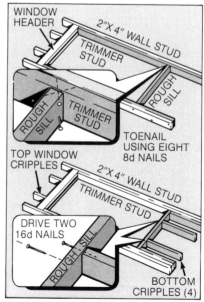

WINDOW HEADER

2"X 4" WALL STUD

TRIMMER STUD

ROUGH SILL

ROUGH SILL

TRIMMER STUD

TOENAIL USING EIGHT 8d NAILS

TOP WINDOW CRIPPLES

2"X 4" WALL STUD

TRIMMER STUD

DRIVE TWO 16d NAILS

ROUGH SILL

BOTTOM CRIPPLES (4)

Figure 43-34. Installing a rough windowsill and cripple studs. Cripple studs follow the stud layout (usually 16" O.C.).

made up, the entire wall can be nailed together on the subfloor. Place top and bottom plates at a distance slightly greater than the length of the studs. Position the corners and openings between the plates according to the plate layout. Place studs in position with crown side up. Spike the plates into the studs, cripples, and trimmers. See Figure 43–35.

On long walls the breaks in the plates should occur over a stud or cripple. If a 4″ × 12″ header without cripples is used, the breaks in the plates should occur over a stud or over the header. See Figure 43–36.

## Placing the Double Top Plate

The double top plate may be placed while the wall is still lying flat on the subfloor or after all the walls have been raised. The top-most plates are nailed so that they overlap the plates below

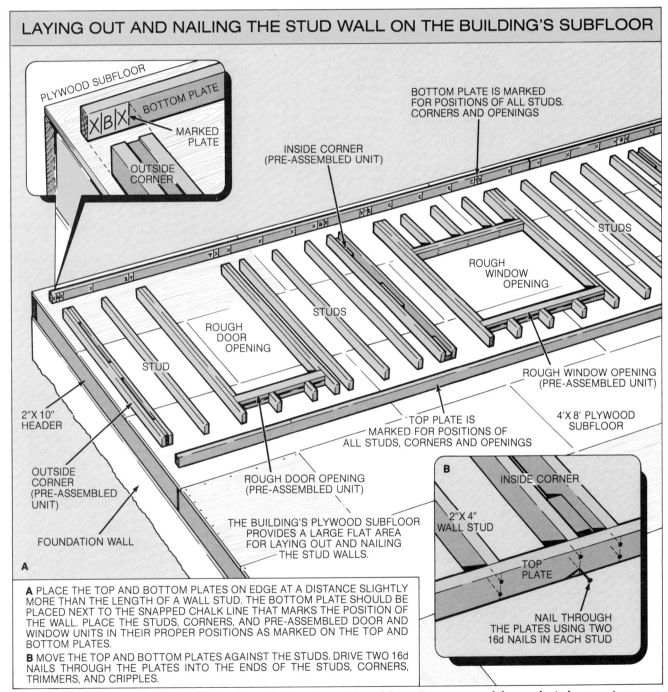

### LAYING OUT AND NAILING THE STUD WALL ON THE BUILDING'S SUBFLOOR

PLYWOOD SUBFLOOR
BOTTOM PLATE
X|B|X| MARKED PLATE
OUTSIDE CORNER

BOTTOM PLATE IS MARKED FOR POSITIONS OF ALL STUDS. CORNERS AND OPENINGS

INSIDE CORNER (PRE-ASSEMBLED UNIT)

BOTTOM PLATE IS MARKED FOR POSITIONS OF ALL STUDS. CORNERS AND OPENINGS

STUDS

ROUGH WINDOW OPENING

STUDS

ROUGH DOOR OPENING

STUD

2″X 10″ HEADER

OUTSIDE CORNER (PRE-ASSEMBLED UNIT)

FOUNDATION WALL

ROUGH DOOR OPENING (PRE-ASSEMBLED UNIT)

THE BUILDING'S PLYWOOD SUBFLOOR PROVIDES A LARGE FLAT AREA FOR LAYING OUT AND NAILING THE STUD WALLS.

TOP PLATE IS MARKED FOR POSITIONS OF ALL STUDS, CORNERS AND OPENINGS

ROUGH WINDOW OPENING (PRE-ASSEMBLED UNIT)

4′X 8′ PLYWOOD SUBFLOOR

**B** INSIDE CORNER

2″X 4″ WALL STUD

TOP PLATE

NAIL THROUGH THE PLATES USING TWO 16d NAILS IN EACH STUD

**A** PLACE THE TOP AND BOTTOM PLATES ON EDGE AT A DISTANCE SLIGHTLY MORE THAN THE LENGTH OF A WALL STUD. THE BOTTOM PLATE SHOULD BE PLACED NEXT TO THE SNAPPED CHALK LINE THAT MARKS THE POSITION OF THE WALL. PLACE THE STUDS, CORNERS, AND PRE-ASSEMBLED DOOR AND WINDOW UNITS IN THEIR PROPER POSITIONS AS MARKED ON THE TOP AND BOTTOM PLATES.

**B** MOVE THE TOP AND BOTTOM PLATES AGAINST THE STUDS. DRIVE TWO 16d NAILS THROUGH THE PLATES INTO THE ENDS OF THE STUDS, CORNERS, TRIMMERS, AND CRIPPLES.

Figure 43–35. Assembling the wall. The wall is nailed together on the subfloor after corners and door and window openings are completed.

Figure 43–36. Joints in plates should occur over a full header or at the center of a stud or cripple.

them at all corners. This helps to tie the walls together. All ends are fastened with two 16d nails. Between the ends, 16d nails are staggered 16″ O.C. The butt joints between the topmost plates should be at least 4′ from any butt joint between the plates below them. See Figure 43–37.

## Squaring the Walls and Placing Braces

A completely framed wall is often squared while it is still lying on the subfloor. In this way, bracing, plywood, or other exterior wall covering can be nailed before the wall is raised. See Figure 43–38.

A let-in diagonal brace may also be placed while the wall is still lying on the subfloor. Lay out and snap lines on the studs to show the location of the brace. See Figure 43–39. The studs

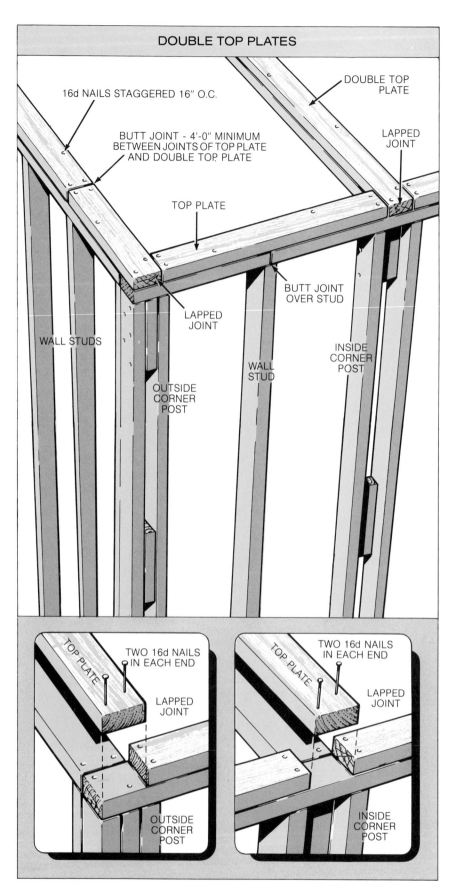

Figure 43–37. The topmost plates of the double top plate overlap the plates below them at all inside corners.

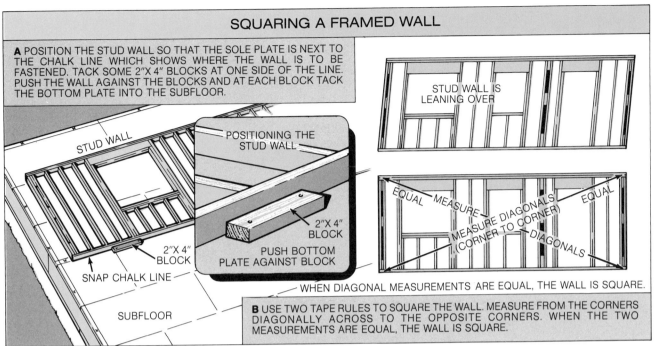

Figure 43–38. A completely framed wall is often squared while it is still lying on the subfloor.

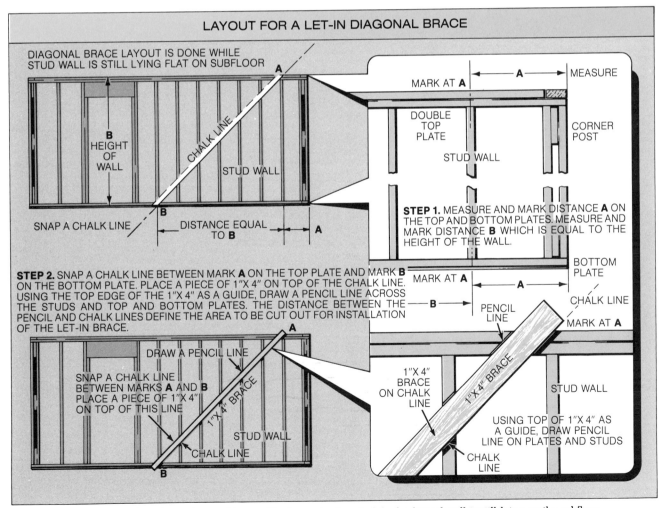

Figure 43–39. Laying out for a let-in diagonal brace. The brace is placed while the framed wall is still lying on the subfloor.

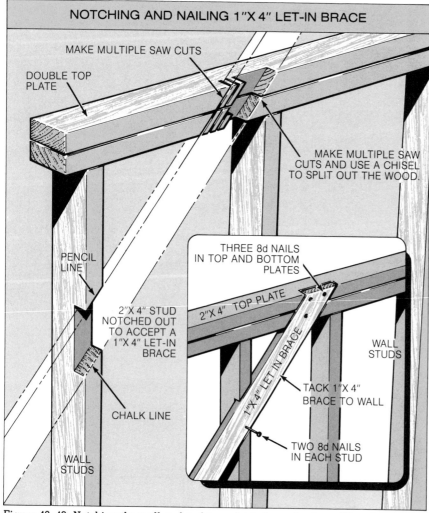

NOTCHING AND NAILING 1"X 4" LET-IN BRACE

MAKE MULTIPLE SAW CUTS

DOUBLE TOP PLATE

MAKE MULTIPLE SAW CUTS AND USE A CHISEL TO SPLIT OUT THE WOOD.

PENCIL LINE

2"X 4" STUD NOTCHED OUT TO ACCEPT A 1"X 4" LET-IN BRACE

CHALK LINE

WALL STUDS

THREE 8d NAILS IN TOP AND BOTTOM PLATES

2"X 4" TOP PLATE

1"X 4" LET-IN BRACE

WALL STUDS

TACK 1"X 4" BRACE TO WALL

TWO 8d NAILS IN EACH STUD

Figure 43–40. Notching the wall and tacking a let-in brace in place. After the wall is raised and the brace adjusted to its final postion, the nails are driven in.

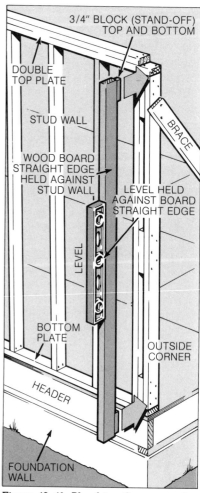

3/4" BLOCK (STAND-OFF) TOP AND BOTTOM

DOUBLE TOP PLATE

STUD WALL

WOOD BOARD STRAIGHT EDGE HELD AGAINST STUD WALL

BRACE

LEVEL HELD AGAINST BOARD STRAIGHT EDGE

LEVEL

BOTTOM PLATE

OUTSIDE CORNER

HEADER

FOUNDATION WALL

Figure 43–41. Plumbing the corner of a wall with a straightedge and level.

may then be notched for the brace. See Figure 43–40. Tack the brace to the studs while the wall is still lying on the subfloor. Tacking instead of nailing allows for some adjustment after the wall is raised. After any necessary adjustment is made, the nails can be securely driven in.

## Placing Fire Blocks

As mentioned earlier in this unit, fire blocks may be placed in the walls to slow the rate of a fire that may occur inside the walls, and also to serve as a nailing base for the edges of plywood or drywall panels.

## Raising the Walls

Most walls can be raised by hand if enough help is available on the job. It is advisable to have one person for every 10′ of wall for the lifting operation.

The order in which walls are framed and raised may vary from job to job. Generally the longer exterior walls are raised first. The shorter exterior walls are then raised, and the corners are nailed together. The order of framing interior partitions depends on the floor layout.

After a wall has been raised, its bottom plates must be nailed securely to the floor. Where the wall rests on a wood subfloor

and joists, 16d nails should be driven through the bottom plate and into the floor joists below the wall.

## Plumbing and Aligning the Walls

Accurate plumbing of the corners is possible only after all the walls are up. Most framing material is not perfectly straight; therefore, walls should never be plumbed by applying a hand level directly to the end stud. Always use a straightedge along with the level. See Figure 43–41. The straightedge can be a piece ripped out of plywood or a straight piece of

2″ × 4″ lumber. Blocks ¾″ thick are nailed to each end. The blocks make it possible to accurately plumb the wall from the bottom plate to the top plate.

Plumbing corners requires two persons working together. One carpenter releases the nails at the bottom end of the corner brace so that the top of the wall can be moved. The other carpenter watches the level. The bottom end of the brace is re-nailed when the level shows a plumb wall.

The tops of the walls are straightened (aligned or *lined up*) after all the corners have been plumbed. Lining up must be done before the floor or ceiling joists are nailed to the tops of the walls. A string is fastened from the top plate at one corner of the wall to the top plate at another corner of the wall. Three small blocks are cut from 1″ × 2″ lumber. One block is placed under each end of the string so that the line is clear of the wall.

Figure 43–43. The bottoms of these bearing partitions have been bolted to the slab. Non-bearing partitions have been fastened with a powder-actuated driver. The two rows of blocking in the wall serve as fire blocks and as a nailing surface for the edges of plasterboard.

The third block is used as a gauge to check the wall at 6′ or 8′ intervals. See Figure 43–42. At each check point a temporary brace is fastened to a wall stud.

When fastening the temporary brace to the wall stud, adjust the wall so that the string is barely touching the gauge block. Nail the other end of the brace to a short 2″ × 4″ block fastened to the subfloor. These temporary braces are not removed until the framing and sheathing for the entire building have been completed.

## Pick-up Framing Operations

Certain smaller framing operations (*pick-up* operations) are usually performed after the walls have been raised. For example, openings are cut in the bottom plates for all door openings at this time. Also, openings are cut in the wall for heating outlets after the walls are raised.

Another pick-up operation is the placement of wall backing where plumbing fixtures (sinks, bathtubs, water closets) will later be fastened to the wall.

## Framing over Concrete Slabs

Often the basement or ground floor of a wood-framed building is a concrete slab. In this case the bottom plates of the walls must be either bolted to the slab or nailed to the slab with a powder-actuated driver. If bolts are used, they must be accurately set into the slab at the time of the concrete pour. Holes for the bolts are laid out and drilled in the bottom plate when the wall is framed. When the wall is raised, it is slipped over the bolts. See Figure 43–43.

Some walls, such as basement

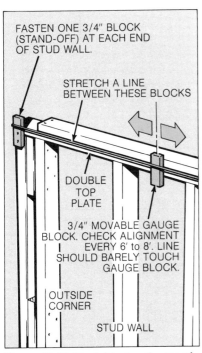

FASTEN ONE 3/4″ BLOCK (STAND-OFF) AT EACH END OF STUD WALL.

STRETCH A LINE BETWEEN THESE BLOCKS

DOUBLE TOP PLATE

3/4″ MOVABLE GAUGE BLOCK. CHECK ALIGNMENT EVERY 6′ to 8′. LINE SHOULD BARELY TOUCH GAUGE BLOCK.

OUTSIDE CORNER

STUD WALL

Figure 43–42. Straightening the top of a wall using a line and ¾″ blocks.

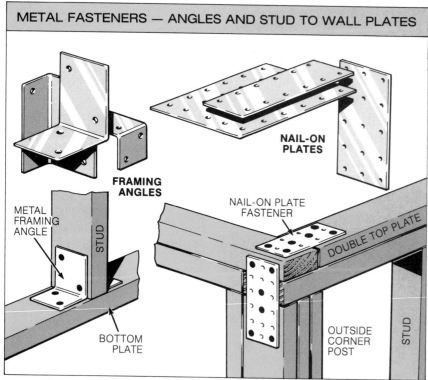

Figure 43–44. Metal framing angles and nail-on plates used with wood framed walls.

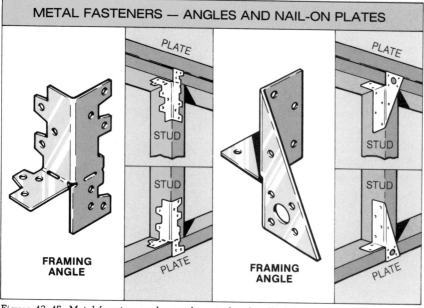

Figure 43–45. Metal framing angles used on stud and wall plate.

or garage walls, have no bottom plate. The top plates of these walls are nailed to the studs, and then the walls are raised. After the walls are lifted into place, the bottom ends of the studs are toe-nailed into the mudsill. The rest of the framing procedure is the same as that for walls nailed on top of a subfloor.

## Metal Fasteners and Devices

Metal fasteners may be used with wood-framed walls. These include flat nail-on plates and framing angles (Figure 43–44), framing anchors (Figure 43–45), wall bracing, and tie-down devices (Figures 43–46 and 43–47). In some areas of the United States the use of metal fasteners is optional and not required by code. In other regions, particularly where there are earthquake and high-wind risk factors, certain types of metal connectors are required by the local building codes.

## Sheathing the Walls

Wall sheathing is the material used for the exterior covering of the outside walls. In the past, nominal 1″ thick boards were nailed to the wall horizontally or at a 45° angle for sheathing. Today plywood and other types of panel products (waferboard, oriented strand board, composite board) are usually used for sheathing. Plywood and non-veneered panels can be applied much more quickly than boards. They add considerable strength to a building and often eliminate the need for diagonal bracing. (To review information on plywood and non-veneered panels, refer to Section 2.)

Generally, the term *wall sheathing* does not include the finished surface of a wall. Siding, shingles, stucco, or brick veneer are placed on top of the sheathing to finish the wall. Exterior finish materials are discussed in Section 12.

**Plywood Sheating.** Plywood is the most widely used sheathing material. Plywood panels that are usually applied to exterior walls range in size from 4′ × 8′ to 4′ × 12′ with thicknesses from 5/16″ to 3/4″. The panels may be placed with the grain running horizon-

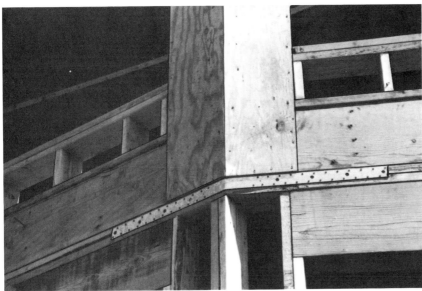

Figure 43–46. Metal strap tying two plates together at the corner of a wood-framed building.

pneumatic stapler is used, the staples should be a minimum of 1¼″ long. They must be spaced 4″ apart on the edges of the panel and 8″ apart at the intermediate studs. See the table in Figure 43–51.

When nailing the panels, leave a ⅛″ gap between the horizontal edges of the panels, and a 1/16″ gap between the vertical edges. These gaps allow for future expansion caused by moisture and prevents panels from buckling.

In larger wood-framed buildings, plywood is often nailed to some of the main interior partitions. The result is called a *shear wall*. Shear walls add considerable strength to the entire building.

Plywood sheathing can be applied when the squared wall is still lying on the subfloor. However, problems can occur after the wall is raised if the floor is

tally or vertically. See Figure 43–48. Local building codes may require blocking along the long edges of horizontally placed panels.

Typical code requirements are: 6d nails with panels ½″ or less in thickness, and 8d nails for panels more than ½″ in thickness. The nails should be spaced 6″

apart along the edges of the panels and 12″ apart at the intermediate studs. See the tables in Figures 43–49 and 43–50. If a

Figure 43–47. Metal fastener attached to corner post and top plates to reinforce the tie.

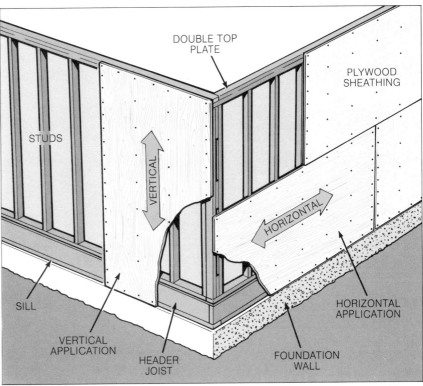

Figure 43–48. Plywood sheathing can be placed with the grain running vertically or horizontally.

| PLYWOOD GRADE | PLYWOOD THICKNESS | MAXIMUM STUD SPACING EXTERIOR COVERING NAILED TO: | | FASTENER TYPE AND SIZE | NAIL SPACING | |
|---|---|---|---|---|---|---|
| | | STUDS | SHEATHING | | PANEL EDGES | INTERMEDIATE STUDS |
| C-D INT<br>C-D INT w/exterior glue, C-C EXT<br><br>or<br><br>STRUCTURAL I<br>C-D INT | 5/16″ | 16″ | 16″* | 6d common smooth, annular or ring-shank, spiral-thread, galvanized box or T-nails or same diameter. Staples also permitted at reduced spacing. | 6″ | 12″ |
| | 3/8″ and 1/2″ 3-ply | 24″ | 16″, 24″* | | 6″ | 12″ |
| | 1/2″ 4-ply and 5-ply | 24″ | 24″ | | 6″ | 12″ |
| *Apply plywood with face grain across studs. | | | | | | |

*American Plywood Association*

Figure 43–49. Recommended thicknesses and nailing schedule for plywood wall sheathing.

| PANEL SPAN RATING | MAXIMUM STUD SPACING | NAIL SIZE | NAIL SPACING PANEL EDGES | NAIL SPACING INTERMEDIATE |
|---|---|---|---|---|
| 12/0, 16/0, 20/0 or wall 16″ O.C. | 16″(a) | 6d for panels 1/2″ thick or less. 8d for thicker panels. | 6″ | 12″ |
| 24/0, 24/16, 32/16 or wall 24″ O.C. | 24″ (b) | 6d for panels 1/2″ thick or less. 8d for thicker panels. | 6″ | 12″ |

(a) Apply plywood panels less than 3/8″ thick with face grain across studs when exterior covering is nailed to sheathing.
(b) Apply 3-ply plywood panels with face grain across studs 24″ O.C. when exterior covering is nailed to sheathing.

*American Plywood Association*

Figure 43–50. Recommended nailing schedule for plywood wall sheathing.

| PANEL THICKNESS | STAPLE LEG LENGTH | SPACING AROUND ENTIRE PERIMETER OF SHEET | SPACING AT INTERMEDIATE MEMBERS |
|---|---|---|---|
| 5/16″ | 1 1/4″ | 4″ | 8″ |
| 3/8″ | 1 3/8″ | 4″ | 8″ |
| 1/2″ | 1 1/2″ | 4″ | 8″ |

*American Plywood Association*

Figure 43–51. Recommended stapling schedule for plywood wall sheathing.

not perfectly straight and level. For this reason some carpenters prefer to place the plywood after the entire building has been framed.

**Non-veneered Panel Sheathing.** Although plywood is still the type of panel employed most often for wall sheathing, non-veneered (reconstituted wood) panels are growing in use. See Figure 43–52. Panels made of waferboard, oriented strand board, and composite board have been approved by most local building codes for wall sheathing. Like plywood, these panels sufficiently resist racking so that no corner bracing is necessary in normal residential and commercial construction. However, under construction conditions where maximum shear strength is called for, conventional veneered plywood panels are still recommended.

The application of non-veneered wall sheathing is similar to that for plywood. Nailing schedules usually call for 6d common nails spaced 6″ O.C. at the panel edges, and 12″ O.C. when nailed into the intermediate studs. Non-veneered panels are usually applied with the long edge of the panel in a vertical position.

Non-wood insulating materials are also available for wall sheathing. These products are discussed in Section 11.

Figure 43–52. Non-veneered waferboard panels are often used to sheath exterior walls.

# UNIT 44

# Ceiling Framing

Ceiling construction begins after all walls have been set in place and fastened to the subfloor, the corners of the walls have been plumbed, and the tops of the walls have been aligned and held in place with temporary bracing. One type of ceiling supports an attic area beneath a sloping (pitched) roof. Another type of ceiling serves as the framework of a flat roof. When a building has two or more floors, the ceiling of a lower story is the floor unit of the story above.

One of the main structural functions of a ceiling frame is to tie together the outside walls of the building. When located under a pitched roof, the ceiling frame also resists the outward pressure placed on the walls by the roof rafters. See Figure 44–1. The tops of interior partitions are fastened to the ceiling frame. In addition to supporting the attic area beneath the roof, the ceiling frame supports the weight of the finish ceiling materials, such as gypsum board or lath and plaster.

## CEILING JOISTS

Joists are the most important framing members of the ceiling. Their size, spacing, and direction of travel are given on the floor plan. The spacing between ceiling joists is usually 16" O.C., although 24" spacing is also used. The size of a ceiling joist is determined by the amount of weight it must carry and the span it covers from one wall to the other. The table in Figure 44–2 gives allowable spans for joists spaced 16" and 24" O.C.

Although it is more convenient to have all the joists running in the same direction, plans sometimes call for different sets of joists running at right angles to each other.

## Interior Support of Joists

One end of a ceiling joist rests on an outside wall. The other end often overlaps over an interior bearing partition or girder. The overlap should be at least 4". Ceiling joists are sometimes butted over the partition or girder. In this case the joists must be cleated with a ¾" thick plywood board, 24" long, or an 18-ga. metal strap, 18" long.

**JOISTS RESIST OUTWARD PRESSURE OF ROOF RAFTERS**

OUTWARD PRESSURE OF ROOF RAFTERS

OUTWARD PRESSURE OF ROOF RAFTERS

ROOF RAFTERS

ROOF RAFTERS

ATTIC

CEILING JOISTS

CEILING JOISTS

INTERIOR WALL

CEILING JOISTS TIE THE OUTSIDE WALLS TOGETHER AND HELP TO RESIST THE OUTWARD PRESSURE OF THE ROOF RAFTERS

OUTSIDE WALL

OUTSIDE WALL

Figure 44–1. A ceiling frame ties together the exterior walls and resists the outward pressure of the roof rafters.

| Species | Grade | Span (feet and inches) | | | | | |
|---|---|---|---|---|---|---|---|
| | | 2 x 4 | | 2 x 6 | | 2 x 8 | |
| | | 16″ oc | 24″ oc | 16″ oc | 24″ oc | 16″ oc | 24″ oc |
| DOUGLAS FIR-LARCH | 2 | 11-6 | 10-0 | 18-1 | 15-7 | 23-10 | 20-7 |
| | 3 | 9-9 | 7-11 | 14-8 | 11-11 | 19-4 | 15-9 |
| DOUGLAS FIR SOUTH | 2 | 10-6 | 9-2 | 16-6 | 14-5 | 21-9 | 19-0 |
| | 3 | 9-5 | 7-9 | 14-2 | 11-7 | 18-9 | 15-3 |
| HEM-FIR | 2 | 10-9 | 9-5 | 16-11 | 13-11 | 22-4 | 18-4 |
| | 3 | 8-7 | 7-0 | 13-1 | 10-8 | 17-2 | 14-0 |
| MOUNTAIN HEMLOCK | 2 | 9-11 | 8-8 | 15-7 | 13-8 | 20-7 | 18-0 |
| | 3 | 8-11 | 7-3 | 13-3 | 10-10 | 17-6 | 14-4 |
| MOUNTAIN HEMLOCK-HEM-FIR | 2 | 9-11 | 8-8 | 15-7 | 13-8 | 20-7 | 18-0 |
| | 3 | 8-7 | 7-0 | 13-1 | 10-8 | 17-2 | 14-0 |
| WESTERN HEMLOCK | 2 | 10-9 | 9-5 | 16-11 | 14-6 | 22-4 | 19-1 |
| | 3 | 9-0 | 7-5 | 13-9 | 11-3 | 18-1 | 14-10 |
| ENGELMANN SPRUCE ALPINE FIR (Engelmann Spruce-Lodgepole Pine) | 2 | 9-11 | 8-8 | 15-6 | 12-8 | 20-7 | 16-8 |
| | 3 | 7-10 | 6-5 | 11-9 | 9-7 | 15-6 | 12-8 |
| LODGEPOLE PINE | 2 | 10-3 | 8-11 | 16-1 | 13-3 | 21-2 | 17-6 |
| | 3 | 8-4 | 6-9 | 12-7 | 10-3 | 16-7 | 13-6 |
| PONDEROSA PINE-SUGAR PINE (Ponderosa Pine-Lodgepole Pine) | 2 | 9-11 | 8-8 | 15-7 | 12-10 | 20-7 | 16-10 |
| | 3 | 8-0 | 6-6 | 12-0 | 9-10 | 15-10 | 12-11 |
| WHITE WOODS (Western Woods) | 2 | 9-7 | 8-5 | 15-1 | 12-6 | 20-2 | 16-5 |
| | 3 | 7-10 | 6-5 | 11-9 | 9-8 | 15-6 | 12-8 |
| IDAHO WHITE PINE | 2 | 10-3 | 8-5 | 15-3 | 12-6 | 20-0 | 16-5 |
| | 3 | 7-10 | 6-5 | 11-9 | 9-8 | 15-6 | 12-8 |
| WESTERN CEDARS | 2 | 9-7 | 8-5 | 15-1 | 13-2 | 20-0 | 17-6 |
| | 3 | 8-3 | 6-9 | 12-7 | 10-3 | 16-7 | 13-7 |

Design Criteria:
Strength—5 lbs. per sq. ft. dead load plus 10 lbs. per sq. ft. live load. No storage above.
Deflection—Limited to span in inches divided by 240 for live load only.

*Western Wood Products Association*

Figure 44–2. Allowable spans for ceiling joists spaced 16″ and 24″ O.C.

Ceiling joists may also butt against the girder and be supported by a ledger strip or joist hangers in the same manner as floor joists.

The structural design of a building may call for the girder to be placed above the ceiling joists. In this case the inside ends of the joists are hung from the beam with special joist hangers.

## Ceiling Joists and Roof Rafters

Whenever possible, the ceiling joists should run in the same direction as the roof rafters. Nailing the outside end of each ceiling joist to the heel of the rafter as well as to the wall plates strengthens the tie between the outside walls of the building. See Figure 44–3.

A building may be designed so that the ceiling joists do not run parallel to the roof rafters. The rafters are therefore pushing out on the walls that are not tied together by the ceiling joist. In this case 2″ × 4″ pieces are added to run in the same direction as the rafters. The top of each ceiling joist should be nailed to the rafters with two 16d nails. See Figure 44–4. The 2″ × 4″ pieces should be spaced no more than 4′ apart. The ends should be secured to the heel of the rafters or to blocking over the outside walls.

## Cutting Ends for Roof Slope

When ceiling joists run in the same direction as the roof rafters, the outside ends must be

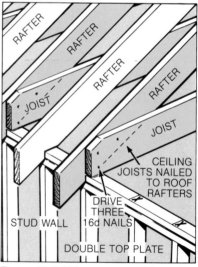

Figure 44–3. Whenever possible, ceiling joists should be nailed to the rafter.

cut to the slope of the roof. (The procedure for laying out the angle for the slope is discussed in Section 10.) Ceiling frames are

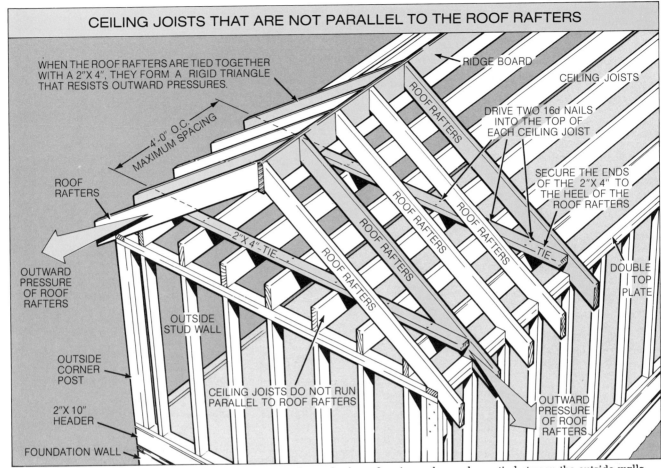

CEILING JOISTS THAT ARE NOT PARALLEL TO THE ROOF RAFTERS

WHEN THE ROOF RAFTERS ARE TIED TOGETHER WITH A 2"X 4", THEY FORM A RIGID TRIANGLE THAT RESISTS OUTWARD PRESSURES.

RIDGE BOARD

CEILING JOISTS

ROOF RAFTERS

4'-0" O.C. MAXIMUM SPACING

DRIVE TWO 16d NAILS INTO THE TOP OF EACH CEILING JOIST

ROOF RAFTERS

SECURE THE ENDS OF THE 2"X 4" TO THE HEEL OF THE ROOF RAFTERS

ROOF RAFTERS

ROOF RAFTERS

ROOF RAFTERS

ROOF RAFTERS

ROOF RAFTERS

TIE

DOUBLE TOP PLATE

2"X 4" TIE

OUTWARD PRESSURE OF ROOF RAFTERS

OUTSIDE STUD WALL

OUTSIDE CORNER POST

CEILING JOISTS DO NOT RUN PARALLEL TO ROOF RAFTERS

OUTWARD PRESSURE OF ROOF RAFTERS

2"X 10" HEADER

FOUNDATION WALL

Figure 44-4. When ceiling joists do not run parallel to the roof rafters, 2 × 4s can be used as a tie between the outside walls.

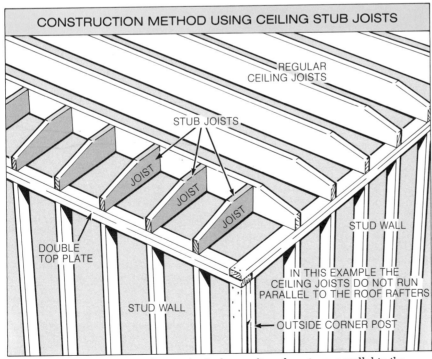

CONSTRUCTION METHOD USING CEILING STUB JOISTS

REGULAR CEILING JOISTS

STUB JOISTS

JOIST

JOIST

JOIST

STUD WALL

DOUBLE TOP PLATE

IN THIS EXAMPLE THE CEILING JOISTS DO NOT RUN PARALLEL TO THE ROOF RAFTERS

STUD WALL

OUTSIDE CORNER POST

Figure 44-5. Stub joists may be required where rafters do not run parallel to the ceiling joists.

sometimes constructed with stub joists. See Figure 44-5. Stub joists are necessary when, in certain sections of the roof, rafters and ceiling joists do not run in the same direction. For example, a low-pitched hip roof requires stub joists in the hip section of the roof.

## Ribbands and Strongbacks

Ceiling joists that do not support a floor above require no header (rim) joists or blocking. Without the additional header joists, however, ceiling joists may twist or bow at the centers of their span. To help prevent this, a 1" × 4" piece called a *ribband* can be nailed at the center of the spans. See Figure 44-6. The ribband is laid flat and fastened to the top

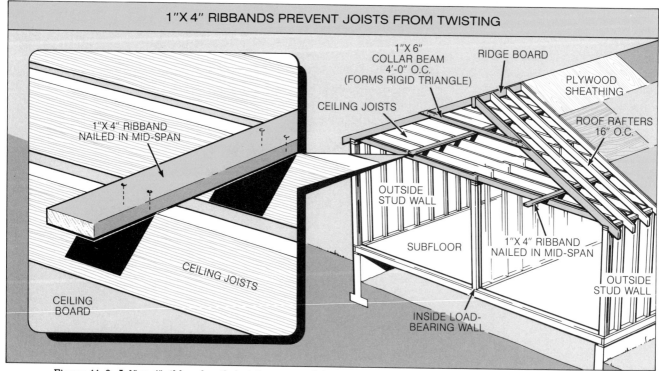

Figure 44–6. A 1″ × 4″ ribband nailed at the center of the joist spans prevents twisting and bowing of the joists.

of each joist with two 8d nails. The end of each ribband is secured to the outside walls of the building.

A more effective method of preventing twisting or bowing of the ceiling joists is to use a *strongback*. A strongback is made of 2″ × 6″ or 2″ × 8″ material nailed to the side of a 2″ × 4″ piece. The 2″ × 4″ piece is fastened with two 16d nails to the top of each ceiling joist as shown in Figure 44–7. The strongbacks are blocked up and supported over the outside walls and interior partitions. Each strongback holds a ceiling joist in line and also helps support the joist at the center of its span.

## LAYING OUT A CEILING FRAME

Ceiling joists should be placed directly above the studs when the spacing between the joists is the same as between the studs. This arrangement makes it eas-

ier to install pipes, flues, or ducts that have to run up the wall and through the roof. However, for buildings with walls that have double top plates, most building codes do not require ceiling joists to line up with the studs below.

If the joists are being placed directly above the studs, they will follow the same layout as the studs below. See Figure 44–8. If the joist layout is different from that of the studs below (for example, if joists are laid out 24″ O.C. over a 16″ O.C. stud layout), mark the first joist at 23¼″ and then at every 24″ O.C. Refer again to Figure 44–8.

It is a good practice to mark the positions of the roof rafters at the time that the ceiling joists are being laid out. If the spacing between the ceiling joists is the same as between the roof rafters, there will be a rafter next to every joist. Often the joists are laid out 16″ O.C. and the roof rafters 24″ O.C. Therefore, every

other rafter can be placed next to a ceiling joist.

## CONSTRUCTING A CEILING FRAME

All the joists for the ceiling frame should be cut to length before they are placed on top of the walls. On houses with pitched roofs, the outside ends of the joists should also be trimmed for the roof slope. This angle must be cut on the *crown* side of the joist. The prepared joists can then be handed up to the carpenters working on top of the walls. They are spread in a flat position along the walls, close to where they will be nailed. Figure 44–9 shows one procedure for constructing the ceiling frame. In this example the joists will lap over an interior partition.

Ceilings may be *furred down* for insulation or sound-deadening purposes. Figure 44–10 shows an example where a header piece the same width as the joist material is nailed against the wall studs. The ends of the joist are then toenailed or fastened with metal anchors to the header piece.

In a *balloon* framing system, the studs are notched for a *ribbon* (a narrow board) that provides the main support for the ends of the joists. See Figure 44–11. The joists are also nailed into the sides of the studs.

## Applying Backing to Walls

Walls running in the same direction as the ceiling joists require *backing* (sometimes called *deadwood*). Backing provides a nailing surface for the edges of the finish ceiling material. Lumber used for backing is usually of 2″ nominal thickness, although 1″ boards are sometimes used.

Figure 44–12 shows backing placed on top of walls. The 2″ × 4″ pieces nailed to the exterior wall project from one side of the wall. The interior wall requires a 2″ × 6″ or 2″ × 8″ piece extending from both sides of the wall. Backing is fastened to the top plates with 16d nails spaced 16″ O.C. It is sometimes also used where joists run at a right angle to the partition. See Figure 44–13.

## Fastening Walls to Ceiling Frame

The tops of walls running in the same direction as the ceiling joists must be securely fastened to the ceiling frame. The method most often used is shown in Figure 44–14. Blocks, 2″ × 4″, spaced 32″ O.C., are laid flat over the top of the partition. The ends of each block are fastened

to the joists with two 16d nails. Two 16d nails are also driven through each block into the top of the wall.

## ATTIC SCUTTLE

The scuttle is an opening framed in the ceiling to provide an entrance into the attic area. The size of the opening is decided by local code requirements and should be indicated in the blueprints. It must be large enough for a person to climb through easily.

The scuttle is framed in the same way as a floor opening. If the opening is no more than 3′ square, it is not necessary to double the joists and headers. Scuttles must be placed away from the lower areas of a sloping roof. The opening may be covered by a piece of plywood resting on stops.

The scuttle opening can be cut out after all the regular ceiling joists have been nailed in place.

## CONSTRUCTING A FLAT-ROOF CEILING

The use of flat roofs has been increasing in residential construction. They are economical to build, and they blend well with

buildings of modern design.

The joists that make up the framework of a flat roof are also the ceiling joists for the living area below. A flat roof should have some pitch (slope), no less than ¼″ per foot, to shed water. To increase the strength of this type of roof in areas of heavy snowfall, heavier joists, spaced more closely together, may be used. In northern states flat roofs are usually required by local

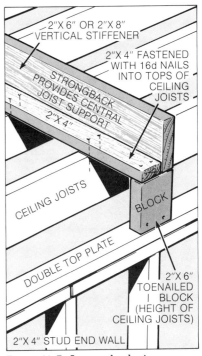

Figure 44–7. A strongback gives central support to the joist span.

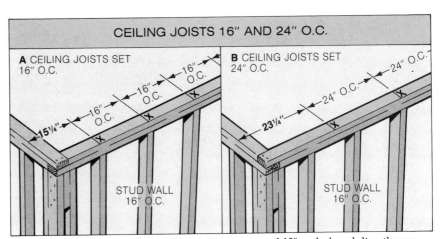

Figure 44–8. In drawing A the ceiling joists are spaced 16″ and placed directly over the studs below. In drawing B the joists are spaced 24″ over studs spaced 16″ O.C.

325

# CONSTRUCTING A TYPICAL CEILING FRAME

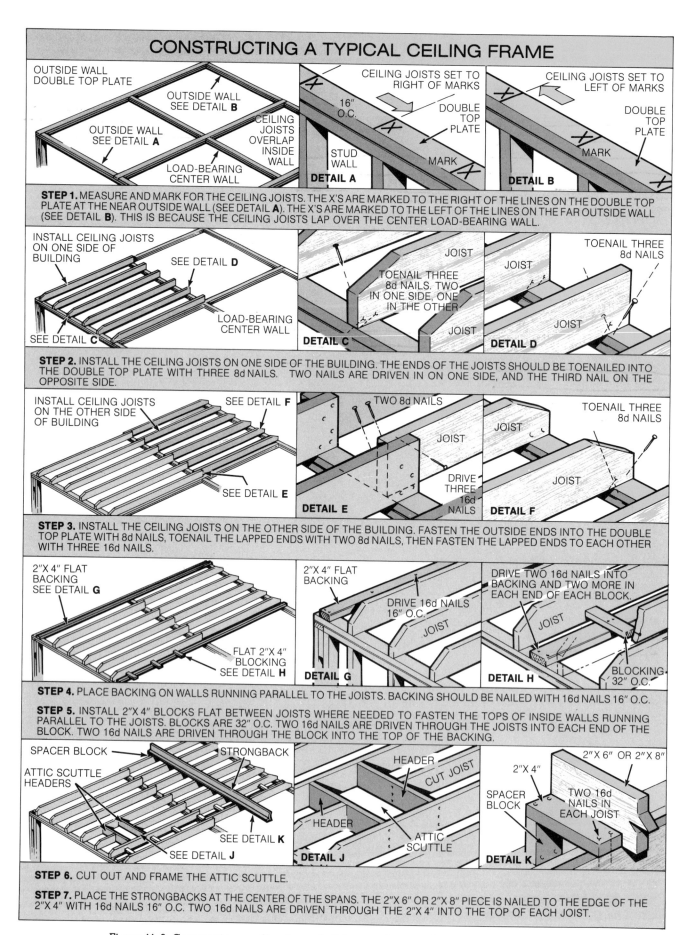

**STEP 1.** MEASURE AND MARK FOR THE CEILING JOISTS. THE X'S ARE MARKED TO THE RIGHT OF THE LINES ON THE DOUBLE TOP PLATE AT THE NEAR OUTSIDE WALL (SEE DETAIL **A**). THE X'S ARE MARKED TO THE LEFT OF THE LINES ON THE FAR OUTSIDE WALL (SEE DETAIL **B**). THIS IS BECAUSE THE CEILING JOISTS LAP OVER THE CENTER LOAD-BEARING WALL.

**STEP 2.** INSTALL THE CEILING JOISTS ON ONE SIDE OF THE BUILDING. THE ENDS OF THE JOISTS SHOULD BE TOENAILED INTO THE DOUBLE TOP PLATE WITH THREE 8d NAILS.  TWO NAILS ARE DRIVEN IN ON ONE SIDE, AND THE THIRD NAIL ON THE OPPOSITE SIDE.

**STEP 3.** INSTALL THE CEILING JOISTS ON THE OTHER SIDE OF THE BUILDING. FASTEN THE OUTSIDE ENDS INTO THE DOUBLE TOP PLATE WITH 8d NAILS, TOENAIL THE LAPPED ENDS WITH TWO 8d NAILS, THEN FASTEN THE LAPPED ENDS TO EACH OTHER WITH THREE 16d NAILS.

**STEP 4.** PLACE BACKING ON WALLS RUNNING PARALLEL TO THE JOISTS. BACKING SHOULD BE NAILED WITH 16d NAILS 16" O.C.

**STEP 5.** INSTALL 2"X 4" BLOCKS FLAT BETWEEN JOISTS WHERE NEEDED TO FASTEN THE TOPS OF INSIDE WALLS RUNNING PARALLEL TO THE JOISTS. BLOCKS ARE 32" O.C. TWO 16d NAILS ARE DRIVEN THROUGH THE JOISTS INTO EACH END OF THE BLOCK. TWO 16d NAILS ARE DRIVEN THROUGH THE BLOCK INTO THE TOP OF THE BACKING.

**STEP 6.** CUT OUT AND FRAME THE ATTIC SCUTTLE.

**STEP 7.** PLACE THE STRONGBACKS AT THE CENTER OF THE SPANS. THE 2"X 6" OR 2"X 8" PIECE IS NAILED TO THE EDGE OF THE 2"X 4" WITH 16d NAILS 16" O.C. TWO 16d NAILS ARE DRIVEN THROUGH THE 2"X 4" INTO THE TOP OF EACH JOIST.

Figure 44–9. Constructing a ceiling frame. In this example, joists lap over an interior partition.

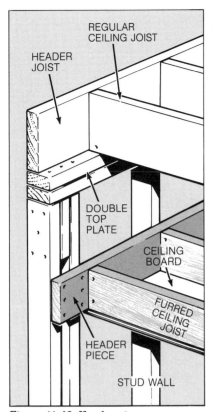

Figure 44-10. Header pieces are nailed against the wall to provide a nailing surface for the joists of the lower ceiling.

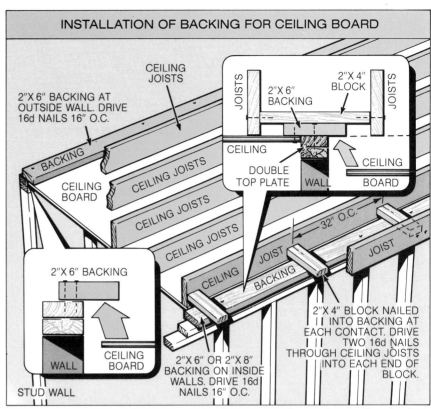

INSTALLATION OF BACKING FOR CEILING BOARD

CEILING JOISTS

2"X 6" BACKING AT OUTSIDE WALL. DRIVE 16d NAILS 16" O.C.

BACKING

CEILING BOARD

CEILING JOISTS

CEILING JOISTS

CEILING JOISTS

CEILING JOIST

2"X 6" BACKING

WALL    CEILING BOARD

STUD WALL

2"X 6" OR 2"X 8" BACKING ON INSIDE WALLS. DRIVE 16d NAILS 16" O.C.

JOISTS    2"X 4" BLOCK    JOISTS

2"X 6" BACKING

CEILING

DOUBLE TOP PLATE    WALL    CEILING BOARD

32" O.C.

JOIST    BACKING    JOIST

CEILING    BACKING

2"X 4" BLOCK NAILED INTO BACKING AT EACH CONTACT. DRIVE TWO 16d NAILS THROUGH CEILING JOISTS INTO EACH END OF BLOCK.

Figure 44-12. Backing is nailed to the top plates to provide a nailing surface for the edges of the finish ceiling material. Note: 2" × 4" backing may also be used.

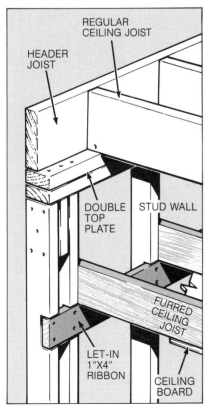

Figure 44-11. A balloon framing system can be used for supporting the joists of the lower ceiling.

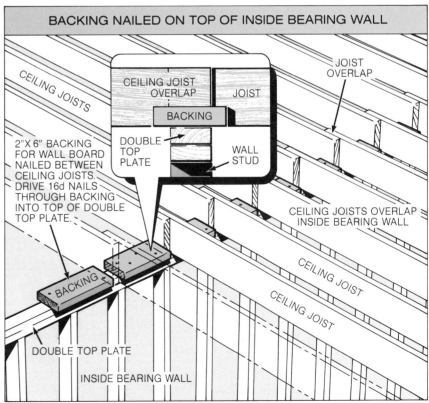

BACKING NAILED ON TOP OF INSIDE BEARING WALL

CEILING JOISTS

CEILING JOIST OVERLAP    JOIST

BACKING

DOUBLE TOP PLATE    WALL STUD

2"X 6" BACKING FOR WALL BOARD NAILED BETWEEN CEILING JOISTS. DRIVE 16d NAILS THROUGH BACKING INTO TOP OF DOUBLE TOP PLATE.

JOIST OVERLAP

CEILING JOISTS OVERLAP INSIDE BEARING WALL

CEILING JOIST

CEILING JOIST

BACKING

DOUBLE TOP PLATE

INSIDE BEARING WALL

Figure 44-13. Backing is sometimes nailed on top of bearing partitions.

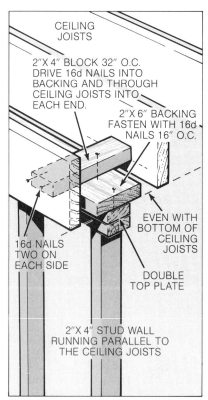

Figure 44–14. Blocking is placed between joists to secure the tops of walls running parallel to the ceiling joists.

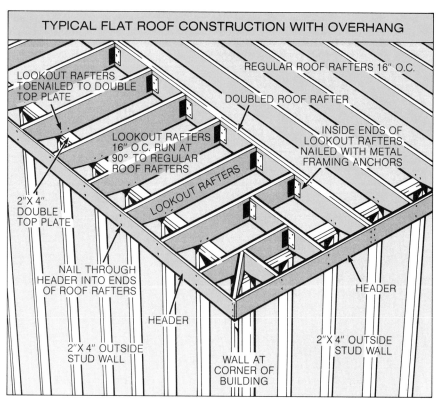

Figure 44–15. A typical flat roof with overhang. Since joists for a flat roof must support the combined load of the ceiling and roof materials, 2″ × 10″ or larger framing members are used.

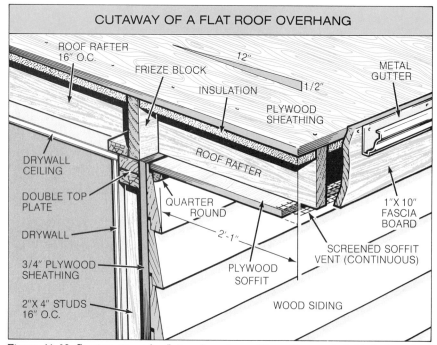

Figure 44–16. Section view of a flat roof overhang. The ceiling material is lath and plaster or gypsum board. The roof is sheathed with plywood over insulation board.

ern states it is usually 20 pounds per square foot. (See Section 10 for a discussion of live loads.)

Figure 44–15 shows a section of the joist framework for a flat roof. The joists are tied together with a header *(band)* where they extend past the sides of the building. The shorter, cantilevered joists are called *lookout rafters.* One end of each lookout raffer is fastened to a regular joist, which will be doubled after all the lookouts have been nailed. A metal framing anchor should also be used. Since the joists for a flat roof must support the combined load of the ceiling and roof materials, 2″ × 10″ or larger pieces are usually used, spaced 16″ O.C.

Figure 44–16 shows a section view of a typical flat roof overhang. The ceiling material is lath and plaster or gypsum board. The roof is sheathed with plywood over insulation board.

building codes to withstand a *live load* (snow and wind load) of 40 pounds per square foot. In cen-

tral states the requirement is usually 30 pounds per square foot, and in southern and west-

# UNIT 45

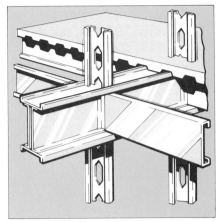

# Metal Framing Systems

The use of metal framing systems continues to grow in residential and other light construction. Many one-family dwellings, apartment buildings, and small commercial buildings are framed entirely in metal today. See Figure 45–1. Many other light construction buildings are framed in a combination of wood and metal.

The material used most often

Inryco, Inc.

Figure 45–1. Four-story building with light-gauge steel framework.

Figure 45–2. Steel joists on top of a wood-framed wall. Steel joists are available in most of the same standard sizes as wood joists.

for metal framing is light-gauge galvanized steel. Aluminum is also used. Both of these products have many features that are an advantage in light framed construction. They will not shrink, swell, twist, or warp. Termites cannot affect them, nor are they susceptible to dry rot. Also, when combined with proper covering material, they have a high fire-resistance rating.

## METAL FLOOR SYSTEMS

Steel floor joists are available in most of the same standard sizes as wood joists. They may rest on top of a concrete foundation wall or a wood-framed (Figure 45–2) or metal-framed wall. End clips, joist hangers, nails, screws, and other devices are used for fastening metal joists to walls of various types. Figure 45–3 shows how metal joists are fastened to a concrete foundation wall. Figure 45–4 shows how

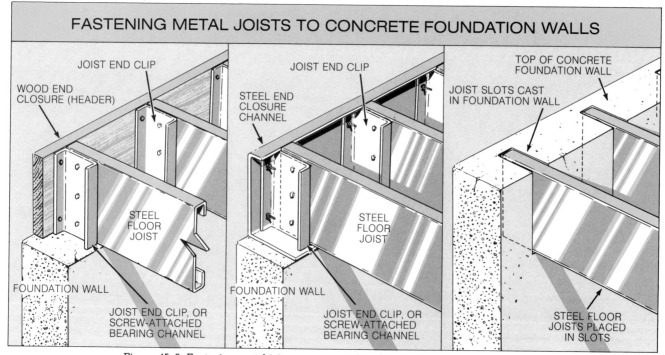

Figure 45–3. Fastening metal joists to concrete foundation walls. End clips are used.

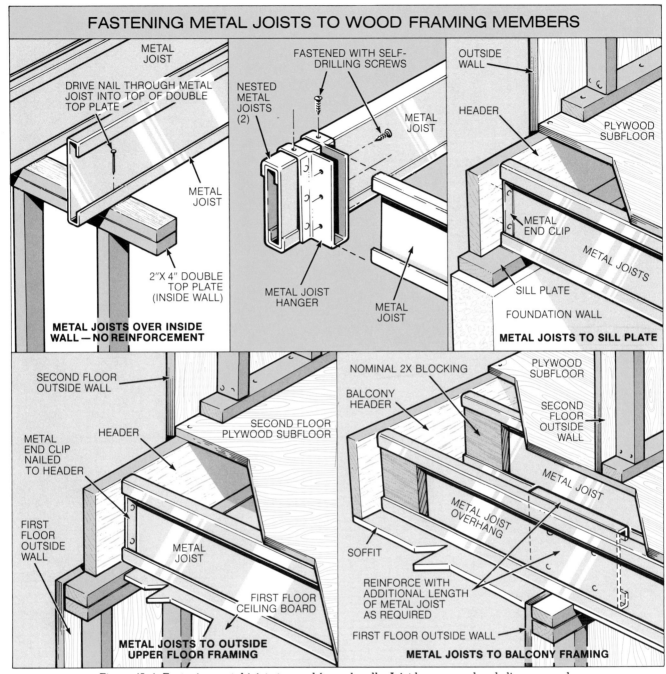

**FASTENING METAL JOISTS TO WOOD FRAMING MEMBERS**

METAL JOIST

DRIVE NAIL THROUGH METAL JOIST INTO TOP OF DOUBLE TOP PLATE

METAL JOIST

2" X 4" DOUBLE TOP PLATE (INSIDE WALL)

**METAL JOISTS OVER INSIDE WALL — NO REINFORCEMENT**

FASTENED WITH SELF-DRILLING SCREWS

NESTED METAL JOISTS (2)

METAL JOIST

METAL JOIST HANGER

METAL JOIST

OUTSIDE WALL

HEADER

PLYWOOD SUBFLOOR

METAL END CLIP

METAL JOISTS

SILL PLATE

FOUNDATION WALL

**METAL JOISTS TO SILL PLATE**

SECOND FLOOR OUTSIDE WALL

HEADER

SECOND FLOOR PLYWOOD SUBFLOOR

METAL END CLIP NAILED TO HEADER

FIRST FLOOR OUTSIDE WALL

METAL JOIST

FIRST FLOOR CEILING BOARD

**METAL JOISTS TO OUTSIDE UPPER FLOOR FRAMING**

NOMINAL 2X BLOCKING

BALCONY HEADER

PLYWOOD SUBFLOOR

SECOND FLOOR OUTSIDE WALL

METAL JOIST

METAL JOIST OVERHANG

SOFFIT

REINFORCE WITH ADDITIONAL LENGTH OF METAL JOIST AS REQUIRED

FIRST FLOOR OUTSIDE WALL

**METAL JOISTS TO BALCONY FRAMING**

Figure 45–4. Fastening metal joists to wood-framed walls. Joist hangers and end clips are used.

they are fastened to a wood-framed wall.

After the joists are set, the floor material can be placed. Often the floor material is plywood panels fastened with self-drilling flat-head screws. In addition, a construction adhesive is usually

recommended. In another type of floor system common in light framed commercial buildings, lightweight corrugated steel decking is placed over the joists, and lightweight concrete fill is placed over the steel decking (Figure 45–5).

## METAL-FRAMED WALLS

Metal-framed walls are similar in design to wood-framed walls. They have channel-shaped studs and cripples, which fasten to tracks that serve as top and bottom plates. Figure 45–6 shows a

331

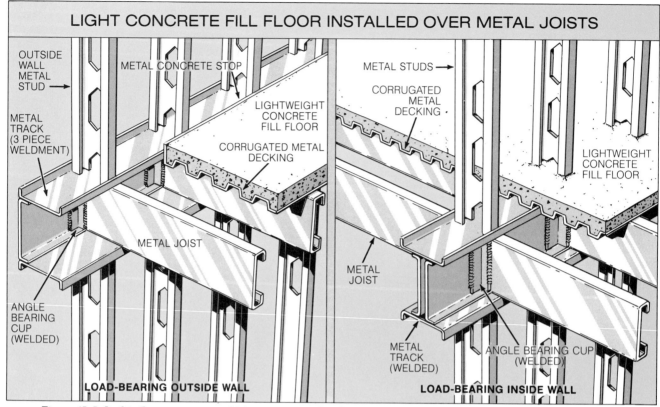

### LIGHT CONCRETE FILL FLOOR INSTALLED OVER METAL JOISTS

OUTSIDE WALL METAL STUD →

METAL CONCRETE STOP

LIGHTWEIGHT CONCRETE FILL FLOOR

CORRUGATED METAL DECKING

METAL TRACK (3 PIECE WELDMENT)

ANGLE BEARING CUP (WELDED)

METAL JOIST

**LOAD-BEARING OUTSIDE WALL**

METAL STUDS →

CORRUGATED METAL DECKING

LIGHTWEIGHT CONCRETE FILL FLOOR

METAL JOIST

METAL TRACK (WELDED)

ANGLE BEARING CUP (WELDED)

**LOAD-BEARING INSIDE WALL**

Figure 45–5. In this floor system, metal joists are covered with corrugated steel decking and lightweight concrete fill.

*United States Steel Corporation*

Figure 45–6. Laying out the position of the studs on the runner tracks.

track being marked for stud positions. The tops of the door and window openings are supported by channel-shaped lintels (headers).

Figure 45–7 shows the major

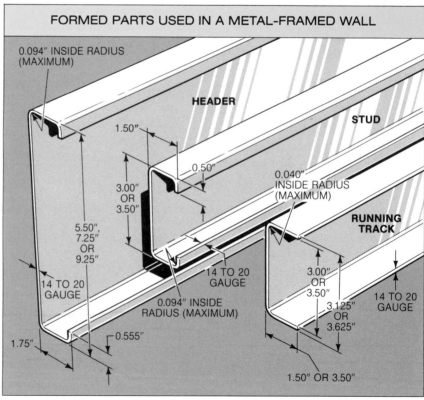

### FORMED PARTS USED IN A METAL-FRAMED WALL

0.094″ INSIDE RADIUS (MAXIMUM)

**HEADER**

**STUD**

1.50″

0.50″

3.00″ OR 3.50″

5.50″, 7.25″ OR 9.25″

14 TO 20 GAUGE

1.75″

0.555″

0.094″ INSIDE RADIUS (MAXIMUM)

14 TO 20 GAUGE

0.040″ INSIDE RADIUS (MAXIMUM)

**RUNNING TRACK**

3.00″ OR 3.50″

3.125″ OR 3.625″

14 TO 20 GAUGE

1.50″ OR 3.50″

Figure 45–7. The major parts of a steel-framed wall are shaped from hot-dipped, galvanized steel or primed sheet metal or from aluminum.

Figure 45–8. Metal walls are often assembled and raised on the job site as shown here for the construction of an apartment building. Note the corrugated steel decking that will later be covered with lightweight concrete.

Figure 45–10. Self-drilling screws are used to fasten metal framing members. One end of a lintel (header) is shown here being fastened to a stud.

Figure 45–9. Metal walls are often prefabricated in a shop. The studs are shown here being welded to the runner tracks that act as top and bottom plates for the wall.

parts of a steel-framed wall. These parts are available in various sizes. Thicknesses range from 14-ga. to 20-ga. steel. The most popular stud widths are 6", 3⅝", and 4". The most popular joist widths are 6", 8", and 10". Metal walls can be assembled at the job site (Figure 45–8) or prefabricated in a shop (Figure 45–9). The structural members can be welded together or joined with self-drilling screws (Figure 45–10). Figure 45–11 shows a metal-framed wall being raised

333

Figure 45–11. A metal-framed wall being raised into place.

Figure 45–12. Metal roofs are constructed of prefabricated trusses.

into place.

Sheathing materials can be applied to a metal wall with self-drilling screws. Drywall is usually used for interior wall finish.

Metal roofs are constructed of prefabricated trusses. See Figure 45–12. They are assembled in a shop or plant, not on the job site.

# SECTION 10

# Roof Frame Construction

The construction of the roof is the last major framing operation for a building. A properly designed roof is strong enough to support heavy snow loads and withstand pressures from strong winds. In addition, it should shed water quickly to reduce the possibility of water leakage. A roof should also have a pleasing appearance, in harmony with the design of the rest of the building.

# UNIT 46

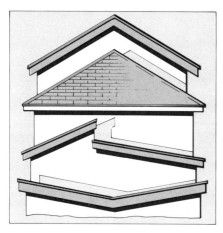

# Basic Roof Types and Roof Theory

## BASIC ROOF TYPES

The basic types of sloping (pitched) roofs are the *shed, gable,* and *hip* roof. See Figure 46–1. These are traditional roof designs that have proven practical over hundreds of years and are still used in all types of construction. (Flat roofs, discussed in Unit 44 of Section 9, are usually used only in buildings of modern design.)

The shed roof, also known as a *lean-to,* is the simplest to construct, as it slopes in only one direction. It is sometimes used as the main roof on a building, but it is more often employed for small sheds, covered porches, and additons to a building.

The gable roof has a ridge at the center and slopes in two directions. It is easy to build and is probably used more often than any other type of roof.

The hip roof has four sloping sides. It is the strongest type of roof, because it is braced by four hip rafters. These hip rafters run at a 45° angle from the corners of the building to the ridge. A hip roof is more difficult to construct than a gable roof.

Two other types of roofs are the *gambrel* and *mansard* roofs.

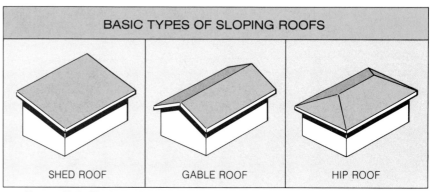

Figure 46–1. Shed, gable, and hip roofs are the basic types of sloping roofs.

See Figure 46–2. Both have the advantage of providing additional living space directly underneath. The gambrel roof is similar to the gable roof except that the slope of the gambrel is broken near the center of the roof, making a double slope on each side. The mansard roof is similar to the hip roof except that it has a double slope on each of its four sides. Compare Figure 46–1 with Figure 46–2.

From the three basic types of roofs (shed, gable, and hip), several other styles of roofs have been designed, such as the *butterfly,* the *monitor,* and the *continuous-slope gable* roof. See Figure 46–3. Also, the basic types can be combined in var-

ious ways, producing *intersecting* roofs. See Figure 46–4.

## PRINCIPLES OF ROOF LAYOUT

Any roof that slopes in two or more directions is based upon the shape of two or more right triangles. A shed roof, which slopes in one direction, is based on the shape of one right triangle. Figure 46–5 shows how the end framework of a gable roof follows the shape of two right triangles placed together.

Carpenters must have the following information to lay out a sloping roof of any type:

1. *Distance of total span.* The

## GAMBREL AND MANSARD ROOFS

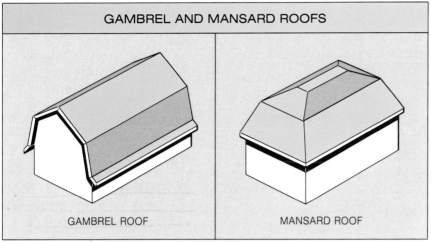

GAMBREL ROOF

MANSARD ROOF

Figure 46–2. Gambrel and mansard roofs are similar to gable and hip roofs except that they have double slopes on each side.

overall width of the building is the total span. This information is found in the floor or roof plan sections of the blueprints.

2. *Distance of total run.* One-half the total span is the total run.

3. *Height of total rise.* The ac-

## EXAMPLES OF SHED AND GABLE ROOF VARIATIONS

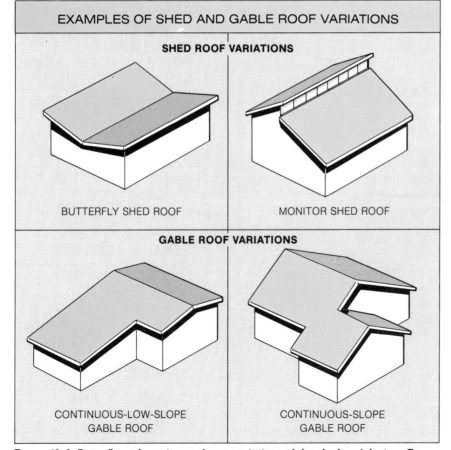

**SHED ROOF VARIATIONS**

BUTTERFLY SHED ROOF

MONITOR SHED ROOF

**GABLE ROOF VARIATIONS**

CONTINUOUS-LOW-SLOPE GABLE ROOF

CONTINUOUS-SLOPE GABLE ROOF

Figure 46–3. Butterfly and monitor roofs are variations of the shed roof design. Continuous-slope gable roofs are a variation of the gable roof design.

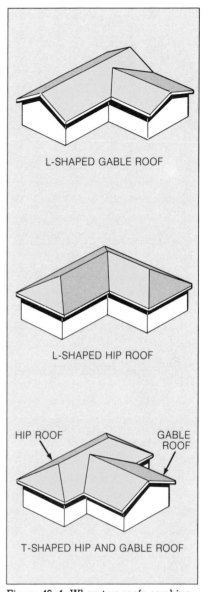

L-SHAPED GABLE ROOF

L-SHAPED HIP ROOF

HIP ROOF    GABLE ROOF

T-SHAPED HIP AND GABLE ROOF

Figure 46–4. When two roofs combine, they form an intersecting roof. Among the many possible variations are the L-shaped hip and L-shaped gable intersecting roofs, and the T-shaped hip-and-gable intersecting roof.

tual height of the roof is the total rise. It is measured from the top wall plate to the ridge of the roof. It is the altitude of the right triangle formed by the roof.

## Roof Pitch and Unit Rise

*Pitch* refers to the angle (slope) of the roof. The amount of pitch is determined by the *unit rise.* In the blueprints, usually on an elevation or section-view drawing, a small triangle is shown with the unit rise and the *unit run* of the roof. The unit run, indicated at the base of the triangle, is always 12″. The unit rise is indicated at the side of the triangle.

The unit rise is the number of inches that the rafter rises vertically for every foot of unit run. See Figure 46–6. As the unit rise of the roof increases, the slope of the roof becomes steeper. Unit rise is specified on the vertical leg of the roof triangle on elevation drawings.

## Total Rise

For carpenters to set the roof ridge to its correct height before nailing the rafters, they must know the *total rise*. (Other methods of setting up the ridge are explained in later units.) The total rise is calculated by multiplying the number of feet in the *total run* times the *unit rise*.

Figure 46–7 shows two examples of how to calculate the total rise of a roof.

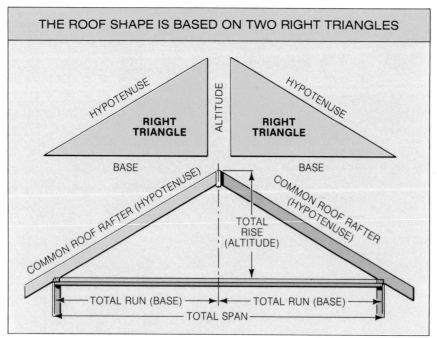

THE ROOF SHAPE IS BASED ON TWO RIGHT TRIANGLES

Figure 46–5. The shape of a gable roof is based on two right triangles. The common rafter forms the third side (hypotenuse) of each triangle.

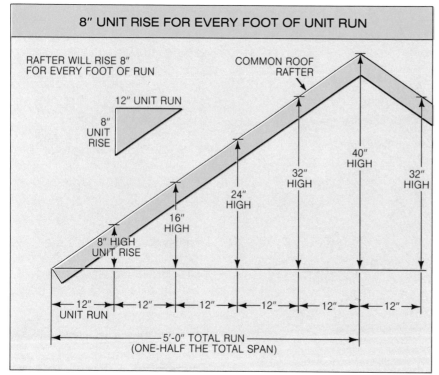

8″ UNIT RISE FOR EVERY FOOT OF UNIT RUN

Figure 46–6. The unit rise of a roof is the number of inches the rafter will rise vertically for every foot of run.

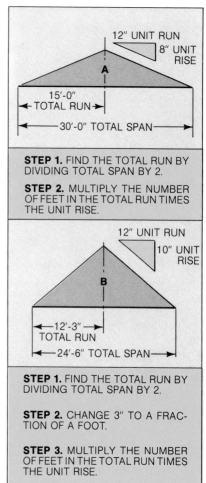

STEP 1. FIND THE TOTAL RUN BY DIVIDING TOTAL SPAN BY 2.

STEP 2. MULTIPLY THE NUMBER OF FEET IN THE TOTAL RUN TIMES THE UNIT RISE.

STEP 1. FIND THE TOTAL RUN BY DIVIDING TOTAL SPAN BY 2.

STEP 2. CHANGE 3″ TO A FRACTION OF A FOOT.

STEP 3. MULTIPLY THE NUMBER OF FEET IN THE TOTAL RUN TIMES THE UNIT RISE.

Figure 46–7. The total rise of a roof is based on the unit rise and total run.

**Example 1.** Find the total rise of a roof with a 30′ span and an 8″ unit rise.

*Step 1.* Find total run by dividing total span by 2.

$$\frac{30'}{2} = 15'\text{-}0'' \text{ total run}$$

*Step 2.* Multiply number of feet in total run times unit rise.

$$15 \times 8'' = 120''$$
$$120'' = 10'\text{-}0'' \text{ total rise}$$

**Example 2.** Find the total rise of a roof with a 24′-6″ span and a 10″ unit rise.

*Step 1.* Find total run by dividing total span by 2.

$$\frac{24'\text{-}6''}{2} = 12'\text{-}3'' \text{ total run}$$

*Step 2.* Change 3″ to a fraction of a foot.

$$\frac{3}{12} = \frac{1}{4}$$

$$12'\text{-}3'' = 12\frac{1}{4}'$$

*Step 3.* Multiply the number of feet in the total run times the unit rise.

$$12\frac{1}{4}' \times 10''$$

Convert $12\frac{1}{4}$ to $\frac{49}{4}$

$$\frac{49}{\underset{2}{4}} \times \overset{5}{10} = \frac{245}{2} = 122\frac{1}{2}$$

$$122\frac{1}{2}'' = 10'\text{-}2\frac{1}{2}'' \text{ total rise}$$

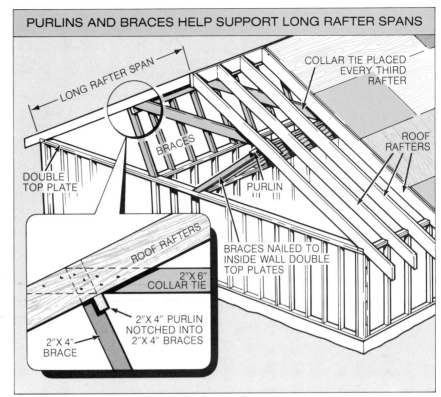

Figure 46–8. Purlins and braces help support the spans of longer rafters.

## STRUCTURAL FACTORS IN ROOF DESIGN

A roof must withstand a great deal of weight and stress. To guarantee structural strength, the *dead load* and *live load* that a roof will bear must be considered in roof design and construction. Rafters, ceiling joists, and devices such as *collar ties* and *purlins* are all factors in roof strength. See Figure 46–8.

### Dead Load

The dead load is the weight of the materials used to construct the roof. Roof rafters, sheathing, insulation, and the finish covering (such as shingles or built-up roofing) are included in the dead load.

### Live Load

The live load is the weight and pressure of wind and snow to which the roof will be subjected.

In most parts of the United States, the combined wind and snow load for a sloped roof will not exceed 30 pounds per square foot. However, wind and snow loads differ from one part of the country to another, and local building codes reflect this difference.

Flat roofs do not shed snow as easily as sloped roofs. Therefore, in cold climates, flat roofs carry heavier snow loads. In the northern states, a flat roof is usually required to withstand a live load of 40 pounds per square foot. (See Unit 44 of Section 9 for a discussion of flat roofs.)

### Allowable Span for Rafters

Dead and live loads have a direct effect on the *allowable span* of rafters used in a roof. The allowable span is the distance from the ridge to the outside wall plates. Some examples of allow-

| RAFTER SIZE | SPACING CENTER TO CENTER | MAXIMUM ALLOWABLE SPAN, FEET AND INCHES | |
|---|---|---|---|
| | | SLOPE OF LESS THAN 4 IN 12 | SLOPE OF 4 IN 12 TO 12 IN 12 |
| 2" × 4" | 12" | 9'6" | 10.'0" |
| | 16" | 8'0" | 9'0" |
| | 24" | 6'6" | 7'6" |
| | 32" | 6'0" | 6'6" |
| 2" × 6" | 12" | 16'6" | 17'6" |
| | 16" | 14'6" | 15'6" |
| | 24" | 12'0" | 12'6" |
| | 32" | 10'6" | 11'0" |

Figure 46–9. Allowable spans for two different rafter sizes. Rafter spacing and roof slope determine the allowable span. For example, the 2" × 4" rafters spaced 16" O.C. have an allowable span of 8'-0" if the unit rise of the roof is less than 4". The allowable span is 9'-0" if the unit rise is 4" or more.

able spans for roof rafters are given in the table in Figure 46–9. This table shows that rafters for a low-pitched roof must be able to support greater weight (live loads) than rafters for a steeper roof.

## Ceiling Joists

Ceiling joists are an important structural factor in any roof system. They hold the tops of the walls in place, and they prevent the weight of the roof from pushing the walls apart. Ceiling joists usually run in the same direction as the rafters. Wherever possible, the seat end of a rafter should be nailed to the side of a ceiling joist as well as to the top wall plates. Ceiling joists are usually spaced 16" O.C., and rafters are usually spaced 24" O.C. Under these conditions rafters can be nailed only into every third ceiling joist.

The Uniform Building Code recommends the following nailing schedule for ceiling joists and rafters at the outside wall plates:

Ceiling joists to plate, toenail: three 8d nails

Ceiling joists to parallel rafter, facenail: three 16d nails

Rafter to plate, toenail: three 8d nails

The material used for ceiling joists is often wider than the rafter stock. In this case, a slope must be cut at the end of the joist. A steel square can be used to mark the angle of the slope, as shown in Figure 46–10. (See Figure 9–16 in Section 3 for more information on the steel square.)

## Collar Ties and Purlins

A way to strengthen the roof is to put collar ties (also known as collar beams) at every second or third pair of rafters. Refer again to Figure 46–8. The collar ties should be placed in the upper third area of the attic space and fastened at each end with four

common nails.

Longer rafters can be supported by purlins, which are horizontal timbers placed beneath the rafters at an intermediate point between the ridge of the roof and the outside wall. Refer again to Figure 46–8. Purlins are supported by braces that extend to the nearest partition.

## Rafter Anchors

Various metal rafter anchors are used in roof construction today. See Figure 46–11. These devices help tie the rafter to the supporting wall. They are nailed to the rafter and into the top plates or studs below.

## ROOF SHEATHING

Sheathing should be nailed to the roof as soon as the framing

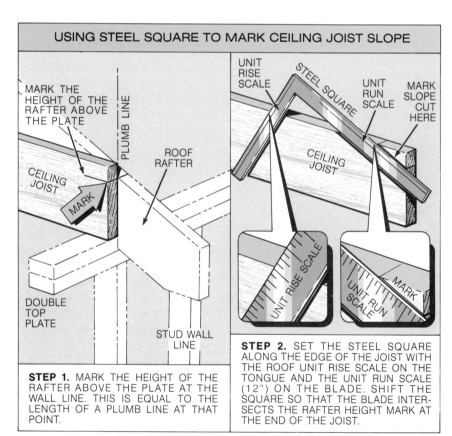

USING STEEL SQUARE TO MARK CEILING JOIST SLOPE

MARK THE HEIGHT OF THE RAFTER ABOVE THE PLATE

PLUMB LINE

ROOF RAFTER

CEILING JOIST

MARK

DOUBLE TOP PLATE

STUD WALL LINE

STEP 1. MARK THE HEIGHT OF THE RAFTER ABOVE THE PLATE AT THE WALL LINE. THIS IS EQUAL TO THE LENGTH OF A PLUMB LINE AT THAT POINT.

UNIT RISE SCALE

STEEL SQUARE

UNIT RUN SCALE

MARK SLOPE CUT HERE

CEILING JOIST

UNIT RISE SCALE

UNIT RUN SCALE

MARK

STEP 2. SET THE STEEL SQUARE ALONG THE EDGE OF THE JOIST WITH THE ROOF UNIT RISE SCALE ON THE TONGUE AND THE UNIT RUN SCALE (12") ON THE BLADE. SHIFT THE SQUARE SO THAT THE BLADE INTERSECTS THE RAFTER HEIGHT MARK AT THE END OF THE JOIST.

Figure 46–10. Laying out the angle for the ceiling joist slope cut.

## TYPICAL FRAMING METHODS USING METAL RAFTER ANCHORS

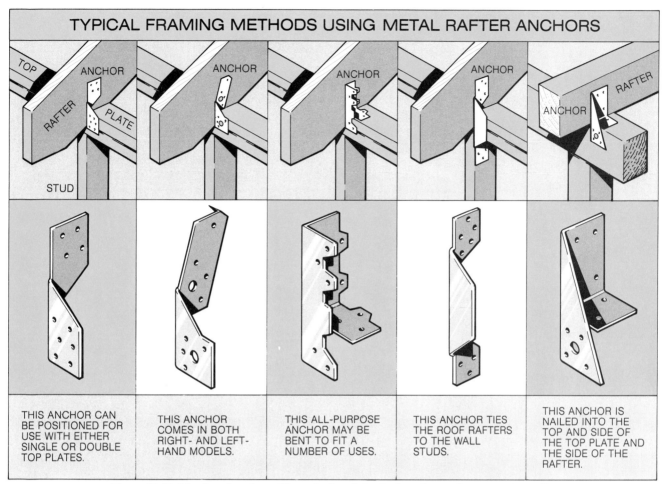

| THIS ANCHOR CAN BE POSITIONED FOR USE WITH EITHER SINGLE OR DOUBLE TOP PLATES. | THIS ANCHOR COMES IN BOTH RIGHT- AND LEFT-HAND MODELS. | THIS ALL-PURPOSE ANCHOR MAY BE BENT TO FIT A NUMBER OF USES. | THIS ANCHOR TIES THE ROOF RAFTERS TO THE WALL STUDS. | THIS ANCHOR IS NAILED INTO THE TOP AND SIDE OF THE TOP PLATE AND THE SIDE OF THE RAFTER. |

Figure 46–11. Metal roof anchors. These devices help tie the rafter to the supporting wall.

is completed. Sheathing serves as a base for the finish roof material and also adds strength to the roof structure. See Figure 46–12.

Plywood panels are still the most widely used material for roof sheathing. However, the use of non-veneered panel products such as particleboard, waferboard, oriented strandboard, and composite board is growing.

Another type of sheathing material comes under the category of structural insulation panels. This type of sheathing is generally used only with exposed-beam roofs. Still another type of sheathing, *spaced-board* sheathing, is used as a base for certain types of roof shingles. (See Section 12.)

When nailing panels to the roof, be sure that the first row of panels is in a straight line. Since the end joints of the panels should be staggered, start the second row with a half-panel.

Plywood panels 3/8″ or 1/2″ thick are usually used for sheathing, depending on the rafter spacing.

*American Plywood Association*                              *Duo-Fast Corporation*

Figure 46–12. Plywood panels are the most widely used material for roof sheathing. A pneumatic coil nailer (at right) is sometimes used to fasten the panels. Sheathing over an open soffit should be exterior grade material, since it will be exposed to the weather.

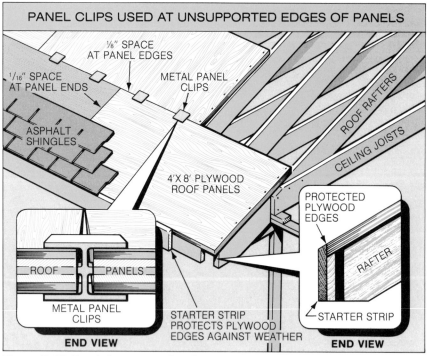

PANEL CLIPS USED AT UNSUPPORTED EDGES OF PANELS

⅛" SPACE
AT PANEL EDGES

1/16" SPACE
AT PANEL ENDS

METAL PANEL
CLIPS

ASPHALT
SHINGLES

ROOF RAFTERS

CEILING JOISTS

4' X 8' PLYWOOD
ROOF PANELS

PROTECTED
PLYWOOD
EDGES

ROOF      PANELS

RAFTER

METAL PANEL
CLIPS

STARTER STRIP
PROTECTS PLYWOOD
EDGES AGAINST WEATHER

STARTER STRIP

END VIEW

END VIEW

Figure 46–13. Panel clips are used at the unsupported edges of roof sheathing panels. These clips eliminate the need for blocking.

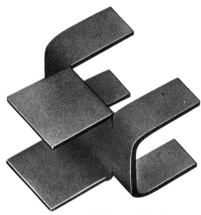

*The Panel-Clip Company*

Figure 46–14. This panel clip is made of 18-ga. galvanized steel and is available in sizes accommodating ⅜", ½", ⅝", and ¾" panel thicknesses.

The recommended thickness for waferboard is ½". Both plywood and waferboard panels require 6d common nails (smooth, ring-shank, or spiral-thread) spaced 6" O.C. along the panel edges, and 12" O.C. at the intermediate rafters.

Staples may also be used with plywood panels. They should be 1⅜" long for ⅜" panels, and 1½" long for ½" panels. Staples are driven in 4" apart at the panel edges, and 8" apart at the intermediate rafters.

Sometimes rafters may be spaced farther apart than the allowable span (distance between rafters) for the thickness of the plywood. In this case, blocking should be nailed between the rafters, or panels with tongue-and-groove edges should be used. A recent development, the *panel clip,* can be used instead of blocking or tongue-and-groove panels. See Figures 46–13 and 46–14.

When placing sheathing on a steeply pitched roof, the carpenters should, for safety, nail 2" × 4" toeholds to the roof as they work toward the top of the ridge.

# UNIT 47

# Gable, Gambrel, and Shed Roofs

## GABLE ROOF

Next to the shed roof, which has only one slope, the gable roof is the simplest type of sloping roof to build because it slopes in only two directions. See Figure 47–1. The basic structural members of the gable roof are the *ridge board,* the *common rafters,* and the *gable studs.* See Figure 47–2.

The ridge board is placed at the peak of the roof. It provides a nailing surface for the top ends of the common rafters. The common rafters extend from the top wall plates to the ridge. The gable studs are upright framing members that provide a nailing surface for siding and sheathing at the gable ends of the roof.

## Common Rafters

All common rafters for a gable roof are the same length. They can be pre-cut before the roof is assembled. Today most common rafters include an *overhang,* as shown in Figure 47–3. The overhang is the part of the rafter that extends past the building line. The *run* of the overhang is the horizontal distance from the building line to the tail cut on the rafter.

*Plumb cuts* are made at the ridge, heel, and tail of the common rafter. A level *seat cut* is made where the rafter rests on the top wall plates. The notch formed by the seat and heel plumb line is often called the *bird's mouth.* See Figure 47–4.

The length of the seat cut should be the width of the wall plates. (For wall plates made of 2″ × 4″ material, the seat cut would be 3½″.) At least 2″ of stock should remain above the

*American Plywood Association*

Figure 47–1. A gable roof slopes in only two directions.

## GABLE ROOF FRAMEWORK

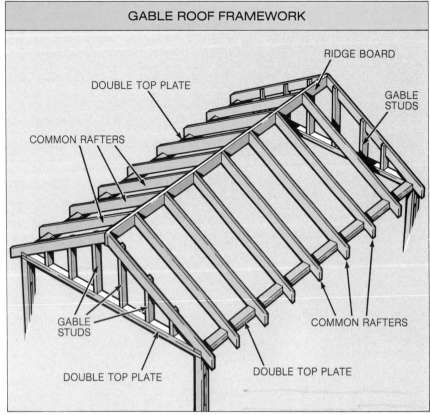

Figure 47–2. Framework of a gable roof includes common rafters, ridge board, and gable studs.

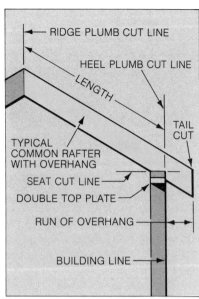

Figure 47–3. Typical common rafter with an overhang. Note plumb cuts at the ridge, heel, and tail of the rafter. The seat cut allows a rafter to rest on the double top plate.

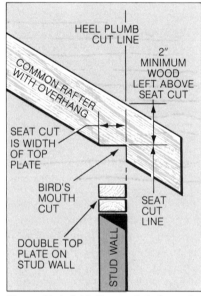

Figure 47–4. A bird's mouth is formed by the heel plumb line and seat line.

seat cut. The procedure for marking these cuts is explained later in this unit under the heading, "Laying out Common Rafters." Usually layout is not done until the length of the rafters is calculated.

**Calculating Length of Common Rafters.** The length of a common rafter is based on the unit rise and total run of the roof. The unit rise and total run are obtained from the blueprints. Three different procedures can be followed to calculate common rafter length. One procedure uses a steel square printed with a rafter table. Another procedure uses a book of rafter tables. In still another procedure (the *step-off* method), rafter layout is combined with calculating length.

*Steel Square Rafter Table.* Steel squares are available with a raf-

ter table printed on their face side. See Figure 47–5. (For more information on the steel square, see Figure 9–16 in Section 3.) The rafter table makes it possible to find the lengths of all types of rafters for pitched roofs with unit rises ranging from 2" to 18". The method for using the steel square rafter tables is given in the following two examples.

**Example 1.** The roof has a 7" unit rise and a 16' span. See Figure 47–6.
Look at the first line of the rafter table on the steel square to find the "length of common rafters per foot of run." Since the roof in this example has a 7" unit rise, locate the number 7 at the top of the square. Directly beneath the number 7 is the figure 13.89. This means that a common rafter with a 7" unit rise will be 13.89" long for every foot of run.

In order to find the length of the rafter, multiply 13.89" by the *number of feet in the total run.* (The total run is always one-half the span.) The total run for a roof with a 16' span is 8'; therefore, multiply 13.89" by 8 to find the rafter length. The mathematical steps in this procedure are:

Step 1. Multiply the number of feet in the total run (8) by the length of the common rafter per foot of run (13.89″).

$$13.89″ \times 8 = 111.12″$$

Step 2. To change .12 of an inch to a fraction of an inch, multiply by 16.

$$.12 \times 16 = 1.92, \text{ or } 1/8″$$

The number 1 to the left of the decimal point represents $1/16″$. The number .92 to the right of the decimal represents $92/100$ of $1/16″$. For practical purposes, 1.92 is calculated as being equal to $1/8″$. (As a general rule in this kind of calculation, if the number to the right of the decimal is more than 5, add $1/16″$ to the figure on the left side of the decimal.)

The result of Steps 1 and 2 is a total common rafter length of $111\frac{1}{8}″$, or $9′\text{-}3\frac{1}{8}″$.

**Example 2.** A roof has a 6″ unit rise and a 25′ span. The total run of the roof is 12′-6″.

The mathematical steps to find the rafter length are:

Step 1. Change 6″ to a fraction of a foot by placing the number 6 over the number 12.

$$\frac{6}{12} = \frac{1}{2}$$

$$\frac{1′}{2} = 6″$$

Step 2. Change the fraction to a decimal by dividing the bottom number (denominator) into the top number (numerator).

$$1 \div 2 = .5$$
$$.5′ = 6″$$

Step 3. Multiply the total run (12.5) by the length of the common rafter per foot of run (13.42″)

$$12.5 \times 13.42 = 167.75″$$

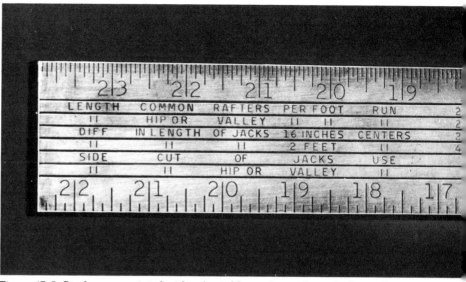

Figure 47–5. Steel square printed with rafter table can be used to calculate rafter lengths for pitched roofs with unit rises of 2″ to 18″.

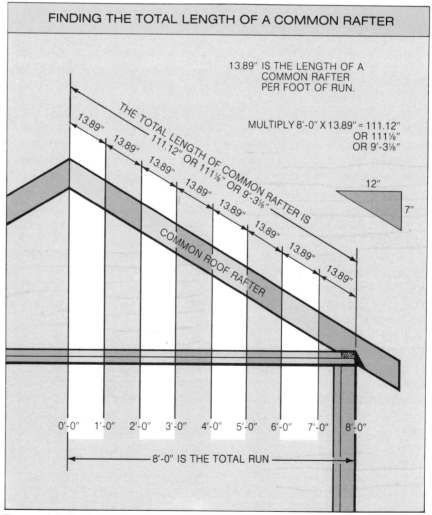

Figure 47–6. Total run times the length of the common rafter per foot of run equals the total length of the common rafter. In this example, 8 × 13.89″ equals 111.2″, or 111 ⅛″.

345

Step 4. To change .75″ to a fraction of an inch, multiply by 16 (for an answer expressed in sixteenths of an inch).

$$.75 \times 16 = 12.$$

$$\frac{12}{16} = \frac{3}{4}$$

The result of these steps is a total common rafter length of 167¾″, or 13′-11¾″.

*Book of Rafter Tables.* Most carpenters prefer to use a book of rafter tables for calculating common rafter length. Such books contain the rafter measurements for a wider range of roof spans and unit rises than the steel square rafter table covers.

*Shortening a Common Rafter.* Rafter length found by any of the methods discussed here is the measurement from the *heel plumb line* to the *center of the ridge.* This is known as the *theoretical* length of the rafter. Since a ridge board, usually 1½″ thick, is placed between the rafters, one-half of the ridge board (¾″) must be deducted from each rafter. This calculation is known as *shortening the rafter.* It is done at the time the rafters are laid out. The actual length (as opposed to the theoretical length) of a rafter is the distance from the heel plumb line to the shortened ridge plumb line. See Figure 47–7.

**Laying Out Common Rafters.**
Before the rafters can be cut, the angles of the cuts must be marked. Layout consists of marking the plumb cuts at the ridge, heel, and tail of the rafter, and the seat cut where the rafter will rest on the wall. The angles are laid out with a steel square. A pair of square gauges is useful in

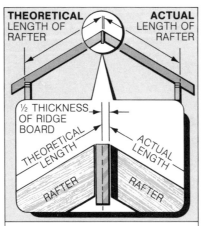

THE **THEORETICAL** RAFTER LENGTH IS THE DISTANCE FROM THE HEEL PLUMB LINE TO THE CENTER LINE OF THE RIDGE BOARD.

THE **ACTUAL** RAFTER LENGTH IS THE DISTANCE FROM THE HEEL PLUMB LINE TO THE CENTER LINE OF THE RIDGE BOARD, **LESS** ONE-HALF THE RIDGE BOARD THICKNESS.

Figure 47–7. The actual (versus theoretical) length of a common rafter is found by deducting one-half the ridge thickness from the theoretical length.

this procedure. One square gauge is secured to the tongue of the square next to the number that is the same as the unit rise. The other gauge is secured to the blade of the square next to the number that is the same as the unit run, which is always 12″. When the square is placed on the rafter stock, the plumb cut can be marked along the tongue (unit rise) side of the square. The seat cut can be marked along the blade (unit run) side of the square. See Figure 47–8.

Rafter layout also includes marking off the required overhang, and making the shortening calculation explained earlier in this unit. A procedure for laying out a common rafter is shown in Figure 47–9.

**Step-off Method for Calculating Length and Laying Out Common Rafters.** The step-off method for rafter layout is an old but still practiced method. It com-

bines the procedure of laying out the rafters with a procedure of *stepping off* the length of the rafter. Figure 47–10 shows this method.

## Constructing a Gable Roof

The major part of gable roof construction is the setting in place of the common rafters. See Figures 47–11 and 47–12. The most efficient method is to pre-cut all common rafters, then fasten them to the ridge board and the wall plates in one continuous operation.

**Marking Wall Plates.** The rafter locations should be marked on the top wall plates when the positions of the ceiling joists are laid out. Proper roof layout will assure that the rafters and joists tie into each other wherever possible. The roof rafters are often spaced 24″ O.C. and the ceiling joists 16″ O.C. An example of this type of layout is shown in Figure 47–13.

**Cutting and Marking a Ridge Board.** The ridge board, like the common rafters, should be pre-cut. The rafter locations are then copied on the ridge board from the markings on the wall plates.

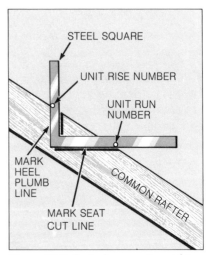

Figure 47–8. A steel square is used to lay out the plumb and seat cuts.

## LAYING OUT A COMMON RAFTER FOR A GABLE ROOF

**STEP 1.** PLACE THE STEEL SQUARE AT ONE END OF THE RAFTER. MARK A LINE ALONG THE TONGUE (UNIT RISE SIDE) OF THE SQUARE. THIS WILL BE THE **FIRST** RIDGE PLUMB LINE.

MARK THE **FIRST** RIDGE PLUMB LINE

2"X 6" RAFTER STOCK

COMMON RAFTER

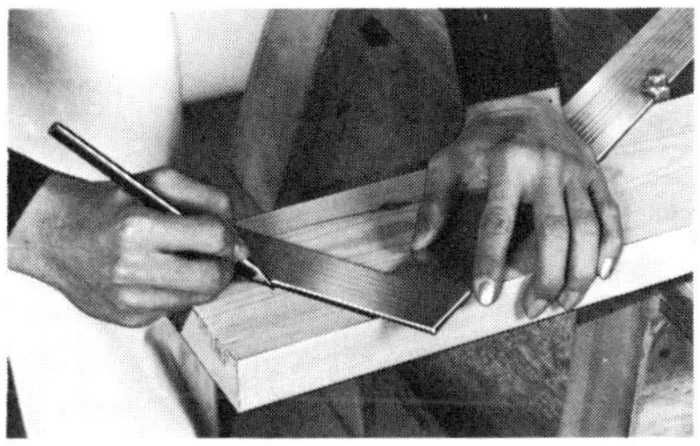

**STEP 2.** SHORTEN THE RAFTER BY ONE-HALF THE THICKNESS OF THE RIDGE BOARD. MEASURE AT A RIGHT ANGLE TO THE FIRST RIDGE PLUMB LINE, AND MARK A **SECOND** RIDGE PLUMB LINE.

MEASURE ONE-HALF THE RIDGE BOARD THICKNESS AND MARK THE **SECOND** RIDGE PLUMB LINE

**SECOND** RIDGE PLUMB LINE

COMMON RAFTER

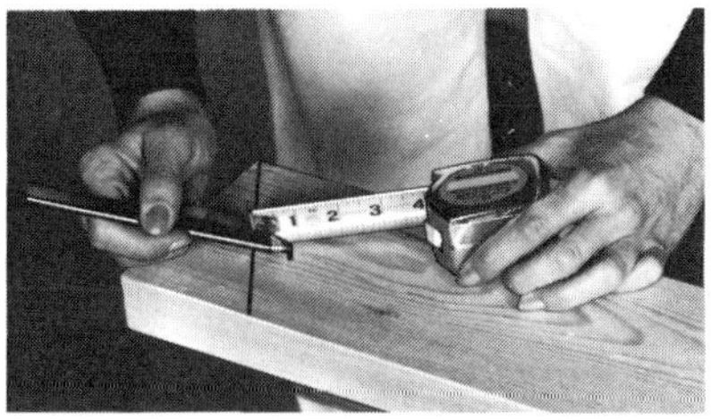

**STEP 3.** MEASURE THE TOTAL LENGTH OF THE RAFTER FROM THE **FIRST** RIDGE PLUMB LINE, AND MAKE A MARK. PLACE THE TONGUE OF THE SQUARE ALONG SIDE THIS MARK, AND DRAW THE HEEL PLUMB LINE.

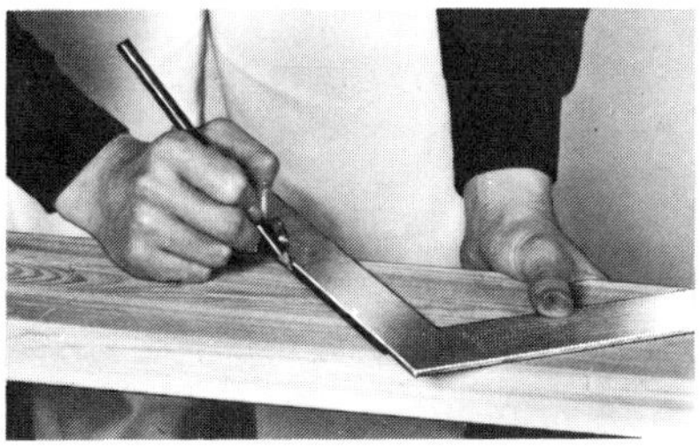

RAFTER LENGTH

MEASURE THE RAFTER LENGTH FROM THE **FIRST** RIDGE PLUMB LINE

MEASURE AND MARK THE HEEL PLUMB LINE

**STEP 4.** SLIDE THE SQUARE TO THE LEFT, AND DRAW A LINE ALONG THE EDGE OF THE BLADE. THIS LINE IS THE SEAT CUT LINE. (THE LENGTH OF THE SEAT CUT EQUALS THE WIDTH OF THE TOP PLATE.)

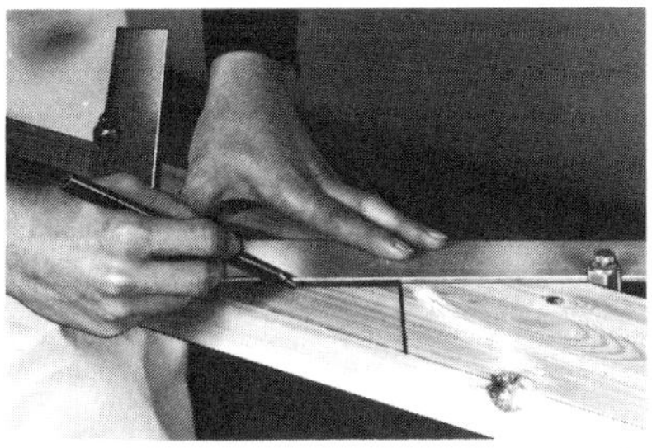

HEEL PLUMB LINE

MARK THE SEAT CUT LINE

3½" FOR A 2"X 4" TOP PLATE

**STEP 5.** SLIDE THE SQUARE TO THE RIGHT SO THAT ONE SIDE LINES UP WITH THE HEEL PLUMB LINE. MEASURE AND MARK THE DISTANCE OF THE OVER-HANG, WHICH IN THIS CASE IS 10". BE SURE THAT YOU ARE USING FIGURES ON THE **SAME** SIDE OF THE SQUARE.

10"

CUT OUT (BIRD'S MOUTH)

MEASURE AND MARK FOR THE LENGTH OF THE OVERHANG (10")

**STEP 6.** SLIDE THE SQUARE TO THE RIGHT. PLACE THE TONGUE NEXT TO THE 10" MARK AND DRAW A LINE ALONG THE EDGE OF THE TONGUE. THIS LINE IS THE TAIL CUT LINE.

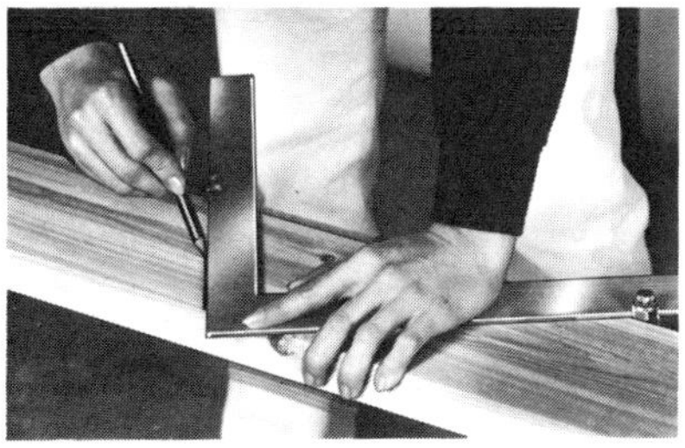

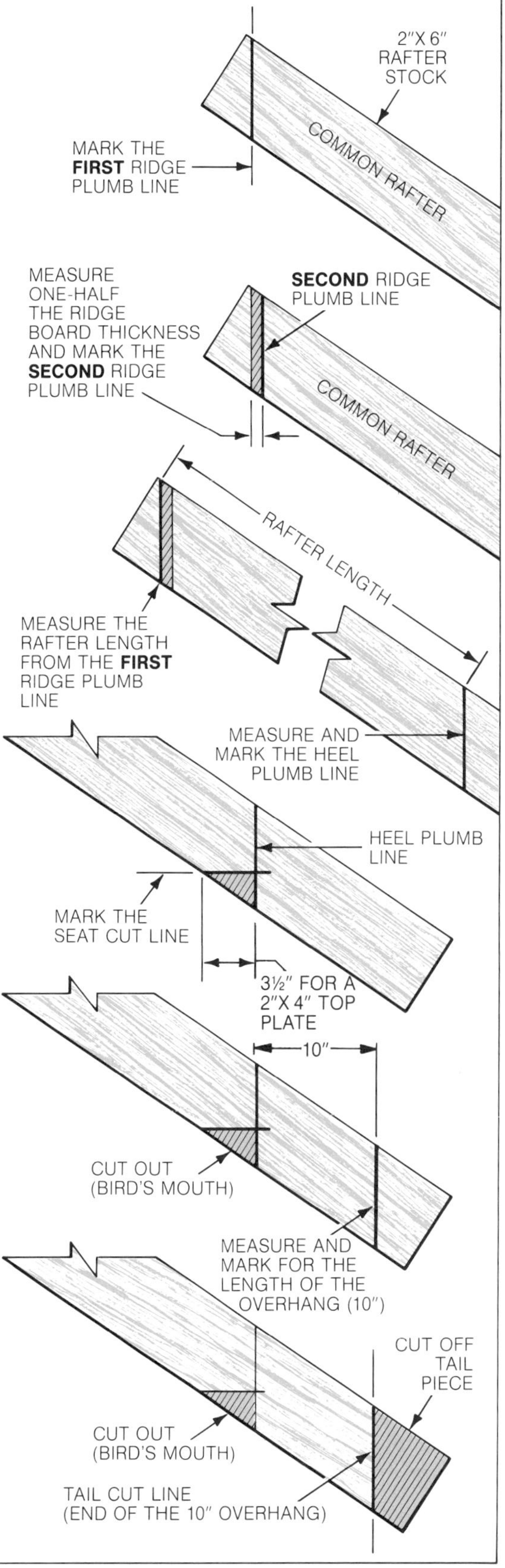

CUT OFF TAIL PIECE

CUT OUT (BIRD'S MOUTH)

TAIL CUT LINE (END OF THE 10" OVERHANG)

Figure 47–9. Laying out a common rafter for a gable roof. The roof in this example has an 8" unit rise and a 10" overhang.

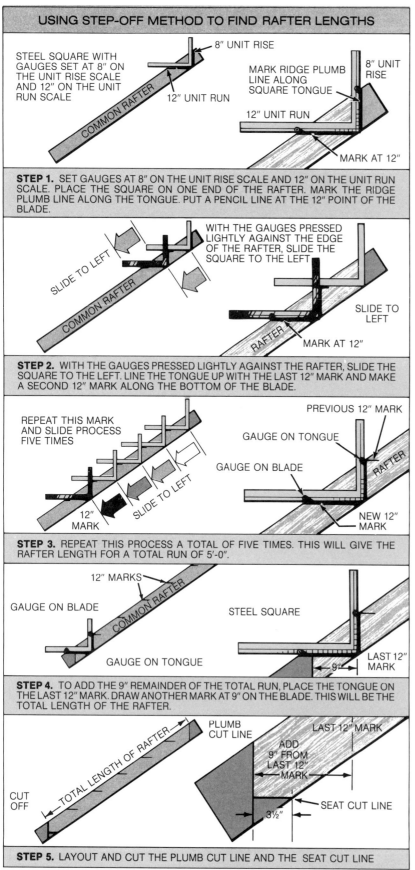

**USING STEP-OFF METHOD TO FIND RAFTER LENGTHS**

STEEL SQUARE WITH GAUGES SET AT 8" ON THE UNIT RISE SCALE AND 12" ON THE UNIT RUN SCALE

8" UNIT RISE

COMMON RAFTER

12" UNIT RUN

MARK RIDGE PLUMB LINE ALONG SQUARE TONGUE

8" UNIT RISE

12" UNIT RUN

MARK AT 12"

**STEP 1.** SET GAUGES AT 8" ON THE UNIT RISE SCALE AND 12" ON THE UNIT RUN SCALE. PLACE THE SQUARE ON ONE END OF THE RAFTER. MARK THE RIDGE PLUMB LINE ALONG THE TONGUE. PUT A PENCIL LINE AT THE 12" POINT OF THE BLADE.

WITH THE GAUGES PRESSED LIGHTLY AGAINST THE EDGE OF THE RAFTER, SLIDE THE SQUARE TO THE LEFT

SLIDE TO LEFT

COMMON RAFTER

SLIDE TO LEFT

RAFTER

MARK AT 12"

**STEP 2.** WITH THE GAUGES PRESSED LIGHTLY AGAINST THE RAFTER, SLIDE THE SQUARE TO THE LEFT. LINE THE TONGUE UP WITH THE LAST 12" MARK AND MAKE A SECOND 12" MARK ALONG THE BOTTOM OF THE BLADE.

REPEAT THIS MARK AND SLIDE PROCESS FIVE TIMES

12" MARK

SLIDE TO LEFT

PREVIOUS 12" MARK

GAUGE ON TONGUE

GAUGE ON BLADE

RAFTER

NEW 12" MARK

**STEP 3.** REPEAT THIS PROCESS A TOTAL OF FIVE TIMES. THIS WILL GIVE THE RAFTER LENGTH FOR A TOTAL RUN OF 5'-0".

12" MARKS

GAUGE ON BLADE

COMMON RAFTER

GAUGE ON TONGUE

STEEL SQUARE

9"

LAST 12" MARK

**STEP 4.** TO ADD THE 9" REMAINDER OF THE TOTAL RUN, PLACE THE TONGUE ON THE LAST 12" MARK. DRAW ANOTHER MARK AT 9" ON THE BLADE. THIS WILL BE THE TOTAL LENGTH OF THE RAFTER.

PLUMB CUT LINE

TOTAL LENGTH OF RAFTER

LAST 12" MARK

ADD 9" FROM LAST 12" MARK

CUT OFF

SEAT CUT LINE

3½"

**STEP 5.** LAYOUT AND CUT THE PLUMB CUT LINE AND THE SEAT CUT LINE

Figure 47–10. Step-off method for calculating common rafter length. In this example the roof has an 8" unit rise and a total run of 5'-9".

See Figure 47–14. The ridge board should be the length of the building plus the overhang at the gable ends.

The material used for the ridge board is usually wider than the rafter stock. For example, a ridge board of 2" x 8" stock would be used with rafters of 2" x 6" stock. Some buildings are long enough to require more than one piece of ridge material. The breaks between these ridge pieces should occur at the center of a rafter.

**Cutting Common Rafters.** One pair of rafters should be cut and checked for accuracy before the other rafters are cut. To check the first pair for accuracy, set them in position with a 1½" piece of wood fitted between them. If the rafters are the correct length, they should fit the building. If, however, the building walls are out of line, adjustments will have to be made on the rafters.

After the first pair of rafters is checked for accuracy (and adjusted if necessary), one of the pair can be used as a pattern for marking all the other rafters. Cutting is usually done with an electric handsaw or a radial-arm saw. See Figure 47–15.

**Placing a Ridge Board and Common Rafters.** Several different methods exist for setting up the ridge board and attaching the rafters to it. When only a few carpenters are present on the job site, the most convenient procedure is to set the ridge board to its required height (total rise) and hold it in place with temporary props. See Figure 47–16. The rafters can then be nailed to the ridge board and the top wall plates. A faster system that can be used when a large crew is present is shown in Figure 47–17.

Plywood panels should be laid

*Southern Forest Products Association*

Figure 47-11. Carpenters completing a gable roof. The ridge board and most of the common rafters are in place.

*Southern Forest Products Association*

Figure 47-12. Looking up at a framed section of a gable roof from the floor below. Collar ties have been placed at every third pair of rafters. Note the ceiling joists lapped over a partition.

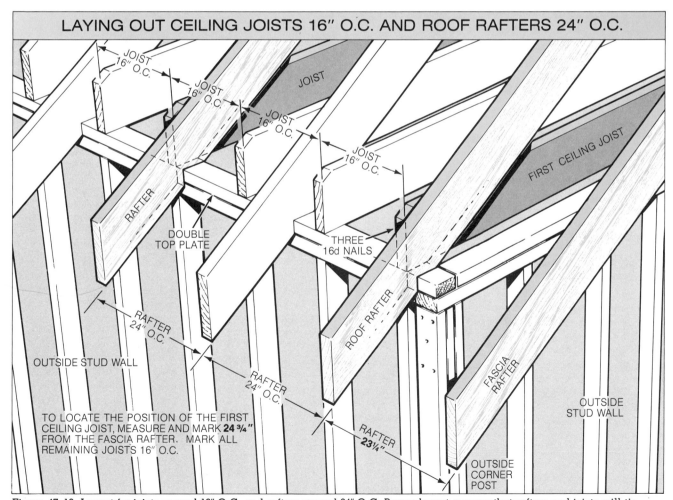

## LAYING OUT CEILING JOISTS 16″ O.C. AND ROOF RAFTERS 24″ O.C.

JOIST 16″ O.C.

JOIST 16″ O.C.

JOIST 16″ O.C.

JOIST

JOIST 16″ O.C.

FIRST CEILING JOIST

RAFTER

DOUBLE TOP PLATE

THREE 16d NAILS

ROOF RAFTER

RAFTER 24″ O.C.

OUTSIDE STUD WALL

RAFTER 24″ O.C.

FASCIA RAFTER

OUTSIDE STUD WALL

TO LOCATE THE POSITION OF THE FIRST CEILING JOIST, MEASURE AND MARK **24 ¾″** FROM THE FASCIA RAFTER.  MARK ALL REMAINING JOISTS 16″ O.C.

RAFTER **23¼″**

OUTSIDE CORNER POST

Figure 47-13. Layout for joists spaced 16″ O.C. and rafters spaced 24″ O.C. Proper layout assures that rafters and joists will tie into each other wherever possible.

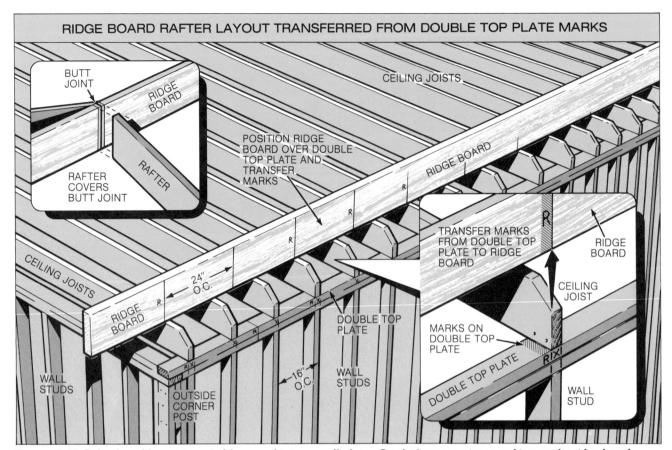

**RIDGE BOARD RAFTER LAYOUT TRANSFERRED FROM DOUBLE TOP PLATE MARKS**

Figure 47–14. Ridge board layout is copied from markings on wall plates. Breaks between pieces making up the ridge board always occur at the center of a rafter.

*Southern Forest Products Association*

Figure 47–15. Carpenters using one rafter as a template to mark off other rafters to be cut with an electric hand-saw.

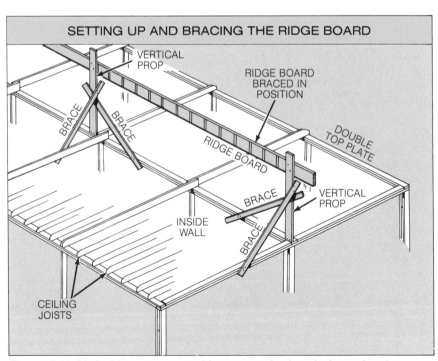

**SETTING UP AND BRACING THE RIDGE BOARD**

Figure 47–16. Method for setting up and bracing a ridge board when only a few carpenters are present on the job site.

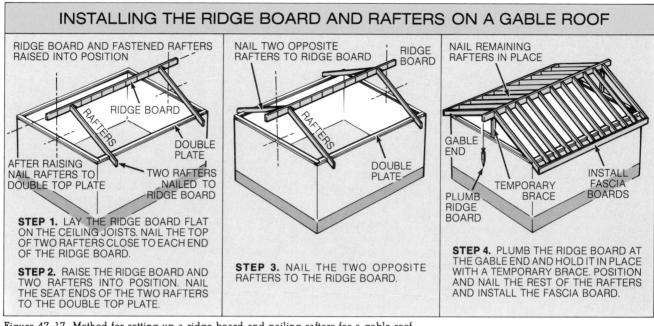

INSTALLING THE RIDGE BOARD AND RAFTERS ON A GABLE ROOF

**STEP 1.** LAY THE RIDGE BOARD FLAT ON THE CEILING JOISTS. NAIL THE TOP OF TWO RAFTERS CLOSE TO EACH END OF THE RIDGE BOARD.

**STEP 2.** RAISE THE RIDGE BOARD AND TWO RAFTERS INTO POSITION. NAIL THE SEAT ENDS OF THE TWO RAFTERS TO THE DOUBLE TOP PLATE.

**STEP 3.** NAIL THE TWO OPPOSITE RAFTERS TO THE RIDGE BOARD.

**STEP 4.** PLUMB THE RIDGE BOARD AT THE GABLE END AND HOLD IT IN PLACE WITH A TEMPORARY BRACE. POSITION AND NAIL THE REST OF THE RAFTERS AND INSTALL THE FASCIA BOARD.

Figure 47–17. Method for setting up a ridge board and nailing rafters for a gable roof.

on top of the ceiling joists where the framing will take place. The panels provide safe and comfortable footing for the carpenters. They also provide a place to put tools and materials.

**Cutting a Common Rafter Overhang.** The overhang of the common rafters can be laid out and cut before the rafters are set in place. However, many carpenters prefer to cut the overhang after the rafters are fastened to the ridge board and wall plates. A line is snapped from one end of the building to the other, and the tail plumb lines are marked with a sliding T-bevel (also called a bevel square). See Figure 47–18. The rafters are then cut with an electric handsaw. This method guarantees that the line of the overhang will be perfectly straight, even if the building is not.

**Framing a Gable End Overhang.** Over each gable end of the building, another overhang is framed. The main framing members of the gable end overhang

are the *fascia rafters,* also referred to as *barge rafters.* They are tied to the ridge board at one end and to the *fascia board* at their lower end. Fascia boards are often nailed to the tail ends of the common rafters to serve as a finish piece at the edge of the roof. By extending past the gable ends of the house, they also help to support the fascia rafters. Figures 47–19 and 47–20 show different methods used to frame gable end overhangs.

**Cutting and Placing Gable Studs.** At each gable end, vertical members called *gable studs* are placed. They decrease in length from the ridge section toward the outside walls. A method for finding the *common length difference* for gable studs is shown in Figure 47–21. Another method, using a steel square, is shown in Figure 47–22. Gable studs also require an angle cut where they fit beneath a top plate or rafter. The steel square can be used to lay out this angle. See Figure 47–23.

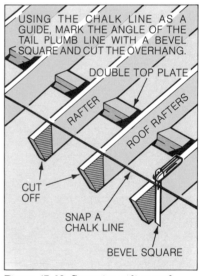

Figure 47–18. Snapping a line and marking plumb cuts for a gable end overhang.

**Placing Purlins and Collar Ties.** If purlins are required, they are nailed beneath the rafters after the roof framing is completed. The rafters should be lined up with a string while the posts are being fitted between the purlin and the supporting partition. The collar ties are also installed at this time. (Purlins and collar ties are discussed in the preceding unit.)

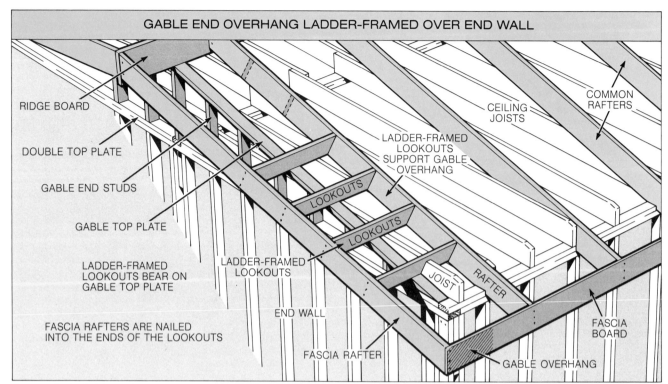

**GABLE END OVERHANG LADDER-FRAMED OVER END WALL**

RIDGE BOARD

DOUBLE TOP PLATE

GABLE END STUDS

GABLE TOP PLATE

LADDER-FRAMED LOOKOUTS BEAR ON GABLE TOP PLATE

FASCIA RAFTERS ARE NAILED INTO THE ENDS OF THE LOOKOUTS

LADDER-FRAMED LOOKOUTS

END WALL

FASCIA RAFTER

LOOKOUTS

LOOKOUTS

LADDER-FRAMED LOOKOUTS SUPPORT GABLE OVERHANG

CEILING JOISTS

COMMON RAFTERS

JOIST

RAFTER

FASCIA BOARD

GABLE OVERHANG

Figure 47–19. Gable end overhang with the end wall framed under the overhang. A fascia rafter is nailed to the ridge board and to the fascia board. Blocking rests on the end wall and is nailed between the fascia rafter and the rafter next to it. This section of the roof will be further strengthened when the roof sheathing is nailed to it.

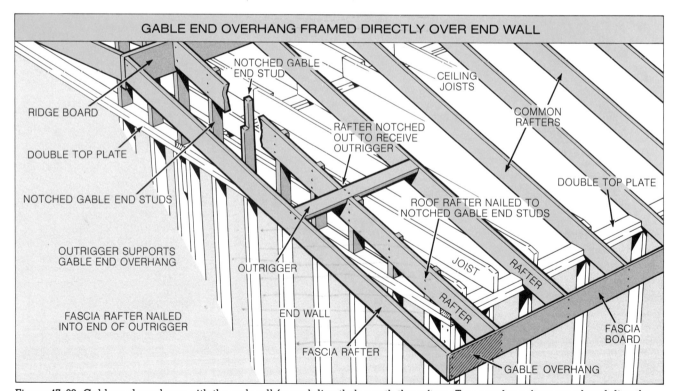

**GABLE END OVERHANG FRAMED DIRECTLY OVER END WALL**

RIDGE BOARD

DOUBLE TOP PLATE

NOTCHED GABLE END STUDS

OUTRIGGER SUPPORTS GABLE END OVERHANG

FASCIA RAFTER NAILED INTO END OF OUTRIGGER

NOTCHED GABLE END STUD

RAFTER NOTCHED OUT TO RECEIVE OUTRIGGER

OUTRIGGER

END WALL

FASCIA RAFTER

ROOF RAFTER NAILED TO NOTCHED GABLE END STUDS

CEILING JOISTS

COMMON RAFTERS

DOUBLE TOP PLATE

JOIST

RAFTER

RAFTER

FASCIA BOARD

GABLE OVERHANG

Figure 47–20. Gable end overhang with the end wall framed directly beneath the rafters. Two regular rafters are placed directly over the gable ends of the building. The fascia rafters are placed between the ridge board and the fascia boards. The gable studs are notched to fit against the rafter above.

## GAMBREL ROOF

The gambrel roof is basically a gable roof with a double slope on each side. This double slope is created by two common rafters that meet between the top of the wall and the ridge. Where these rafters meet, they must be supported either by a wall or by purlins and posts.

Figure 47–24 shows a section view of a gambrel roof. The meeting points of the upper and lower rafters in this gambrel roof are supported by walls.

### Calculating Length of Common Rafters

The procedures used to calculate the length of common rafters for a gable roof also apply to common rafters for a gambrel roof. The length of rafters at the top portion of the roof is based on the total run at the top, and the length of the rafters at the bottom portion of the roof is based on the total run at the bottom. Refer again to Figure 47–24.

### Laying out Common Rafters

A procedure is shown in Figure 47–25 for laying out the common rafters for the gambrel roof shown in Figure 47–24. The angles for plumb cuts are marked at the ridge, tail, and heel of the common rafters. A seat cut is marked where the rafter will rest on the wall. Finally, the overhang cut is marked.

### Constructing a Gambrel Roof

If the attic space provided by a gambrel roof is to be used for a living area, the subfloor should be placed before the roof framing begins. The two outside rows of subfloor panels should be held back or cut so that the seat of the rafters can be easily nailed to the top of the wall plates. One recommended procedure for framing a gambrel roof is shown in Figure 47–26.

### SHED ROOF

The shed roof has only one slope. See Figure 47–27. The common rafters for a shed roof are marked on each end for seat cuts where they will rest on the two opposite walls of the build-

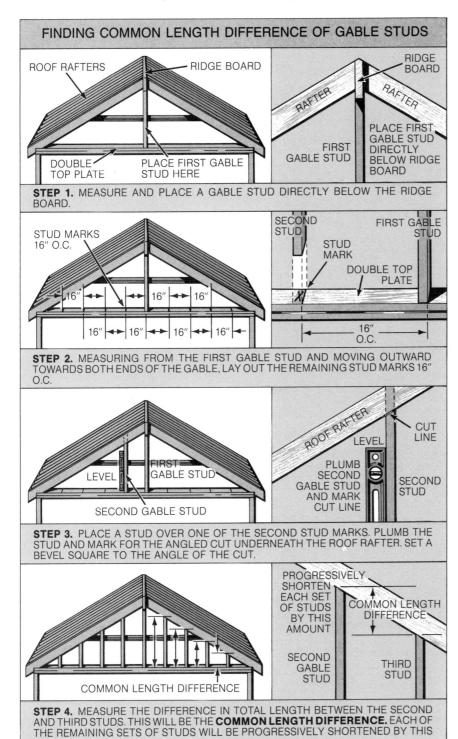

Figure 47-21. Gable studs decrease in length from the ridge section toward the outside walls.

## FINDING COMMON LENGTH DIFFERENCE WITH STEEL SQUARE

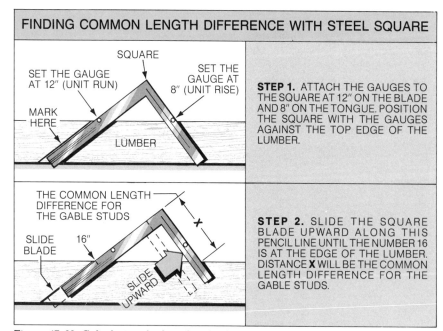

SET THE GAUGE AT 12" (UNIT RUN)

SQUARE

SET THE GAUGE AT 8" (UNIT RISE)

MARK HERE

LUMBER

**STEP 1.** ATTACH THE GAUGES TO THE SQUARE AT 12" ON THE BLADE AND 8" ON THE TONGUE. POSITION THE SQUARE WITH THE GAUGES AGAINST THE TOP EDGE OF THE LUMBER.

THE COMMON LENGTH DIFFERENCE FOR THE GABLE STUDS

SLIDE BLADE

16"

X

SLIDE UPWARD

**STEP 2.** SLIDE THE SQUARE BLADE UPWARD ALONG THIS PENCIL LINE UNTIL THE NUMBER 16 IS AT THE EDGE OF THE LUMBER. DISTANCE **X** WILL BE THE COMMON LENGTH DIFFERENCE FOR THE GABLE STUDS.

Figure 47–22. Calculating the lengths of gable studs with a steel square.

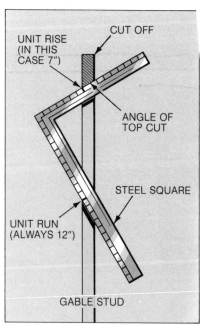

CUT OFF

UNIT RISE (IN THIS CASE 7")

ANGLE OF TOP CUT

STEEL SQUARE

UNIT RUN (ALWAYS 12")

GABLE STUD

Figure 47–23. Using steel square to find the angle of the top cut on gable studs. Combine the unit rise and 12". Mark the cut along the unit rise side.

## CONSTRUCTION DETAILS AND DIMENSIONS FOR A GAMBREL ROOF

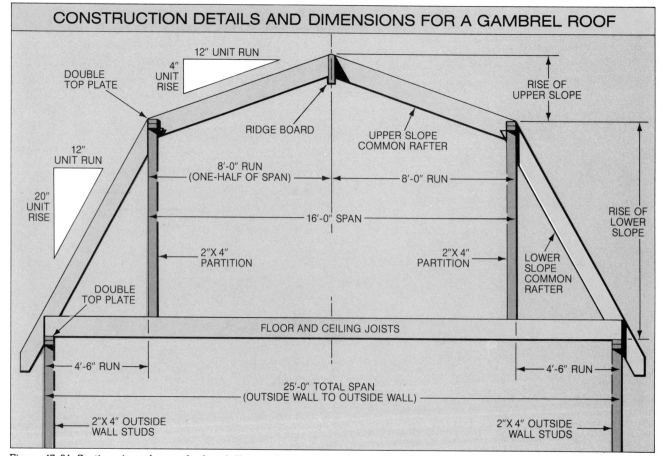

Figure 47–24. Section view of a gambrel roof. Note walls supporting the meeting points of the upper and lower rafters. The length of the rafters for the lower portion of the roof is based on a 20" unit rise and a 4'-6" run. The length of the rafters for the upper section is based on a 4" unit rise and an 8'-0" run.

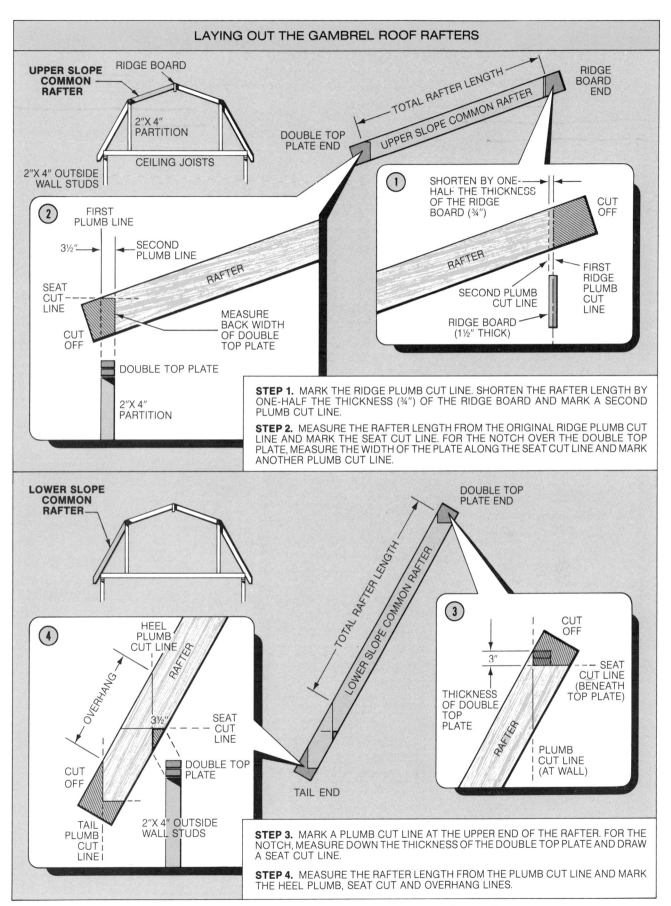

LAYING OUT THE GAMBREL ROOF RAFTERS

UPPER SLOPE COMMON RAFTER

RIDGE BOARD

2"X 4" PARTITION

CEILING JOISTS

2"X 4" OUTSIDE WALL STUDS

TOTAL RAFTER LENGTH

DOUBLE TOP PLATE END

UPPER SLOPE COMMON RAFTER

RIDGE BOARD END

① SHORTEN BY ONE-HALF THE THICKNESS OF THE RIDGE BOARD (¾")

RAFTER

CUT OFF

SECOND PLUMB CUT LINE

RIDGE BOARD (1½" THICK)

FIRST RIDGE PLUMB CUT LINE

② FIRST PLUMB LINE

3½"

SECOND PLUMB LINE

RAFTER

SEAT CUT LINE

CUT OFF

MEASURE BACK WIDTH OF DOUBLE TOP PLATE

DOUBLE TOP PLATE

2"X 4" PARTITION

**STEP 1.** MARK THE RIDGE PLUMB CUT LINE. SHORTEN THE RAFTER LENGTH BY ONE-HALF THE THICKNESS (¾") OF THE RIDGE BOARD AND MARK A SECOND PLUMB CUT LINE.

**STEP 2.** MEASURE THE RAFTER LENGTH FROM THE ORIGINAL RIDGE PLUMB CUT LINE AND MARK THE SEAT CUT LINE. FOR THE NOTCH OVER THE DOUBLE TOP PLATE, MEASURE THE WIDTH OF THE PLATE ALONG THE SEAT CUT LINE AND MARK ANOTHER PLUMB CUT LINE.

LOWER SLOPE COMMON RAFTER

DOUBLE TOP PLATE END

TOTAL RAFTER LENGTH

LOWER SLOPE COMMON RAFTER

③ CUT OFF

3"

SEAT CUT LINE (BENEATH TOP PLATE)

THICKNESS OF DOUBLE TOP PLATE

RAFTER

PLUMB CUT LINE (AT WALL)

④ HEEL PLUMB CUT LINE

RAFTER

OVERHANG

3½"

SEAT CUT LINE

DOUBLE TOP PLATE

CUT OFF

TAIL PLUMB CUT LINE

2"X 4" OUTSIDE WALL STUDS

TAIL END

**STEP 3.** MARK A PLUMB CUT LINE AT THE UPPER END OF THE RAFTER. FOR THE NOTCH, MEASURE DOWN THE THICKNESS OF THE DOUBLE TOP PLATE AND DRAW A SEAT CUT LINE.

**STEP 4.** MEASURE THE RAFTER LENGTH FROM THE PLUMB CUT LINE AND MARK THE HEEL PLUMB, SEAT CUT AND OVERHANG LINES.

Figure 47–25. Laying out the cuts for common rafters for the gambrel roof in Figure 47–24.

ing. They are also marked for overhang cuts on each end. The length of the common rafters is based on the unit rise of the roof and the total run. <u>The total run for a shed roof is the width of the building minus one wall width.</u> A procedure for laying out a shed roof is shown in Figure 47–28.

## DORMERS

Dormers add space, light, and ventilation to an attic area. They require a roof with a steep slope and a high ridge. Gambrel roofs are particularly well suited for dormer construction.

<u>Most dormers are of gable or shed design. See Figures 47–29 and 47–30.</u> The construction of

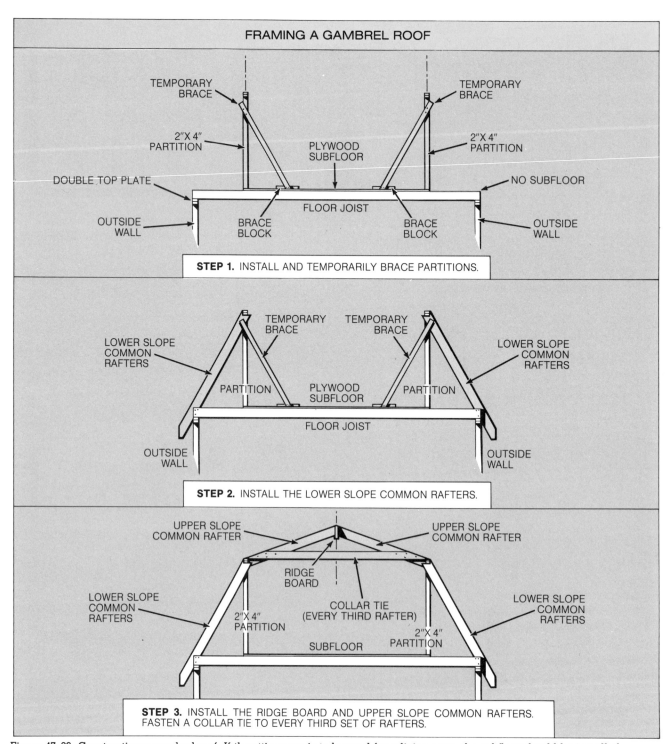

Figure 47–26. Constructing a gambrel roof. If the attic space is to be used for a living area, the subfloor should be installed before the roof is constructed.

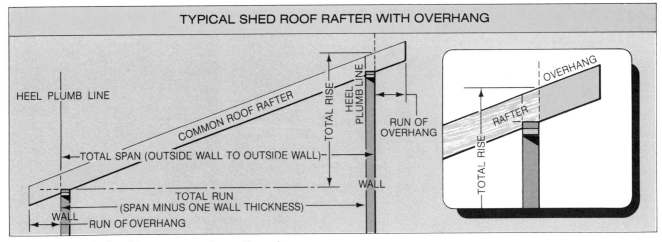

Figure 47–27. A shed roof has common rafters with overhangs.

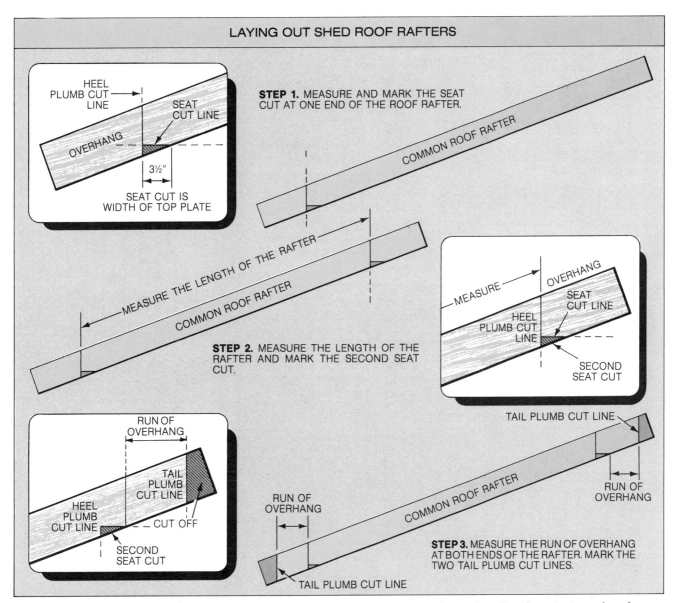

Figure 47–28. Laying out common rafters for a shed roof. The length of the common rafters is based on the unit rise and total run of the roof.

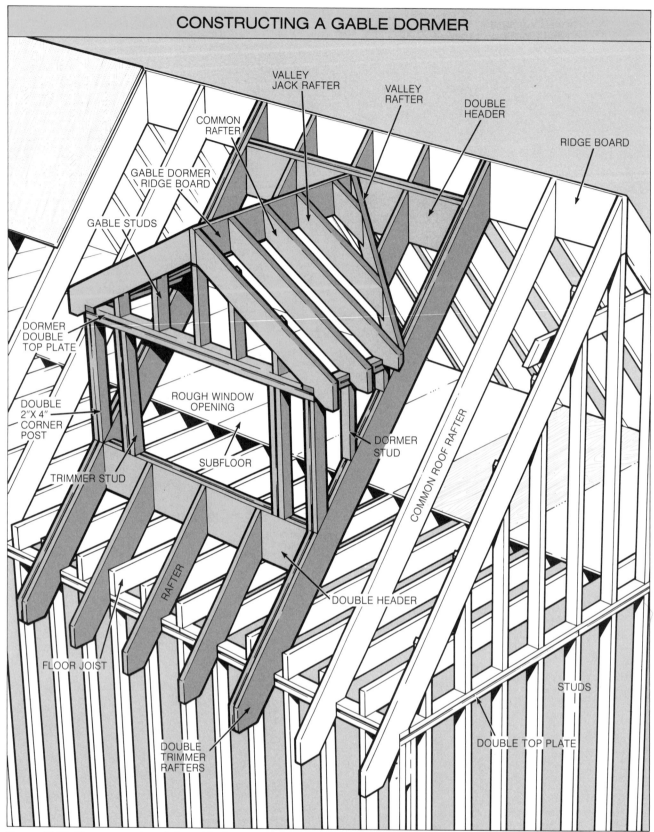

## CONSTRUCTING A GABLE DORMER

VALLEY
JACK RAFTER

VALLEY
RAFTER

DOUBLE
HEADER

COMMON
RAFTER

RIDGE BOARD

GABLE DORMER
RIDGE BOARD

GABLE STUDS

DORMER
DOUBLE
TOP PLATE

ROUGH WINDOW
OPENING

DORMER
STUD

DOUBLE
2" X 4"
CORNER
POST

COMMON ROOF RAFTER

SUBFLOOR

TRIMMER STUD

RAFTER

DOUBLE HEADER

FLOOR JOIST

STUDS

DOUBLE TOP PLATE

DOUBLE
TRIMMER
RAFTERS

Figure 47–29. The construction of the gable dormer is similar to that of a gable intersecting roof. Note that the dormer consists of a level ridge board, common rafters, valley rafters, and valley jack rafters.

the gable dormer is similar to that of a gable intersecting roof. The gable dormer consists of a level ridge board, common rafters, valley rafters, and valley jack rafters. The front wall of a shed dormer is usually directly over the exterior wall of the building. The rafters extend from

358

## CONSTRUCTING A SHED DORMER

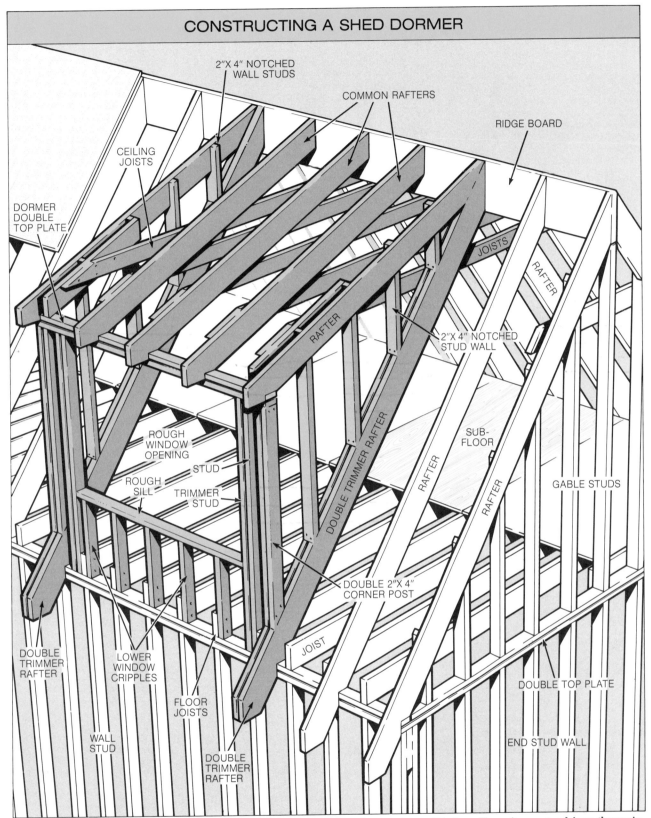

Figure 47–30. The front wall of a shed dormer is usually directly over the exterior wall below. The rafters extend from the main ridge. They must be pitched enough to shed water and snow.

the main ridge. They must be pitched enough to shed water and snow.

When an opening is framed for a dormer in a roof, the rafters on both sides of the opening are doubled. Also, double headers are placed at the top and bottom of the opening.

359

# UNIT 48

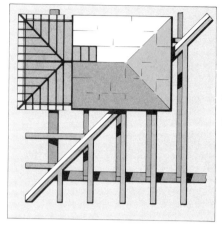

# Hip Roof

A hip roof has four sloping sides. Four *hip rafters* run at a 45° angle from the corners of the building to the ridge. *Hip jack rafters* frame the space between the hip rafters and the tops of the outside walls. See Figure 48–1.

Common rafters for hip roofs are the same as for gable roofs. For a description of their layout and the procedure for calculating their length, see the preceding unit.

## HIP RAFTERS

Because a hip rafter travels at a diagonal (a 45° angle) to reach the ridge, it is longer than a common rafter. It differs from the common rafter in other ways, too. In addition to plumb cuts at the ridge, heel, and tail, the hip rafter requires side cuts where it meets the ridge. Side cuts are also necessary at the tail in order for the overhang of the hip raf-

ters to be in line with the overhang of the common rafters. See Figure 48–2. The procedure for marking these cuts on the rafters is explained later in this unit under the heading, "Laying out Hip Rafters." Usually layout is not done until the length of the rafters is calculated.

The unit run of a hip rafter is 17″, compared with 12″ for a common rafter. Since the hip rafter runs at a 45° angle to the

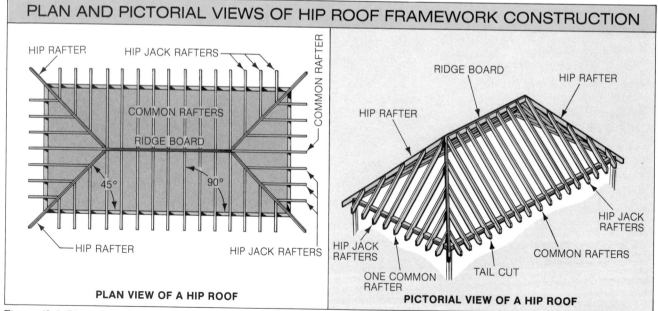

## PLAN AND PICTORIAL VIEWS OF HIP ROOF FRAMEWORK CONSTRUCTION

HIP RAFTER    HIP JACK RAFTERS    COMMON RAFTER

COMMON RAFTERS

RIDGE BOARD

45°    90°

HIP RAFTER    HIP JACK RAFTERS

**PLAN VIEW OF A HIP ROOF**

RIDGE BOARD    HIP RAFTER

HIP RAFTER

HIP JACK RAFTERS

HIP JACK RAFTERS

ONE COMMON RAFTER    TAIL CUT    COMMON RAFTERS

**PICTORIAL VIEW OF A HIP ROOF**

Figure 48–1. Plan and pictorial views of a hip roof. Four hip rafters run at a 45° angle from the outside walls. Hip jack rafters frame the space between the hip rafters and the top of the exterior walls.

common rafter, its unit run is figured by taking the diagonal of a 12″ square. The diagonal of a 12″ square is 16.97″, which rounds off to 17″. A hip rafter must run 17″ in order to reach the same height that a common rafter reaches in 12″. See Figure 48–3.

## Calculating Length of Hip Rafters

Hip rafter lengths, like common rafter lengths, can be calculated by three methods: the steel square rafter table, a book of rafter tables, or the step-off method that combines laying out with calculating length.

**Steel Square Rafter Table.** The steel square printed with a rafter table shown in the preceding unit (and in Figure 9–16 in Section 3) can be used for calculating hip rafters by the same procedure used for common rafters, except that the second line of the table should be used rather than the first. This line gives the *length of hip or valley per foot run.* The mathematical procedure is shown in the following example.

**Example.** The roof has a 6″ unit rise and a 28′ span.

Step 1. Divide the total span (28′) by 2 in order to find the total run.

$$28 \div 2 = 14$$

Step 2. Look under the number 6 at the top of the square. On the second line the number 18 is given as the length of hip or valley per foot of run.

Step 3. Multiply the length of hip per foot of run (18″) by the number of feet in the total run (14).

$$18 \times 14 = 252″ \text{ or } 21′0″$$

**Book of Rafter Tables.** Hip raf-

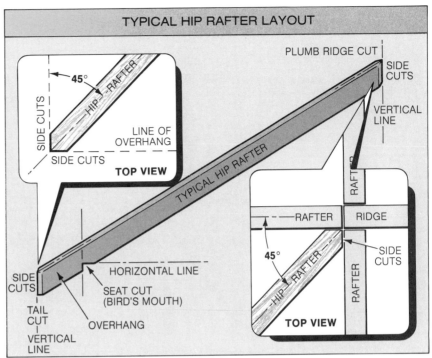

Figure 48–2. A hip rafter. Side cuts are required at the ridge and tail.

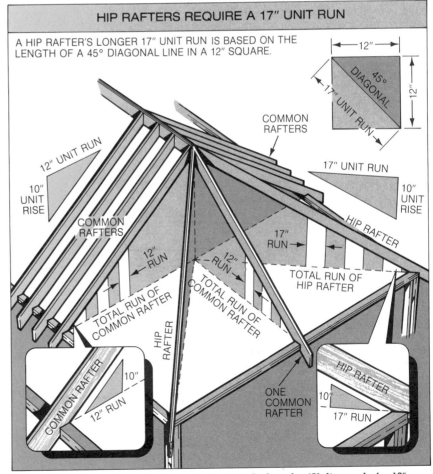

Figure 48–3. The unit run of a hip rafter is 17″, which is the 45° diagonal of a 12″ square.

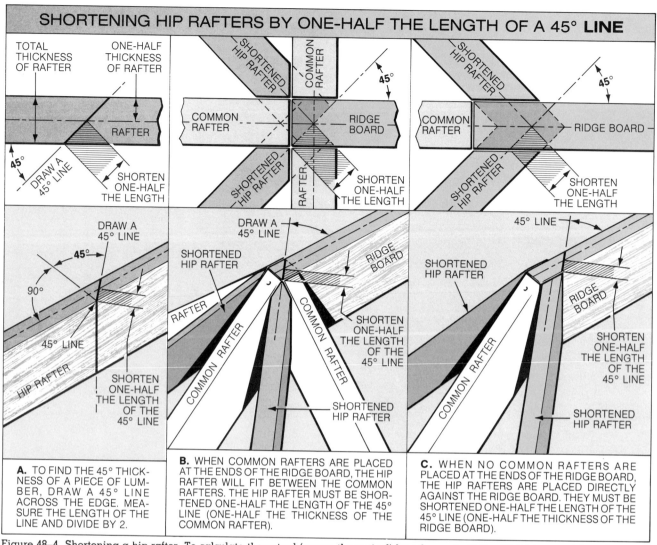

**A.** TO FIND THE 45° THICKNESS OF A PIECE OF LUMBER, DRAW A 45° LINE ACROSS THE EDGE. MEASURE THE LENGTH OF THE LINE AND DIVIDE BY 2.

**B.** WHEN COMMON RAFTERS ARE PLACED AT THE ENDS OF THE RIDGE BOARD, THE HIP RAFTER WILL FIT BETWEEN THE COMMON RAFTERS. THE HIP RAFTER MUST BE SHORTENED ONE-HALF THE LENGTH OF THE 45° LINE (ONE-HALF THE THICKNESS OF THE COMMON RAFTER).

**C.** WHEN NO COMMON RAFTERS ARE PLACED AT THE ENDS OF THE RIDGE BOARD, THE HIP RAFTERS ARE PLACED DIRECTLY AGAINST THE RIDGE BOARD. THEY MUST BE SHORTENED ONE-HALF THE LENGTH OF THE 45° LINE (ONE-HALF THE THICKNESS OF THE RIDGE BOARD).

Figure 48–4. Shortening a hip rafter. To calculate the actual (versus theoretical) length, one-half the 45° thickness of the ridge piece that fits between the rafters is deducted from the theoretical length.

ter lengths can also be calculated by using a book of rafter tables as described in the preceding unit.

**Shortening a Hip Rafter.** All of the methods discussed in this text for calculating hip rafter length give the theoretical length rather than the actual length. The theoretical length is the distance from the heel plumb line to the center of the ridge. For the actual length, one-half the 45° thickness of the ridge piece that fits between the rafters must be deducted. Hip roofs may or may not be framed with common rafters at the ends of the ridge. If common rafters are used, one-half the 45° thickness of the

common rafter must be deducted. See Figure 48–4.

## Laying out Hip Rafters

The layout procedure begins with marking the ridge plumb and side cuts. Next, the seat and heel plumb cuts are marked, then the lines showing the overhang. A procedure for laying out hip rafters is shown in Figure 48–5.

**Overhang.** Since the hip rafter runs at a diagonal, its overhang is longer than the common rafter overhang. (Refer to Figure 48–2.) The run of the hip rafter overhang is 1.42″ for every 1″ of common rafter overhang. There-

fore, to find the run of the hip rafter overhang, multiply 1.42″ by the run of the common rafter overhang. See Figure 48–6.

The steel square can also be used to calculate the run of the hip rafter overhang. Take a diagonal measurement from two points that are equal to the sides of a square formed by the common rafter overhang. See Figure 48–7.

**Plumb and Seat Cuts.** The 17″ unit run of the hip rafter makes the angles of the plumb and seat cuts different from those of a common rafter. To mark these cuts, use the unit rise measurement on the tongue of the steel

# LAYING OUT AN 8" UNIT RISE HIP RAFTER

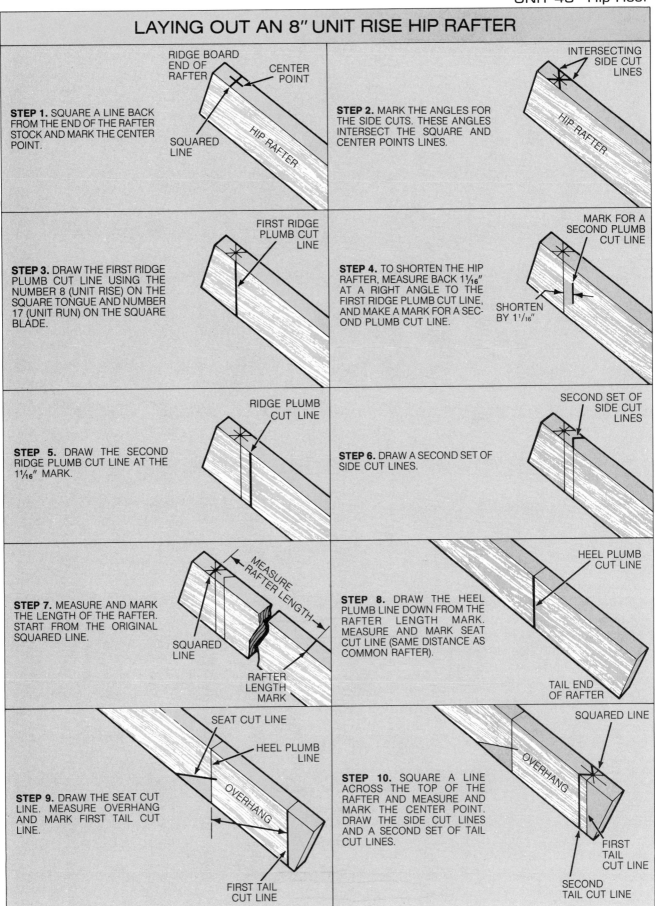

**STEP 1.** SQUARE A LINE BACK FROM THE END OF THE RAFTER STOCK AND MARK THE CENTER POINT.

RIDGE BOARD END OF RAFTER

CENTER POINT

SQUARED LINE

HIP RAFTER

**STEP 2.** MARK THE ANGLES FOR THE SIDE CUTS. THESE ANGLES INTERSECT THE SQUARE AND CENTER POINTS LINES.

INTERSECTING SIDE CUT LINES

HIP RAFTER

**STEP 3.** DRAW THE FIRST RIDGE PLUMB CUT LINE USING THE NUMBER 8 (UNIT RISE) ON THE SQUARE TONGUE AND NUMBER 17 (UNIT RUN) ON THE SQUARE BLADE.

FIRST RIDGE PLUMB CUT LINE

**STEP 4.** TO SHORTEN THE HIP RAFTER, MEASURE BACK 1 1/16" AT A RIGHT ANGLE TO THE FIRST RIDGE PLUMB CUT LINE, AND MAKE A MARK FOR A SECOND PLUMB CUT LINE.

MARK FOR A SECOND PLUMB CUT LINE

SHORTEN BY 1 1/16"

**STEP 5.** DRAW THE SECOND RIDGE PLUMB CUT LINE AT THE 1 1/16" MARK.

RIDGE PLUMB CUT LINE

**STEP 6.** DRAW A SECOND SET OF SIDE CUT LINES.

SECOND SET OF SIDE CUT LINES

**STEP 7.** MEASURE AND MARK THE LENGTH OF THE RAFTER. START FROM THE ORIGINAL SQUARED LINE.

MEASURE RAFTER LENGTH

SQUARED LINE

RAFTER LENGTH MARK

**STEP 8.** DRAW THE HEEL PLUMB LINE DOWN FROM THE RAFTER LENGTH MARK. MEASURE AND MARK SEAT CUT LINE (SAME DISTANCE AS COMMON RAFTER).

HEEL PLUMB CUT LINE

TAIL END OF RAFTER

**STEP 9.** DRAW THE SEAT CUT LINE. MEASURE OVERHANG AND MARK FIRST TAIL CUT LINE.

SEAT CUT LINE

HEEL PLUMB LINE

OVERHANG

FIRST TAIL CUT LINE

**STEP 10.** SQUARE A LINE ACROSS THE TOP OF THE RAFTER AND MEASURE AND MARK THE CENTER POINT. DRAW THE SIDE CUT LINES AND A SECOND SET OF TAIL CUT LINES.

SQUARED LINE

OVERHANG

FIRST TAIL CUT LINE

SECOND TAIL CUT LINE

Figure 48–5. Laying out a hip rafter. The angles for the side cuts and the ridge plumb cuts are marked first.

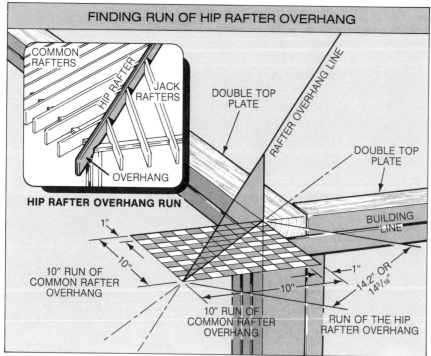

FINDING RUN OF HIP RAFTER OVERHANG

Figure 48–6. A hip rafter overhang runs 1.42″ for every 1″ of common rafter run. In this example the run of the common rafter is 10″. To find the run of the hip overhang, multiply 1.42″ by 10. The result is 14.2″, or 14 3/16″.

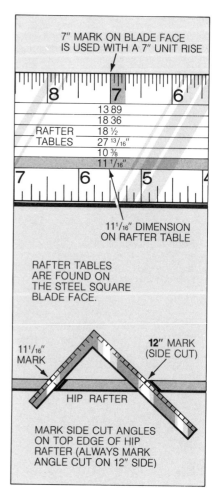

Figure 48–8. Finding the side cut angle for a hip rafter. In this example the roof has a 7″ unit rise.

square and the 17″ measurement on the blade. Mark the plumb cut along the tongue and the seat cut along the blade.

When positioning the square to mark the seat cut of the hip rafter, first check the amount of stock left above the seat cut at the heel plumb line of the common rafter for that roof. Measure and mark the same amount at the heel plumb line of the hip rafter and draw the seat cut line. The length of the hip rafter seat

cut is greater than that of the common rafter because it rests diagonally at the corner of the building.

**Side Cuts.** Some carpenters use a 45° angle for all the side cuts on a hip rafter. However, this may result in a poor fit against the ridge, as the side cut angle is different for every unit rise. The rafter table on the steel square can be used to find the exact angle for the side cut of any hip rafter. This procedure is shown in Figure 48–8.

## Step-off Method for Calculating Length and Laying out Hip Rafters

The step-off method described in the preceding unit for the common rafter can also be used for the hip rafter. The steel square is set up with the unit rise measurement on the tongue and the 17″ (rather than 12″) measurement on the blade.

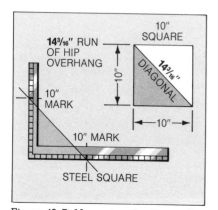

Figure 48–7. Measuring the diagonal on the steel square to find the run of a hip rafter overhang. In this example the run of the roof overhang is 10″.

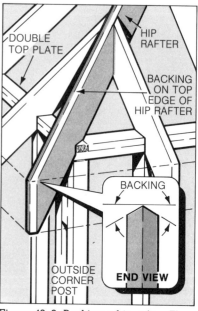

Figure 48–9. Backing a hip rafter. The top edges of the rafter are chamfered.

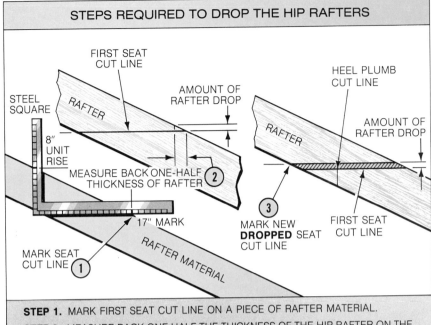

STEPS REQUIRED TO DROP THE HIP RAFTERS

**STEP 1.** MARK FIRST SEAT CUT LINE ON A PIECE OF RAFTER MATERIAL.

**STEP 2.** MEASURE BACK ONE-HALF THE THICKNESS OF THE HIP RAFTER ON THE SEAT CUT LINE. MEASURE AT A RIGHT ANGLE FROM THIS POINT TO THE EDGE OF THE RAFTER. THIS WILL GIVE THE AMOUNT OF HIP RAFTER DROP.

**STEP 3.** FROM A LAID OUT HIP RAFTER, MEASURE THE AMOUNT OF DROP FROM THE FIRST SEAT CUT LINE AND MARK A NEW **DROPPED** SEAT CUT LINE.

Figure 48–10. Dropping a hip rafter. This procedure is faster than backing a rafter and accomplishes the same purpose.

## Backing or Dropping Hip Rafters

Chamfering the top edges of the hip rafter is called *backing* the rafters. See Figure 48–9. Backing prevents the roof sheathing from being higher where it covers the hip rafters than where it covers the common and jack rafters.

Another method to accomplish the same end is *dropping* the hip rafter. This procedure is shown in Figure 48–10. The seat cut is enlarged, causing the rafter to drop. Consequently, the sheathing rests on the top corners of the rafter and is in line with the roof. Most carpenters use the dropping method because it is faster than the backing method.

## HIP JACK RAFTERS

*Hip jack rafters* frame the space between the hip rafters and the wall plates. (Refer to Figure 48–1.) They run in pairs and are the same distance apart as the common rafters. If the common rafters are spaced 24″ O.C., so are the hip jack rafters.

## Common Length Difference

The hip jack rafters decrease in length as they get closer to the end of the building. They have a *common length difference* as long as they are the same distance from each other. When the length of one of the hip jack rafters is known, the length of the others can be found by subtracting or adding the common length difference.

The common length difference for any roof can be easily calculated by using the steel square rafter table. It is expressed as the "difference in length of jacks." The measurements on the third line of the rafter table are for jacks spaced 16″ O.C. The measurements on the fourth line are for jacks spaced 2′ (24″) O.C.

## Laying out Hip Jack Rafters

Layout for hip jack rafters placement may begin from a common rafter at the end of the ridge. The procedure for this is shown in Figure 48–11. Layout may also begin from a common rafter located at some point other than the end of the ridge. This procedure is shown in Figure 48–12. In a third hip jack layout method, layout begins at the corner of the building. See Figure 48–13.

The hip jack rafter has plumb cuts where it fastens to the hip, as well as at the heel and tail. A seat cut is made where it rests on the plate. The plumb and seat cuts can be marked by combining the unit rise and 12″ on the steel square, as is done for the common rafter. The plumb cut is marked on the unit rise side of the square and the seat cut is marked on the 12″ side. The bird's mouth and overhang are marked the same way they are marked for the common rafter.

Hip jack rafters require a single side cut where they fasten to the hip. The steel square can be used to find the angle of the side cut. See Figure 48–14. A procedure for laying out the cuts on a hip jack rafter is shown in Figure 48–15.

## CONSTRUCTING A HIP ROOF

As with all other types of roof, the main framing members for the hip roof are pre-cut. The ridge board for a hip roof must be pre-cut to its exact length before it can be assembled. A procedure for finding the theoretical length of the ridge is shown in Figure 48–16. The actual length is affected by the framing method used at the end of the ridge. See Figure 48–17.

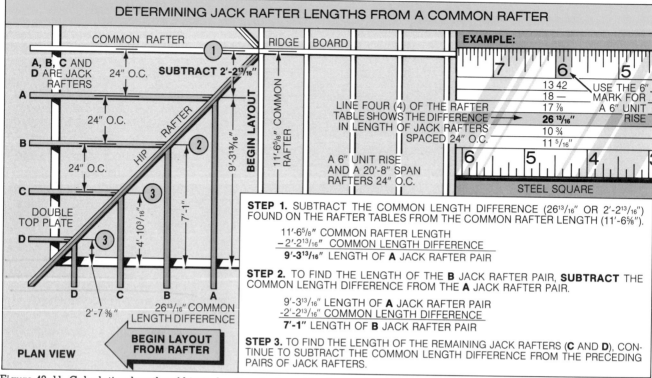

**DETERMINING JACK RAFTER LENGTHS FROM A COMMON RAFTER**

A, B, C AND D ARE JACK RAFTERS

COMMON RAFTER

SUBTRACT 2'-2¹³/₁₆"

24" O.C.

24" O.C.

24" O.C.

DOUBLE TOP PLATE

HIP RAFTER

9'-3¹³/₁₆"

7'-1"

4'-10³/₁₆"

BEGIN LAYOUT

11'-6⅝" COMMON RAFTER

RIDGE BOARD

**EXAMPLE:**

13 42
18 —
17 ⅞
**26 ¹³/₁₆"**
10 ¾
11 ⁵/₁₆"

USE THE 6" MARK FOR A 6" UNIT RISE

STEEL SQUARE

LINE FOUR (4) OF THE RAFTER TABLE SHOWS THE DIFFERENCE IN LENGTH OF JACK RAFTERS SPACED 24" O.C.

A 6" UNIT RISE AND A 20'-8" SPAN RAFTERS 24" O.C.

**STEP 1.** SUBTRACT THE COMMON LENGTH DIFFERENCE (26¹³/₁₆" OR 2'-2¹³/₁₆") FOUND ON THE RAFTER TABLES FROM THE COMMON RAFTER LENGTH (11'-6⅝").

11'-6⅝" COMMON RAFTER LENGTH
−2'-2¹³/₁₆" COMMON LENGTH DIFFERENCE
**9'-3¹³/₁₆"** LENGTH OF **A** JACK RAFTER PAIR

**STEP 2.** TO FIND THE LENGTH OF THE **B** JACK RAFTER PAIR, **SUBTRACT** THE COMMON LENGTH DIFFERENCE FROM THE **A** JACK RAFTER PAIR.

9'-3¹³/₁₆" LENGTH OF **A** JACK RAFTER PAIR
−2'-2¹³/₁₆" COMMON LENGTH DIFFERENCE
**7'-1"** LENGTH OF **B** JACK RAFTER PAIR

**STEP 3.** TO FIND THE LENGTH OF THE REMAINING JACK RAFTERS (**C** AND **D**), CONTINUE TO SUBTRACT THE COMMON LENGTH DIFFERENCE FROM THE PRECEDING PAIRS OF JACK RAFTERS.

2'-7⅜" 26¹³/₁₆" COMMON LENGTH DIFFERENCE

**BEGIN LAYOUT FROM RAFTER**

PLAN VIEW

Figure 48–11. Calculating lengths of hip jack rafters for layout beginning from common rafter at end of ridge. In this example, the roof has a 6" unit rise and a 20'-8" span. The length of the common rafter is 11'-6 ⅝". The jacks are spaced 24" O.C. The common length difference is 26 ¹³/₁₆".

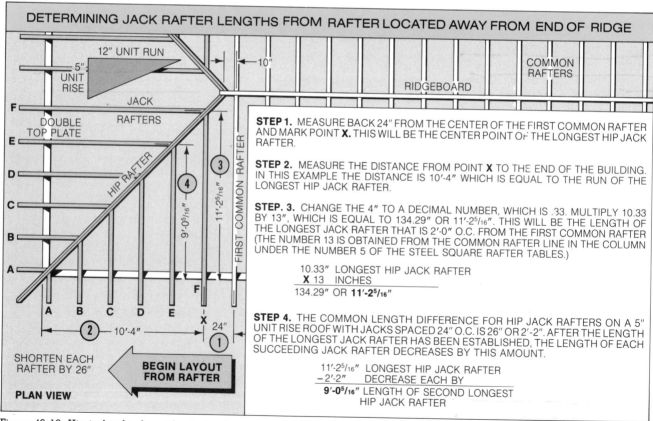

**DETERMINING JACK RAFTER LENGTHS FROM RAFTER LOCATED AWAY FROM END OF RIDGE**

12" UNIT RUN

5" UNIT RISE

JACK RAFTERS

DOUBLE TOP PLATE

HIP RAFTER

10"

RIDGEBOARD

COMMON RAFTERS

11'-2⁵/₁₆"

9'-0⁵/₁₆"

FIRST COMMON RAFTER

10'-4"

24"

SHORTEN EACH RAFTER BY 26"

**BEGIN LAYOUT FROM RAFTER**

PLAN VIEW

**STEP 1.** MEASURE BACK 24" FROM THE CENTER OF THE FIRST COMMON RAFTER AND MARK POINT **X.** THIS WILL BE THE CENTER POINT OF THE LONGEST HIP JACK RAFTER.

**STEP 2.** MEASURE THE DISTANCE FROM POINT **X** TO THE END OF THE BUILDING. IN THIS EXAMPLE THE DISTANCE IS 10'-4" WHICH IS EQUAL TO THE RUN OF THE LONGEST HIP JACK RAFTER.

**STEP. 3.** CHANGE THE 4" TO A DECIMAL NUMBER, WHICH IS .33. MULTIPLY 10.33 BY 13", WHICH IS EQUAL TO 134.29" OR 11'-2⁵/₁₆". THIS WILL BE THE LENGTH OF THE LONGEST JACK RAFTER THAT IS 2'-0" O.C. FROM THE FIRST COMMON RAFTER (THE NUMBER 13 IS OBTAINED FROM THE COMMON RAFTER LINE IN THE COLUMN UNDER THE NUMBER 5 OF THE STEEL SQUARE RAFTER TABLES.)

10.33" LONGEST HIP JACK RAFTER
**X** 13 INCHES
134.29" OR **11'-2⁵/₁₆"**

**STEP 4.** THE COMMON LENGTH DIFFERENCE FOR HIP JACK RAFTERS ON A 5" UNIT RISE ROOF WITH JACKS SPACED 24" O.C. IS 26" OR 2'-2". AFTER THE LENGTH OF THE LONGEST JACK RAFTER HAS BEEN ESTABLISHED, THE LENGTH OF EACH SUCCEEDING JACK RAFTER DECREASES BY THIS AMOUNT.

11'-2⁵/₁₆" LONGEST HIP JACK RAFTER
−2'-2" DECREASE EACH BY
**9'-0⁵/₁₆"** LENGTH OF SECOND LONGEST HIP JACK RAFTER

Figure 48–12. Hip jack rafter layout beginning from common rafter located away from end of ridge. In this example, the roof has a 5" unit rise and 23'-0" span. The jacks are spaced 24" O.C. The common length difference is 26".

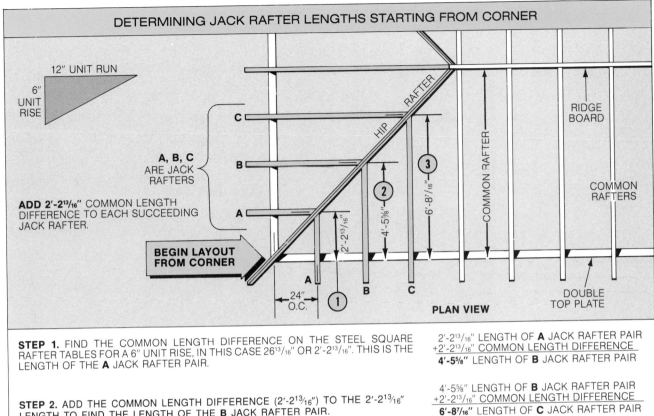

## DETERMINING JACK RAFTER LENGTHS STARTING FROM CORNER

12" UNIT RUN

6" UNIT RISE

C
B
A, B, C
ARE JACK
RAFTERS

**ADD 2'-2¹³⁄₁₆"** COMMON LENGTH
DIFFERENCE TO EACH SUCCEEDING
JACK RAFTER.

**BEGIN LAYOUT
FROM CORNER**

HIP RAFTER

③ 6'-8⁷⁄₁₆"
② 4'-5⁵⁄₈"
① 2'-2¹³⁄₁₆"

A
24"
O.C.

A   B   C

COMMON RAFTER

RIDGE BOARD

COMMON RAFTERS

DOUBLE TOP PLATE

**PLAN VIEW**

**STEP 1.** FIND THE COMMON LENGTH DIFFERENCE ON THE STEEL SQUARE RAFTER TABLES FOR A 6" UNIT RISE, IN THIS CASE 26¹³⁄₁₆" OR 2'-2¹³⁄₁₆". THIS IS THE LENGTH OF THE **A** JACK RAFTER PAIR.

**STEP 2.** ADD THE COMMON LENGTH DIFFERENCE (2'-2¹³⁄₁₆") TO THE 2'-2¹³⁄₁₆" LENGTH TO FIND THE LENGTH OF THE **B** JACK RAFTER PAIR.

**STEP 3.** ADD THE COMMON LENGTH DIFFERENCE (2'-2¹³⁄₁₆") TO THE LENGTH OF THE B JACK RAFTER PAIR (4'-5⁵⁄₈") TO FIND THE LENGTH OF THE **C** JACK RAFTER PAIR.

```
  2'-2¹³⁄₁₆" LENGTH OF A JACK RAFTER PAIR
+ 2'-2¹³⁄₁₆" COMMON LENGTH DIFFERENCE
─────────────────────────────────────
  4'-5⁵⁄₈"  LENGTH OF B JACK RAFTER PAIR
```

```
  4'-5⁵⁄₈"  LENGTH OF B JACK RAFTER PAIR
+ 2'-2¹³⁄₁₆" COMMON LENGTH DIFFERENCE
─────────────────────────────────────
  6'-8⁷⁄₁₆" LENGTH OF C JACK RAFTER PAIR
```

Figure 48–13. Hip jack rafter layout beginning from corner of building. In this example the roof has a 6" unit rise. The jacks are spaced 24" O.C. The 26 ¹³⁄₁₆" or 2'-2 ¹³⁄₁₆" common length difference shown in the rafter table is the length of the first jack placed 24" O.C. from the corner of the building. The length of each succeeding jack increases by that amount.

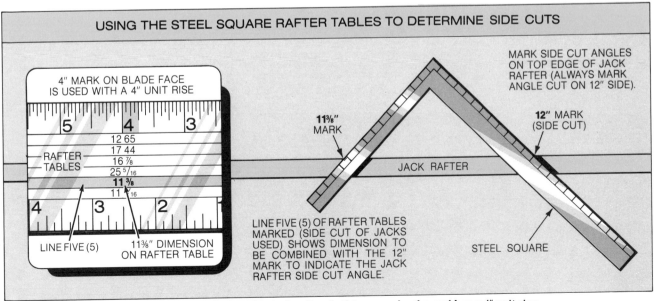

## USING THE STEEL SQUARE RAFTER TABLES TO DETERMINE SIDE CUTS

4" MARK ON BLADE FACE
IS USED WITH A 4" UNIT RISE

5   4   3

RAFTER TABLES

```
12 65
17 44
16 ⅞
25 ⁵⁄₁₆
11 ⅜
11 ⁵⁄₁₆
```

4   3   2

LINE FIVE (5)          11⅜" DIMENSION
                        ON RAFTER TABLE

MARK SIDE CUT ANGLES ON TOP EDGE OF JACK RAFTER (ALWAYS MARK ANGLE CUT ON 12" SIDE).

11⅜" MARK

12" MARK (SIDE CUT)

JACK RAFTER

STEEL SQUARE

LINE FIVE (5) OF RAFTER TABLES MARKED (SIDE CUT OF JACKS USED) SHOWS DIMENSION TO BE COMBINED WITH THE 12" MARK TO INDICATE THE JACK RAFTER SIDE CUT ANGLE.

Figure 48–14. Calculating the side cut angle for a hip jack rafter. In this example, the roof has a 4" unit rise.

# LAYING OUT HIP JACK RAFTERS

**STEP 1.** DRAW A SQUARED LINE ACROSS THE TOP EDGE OF THE RAFTER MATERIAL CLOSE TO THE END. MEASURE AND MARK THE CENTER POINT OF THE TOP EDGE (¾" ON NOMINAL 2"X 10" MATERIAL).

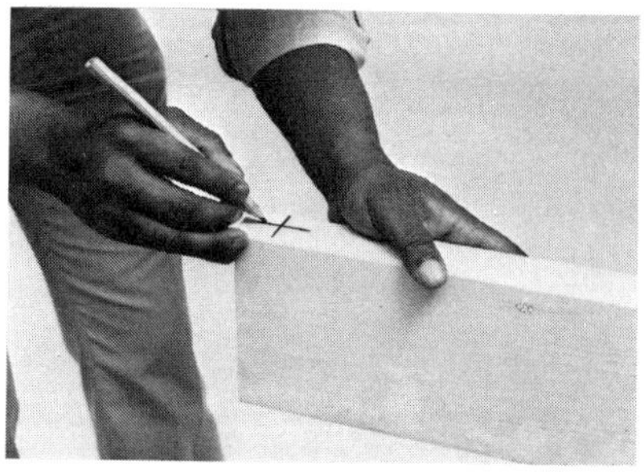

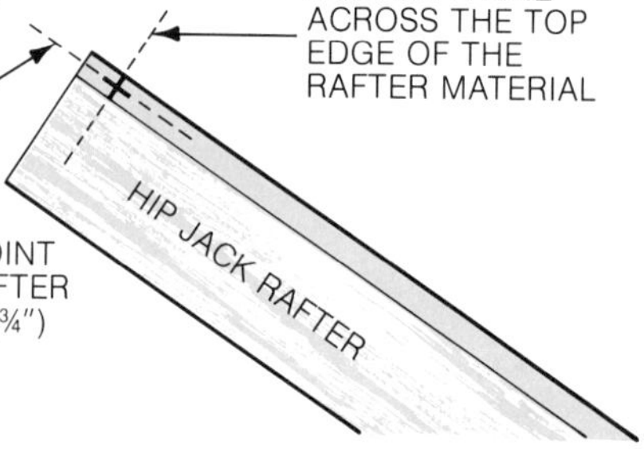

TOP END OF RAFTER

SQUARE A LINE ACROSS THE TOP EDGE OF THE RAFTER MATERIAL

MARK THE CENTER POINT OF THE RAFTER MATERIAL (¾")

HIP JACK RAFTER

**STEP 2.** USING A STEEL SQUARE MEASURE AND MARK THE ANGLE FOR THE SIDE CUT LINE ON THE TOP EDGE OF THE RAFTER MATERIAL.

MARK THE ANGLE FOR THE SIDE CUT LINE

ANGLE OF SIDE CUT LINE

HIP JACK RAFTER

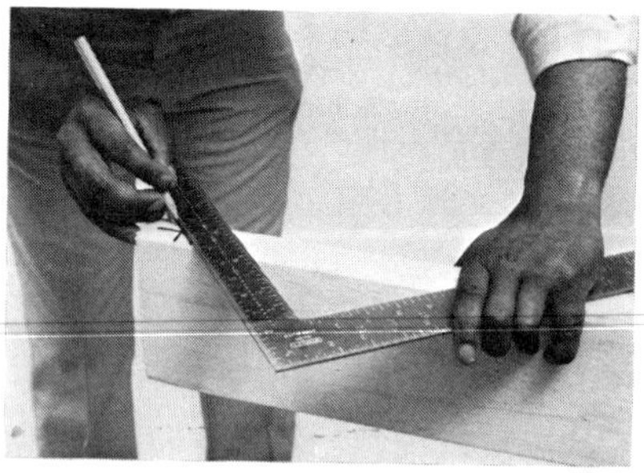

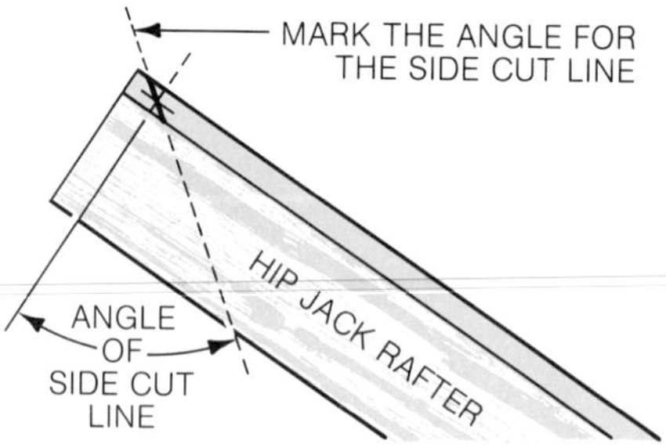

**STEP 3.** USING A STEEL SQUARE MARK THE **FIRST** TOP PLUMB CUT LINE ON THE SIDE OF THE RAFTER MATERIAL.

MARK THE **FIRST** TOP PLUMB CUT LINE

HIP JACK RAFTER

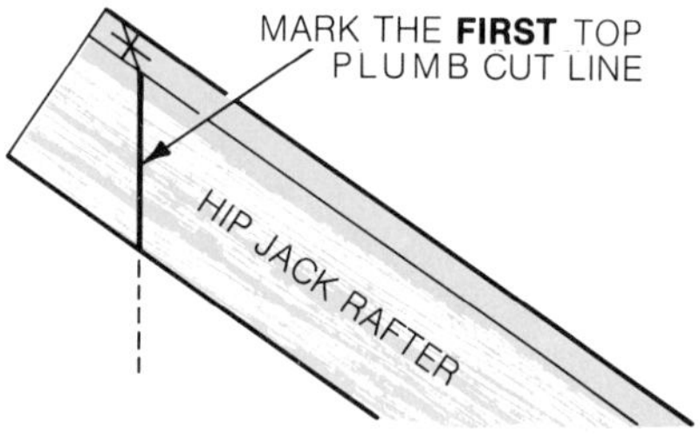

**STEP 4.** SHORTEN THE HIP JACK RAFTER BY ONE-HALF THE 45° THICKNESS OF THE RAFTER. DRAW A **SECOND** TOP PLUMB CUT LINE ON THE SIDE OF THE RAFTER MATERIAL.

SHORTEN BY ONE-HALF THE 45° THICKNESS OF THE HIP JACK RAFTER AND MAKE A **SECOND** RIDGE PLUMB CUT LINE

HIP JACK RAFTER

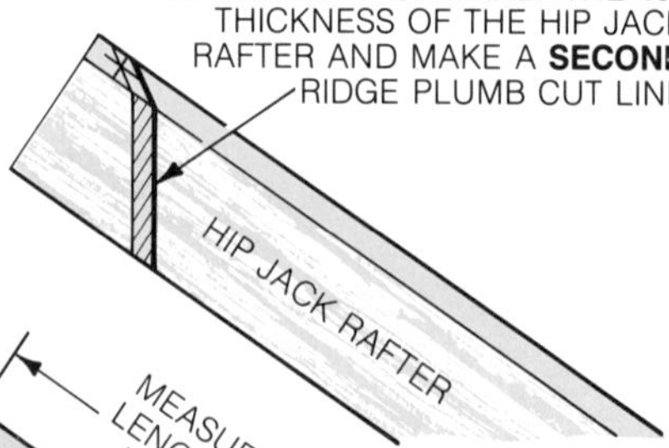

**STEP 5.** MEASURE THE LENGTH OF THE HIP JACK RAFTER FROM THE SQUARED LINE ON THE TOP EDGE. DRAW A HEEL PLUMB CUT LINE AND LAY OUT THE BIRD'S MOUTH BY MARKING THE SEAT CUT LINE. MEASURE THE LENGTH OF THE RAFTER OVERHANG FROM THE HEEL PLUMB CUT LINE AND MARK THE TAIL PLUMB CUT LINE.

MEASURE AND MARK LENGTH OF THE THE HIP JACK RAFTER

MARK THE HEEL PLUMB CUT LINE

MEASURE THE LENGTH FROM THE SQUARED LINE ON THE TOP EDGE OF THE HIP JACK RAFTER.

MARK THE SEAT CUT LINE

MARK THE TAIL PLUMB CUT LINE

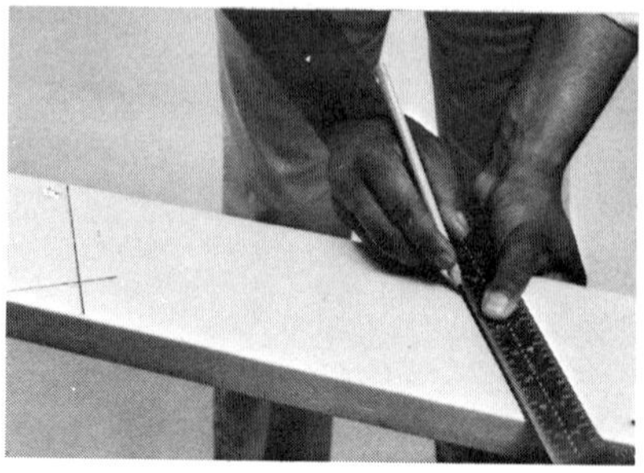

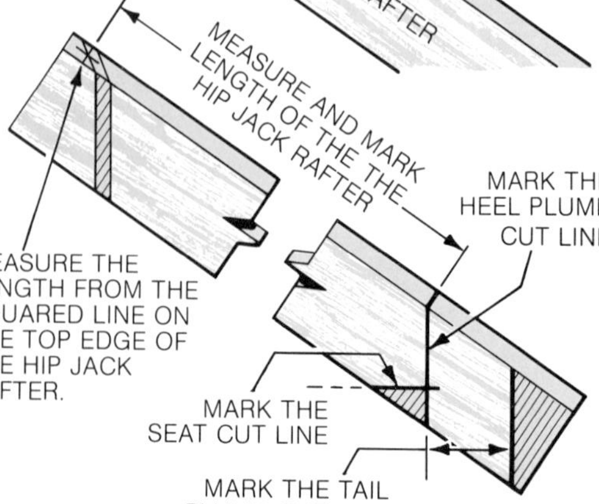

Figure 48–15. Laying out a hip jack rafter. It has plumb cuts where it fastens to the hip as well as cuts at the heel and tail.

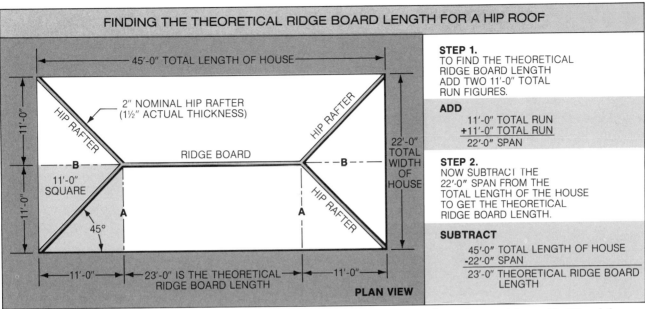

FINDING THE THEORETICAL RIDGE BOARD LENGTH FOR A HIP ROOF

**STEP 1.**
TO FIND THE THEORETICAL RIDGE BOARD LENGTH ADD TWO 11'-0" TOTAL RUN FIGURES.

**ADD**
11'-0" TOTAL RUN
+11'-0" TOTAL RUN
22'-0" SPAN

**STEP 2.**
NOW SUBTRACT THE 22'-0" SPAN FROM THE TOTAL LENGTH OF THE HOUSE TO GET THE THEORETICAL RIDGE BOARD LENGTH.

**SUBTRACT**
45'-0" TOTAL LENGTH OF HOUSE
-22'-0" SPAN
23'-0" THEORETICAL RIDGE BOARD LENGTH

Figure 48–16. Finding the theoretical length of the ridge of a hip roof. In this example, the total span of the roof is 22' and the length of the building is 45'.

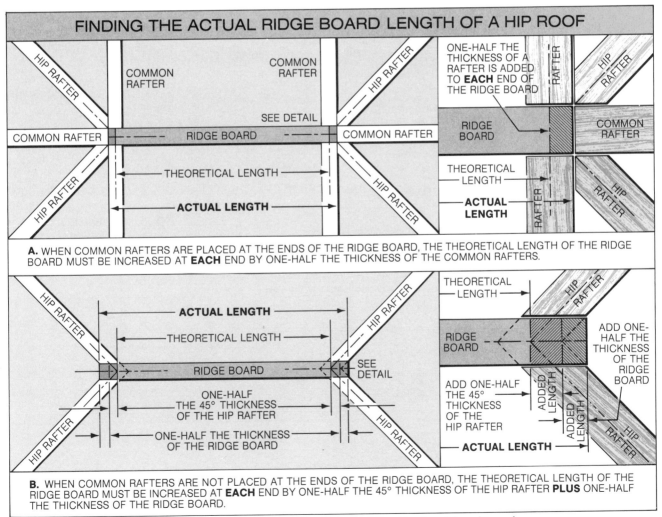

FINDING THE ACTUAL RIDGE BOARD LENGTH OF A HIP ROOF

**A.** WHEN COMMON RAFTERS ARE PLACED AT THE ENDS OF THE RIDGE BOARD, THE THEORETICAL LENGTH OF THE RIDGE BOARD MUST BE INCREASED AT **EACH** END BY ONE-HALF THE THICKNESS OF THE COMMON RAFTERS.

**B.** WHEN COMMON RAFTERS ARE NOT PLACED AT THE ENDS OF THE RIDGE BOARD, THE THEORETICAL LENGTH OF THE RIDGE BOARD MUST BE INCREASED AT **EACH** END BY ONE-HALF THE 45° THICKNESS OF THE HIP RAFTER **PLUS** ONE-HALF THE THICKNESS OF THE RIDGE BOARD.

Figure 48–17. Finding the actual length of the ridge board for a hip roof.

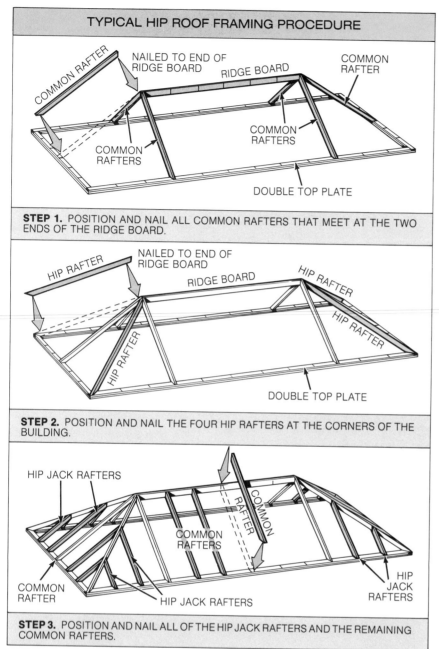

## TYPICAL HIP ROOF FRAMING PROCEDURE

COMMON RAFTER

NAILED TO END OF RIDGE BOARD

RIDGE BOARD

COMMON RAFTER

COMMON RAFTERS

COMMON RAFTERS

DOUBLE TOP PLATE

**STEP 1.** POSITION AND NAIL ALL COMMON RAFTERS THAT MEET AT THE TWO ENDS OF THE RIDGE BOARD.

HIP RAFTER

NAILED TO END OF RIDGE BOARD

RIDGE BOARD

HIP RAFTER

HIP RAFTER

HIP RAFTER

DOUBLE TOP PLATE

**STEP 2.** POSITION AND NAIL THE FOUR HIP RAFTERS AT THE CORNERS OF THE BUILDING.

HIP JACK RAFTERS

COMMON RAFTER

COMMON RAFTERS

COMMON RAFTER

HIP JACK RAFTERS

HIP JACK RAFTERS

**STEP 3.** POSITION AND NAIL ALL OF THE HIP JACK RAFTERS AND THE REMAINING COMMON RAFTERS.

Figure 48–18. Constructing a hip roof. The main framing members should be precut before construction begins.

Figure 48–19. Hip roof under construction. Carpenter at right is nailing hip jack rafters to a hip rafter. Carpenter at left is placing common rafters.

After the ridge board is cut to length, the layout markings for placing the rafters can be transferred from the top wall plates to the ridge board.

The conditions that require the use of purlins, braces, and collar ties are described in Unit 46. Collar ties may be placed at every second or third pair of rafters. Purlins may be used to support longer rafters. They are placed beneath the rafters at an intermediate point between the ridge of the roof and the outside wall. Braces extending to the nearest partition support the purlins.

Some general construction procedures for erecting the hip roof are shown in Figures 48–18 and 48–19.

# UNIT 49

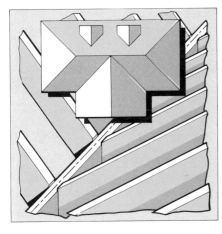

# Intersecting Roof

An intersecting roof, also known as a *combination* roof, consists of two or more sections sloping in different directions. A valley is formed where the different sloping sections come together. See Figure 49–1.

The two sections of an intersecting roof may or may not be the same width. If they are the same width, the roof is said to have *equal spans*. If they are not the same width, the roof is said to have *unequal spans*.

## INTERSECTING ROOF WITH EQUAL SPANS

In a roof with equal spans, the height (total rise) is the same for each of the two ridges. See Figure 49–2. Where the slopes of the roof meet to form a valley between the two sections, a pair of *valley rafters* is placed. These rafters go from the inside corners formed by the two sections of the building to the corners formed by the intersecting ridges. *Valley jack rafters* run from the valley rafters to both ridges. *Hip-valley cripple jack rafters* are placed between the valley and hip rafters.

## INTERSECTING ROOF WITH UNEQUAL SPANS

An intersecting roof with unequal spans requires a *supporting* valley rafter to run from the inside corner formed by the two sections of the building to the main ridge. See Figure 49–3. A *shortened* valley rafter runs from the other inside corner of the building to the supporting valley rafter. Like an intersecting roof with equal spans, one with unequal spans also requires valley jack rafters and hip-valley cripple jack rafters. In addition, a *valley crip-*

*ple jack rafter* is placed between the supporting and shortened valley rafters.

## VALLEY RAFTERS

Valley rafters run at a 45° angle to the outside walls of the building. This places them parallel to the hip rafters. Consequently, they are the same length as the hip rafters.

### Laying out Valley Rafters

The layout of the valley rafter is almost identical to that of the hip

Figure 49–1. In this intersecting roof the hip section over the garage intersects with the main gable roof.

371

## INTERSECTING ROOF SECTIONS WITH EQUAL SPANS

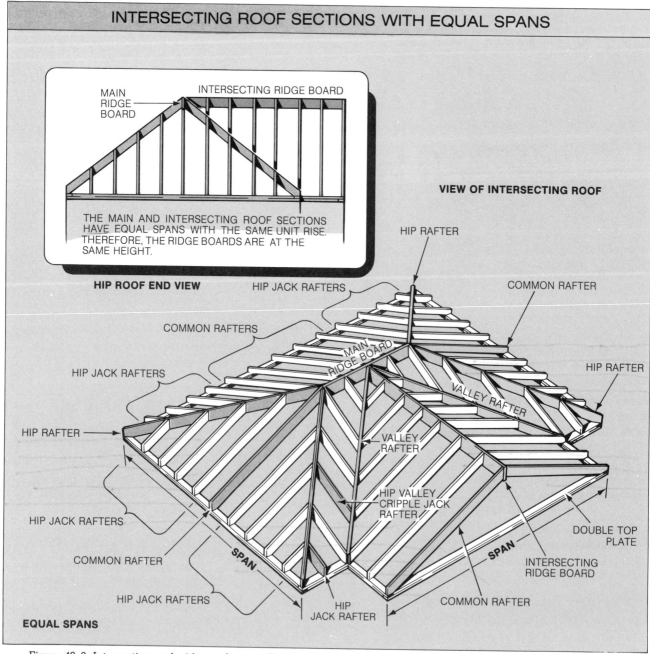

MAIN RIDGE BOARD

INTERSECTING RIDGE BOARD

THE MAIN AND INTERSECTING ROOF SECTIONS HAVE EQUAL SPANS WITH THE SAME UNIT RISE. THEREFORE, THE RIDGE BOARDS ARE AT THE SAME HEIGHT.

**HIP ROOF END VIEW**

**VIEW OF INTERSECTING ROOF**

HIP RAFTER

COMMON RAFTER

HIP JACK RAFTERS

COMMON RAFTERS

HIP JACK RAFTERS

MAIN RIDGE BOARD

HIP RAFTER

VALLEY RAFTER

HIP RAFTER

VALLEY RAFTER

HIP VALLEY CRIPPLE JACK RAFTER

HIP JACK RAFTERS

COMMON RAFTER

SPAN

SPAN

DOUBLE TOP PLATE

INTERSECTING RIDGE BOARD

HIP JACK RAFTERS

HIP JACK RAFTER

COMMON RAFTER

**EQUAL SPANS**

Figure 49–2. Intersecting roof with equal spans. Both sections are the same width, and both ridges are the same height.

rafter. The unit rise measurement and the 17″ measurement on the steel square give the angles for the plumb and seat cuts. The angles of the side cuts are the same as for the hip rafter. The only difference in layout occurs at the seat and tail of the valley rafter. Side cuts must be angled back at the heel plumb line to allow the valley rafter to drop down

into the inside corner of the building. See Figure 49–4. Side cuts are also required at the tail of the overhang so that the corner formed by the valley will line up with the rest of the roof overhang. See Figure 49–5.

The angle of the side cuts at the heel and tail is the same as the angle where the rafter connects with the ridge. The com-

plete layout procedure for a valley rafter is shown in Figure 49–6. Note that, unlike a hip rafter, the valley rafter requires no backing or dropping.

## Supporting and Shortened Valley Rafters

An intersecting roof with unequal spans requires two types of val-

372

## INTERSECTING ROOF SECTIONS WITH UNEQUAL SPANS

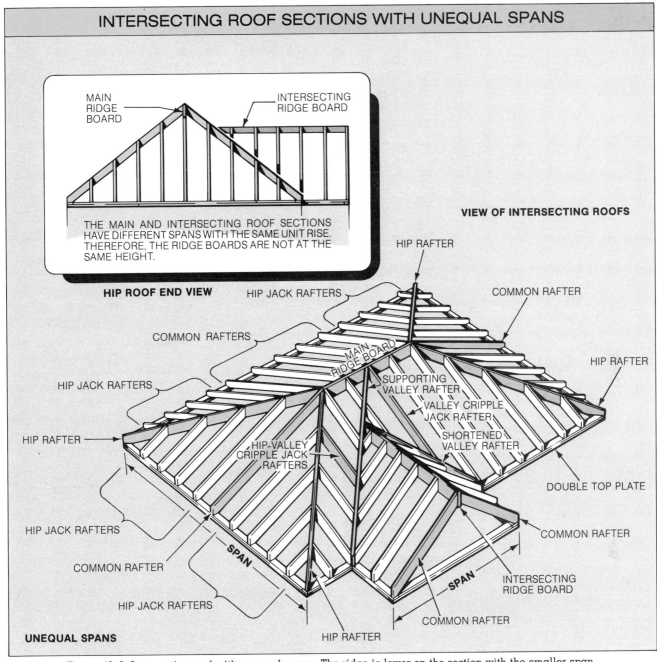

Figure 49-3. Intersecting roof with unequal spans. The ridge is lower on the section with the smaller span.

ley rafters: supporting and short-ened. See Figure 49–7. The supporting valley rafter extends to the main ridge. It has a single side cut where it fits against the ridge. The shortened valley rafter runs at a 90° angle to the supporting valley. It has a square cut where it butts against the longer rafter. Its length is based on the run of the narrower roof. See

Figure 49–8. The layout procedure for a shortened valley rafter is shown in Figure 49–9.

## VALLEY JACK RAFTERS

Valley jack rafters bridge the area between the valley rafters and the ridges of an intersecting roof (refer again to Figures 49–2,

49–3, and 49–7). Spacing between valley jack rafters is the same as the spacing between common rafters in the roof.

## Calculating Length of Valley Jack Rafters

Valley jack rafters decrease in length as they get closer to the

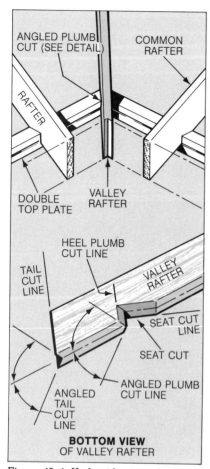

Figure 49–4. Underside view of the angled cuts at the seat of a valley rafter.

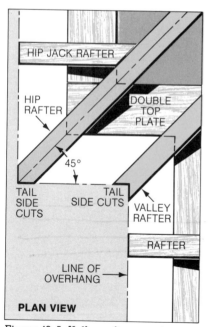

Figure 49–5. Valley rafters require side cuts at the seat and tail of the overhang.

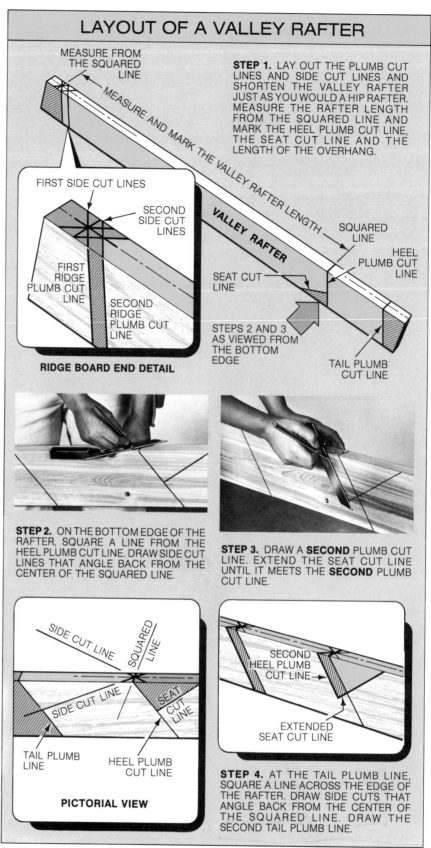

# LAYOUT OF A VALLEY RAFTER

**STEP 1.** LAY OUT THE PLUMB CUT LINES AND SIDE CUT LINES AND SHORTEN THE VALLEY RAFTER JUST AS YOU WOULD A HIP RAFTER. MEASURE THE RAFTER LENGTH FROM THE SQUARED LINE AND MARK THE HEEL PLUMB CUT LINE, THE SEAT CUT LINE AND THE LENGTH OF THE OVERHANG.

**STEP 2.** ON THE BOTTOM EDGE OF THE RAFTER, SQUARE A LINE FROM THE HEEL PLUMB CUT LINE. DRAW SIDE CUT LINES THAT ANGLE BACK FROM THE CENTER OF THE SQUARED LINE.

**STEP 3.** DRAW A **SECOND** PLUMB CUT LINE. EXTEND THE SEAT CUT LINE UNTIL IT MEETS THE **SECOND** PLUMB CUT LINE.

**STEP 4.** AT THE TAIL PLUMB LINE, SQUARE A LINE ACROSS THE EDGE OF THE RAFTER. DRAW SIDE CUTS THAT ANGLE BACK FROM THE CENTER OF THE SQUARED LINE. DRAW THE SECOND TAIL PLUMB LINE.

Figure 49–6. Layout of a valley rafter. Unlike a hip rafter, a valley rafter requires no backing or dropping.

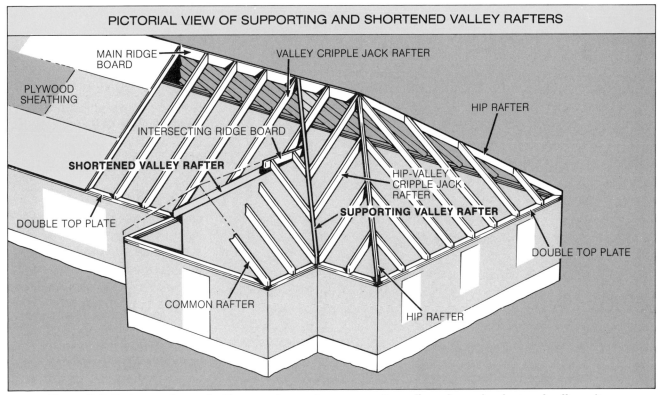

PICTORIAL VIEW OF SUPPORTING AND SHORTENED VALLEY RAFTERS

MAIN RIDGE BOARD

VALLEY CRIPPLE JACK RAFTER

PLYWOOD SHEATHING

INTERSECTING RIDGE BOARD

HIP RAFTER

SHORTENED VALLEY RAFTER

HIP-VALLEY CRIPPLE JACK RAFTER

SUPPORTING VALLEY RAFTER

DOUBLE TOP PLATE

DOUBLE TOP PLATE

COMMON RAFTER

HIP RAFTER

Figure 49–7. An intersecting roof with unequal spans has a supporting valley rafter and a shortened valley rafter.

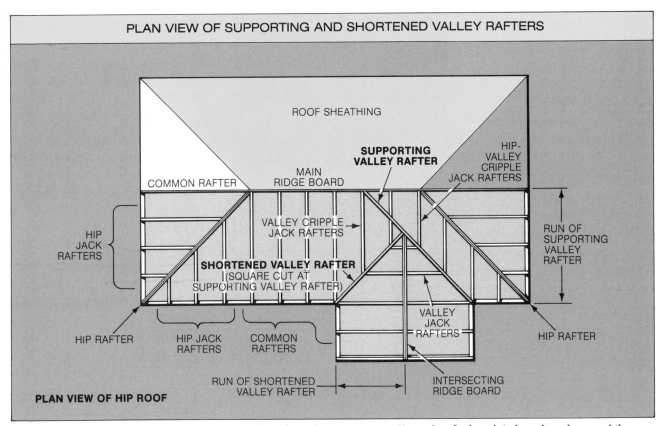

PLAN VIEW OF SUPPORTING AND SHORTENED VALLEY RAFTERS

ROOF SHEATHING

COMMON RAFTER

MAIN RIDGE BOARD

SUPPORTING VALLEY RAFTER

HIP-VALLEY CRIPPLE JACK RAFTERS

HIP JACK RAFTERS

VALLEY CRIPPLE JACK RAFTERS

RUN OF SUPPORTING VALLEY RAFTER

SHORTENED VALLEY RAFTER
(SQUARE CUT AT SUPPORTING VALLEY RAFTER)

HIP RAFTER

HIP JACK RAFTERS

COMMON RAFTERS

VALLEY JACK RAFTERS

HIP RAFTER

RUN OF SHORTENED VALLEY RAFTER

INTERSECTING RIDGE BOARD

PLAN VIEW OF HIP ROOF

Figure 49–8. A shortened valley rafter runs at a 90° angle to the supporting valley rafter. Its length is based on the run of the narrower roof section.

375

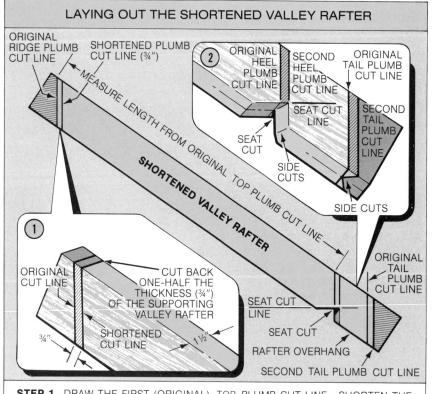

## LAYING OUT THE SHORTENED VALLEY RAFTER

**STEP 1.** DRAW THE FIRST (ORIGINAL) TOP PLUMB CUT LINE. SHORTEN THE VALLEY RAFTER BY ONE-HALF THE THICKNESS (¾″) OF THE SUPPORTING VALLEY RAFTER.

**STEP 2.** MEASURE THE RAFTER LENGTH FROM THE ORIGINAL RIDGE PLUMB CUT LINE. LAY OUT THE SEAT CUT AND OVERHANG.

Figure 49–9. Laying out a shortened valley rafter. It has a square cut where it butts against the supporting valley rafter.

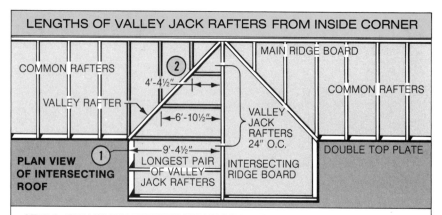

## LENGTHS OF VALLEY JACK RAFTERS FROM INSIDE CORNER

**STEP 1.** THE LENGTH OF THE FIRST PAIR OF VALLEY JACK RAFTERS (9′-4½″) IS THE SAME AS THE LENGTH OF THE COMMON RAFTERS. TO FIND THE LENGTH OF THE SECOND PAIR OF VALLEY JACK RAFTERS, SUBTRACT THE COMMON LENGTH DIFFERENCE 2′-6″ (30″) (FOUND ON LINE FOUR OF THE STEEL SQUARE RAFTER TABLES) FROM 9′-4½″.

> 9′-4½″ LENGTH OF FIRST PAIR OF VALLEY JACK RAFTERS
> − 2′-6″ COMMON LENGTH DIFFERENCE
> 6′-10½″ LENGTH OF SECOND PAIR OF VALLEY JACK RAFTERS

**STEP 2.** TO FIND THE LENGTH OF THE REMAINING VALLEY JACK RAFTERS, CONTINUE TO SUBTRACT THE COMMON LENGTH DIFFERENCE FROM EACH SUCCEEDING PAIR.

Figure 49-10. Calculating lengths of valley jack rafters when the longest jack begins at the inside corner of the building. In this example, the roof has a 9″ unit rise. The rafter spacing is 24″ O.C.

top of the roof. They have a *common length difference* as long as they are placed the same distance apart. The common length differences are the same as those for hip jack rafters. The third and fourth lines of a steel square rafter table give these common length differences.

The best procedure for calculating the lengths of valley jack rafters depends on how the rafters are positioned on the roof. Figure 49–10 shows the procedure to use when the spacing begins from the inside corner of the building. Figure 49–11 shows the procedure to use when spacing begins from a common rafter positioned away from the inside corner. Figure 49–12 shows the procedure to use when spacing begins from the center point of the intersecting ridges.

## Laying out Valley Jack Rafters

The angles for all the cuts on the valley jack rafters are the same as those for the hip jack rafters. For the plumb cut, use the unit rise measurement and the 12″ measurement on the steel square, and mark on the unit rise side. For the side cut, use the number found on the fifth line of the rafter table and the number 12. Mark on the side of the 12.

Valley jack rafters require a square cut where they are nailed against the ridge, and a side cut where they meet the valley rafter. The procedure for laying out a valley jack rafter is shown in Figure 49–13.

## HIP-VALLEY CRIPPLE JACK RAFTERS

When the hip and valley rafters are placed close together, the space between them is framed

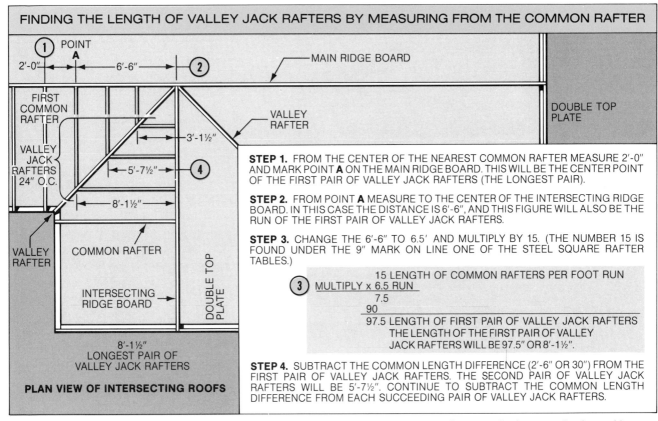

Figure 49–11. Calculating lengths of valley jack rafters when the first jack falls past the inside corner. In this example, the roof has a 9″ unit rise. The rafter spacing is 24″ O.C.

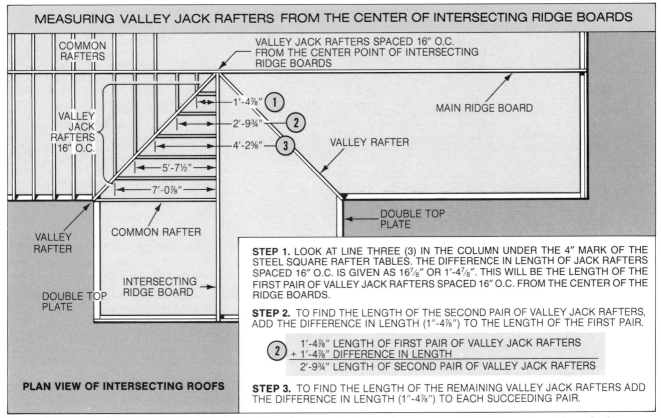

Figure 49–12. Calculating lengths of valley jack rafters when spacing begins from center point of intersecting ridges. In this example, the roof has a 4″ unit rise. The rafter spacing is 16″ O.C.

# LAYOUT OF A TYPICAL VALLEY JACK RAFTER

**STEP 1.** SQUARE A LINE ACROSS THE TOP EDGE OF THE RAFTER. DRAW THE **FIRST** RIDGE PLUMB CUT LINE.

**STEP 2.** SHORTEN THE VALLEY JACK RAFTER BY ONE-HALF THE THICKNESS OF THE RIDGE BOARD. DRAW THE **SECOND** RIDGE PLUMB CUT LINE.

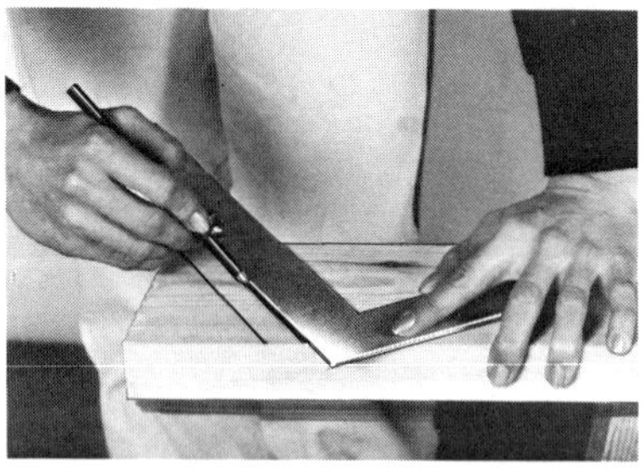

**STEP 3.** MEASURE THE LENGTH OF THE RAFTER FROM THE **FIRST** PLUMB CUT LINE. SQUARE A LINE ACROSS THE TOP EDGE OF THE RAFTER AND MARK THE CENTER POINT. MARK THE SIDE CUT ANGLE LINES.

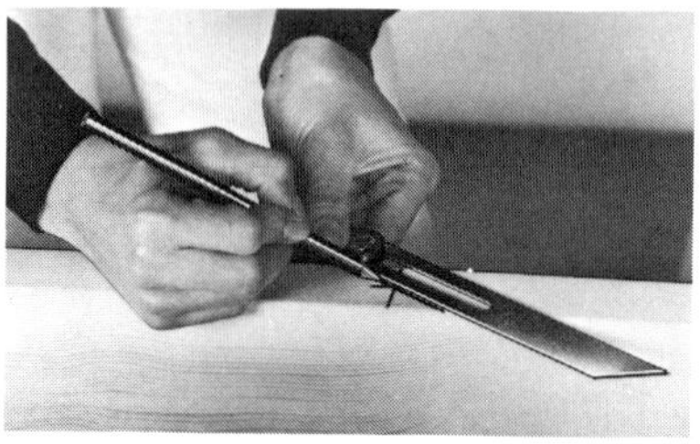

**STEP 4.** DRAW A PLUMB CUT LINE FROM THE SIDE CUT ANGLE LINES.

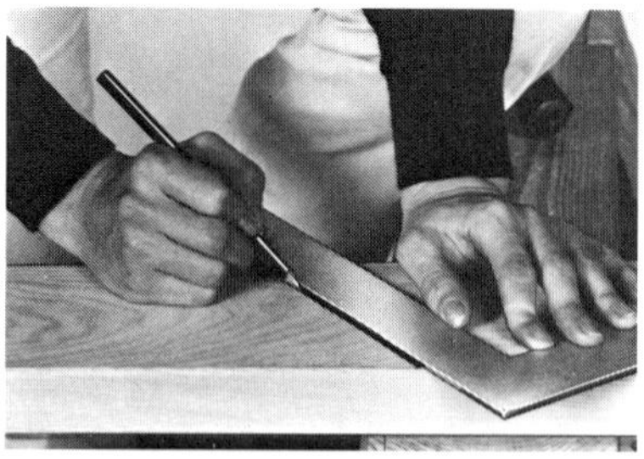

**STEP 5.** SHORTEN THE VALLEY JACK RAFTER BY ONE-HALF THE 45° THICKNESS OF THE VALLEY RAFTER.

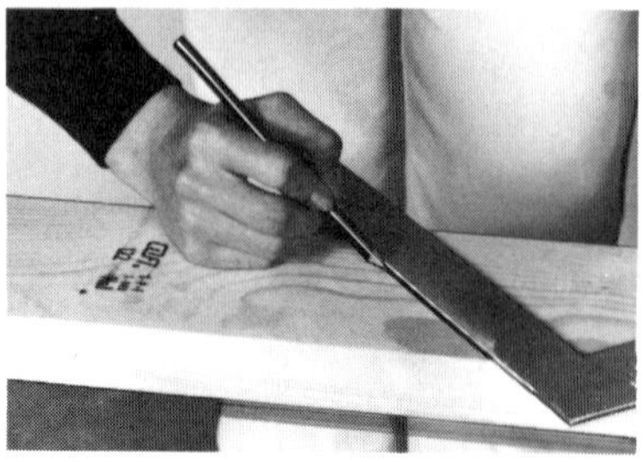

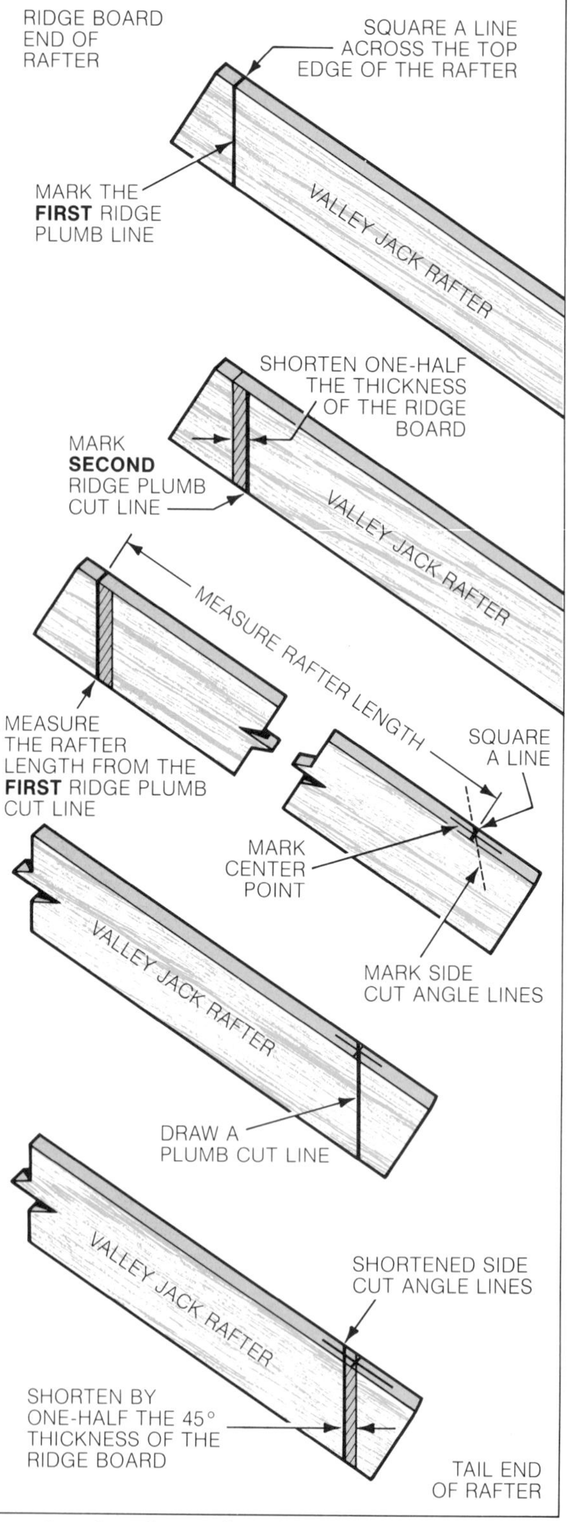

RIDGE BOARD END OF RAFTER

SQUARE A LINE ACROSS THE TOP EDGE OF THE RAFTER

MARK THE **FIRST** RIDGE PLUMB LINE

VALLEY JACK RAFTER

SHORTEN ONE-HALF THE THICKNESS OF THE RIDGE BOARD

MARK **SECOND** RIDGE PLUMB CUT LINE

VALLEY JACK RAFTER

MEASURE RAFTER LENGTH

MEASURE THE RAFTER LENGTH FROM THE **FIRST** RIDGE PLUMB CUT LINE

SQUARE A LINE

MARK CENTER POINT

MARK SIDE CUT ANGLE LINES

VALLEY JACK RAFTER

DRAW A PLUMB CUT LINE

VALLEY JACK RAFTER

SHORTENED SIDE CUT ANGLE LINES

SHORTEN BY ONE-HALF THE 45° THICKNESS OF THE RIDGE BOARD

TAIL END OF RAFTER

Figure 49–13. Layout of a valley jack rafter. It has a square cut where it fits against the ridge, and a side cut where it meets the valley rafter.

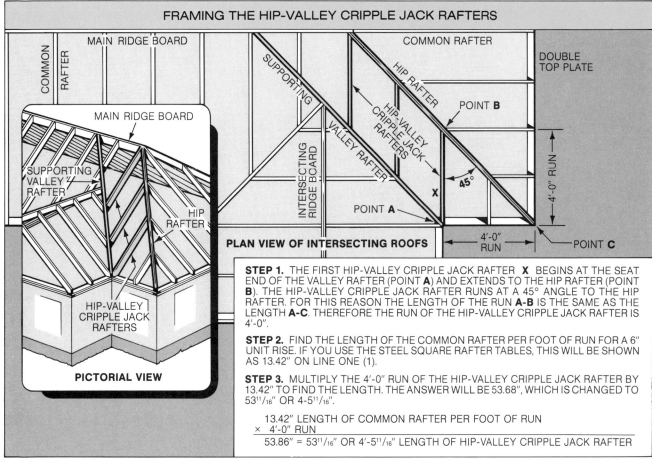

## FRAMING THE HIP-VALLEY CRIPPLE JACK RAFTERS

**PLAN VIEW OF INTERSECTING ROOFS**

**PICTORIAL VIEW**

**STEP 1.** THE FIRST HIP-VALLEY CRIPPLE JACK RAFTER **X** BEGINS AT THE SEAT END OF THE VALLEY RAFTER (POINT **A**) AND EXTENDS TO THE HIP RAFTER (POINT **B**). THE HIP-VALLEY CRIPPLE JACK RAFTER RUNS AT A 45° ANGLE TO THE HIP RAFTER. FOR THIS REASON THE LENGTH OF THE RUN **A-B** IS THE SAME AS THE LENGTH **A-C**. THEREFORE THE RUN OF THE HIP-VALLEY CRIPPLE JACK RAFTER IS 4'-0".

**STEP 2.** FIND THE LENGTH OF THE COMMON RAFTER PER FOOT OF RUN FOR A 6" UNIT RISE. IF YOU USE THE STEEL SQUARE RAFTER TABLES, THIS WILL BE SHOWN AS 13.42" ON LINE ONE (1).

**STEP 3.** MULTIPLY THE 4'-0" RUN OF THE HIP-VALLEY CRIPPLE JACK RAFTER BY 13.42" TO FIND THE LENGTH. THE ANSWER WILL BE 53.68", WHICH IS CHANGED TO 53$^{11}$/$_{16}$" OR 4-5$^{11}$/$_{16}$".

13.42" LENGTH OF COMMON RAFTER PER FOOT OF RUN
× 4'-0" RUN
53.86" = 53$^{11}$/$_{16}$" OR 4'-5$^{11}$/$_{16}$" LENGTH OF HIP-VALLEY CRIPPLE JACK RAFTER

Figure 49-14. The steel square rafter tables may be used to calculate the length of hip-valley cripple jack rafters. In this example, the roof has a 6" unit rise. Additional information required is the distance from the end of the main section to the intersecting roof section. This distance, shown as points A and C on the drawing, is 4'-0".

with hip-valley cripple jack rafters. All these rafters are the same size. See Figure 49–14. (Also refer to Figures 49–2 and 49–3.)

## Calculating Length of Hip-Valley Cripple Jack Rafters

The steel square rafter table can be used to find the lengths of hip-valley cripple jack rafters. First, find the distance from the end of the main roof section to the intersecting roof section. Multiply the number of feet in this distance by the number on the common rafter line under the unit rise number at the top of the square. This procedure is shown in Figure 49–14.

A book of rafter tables (refer to Unit 47) can also be used to find the lengths of hip-valley cripple jack rafters.

## Laying out Hip-Valley Cripple Jack Rafters

Because the hip-valley cripple jack rafter fits between the hip and the valley rafter, it requires a plumb cut and side cut at each end. A layout procedure is shown in Figure 49–15.

## VALLEY CRIPPLE JACK RAFTERS

Valley cripple jack rafters are used only on intersecting roofs

with unequal spans. They are placed between the shortened and the supporting valley rafters. Their purpose is to bridge the space in the main roof section between the supporting and shortened valley rafters. (Refer again to Figure 49–3.)

## Calculating Length of Valley Cripple Jack Rafters

The run of a valley cripple jack rafter is always twice the run of the valley jack rafter that it meets at the shortened valley rafter. For this reason, the length of a valley cripple jack rafter is also twice the length of that valley jack rafter. See Figure 49–16.

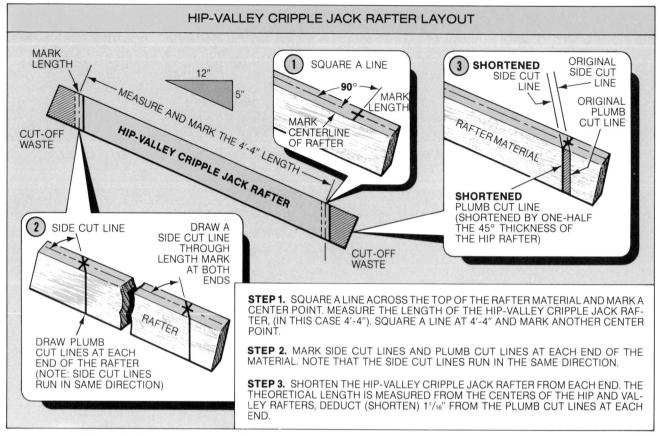

## HIP-VALLEY CRIPPLE JACK RAFTER LAYOUT

MARK LENGTH

MEASURE AND MARK THE 4'-4" LENGTH

12"

5"

HIP-VALLEY CRIPPLE JACK RAFTER

CUT-OFF WASTE

**1** SQUARE A LINE

90°

MARK LENGTH

MARK CENTERLINE OF RAFTER

**3 SHORTENED** SIDE CUT LINE

ORIGINAL SIDE CUT LINE

ORIGINAL PLUMB CUT LINE

RAFTER MATERIAL

**SHORTENED** PLUMB CUT LINE (SHORTENED BY ONE-HALF THE 45° THICKNESS OF THE HIP RAFTER)

**2** SIDE CUT LINE

DRAW A SIDE CUT LINE THROUGH LENGTH MARK AT BOTH ENDS

RAFTER

DRAW PLUMB CUT LINES AT EACH END OF THE RAFTER (NOTE: SIDE CUT LINES RUN IN SAME DIRECTION)

CUT-OFF WASTE

**STEP 1.** SQUARE A LINE ACROSS THE TOP OF THE RAFTER MATERIAL AND MARK A CENTER POINT. MEASURE THE LENGTH OF THE HIP-VALLEY CRIPPLE JACK RAFTER, (IN THIS CASE 4'-4"). SQUARE A LINE AT 4'-4" AND MARK ANOTHER CENTER POINT.

**STEP 2.** MARK SIDE CUT LINES AND PLUMB CUT LINES AT EACH END OF THE MATERIAL. NOTE THAT THE SIDE CUT LINES RUN IN THE SAME DIRECTION.

**STEP 3.** SHORTEN THE HIP-VALLEY CRIPPLE JACK RAFTER FROM EACH END. THE THEORETICAL LENGTH IS MEASURED FROM THE CENTERS OF THE HIP AND VALLEY RAFTERS, DEDUCT (SHORTEN) 1¹/₁₆" FROM THE PLUMB CUT LINES AT EACH END.

Figure 49–15. Layout of a hip-valley cripple jack rafter. It has a plumb cut and side cut at each end.

## Laying out Valley Cripple Jack Rafters

The angles for the plumb cuts and side cuts on valley cripple jack rafters are found by using the same steel square method described for laying out other types of jack rafters. A procedure for laying out the valley cripple jack rafter is shown in Figure 49–17.

## CONSTRUCTING AN INTERSECTING ROOF

Construction of an intersecting roof usually begins with setting up the ridge of the main roof. This ridge is supported by a pair of common or hip rafters at each end, as described in earlier units for gable and hip roofs (Units 47 and 48).

## Locating Point of Intersection

After the main ridge has been set in place, the ridge of the intersecting portion of the roof can be erected. First, however, the correct point of intersection must be marked on the main ridge. Ways to locate the point of intersection for three different types of intersecting roofs are shown in Figures 49–18, 49–19, and 49–20.

## Calculating Intersecting Ridge Lengths

Whenever possible, intersecting ridges should be cut to their exact lengths before they are set in place. Ways to calculate ridge lengths for three different types of intersecting roofs are shown in Figures 49–21, 49–22, and 49–23.

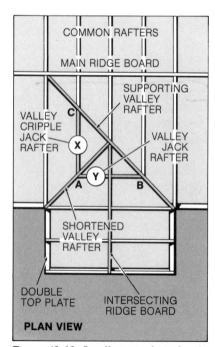

COMMON RAFTERS

MAIN RIDGE BOARD

SUPPORTING VALLEY RAFTER

VALLEY CRIPPLE JACK RAFTER

C

X

VALLEY JACK RAFTER

A Y B

SHORTENED VALLEY RAFTER

DOUBLE TOP PLATE

INTERSECTING RIDGE BOARD

**PLAN VIEW**

Figure 49–16. A valley cripple jack rafter (X) is always twice the length of the valley jack rafter (Y). Note that the valley cripple jack rafter and the valley jack rafter meet at the same point (A) on the shortened valley rafter.

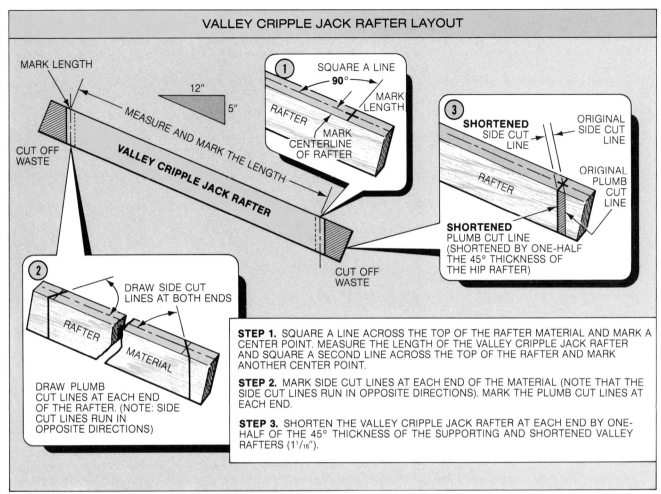

## VALLEY CRIPPLE JACK RAFTER LAYOUT

**STEP 1.** SQUARE A LINE ACROSS THE TOP OF THE RAFTER MATERIAL AND MARK A CENTER POINT. MEASURE THE LENGTH OF THE VALLEY CRIPPLE JACK RAFTER AND SQUARE A SECOND LINE ACROSS THE TOP OF THE RAFTER AND MARK ANOTHER CENTER POINT.

**STEP 2.** MARK SIDE CUT LINES AT EACH END OF THE MATERIAL (NOTE THAT THE SIDE CUT LINES RUN IN OPPOSITE DIRECTIONS). MARK THE PLUMB CUT LINES AT EACH END.

**STEP 3.** SHORTEN THE VALLEY CRIPPLE JACK RAFTER AT EACH END BY ONE-HALF OF THE 45° THICKNESS OF THE SUPPORTING AND SHORTENED VALLEY RAFTERS (1 1/16″).

Figure 49–17. Layout of a valley cripple jack rafter. Its run is twice the run of the valley jack rafter that it meets at the shortened valley rafter.

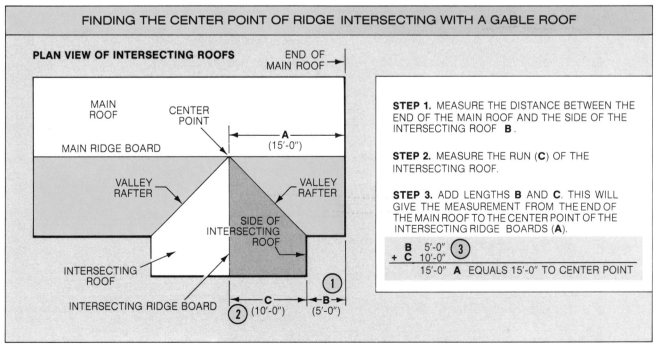

## FINDING THE CENTER POINT OF RIDGE INTERSECTING WITH A GABLE ROOF

**PLAN VIEW OF INTERSECTING ROOFS**

**STEP 1.** MEASURE THE DISTANCE BETWEEN THE END OF THE MAIN ROOF AND THE SIDE OF THE INTERSECTING ROOF **B**.

**STEP 2.** MEASURE THE RUN (**C**) OF THE INTERSECTING ROOF.

**STEP 3.** ADD LENGTHS **B** AND **C**. THIS WILL GIVE THE MEASUREMENT FROM THE END OF THE MAIN ROOF TO THE CENTER POINT OF THE INTERSECTING RIDGE BOARDS (**A**).

```
  B    5'-0"     ③
+ C   10'-0"
      15'-0"   A  EQUALS 15'-0" TO CENTER POINT
```

Figure 49–18. Locating the center point of a ridge intersecting a gable roof.

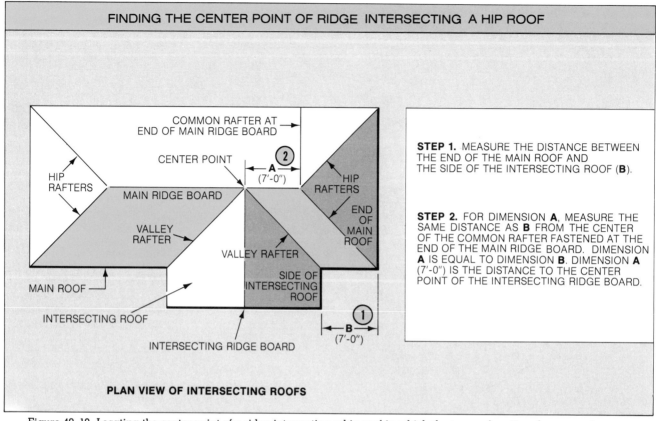

Figure 49–19. Locating the center point of a ridge intersecting a hip roof in which the two roof sections have equal spans.

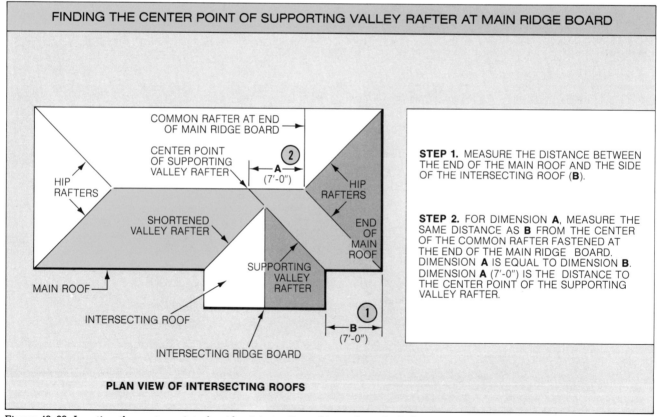

Figure 49–20. Locating the center point of a ridge intersecting a hip roof in which the two roof sections have unequal spans. The intersecting ridge fits into the corner formed by the supporting and shortened valley rafters.

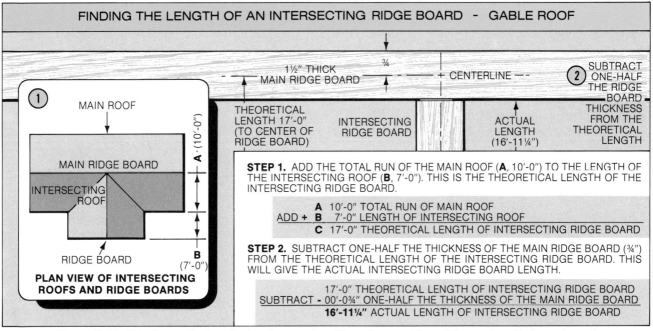

### FINDING THE LENGTH OF AN INTERSECTING RIDGE BOARD - GABLE ROOF

**STEP 1.** ADD THE TOTAL RUN OF THE MAIN ROOF (**A**, 10'-0") TO THE LENGTH OF THE INTERSECTING ROOF (**B**, 7'-0"). THIS IS THE THEORETICAL LENGTH OF THE INTERSECTING RIDGE BOARD.

| | | |
|---|---|---|
| | **A** | 10'-0" TOTAL RUN OF MAIN ROOF |
| ADD + | **B** | 7'-0" LENGTH OF INTERSECTING ROOF |
| | **C** | 17'-0" THEORETICAL LENGTH OF INTERSECTING RIDGE BOARD |

**STEP 2.** SUBTRACT ONE-HALF THE THICKNESS OF THE MAIN RIDGE BOARD (¾") FROM THE THEORETICAL LENGTH OF THE INTERSECTING RIDGE BOARD. THIS WILL GIVE THE ACTUAL INTERSECTING RIDGE BOARD LENGTH.

| | |
|---|---|
| | 17'-0" THEORETICAL LENGTH OF INTERSECTING RIDGE BOARD |
| SUBTRACT - | 00'-0¾" ONE-HALF THE THICKNESS OF THE MAIN RIDGE BOARD |
| | **16'-11¼"** ACTUAL LENGTH OF INTERSECTING RIDGE BOARD |

Figure 49–21. Calculating the length of an intersecting ridge when a gable roof intersects with the main roof.

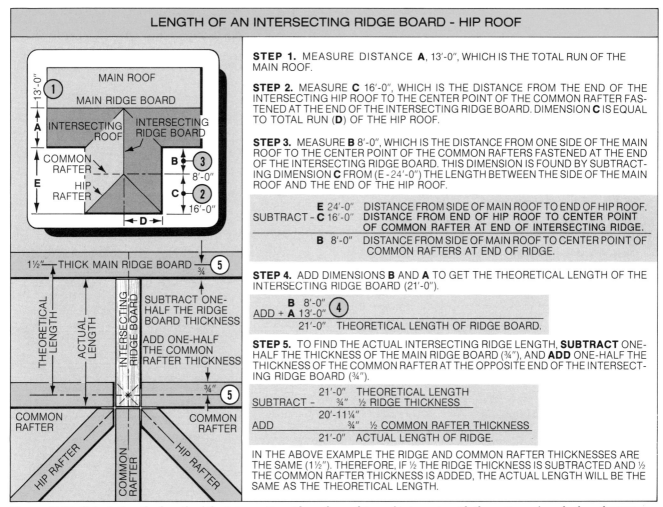

### LENGTH OF AN INTERSECTING RIDGE BOARD - HIP ROOF

**STEP 1.** MEASURE DISTANCE **A**, 13'-0", WHICH IS THE TOTAL RUN OF THE MAIN ROOF.

**STEP 2.** MEASURE **C** 16'-0", WHICH IS THE DISTANCE FROM THE END OF THE INTERSECTING HIP ROOF TO THE CENTER POINT OF THE COMMON RAFTER FASTENED AT THE END OF THE INTERSECTING RIDGE BOARD. DIMENSION **C** IS EQUAL TO TOTAL RUN (**D**) OF THE HIP ROOF.

**STEP 3.** MEASURE **B** 8'-0", WHICH IS THE DISTANCE FROM ONE SIDE OF THE MAIN ROOF TO THE CENTER POINT OF THE COMMON RAFTERS FASTENED AT THE END OF THE INTERSECTING RIDGE BOARD. THIS DIMENSION IS FOUND BY SUBTRACTING DIMENSION **C** FROM (E - 24'-0") THE LENGTH BETWEEN THE SIDE OF THE MAIN ROOF AND THE END OF THE HIP ROOF.

| | | |
|---|---|---|
| | **E** 24'-0" | DISTANCE FROM SIDE OF MAIN ROOF TO END OF HIP ROOF. |
| SUBTRACT - | **C** 16'-0" | DISTANCE FROM END OF HIP ROOF TO CENTER POINT OF COMMON RAFTER AT END OF INTERSECTING RIDGE. |
| | **B** 8'-0" | DISTANCE FROM SIDE OF MAIN ROOF TO CENTER POINT OF COMMON RAFTERS AT END OF RIDGE. |

**STEP 4.** ADD DIMENSIONS **B** AND **A** TO GET THE THEORETICAL LENGTH OF THE INTERSECTING RIDGE BOARD (21'-0").

| | |
|---|---|
| | **B** 8'-0" |
| ADD + | **A** 13'-0" |
| | 21'-0" THEORETICAL LENGTH OF RIDGE BOARD. |

**STEP 5.** TO FIND THE ACTUAL INTERSECTING RIDGE LENGTH, **SUBTRACT** ONE-HALF THE THICKNESS OF THE MAIN RIDGE BOARD (¾"), AND **ADD** ONE-HALF THE THICKNESS OF THE COMMON RAFTER AT THE OPPOSITE END OF THE INTERSECTING RIDGE BOARD (¾").

| | |
|---|---|
| | 21'-0" THEORETICAL LENGTH |
| SUBTRACT - | ¾" ½ RIDGE THICKNESS |
| | 20'-11¼" |
| ADD | ¾" ½ COMMON RAFTER THICKNESS |
| | 21'-0" ACTUAL LENGTH OF RIDGE. |

IN THE ABOVE EXAMPLE THE RIDGE AND COMMON RAFTER THICKNESSES ARE THE SAME (1½"). THEREFORE, IF ½ THE RIDGE THICKNESS IS SUBTRACTED AND ½ THE COMMON RAFTER THICKNESS IS ADDED, THE ACTUAL LENGTH WILL BE THE SAME AS THE THEORETICAL LENGTH.

Figure 49–22. Calculating the length of the intersecting ridge when a hip roof intersects with the main roof, and when the two sections have equal spans.

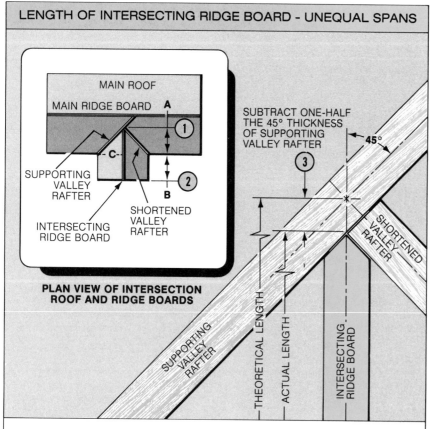

## LENGTH OF INTERSECTING RIDGE BOARD - UNEQUAL SPANS

**STEP 1.** MEASURE DISTANCE **A**, WHICH IS THE SAME TOTAL AS RUN **C** OF THE INTERSECTING ROOF.

**STEP 2.** ADD DIMENSION **A** TO DIMENSION **B**, WHICH IS THE LENGTH OF THE INTERSECTING ROOF. THIS WILL GIVE THE THEORETICAL LENGTH OF THE INTERSECTING RIDGE BOARD.

**STEP 3.** TO FIND THE ACTUAL LENGTH OF THE INTERSECTING RIDGE BOARD, SUBTRACT ONE-HALF THE 45° THICKNESS OF THE VALLEY RAFTERS.

Figure 49-23. Calculating the length of an intersecting ridge when a gable roof intersects with the main roof, and when the two sections have unequal spans.

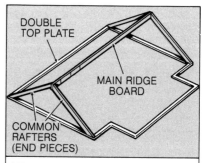

**STEP 1.** INSTALL THE MAIN RIDGE BOARD AND THE FOUR SUPPORTING END COMMON RAFTERS.

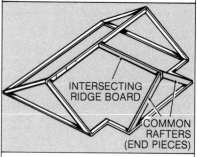

**STEP 2.** INSTALL THE INTERSECTING RIDGE BOARD. NAIL THE INTERSECTING END TO THE MAIN RIDGE BOARD AND PLACE TWO COMMON RAFTERS AT THE OTHER END.

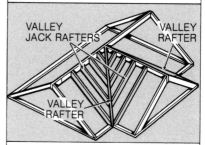

**STEP 3.** INSTALL THE TWO VALLEY RAFTERS RUNNING FROM THE MAIN RIDGE BOARD TO THE INSIDE BUILDING CORNER.

**STEP 4.** INSTALL THE VALLEY JACK RAFTERS.

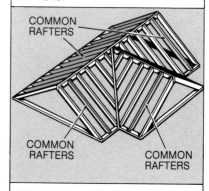

**STEP 5.** INSTALL ALL REMAINING COMMON RAFTERS.

Figure 49-24. Framing an intersecting roof with equal spans. In this example, both sections of the intersecting roof are gable roofs.

## Framing Intersecting Roofs with Equal Spans

A procedure for framing an intersecting roof with equal spans is shown in Figure 49-24. In this example both sections of the intersecting roof are gable roofs.

## Framing Intersecting Roofs with Unequal Spans

The framing procedure for an intersecting roof with unequal spans differs somewhat from the procedure for a roof with equal

spans. In a roof with unequal spans, the ridge of the smaller roof section is lower than the main ridge. It is fastened to the intersecting point of the shortened valley rafter and the supporting valley rafter. One method for framing this type of roof is shown in Figure 49-25.

## Blind Valley Construction of Intersecting Roof

Blind valley construction is a method of building intersecting roofs without valley rafters. See Figures 49-26 and 49-27. The

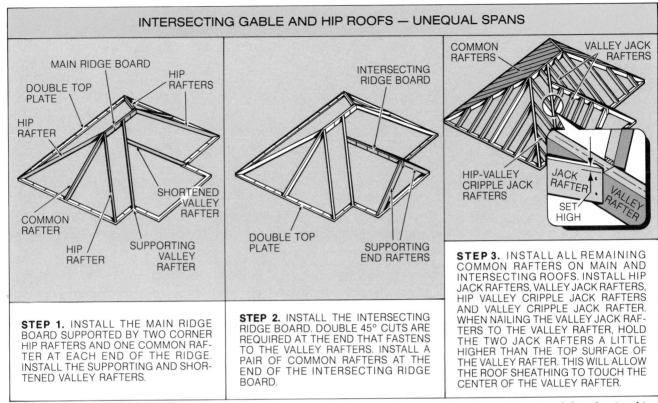

## INTERSECTING GABLE AND HIP ROOFS — UNEQUAL SPANS

MAIN RIDGE BOARD
HIP RAFTERS
DOUBLE TOP PLATE
HIP RAFTER
COMMON RAFTER
HIP RAFTER
SUPPORTING VALLEY RAFTER
SHORTENED VALLEY RAFTER

INTERSECTING RIDGE BOARD
DOUBLE TOP PLATE
SUPPORTING END RAFTERS

COMMON RAFTERS
VALLEY JACK RAFTERS
HIP-VALLEY CRIPPLE JACK RAFTERS
JACK RAFTER
VALLEY RAFTER
SET HIGH

**STEP 1.** INSTALL THE MAIN RIDGE BOARD SUPPORTED BY TWO CORNER HIP RAFTERS AND ONE COMMON RAFTER AT EACH END OF THE RIDGE. INSTALL THE SUPPORTING AND SHORTENED VALLEY RAFTERS.

**STEP 2.** INSTALL THE INTERSECTING RIDGE BOARD. DOUBLE 45° CUTS ARE REQUIRED AT THE END THAT FASTENS TO THE VALLEY RAFTERS. INSTALL A PAIR OF COMMON RAFTERS AT THE END OF THE INTERSECTING RIDGE BOARD.

**STEP 3.** INSTALL ALL REMAINING COMMON RAFTERS ON MAIN AND INTERSECTING ROOFS. INSTALL HIP JACK RAFTERS, VALLEY JACK RAFTERS, HIP VALLEY CRIPPLE JACK RAFTERS AND VALLEY CRIPPLE JACK RAFTER. WHEN NAILING THE VALLEY JACK RAFTERS TO THE VALLEY RAFTER, HOLD THE TWO JACK RAFTERS A LITTLE HIGHER THAN THE TOP SURFACE OF THE VALLEY RAFTER. THIS WILL ALLOW THE ROOF SHEATHING TO TOUCH THE CENTER OF THE VALLEY RAFTER.

Figure 49–25. Framing an intersecting roof with unequal spans. In this example, one section is a gable roof and the other is a hip roof.

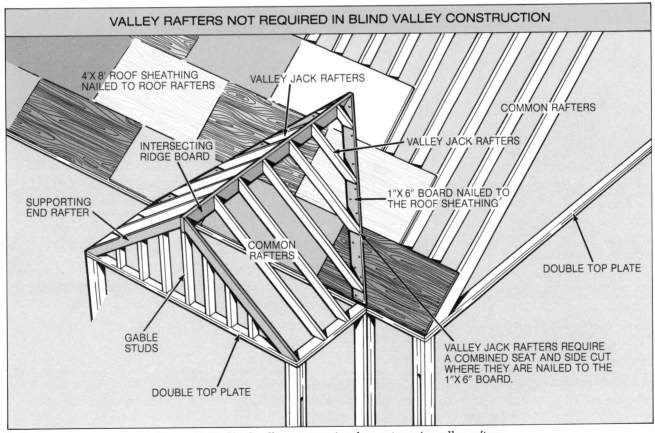

## VALLEY RAFTERS NOT REQUIRED IN BLIND VALLEY CONSTRUCTION

4'X 8' ROOF SHEATHING NAILED TO ROOF RAFTERS
VALLEY JACK RAFTERS
COMMON RAFTERS
INTERSECTING RIDGE BOARD
VALLEY JACK RAFTERS
1"X 6" BOARD NAILED TO THE ROOF SHEATHING
SUPPORTING END RAFTER
COMMON RAFTERS
GABLE STUDS
DOUBLE TOP PLATE
DOUBLE TOP PLATE
VALLEY JACK RAFTERS REQUIRE A COMBINED SEAT AND SIDE CUT WHERE THEY ARE NAILED TO THE 1"X 6" BOARD.

Figure 49–26. Blind valley construction does not require valley rafters.

Figure 49–27. Blind valley construction on a roof with spaced sheathing.

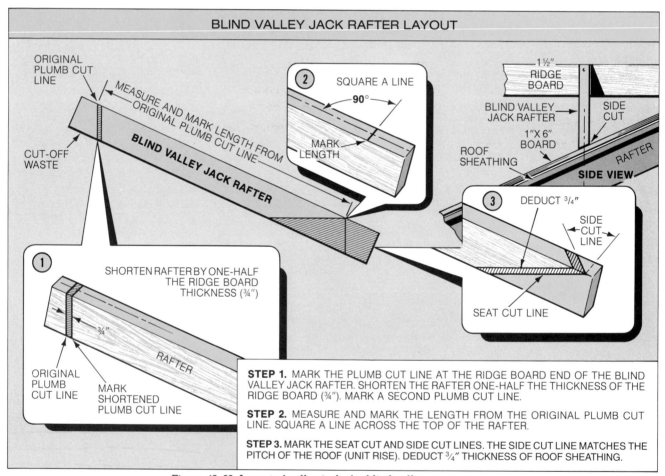

**BLIND VALLEY JACK RAFTER LAYOUT**

ORIGINAL PLUMB CUT LINE

MEASURE AND MARK LENGTH FROM ORIGINAL PLUMB CUT LINE

CUT-OFF WASTE

BLIND VALLEY JACK RAFTER

② SQUARE A LINE

90°

MARK LENGTH

1½"
RIDGE BOARD

BLIND VALLEY JACK RAFTER

SIDE CUT

1"X 6" BOARD

ROOF SHEATHING

RAFTER

**SIDE VIEW**

③ DEDUCT ¾"

SIDE CUT LINE

SEAT CUT LINE

① SHORTEN RAFTER BY ONE-HALF THE RIDGE BOARD THICKNESS (¾")

¾"

RAFTER

ORIGINAL PLUMB CUT LINE

MARK SHORTENED PLUMB CUT LINE

**STEP 1.** MARK THE PLUMB CUT LINE AT THE RIDGE BOARD END OF THE BLIND VALLEY JACK RAFTER. SHORTEN THE RAFTER ONE-HALF THE THICKNESS OF THE RIDGE BOARD (¾"). MARK A SECOND PLUMB CUT LINE.

**STEP 2.** MEASURE AND MARK THE LENGTH FROM THE ORIGINAL PLUMB CUT LINE. SQUARE A LINE ACROSS THE TOP OF THE RAFTER.

**STEP 3.** MARK THE SEAT CUT AND SIDE CUT LINES. THE SIDE CUT LINE MATCHES THE PITCH OF THE ROOF (UNIT RISE). DEDUCT ¾" THICKNESS OF ROOF SHEATHING.

Figure 49–28. Layout of valley jacks for blind valley construction.

main roof is sheathed, and the intersecting section is built on top of the sheathing.

Boards (1" × 6") are fastened to the top of the sheathing as a base for nailing the valley jacks.

The roof section consists of common rafters and valley jack rafters. The length of the longest set of jacks is found by subtracting the common length difference from the common rafter. The val-

ley jacks require a seat cut combined with a side cut where they fasten to the 1" × 6" board. The layout for the valley jack cuts is shown in Figure 49–28.

386

# UNIT 50

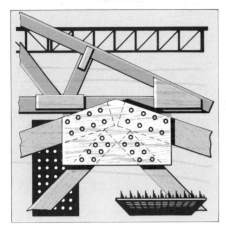

# Trussed Roof

A roof truss, also known as a *trussed rafter,* is a framed unit built to the shape of a roof. It is assembled before it is lifted and set into place on top of the outside walls of a building. See Figures 50–1 and 50–2. The earliest wood truss systems were developed hundreds of years ago in medieval Europe.

Until recently roof trusses were primarily used to cover wide spans in industrial buildings such as warehouses and factories. See Figure 50–3. Now, however, they are in widespread use in residential and other light construction. More than 60% of all new light frame construction is built with some type of truss system. This development is due mainly to the growing use of mass-production framing methods, particularly in tract house construction.

Under certain conditions a truss system offers many advantages over rafter framing. Truss systems require less labor and material. They are usually designed to cover a span of 20′ to 32′, or even more, without requiring any in-between support by wall or girder. See Figure 50–4. This allows greater choice in planning the locations of interior partitions.

Trusses can be built on the job site. However, when large quantities of trusses are required, such as for housing tracts, it is more economical to order them from a truss manufacturer.

*Gang-Nail Systems, Inc.*

Figure 50–1. Roof trusses placed over wood-framed walls. A floor truss system is used at the second floor level. The ground floor features concrete slab-at-grade construction.

*Gang-Nail Systems, Inc.*

Figure 50–2. A trussed roof over masonry walls.

Southern Pine Council

Figure 50-3. Trusses constructed of southern pine lumber are set in place by crane. They can span long distances over warehouses, factories, assembly halls, and other types of commercial construction.

Trusses can be used to greatest advantage on rectangular buildings with simple roof designs. However, truss manufacturers also offer trusses for more complicated roof designs, such as hip roofs and intersecting roofs. See Figures 50–5 and 50–6.

Since prefabricated trusses must be transported from the manufacturer to the job site, the manufacturer must be located within a reasonable distance from the job site for a prefabricated trussed roof to be practical.

## COMPONENTS OF A ROOF TRUSS

The basic components of a roof truss are the top and bottom chords and the web members. See Figure 50–7. The top chords serve as roof rafters. The bottom chords act as ceiling joists. The web members run between the top and bottom chords. The truss parts are usually made of 2″ × 4″ or 2″ × 6″ material and are tied together with metal or plywood gusset plates.

## TYPES OF ROOF TRUSSES

Roof trusses come in a variety of shapes. See Figure 50–8. The

ones most commonly used for residential and other light framing are the *king-post* truss, the *W-type* (or *fink* truss), and the *scissors* truss.

### King-post Truss

The simplest type of truss used in frame construction is the king-post truss. It consists of top and bottom chords and a vertical post at the center. Used mainly for

garage roofs or small homes, king-post trusses can be placed over spans up to 22′.

## W-type (Fink Truss)

The most widely used truss in light frame construction is the W-type (fink) truss. It consists of top and bottom chords tied together with web members. The W-type truss provides uniform load-carrying capacity and can be placed over spans up to 50′.

## Scissors Truss

The scissors truss is used for buildings with sloping ceilings. Many residential, church, and commercial buildings require this type of truss. Generally the slope of the bottom chord of a scissors truss equals one-half the slope of the top chord. Scissors trusses can be placed over spans up to 50′.

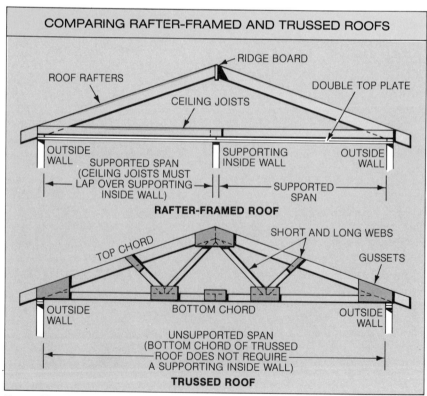

Figure 50–4. Both of these roofs have the same span. The ceiling joists must lap over an interior partition in the rafter-framed roof. The bottom chord of the trussed roof requires no central support.

## PRE-FABRICATED TRUSSES USED IN CONSTRUCTION OF INTERSECTING ROOF

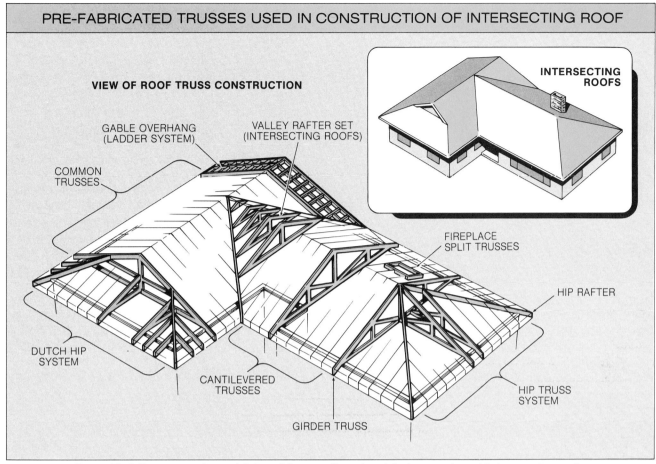

VIEW OF ROOF TRUSS CONSTRUCTION

INTERSECTING ROOFS

GABLE OVERHANG (LADDER SYSTEM)

VALLEY RAFTER SET (INTERSECTING ROOFS)

COMMON TRUSSES

FIREPLACE SPLIT TRUSSES

HIP RAFTER

DUTCH HIP SYSTEM

CANTILEVERED TRUSSES

GIRDER TRUSS

HIP TRUSS SYSTEM

Figure 50–5. Trusses can be prefabricated for complicated roof designs, such as this intersecting roof.

## PRINCIPLES OF ROOF TRUSS DESIGN

A roof truss is an engineered structural frame resting on two outside walls of a building. The load carried by the truss is transferred to these outside walls.

### Weight and Stress

Truss design must accommodate the weight of the materials used for the truss framework, sheathing, finish roofing materials, and the finish ceiling materials. In addition, local snow and wind conditions must be considered. The flatter the slope of a truss, for example, the heavier the snow load it may have to support. In this event, the truss should be constructed of heavier 2″ × 6″ material rather than 2″ × 4″ material.

### Tension and Compression

Each part of a truss is in a state of either tension or compression. The parts in a state of tension are subjected to a *pulling-apart* force. The parts in a state of compression are subjected to a *pushing-together* force. The balance of tension and compression gives the truss its ability to carry heavy loads and to cover wide spans. See Figure 50–9.

Web members must be fastened at certain points along the top and bottom chords in order to handle the stress and weight placed upon the truss. A typical layout for a W-type (fink) truss is shown in Figure 50–10.

### Truss Connectors

A properly constructed truss has tight-fitting joints. The ends of the truss members are usually tied together with some kind of metal device. Often a simple prepunched nailing plate is used.

*Gang-Nail Systems, Inc.*
Figure 50–6. Valley jack trusses are used on this intersecting roof.

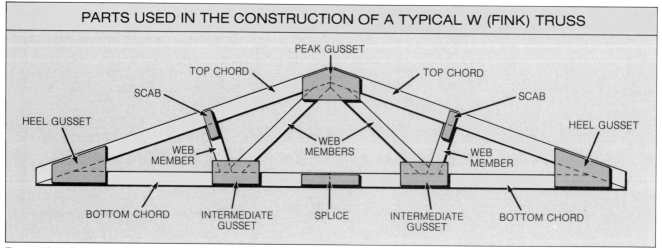

**PARTS USED IN THE CONSTRUCTION OF A TYPICAL W (FINK) TRUSS**

Figure 50–7. A truss is made up of top chords, bottom chords, and web members tied to the chords with gusset plates. Gussets shown in this drawing are made of plywood.

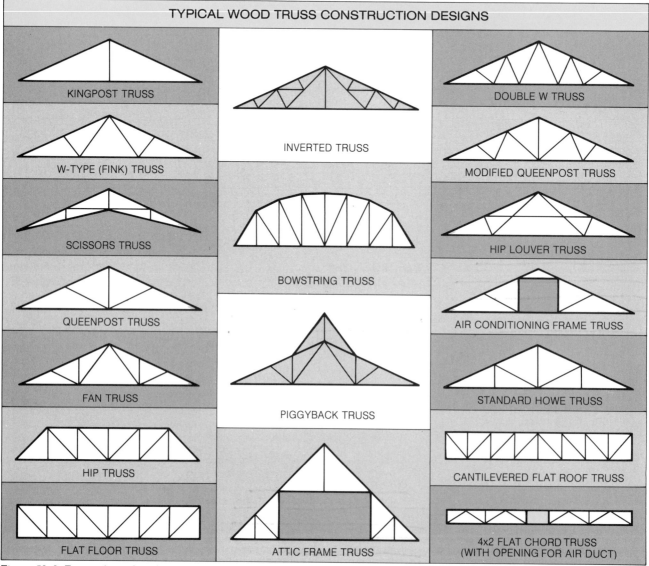

**TYPICAL WOOD TRUSS CONSTRUCTION DESIGNS**

KINGPOST TRUSS

W-TYPE (FINK) TRUSS

SCISSORS TRUSS

QUEENPOST TRUSS

FAN TRUSS

HIP TRUSS

FLAT FLOOR TRUSS

INVERTED TRUSS

BOWSTRING TRUSS

PIGGYBACK TRUSS

ATTIC FRAME TRUSS

DOUBLE W TRUSS

MODIFIED QUEENPOST TRUSS

HIP LOUVER TRUSS

AIR CONDITIONING FRAME TRUSS

STANDARD HOWE TRUSS

CANTILEVERED FLAT ROOF TRUSS

4x2 FLAT CHORD TRUSS
(WITH OPENING FOR AIR DUCT)

Figure 50–8. Types of wood roof trusses. The king-post, W-type (fink), and scissors truss are most commonly used in light construction.

## TRUSS ROOF TRANSFERS THE ROOF LOAD TO THE WALLS

**STEP 1.** THE ENDS OF THE TWO TOP CHORDS (**A-B** AND **A-C**) ARE BEING PUSHED TOGETHER (COMPRESSED). THE BOTTOM CHORD HOLDS THE LOWER ENDS (**B** AND **C**) OF THE TOP CHORDS FROM PUSHING OUT, THEREFORE THE BOTTOM CHORD IS IN A PULLING APART STATE (TENSION). BECAUSE THE LOWER ENDS OF THE TOP CHORDS CANNOT PULL APART, THE PEAK OF THE TRUSS (**A**) CANNOT DROP DOWN.

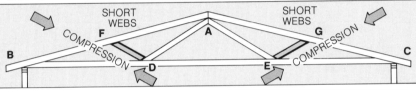

**STEP 2.** THE LONG WEBS ARE SECURED TO THE PEAK OF THE TRUSS (**A**) AND ALSO FASTENED TO THE BOTTOM CHORD AT POINTS **D** AND **E**. THIS GIVES THE BOTTOM CHORD SUPPORT ALONG THE OUTSIDE WALL SPAN. THE WEIGHT OF THE BOTTOM CHORD HAS A PULLING APART EFFECT (TENSION) ON THE LONG WEBS.

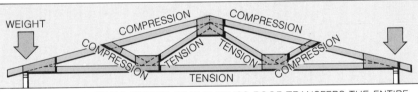

**STEP 3.** THE SHORT WEBS RUN FROM THE INTERMEDIATE POINTS **F AND G** OF THE TOP CHORD TO POINTS **D** AND **E** OF THE BOTTOM CHORD. THEIR PURPOSE IS TO PROVIDE SUPPORT TO THE TOP CHORD. THIS EXERTS A DOWNWARD, PUSHING TOGETHER FORCE (COMPRESSION) ON THE SHORT WEB.

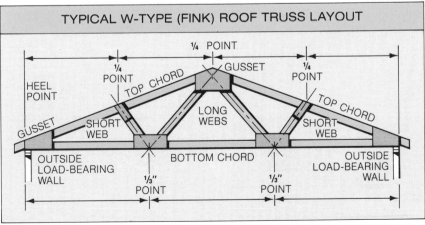

**STEP 4.** THE OVERALL DESIGN OF THE TRUSS ROOF TRANSFERS THE ENTIRE LOAD (ROOF WEIGHT, SNOW LOAD, WIND LOAD, ETC.) DOWN THROUGH THE OUTSIDE WALLS TO THE FOUNDATION.

Figure 50–9. How a truss works. Each part is in a state of either tension or compression.

## TYPICAL W-TYPE (FINK) ROOF TRUSS LAYOUT

Figure 50–10. Layout for a W-type (fink) truss. The points at which the lower ends of the web members fasten to the bottom chord divide the bottom chord into three equal parts. Each short web meets the top chord at a point that is one-fourth the horizontal distance of the bottom chord.

See Figure 50–11. A more effective device is the *clinch-nail* connector, which features plates on both sides of the splice. See Figure 50–12. Another metal connector is a plate with long teeth that is pressed into place by hydraulic pressing equipment. See Figure 50–13. Split-ring connectors are used for heavy trusses. Circular grooves are cut into the chords and webs to bolt the split rings together. See Figure 50–14.

Although used less frequently today, glued and nailed plywood *gusset plates* are still considered an acceptable method for tying a truss together. See Figure 50–15.

## PREFABRICATED ROOF TRUSS

A truss manufacturer has the equipment to construct trusses at a much lower cost than job-built trusses. The blueprints for the building should provide the truss manufacturer with all the necessary information. See Figure 50–16. For example, the engineer designing the truss must know the slope of the roof, the span the truss must cover, and the live load, dead load, and wind stress factors the truss must withstand. Based on this information, the engineer determines the truss design, the size and type of lumber, and the kind of connector plates to be used.

Trusses are assembled on large tables at the manufacturing plant. One type of assembly table is shown in Figure 50–17. The metal connector plates are placed on both sides of the truss. The truss passes between rollers that compress the connector plates and force the prongs into the wood. When the trusses are completed, they are transported to the job site by truck. See Figure 50–18.

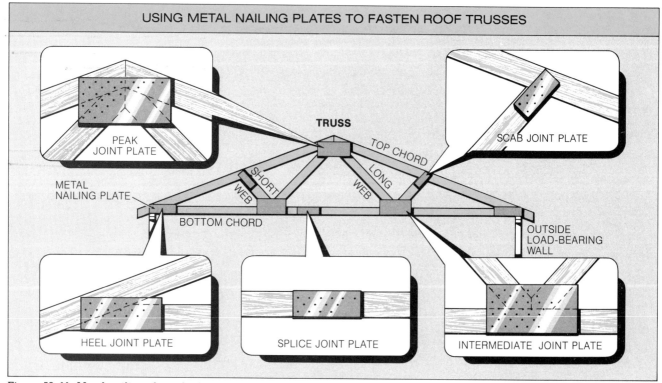

USING METAL NAILING PLATES TO FASTEN ROOF TRUSSES

PEAK JOINT PLATE

SCAB JOINT PLATE

TRUSS

TOP CHORD

METAL NAILING PLATE

SHORT WEB

LONG WEB

BOTTOM CHORD

OUTSIDE LOAD-BEARING WALL

HEEL JOINT PLATE

SPLICE JOINT PLATE

INTERMEDIATE JOINT PLATE

Figure 50–11. Metal nailing plates for fastening truss members. These are flat pieces usually manufactured from 20-ga. zinc-coated or galvanized steel. The holes for the nails are pre-punched.

*Bostich Division of Texron, Inc.*

Figure 50–12. Fastening clinch-nail connector plates at truss joints with a coil-fed pneumatic nailer.

## JOB-BUILT ROOF TRUSS

When no truss manufacturer is located a reasonable distance from the construction site, trusses must be built on the job.

*Weyerhauser Company*

Figure 50–13. Pressed connector plates tie the webs of these roof trusses to the chords.

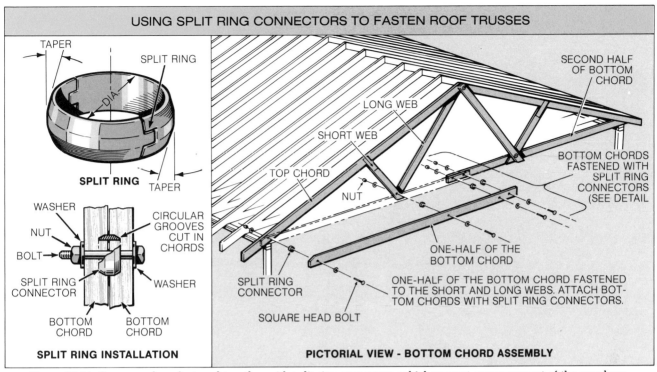

Figure 50–14. Truss members fastened together with split-ring connectors, which prevent any movement of the members.

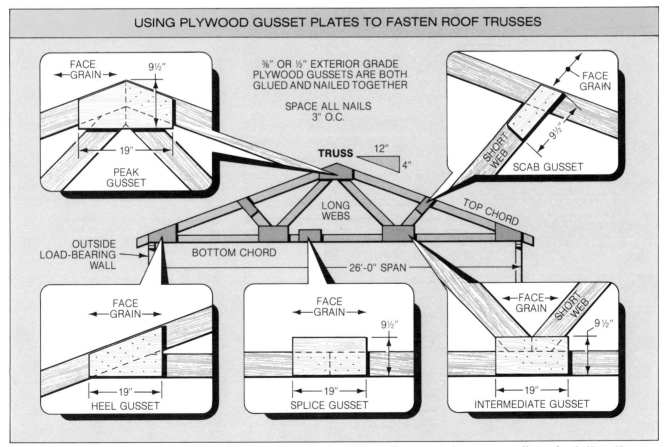

Figure 50–15. Truss members fastened together with plywood gusset plates. The gusset plates are usually made of ⅜" or ½" exterior grade plywood. For ⅜" plywood, 4d nails should be used. For ½" or thicker plywood 6d nails should be used. The nails are spaced 3" O.C. and positioned ¾" in from the edge of the gusset. Glue is usually recommended in addition to nailing.

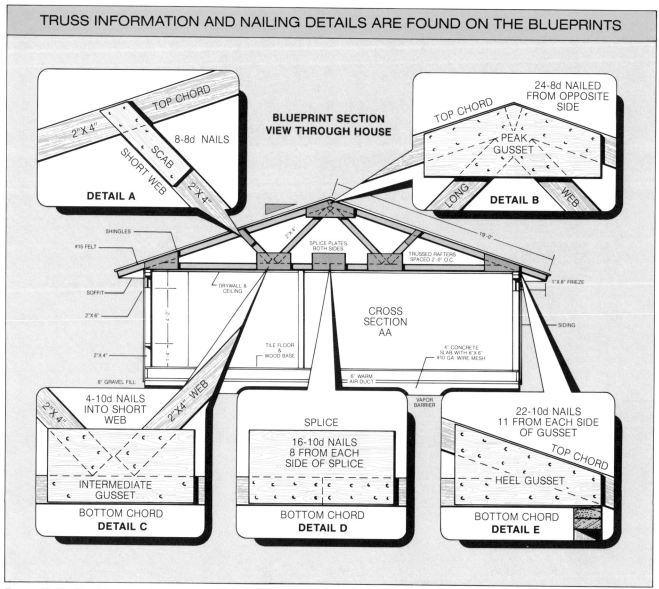

**TRUSS INFORMATION AND NAILING DETAILS ARE FOUND ON THE BLUEPRINTS**

TOP CHORD
2" X 4"
SCAB
SHORT WEB
8-8d NAILS
2" X 4"
**DETAIL A**

BLUEPRINT SECTION
VIEW THROUGH HOUSE

24-8d NAILED
FROM OPPOSITE
SIDE
TOP CHORD
PEAK
GUSSET
LONG
WEB
**DETAIL B**

SHINGLES
#15 FELT
SOFFIT
2" X 6"
2" X 4"
6" GRAVEL FILL
2'4"
SPLICE PLATES
BOTH SIDES
TRUSSED RAFTERS
SPACED 2'-0" O.C.
19'-0"
1" X 8" FRIEZE
DRYWALL & CEILING
CROSS
SECTION
AA
SIDING
4'-2"
1'-4"
TILE FLOOR & WOOD BASE
4" CONCRETE SLAB WITH 6"X 6" #10 GA. WIRE MESH
6" WARM AIR DUCT
VAPOR BARRIER

4-10d NAILS
INTO SHORT
WEB
2" X 4"
2" X 4" WEB
INTERMEDIATE
GUSSET
BOTTOM CHORD
**DETAIL C**

SPLICE
16-10d NAILS
8 FROM EACH
SIDE OF SPLICE
BOTTOM CHORD
**DETAIL D**

22-10d NAILS
11 FROM EACH SIDE
OF GUSSET
TOP CHORD
HEEL GUSSET
BOTTOM CHORD
**DETAIL E**

Figure 50–16. This blueprint drawing for a W-type truss gives all information needed for installation, including details for all connections.

*Gang-Nail Systems, Inc.*

Figure 50–17. A truss assembly table with equipment for compressing the connector plates.

Figure 50–18. Prefabricated trusses are delivered to the job site by truck.

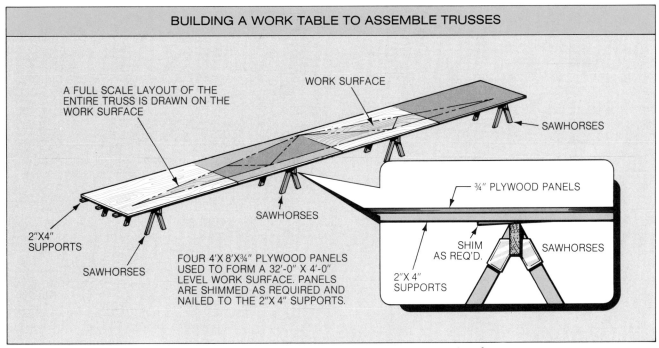

BUILDING A WORK TABLE TO ASSEMBLE TRUSSES

A FULL SCALE LAYOUT OF THE ENTIRE TRUSS IS DRAWN ON THE WORK SURFACE

WORK SURFACE

SAWHORSES

¾" PLYWOOD PANELS

2"X4" SUPPORTS

SAWHORSES

SAWHORSES

SHIM AS REQ'D.

SAWHORSES

2"X 4" SUPPORTS

FOUR 4'X 8'X¾" PLYWOOD PANELS USED TO FORM A 32'-0" X 4'-0" LEVEL WORK SURFACE. PANELS ARE SHIMMED AS REQUIRED AND NAILED TO THE 2"X 4" SUPPORTS.

Figure 50–19. Assembly table for constructing trusses on the job.

Also, if a small number of trusses is required, it may be cheaper to build them on the job than to obtain them from a truss manufacturer.

An assembly table (work table) should be set up where the truss parts can be laid out and clamped before nailing. Portable assembly units are also available. They can be adjusted for different truss sizes and designs.

Job-built trusses are usually fastened together with flat metal nailing plates. Less frequently, plywood gussets are used.

## Constructing a Job-built Truss

An assembly table, or work table, for a job-built truss can be made by placing 2" × 4" pieces across sawhorses and nailing plywood panels to the top of the 2" × 4" pieces. See Figure 50–19. Shims are placed under the panels to level and straighten the work table.

The entire truss is laid out on this work table. The exact length of each truss member and the angles of the cuts are marked. The top and bottom chords and the webs are measured and cut according to this layout. After one set of truss members is completed, it can be used as a pattern for cutting the pieces for all the trusses required. A procedure for laying out a W-type truss on a work table is shown in Figure 50–20. The work table also serves as a platform for assembling the trusses.

A tight fit between truss members is important. See Figure 50–21. To ensure tight fits, cuts must be accurate. Also, truss members should be toenailed or clamped together, then metal or plywood gusset plates should be nailed at the joints between the truss members. The truss is then turned over and gusset plates are nailed to the other side.

*Guide blocks, wedge blocks,* and *wedges* make assembly more efficient. See Figure 50–22. Guide blocks hold the web members in their proper po-

sition. Wedges are driven between the wedge blocks and the truss members, forcing a very tight fit between members.

The bottom chord of a truss tends to sag at the center after it has been set in place. To prevent sagging, the bottom chord should be arched a small amount when the truss is being constructed. The crown-like result is known as *camber.*

## INSTALLING ROOF TRUSSES

Trusses are usually spaced 2' O.C. They must be lifted into place, fastened to the walls, and braced. Small trusses can be placed by hand as shown in Figure 50–23. Carpenters are required on the two opposite walls to fasten the ends of the trusses. One or two workers on the floor below can push the truss to an upright position.

It is always more efficient to use mechanical equipment such as forklifts and cranes to install

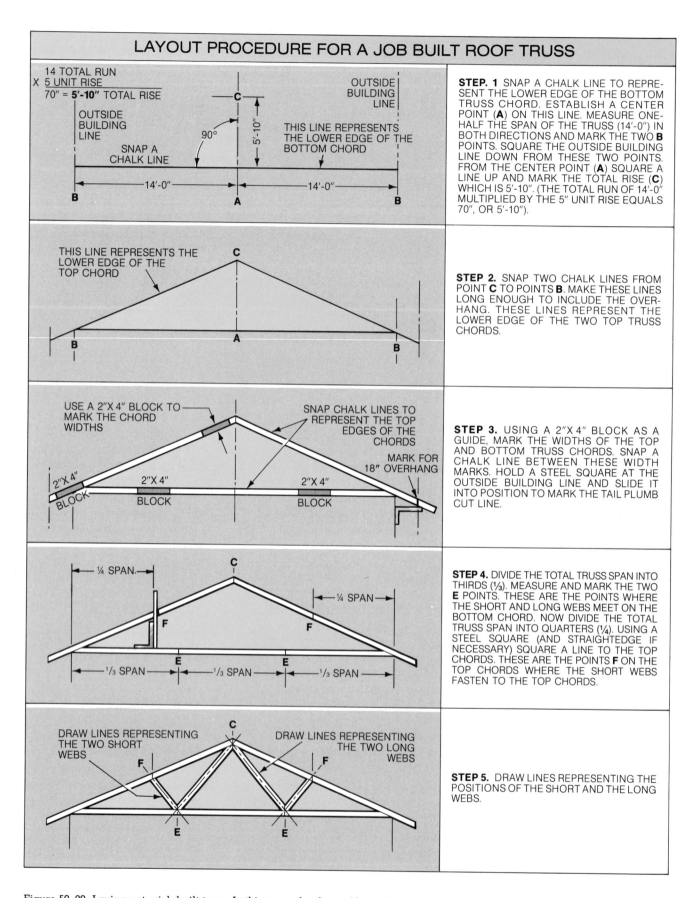

## LAYOUT PROCEDURE FOR A JOB BUILT ROOF TRUSS

**STEP. 1** SNAP A CHALK LINE TO REPRESENT THE LOWER EDGE OF THE BOTTOM TRUSS CHORD. ESTABLISH A CENTER POINT (**A**) ON THIS LINE. MEASURE ONE-HALF THE SPAN OF THE TRUSS (14'-0") IN BOTH DIRECTIONS AND MARK THE TWO **B** POINTS. SQUARE THE OUTSIDE BUILDING LINE DOWN FROM THESE TWO POINTS. FROM THE CENTER POINT (**A**) SQUARE A LINE UP AND MARK THE TOTAL RISE (**C**) WHICH IS 5'-10". (THE TOTAL RUN OF 14'-0" MULTIPLIED BY THE 5" UNIT RISE EQUALS 70", OR 5'-10").

**STEP 2.** SNAP TWO CHALK LINES FROM POINT **C** TO POINTS **B**. MAKE THESE LINES LONG ENOUGH TO INCLUDE THE OVERHANG. THESE LINES REPRESENT THE LOWER EDGE OF THE TWO TOP TRUSS CHORDS.

**STEP 3.** USING A 2"X 4" BLOCK AS A GUIDE, MARK THE WIDTHS OF THE TOP AND BOTTOM TRUSS CHORDS. SNAP A CHALK LINE BETWEEN THESE WIDTH MARKS. HOLD A STEEL SQUARE AT THE OUTSIDE BUILDING LINE AND SLIDE IT INTO POSITION TO MARK THE TAIL PLUMB CUT LINE.

**STEP 4.** DIVIDE THE TOTAL TRUSS SPAN INTO THIRDS (⅓). MEASURE AND MARK THE TWO **E** POINTS. THESE ARE THE POINTS WHERE THE SHORT AND LONG WEBS MEET ON THE BOTTOM CHORD. NOW DIVIDE THE TOTAL TRUSS SPAN INTO QUARTERS (¼). USING A STEEL SQUARE (AND STRAIGHTEDGE IF NECESSARY) SQUARE A LINE TO THE TOP CHORDS. THESE ARE THE POINTS **F** ON THE TOP CHORDS WHERE THE SHORT WEBS FASTEN TO THE TOP CHORDS.

**STEP 5.** DRAW LINES REPRESENTING THE POSITIONS OF THE SHORT AND THE LONG WEBS.

Figure 50–20. Laying out a job-built truss. In this example, the roof has a 5" unit rise and an 18" overhang. The span of the roof is 28'. The truss is constructed of 2" × 4" material.

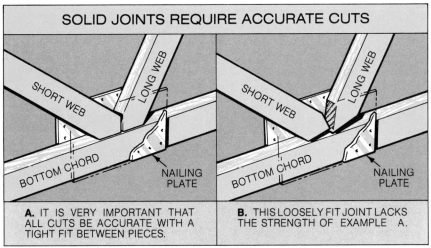

## SOLID JOINTS REQUIRE ACCURATE CUTS

**A.** IT IS VERY IMPORTANT THAT ALL CUTS BE ACCURATE WITH A TIGHT FIT BETWEEN PIECES.

**B.** THIS LOOSELY FIT JOINT LACKS THE STRENGTH OF EXAMPLE A.

Figure 50–21. Tight joints are required between truss members.

trusses. See Figures 50–24 and 50–25.

### Temporary Bracing

After the truss bundles have been set on the walls, they are moved individually into position, nailed down, and temporarily braced. See Figure 50–26. Without temporary bracing the truss may topple over, causing damage to the truss and possible injury to the workers installing the truss. A recommended procedure

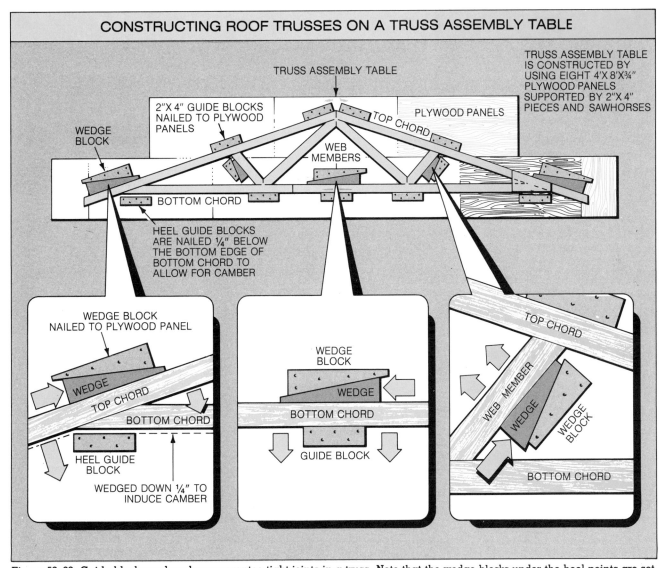

## CONSTRUCTING ROOF TRUSSES ON A TRUSS ASSEMBLY TABLE

Figure 50–22. Guide blocks and wedges guarantee tight joints in a truss. Note that the wedge blocks under the heel points are set ¼" down from the ends of the bottom chord. When the other truss members are wedged in place, the wedges above the heel points are tightened. This tightening causes the two ends of the truss to bend downward, producing the desired camber in both top and bottom chords.

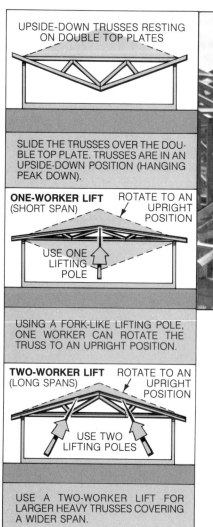

UPSIDE-DOWN TRUSSES RESTING ON DOUBLE TOP PLATES

SLIDE THE TRUSSES OVER THE DOUBLE TOP PLATE. TRUSSES ARE IN AN UPSIDE-DOWN POSITION (HANGING PEAK DOWN).

**ONE-WORKER LIFT** (SHORT SPAN)
ROTATE TO AN UPRIGHT POSITION
USE ONE LIFTING POLE

USING A FORK-LIKE LIFTING POLE, ONE WORKER CAN ROTATE THE TRUSS TO AN UPRIGHT POSITION.

**TWO-WORKER LIFT** (LONG SPANS)
ROTATE TO AN UPRIGHT POSITION
USE TWO LIFTING POLES

USE A TWO-WORKER LIFT FOR LARGER HEAVY TRUSSES COVERING A WIDER SPAN.

Figure 50–23. Small trusses can be placed manually.

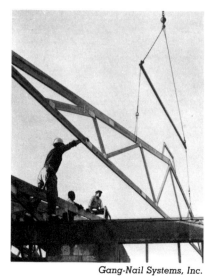

*Gang-Nail Systems, Inc.*

Figure 50–24. Light crane being used to place roof trusses over masonry walls.

Figure 50–25. Scissors trusses being placed by crane. The scissors design provides an arched ceiling in the interior of the building.

for bracing trusses as they are being set in place is shown in Figure 50–27.

## Permanent Bracing

The temporary bracing is removed as the roof sheathing is

Figure 50–26. Carpenter nailing top of truss to a temporary 2″ × 4″ brace.

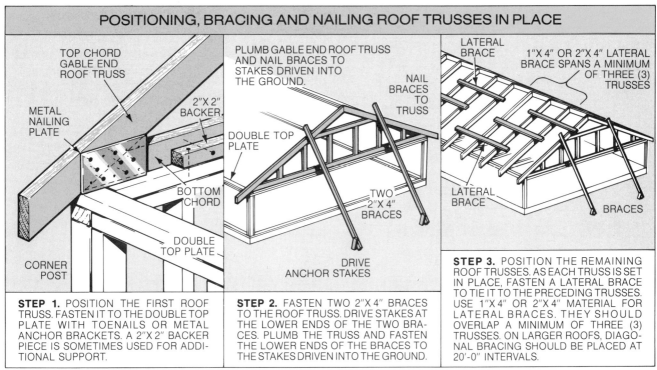

### POSITIONING, BRACING AND NAILING ROOF TRUSSES IN PLACE

TOP CHORD GABLE END ROOF TRUSS

METAL NAILING PLATE

2″X 2″ BACKER

BOTTOM CHORD

DOUBLE TOP PLATE

CORNER POST

PLUMB GABLE END ROOF TRUSS AND NAIL BRACES TO STAKES DRIVEN INTO THE GROUND.

DOUBLE TOP PLATE

NAIL BRACES TO TRUSS

TWO 2″X 4″ BRACES

DRIVE ANCHOR STAKES

LATERAL BRACE

1″X 4″ OR 2″X 4″ LATERAL BRACE SPANS A MINIMUM OF THREE (3) TRUSSES

LATERAL BRACE

BRACES

**STEP 1.** POSITION THE FIRST ROOF TRUSS. FASTEN IT TO THE DOUBLE TOP PLATE WITH TOENAILS OR METAL ANCHOR BRACKETS. A 2″X 2″ BACKER PIECE IS SOMETIMES USED FOR ADDITIONAL SUPPORT.

**STEP 2.** FASTEN TWO 2″X 4″ BRACES TO THE ROOF TRUSS. DRIVE STAKES AT THE LOWER ENDS OF THE TWO BRACES. PLUMB THE TRUSS AND FASTEN THE LOWER ENDS OF THE BRACES TO THE STAKES DRIVEN INTO THE GROUND.

**STEP 3.** POSITION THE REMAINING ROOF TRUSSES. AS EACH TRUSS IS SET IN PLACE, FASTEN A LATERAL BRACE TO TIE IT TO THE PRECEDING TRUSSES. USE 1″X 4″ OR 2″X 4″ MATERIAL FOR LATERAL BRACES. THEY SHOULD OVERLAP A MINIMUM OF THREE (3) TRUSSES. ON LARGER ROOFS, DIAGONAL BRACING SHOULD BE PLACED AT 20′-0″ INTERVALS.

Figure 50–27. Installing roof trusses. After the trusses are temporarily braced, the sheathing is applied.

Figure 50–28. Plywood sheathing has been placed on top of this trussed roof.

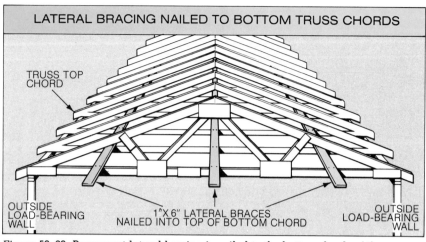

LATERAL BRACING NAILED TO BOTTOM TRUSS CHORDS

TRUSS TOP
CHORD

OUTSIDE
LOAD-BEARING
WALL

1"X 6" LATERAL BRACES
NAILED INTO TOP OF BOTTOM CHORD

OUTSIDE
LOAD-BEARING
WALL

Figure 50–29. Permanent lateral bracing is nailed to the bottom chords of the trusses. The braces are tied to the end walls and spaced 10' O.C.

nailed. Properly nailed plywood sheathing is considered sufficient to tie together the top chords of the trusses. See Figure 50–28. Permanent lateral bracing of 1" × 4" material is recommended at the bottom chords. See Figure 50–29.

## Metal Truss Anchors

The same types of metal anchors used to tie regular rafters to the outside walls are equally effective for fastening the ends of the truss. (These anchors are shown in Unit 46.)

# SECTION 11

# Energy Conservation: Insulation and Construction Methods

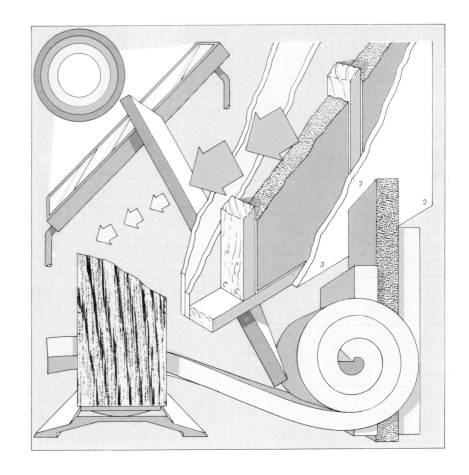

Between 1950 and 1970 energy fuel consumption doubled in the United States, and it continues to increase at an alarming pace. Obviously, the natural resources that give us energy cannot indefinitely supply our needs unless the rate of consumption is slowed. Insulation and improved construction methods provide an important measure of energy conservation.

# UNIT 51

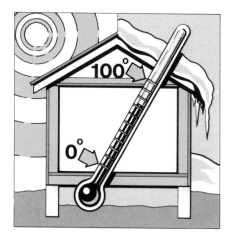

# Temperature Control, Condensation, and Ventilation

The two most important factors affecting the temperature inside a building are the degree of *heat flow* in and out of the structure and the amount of moisture resulting from *condensation*. These conditions can be controlled by insulation and by proper building design. When these conditions are well controlled, less fuel is required for the furnace to warm the building, and less electricity is required for air conditioners to cool the building.

## HEAT TRANSFER

Heat always moves toward coldness. Consequently, in the winter the warm air inside a building escapes through the framework of the building, moving to the cooler air outside. In the summer the warm air outside a building flows toward the cooler air inside. See Figure 51–1. The flow of heat is called *heat transfer*.

All construction materials used for the outside walls of buildings (wood, brick, concrete, and masonry) prevent a certain amount of heat flow through the walls. However, a much greater amount of heat flow can be prevented by adding *thermal insulation*. (The word *thermal* de-

scribes materials that have a high heat-flow resistance.) Insulation is usually installed by carpenters. It is placed in the walls, floors, and ceilings that surround the living areas of a building. In a one-story house with an unheated attic and crawl space, for example, insulation should be placed in the exterior walls, floor, and ceiling. See Figure 51–2. In a one-and-one-half-story house with a full basement, insulation should be placed in the exterior walls of all the living areas, including the basement. It should also be placed along the rafter

sections and in the walls and ceiling that enclose the living space in the attic. See Figure 51–3.

## Methods of Heat Transfer

The three methods by which heat transfer occurs are *conduction, convection,* and *radiation*.

**Conduction.** Conduction is the movement of heat through a solid substance. The heat passes from one molecule to another. When heat is applied to

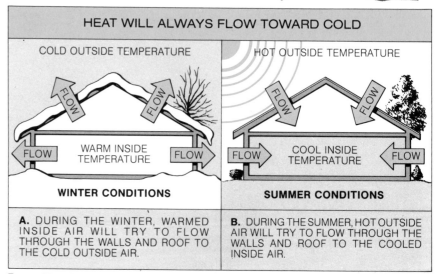

**HEAT WILL ALWAYS FLOW TOWARD COLD**

COLD OUTSIDE TEMPERATURE

FLOW FLOW

WARM INSIDE TEMPERATURE

FLOW FLOW

**WINTER CONDITIONS**

**A.** DURING THE WINTER, WARMED INSIDE AIR WILL TRY TO FLOW THROUGH THE WALLS AND ROOF TO THE COLD OUTSIDE AIR.

HOT OUTSIDE TEMPERATURE

FLOW FLOW

COOL INSIDE TEMPERATURE

FLOW FLOW

**SUMMER CONDITIONS**

**B.** DURING THE SUMMER, HOT OUTSIDE AIR WILL TRY TO FLOW THROUGH THE WALLS AND ROOF TO THE COOLED INSIDE AIR.

Figure 51-1. Heat always moves toward cold. The flow of heat is called heat transfer.

## INSULATING THE CEILING, OUTSIDE WALLS AND FLOOR

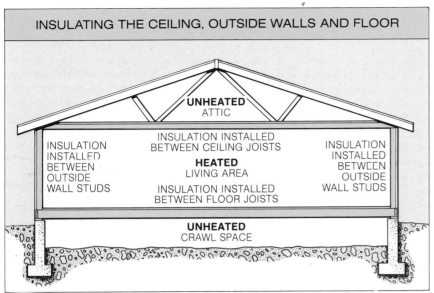

Figure 51-2. In a one-story house with an unheated attic and crawl-space area, insulation should be placed in the exterior walls, floor, and ceiling.

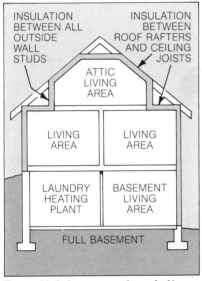

Figure 51-3. In a one-and-one-half-story house with a full basement, insulation should be placed in the exterior walls of all living areas. It should also be placed along the rafter sections and in the walls and ceilings that enclose the living space in the attic.

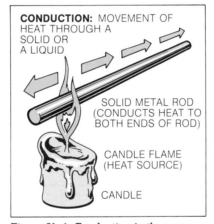

Figure 51-4. Conduction is the movement of heat through a solid substance. For example, the heat from a flame is conducted from one end of a metal rod to the other.

one end of a metal rod, for example, the other end of the rod eventually becomes hot by means of conduction. See Figure 51-4. Dense materials such as metals are called *conductors* because they transfer large quanti-

ties of heat quickly. Less dense materials such as wood or plastic are called *insulators,* because they do not transfer heat quickly. Insulators are therefore used to retard heat flow.

**Convection.** Convection is the movement of heat through the circulatory motion of air or liquid. An example of heat transfer by convection is air heated in a furnace and blown to different

areas of a house. See Figure 51-5. When air is heated, it expands, becomes lighter, and rises. Colder air then takes the place of warmer air. As a result, a convection loop is established that circulates the heat throughout the house.

Another example of convection is the way a pipe is warmed by hot water moving through it. In convection, heat is transferred by the movement of the molecules in a fluid (air or liquid) substance.

**Radiation.** Radiation is the direct transmission of heat by invisible waves similar to light waves. An example of radiation is the way radiant heat waves travel through the vacuum of space and warm the surface of the earth. Another example of radiation is the heat that a person feels when standing near a fire. Figure 51-6 shows how radiant heat can be transmitted by a mechanical heat source such as a radiator. The heat from the radiator strikes and warms objects such as the walls, ceilings, and furniture. In addition, some of this heat transferred by radiation is reflected and warms the air by convection.

Figure 51-5. Convection is the movement of heat through a fluid substance such as air or water. For example, air warmed at a source such as a furnace or wall heater rises. It is then replaced by colder air, which in turn is heated and also rises. A convection loop circulates the heat throughout the area.

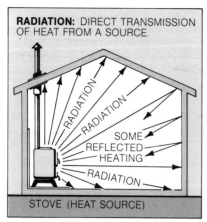

Figure 51-6. Radiation is the direct transmission of heat waves from a heat source. For example, heat from a radiator strikes and warms objects such as the walls, ceilings, and furniture. Some of the heat is reflected and warms the air through convection (as shown by the arrows).

## Insulation as Protection against Heat Transfer

Materials used for insulation should protect a building from losing heat or coolness by any of the methods of heat transfer. The insulation material should be a poor conductor of heat in order to prevent heat transfer by conduction. It should stop or slow the flow of air from warm to cold areas of the building in order to prevent heat transfer by convection. Finally, it should reflect heat rather than absorb it in order to prevent heat transfer by radiation.

## Measuring Heat Transfer and Resistance

Many different kinds of insulation are available. A knowledge of how various materials used for insulation resist the passage of heat enables the architect or builder to select the materials best suited for insulating a particular building.

Technical information about each type of insulation is ex-

pressed in letters that represent certain *factors:* conductivity (k), conductance (C), transmittance (U). A *value* of resistance (R), is based upon these factors.

Heat is measured by British thermal units (Btu). One Btu is the amount of heat required to raise the temperature of one pound of water 1° F (Fahrenheit). See Figure 51-7. Heating equipment used in a building, such as furnaces and boilers, is rated by its output. When deciding on such equipment, engineers must calculate the amount of space to be heated in addition to the expected heat loss in the building.

**Conductivity (k) and Conductance (C) Factors.** The best materials for insulation have a low k or C factor. The *k factor* (conductivity) measures the amount of heat that travels through homogenous material. See Figure 51-8. Concrete, wood, and poly-

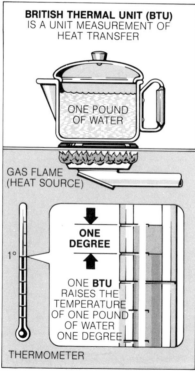

Figure 51-7. The British thermal unit (Btu) is the unit measurement of heat transfer. One Btu is the amount of heat needed to raise the temperature of one pound of water 1°F.

urethane foam are examples of homogenous materials—they have the same composition throughout. The k factor is a decimal measurement of how many Btu per hour pass through one square foot of material that is 1″ thick and has a temperature difference of 1°F between its inside and outside surfaces. Wood has a lower k factor than concrete, and polyurethane foam has a lower k factor than wood.

The *C factor* (conductance) is based on the same principle as the k factor; however, it applies to materials of any thickness that are not homogenous or that have air cavities, such as hollow concrete blocks.

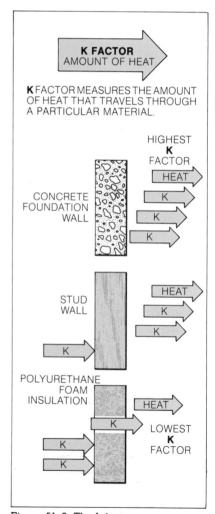

Figure 51-8. The k factor represents the amount of heat that travels through a homogenous material.

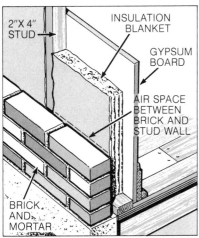

Figure 51–9. The U factor for a brick-veneered wood stud wall with insulation expresses heat transfer through the combined materials plus the air space.

**Transmittance (U) Factor.** The *U factor* (transmittance) is a decimal measurement of how many Btu per hour pass through one square foot of the *combination of materials* that make up the floor, ceiling, wall, or roof area of a building when a 1° difference in temperature exists between the inside and outside of the building. See Figure 51–9. The U factor differs from k and C factors in that it expresses heat transfer through combined materials plus any air space that may exist rather than through a single material.

**Resistance (R) Value.** The *R value* (resistance) represents the ability of a material to resist heat flow. Most insulation products are labeled with an R value on the wrapper. The total heat-flow resistance of a wall includes the R value of the material used to construct the wall (wood, brick, concrete, hollow concrete blocks), plus the R value of the insulation material placed in the wall. An air space, such as the kind found between a wood frame and brick veneer wall, adds to the total R value of the wall.

The total R value needed for a building depends on regional weather conditions, since buildings in colder climates require more insulation. Local building codes often specify insulation requirements for the area. Figure 51–10 gives examples of the R values (based on k and C factors) of a number of different building materials.

## CONDENSATION

Moisture is always present in the air. Usually it is invisible. When air temperature increases, the moisture content of the air increases.

| MATERIAL OR PRODUCT | CONDUCTIVITY k | CONDUCTANCE C | RESISTANCE | |
|---|---|---|---|---|
| | | | PER 1" THICK-NESS (1/k) | PER THICKNESS SHOWN (1/C) |
| CONCRETE | 12.0 | | 0.08 | |
| FACE BRICK | 9.0 | | 0.11 | |
| HOLLOW CONCRETE BLOCK, 8" | | 0.90 | | 1.11 |
| STUCCO | 5.0 | | 0.20 | |
| METAL LATH & PLASTER, ¾" | | 7.70 | | 0.13 |
| GYPSUM BOARD, ½" | | 2.22 | | 0.45 |
| PLYWOOD, ½" | | 1.60 | | 0.62 |
| PINE, FIR, OTHER SOFTWOODS | 0.80 | | 1.25 | |
| OAK, MAPLE, OTHER HARDWOODS | 1.10 | | 0.91 | |
| ASPHALT SHINGLES | | 2.27 | | 0.44 |
| BUILT-UP ROOFING, ⅜" | | 3.00 | | 0.33 |
| WOOD SHINGLES | | 1.06 | | 0.94 |
| STRUCTURAL INSULATION BOARD, ½" | | 0.76 | | 1.32 |
| MINERAL WOOL BATTS, 3"—4" | | 0.09 | | 11.00 ⎫ |
| 5"—6" | | 0.05 | | 19.00 ⎪ INSULATION |
| 6½"—7" | | 0.05 | | 22.00 ⎬ MATERIALS |
| 8½"—9" | | 0.03 | | 30.00 ⎭ |
| EXPANDED POLYSTYRENE, 1" | | 0.28 | | 3.57 |
| INSIDE SURFACE AIR FILM[2] | | | | 0.68 |
| OUTSIDE SURFACE AIR FILM[3] | | | | 0.17 |
| AIR SPACE ¾", NONREFLECTIVE[4] | | 1.0 | | 1.01 |
| AIR SPACE ¾", REFLECTIVE[4] | | | | 3.48 |

[1]At 75°F mean temperature, from ASHRAE 1977 Fundamentals Handbook.
[2]Heat flow horizontal, still air.
[3]Heat flow any direction, 15 mph wind.
[4]Heat flow horizontal.

Figure 51–10. R values of some common building materials based on their conductivity (k) or conductance (C). Note how much higher the R values are for insulation materials than for other materials.

Moisture in the form of water vapor is produced from many sources within a building. Activities of an average family of four in a heated building generate about 22.5 pounds of water vapor within a 24-hour period. See Figure 51–11.

When the warm, moisture-laden air within a building comes into contact with the cooler outside surfaces of the building, it is unable to hold all of its moisture. Some of the moisture comes out of the air and forms drops of water on the cooler surfaces. This process is known as *condensation.*

The temperature at which condensation occurs is the *dew point.* Moisture problems because of condensation are greater in colder climates, where houses require more heating during the winter.

Condensation that takes place on the face of a wall is called *surface condensation.* This type of condensation is easily visible. Often, however, condensation occurs inside a wall. This *concealed condensation* can cause many problems. Over a period of time, it can cause serious decay in the framing members. Blistering may occur on the outside paint surfaces of the building. Plaster or gypsum board inside the building may begin to crack and crumble.

In the past houses were not built to resist the effects of weather as effectively as they are today. Construction allowed more air leakage, which in turn permitted more of the interior moisture to move out of the building. Modern homes are generally smaller and more tightly constructed than older ones. They are also better insulated, with very little air leakage. As a result, air moisture and condensation pose a greater problem in new buildings than in older buildings.

## Moisture Control

Condensation in the floors, walls, ceiling, and roof of a building can be controlled by means of vapor barriers and ventilation.

**Vapor Barriers.** A *vapor barrier* is a thin, covering material through which water cannot easily pass. Many insulation products have a vapor barrier on one surface. These products should be placed with the vapor barrier facing the *warm side* (inside surface) of the wall. The vapor barrier prevents damage to the insulation material from the moisture that collects inside a wall. If the insulation product does not have a vapor barrier, a separate vapor barrier should be attached to the warm side.

Information about a vapor barrier material often refers to its *perm rating.* The perm rating indicates *permeability* (or permeance). It is based on a formula that measures the water vapor flow through the thickness of single or combined materials. Most building materials are permeable to some degree. A few materials, such as metals and glass, are completely *impermeable*, meaning that they allow no vapor to pass through. An impermeable material has a perm rating of 0.00. Any material that has a rating of 1 or less qualifies as a vapor barrier. However, construction conditions today often require a vapor barrier with a rating of 0.5 or less.

Polyethylene film and aluminum foil have very low perm ratings and are often used as vapor barriers.

**Ventilation.** Vapor barriers and insulation can effectively control condensation in the walls, floor, and ceiling of inhabited parts of a house. However, ventilation is needed in closed-off areas, such as the attic space above the ceiling and below the roof, and the area between the floor and ground in a house with a crawl-space foundation. Ventilation is provided by openings that permit warm air to escape, as well as by the circulation of air in the enclosed areas.

*Ventilation for Attic Roofs.* Condensation occurs beneath the roof when water vapor is unable to escape through the roofing material. Wood shingles or shakes do not hold back vapor movement. Asphalt shingles or built-up roofs are highly resistant. Several venting methods are possible for attic areas of wood-framed buildings. Most have serious shortcomings when used alone but are effective when combined.

1. *Gable end louvers* are probably the oldest ventilation system used for gable roofs. See Figure 51–12. Slanted boards in the louver prevent rainwater from entering. A screen covers the inside of the louver to keep out insects. Alone, gable end louvers are considered less efficient than other systems. They only work well when a breeze is blowing at a right angle to the louvers.

2. *Soffit vents* are located beneath the roof cornice. They may consist of a series of small openings or one continuous slot. See Figure 51–13. With soffit vents, most of the ventilation occurs

| SOURCE OR FUNCTION | MOISTURE GENERATED (LBS.) |
|---|---|
| BREATHING AND PERSPIRING | 13 |
| COOKING | 5 |
| BATHING | 1 |
| DISHWASHING | 3.5* |

*As much as 30 lbs. or more of moisture can be added by clothes washing and inside drying.

Figure 51–11. In a heated building, an average family of four generates about 22.5 pounds of vapor in a 24-hour period.

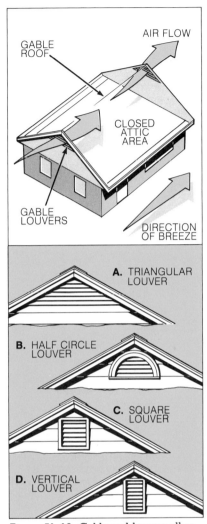

Figure 51–12. Gable end louvers allow air flow through an attic.

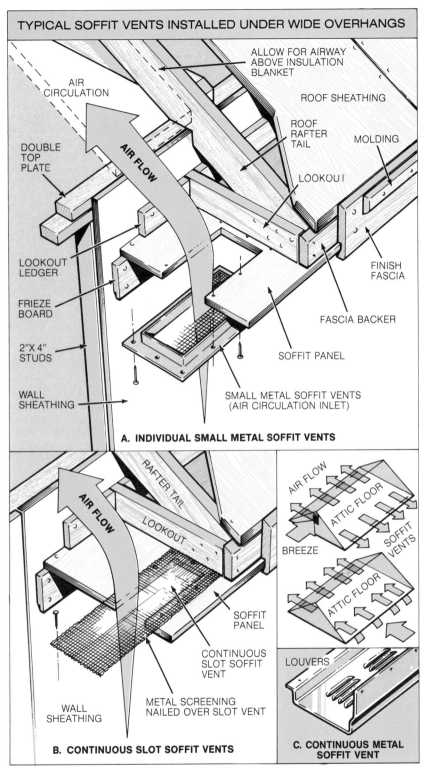

Figure 51–13. Types of soffit vents. Most of the ventilation through soffit vents occurs over the attic floor.

over the attic floor. There is little air movement beneath the roof sheathing.

3. *Roof vents* are located on top of the roof. See Figure 51–14. They allow warm air to escape, but do not allow much air flow in the attic space.

4. *Continuous-ridge vents* run along the ridge of the roof. See Figure 51–15. They are unique in that their design provides some air flow resulting from temperature differences alone. General air circulation is limited, however.

Common combinations of these venting systems are *roof-and-soffit, gable-louver-and-sof-fit,* and *ridge-and-soffit.* See Figure 51–16.

**Ventilation for Crawl-space Foundation.** Moisture rising by capillary action from the ground can cause condensation that results in stain and decay beneath the floor unit of a building with a crawl-space foundation. Vapor

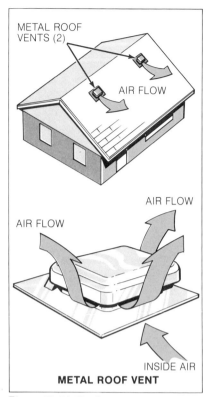

Figure 51-14. Roof vents are effective in allowing warm air to escape but do not allow much air flow into the attic space.

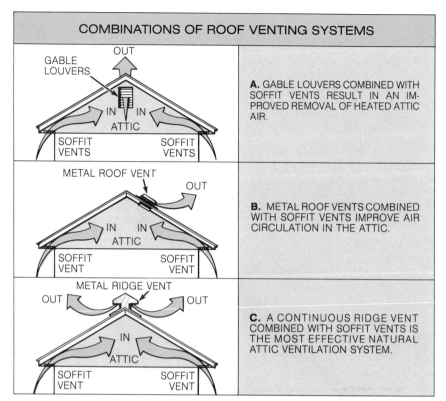

Figure 51-16. Common combinations of roof venting systems.

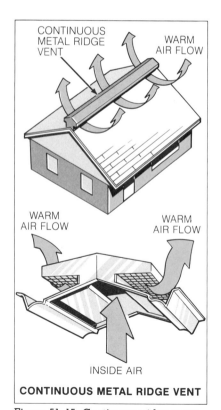

Figure 51-15. Continuous-ridge vents run along the entire ridge of the roof.

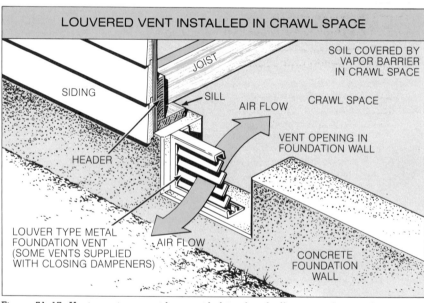

Figure 51-17. Vent openings must be provided in closed-off areas under crawl-space foundations. A variety of vent designs are available.

barriers placed over the ground help reduce moisture. In addition, vent openings should be provided, as shown in Figure 51–17. There should be at least two such openings, with a total net free area of not less that 1/1,500 of the crawl-space area. Foundation vents should be placed as high as possible and in opposite walls. Various styles of vent coverings are available.

408

# UNIT 52

# Thermal Insulation and Other Insulating Methods

Studies by government agencies show that 20% of the energy used in an average American home is used for heating water, and 10% is used for lighting rooms, cooking food, and powering appliances such as dishwashers, washing machines, and dryers. *The remaining 70% of the energy used in an average home is for heating and cooling the building.*

Today the amount of energy required to heat and cool a building has been greatly reduced by insulation and by improved construction methods.

## THERMAL INSULATION

Thermal insulation materials are placed in floors, walls, ceilings, and roofs to resist heat flow out of the building during cold weather and heat flow into the building during hot weather. Insulation products used in construction are made of mineral fibers, fiberglass, mineral wool, foam, or other organic materials.

They are classified as *loose fill, flexible, rigid, reflective,* or *foamed-in-place.* All of these products are very light in weight, since they consist largely of air trapped in and around the tiny cells that make up the material.

Insulation is rated by its R value, which indicates its resistance to heat flow. Buildings in colder climates require insulation with a higher R value. The map in Figure 52–1 shows the R values required for insulation in different areas of the United States. The chart in Figure 52–2 lists some of the more widely used insulation products with their R values.

In addition to resisting heat flow, most types of thermal insulation reduce sound transmission. Many types also have high fire-resistive ratings.

### Loose Fill Insulation

Loose fill insulation products come in large bags. The materials are either poured directly from the bag or blown in place with a pressurized hose. See Figure 52–3. Loose fill products may be composed of rock or fiberglass wool, wood fiber, shredded redwood bark, cord, vermicu-

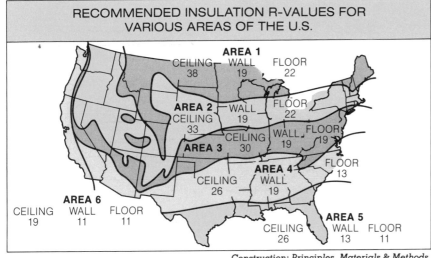

RECOMMENDED INSULATION R-VALUES FOR VARIOUS AREAS OF THE U.S.

AREA 1
CEILING 38 WALL 19 FLOOR 22

AREA 2
CEILING 33 WALL 19 FLOOR 22

AREA 3
CEILING 30 WALL 19 FLOOR 19

AREA 4
CEILING 26 WALL 19 FLOOR 13

AREA 5
CEILING 26 WALL 13 FLOOR 11

AREA 6
CEILING 19 WALL 11 FLOOR 11

*Construction: Principles, Materials & Methods*
*Institute of Financial Education*

Figure 52–1. Recommended R values for insulation used in different climate areas of the United States. Note the difference between the western and northeastern areas.

| PRODUCT | R PER 1" THICKNESS* |
|---|---|
| **LOOSE FILL** | |
| MINERAL FIBER (ROCK, SLAG OR GLASS) | 2.20–3.00 |
| CELLULOSE | 3.70 |
| PERLITE | 2.70 |
| VERMICULITE | 2.13 |
| **FLEXIBLE** | |
| MINERAL WOOL BATTS | 3.10-3.70 |
| **RIGID** | |
| CELLULAR GLASS | 2.63 |
| EXPANDED POLYSTYRENE (EXTRUDED) | 5.00 |
| EXPANDED POLYSTYRENE (MOLDED) | 3.57 |
| EXPANDED POLYURETHANE | 6.25 |
| MINERAL FIBERBOARD | 3.45 |
| POLYISOCYANURATE | 7.20 |
| **REFLECTIVE** | |
| ALUMINUM FOIL** | 3.48 |
| **FOAMED-IN-PLACE** | |
| UREA FORMALDEHYDE | 4.20 |
| POLYURETHANE | 6.25 |

*All values are for 75° F mean temperature.
**Thickness of foil not a factor; ¾" air space on room side.

Figure 52–2. R values of common insulation products.

Owens-Corning Fiberglas Corp.

Figure 52–3. A pressurized hose is used to blow loose fill insulation between the ceiling joists.

Manville Building Materials Corporation

Figure 52–4. Flexible batt (at right) and blanket (at left) insulation differ only in length.

lite, or perlite.

Loose fill insulation is often used for attics. It is placed directly on top of the ceiling below. It can also be blown or packed into the side walls of older buildings that were not insulated when constructed.

## Flexible Insulation

Flexible insulation comes in *blanket* or *batt* form. The two are alike in composition and appearance but differ in length. Batts are usually 48" long, whereas blankets are longer. See Figure 52–4. Both types are available in widths suitable for 16" O.C. and 24" O.C. stud and joist spacings. Thicknesses used in walls, floors, and ceilings usually range from 3" to 7" or more.

The more widely used flexible insulation products are made of mineral wools that are resistant to fire, moisture, and vermin. The type used most often is *faced* with a vapor barrier on one side. The facing material is generally asphalt-saturated paper or aluminum foil. This insulation has tabs extending from the sides for stapling to joists or studs. Another type of flexible insulation is *unfaced* and depends on friction to stay in place.

Flexible batt or blanket insulation is used extensively in walls, ceilings, and floors. Figure 52–5 shows a procedure for installing blanket insulation in a floor over a crawl-space foundation. First, chicken wire is secured to the

Owens-Corning Fiberglas Corporation

Figure 52-6. These batts are faced with kraft paper. Flanges extend on each side for stapling into the ceiling joists.

floor joists. Then the insulation is placed on top of the chicken wire. Figure 52–6 shows batt insulation being placed between ceiling joists.

## Rigid Insulation

Rigid insulation is available in panels or tiles. Many types can be placed on walls, floors, roofs, and ceilings. They are made of wood and vegetable fibers, polyurethane and polystyrene foams, fiberglass, and other materials. There are many brand names by which rigid insulation is known. Most of them can be categorized as structural or nonstructural.

**Structural Rigid Insulation.** Structural insulation is used most often on framed exterior walls. It comes in the form of wood or cane fiber panels. Many of these products are strong enough to meet code requirements for outside wall sheathing, and no corner braces are needed in the walls. Panels ½″ thick are applied with 6d nails. Thicker panels are applied with 8d nails. Nails should be spaced 3″ apart at the edges, and 6″ apart at the intermediate studs. Staples may be used instead of nails.

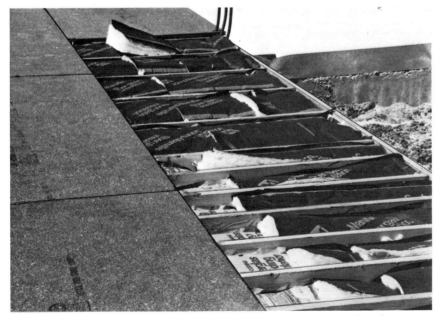

Figure 52–5. In one method of insulating the floor over a crawl-space foundation, chicken wire is secured to the floor joists. Then flexible blanket insulation is placed on top of the chicken wire. Finally, the subfloor is nailed over the insulation.

Structural insulation for walls is categorized as regular density, intermediate density, or nail-based. Panels are 2' x 8', 4' x 8', or 4' x 9', in thicknesses of ½" or 25/32". The 2' x 8' panels are also available with tongue-and-groove or shiplap joints.

Structural insulation for roofs is available in various types. One type is designed primarily for flat or sloping roofs with exposed beams. The exposed underside of the material is finished off in various designs and textures. Two other types are made of lightweight aggregate or wood fiber materials. They come in panels of different thicknesses and sizes with butt joints or tongue-and-groove edges. Recently a waferboard–polyurethane foam type has been introduced to the market.

**Nonstructural Rigid Insulation.** Nonstructural rigid insulation is manufactured from such materials as urethane foam, polystyrene foam, expanded perlite, cellular glass, and glass fibers. Panels of different sizes are

Dow Chemical U.S.A.

Figure 52–8. Exterior of a full-basement foundation completely covered with rigid insulation.

Dow Chemical U.S.A.

Figure 52–9. When rigid insulation sheets are installed on foundation walls, a protective coating is recommended on the above-grade portion of the wall.

Dow Chemical U.S.A.

Figure 52–7. Rigid (styrofoam) insulation is applied to a concrete-block foundation wall with mastic. Concrete nails hold the sheet in place while the mastic is setting up.

used to insulate foundations, walls, and roofs.

Full-basement foundation systems that are constructed of concrete or hollow concrete blocks are often insulated with nonstructural rigid panels attached to the outside walls with an adhesive. Concrete nails hold the panels in place until the mastic sets up. See Figures 52–7 and 52–8. A

protective coating is recommended for the panels on the above-grade portion of the foundation wall. See Figure 52–9. Rigid panels are also used around the perimeters of a slab-type foundation system.

To insulate exterior walls, nonstructural rigid panels are often nailed against the studs and then covered with siding or a masonry

veneer. See Figure 52–10. Aluminum-faced fiberglass sheathing is also available for exterior walls. It is applied to the walls before the walls are raised.

Roofs are often insulated with special types of nonstructural rigid panels that are combined with wood decks and covered with solid or spaced sheathing. Figure 52–11 shows nonstructural rigid panels being applied to a roof deck, then covered with spaced sheathing, and then with roof tiles.

Figure 52–12 shows the finish material for a built-up roof being applied to urethane foam insulation panels that have been applied to the roof deck.

## Reflective Insulation

Reflective insulation usually comes in the form of aluminum

foil sheets. It protects against radiant heat loss by reflecting heat rather than absorbing it. Often it is used with other insulation materials to serve as a vapor barrier. For example, flexible batts or blankets may be faced with reflective foil on one side. Gypsum wallboard and some types of rigid insulation panels may also have reflective foil on one side.

## Foamed-in-place Insulation

Foamed-in-place insulation is a recent development. Plastic chemical foam is poured or blown into the wall cavities. It then expands and completely fills the cavity. Some types of chemical foams can also be sprayed on the inside surface of an open wall. See Figure 52–13. Chemi-

*Simplex Products Division*

Figure 52–11. Rigid nonstructural panels are applied to a roof deck, then covered with spaced sheathing, and then with roof tiles.

cal foam has a very high R value. It is particularly well suited for insulating older buildings that were not insulated when constructed.

Before foam is blown into a wood-framed wall cavity, a course of shingles or siding near

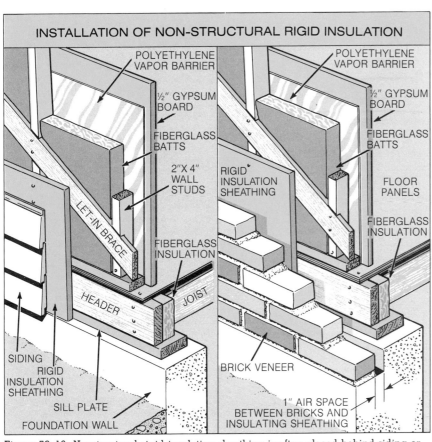

INSTALLATION OF NON-STRUCTURAL RIGID INSULATION

POLYETHYLENE VAPOR BARRIER

POLYETHYLENE VAPOR BARRIER

½" GYPSUM BOARD

½" GYPSUM BOARD

FIBERGLASS BATTS

FIBERGLASS BATTS

2" X 4" WALL STUDS

RIGID INSULATION SHEATHING

FLOOR PANELS

LET-IN BRACE

FIBERGLASS INSULATION

FIBERGLASS INSULATION

HEADER

JOIST

SIDING

RIGID INSULATION SHEATHING

SILL PLATE

FOUNDATION WALL

BRICK VENEER

1" AIR SPACE BETWEEN BRICKS AND INSULATING SHEATHING

Figure 52–10. Nonstructural rigid insulation sheathing is often placed behind siding or masonry veneer.

413

Mobay Chemical Corporation

Figure 52-12. Urethane foam insulation panels have been placed over the wood deck. Roofers are applying the finish materials for a built-up roof.

the top and bottom of the wall is removed. Small holes are then drilled through the sheathing and between the studs. The foam is blown in with a pressurized air hose. The holes are plugged up after the foam is in place. Foam can also be blown into the cavities of masonry walls if individual blocks are removed and then replaced. See Figure 52-14.

## HEAT LOSS THROUGH DOORS AND WINDOWS

In a typical house, more heat is lost through windows and exterior doors than through any other part of the building. Studies made by the U.S. Department of Housing and Urban Development show that in many areas of the country 70% of the total heating load is used to replace heat lost through doors and windows.

Most of this type of heat loss occurs as a result of *infiltration*, which is air leakage through cracks around the window and door frames. The second major cause of such heat loss is due to heat *transmission* (transfer) through the door or window material.

## Preventing Infiltration

Most heat loss caused by infiltration can be eliminated by sealing

CPR Division, The Upjohn Company

Figure 52-13. Some types of chemical foam can be sprayed on the inside surface of an open wall.

Mobay Chemical Corporation

Figure 52-14. An air gun is used to place chemical foam insulation in a masonry wall cavity.

Figure 52–15. A caulking gun with car-
tridge is used to apply caulking between the window frame and the siding.

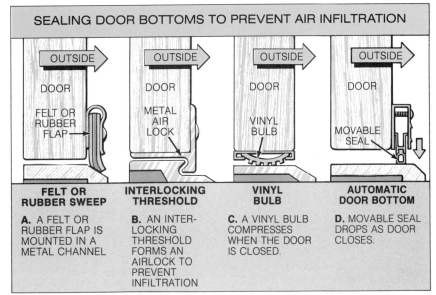

## SEALING DOOR BOTTOMS TO PREVENT AIR INFILTRATION

| FELT OR RUBBER SWEEP | INTERLOCKING THRESHOLD | VINYL BULB | AUTOMATIC DOOR BOTTOM |
|---|---|---|---|
| **A.** A FELT OR RUBBER FLAP IS MOUNTED IN A METAL CHANNEL | **B.** AN INTER-LOCKING THRESHOLD FORMS AN AIRLOCK TO PREVENT INFILTRATION | **C.** A VINYL BULB COMPRESSES WHEN THE DOOR IS CLOSED. | **D.** MOVABLE SEAL DROPS AS DOOR CLOSES. |

Figure 52–17. Methods for sealing door bottoms to prevent air infiltration.

any cracks around the frames of doors and windows. Sealing is done by caulking, weather-strip-ping, and installing thresholds under the doors.

**Caulking.** The best method for sealing cracks between a door or window frame and the outside wall is caulking. Caulking mate-rial used for construction pur-poses comes in cartridge-type tubes and is applied with a caulking gun. See Figure 52–15.

**Weatherstripping.** To prevent leakage between a door or win-dow and its frame, weatherstrip-ping materials are applied. Com-mon types of weatherstripping are adhesive-backed foam rub-ber, wood-backed foam rubber, rolled vinyl, and V-strip. Figure 52–16 shows each type and ex-plains its applications.

**Sealing Door Bottoms.** The space at the bottom of a door is a major cause of air leakage. Figure 52–17 shows how this space can be sealed off with a sweep, an interlocking threshold, an automatic sweep, or a vinyl bulb threshold.

## Preventing Transmission

Heat loss caused by transmis-sion (heat transfer) can be re-duced by using insulated doors and windows or by adding storm doors and windows, or both.

**Doors.** Both wood and metal-faced entry doors are available with plastic foam cores that allow very little heat transmission. See Figure 52–18. An older method of preventing transmission is the *storm door.* This is an additional door (wood or metal) that is hung on the outside of the door frame and can be removed during warmer weather. Metal-framed

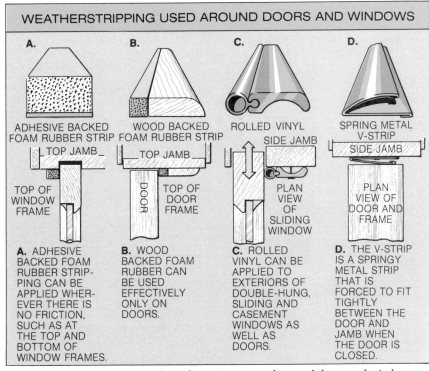

## WEATHERSTRIPPING USED AROUND DOORS AND WINDOWS

**A.** ADHESIVE BACKED FOAM RUBBER STRIP

**B.** WOOD BACKED FOAM RUBBER STRIP

**C.** ROLLED VINYL

**D.** SPRING METAL V-STRIP

**A.** ADHESIVE BACKED FOAM RUBBER STRIP-PING CAN BE APPLIED WHER-EVER THERE IS NO FRICTION, SUCH AS AT THE TOP AND BOTTOM OF WINDOW FRAMES.

**B.** WOOD BACKED FOAM RUBBER CAN BE USED EFFECTIVELY ONLY ON DOORS.

**C.** ROLLED VINYL CAN BE APPLIED TO EXTERIORS OF DOUBLE-HUNG, SLIDING AND CASEMENT WINDOWS AS WELL AS DOORS.

**D.** THE V-STRIP IS A SPRINGY METAL STRIP THAT IS FORCED TO FIT TIGHTLY BETWEEN THE DOOR AND JAMB WHEN THE DOOR IS CLOSED.

Figure 52–16. Common types of weatherstripping used around doors and windows.

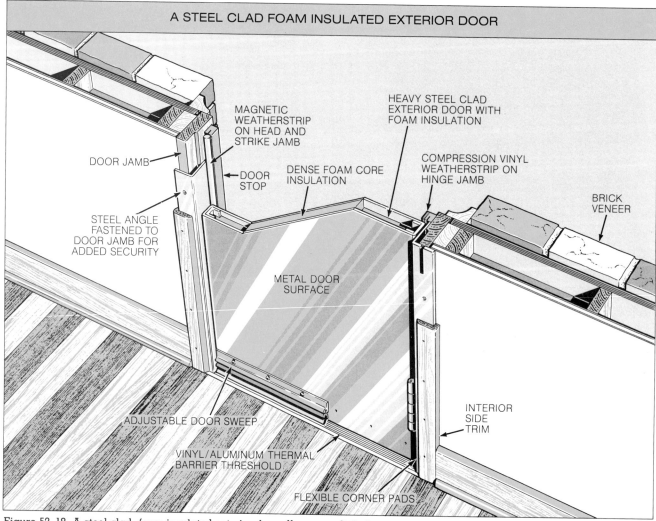

A STEEL CLAD FOAM INSULATED EXTERIOR DOOR

MAGNETIC WEATHERSTRIP ON HEAD AND STRIKE JAMB

HEAVY STEEL CLAD EXTERIOR DOOR WITH FOAM INSULATION

DOOR JAMB

DOOR STOP

DENSE FOAM CORE INSULATION

COMPRESSION VINYL WEATHERSTRIP ON HINGE JAMB

BRICK VENEER

STEEL ANGLE FASTENED TO DOOR JAMB FOR ADDED SECURITY

METAL DOOR SURFACE

ADJUSTABLE DOOR SWEEP

INTERIOR SIDE TRIM

VINYL/ALUMINUM THERMAL BARRIER THRESHOLD

FLEXIBLE CORNER PADS

Figure 52–18. A steel-clad, foam-insulated exterior door allows very little heat transmission. Note weatherstripping and threshold.

storm doors with removable glass panels are also available. The glass can be taken out and replaced with screening material during the summer.

**Windows.** Glass is a poor insulator. One square foot of ¼" clear glass conducts six to ten times more Btu per hour than a square foot section of framed wall. An effective way of reducing heat transmission through glass windows is by *multiple glazing.* The windows are made up with two or three layers of glass set into the window sash. See Figure 52–19. The air spaces between the layers of

*Continental Aluminum Products Company*
Figure 52–19. This triple-glazed window has special ½" sealed insulated glass and dead air spaces to provide a highly insulated unit.

glass greatly reduce the heat flow.

Most older homes in cold climates have *storm windows.* These are attached to the outside of the window frame. They are usually installed in the fall and taken down in the spring.

## BUILDING SYSTEMS FOR IMPROVED INSULATION

Building design and construction methods are continually being modified for energy-saving efficiency. Some improvements are (1) the use of 2" x 6" material in-

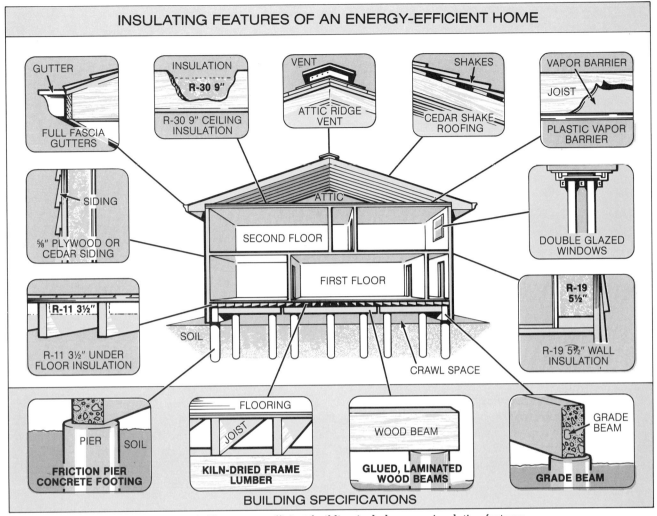

Figure 52–20. This energy-efficient building includes many insulating features.

stead of 2″ x 4″ material for exterior wall frames, (2) berms, and (3) the Plen-Wood system.

## Framing Methods

In the past, 2″ x 4″ material was used for framing exterior walls in most one-story residential construction. Increasingly, however, 2″ x 6″ lumber is being used instead. The 2″ x 6″ lumber allows the use of thicker insulation materials with higher R values inside the walls. Figure 52–20 shows a cross section of an energy-efficient house utilizing this

idea along with other energy-saving methods.

## Berms

Another way to help insulate a building is to build earth slopes called *berms* against one or more walls of the house. The earth reduces heat flow in or out of the building.

## Plen-Wood System

The Plen-Wood system is a recently developed energy-saving method that takes its name from

the word *plenum*, which refers to a space under the first floor of a building where air is heated or cooled and then distributed to the floor above. At present, the Plen-Wood system is recommended only for single-story houses. It can be incorporated in most crawl-space foundation designs.

**Operation.** In the Plen-Wood system, a downflow furnace delivers hot air to the underfloor plenum area. The hot air in the plenum then moves up into the living area through conventional floor or baseboard registers. See Figure 52–21. The Plen-Wood

417

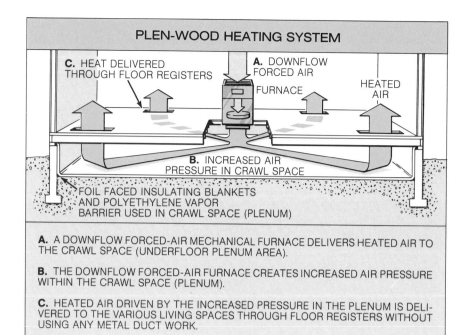

PLEN-WOOD HEATING SYSTEM

**C.** HEAT DELIVERED THROUGH FLOOR REGISTERS

**A.** DOWNFLOW FORCED AIR

FURNACE

HEATED AIR

**B.** INCREASED AIR PRESSURE IN CRAWL SPACE

FOIL FACED INSULATING BLANKETS AND POLYETHYLENE VAPOR BARRIER USED IN CRAWL SPACE (PLENUM)

**A.** A DOWNFLOW FORCED-AIR MECHANICAL FURNACE DELIVERS HEATED AIR TO THE CRAWL SPACE (UNDERFLOOR PLENUM AREA).

**B.** THE DOWNFLOW FORCED-AIR FURNACE CREATES INCREASED AIR PRESSURE WITHIN THE CRAWL SPACE (PLENUM).

**C.** HEATED AIR DRIVEN BY THE INCREASED PRESSURE IN THE PLENUM IS DELIVERED TO THE VARIOUS LIVING SPACES THROUGH FLOOR REGISTERS WITHOUT USING ANY METAL DUCT WORK.

Figure 52–21. The Plen-Wood heating system eliminates the expense of installing ducts for delivering heat.

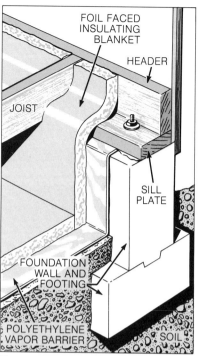

FOIL FACED INSULATING BLANKET

HEADER

JOIST

SILL PLATE

FOUNDATION WALL AND FOOTING

POLYETHYLENE VAPOR BARRIER

SOIL

Figure 52–22. In the Plen-Wood heating system, the plenum of the building is sealed off with a vapor barrier and insulating blankets.

system eliminates the expense of installing ducts for delivering heat.

**Insulation and Sealing.** The underfloor plenum area must be airtight for the Plen-Wood system to operate correctly. A polyethyl- ene vapor barrier over the ground is necessary to keep the plenum dry. A separate piece should also be placed on the foundation wall to seal off any joints. Foil-faced blankets with a minimum R value of 11 should be fastened at the top and draped along the sidewall from the subfloor to grade. See Figure 52–22. The foil face should be turned inward to act as a vapor barrier and also provide additional reflective insulation.

# UNIT 53

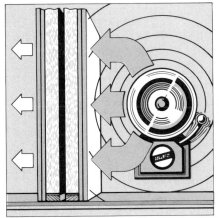

# Sound Control

Noise disturbance can be a serious problem in homes and offices. High noise levels interfere with rest and relaxation at home. They contribute to inefficiency and fatigue at work. Constant exposure to extremely high noise levels can cause ear injury and hearing loss.

Noise problems can be reduced through the use of insulation materials and special sound-reducing construction methods for walls and ceilings. Many of the materials used for sound insulation are the same as or similar to the materials used for thermal insulation.

## SOUND TRANSMISSION AND SOUND LEVELS

Sound travels in different ways and at various levels. These factors must be considered in designing an effective sound control system.

## Airborne Sound Transmission

Airborne noise such as speech or music is conducted through the air as pressure waves that strike against the wall surface. The sound is then transmitted through the air in the wall cavity to the opposite wall surface,

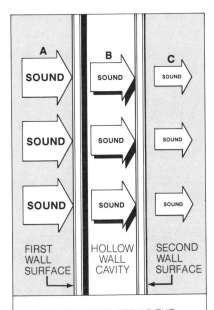

**A.** SOUND WAVES STRIKE THE FIRST WALL SURFACE, MAKING IT VIBRATE.

**B.** SOUND WAVES PASS THROUGH THE WALL CAVITY AND STRIKE THE SECOND WALL SURFACE.

**C.** THE SECOND WALL SURFACE VIBRATES AND TRANSMITS THE SOUND INTO THE ADJOINING ROOM.

Figure 53–1. Airborne sound transmission is conducted through the air as pressure waves.

causing it to vibrate and transmit the sound to the adjoining room. See Figure 53–1. Airborne sound travels through floors and ceilings in the same way that it travels through walls.

Resistance of a floor, wall, or ceiling system to airborne sound transmission is indicated by its *sound transmission class* (STC) rating. See Figure 53–2. A high STC number indicates more resistance than a low number. Acceptable STC ratings range from approximately 39 to 65.

| | |
|---|---|
| 25 | NORMAL SPEECH CAN BE UNDERSTOOD QUITE EASILY |
| 30 | LOUD SPEECH CAN BE UNDERSTOOD FAIRLY WELL |
| 35 | LOUD SPEECH AUDIBLE BUT NOT INTELLIGIBLE |
| 42 | LOUD SPEECH AUDIBLE AS A MURMUR |
| 45 | MUST STRAIN TO HEAR LOUD SPEECH |
| 48 | SOME LOUD SPEECH BARELY AUDIBLE |
| 50 | LOUD SPEECH NOT AUDIBLE |

Figure 53–2. Common sound transmission class (STC) ratings.

419

## Structure-borne Sound Transmission

Structure-borne sound transmission is produced when a part of the building structure, such as the floor, is set into vibration by a direct impact. Walking on the floor or dropping objects on it, and striking the surface of the wall or ceiling are examples of the way structure-borne sound transmission can be produced by impact. See Figure 53–3.

Resistance of any part of the building structure to structure-borne sound transmission is indicated by its *impact insulation class* (IIC) rating. A high number indicates more resistance than a low number. An IIC rating of 55 is needed for good control of impact noise. A rating of 60 or

| | DECIBELS | THRESHOLD OF FEELING |
|---|---|---|
| DEAFENING | 120 — — 110 — — 100 — | THUNDER, ARTILLERY NEARBY RIVETER ELEVATED TRAIN BOILER FACTORY |
| VERY LOUD | — 90 — — 80 — | LOUD STREET NOISE NOISY FACTORY TRUCK UNMUFFLED POLICE WHISTLE |
| LOUD | — 70 — — 60 — | NOISY OFFICE AVERAGE STREET NOISE AVERAGE RADIO AVERAGE FACTORY |
| MODERATE | — 50 — — 40 — | NOISY HOME AVERAGE OFFICE AVERAGE CONVERSATION QUIET RADIO |
| FAINT | — 30 — — 20 — | QUIET HOME OR PRIVATE OFFICE AVERAGE AUDITORIUM QUIET CONVERSATION |
| VERY FAINT | — 10 — — 0 — | RUSTLE OF LEAVES WHISPER SOUND PROOF ROOM THRESHOLD OF AUDIBILITY |

Figure 53–4. The intensity of sound is expressed in decibels.

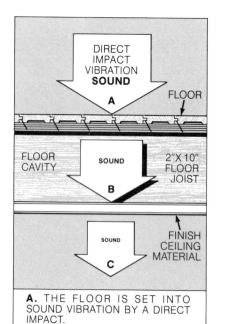

A. THE FLOOR IS SET INTO SOUND VIBRATION BY A DIRECT IMPACT.

B. REDUCED SOUND WAVES PASS THROUGH THE FLOOR CAVITY AND STRIKE THE CEILING MATERIAL BELOW.

C. THE CEILING MATERIAL SURFACE VIBRATES AND TRANSMITS THE SOUND INTO THE ROOM BELOW.

Figure 53–3. Impact produces structure-borne sound transmission.

more is needed for maximum privacy. A rating of less than 50 is considered unacceptable for sound control.

Structure-borne transmission is also caused by appliances such as vacuum cleaners, washers, dryers, garbage disposals, fans, and compressors and by pipes and ducts.

## Sound Levels

The intensity of sound is expressed in *decibels* (db). One decibel is equal to the smallest change in sound intensity that can be detected by the average human ear. Figure 53–4 shows a decibel scale ranging from zero to 120.

## SOUND CONTROL FOR WALLS

A typical interior wood stud wall, with a single layer of ⅝" gypsum wallboard on each side, has an STC rating of 34. See Figure 53–5. According to the chart in Figure 53–2 this type of wall is effective against quiet conversation on the other side of the wall, but not against loud speech.

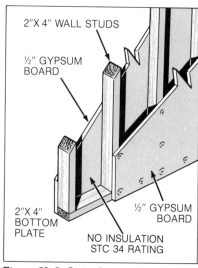

Figure 53–5. A single wood stud wall with a single layer of ½" gypsum board on each side and no insulation between the studs has a STC rating of 34.

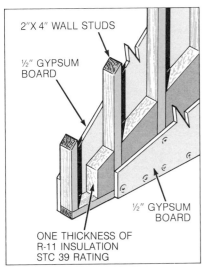

Figure 53–6. The STC rating of a single wood stud wall with ½" gypsum board on each side can be increased from 34 to 39 by placing one thickness of insulation material between the studs.

Three methods of reducing sound transmission are: (1) to provide cavity absorption, (2) to increase mass, and (3) to break the sound vibration path. Best results are obtained when these methods are combined. Staggered-stud or double-stud walls also reduce sound transmission.

## Cavity Absorption

One way to reduce sound transmission through a wall is to fill the wall cavity with sound-absorbing materials. Fiberglass is often used for this purpose. The STC rating of a single wood stud wall with ½" gypsum wallboard on each side can be increased from 34 to 39 by placing one thickness of insulation material between the studs. See Figure 53–6.

## Increasing Mass

Heavy materials block sound better than light materials. An ex-

ample of increasing mass is adding another layer of gypsum wallboard to one or both sides of a partition. Doubling the gypsum wallboard on both sides increases the STC rating of a single wood stud wall (with one thickness of insulation material between the studs) from 39 to 45. See Figure 53–7.

## Breaking Vibration Path

Most sounds traveling through walls are a result of vibration from one face of the wall to the other through structural members such as wall studs. Metal is a more resilient material in relation to sound transmission than wood. Therefore, walls constructed with metal studs transmit less sound vibration than walls constructed with wood studs. An effective way of improving the STC rating of wood stud walls is to place a metal resilient channel between the gypsum wallboard and the stud to break the vibration path. A single wood stud wall with one thick-

ness of insulation material between the studs, double layers of gypsum wallboard on both sides, and a resilient channel has an STC rating of about 56. See Figure 53–8.

## Staggered-stud and Double-stud Walls

In the staggered-stud system, a 2" x 6" plate is used with 2" x 4" wall studs. See Figure 53–9. In the double-stud system, two separate wood stud walls are built with a space between them for insulation. See Figure 53–10. With insulation material between the studs and double layers of gypsum wallboard on both sides of the wall, a staggered-stud wall has an STC rating of about 55, and a double-stud wall has an STC rating of about 63.

## FLOOR-CEILING SOUND CONTROL

Probably the most disturbing household noises are footsteps

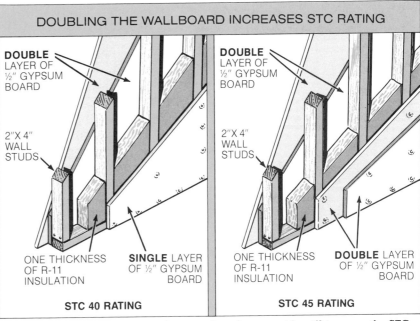

Figure 53–7. Doubling the layer of gypsum board on one side will increase the STC rating of single wood stud wall with one thickness of insulation material from 39 to 40. Doubling the gypsum board on both sides will increase it to 45.

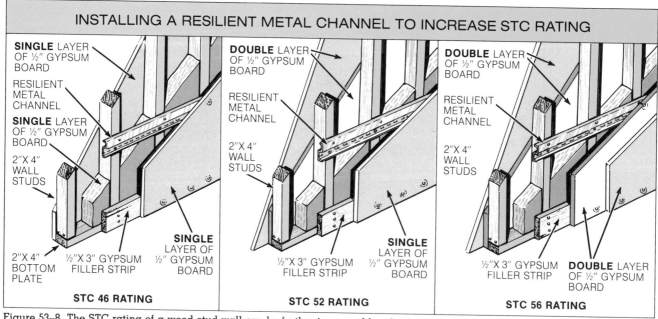

Figure 53–8. The STC rating of a wood stud wall can be further increased by placing a resilient channel on one side of the wall.

and vibrations from the floor above. These sounds are also more difficult than other sounds to eliminate. Several floor-ceiling sound control methods have been developed. See Figure 53–11. They all require carpets and pads over the subfloor. Mineral insulation is placed between the joists. The ceiling below consists of gypsum board fastened to resilient channels.

## SOUND CONTROL BY BUILDING DESIGN AND CONSTRUCTION PRACTICES

The location of doors and windows in a building is important to sound control. Building design should take this factor into account. When possible, windows should not face noisy areas. They should be separated to reduce cross-talk. Doors opening on opposite sides of a hallway should be staggered. See Figure 53–12. In some cases, sound control requires that window area

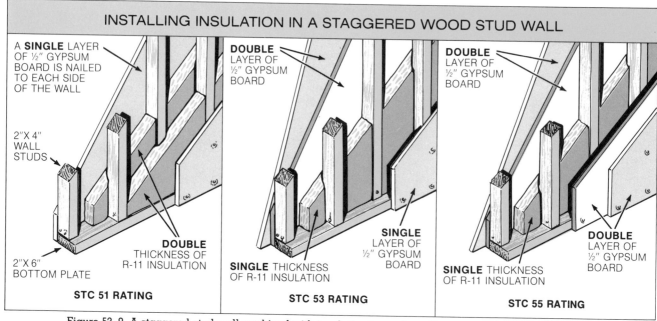

Figure 53–9. A staggered-stud wall combined with insulation is another effective method of sound control.

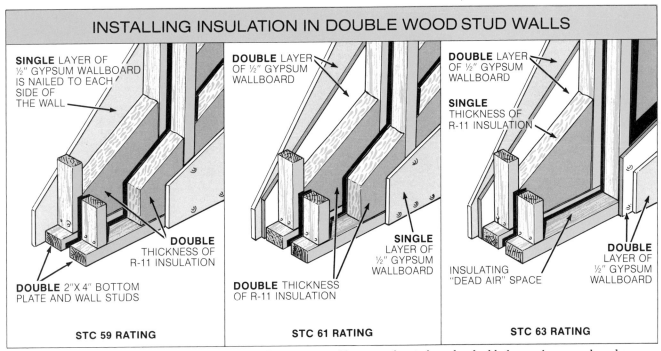

## INSTALLING INSULATION IN DOUBLE WOOD STUD WALLS

**SINGLE** LAYER OF ½" GYPSUM WALLBOARD IS NAILED TO EACH SIDE OF THE WALL

**DOUBLE** THICKNESS OF R-11 INSULATION

**DOUBLE** 2"X 4" BOTTOM PLATE AND WALL STUDS

**STC 59 RATING**

**DOUBLE** LAYER OF ½" GYPSUM WALLBOARD

**DOUBLE** THICKNESS OF R-11 INSULATION

**SINGLE** LAYER OF ½" GYPSUM WALLBOARD

**STC 61 RATING**

**DOUBLE** LAYER OF ½" GYPSUM WALLBOARD

**SINGLE** THICKNESS OF R-11 INSULATION

INSULATING "DEAD AIR" SPACE

**DOUBLE** LAYER OF ½" GYPSUM WALLBOARD

**STC 63 RATING**

Figure 53–10. A doubled-stud wall that includes insulation material between the studs and a double layer of gypsum board on both sides provides extremely effective sound control (STC rating of 63).

be reduced from the original plan.

Several precautions should be taken in construction to reduce sound transmission. For example, all perimeters where walls butt against each other or against floors and ceilings should be properly sealed. A nonhardening, permanently resilient caulking material such as butyl-rubber-based compound is

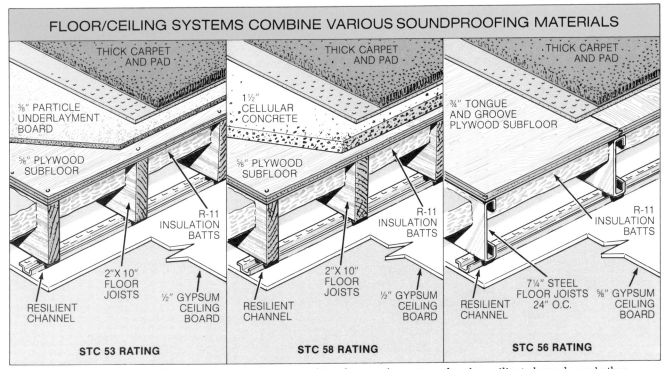

## FLOOR/CEILING SYSTEMS COMBINE VARIOUS SOUNDPROOFING MATERIALS

THICK CARPET AND PAD

⅜" PARTICLE UNDERLAYMENT BOARD

⅝" PLYWOOD SUBFLOOR

R-11 INSULATION BATTS

2"X 10" FLOOR JOISTS

½" GYPSUM CEILING BOARD

RESILIENT CHANNEL

**STC 53 RATING**

THICK CARPET AND PAD

1½" CELLULAR CONCRETE

⅝" PLYWOOD SUBFLOOR

R-11 INSULATION BATTS

2"X 10" FLOOR JOISTS

½" GYPSUM CEILING BOARD

RESILIENT CHANNEL

**STC 58 RATING**

THICK CARPET AND PAD

¾" TONGUE AND GROOVE PLYWOOD SUBFLOOR

R-11 INSULATION BATTS

7¼" STEEL FLOOR JOISTS 24" O.C.

⅝" GYPSUM CEILING BOARD

RESILIENT CHANNEL

**STC 56 RATING**

Figure 53–11. Effective floor-ceiling noise control systems combine the use of carpets and pads, resilient channels, and other sound-deadening materials.

recommended for both sides of a partition at the top and bottom plates. Joint compound and tape applied on multiple-layered and staggered wallboard is also an effective seal.

The use of insulated glass in windows and proper weather-stripping around windows and doors helps reduce transmission of outside sound into the building. Solid wood-core or metal doors should be used. Sliding doors should be avoided.

Proper sealing around electrical and plumbing installations is also important to sound control. Cut tight-fitting holes for electrical boxes, and apply elastic caulking around the outlets before the plates are installed. See Figure 53–13. (Do not place light switches or outlets back to back.) Seal around surface-mounted ceiling fixtures in gypsum ceilings to make them airtight. See Figure 53–14.

Apply elastic caulking or other resilient material around all openings made for pipes in the wall plates as well as in the wall covering. See Figure 53–15. Seal all joints between subfloor panels. See Figure 53–16.

Noise control can also be reduced by enclosing noisy equipment when possible. Also, if cabinets are placed on opposite sides of a wall, they should not

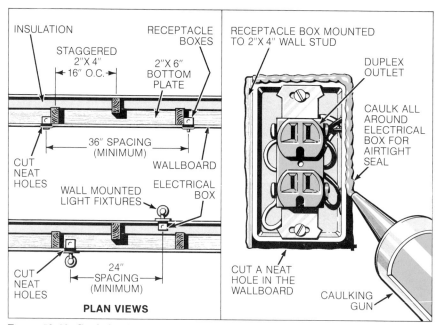

Figure 53–13. Cut holes for electrical outlets neatly to reduce noise leaks. Place elastic caulking around the outlets before installing cover plates.

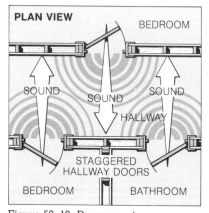

Figure 53–12. Doors opening on opposite sides of a hallway should be staggered whenever possible.

be placed back to back unless they are surface-mounted. See Figure 53–17.

## ROOM ACOUSTICS

Airborne and structure-borne noises enter a room through the walls, floors, and ceilings. However, many disturbing noises are also created within the room. Such sounds include voices, the sound of operating appliances, and the sound produced by walking on non-carpeted floors. In offices, shops, and factories, high noise levels are created by machinery and other types of working equipment.

High levels of sound created within an enclosed space, such as a room, are caused by *reflection*. For example, the sound waves created by a typewriter in use travel in all directions until they strike against an obstacle such as a wall or ceiling. The sound waves then bounce off these reflective surfaces, creating a greater noise disturbance than the original sound. See Figure 53–18.

## Absorbing Sound

Ordinary smooth, hard building surfaces reflect up to 98% of the sound that strikes them. This reflection can be greatly reduced by covering some of the room surfaces with acoustical materials that are designed to absorb the sound. When properly applied, such materials reduce

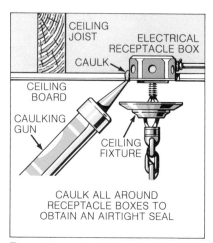

Figure 53–14. Surface-mount ceiling fixtures placed in gypsum ceilings. Seal openings around the fixtures to make them airtight.

sound reflection as much as 50%.

**Acoustical Tile.** Sound-absorbing material usually is in the form of acoustical tile. The tiles are manufactured from wood fiber or similar materials and are available in different sizes of square or rectangular shapes. They are generally applied to ceiling areas directly below the joists or are used in suspended ceilings.

Acoustical tile is designed with numerous tiny sound traps in the tile surface. These sound traps consist of drilled or punched holes or fissures, or a combina-

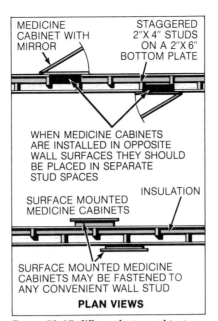

WHEN MEDICINE CABINETS ARE INSTALLED IN OPPOSITE WALL SURFACES THEY SHOULD BE PLACED IN SEPARATE STUD SPACES

SURFACE MOUNTED MEDICINE CABINETS MAY BE FASTENED TO ANY CONVENIENT WALL STUD

**PLAN VIEWS**

Figure 53-17. When placing cabinets on opposite sides of a wall, separate the cabinets or surface-mount them as shown.

tion of both. When sound strikes the tile, it is trapped in the holes or fissures. See Figure 53-19. Installation of acoustical tile is

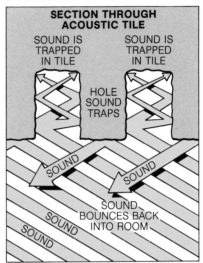

Figure 53-19. Sound-absorbing materials such as acoustical tile have holes or fissures, or a combination of both, to trap sound.

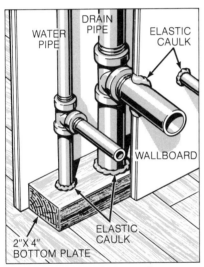

Figure 53-15. Use elastic caulking or other resilient materials around all openings made for pipes in bottom plates or the wall covering.

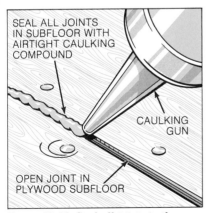

Figure 53-16. Seal all joints in the subfloor with caulking compound to make them airtight.

**SOUND WAVES ARE REFLECTED WITHIN AN ENCLOSED AREA**

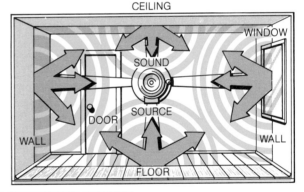

**A.** SOUND WAVES TRAVEL RADIALLY IN ALL DIRECTIONS. THEY ARE REFLECTED BY THE HARD SURFACES OF THE ROOM (WALLS, FLOOR AND CEILING).

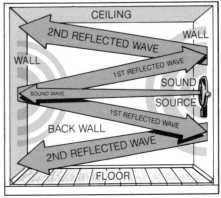

**B.** THE SOUND WAVES FROM THE ORIGINAL SOUND SOURCE CONTINUE TO BE REFLECTED BACK AND FORTH IN WHAT IS CALLED MULTIPLE REFLECTION. THIS CAN CAUSE A HIGHER NOISE LEVEL THAN THE ORIGINAL SOUND SOURCE.

Figure 53-18. Reflection creates high levels of sound within an enclosed area.

discussed in Section 13.

**Acoustical Plaster.** Another acoustical product is a plaster that is usually sprayed directly against the ceiling. Acoustical plaster is generally a mixture of ground gypsum combined with chemical ingredients that cause tiny air bubbles to form in the material. The air bubbles trap and absorb sound.

# UNIT 54

# Solar Energy

Solar heating is an important alternate method of energy-saving. Many buildings are being constructed to accommodate solar heating. Solar heating systems are also being installed in many older structures.

Two basic types of solar heating are used. One is the *passive* method, which does not rely on any mechanical means. The other is the *active* method, which requires a sophisticated system of collection, transport, and storage. Both methods rely on direct energy from the sun's rays, which produce the energy that is collected and distributed within the building.

Solar heating systems as they exist today cannot entirely replace other types of heating systems within a building. During prolonged cloudy or stormy weather, the conventional gas, electric, or wood-burning methods of heating are required. However, a solar heating system can significantly reduce heating costs in a building and cut down on the use of fossil fuels.

## PASSIVE SOLAR HEATING

In a passive solar heating system, the building structure is designed to both collect and store solar energy. The south-facing side of the building must be provided with large areas of glass or plastic glazing through which sunlight can pass into the building. See Figures 54–1 and 54–2. Once collected, the heat is absorbed and stored by thick masonry materials or by water. In the post-and-beam building in Figures 54–1 and 54–2, heat is absorbed into and stored in the tile floor, the brick wall, and a rockbed under the floor. The sloped window areas can be shaded during summer months.

Principles of passive solar heating date back 2,500 years. Using these principles, the Greeks built entire cities facing

*Deck House, Inc., Acton, MA*

Figure 54–1. A contemporary house with passive solar heating. This post-and-beam building relies on a broad expanse of south-facing windows to collect winter sunlight for interior heating purposes. Berms, shown at the lower wall at the right, are another important temperature control feature.

427

Figure 54–2. Interior view of the conservatory collecting area of the house shown in Figure 54–1. The sloped window areas combined with the lower sections allow for maximum collection of winter sunlight. (The sloped area can be shaded during summer months.) Sliding doors, placed strategically in the conservatory, can be opened to allow warm air to flow into other living areas of the house. Heat for later use (after the sun goes down) is absorbed into and stored in the tile floor and brick wall, as well as in a rockbed under the floor.

*Deck House, Inc., Acton, MA*

## Collectors

Collectors are the large glass or plastic areas through which sunlight enters the building. They should face within 30° of true south and should not be shaded by trees or other buildings.

## Absorber

The absorber is the hard, darkened surface of the storage material. This surface should be in the direct path of the sunlight that enters through the collectors. The sunlight is absorbed as heat.

## Storage

The storage material is usually masonry (concrete, concrete block, terra cotta tile) or water. It is directly behind or below the absorber. (The absorber for masonry storage material is simply the exposed surface of the storage material.) The heat absorbed from sunlight by the absorber is retained in the storage material.

## Distribution

The method of distribution of solar heat is conduction, convection, and radiation. Sometimes mechanical devices such as fans, ducts, and blowers are used to help circulate the heat through the building.

## Control

The control element is some type of movable insulation such as screens, drapes, or aluminized shades that can be applied at nighttime to the collectors to prevent heat loss back through the collectors. The movable insulation is also applied to the collectors during summer to keep the building cool.

south to take full advantage of the sun's rays. In this way they were able to cope with the "energy crisis" they experienced when overuse of wood fuel had stripped the countryside of its trees.

Five elements are necessary for a passive solar heating system (Figure 54–3):

1. *Collectors* through which sunlight enters the building.
2. An *absorber* to take in the heat.
3. *Storage* materials to retain the heat.
4. A method of *distribution* of the heat.
5. A *control* element to prevent heat loss back through the collector.

The three main types of passive solar heating system are *direct gain, indirect gain,* and *isolated gain.* See Figure 54–4.

## Direct Gain

The simplest passive method is the direct gain method. Sunlight enters through large, south-facing windows (collectors) and strikes the walls and floors. The walls and floors should be 4″ to 8″ thick masonry material such as concrete, concrete blocks, or brick. The surfaces should be a dark color in order to absorb the sun's heat, which is then stored in the masonry. At night, as the room cools, the heat stored in the masonry will radiate into the room. This is called a *time-lag* heating process. To control heat loss, movable insulation is pulled down to cover the collector area at night. During the summer the movable insulation covers the collector area during the day to prevent the sun from overheating the building.

## Indirect Gain

The best known passive systems are two indirect gain systems: the *Thrombe wall* and the *water wall.* In these systems, the solar radiation is intercepted by an absorber-and-storage unit that separates the south-facing glass from the room.

The Thrombe wall system uses an 8″ to 16″ masonry wall. A single or double layer of glass or plastic glazing is mounted about 4″ in front of the wall surface between the masonry wall and the glazing. The dark outside surface of the wall then absorbs the heat, which is stored in the wall mass.

The water wall system uses water-filled containers instead of a masonry wall. Water walls can be built in a number of ways. One way is to use tall, hard plastic tubes as shown in Figure 54–4. A water wall will actually absorb and store more heat than a masonry wall of equal volume.

## Isolated Gain

The isolated gain method requires a separate space such as

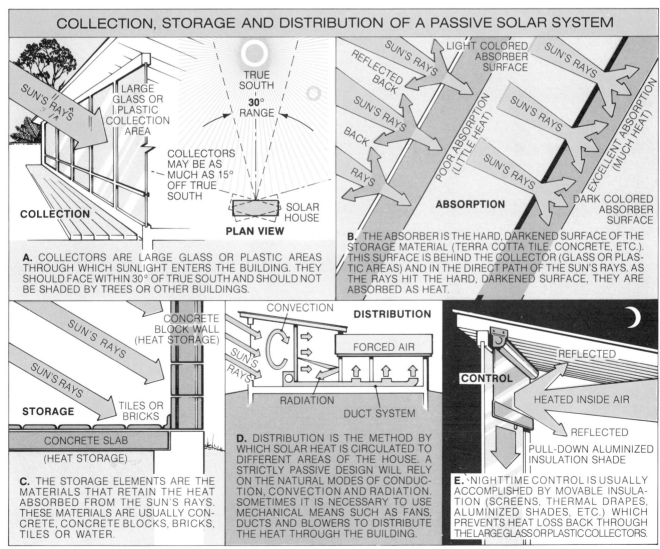

Figure 54–3. Five basic elements are necessary for a passive solar heating system.

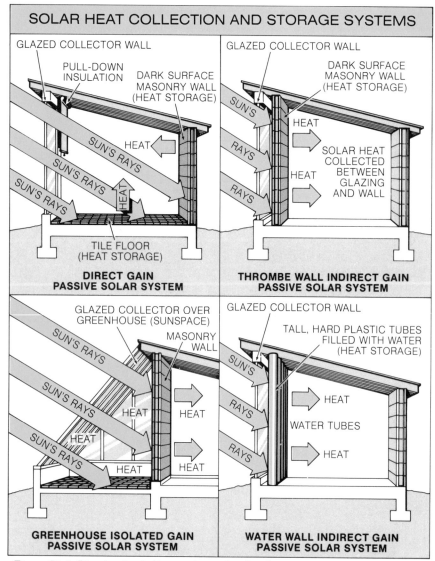

SOLAR HEAT COLLECTION AND STORAGE SYSTEMS

GLAZED COLLECTOR WALL
PULL-DOWN INSULATION
DARK SURFACE MASONRY WALL (HEAT STORAGE)
SUN'S RAYS
HEAT
TILE FLOOR (HEAT STORAGE)

**DIRECT GAIN PASSIVE SOLAR SYSTEM**

GLAZED COLLECTOR WALL
DARK SURFACE MASONRY WALL (HEAT STORAGE)
SUN'S RAYS
HEAT
SOLAR HEAT COLLECTED BETWEEN GLAZING AND WALL

**THROMBE WALL INDIRECT GAIN PASSIVE SOLAR SYSTEM**

GLAZED COLLECTOR OVER GREENHOUSE (SUNSPACE)
MASONRY WALL
SUN'S RAYS
HEAT

**GREENHOUSE ISOLATED GAIN PASSIVE SOLAR SYSTEM**

GLAZED COLLECTOR WALL
TALL, HARD PLASTIC TUBES FILLED WITH WATER (HEAT STORAGE)
SUN'S RAYS
HEAT
WATER TUBES
HEAT

**WATER WALL INDIRECT GAIN PASSIVE SOLAR SYSTEM**

Figure 54–4. Direct gain, indirect gain, and isolated gain passive solar heating systems. One indirect gain system uses a Thrombe wall, and the other uses a water wall.

a greenhouse (also called a sunspace, solar room, solarium, or atrium) to capture the solar radiation.

The greenhouse can be built as part of a new buiding or added to an older building. Solar heat is collected through the greenhouse glazing. It can be absorbed and stored in various ways. A masonry wall or water-filled containers can be used for storage. Several methods can be used to transfer the heat collected in the greenhouse to the living space. If a masonry wall is

used for heat storage, a time-lag heating process can take place. Ceiling and floor-level vents may allow for a natural convective loop of warm and cold air. A low-horsepower fan or blower can also be used.

## ACTIVE SOLAR HEATING

Unlike a passive system, an active solar heating system does not require a specially designed building. An active system is a mechanical method that can be

installed in most buildings of new or old design. It heats space or water, or both. Some of these systems can provide up to 80% of the winter space-heating requirements and 60% of the water-heating requirements of a well-insulated house.

The basic elements of an active solar heating system are (Figure 54–5):
1. Collectors.
2. A thermal storage unit.
3. An air mover.
4. An auxiliary heating system.

## Collectors

Collectors are usually rectangular-shaped containers with transparent glass or plastic glazing to allow sunlight to enter, an absorber plate to absorb the heat from the sunlight, and insulation to reduce heat loss from the collector. See Figure 54–6.

Usually the collectors are roof-mounted and face south. See Figure 54–7. In new construction they can be installed between the roof rafters. When placed on older homes, they are usually mounted on wood strips or metal brackets fastened to the top of the existing roofs. In another method of active solar heating, the collectors are installed on a structure remote from the building to be heated. This method is an advantage when the roof of a building does not lend itself to southern exposure for the collectors. See Figure 54–8.

## Air Movement after Entering Collectors

Air is introduced into one end of the collectors through a pipe or duct. It is heated as it passes along the surface of the absorber plate within the collector. The hot air exits from the opposite end of

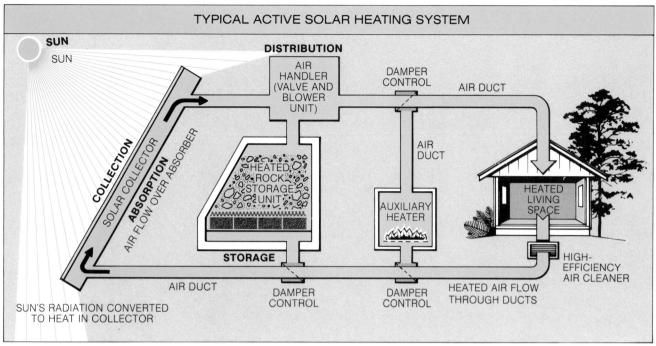

Figure 54–5. Schematic drawing for an active solar heating system used for space heating.

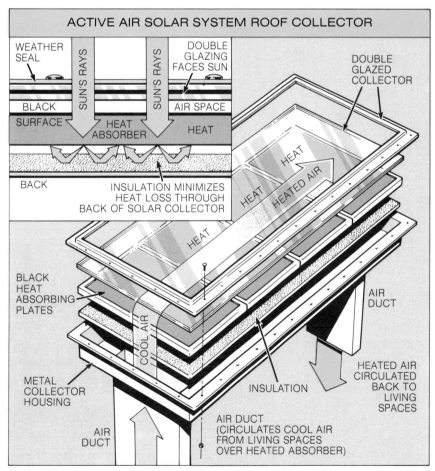

the collectors, where it is carried directly into the building space to be heated. If heat is not required in the building space at the particular time, it can be directed to a heat storage unit.

## Thermal Storage Unit

A method of storing heat is necessary, since much of a building's heating needs occur in the evening after the sun has set and solar radiation is no longer available. A thermal storage unit for active systems usually consists of small rocks, 1″ to 1½″ in diameter, housed in a wood or concrete box. See Figure 54–9. The size of the rock storage area depends on the size of the collector area. Most manufacturers recommend approximately ½ to 1 cubic foot of rock for every square foot of the collector.

The thermal storage unit requires an open area at the bottom. This open area is called a *plenum*. A plenum can be created with hollow concrete blocks

Figure 54–6. A typical collector used with an active solar heating system. The sun's rays enter through the insulated glass. The black absorber plate absorbs the heat transmitted through the glass. Insulation behind the absorber minimizes heat loss.

*Acorn Structures, Inc.*©

Figure 54–7. Solar collectors on the roof face south for maximum exposure to the sun. A glassed-in solarium at the first floor level at the front of the house permits passive solar collection in addition to the active system on the roof.

*Research Products Corporation*

Figure 54–8. A remote solar collector unit is necessary when a roof-mounted system is not practical.

## STORING SOLAR HEAT IN A WELL-INSULATED THERMAL ROCK STORAGE UNIT

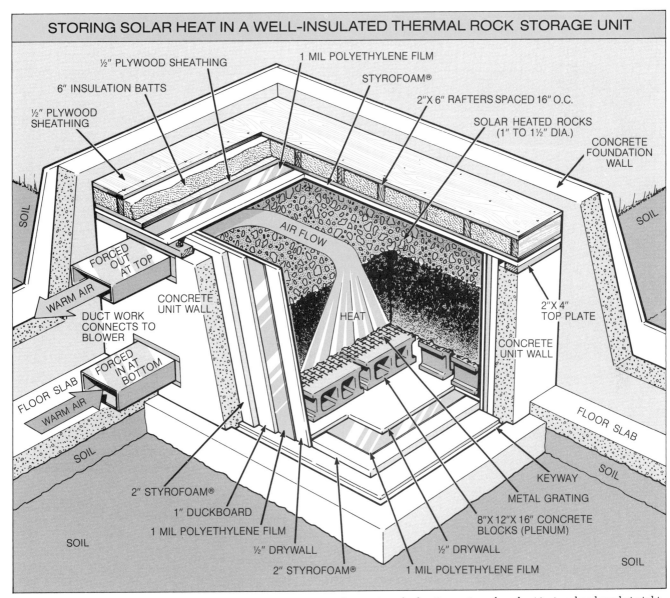

½" PLYWOOD SHEATHING

6" INSULATION BATTS

½" PLYWOOD SHEATHING

1 MIL POLYETHYLENE FILM

STYROFOAM®

2"X 6" RAFTERS SPACED 16" O.C.

SOLAR HEATED ROCKS (1" TO 1½" DIA.)

CONCRETE FOUNDATION WALL

SOIL

SOIL

AIR FLOW

FORCED OUT AT TOP

WARM AIR

DUCT WORK CONNECTS TO BLOWER

CONCRETE UNIT WALL

HEAT

2"X 4" TOP PLATE

CONCRETE UNIT WALL

FORCED IN AT BOTTOM

FLOOR SLAB

WARM AIR

FLOOR SLAB

SOIL

SOIL

2" STYROFOAM®

1" DUCKBOARD

1 MIL POLYETHYLENE FILM

½" DRYWALL

2" STYROFOAM®

½" DRYWALL

1 MIL POLYETHYLENE FILM

KEYWAY

METAL GRATING

8"X 12"X 16" CONCRETE BLOCKS (PLENUM)

SOIL

SOIL

SOIL

Figure 54-9. Warm air can be directed to a thermal rock storage unit of an active solar heating system where heat is stored and used at night.

covered with a steel grate, with allowance made for another air space at the top of the enclosure. The entire thermal storage unit must be well sealed and insulated.

## Air Mover

An air mover unit (air handler) includes a blower assembly and automatically controlled dampers. It is usually positioned next to the rock storage enclosure. Through control of the automatic dampers, the solar-heated air

can be moved directly to the building space to be heated, or the heated air can be directed into the rock-filled enclosure for storage. Later the air mover can be used to direct the hot air from the storage area into the space to be heated.

## Auxiliary Heating Systems

All buildings equipped with an active solar heating system should have an auxiliary heating

system for use on days when there is not adequate sunshine to operate the solar system. The auxiliary system may be powered by natural gas, propane, fuel oil, or electricity.

## Automatic Controls

The more sophisticated active solar heating systems feature a fully automatic control system. It coordinates the total operation of the system in conjunction with thermostats set in the living area of the building. As a result, four

433

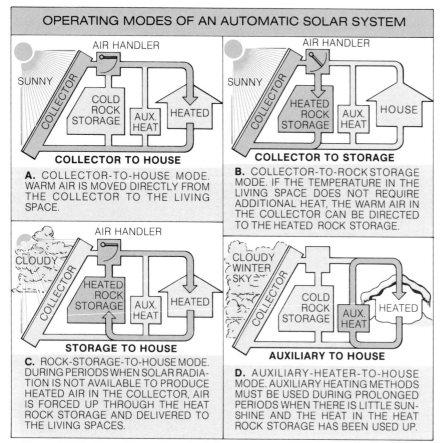

Figure 54–10. Operating modes of a fully automatic active solar heating system.

operating modes are possible (Figure 54–10):

1. Collectors-to-house mode.
2. Collectors-to-storage mode.
3. Storage-to-house mode.
4. Auxiliary-to-house mode.

In the collectors-to-house mode, warm air is moved directly from the collectors to the living area.

In the collectors-to-storage mode, the warm air from the collectors is directed to the rock storage unit when the temperature of the living area is high enough that additional heat is not required.

In the storage-to-house mode, warm air is forced up through the heated rock storage unit and delivered to the living area when solar radiation is not available to produce heated air in the collectors.

In the auxiliary-to-house mode, mechanical heating methods are used when solar radiation is not available to produce heated air in the collectors and stored heat has been used up.

# SECTION 12

# Exterior Finish

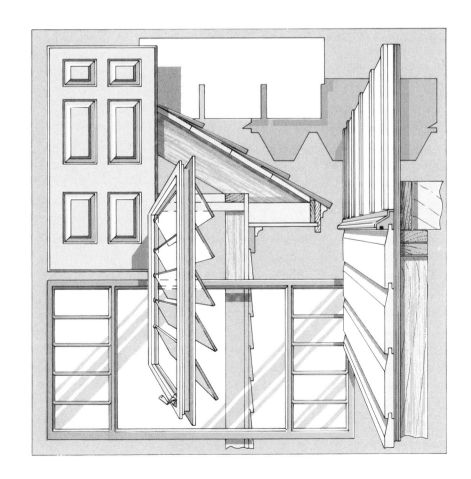

The exterior finish of a building includes the door and window units and all the materials that cover the roof and exterior walls. Wood, metal, or plastic products may be used for exterior finish. The materials chosen for a particular building should be in keeping with its general design. They must also give weathertight protection to the roof and exterior walls.

# UNIT 55

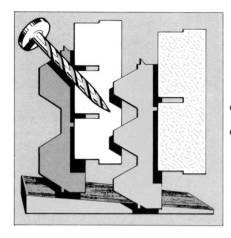

# Roof Finish

Roof finish includes overhang work and cornice work as well as the finish materials covering the roof. Finish work can begin immediately after the roof has been sheathed and after all items that will project from the roof have been placed. (Items that project from the roof include chimneys, vent pipes, flues, and sometimes electrical conduits.) Completing the overhangs and cornices is the first step in finishing the roof.

## ROOF OVERHANGS AND CORNICE WORK

The roof overhangs *(eaves)* are the portions of the roof that project past the side walls of the building. The cornice is the area beneath the overhangs. Several basic designs are used for finishing off the roof overhangs and cornices. Most of these designs come under the category of *open cornice* or *closed cornice.* They not only add to the attractiveness of a building but also perform a practical function. They help protect the side walls of the building from rain and snow, and wider overhangs also shade windows from the hot summer sun.

## Open Cornice

In open cornice construction the undersides of the rafters and roof sheathing are exposed. A 1″ or 2″ thick trim piece *(fascia board)* is usually nailed to the tail ends of the rafters. See Figure 55–1. Most spaces between the rafters are blocked off. Some spaces are left open (and screened) to allow attic ventilation. Usually a *frieze board* is nailed to the wall below the rafters. Sometimes the frieze board is notched between the rafters and molding is nailed over it. See Figures 55–2 and 55–3.

## Closed Cornice

In closed cornice construction the bottom of the roof overhang is closed off. The two most common types of closed cornices are the *flat box cornice* and the *sloped box cornice.*

The flat box cornice requires framing pieces called *lookouts* that are toenailed to the wall and facenailed to the ends of the rafters. The lookouts provide a nailing base for the *soffit,* which is the material that is fastened to the underside of the cornice. A typical flat box cornice is shown

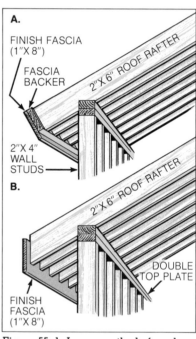

**A.**

FINISH FASCIA (1″ X 8″)

FASCIA BACKER

2″ X 6″ ROOF RAFTER

2″ X 4″ WALL STUDS

**B.**

2″ X 6″ ROOF RAFTER

DOUBLE TOP PLATE

FINISH FASCIA (1″ X 8″)

Figure 55–1. In one method of applying fascia, a subfascia is nailed below the finish fascia (A). In another method, the finish fascia is nailed directly to the rafter tails (B).

in Figure 55–4.

For a sloped box cornice the soffit material is nailed directly to the underside of the rafters. See Figure 55–5. This design may be found on houses with wide overhangs.

The construction of a *gable end overhang,* also known as the *rake section,* is explained in

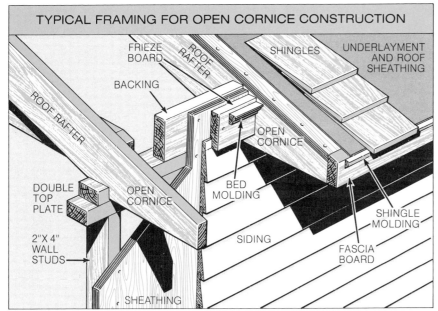

TYPICAL FRAMING FOR OPEN CORNICE CONSTRUCTION

Figure 55–2. Example of open cornice construction. The frieze board is cut between the rafters and the molding is nailed over the frieze board.

Figure 55–3. View from beneath an open cornice.

Section 10. The basic trim pieces in the rake section normally are the fascia and soffit material. Figure 55–6 shows the finished rake section for a flat box cornice. Figure 55–7 shows the finished rake section for a sloped box cornice.

## Cornice Soffit Systems

Wood products used for cornice soffits include plywood, hard-board, and fiberboard panels. See Figure 55–8. For a more rustic effect, different types of siding board patterns may be applied.

Cornice trim and soffit systems are also available in aluminum, in a variety of pre-finished colors and designs. See Figure 55–9.

The manufacturer's directions should be carefully followed when installing these systems.

## Gutters and Downspouts

Gutters and downspouts are fastened to the fascia to receive water runoff from the roof. Refer again to Figure 55–9. They may be made of aluminum, plastic, or galvanized steel. In the past gutters (but not downspouts) were made of wood, and occasionally they still are.

## COVERING THE ROOF

The main purpose of any type of finish roof covering is to shed water. In addition, the covering should help protect the building from sun, wind, and dust infiltration.

Many different kinds of roofing materials are available. The type best suited for a particular building depends on the roof slope, the overall design of the structure, and local code regulations. The color, texture, and pattern of

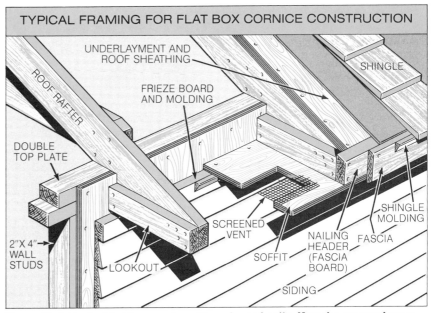

TYPICAL FRAMING FOR FLAT BOX CORNICE CONSTRUCTION

Figure 55–4. Typical flat box cornice with a plywood soffit. Note the screened opening vent for roof ventilation.

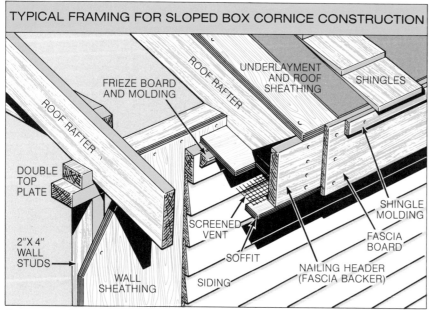

**TYPICAL FRAMING FOR SLOPED BOX CORNICE CONSTRUCTION**

Figure 55–5. Typical sloped box cornice. No lookouts are required. The soffit material is nailed directly to the underside of the rafters.

*The Upson Company*

Figure 55–8. Fiberboard cornice soffit. Note the vent openings.

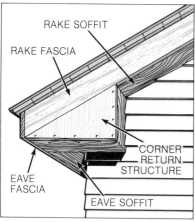

Figure 55–6. A flat box cornice requires a cornice return at the corner juncture of the rake and eave soffits.

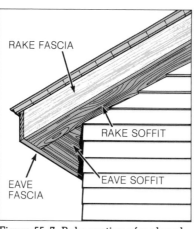

Figure 55–7. Rake section of a sloped box cornice.

*The Aluminum Association, Inc.*

Figure 55–9. Aluminum soffits finish off the overhangs of this building. Gutters and downspouts are also aluminum.

a roofing material can add greatly to the general attractiveness of the building.

Some types of roofing materials are usually applied by carpenters, while other types are applied by specialists called *roofers.* For example, *built-up roof coverings* (discussed at the end of this unit) are applied by roofers.

Within the carpentry trade, a

trend has occurred toward specialization of roof application. In many areas of the country, roofing operations are performed by carpenters who do only roofing work. All carpenters, however, should understand roofing application and be able to make repairs when necessary.

*Shingles,* which are usually applied by carpenters, are made of asphalt, mineral fiber, cedar wood, or fiberglass. *Shakes,* which are similar to cedar shin-

gles but are split rather than sawed from cedar logs, are applied when a more rustic effect is desired.

Key terms related to shingle application are (Figure 55–10):

1. *Shingle Width:* the total measurement across the top of either a strip type or individual type of shingle.

2. *Toplap:* the area that one shingle overlaps a shingle in the course (row) below it.

3. *Sidelap:* the area that one shingle overlaps a shingle next to it (in the same course).

4. *Headlap:* the area that one shingle overlaps a shingle two courses below it. Headlap is measured from the bottom edge of an overlapping shingle to the nearest top edge of an over-lapped shingle.

5. *Exposure:* the area that is exposed (not overlapped) in a shingle.

For the best protection against leakage, shingles (or shakes) should be applied only on roofs with a unit rise of 4" or more. A lesser slope creates slower water runoff, which increases the possibility of leakage as a result of windblown rain or snow being driven underneath the butt ends of the shingles.

## Preparing the Roof for Finish Covering

Before any roofing finish material can be applied, the roof must be sheathed, underlayment installed (except under wood shingles), and flashing placed wherever conditions require it.

**Sheathing.** Sheathing is usually plywood or a non-veneered panel such as particleboard or waferboard. Post-and-beam roofs (discussed in Section 15) often have 2" tongue-and-groove planks. Spaced boards of 1" × 3", 1" × 4", or 1" × 6" nominal width are frequently used as sheathing under wood shakes or shingles. Other sheathing materials include the various types of rigid roof insulation products discussed in Section 11.

**Underlayment.** Roof underlayment has a number of important

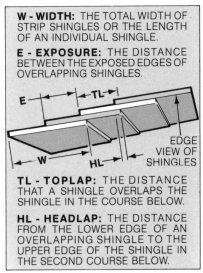

W - WIDTH: THE TOTAL WIDTH OF STRIP SHINGLES OR THE LENGTH OF AN INDIVIDUAL SHINGLE.

E - EXPOSURE: THE DISTANCE BETWEEN THE EXPOSED EDGES OF OVERLAPPING SHINGLES.

TL - TOPLAP: THE DISTANCE THAT A SHINGLE OVERLAPS THE SHINGLE IN THE COURSE BELOW.

HL - HEADLAP: THE DISTANCE FROM THE LOWER EDGE OF AN OVERLAPPING SHINGLE TO THE UPPER EDGE OF THE SHINGLE IN THE SECOND COURSE BELOW.

Figure 55–10. Key words used in directions for placing shingles.

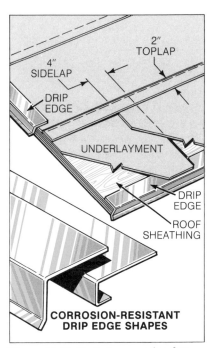

Figure 55–11. Underlayment for shingles placed over roof sheathing. Note the 2" toplap (overlapping) and 4" side-lap.

functions. It protects the sheathing from moisture until the shingles are placed. It provides additional weather protection from wind-driven rainwater that may penetrate underneath the shingles. It also prevents direct

contact between the shingles and the wood sheathing, which is important when asphalt shingles are used. Wood resins in the sheathing can cause chemical reactions that are damaging to asphalt shingles.

Underlayment is always necessary under asphalt, asbestos, and fiberglass shingles. It is not always necessary under wood shingles. Asphalt-saturated felt (tar paper) is used for underlayment. It should be spread over the entire roof surface as soon as the sheathing has been completed. A 2" overlapping *(toplap)* is required at all horizontal joints and 4" overlapping *(sidelap)* at the end joints. See Figure 55–11.

**Flashing.** The shape and construction of different types of roofs can create different types of water leakage problems. Water leakage can be prevented by placing flashing materials in and around the vulnerable areas of the roof. One vulnerable area is around a chimney or other type of stack that projects above the roof surface.

Flashing materials used on roofs may be asphalt-saturated felt, metal, or plastic. The felt types of flashing are generally used at the ridges, hips, and valleys. However, metal flashing made of aluminum, galvanized steel, or copper is considered superior to felt. Flashing products made of plastics such as vinyl are a recent innovation.

## Asphalt Shingles

Most homes in the United States are covered with some sort of asphalt shingles. They are less expensive than other types of shingles, easy to apply, and have a life expectancy of 15 to 25 years.

Asphalt shingles are made of

an asphalt-saturated felt coated with mineral granules. They are available in a wide range of styles, textures, and colors. Light-colored shingles reflect heat, keep the attic cooler, and make a house look larger than it is. Dark-colored shingles absorb heat and make a house look lower and smaller than it is.

Asphalt shingles are available in several different designs of the *strip* type (Figure 55–12) and of the *individual* type (Figure 55–13). The individual type is used less frequently today than in the past.

**Preparing the Roof for Asphalt Shingles.** An asphalt-shingle roof requires felt underlayment and usually some type of flashing. *Open valley* or *closed valley* flashing may be used. See Figure 55–14. In open valley flashing, a strip of mineral-surfaced roofing material is placed face down at the valley. A second piece is cemented face up over the first strip. The shingles are applied over the flashing. In closed valley flashing, a piece of mineral-surfaced roofing material is placed face up in the valley. The shingle strips are then placed to cross the valley.

Flashing is necessary where a sloping roof meets a vertical wall (Figure 55–15) and where stacks project above the roof line. Figure 55–16 shows a flashing procedure around a stack.

In cold-weather areas, flashing is also recommended along the eaves of the roof. See Figure 55–17. A strip of 50-pound asphalt-saturated felt is applied over the regular layer of underlayment at the roof overhang. Flashing for the eaves should extend 12″ to 24″ inside the interior wall line of the building. This protects the roof boards from damage caused by ice dams that may form in this area. A *drip edge* at the top of the fascia is a

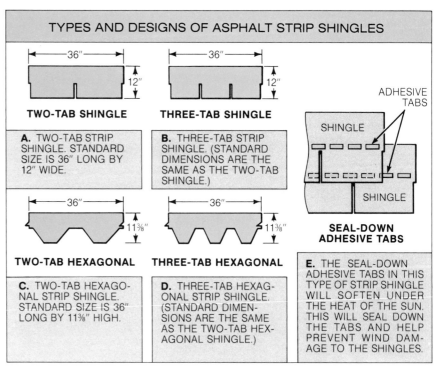

Figure 55–12. Types of asphalt strip shingles.

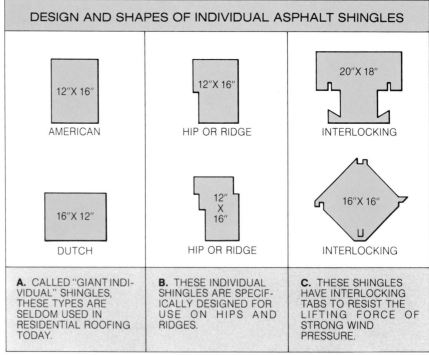

Figure 55–13. Types of individual asphalt shingles.

metal piece that protects the edges of the roof deck and also helps prevent leakage.

**Applying Asphalt Shingles.** Special noncorrosive, hot-dipped galvanized steel or aluminum nails are manufactured for asphalt shingles. See Figure 55–18. These nails have flat heads 3/8″ to 7/16″ in diameter and sharp points. Staples are also commonly used to fasten asphalt shingles. See Figure 55–19.

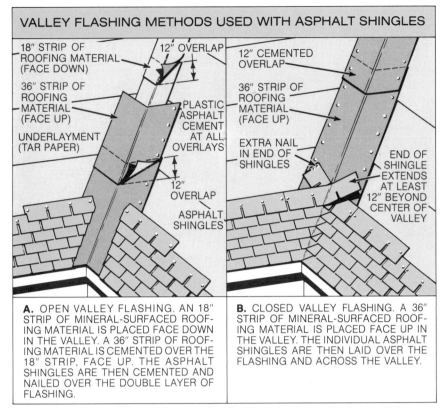

VALLEY FLASHING METHODS USED WITH ASPHALT SHINGLES

18" STRIP OF ROOFING MATERIAL (FACE DOWN)

12" OVERLAP

12" CEMENTED OVERLAP

36" STRIP OF ROOFING MATERIAL (FACE UP)

36" STRIP OF ROOFING MATERIAL (FACE UP)

PLASTIC ASPHALT CEMENT AT ALL OVERLAYS

UNDERLAYMENT (TAR PAPER)

EXTRA NAIL IN END OF SHINGLES

END OF SHINGLE EXTENDS AT LEAST 12" BEYOND CENTER OF VALLEY

12" OVERLAP

ASPHALT SHINGLES

**A.** OPEN VALLEY FLASHING. AN 18" STRIP OF MINERAL-SURFACED ROOF-ING MATERIAL IS PLACED FACE DOWN IN THE VALLEY. A 36" STRIP OF ROOF-ING MATERIAL IS CEMENTED OVER THE 18" STRIP, FACE UP. THE ASPHALT SHINGLES ARE THEN CEMENTED AND NAILED OVER THE DOUBLE LAYER OF FLASHING.

**B.** CLOSED VALLEY FLASHING. A 36" STRIP OF MINERAL-SURFACED ROOF-ING MATERIAL IS PLACED FACE UP IN THE VALLEY. THE INDIVIDUAL ASPHALT SHINGLES ARE THEN LAID OVER THE FLASHING AND ACROSS THE VALLEY.

Figure 55–14. Open valley or closed valley flashing may be used on asphalt-shingled roofs.

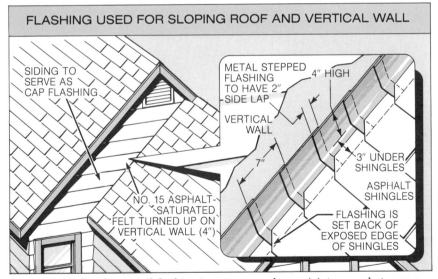

FLASHING USED FOR SLOPING ROOF AND VERTICAL WALL

SIDING TO SERVE AS CAP FLASHING

METAL STEPPED FLASHING TO HAVE 2" SIDE LAP

4" HIGH

VERTICAL WALL

7"

3" UNDER SHINGLES

ASPHALT SHINGLES

NO. 15 ASPHALT-SATURATED FELT TURNED UP ON VERTICAL WALL (4")

FLASHING IS SET BACK OF EXPOSED EDGE OF SHINGLES

Figure 55–15. Metal, stepped flashing is necessary where a joint occurs between a sloping roof and a vertical wall.

Figure 55–20 shows a procedure for placing asphalt strip shingles. A starter strip is placed beneath the first course of shingles. The first course starts with a full shingle. To stagger the cut-outs on the next course, the first tab of the first shingle must be cut. For example, if 6" is cut off the first shingle, the first shingle is 30" wide (36" – 6" = 30"). An additional 6" is cut off of the first shingle of the next course, resulting in a width of 24" (36" – 12" = 24"). When the first shingle of a course has been reduced to a 6" width, the next course begins with a full shingle.

A common procedure for finishing off the ridge and hips is the *Boston Method.* See Figure 55–21. In this method, shingles specially formed for the ridge and hips are overlapped with a 5" exposure.

## Wood Shingles and Shakes

Wood shingles and shakes are among the oldest types of roof covering. They are applied today when a rustic architectural effect is desired. In some areas, however, local fire codes prohibit this type of roof covering.

Most shingles and shakes are produced from western red cedar trees, which are slow-growing coniferous trees found in the Pacific Northwest. Western red cedar has exceptional strength in proportion to its weight. It also has little expansion and contraction with changes in its moisture content. These factors, along with its resistance to decay, make western red cedar a superior wood for roofing or any other exterior use. For example, wood shingles and shakes may be used for a sidewall finish (discussed in Unit 57).

Shingles have a smoother finish than shakes. See Figure 55–22. They are cut from cedar blocks by a shingle-cutting machine. Most wood shingles are produced in random widths ranging from 3" to 14". The standard lengths are 16", 18", and 24". The shingles are tapered to be thicker at the exposed butt end than at the concealed end.

Different grades of wood shingles are available. No. 1 (Blue Label) Grade shingles, recommended for roofs (and sidewalls), are cut from clear heartwood. They are also *edgegrained,*

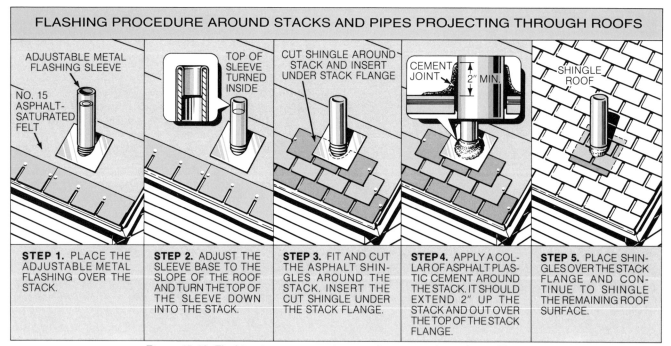

FLASHING PROCEDURE AROUND STACKS AND PIPES PROJECTING THROUGH ROOFS

**STEP 1.** PLACE THE ADJUSTABLE METAL FLASHING OVER THE STACK.

**STEP 2.** ADJUST THE SLEEVE BASE TO THE SLOPE OF THE ROOF AND TURN THE TOP OF THE SLEEVE DOWN INTO THE STACK.

**STEP 3.** FIT AND CUT THE ASPHALT SHINGLES AROUND THE STACK. INSERT THE CUT SHINGLE UNDER THE STACK FLANGE.

**STEP 4.** APPLY A COLLAR OF ASPHALT PLASTIC CEMENT AROUND THE STACK. IT SHOULD EXTEND 2" UP THE STACK AND OUT OVER THE TOP OF THE STACK FLANGE.

**STEP 5.** PLACE SHINGLES OVER THE STACK FLANGE AND CONTINUE TO SHINGLE THE REMAINING ROOF SURFACE.

Figure 55–16. Flashing is necessary where stacks project above the roof line.

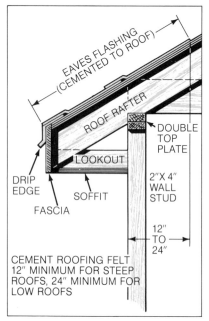

Figure 55–17. In cold-weather areas, flashing is necessary along the eaves of the roof.

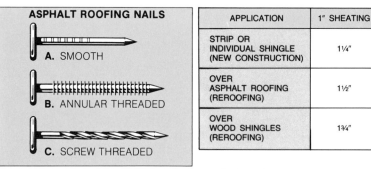

ASPHALT ROOFING NAILS

A. SMOOTH

B. ANNULAR THREADED

C. SCREW THREADED

| APPLICATION | 1" SHEATING | 3/8" PLYWOOD |
|---|---|---|
| STRIP OR INDIVIDUAL SHINGLE (NEW CONSTRUCTION) | 1¼" | 7/8" |
| OVER ASPHALT ROOFING (REROOFING) | 1½" | 1" |
| OVER WOOD SHINGLES (REROOFING) | 1¾" | — |

Figure 55–18. Types of nails for applying asphalt roofing and recommended nail lengths.

meaning that the grain runs in the direction of the long dimension of the shingle.

Shakes are very similar to shingles, but are split rather than sawed from cedar logs. Splitting

*Duo–Fast Corporation*

Figure 55–19. Asphalt shingles may be fastened with a pneumatically powered roofing stapler.

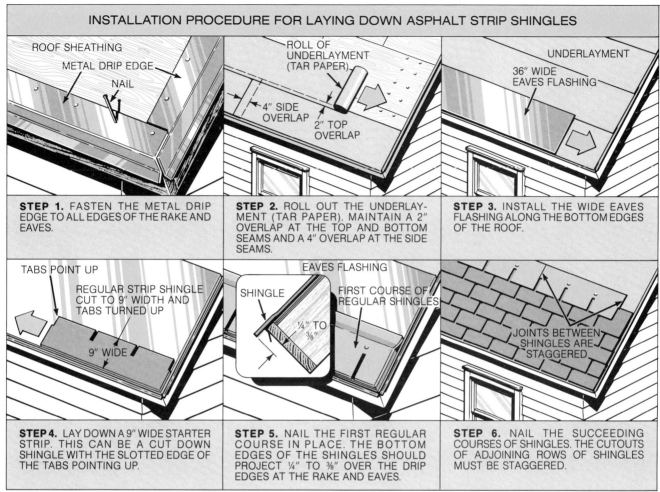

## INSTALLATION PROCEDURE FOR LAYING DOWN ASPHALT STRIP SHINGLES

ROOF SHEATHING
METAL DRIP EDGE
NAIL

**STEP 1.** FASTEN THE METAL DRIP EDGE TO ALL EDGES OF THE RAKE AND EAVES.

ROLL OF UNDERLAYMENT (TAR PAPER)
4" SIDE OVERLAP
2" TOP OVERLAP

**STEP 2.** ROLL OUT THE UNDERLAYMENT (TAR PAPER). MAINTAIN A 2" OVERLAP AT THE TOP AND BOTTOM SEAMS AND A 4" OVERLAP AT THE SIDE SEAMS.

UNDERLAYMENT
36" WIDE EAVES FLASHING

**STEP 3.** INSTALL THE WIDE EAVES FLASHING ALONG THE BOTTOM EDGES OF THE ROOF.

TABS POINT UP
REGULAR STRIP SHINGLE CUT TO 9" WIDTH AND TABS TURNED UP
9" WIDE

**STEP 4.** LAY DOWN A 9" WIDE STARTER STRIP. THIS CAN BE A CUT DOWN SHINGLE WITH THE SLOTTED EDGE OF THE TABS POINTING UP.

EAVES FLASHING
SHINGLE
FIRST COURSE OF REGULAR SHINGLES
¼" TO ⅜"

**STEP 5.** NAIL THE FIRST REGULAR COURSE IN PLACE. THE BOTTOM EDGES OF THE SHINGLES SHOULD PROJECT ¼" TO ⅜" OVER THE DRIP EDGES AT THE RAKE AND EAVES.

JOINTS BETWEEN SHINGLES ARE STAGGERED

**STEP 6.** NAIL THE SUCCEEDING COURSES OF SHINGLES. THE CUTOUTS OF ADJOINING ROWS OF SHINGLES MUST BE STAGGERED.

Figure 55–20. Procedure for placing asphalt strip shingles. In this example, three-tab strip shingles are used.

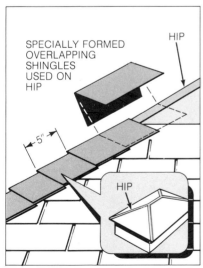

SPECIALLY FORMED OVERLAPPING SHINGLES USED ON HIP
HIP
5"
HIP

Figure 55–21. Boston method for finishing off the ridge and hips of an asphalt-shingled roof. Specially formed overlapping shingles are used.

SHAKE

SHINGLE

*Red Cedar Shingle & Handsplit Shake Bureau*

Figure 55–22. Shingles have a smoother finish than shakes and are tapered from the butt to the concealed end.

Figure 55–23. Spaced sheathing on a roof to be covered with wood cedar shakes or shingles. Note the solid sheathing over the eaves.

thing are $1'' \times 3''$, $1'' \times 4''$, and $1'' \times 6''$. They are spaced on centers equal to the exposure area of the shingles or shakes that will be applied. (The exposure is the distance between the exposed edges of overlapping shakes or shingles. Refer to Figure 55–10.)

In more severe winter climates, where wind-driven snow may be a problem, solid sheathing is a recommended as a nailing base.

Felt underlayment is not required under wood shingles except over the eave which may be subject to infiltration by wind-driven rain or snow.

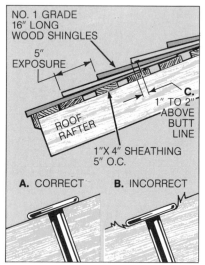

Figure 55–24. When nailing shingles or shakes, drive the nails until heads rest on the surface and no further (A). If nails are driven into the surface, they will have less holding power (B). Use only two nails to a shake, placed 1″ from each edge, and from 1″ to 2″ above the butt line of the following course (C).

produced from 100% heartwood, free of bark and sapwood. Shakes come in random widths starting from 4″, and in standard lengths of 18″, 24″, and 32″.

**Preparing the Roof for Wood Shingles or Shakes.** Either spaced or solid sheathing is used as a base for nailing shingles and shakes. Spaced sheathing, except over the eaves, is recommended in milder climates. See Figure 55–23. Wood shingles and shakes tend to absorb some rainwater. For this reason, it is helpful if air is able to circulate underneath the shakes or shingles. The circulation of air will prevent uneven drying, which can eventually cause splits to develop.

Boards used for spaced shea-

**Applying Wood Shingles.** Hot-dipped, zinc-coated nails are recommended for wood shingles and shakes. Aluminum and stainless steel nails are also acceptable. The nails must be long enough to go through the shingle or shake and penetrate the sheathing at least 1/2″. Proper nailing procedure is shown in Figure 55–24. The nail heads should rest on the surface. If the nail is driven further, it will have less holding power. Two nails are used per shingle, each 1″ from the edge and 1″ to 2″ above the butt line of the following course of shingles.

The recommended exposure for wood shingles depends on the size of the shingle and the slope (pitch) of the roof. See the table in Figure 55–25. A proce-

produces a rougher and more rustic appearance than that of shingles. Three types of shakes are produced for exterior finish work. *Tapersplit* shakes are cut at a taper. *Straightsplit* shakes do not have a taper. *Handsplit* and *resawn* shakes are split and then resawn at a taper.

There is only one grade (No. 1) of cedar shakes. They are

| SHINGLE LENGTH | SHINGLE THICKNESS (GREEN) | MAXIMUM EXPOSURE | |
|---|---|---|---|
| | | SLOPE LESS THAN 4 IN 12 | SLOPE 5 IN 12 AND OVER |
| 16″ | 5 BUTTS IN 2″ | 3¾″ | 5″ |
| 18″ | 5 BUTTS in 2¼″ | 4¼″ | 5½″ |
| 24″ | 4 BUTTS IN 2″ | 5¾″ | 7½″ |

Figure 55–25. Recommended exposure of wood shingles. Note that greater exposure areas are allowed for roofs with steeper slopes.

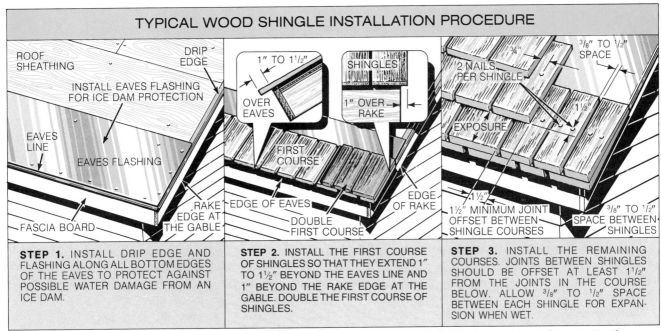

## TYPICAL WOOD SHINGLE INSTALLATION PROCEDURE

**STEP 1.** INSTALL DRIP EDGE AND FLASHING ALONG ALL BOTTOM EDGES OF THE EAVES TO PROTECT AGAINST POSSIBLE WATER DAMAGE FROM AN ICE DAM.

**STEP 2.** INSTALL THE FIRST COURSE OF SHINGLES SO THAT THEY EXTEND 1″ TO 1½″ BEYOND THE EAVES LINE AND 1″ BEYOND THE RAKE EDGE AT THE GABLE. DOUBLE THE FIRST COURSE OF SHINGLES.

**STEP 3.** INSTALL THE REMAINING COURSES. JOINTS BETWEEN SHINGLES SHOULD BE OFFSET AT LEAST 1½″ FROM THE JOINTS IN THE COURSE BELOW. ALLOW ³/₈″ TO ½″ SPACE BETWEEN EACH SHINGLE FOR EXPANSION WHEN WET.

Figure 55–26. Applying wood shingles. Hot-dipped, zinc-coated nails are recommended. Aluminum and stainless steel nails are also acceptable.

dure for placing wood shingles is shown in Figure 55–26.

**Applying Wood Shakes.** The procedure for placing wood shakes is similar but not identical to that for shingles. Shakes are much thicker than shingles; therefore, longer nails should be used. Shakes are also longer than shingles; therefore, they have a greater exposure. Common exposures are 7½″ for 18″ shakes, 10″ for 24″ shakes, and

13″ for 32″ shakes.

The rough and uneven surfaces of shakes increase the possibility of infiltration by wind-driven rain and snow. For this reason, it is necessary to place a strip of underlayment between each course as the shakes are being applied. See Figure 55–27. A 36″ wide strip of 15-pound (minimum) asphalt-saturated felt is laid over the eave line. The bottom course of shakes is doubled. After each

course of shakes, an 18″ wide strip of 15-pound felt is placed over the top portion of the shake, extending onto the sheathing. The bottom edge of the felt placed on top of the shake is positioned from the butt a distance equal to twice the weather exposure of the shake. To allow for possible expansion, the shakes are spaced about ½″ apart. The joints between the shakes are offset at least 1½″ from adjacent courses.

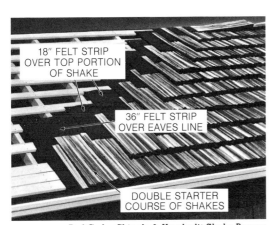

*Red Cedar Shingle & Handsplit Shake Bureau*

*Red Cedar Shingle & Handsplit Shake Bureau*

Figure 55–27. Because shakes have rough and uneven surfaces, underlayment must be placed between each course. Note the spaced sheathing.

*Red Cedar Shingle & Handsplit Shake Bureau*
Figure 55–28. Hip sections of a roof are finished off with a hip cap.

*Red Cedar Shingle & Handsplit Shake Bureau*
Figure 55–30. Ridge and hip caps are also available in factory-manufactured units.

*Red Cedar Shingle & Handsplit Shake Bureau*
Figure 55–33. Flashing and counterflashing must be placed on a shaked roof where it meets a brick wall.

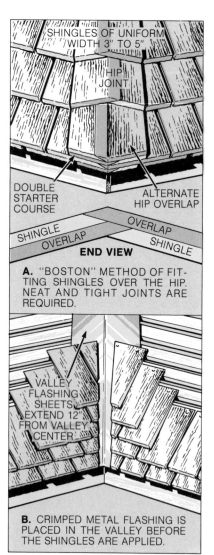

Figure 55–29. Finishing over hip section and placing flashing in valley.

**Ridges, Hips, and Valleys.** The same methods are used with wood shingles as with wood shakes for finishing off ridges,

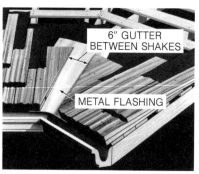

*Red Cedar Shingle & Handsplit Shake Bureau*
Figure 55–31. The open valley method is considered the most practical for finishing off a valley.

*Red Cedar Shingle & Handsplit Shake Bureau*
Figure 55–32. Flashing and counterflashing must be placed around a chimney on a shaked roof.

hips, and valleys.

The hip sections of the roof are finished off with a hip cap. See Figure 55–28. A strip of 15-pound roofing felt, at least 8″ wide, must first be applied, and the starter course area of the hip cap is doubled. See Figure 55–29. The joints between the shakes are beveled. The direc-

tions of the bevels are alternated with each course.

The ridge of the roof is finished off with a ridge cap. See Figure 55–30. The general procedure for placing a ridge cap is the same as for placing a hip cap.

Many roof leaks occur at points where water is channeled for running off the roof, such as the valleys between intersecting roofs. The *open valley* method is considered the most practical method of finishing off a valley. See Figure 55–31. The first step is to apply 15-pound roofing felt directly over the sheathing. Sheets of metal valley flashing at least 20″ wide, and with a 4″ to 6″ headlap, are then nailed down. The metal flashing should be 26-ga. or heavier galvanized iron. Shingles (or shakes) placed in the valley should be fitted so that they run in a line parallel with the valley. They should also be held back so that a 6″ gutter is formed between the lines of shakes.

Another area vulnerable to leakage is where a roof meets a vertical wall or chimney. Flashing and counterflashing must be placed around the chimney or where the roof meets the wall. See Figures 55–32 and 55–33. The flashing should extend 6″ under the shingles. The counterflashing should be inserted ¾″

into the mortar joints of the brick and should fold over the flashing.

## Tile Roofing

Tile is one of the older types of finish covering used on sloping roofs. It has grown increasingly popular because of its fireproofing qualities. See Figure 55–34. Clay or concrete tile roofing is available in a variety of styles. Clay tile is manufactured by baking plates of molded clay into tile. Clay tile is the lighter of the two; however, concrete tile is considered more durable and is used more often than clay tile. Concrete tile is composed of portland cement, sand, and water.

*Monier Roof Tile Co.*

Figure 55–34. Tile roofing provides an attractive and fireproof finish roof covering.

Field tile is generally classified as flat tile or roll tile. *Flat tile* is flat in cross section. *Roll tile* is curved in cross section. *Accessory tiles* are specially designed tiles used for different intersecting points on the roof. See Figure 55–35.

**Field Tile.** A common dimension for field tile is 13″ wide and 16½″ long. These dimensions may vary slightly with different manufacturers. A typical field tile has head lugs (anchors) on its underside toward the top of the tile. Transverse bars at the lower end of the underside act as weather checks. Interlocking ribs are provided on the long edges of the tile to act as a waterlock and control latitudinal movement. One nail hole is located near the top of flat tile. Two nail holes are located near the top of roll tile.

**Accessory Tiles.** *Ridge and hip tiles* are used to cover ridges and normal hips. *Hip starter tiles* are used to start tile at the eaves of hip roofs. *Rake tiles* are used to finish off the gable ends of a roof. *Three-way apex tiles* are used to cap the intersection of the end of the roof ridge and hip. *Four-way apex tiles* are used to cap the peak of a pyramid-shaped roof.

**Preparation for Tiles.** Tiles are placed on spaced or solid sheathing that conforms to building code requirements for the loads involved. Spaced sheathing normally requires minimum 1″ × 6″ boards spanning a maximum of 24″ between rafters. Underlayment material must be at least No. 30 asphalt-saturated felt installed with a minimum 2″ head lap and 6″ side lap. Flashing for eaves may be required in cold weather areas.

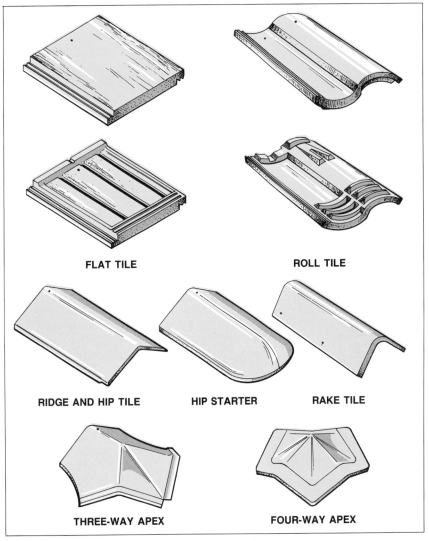

FLAT TILE        ROLL TILE

RIDGE AND HIP TILE      HIP STARTER      RAKE TILE

THREE-WAY APEX        FOUR-WAY APEX

Figure 55–35. Tile roofing is available in a variety of field tiles and accessories.

447

**Battens.** The use of battens under tiles is a frequent building code requirement for installation over solid sheathing. Battens provide positive anchorage for the tiles. The battens are 1″ × 2″ boards nailed to chalklines snapped after placing the underlayment. The head lugs of each tile lap over and rest on the batten. A 1/2″ to 1″ break every 4′-0″ in each row of battens provides for water drainage. An alternate method is to place 1/4″ shims to create a continuous space beneath the battens. The layout and placement of battens is critical because their placement governs the final position of the tiles.

Battens are spaced 13″ to 13 1/2″ O.C. for 16 1/2″ long tiles. This provides a 3″ to 3 1/2″ minimum headlap of the tiles. The line snapped for the upper edge of the top batten is 2″ from the ridge to allow room for a 2″ × 3″ nailer piece for the ridge tiles. The line snapped for the upper edge of the first batten measured from the eave is 15″ to allow for a 1″ tile overhang past the eave. To determine batten spacing between top and bottom lines, the distance between lines is measured and then divided by 13. The answer gives the number of tile courses required. The exact layout of the battens is then found by dividing the number of required courses into the distance between the top and bottom batten lines. See Figure 55–36.

**Flashing.** No. 28 gauge corrosion-resistant metal flashing is required at all valleys, chimneys, skylights, and where the roof butts up against a straight wall. Valley flashing must extend at least 11″ in both directions from the centerline with a splash diverter rib not less than 1″ high at the flow line. In open valley construction, the tiles are cut to the angle of the valley and held back 2″ from the diverter strip. In closed valley construction, the tiles butt up against the diverter strip. Pan flashing is used where the tiles butt up against a wall. Counterflashing is also recommended. Flashing material placed around flues and vents is lead, copper, or any other approved material that can be formed to follow the shape of the tile. See Figure 55–37. The sides of chimneys and skylights are flashed in the same manner as straight walls. The top and bottom areas require flexible flashing material.

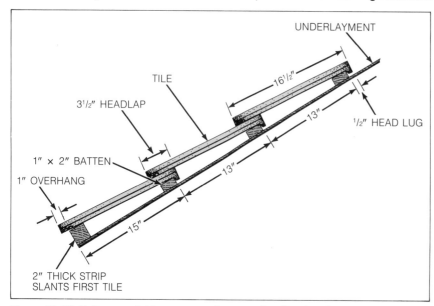

Figure 55-36. Battens are laid out to provide proper headlap for tiles.

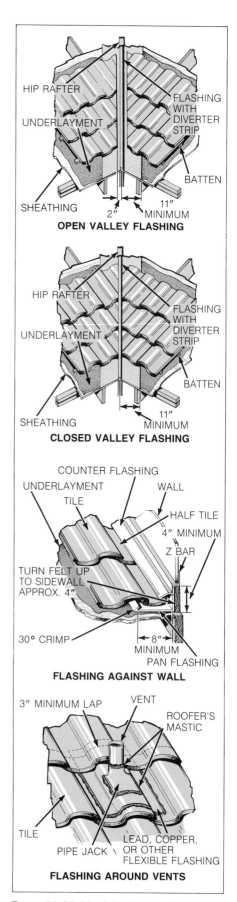

Figure 55-37. Metal flashing is placed in all roof areas where leaks may occur.

**Placing Tiles.** Field tiles are placed in vertical as well as horizontal alignment. Vertical interlocking joints must be free of any foreign matter to assure a correct fit and interlock of the tiles. Tiles are placed from left to right because of the vertical keyway between the tiles. The number of tiles to be fastened with nails depends on the roof slope, whether or not battens are used, and the anticipated wind velocity in the area. Local building codes should be consulted for the tile nailing schedule.

A cant strip used at the eave must be twice the thickness of the battens to properly slant the first course of tiles. For rake tiles used to finish off gable ends, the field tiles are cut and held back 1″ to 2″ from the outside edge of the sheathing.

Ridge and hip tiles are nailed or wired to a 2″ × 3″ nailer strip fastened to the roof. A bead of roofer's mastic is applied where the ridge tiles overlap. The juncture where the ridge and hip tiles rest on the field tiles is weatherproofed with mortar or an approved dry ridge/hip system. Ends of the ridge can be finished off with pieces cut from rake tiles. A starter tile is placed at the lower end of the hip section when starting hip tiles.

## Built-Up Roof Covering

A built-up roof covering is usually used on flat decks. It consists of three, four, or five layers of roofing felt. Each layer is mopped down with hot tar or asphalt. The final layer is coated with gravel embedded in the tar or asphalt. See Figure 55–38.

Built-up roof coverings are installed by roofing companies who employ roofers to specialize in this kind of work. Carpenters are

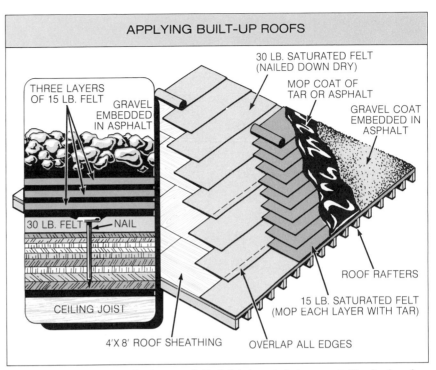

Figure 55–38. Applying a built-up roof. Each felt layer is hot-mopped. The final surface is covered with gravel embedded in asphalt or tar.

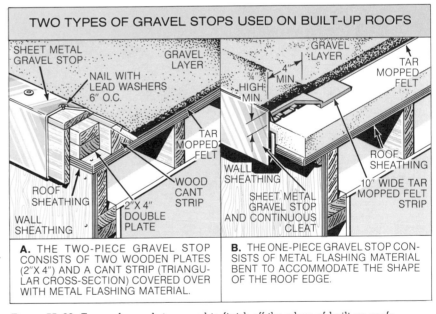

Figure 55–39. Types of gravel stops used to finish off the edges of built-up roofs.

not involved in the application of the built-up roof. They may, however, do certain preparatory work such as nailing on *gravel stops, cant strips,* and flashing. See Figure 55–39. Gravel stops may consist of one or two pieces. A one-piece gravel stop is made of galvanized steel or copper bent to accommodate the shape of the roof edge. A two-piece gravel stop is made of two wood plates (2″ × 4″) and a cant strip covered with metal flashing material.

# UNIT 56

# Exterior Door and Window Frames

Window units and exterior door units are usually assembled at a factory or mill-cabinet shop. They are delivered to the construction site and set in place by carpenters on the job. Frames can be obtained unassembled. Usually, however, preassembled frames are ordered with window sashes and doors already fitted in them. See Figure 56–1. Weatherstripping is often included, and the

Figure 56–1. Preassembled window units are set in place by carpenters.

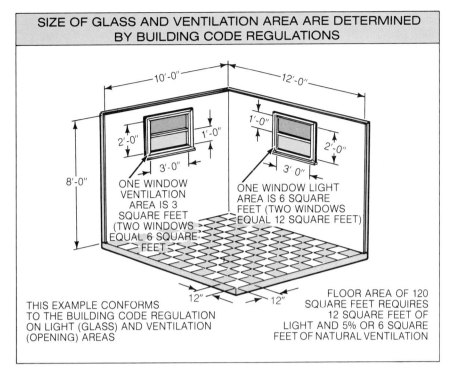

SIZE OF GLASS AND VENTILATION AREA ARE DETERMINED BY BUILDING CODE REGULATIONS

ONE WINDOW VENTILATION AREA IS 3 SQUARE FEET (TWO WINDOWS EQUAL 6 SQUARE FEET

ONE WINDOW LIGHT AREA IS 6 SQUARE FEET (TWO WINDOWS EQUAL 12 SQUARE FEET)

THIS EXAMPLE CONFORMS TO THE BUILDING CODE REGULATION ON LIGHT (GLASS) AND VENTILATION (OPENING) AREAS

FLOOR AREA OF 120 SQUARE FEET REQUIRES 12 SQUARE FEET OF LIGHT AND 5% OR 6 SQUARE FEET OF NATURAL VENTILATION

Figure 56–2. In this example, the two windows provide 12 square feet of glass area, which is 10% (1/10) of the 120' floor area. They also provide 6 square feet of natural ventilation, which is 5% (1/20) of the floor area.

exterior casing may also be attached.

Information regarding the door and window units is provided on the floor plans, elevation plans, and details of the blueprints.

Most blueprints also include a door and window schedule specifying the types of doors and windows to be installed (Refer to Section 6.)

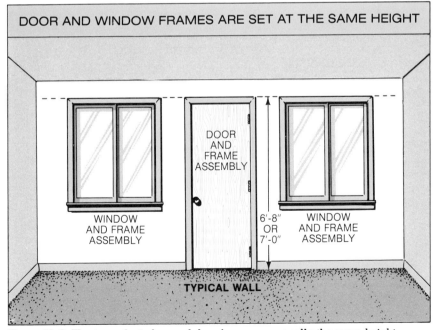

DOOR AND WINDOW FRAMES ARE SET AT THE SAME HEIGHT

DOOR AND FRAME ASSEMBLY

WINDOW AND FRAME ASSEMBLY

6'-8" OR 7'-0"

WINDOW AND FRAME ASSEMBLY

TYPICAL WALL

Figure 56–3. The tops of window and door frames are usually the same height.

## WINDOW UNITS

Windows allow air and light into a building. Local building codes often specify the minimum glass and venting area required for inhabited rooms. For example, the Uniform Building Code recommends that "habitable rooms within a dwelling unit shall be provided with natural light by means of exterior glazed openings with an area not less than one-tenth of the floor area of such rooms with a minimum of 10 square feet." It further states that these rooms must be provided with natural ventilation "by means of openable exterior openings with an area of not less than one-twentieth of the floor area of such rooms with minimum of 5 square feet." See Figure 56–2. Openings for natural ventilation are not required if a complete mechanical ventilation system exists in the structure.

Although windows come in many different sizes, their tops will usually line up with the door heights of the building. See Figure 56–3. Door heights are usually 6'-8" in residences and 7'-0" in public buildings.

Wood (Figure 56–4) is the oldest type of material used for window units. Today aluminum and steel (Figure 56–5) are also widely used. Aluminum windows can be obtained in natural color or anodized with a black, brown, or gold finish. Vinyl-clad windows are also available. See Figure 56–6.

### Components of a Window Unit

The wood or metal framework that holds the glass is the window *sash*. A complete window unit is made up of one or more sashes that fit into the window frame. See Figure 56–7. A double-hung window has a top and

*Hurd Millwork Company*

Figure 56–4. Interior view of a preassembled double-hung window unit made of wood.

DOUBLE-HUNG
WINDOW

SLIDER
WINDOW

*Air Master Corporation*

Figure 56–5. Examples of preassembled aluminum window units. The ones shown here are replacement units that are used for older homes.

*Weather Shield Mfg., Inc.*

Figure 56–6. Vinyl-clad window units are popular. The frame and sash have a wood core covered with pre-finished rigid vinyl.

bottom sash. These sashes move vertically. The glass area of the window shown in Figure 56–7 is separated into several panes of glass. Each pane of glass is called a *light.* The pieces that separate the lights are *muntins.*

A window frame is made up of a top piece and two side pieces. These pieces are equal to the finished thickness of the wall. The *sill* is a slanted piece at the bottom of the window frame. The space between the frame and the wall is covered by the *casing.* An *apron* is placed below another finish piece called a *stool.*

## Types of Windows

The types of windows chosen for a particular building should harmonize with the building's gen-eral design. Some windows look better with traditional building de-signs, while others blend well with more modern structures.

Operating and locking hard-ware varies according to the type of window. *Fixed-sash* windows, however, require no hardware, since they are stationary win-dows that do not open. They are often used in combination with movable windows. See Figure 56–8.

**Double-hung Windows.** Double-hung windows have upper and lower sash sections that move up and down in tracks provided in the window frame. See Figure 56–9. Some type of balancing device is necessary to hold the sash in an open vertical position.

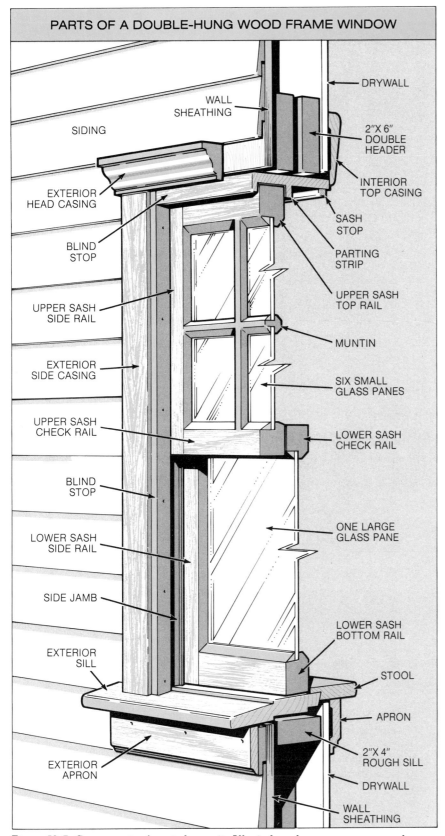

PARTS OF A DOUBLE-HUNG WOOD FRAME WINDOW

Figure 56–7. Components of a window unit. All windows have one or more sashes that fit into the window frame. A double-hung unit with two sashes is shown here.

*Hurd Millwork Company*

Figure 56–8. The three center units of this bow window combination are fixed-sash types. They cannot be opened. Casement windows are shown at each end of the fixed-sash units.

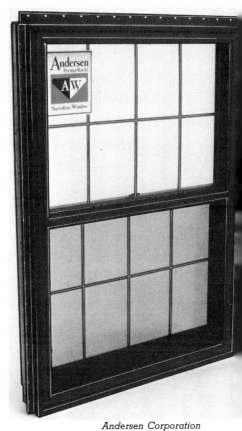

*Andersen Corporation*

Figure 56–9. A double-hung window unit has upper and lower sash sections that move up and down in tracks provided in the window frame. Each sash in this model contains eight lights.

See Figure 56–10. Spring-type devices are usually installed. With lightweight windows, a compressible weatherstripping is sometimes used for this purpose. The old method was to use a cord and counterweight inside the wall.

Typical hardware used with double-hung windows includes a sash lift and a sash lock. Ventilation offered by double-hung win-

## BALANCING DEVICES USED ON DOUBLE-HUNG WINDOWS

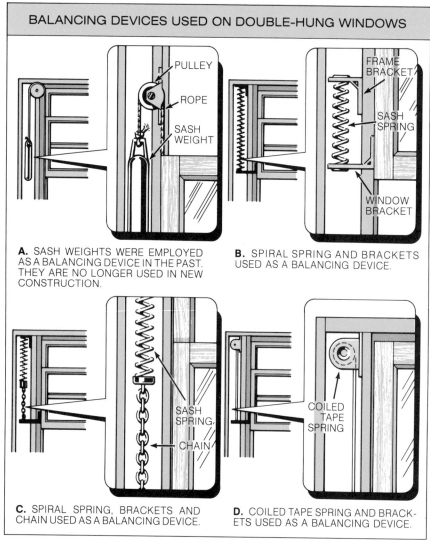

**A.** SASH WEIGHTS WERE EMPLOYED AS A BALANCING DEVICE IN THE PAST. THEY ARE NO LONGER USED IN NEW CONSTRUCTION.

**B.** SPIRAL SPRING AND BRACKETS USED AS A BALANCING DEVICE.

**C.** SPIRAL SPRING, BRACKETS AND CHAIN USED AS A BALANCING DEVICE.

**D.** COILED TAPE SPRING AND BRACKETS USED AS A BALANCING DEVICE.

Figure 56–10. Double-hung windows require a balancing device to hold them stationary in an open position.

dows is limited to 50% of the window opening.

**Casement Windows.** Casement windows are hinged on one side. See Figure 56–11. They have the same swing action as a door. Ventilation from a casement window is 100% of the window opening. Most casement windows are installed to swing out from the building. However, they can also be installed to swing into the building. In addition to hinges, other hardware required is a crank operator and some type of sash lock. Double or single casement windows are available.

**Horizontal Sliding Windows.** Horizontal sliding windows move along tracks (or guides) located above and below the window. A locking device is the only hardware required. Sliding window units often consist of one movable and one stationary window. A three-sash design consists of a fixed middle window and sliding windows on either side.

**Awning Windows.** Awning windows are hinged at the top and swing out at the bottom. See Figure 56–12. They require

*Hurd Millwork Company*
Figure 56–11. Wood double-casement window unit. In this model, the windows swing out from the room. Note the rotary operators at the bottom of each sash that are used to move the windows in and out. Sash locks are located toward the top of each window.

*Continental Aluminum Products Company*
Figure 56–12. These aluminum sash units feature a series of awning windows.

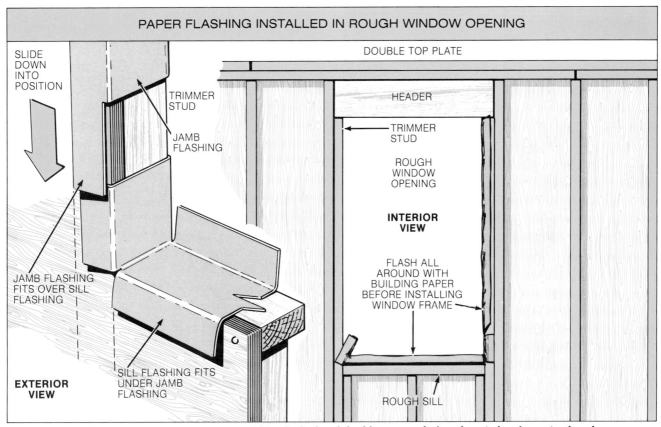

PAPER FLASHING INSTALLED IN ROUGH WINDOW OPENING

SLIDE DOWN INTO POSITION

TRIMMER STUD

JAMB FLASHING

JAMB FLASHING FITS OVER SILL FLASHING

EXTERIOR VIEW

SILL FLASHING FITS UNDER JAMB FLASHING

DOUBLE TOP PLATE

HEADER

TRIMMER STUD

ROUGH WINDOW OPENING

**INTERIOR VIEW**

FLASH ALL AROUND WITH BUILDING PAPER BEFORE INSTALLING WINDOW FRAME

ROUGH SILL

Figure 56–13. Window openings should be flashed with building paper before the window frame is placed.

hinges and a crank operator or push bar. They are often combined with a fixed-sash window. Sometimes a series of awning windows is placed in the same opening.

**Hopper Windows.** Hopper windows are similar to awning windows, but they are hinged at the bottom instead of at the top. They swing in or out. They require a crank operator or push bar and a locking device.

**Jalousie Windows.** Jalousie windows are made up of a series of small glass slats set into metal clips on both sides of the frame. A crank operates the slats. They can be pivoted open up to 90°.

## Installing a Window Unit

Wood or metal window units are usually delivered to the job with the sash already placed in the

frame. Manufacturer's instructions should be followed for installation. Layout and construction of the rough openings for window units are discussed in Section 9. If the opening has been correctly laid out and framed for each window unit, there should be no problem with installation.

It is good practice to flash all four sides of the rough window opening with strips of building paper 10″ to 12″ wide. The paper acts as an added weatherseal around the opening. See Figure 56–13.

Some manufacturers recommend that the window sash be removed before setting the frame to avoid possible damage to the sash. Others suggest that the sash not be removed, enabling faster installation of the unit. Figure 56–14 describes one method of installation. The procedure for trimming out (finishing off) the in-

terior of the window frame is discussed in Section 13.

**Metal Window Units.** The type of metal frame used most often has a flange with nail holes on all four sides. Nails or screws are driven through these holes into the sheathing and framework around the opening. Metal window units are easier to install than wood units. However, care must be taken to place them with bottoms level and sides plumb.

Another type of metal unit is designed to fit into an existing wood frame. These units are often used to replace worn-out wood sashes in older homes. They are attached by driving screws through holes provided in the metal frame into the wood frame. See Figure 56–15.

**Skylights.** Skylights are in effect, windows placed in the roof of a building. They allow added natu-

455

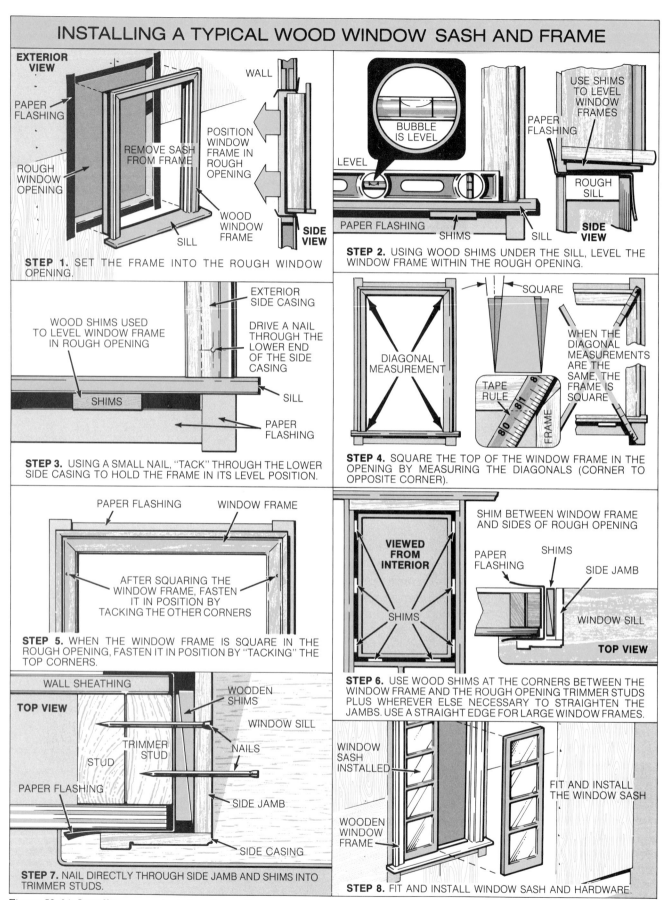

## INSTALLING A TYPICAL WOOD WINDOW SASH AND FRAME

**EXTERIOR VIEW**

PAPER FLASHING

ROUGH WINDOW OPENING

REMOVE SASH FROM FRAME

WALL

POSITION WINDOW FRAME IN ROUGH OPENING

WOOD WINDOW FRAME

SILL

**SIDE VIEW**

**STEP 1.** SET THE FRAME INTO THE ROUGH WINDOW OPENING.

BUBBLE IS LEVEL

LEVEL

USE SHIMS TO LEVEL WINDOW FRAMES

PAPER FLASHING

ROUGH SILL

PAPER FLASHING

SHIMS

SILL

**SIDE VIEW**

**STEP 2.** USING WOOD SHIMS UNDER THE SILL, LEVEL THE WINDOW FRAME WITHIN THE ROUGH OPENING.

WOOD SHIMS USED TO LEVEL WINDOW FRAME IN ROUGH OPENING

EXTERIOR SIDE CASING

DRIVE A NAIL THROUGH THE LOWER END OF THE SIDE CASING

SHIMS

SILL

PAPER FLASHING

**STEP 3.** USING A SMALL NAIL, "TACK" THROUGH THE LOWER SIDE CASING TO HOLD THE FRAME IN ITS LEVEL POSITION.

SQUARE

DIAGONAL MEASUREMENT

WHEN THE DIAGONAL MEASUREMENTS ARE THE SAME, THE FRAME IS SQUARE

TAPE RULE

FRAME

**STEP 4.** SQUARE THE TOP OF THE WINDOW FRAME IN THE OPENING BY MEASURING THE DIAGONALS (CORNER TO OPPOSITE CORNER).

PAPER FLASHING

WINDOW FRAME

AFTER SQUARING THE WINDOW FRAME, FASTEN IT IN POSITION BY TACKING THE OTHER CORNERS

**STEP 5.** WHEN THE WINDOW FRAME IS SQUARE IN THE ROUGH OPENING, FASTEN IT IN POSITION BY "TACKING" THE TOP CORNERS.

**VIEWED FROM INTERIOR**

SHIMS

SHIM BETWEEN WINDOW FRAME AND SIDES OF ROUGH OPENING

PAPER FLASHING

SHIMS

SIDE JAMB

WINDOW SILL

**TOP VIEW**

**STEP 6.** USE WOOD SHIMS AT THE CORNERS BETWEEN THE WINDOW FRAME AND THE ROUGH OPENING TRIMMER STUDS PLUS WHEREVER ELSE NECESSARY TO STRAIGHTEN THE JAMBS. USE A STRAIGHT EDGE FOR LARGE WINDOW FRAMES.

WALL SHEATHING

**TOP VIEW**

WOODEN SHIMS

WINDOW SILL

NAILS

TRIMMER STUD

STUD

PAPER FLASHING

SIDE JAMB

SIDE CASING

**STEP 7.** NAIL DIRECTLY THROUGH SIDE JAMB AND SHIMS INTO TRIMMER STUDS.

WINDOW SASH INSTALLED

WOODEN WINDOW FRAME

FIT AND INSTALL THE WINDOW SASH

**STEP 8.** FIT AND INSTALL WINDOW SASH AND HARDWARE.

Figure 56–14. Installing a wood window frame. Some manufacturers recommend that the sash be removed before installation.

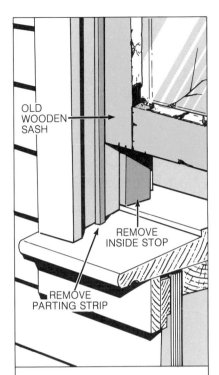

**STEP 1.** REMOVE THE PARTING STRIPS AND THE INSIDE STOP. TAKE OUT THE OLD WOODEN WINDOW SASH.

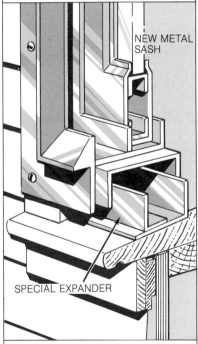

**STEP 2.** SCREW THE NEW METAL FRAME INTO THE OLD FRAME. IF THERE IS A GAP BETWEEN THE BOTTOM OF THE NEW FRAME AND THE SILL, INSTALL A SPECIAL EXPANDER UNIT.

Figure 56–15. Replacing a wood window unit with a metal unit. The metal unit is designed to fit into the existing wood frame in the building.

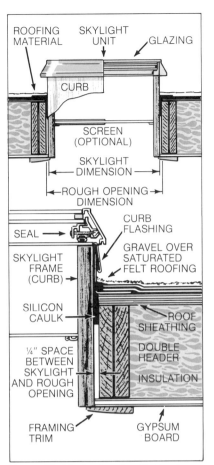

Figure 56–16. Section detail of a skylight on a flat roof.

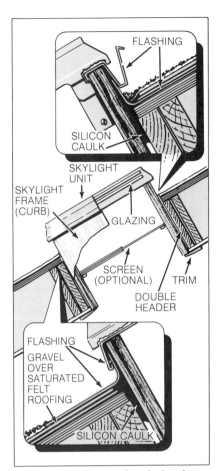

Figure 56–17. Section detail of a skylight on a sloping roof.

ral light in the living space below the roof. They are also an additional means of passive heat collection for the interior of the building.

Skylights come in different shapes and sizes. Smaller types are designed to fit between 16″ and 24″ O.C. rafter spacing. Large skylights require the same framing procedures used for chimneys and flue openings. One or more rafters may have to be cut, and headers must be installed.

Most skylights are made of clear or translucent plastic held in aluminum or wood frames. Insulating glass is also used, but less often. As a rule, the skylight is set on a curb constructed over the opening in the roof. See Figures 56–16 and 56–17.

Skylights can be used instead of dormer windows in gambrel or in steep-pitched gable roofs. See Figures 56–18 and 56–19. They may also be installed over a light shaft in an attic roof. See Figure 56–20.

## EXTERIOR DOOR UNITS

Exterior doors (entry doors) provide passage between the inside and outside of a building. The main-entrance door unit is usually at the front or street side of the building. This may be a single door (Figure 56–21), double doors (Figure 56–22), or a single door flanked by windows (Figure 56–23). The better quality types of main-entrance doors are of solid wood construction, fiberglass, or are metal-faced with an

Roto Frank of America, Inc.        Roto Frank of America, Inc.

Figure 56–18. Interior and exterior views of skylights on a sloping roof. These are made of insulated glass set in an aluminum sash.

Figure 56–21. This single-entrance door has a molded fiberglass surface. The interior core of the door is thick-density, foamed-in-place polyurethane.

Ventarama Skylight Corporation

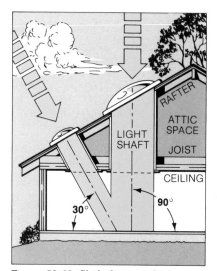

Figure 56–20. Skylights may be installed over a light shaft in an attic roof. The light shaft is framed between the ceiling joist and roof rafters. It is then surfaced with plasterboard or plywood.

National Woodwork Manufacturers Association

Figure 56–22. Exploded view of an exterior entryway system consisting of double panel doors, frames, sills, sidelights, and sidelight grills.

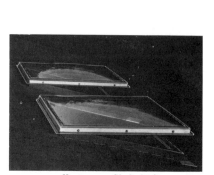

Ventarama Skylight Corporation

Figure 56–19. Interior and exterior views of skylights made of molded aluminum plexiglass set in an aluminum frame.

insulated foam core.

Door frames *(jambs)* may be of wood or metal. Some examples of exterior door styles are shown in Figure 56–24. (Additional information on door design and construction is in Section 13.)

Other entrances to the building are provided by one or more rear-entrance doors. These doors are sometimes referred to as *service doors*. Sliding glass door units are often used for service doors leading to porches, patios, or terraces. See Figure 56–25.

## Installing an Exterior Door Unit

Exterior door units are usually preassembled at a factory or mill-cabinet shop. They include a door *jamb,* outside *casing,* and

*Timber Company, Columbia Door Divison*

Figure 56–23. Solid-core wood entrance door with overhead and side window areas.

*Hurd Millwork Company*

Figure 56–25. Exterior glass sliding doors are frequently used for passage to porches, patios, and terraces. This model has wood frames.

*threshold* (door sill). The jamb is the finish frame for the door opening. The casing is the molding that covers the space between the jamb and the wall. The threshold is a wood, metal, or stone piece that covers the space between the jamb and the bottom of the door opening. Often the doors are pre-hung, meaning that they have already been fitted and hinged to the jambs. Before the jamb is placed in the wall, the opening should be flashed with *building paper.*

(Building paper is a water-resistant paper, also called *sheathing paper,* discussed in the following unit.) A procedure for installing a wood exterior door frame is described in Figure 56–26. (The manner in which the doors are fitted and hung in the frame is discussed in Section 13.) Figure 56–27 shows a wood door frame set in the wall of a building under construction.

**Metal Door Frames.** Metal door frames are delivered to the job site as packaged units and are assembled on the job according to the manufacturer's instructions. When installed, the units are plumbed and aligned in the rough opening. They are fastened to the framework with screws. Doors hung in metal frames may be wood or metal. (Additional information on metal door frames is in Section 13.)

**Wood or Metal Doors in Masonry Walls.** Close coordination with the bricklayers or masons is required for carpenters to install wood or metal frames in masonry (brick or hollow concrete blocks) walls. The frames are set in position when the exterior walls have reached the correct level below the door opening.

*Stanley Door Systems, division of the Stanley Works*

Figure 56–24. Examples of exterior door styles. Exterior doors provide passage between the inside and outside of a building.

# INSTALLING A WOODEN EXTERIOR DOOR FRAME

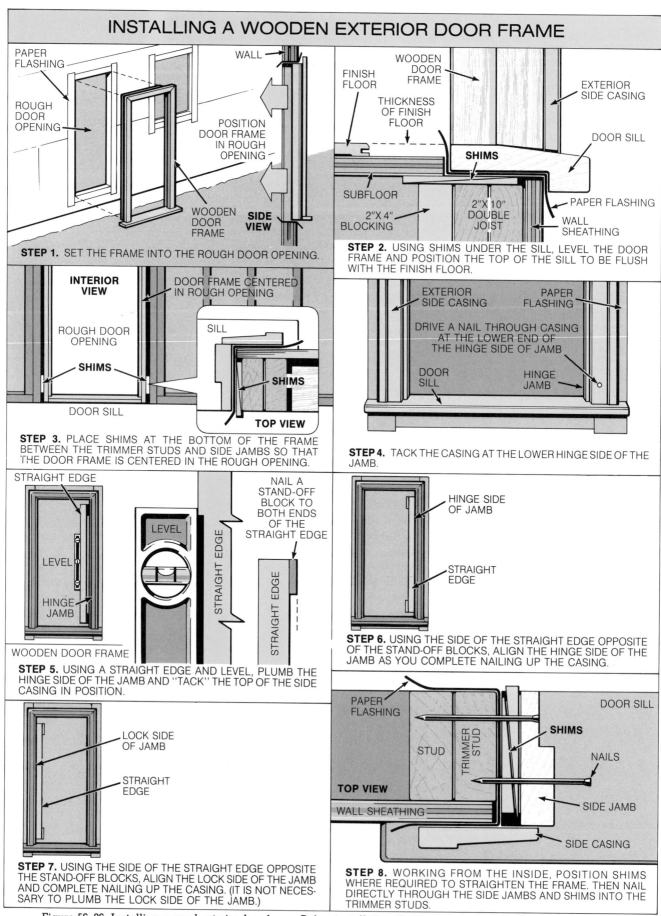

**STEP 1.** SET THE FRAME INTO THE ROUGH DOOR OPENING.

**STEP 2.** USING SHIMS UNDER THE SILL, LEVEL THE DOOR FRAME AND POSITION THE TOP OF THE SILL TO BE FLUSH WITH THE FINISH FLOOR.

**STEP 3.** PLACE SHIMS AT THE BOTTOM OF THE FRAME BETWEEN THE TRIMMER STUDS AND SIDE JAMBS SO THAT THE DOOR FRAME IS CENTERED IN THE ROUGH OPENING.

**STEP 4.** TACK THE CASING AT THE LOWER HINGE SIDE OF THE JAMB.

**STEP 5.** USING A STRAIGHT EDGE AND LEVEL, PLUMB THE HINGE SIDE OF THE JAMB AND "TACK" THE TOP OF THE SIDE CASING IN POSITION.

**STEP 6.** USING THE SIDE OF THE STRAIGHT EDGE OPPOSITE OF THE STAND-OFF BLOCKS, ALIGN THE HINGE SIDE OF THE JAMB AS YOU COMPLETE NAILING UP THE CASING.

**STEP 7.** USING THE SIDE OF THE STRAIGHT EDGE OPPOSITE THE STAND-OFF BLOCKS, ALIGN THE LOCK SIDE OF THE JAMB AND COMPLETE NAILING UP THE CASING. (IT IS NOT NECESSARY TO PLUMB THE LOCK SIDE OF THE JAMB.)

**STEP 8.** WORKING FROM THE INSIDE, POSITION SHIMS WHERE REQUIRED TO STRAIGHTEN THE FRAME. THEN NAIL DIRECTLY THROUGH THE SIDE JAMBS AND SHIMS INTO THE TRIMMER STUDS.

Figure 56-26. Installing a wood exterior door frame. Before installation, the opening is flashed with building paper.

Figure 56–28. Metal door frames plumbed and braced by carpenters during the construction of concrete-block walls.

Figure 56–27. Wood door frame set in the wall of a building under construction. Note paper flashing under the casing, and metal flashing above the drip cap at the top of the door frame.

Figure 56–29. Close-up of a metal door frame tie set in the mortar joint of a concrete-block wall.

They are plumbed, aligned, and securely braced. See Figure 56–28. The walls are then built up on both sides, and the door frame is anchored at the masonry joints. See Figure 56–29. The procedure for installing a wood door in a hollow-block wall is shown in Figure 56–30.

**Sliding Glass Doors.** Sometimes called *patio doors,* sliding glass doors are often constructed with aluminum frames and panels. The unit includes a frame, weatherstripping, and hardware. Wood types are also available. A typical sliding door has nylon rollers at the bottom to enable the door to move over tracks set in the sill. The sill is usually made of heavy extruded aluminum. The top of the door is held in place and guided by a channel or guide.

In two-door units both doors may slide or one door may be stationary. In a typical three-door combination, the center door slides and the side doors have insulating glass that eliminates the need for a storm door. Separate tracks are provided for sliding screen sections used during the summer.

## OVERHEAD GARAGE DOORS

Garage doors must be functional and also present an attractive appearance in keeping with the rest of the building design. See Figure 56–31. The garage doors installed today are usually overhead models. The older type of

461

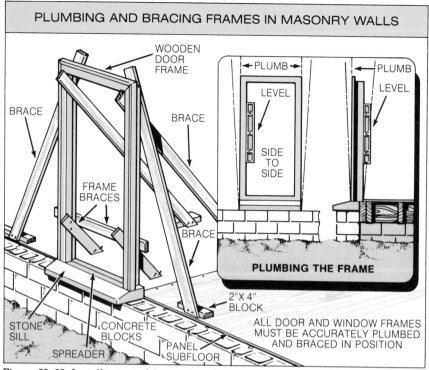

PLUMBING AND BRACING FRAMES IN MASONRY WALLS

WOODEN DOOR FRAME

BRACE

BRACE

FRAME BRACES

BRACE

STONE SILL

CONCRETE BLOCKS

SPREADER

PANEL SUBFLOOR

2" X 4" BLOCK

PLUMB — LEVEL — SIDE TO SIDE

PLUMB — LEVEL

PLUMBING THE FRAME

ALL DOOR AND WINDOW FRAMES MUST BE ACCURATELY PLUMBED AND BRACED IN POSITION

Figure 56–30. Installing wood frames in exterior masonry walls. The frames must be accurately plumbed and braced. Then the masonry walls are constructed on both sides.

hinged garage door is rarely used in new construction.

Overhead garage doors are available in a variety of stock dimensions for single-car or double-car garages. They can also be built to a special size. They are designed to give a flush or paneled appearance. Frames are of wood or metal. Materials used to finish off the doors include plywood, hardboard, particleboard,

aluminum, and fiberglass.

There are three basic types of overhead garage doors. Each type requires special hardware.

The *one-piece swing-up door* depends on springs and counterbalances for its operation. It is fitted with a ⅜" to ½" space on each side to permit air to enter the garage.

The *sectional roll-up door* is more expensive and operates more efficiently than the swing-up door. It is constructed in sections that are hinged together. The door movement is controlled by a track system fastened to the side jambs and suspended from the ceiling of the garage. See Figure 56–32.

The *rolling steel door* is normally used in industrial and commercial buildings. See Figure 56–33.

## Automatic Openers for Garage Doors

For convenience and security, overhead doors can be equipped

*Overhead Door Corporation*
Figure 56–31. Panel-design overhead garage door for a two-car garage.

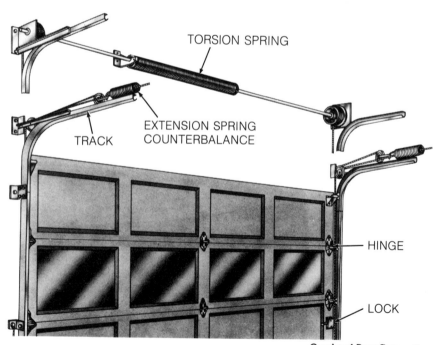

TORSION SPRING

EXTENSION SPRING COUNTERBALANCE

TRACK

HINGE

LOCK

*Overhead Door Corporation*
Figure 56–32. Sectional roll-up garage doors are guided by a track system fastened to the jambs and suspended from the ceiling.

Figure 56–33. Steel overhead rolling doors are used in industrial and commercial buildings.

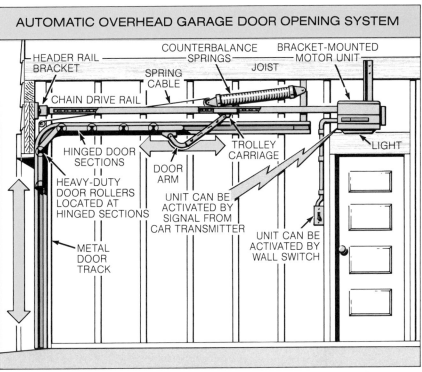

Figure 56–34. Automatic opening systems are widely used with residential overhead garage doors. The system shown here has a chain-and-cable mechanism operated by an electric motor. It is activated by a pushbutton or a transmitter carried in an automobile.

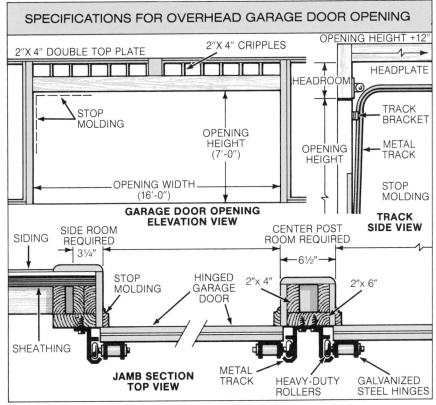

Figure 56–35. Example of architectural specifications for an overhead garage door opening.

with automatic opening systems. See Figure 56–34. An electric motor operates chains or cables that lift and lower the doors. The electric motor may be activated by a radio-controlled unit inside the car or by a key switch.

## Installing an Overhead Garage Door

If the rough opening for an overhead garage door has been constructed to the correct dimensions, there should be no installation problems. Correct layout for a rough opening is based on the size of the door, the jamb thickness, and a number of required clearances. See Figure 56–35.

Manufacturer's instructions for installing the hardware for an overhead garage door should be carefully followed. Typical hardware is shown in Figure 56–36.

463

## TYPICAL HARDWARE USED ON AN OVERHEAD GARAGE DOOR SYSTEM

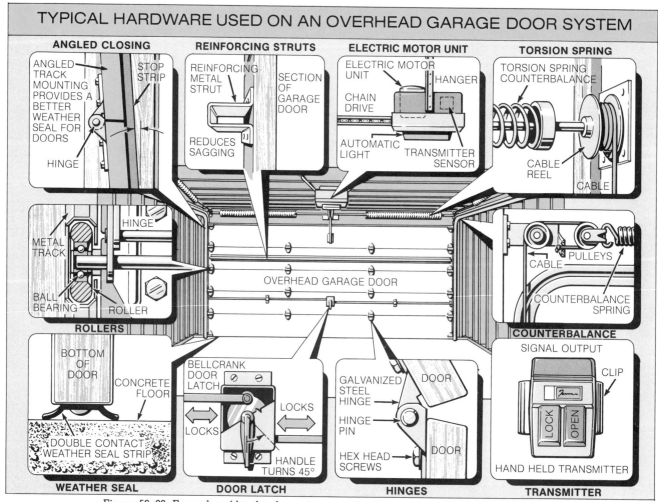

Figure 56–36. Examples of key hardware items used with an overhead garage door system.

Angled tracks provide a weather-tight seal for the door. Torsion spring counterbalance allows the door to open and close smoothly. An extension spring counterbalance device incorporates a safety cable for use in the event of a broken spring.

Other hardware includes heavy-duty rollers, hinges, a double-latch lock, and reinforcing steel struts.

# UNIT 57

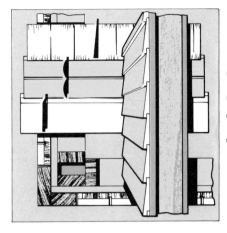

# Exterior Wall Finish, Porches, and Decks

Finish covering for exterior walls (*siding*) is usually placed after the door and window frames have been installed. Materials used for siding include wood, aluminum, vinyl, and asbestos. Exterior wall covering applied by workers in trades other than carpentry include stucco, brick, and stone. Often wood and masonry materials are combined for the exterior wall finish of a building.

Another part of the finish work on the outside of a building is the construction of a porch or deck. This is a popular feature in mild climates that allow many months of outdoor living each year. Porches and decks are made of wood.

## WOOD SIDING

Wood is one of the oldest materials used for exterior wall covering. It is also one of the most attractive and most efficient in protecting the outside surface of a building. For these reasons, wood siding still covers more home exteriors than any other type of siding. See Figure 57–1. Recently, wood-fabricated products such as plywood and hardboard have also become popular. Both wood and wood-fabricated products for sid-

ing are available in boards of various shapes (*patterns*), in panels, or in shingles or shakes.

## Board Siding

Redwood and cedar are highly recommended for wood board siding because of their superior resistance to decay. Western hemlock, cypress, yellow poplar, spruce, and several pine species are also satisfactory. Wood board siding should be a good grade of lumber, free of warp, twist, knots, and pitch pockets.

Allowable moisture content for wood board siding is approximately 10% to 12% in most areas of the United States. In the drier southwestern states, it is 8% to 9%. The siding should be *back-primed* before nailing. In back-priming, the backs of the boards are painted with a sealer (either paint primer or a water-repellent preservative). As the boards are cut for fitting and nailing to the wall, the freshly cut ends are also painted with sealer.

Hardboard, a wood-fabricated product, can also be used for siding. (For a review of its manufacturing process, refer to Section 2.) Hardboard can be manufactured to give the appearance of many different lumber species.

*Georgia-Pacific Corporation*
Figure 57–1. Cedar lap siding produces an attractive exterior finish on this traditionally designed home. Note solar collectors on the main roof.

It can be made with different textures and painted or stained.

### Patterns of Board Siding.
Board siding is available in other shapes (or patterns) than the straight board shape. For example, some panels have a bevel shape, some have tongue-and-groove edges, and some have rabbeted edges. Many other variations are also available. See Figure 57–2. In addition, all of

# WOOD SIDING–AVAILABLE PATTERNS, APPLICATION AND NAILING, GRADES

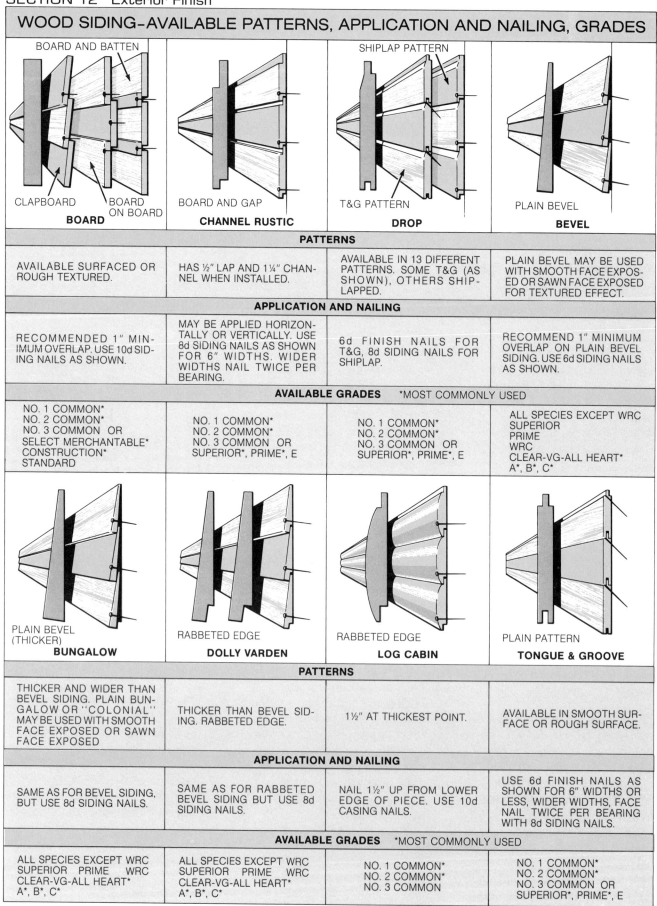

|  | BOARD | CHANNEL RUSTIC | DROP | BEVEL |
|---|---|---|---|---|
| **PATTERNS** | AVAILABLE SURFACED OR ROUGH TEXTURED. | HAS ½″ LAP AND 1¼″ CHANNEL WHEN INSTALLED. | AVAILABLE IN 13 DIFFERENT PATTERNS. SOME T&G (AS SHOWN), OTHERS SHIPLAPPED. | PLAIN BEVEL MAY BE USED WITH SMOOTH FACE EXPOSED OR SAWN FACE EXPOSED FOR TEXTURED EFFECT. |
| **APPLICATION AND NAILING** | RECOMMENDED 1″ MINIMUM OVERLAP. USE 10d SIDING NAILS AS SHOWN. | MAY BE APPLIED HORIZONTALLY OR VERTICALLY. USE 8d SIDING NAILS AS SHOWN FOR 6″ WIDTHS. WIDER WIDTHS NAIL TWICE PER BEARING. | 6d FINISH NAILS FOR T&G, 8d SIDING NAILS FOR SHIPLAP. | RECOMMEND 1″ MINIMUM OVERLAP ON PLAIN BEVEL SIDING. USE 6d SIDING NAILS AS SHOWN. |
| **AVAILABLE GRADES   *MOST COMMONLY USED** | NO. 1 COMMON* NO. 2 COMMON* NO. 3 COMMON OR SELECT MERCHANTABLE* CONSTRUCTION* STANDARD | NO. 1 COMMON* NO. 2 COMMON* NO. 3 COMMON OR SUPERIOR*, PRIME*, E | NO. 1 COMMON* NO. 2 COMMON* NO. 3 COMMON OR SUPERIOR*, PRIME*, E | ALL SPECIES EXCEPT WRC SUPERIOR PRIME WRC CLEAR-VG-ALL HEART* A*, B*, C* |

|  | BUNGALOW | DOLLY VARDEN | LOG CABIN | TONGUE & GROOVE |
|---|---|---|---|---|
| **PATTERNS** | THICKER AND WIDER THAN BEVEL SIDING. PLAIN BUNGALOW OR "COLONIAL" MAY BE USED WITH SMOOTH FACE EXPOSED OR SAWN FACE EXPOSED | THICKER THAN BEVEL SIDING. RABBETED EDGE. | 1½″ AT THICKEST POINT. | AVAILABLE IN SMOOTH SURFACE OR ROUGH SURFACE. |
| **APPLICATION AND NAILING** | SAME AS FOR BEVEL SIDING, BUT USE 8d SIDING NAILS. | SAME AS FOR RABBETED BEVEL SIDING BUT USE 8d SIDING NAILS. | NAIL 1½″ UP FROM LOWER EDGE OF PIECE. USE 10d CASING NAILS. | USE 6d FINISH NAILS AS SHOWN FOR 6″ WIDTHS OR LESS, WIDER WIDTHS, FACE NAIL TWICE PER BEARING WITH 8d SIDING NAILS. |
| **AVAILABLE GRADES   *MOST COMMONLY USED** | ALL SPECIES EXCEPT WRC SUPERIOR  PRIME  WRC CLEAR-VG-ALL HEART* A*, B*, C* | ALL SPECIES EXCEPT WRC SUPERIOR  PRIME  WRC CLEAR-VG-ALL HEART* A*, B*, C* | NO. 1 COMMON* NO. 2 COMMON* NO. 3 COMMON | NO. 1 COMMON* NO. 2 COMMON* NO. 3 COMMON OR SUPERIOR*, PRIME*, E |

*Information from Western Wood Products Association*

Figure 57–2. Wood siding patterns. Each pattern is available in various widths and thicknesses.

| PRODUCT | NOMINAL SIZE | | DRESSED DIMENSIONS | | |
|---|---|---|---|---|---|
| | THICKNESS IN. | WIDTH IN. | THICKNESS IN. | WIDTH IN. | |
| BEVEL SIDING<br>FOR WRC SIZES<br>SEE FOOTNOTE* | 1/2 | 4<br>5<br>6 | 11/22 BUTT, 3/16 TIP | 3 1/2<br>4 1/2<br>5 1/2 | |
| WIDE BEVEL SIDING (COLONIAL OR BUNGA-LOW) | 3/4 | 8<br>10<br>12 | 3/4 BUTT, 3/16 TIP | 7 1/4<br>9 1/4<br>11 1/4 | |
| | | | | FACE | OVERALL |
| RABBETED BEVEL SIDING (DOLLY VARDEN) | 3/4<br>1 | 6<br>8<br>10<br>12 | 5/8 by 3/16<br>11/16 by 13/32 | 5<br>6 3/4<br>8 1/4<br>10 3/4 | 5 1/2<br>7 1/4<br>9 1/2<br>11 1/4 |
| RUSTIC AND DROP SIDING (DRESSED AND MATCHED) | 1 | 6<br>8<br>10<br>12 | 22/32 | 5 1/8<br>6 7/8<br>8 7/8<br>10 7/8 | 5 3/8<br>7 1/8<br>9 1/8<br>11 1/8 |
| RUSTIC AND DROP SIDING (SHIPLAPPED, 3/8-IN. LAP) | 1 | 6<br>8<br>10<br>12 | 22/32 | 5<br>6 3/4<br>8 3/4<br>10 3/4 | 5 3/8<br>7 1/8<br>9 1/8<br>11 1/8 |
| RUSTIC AND DROP SIDING (SHIPLAPPED, 1/2-IN. LAP) | 1 | 6<br>8<br>10<br>12 | 22/32 | 4 15/16<br>6 5/8<br>8 5/8<br>10 5/8 | 5 7/16<br>7 1/8<br>9 1/8<br>11 1/8 |
| LOG CABIN SIDING | 1 1/2 (6/4) | 6<br>8<br>10 | 1 1/2" THICKEST POINT | 4 15/16<br>6 5/8<br>8 5/8 | 5 7/16<br>7 1/8<br>9 1/8 |
| TONGUE & GROOVE (T&G) S2S AND CM | 1 (4/4) | 4<br>6<br>8<br>10<br>12 | 3/4 | 3 1/8<br>5 1/8<br>6 7/8<br>8 7/8<br>10 7/8 | 3 3/8<br>5 3/8<br>7 1/8<br>9 1/8<br>11 1/8 |

*Western Red Cedar Bevel Siding available in 1/2", 5/8", 3/4" nominal thickness. Corresponding surfaced thick edge is 15/32", 9/16" and 3/4". Widths 8" and wider 1/2" off.

*Western Wood Products Association*

Figure 57–3. Nominal and dressed dimensions of wood siding patterns.

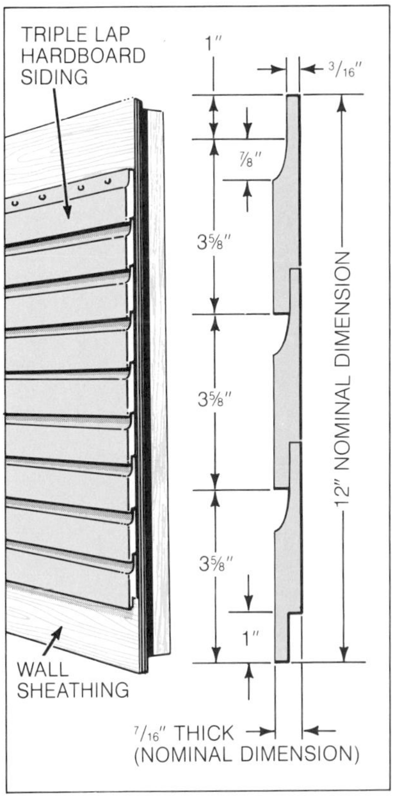


Figure 57–4. Triple-lap hardboard siding allows fast installation.

these patterns are produced in various widths and thicknesses. Figure 57–3 gives the nominal and dressed dimensions of some patterns. A triple-lap design is available (in hardboard) that allows faster installation. See Figure 57–4.

**Preparing the Wall for Board Siding.** In most cases, siding is applied over some type of sheathing material. In mild climates, however, sheathing is not always required, and the siding may be nailed directly to the wall studs. Where structural sheathing or rigid insulation board is used for the nailing base, it is sufficient to nail the siding just into the sheathing material without penetrating the wall studs. When nonstructural sheathing is used, such as gypsum and other types of insulation panels, the siding must be nailed through the sheathing and into the wall studs.

Where board sheathing or no sheathing is used, a layer of *sheathing paper* (building paper) should be placed on the wall. This water-resistant paper helps prevent future air infiltration and water leakage. See Figure 57–5. Sheathing paper must be sufficiently waterproof to resist water penetration from the outside, yet not resistant enough to prevent the escape of water vapor from the inside of the building during cold weather. If water vapor is prevented from escaping, condensation could form inside the walls.

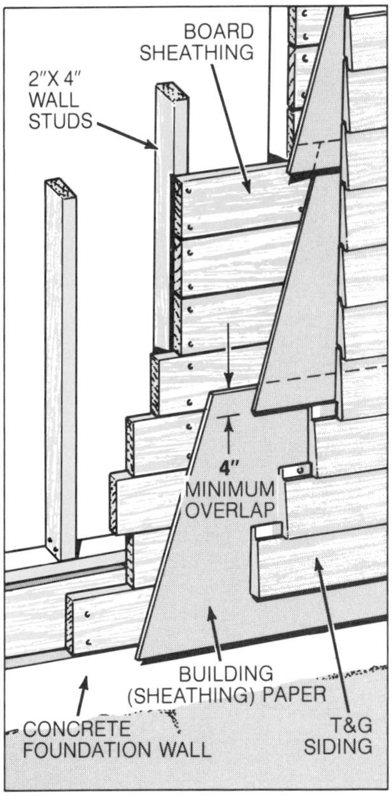


Figure 57–5. Sheathing (building) paper is required over board sheathing.

Figure 57–6. Asphalt-saturated felt sheathing is applied over the studs before the siding material is placed.

American Plywood Association

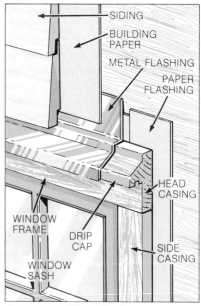

Figure 57–7. Metal flashing is applied over the drip caps of door and window frames.

## SIDING SYSTEM USING TYPICAL METAL TRIM PIECES

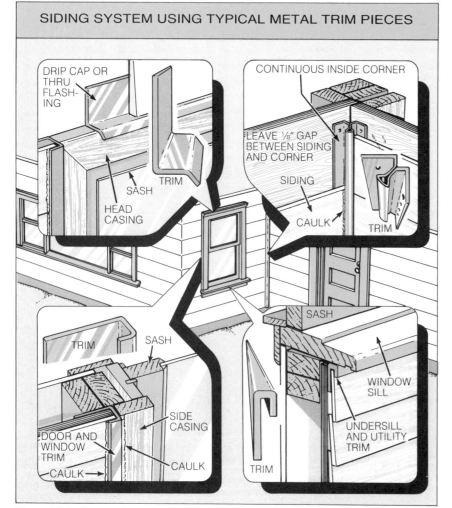

Figure 57–8. Some siding systems use metal trim pieces around the door and window frames.

Asphalt-saturated felt, which has low resistance to water vapor, is generally used as sheathing paper for board sheathing. See Figure 57–6. Panel sheathing (plywood, particleboard, waferboard, etc.) and most types of rigid insulation sheathing do not require sheathing paper.

Flashing is recommended over the *drip caps* of doors and windows to prevent rain leakage. A drip cap is a piece of molding placed over exterior wood frames to direct water away from the door or window. See Figure 57–7. Flashing material on exterior walls is usually galvanized sheet metal. Vinyl flashing is also used.

Other preparations for siding include metal trim pieces placed around the door and window openings. See Figure 57–8.

**Applying Board Siding.** The procedure for applying board siding is similar for all the different patterns. Application is usually horizontal. See Figure 57–9. However, board, shiplap, channel rustic, and tongue-and-groove patterns can also be applied vertically. See Figure 57–10. An at-

468

tractive effect can be created by combining horizontal and vertical applications. See Figure 57–11. Diagonal siding (Figure 57–12) is also sometimes used.

*Horizontal Application.* The first row of boards is applied in a level, perfectly straight line. The bottom of the first row of boards should be at least 1″ below the top of the foundation wall. For bevel siding, a spacer strip equal to the thickness of the thin edge of the siding should be nailed to the wall underneath the bottom

edge. This will slant the first row of siding. See Figure 57–13. Bevel siding should overlap at least 1″. To maintain a uniform overlap, use a notched spacer block as shown in Figure 57–13.

Whenever possible, the rows of siding should be laid out so that they line up with the top of the window drip cap and the bot-

tom of the windowsill. See Figure 57–14. When this layout cannot be followed, the siding boards must be notched around the windows. See Figure 57–15.

The recommended procedure for deciding the siding sequence for a particular wall is to make a siding *story pole*. (A story pole for wall framing is discussed in

*California Redwood Association*

Figure 57–11. Combined vertical and horizontal siding design on a commercial building.

*California Redwood Association*

Figure 57–9. Horizontally applied bevel siding.

*California Redwood Association*

Figure 57–10. Vertically applied channel rustic siding.

*California Redwood Association*

Figure 57–12. Diagonally applied siding gives a contemporary appearance.

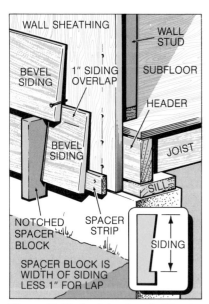

Figure 57-13. Applying bevel siding. Note the spacer strip underneath the bottom edge of the first course.

Section 9.) By laying out the position and height of the window opening and the siding courses on a wood rod (the story pole), the carpenter can determine the layout of the boards. One procedure for laying out a siding story pole is shown in Figure 57-16. In this case, the story pole shows that the boards can be made to line up with the top and bottom of the window.

*Vertical Application.* The layout procedure is somewhat simpler for vertical than for horizontal application. Position the starting corner board so that it is straight and plumb. See Figure 57-17. When vertical siding is placed over walls that are not covered with structural sheathing or insulation, blocks for nailing must be placed between the studs 16″ to 24″ apart. Figure 57-18 shows vertical siding being applied to a wall covered with sheathing.

*Joints.* Fits between the ends of the boards should be *butt* joints. They should be tightly fitted and

staggered as shown in Figure 57-19. Metal butt joint covers are often used with hardboard siding. They enable faster application and provide a weathertight joint. See Figure 57-20.

A joint between vertical and horizontal boards occurs on buildings with gable roofs where vertical siding is applied at the gable ends of the roof, and horizontal boards are placed on the walls below. (Such a design appears in the east and south elevations of the Three-bedroom House Plan in Section 6.) When this type of joint occurs, special steps must be taken to prevent water infiltration where the vertical and horizontal siding meet. One method of preventing water infiltration is to install a drip cap. See Figure 57-21. Another way is to offset the upper wall so that the vertical pieces lap over the top horizontal piece. See Figure 57-22.

*Corners.* Corners must have a neat appearance and be weathertight. Metal corners that give a mitered effect are easy to install. See Figure 57-23. Wooden outside corner boards and in-

side corner strips are often used at corners. Outside corners may be mitered, although this is a more difficult procedure since the cut must be extremely accurate to produce a tight joint. See Figure 57-24.

*Nailing Methods.* Proper nailing is important to the appearance and durability of siding. Noncorrosive nails should be used to avoid unsightly stains or rust streaks. See Figure 57-25. Aluminum alloy, stainless steel, or top quality hot-dipped galvanized nails are recommended for wood siding. Nail shanks may be spiral-grooved or annular-ringed. See Figure 57-26. Although box, casing, and finish nails can be used for siding, special *siding nails* are best. They have a slightly tapered head that can be driven flush with the siding or countersunk without crushing surrounding wood. Generally, 6d nails are used for ½″ siding and 8d nails for ¾″ siding. Although structural sheathing is usually considered an adequate nail base by itself, better fastening is obtained by driving nails through the sheathing and into the wall

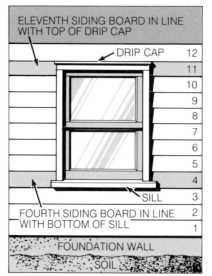

Figure 57-14. In this siding layout, courses line up with the windowsill and drip cap.

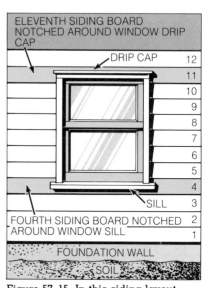

Figure 57-15. In this siding layout, boards are notched around the windowsill and drip cap.

# LAYOUT AND USE OF A STORY POLE MADE FOR 7¼" BEVEL SIDING

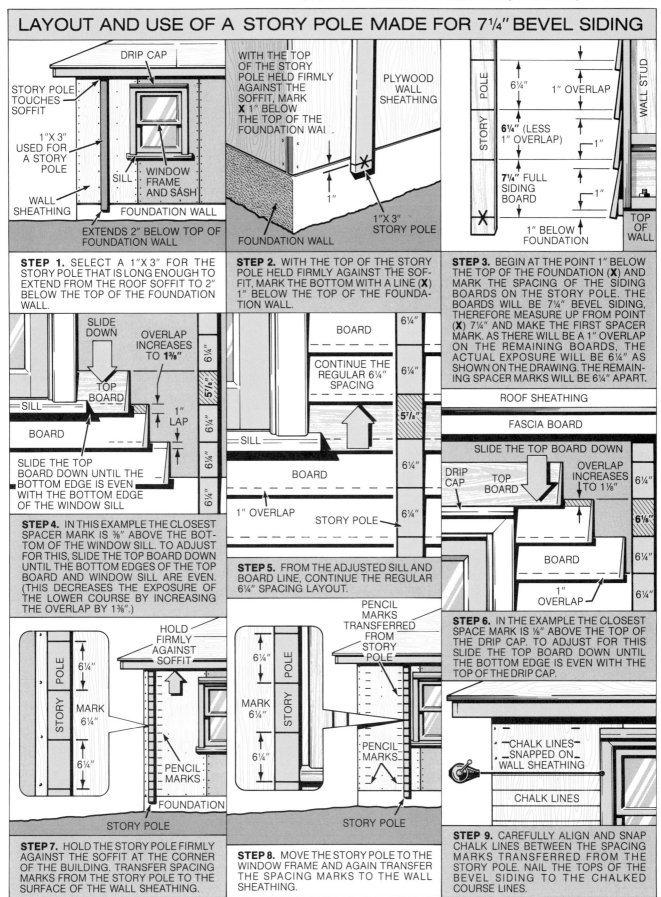

**STEP 1.** SELECT A 1"X 3" FOR THE STORY POLE THAT IS LONG ENOUGH TO EXTEND FROM THE ROOF SOFFIT TO 2" BELOW THE TOP OF THE FOUNDATION WALL.

**STEP 2.** WITH THE TOP OF THE STORY POLE HELD FIRMLY AGAINST THE SOFFIT, MARK THE BOTTOM WITH A LINE (**X**) 1" BELOW THE TOP OF THE FOUNDATION WALL.

**STEP 3.** BEGIN AT THE POINT 1" BELOW THE TOP OF THE FOUNDATION (**X**) AND MARK THE SPACING OF THE SIDING BOARDS ON THE STORY POLE. THE BOARDS WILL BE 7¼" BEVEL SIDING, THEREFORE MEASURE UP FROM POINT (**X**) 7¼" AND MAKE THE FIRST SPACER MARK. AS THERE WILL BE A 1" OVERLAP ON THE REMAINING BOARDS, THE ACTUAL EXPOSURE WILL BE 6¼" AS SHOWN ON THE DRAWING. THE REMAINING SPACER MARKS WILL BE 6¼" APART.

**STEP 4.** IN THIS EXAMPLE THE CLOSEST SPACER MARK IS ⅜" ABOVE THE BOTTOM OF THE WINDOW SILL. TO ADJUST FOR THIS, SLIDE THE TOP BOARD DOWN UNTIL THE BOTTOM EDGES OF THE TOP BOARD AND WINDOW SILL ARE EVEN. (THIS DECREASES THE EXPOSURE OF THE LOWER COURSE BY INCREASING THE OVERLAP BY 1⅜".)

**STEP 5.** FROM THE ADJUSTED SILL AND BOARD LINE, CONTINUE THE REGULAR 6¼" SPACING LAYOUT.

**STEP 6.** IN THE EXAMPLE THE CLOSEST SPACE MARK IS ⅛" ABOVE THE TOP OF THE DRIP CAP. TO ADJUST FOR THIS SLIDE THE TOP BOARD DOWN UNTIL THE BOTTOM EDGE IS EVEN WITH THE TOP OF THE DRIP CAP.

**STEP 7.** HOLD THE STORY POLE FIRMLY AGAINST THE SOFFIT AT THE CORNER OF THE BUILDING. TRANSFER SPACING MARKS FROM THE STORY POLE TO THE SURFACE OF THE WALL SHEATHING.

**STEP 8.** MOVE THE STORY POLE TO THE WINDOW FRAME AND AGAIN TRANSFER THE SPACING MARKS TO THE WALL SHEATHING.

**STEP 9.** CAREFULLY ALIGN AND SNAP CHALK LINES BETWEEN THE SPACING MARKS TRANSFERRED FROM THE STORY POLE. NAIL THE TOPS OF THE BEVEL SIDING TO THE CHALKED COURSE LINES.

Figure 57–16. Layout and use of a siding story pole. Use of a story pole is the best way to decide on the sequence for siding application.

471

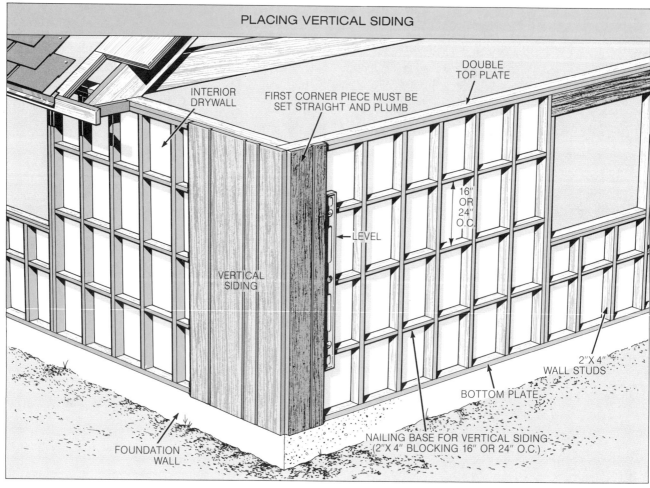

PLACING VERTICAL SIDING

INTERIOR DRYWALL

FIRST CORNER PIECE MUST BE SET STRAIGHT AND PLUMB

DOUBLE TOP PLATE

LEVEL

VERTICAL SIDING

16" OR 24" O.C.

2"X 4" WALL STUDS

BOTTOM PLATE

NAILING BASE FOR VERTICAL SIDING (2"X 4" BLOCKING 16" OR 24" O.C.)

FOUNDATION WALL

Figure 57–17. The first corner piece of vertical siding must be set straight and plumb. In this example, structural sheathing is not being used. Therefore, blocking 16" to 24" O.C. must be placed between the studs as a nailing base for the siding.

Georgia-Pacific Corporation

Figure 57–18. Carpenters placing siding vertically on a wall covered with sheathing.

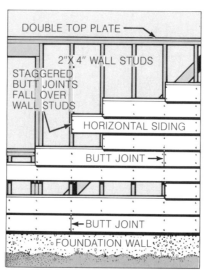

DOUBLE TOP PLATE

2"X 4" WALL STUDS

STAGGERED BUTT JOINTS FALL OVER WALL STUDS

HORIZONTAL SIDING

BUTT JOINT

BUTT JOINT

FOUNDATION WALL

Figure 57–19. Butt joints of horizontal siding should be staggered and fall over a wall stud.

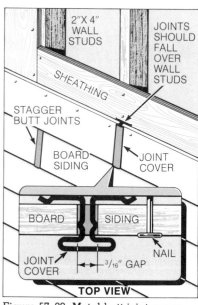

2"X 4" WALL STUDS

JOINTS SHOULD FALL OVER WALL STUDS

SHEATHING

STAGGER BUTT JOINTS

BOARD SIDING

JOINT COVER

BOARD      SIDING

JOINT COVER

NAIL

3/16" GAP

TOP VIEW

Figure 57–20. Metal butt joint covers are often used with hardboard siding.

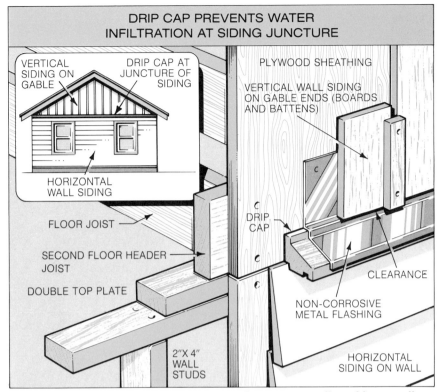

Figure 57–21. A drip cap placed at the juncture between vertically and horizontally placed siding prevents water infiltration.

studs.

If the siding is to be painted or a solid-colored stain is to be used, the nails should be countersunk and the holes puttied. If the finish is a semi-transparent stain, a better appearance is produced by not countersinking the nails.

When placing hardboard siding, drive the nails so that the back of the nail head is in contact with the surface of the siding. When nailing close to the end of a piece, drive the nail into a pre-bored hole approximately three-quarters the diameter of the nail shank. Figures 57–27 and 57–28 show nailing methods for commonly used siding patterns.

Pneumatic nailers and staplers are also used to fasten finish siding. Like nails, staples should be rust-resistant.

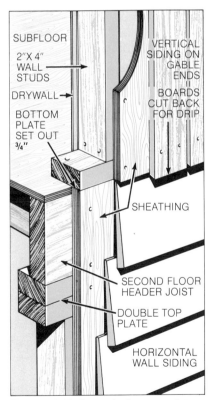

Figure 57–22. Instead of using a drip cap to prevent water infiltration at the juncture between vertical and horizontal siding, the upper wall may be offset so that vertical pieces lap over the top horizontal piece.

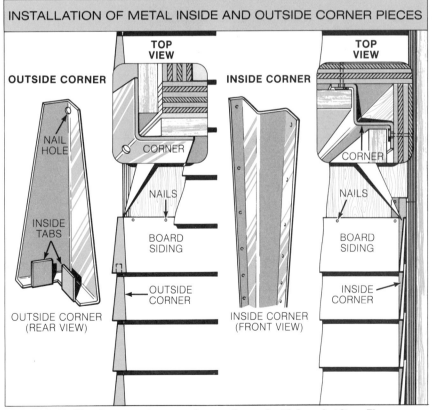

Figure 57–23. Metal corner pieces are frequently used with board siding. They are easy to install and give a mitered effect to the corners.

473

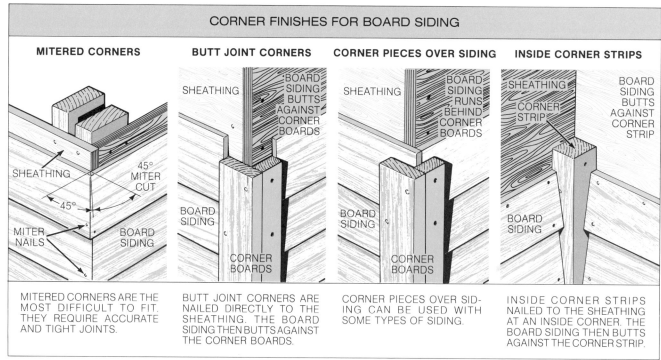

CORNER FINISHES FOR BOARD SIDING

**MITERED CORNERS**

SHEATHING

45° MITER CUT

45°

MITER NAILS

BOARD SIDING

MITERED CORNERS ARE THE MOST DIFFICULT TO FIT. THEY REQUIRE ACCURATE AND TIGHT JOINTS.

**BUTT JOINT CORNERS**

SHEATHING

BOARD SIDING BUTTS AGAINST CORNER BOARDS

BOARD SIDING

CORNER BOARDS

BUTT JOINT CORNERS ARE NAILED DIRECTLY TO THE SHEATHING. THE BOARD SIDING THEN BUTTS AGAINST THE CORNER BOARDS.

**CORNER PIECES OVER SIDING**

SHEATHING

BOARD SIDING RUNS BEHIND CORNER BOARDS

BOARD SIDING

CORNER BOARDS

CORNER PIECES OVER SIDING CAN BE USED WITH SOME TYPES OF SIDING.

**INSIDE CORNER STRIPS**

SHEATHING

CORNER STRIP

BOARD SIDING BUTTS AGAINST CORNER STRIP

BOARD SIDING

INSIDE CORNER STRIPS NAILED TO THE SHEATHING AT AN INSIDE CORNER. THE BOARD SIDING THEN BUTTS AGAINST THE CORNER STRIP.

Figure 57–24. Types of wood corner finishes for board siding.

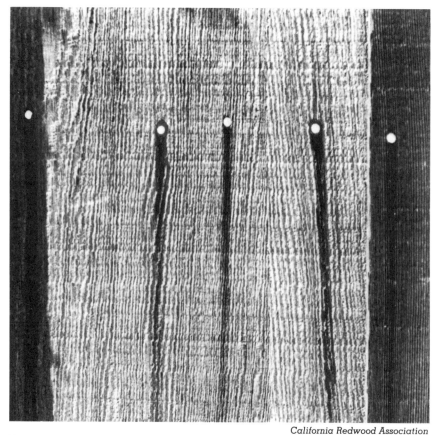

*California Redwood Association*

Figure 57–25. Stains and rust streaks may mar the appearance of siding unless noncorrosive nails are used.

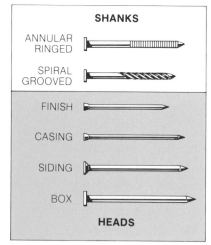

**SHANKS**

ANNULAR RINGED

SPIRAL GROOVED

FINISH

CASING

SIDING

BOX

**HEADS**

Figure 57–26. Nails for exterior trim work. The siding nail is considered best for board siding. The head is slightly tapered and can be driven flush or countersunk without crushing surrounding wood.

## Panel Siding

Panel siding can be installed more quickly than board siding. Plywood or hardboard is usually used for panel siding.

**Plywood Panel Siding.** Plywood

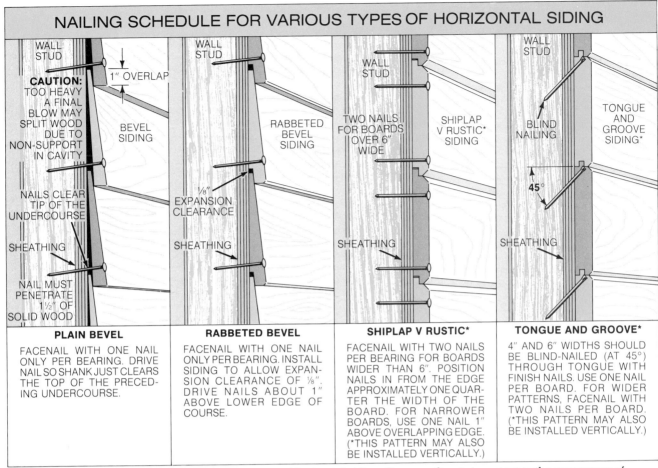

## NAILING SCHEDULE FOR VARIOUS TYPES OF HORIZONTAL SIDING

| PLAIN BEVEL | RABBETED BEVEL | SHIPLAP V RUSTIC* | TONGUE AND GROOVE* |
|---|---|---|---|
| FACENAIL WITH ONE NAIL ONLY PER BEARING. DRIVE NAIL SO SHANK JUST CLEARS THE TOP OF THE PRECEDING UNDERCOURSE. | FACENAIL WITH ONE NAIL ONLY PER BEARING. INSTALL SIDING TO ALLOW EXPANSION CLEARANCE OF ⅛". DRIVE NAILS ABOUT 1" ABOVE LOWER EDGE OF COURSE. | FACENAIL WITH TWO NAILS PER BEARING FOR BOARDS WIDER THAN 6". POSITION NAILS IN FROM THE EDGE APPROXIMATELY ONE QUARTER THE WIDTH OF THE BOARD. FOR NARROWER BOARDS, USE ONE NAIL 1" ABOVE OVERLAPPING EDGE. (*THIS PATTERN MAY ALSO BE INSTALLED VERTICALLY.) | 4" AND 6" WIDTHS SHOULD BE BLIND-NAILED (AT 45°) THROUGH TONGUE WITH FINISH NAILS. USE ONE NAIL PER BOARD. FOR WIDER PATTERNS, FACENAIL WITH TWO NAILS PER BOARD. (*THIS PATTERN MAY ALSO BE INSTALLED VERTICALLY.) |

Figure 57–27. Nailing schedules for four common types of horizontal siding. Proper nailing is important to the appearance of siding.

panels come in a variety of textures and patterns. See Figure 57–29. Most come in a 4' standard width and in lengths of 7', 8', 9', 10', and 12'. Common thicknesses are ¹¹⁄₃₂", ⅜", ¹⁵⁄₃₂", ½", ¹⁹⁄₃₂", and ⅝".

Plywood panels may be applied directly to studs spaced 16" or 24" O.C. or over nonstructural (insulative) sheathing. They are usually applied vertically, but they can be placed horizontally if the horizontal joints are blocked.

Some plywood panel siding has a structural rating. This means that the panels have the same high strength and rack resistance as structural panel sheathing. Therefore, with this type of panel siding it is not necessary to put diagonal bracing inside the framed wall or another layer of structural sheathing un-

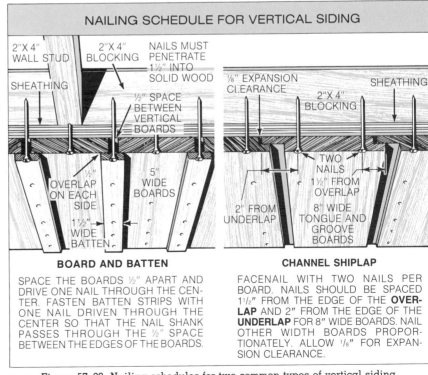

## NAILING SCHEDULE FOR VERTICAL SIDING

| BOARD AND BATTEN | CHANNEL SHIPLAP |
|---|---|
| SPACE THE BOARDS ½" APART AND DRIVE ONE NAIL THROUGH THE CENTER. FASTEN BATTEN STRIPS WITH ONE NAIL DRIVEN THROUGH THE CENTER SO THAT THE NAIL SHANK PASSES THROUGH THE ½" SPACE BETWEEN THE EDGES OF THE BOARDS. | FACENAIL WITH TWO NAILS PER BOARD. NAILS SHOULD BE SPACED 1½" FROM THE EDGE OF THE **OVERLAP** AND 2" FROM THE EDGE OF THE **UNDERLAP** FOR 8" WIDE BOARDS. NAIL OTHER WIDTH BOARDS PROPORTIONATELY. ALLOW ⅛" FOR EXPANSION CLEARANCE. |

Figure 57–28. Nailing schedules for two common types of vertical siding.

derneath the panel siding. See Figure 57–30.

**Hardboard Panel Siding.** Like plywood panels, hardboard panels come in a variety of textures and patterns. Pre-finished panels are also available. Panels are usually 7⁄16″ thick, 4′ wide, and 8′, 9′, or 10′ long.

Hardboard panels do not have the high shear strength of plywood panel siding. They are often applied over structural sheathing. If they are applied over nonstructural sheathing, or directly to the studs, diagonal bracing should be placed in the framed wall.

Hardboard panels must be acclimatized to the temperature conditions around the building they will be covering. For this reason, they should be stored on the job site at least five days before installation.

**Preparing the Wall for Panel Siding.** Walls must be prepared for panel siding in the same manner described for board siding. Place flashing and drip caps over the door and window openings and apply building paper if necessary. Building paper is not required if the vertical joints of the panels have shiplap edges or if the square edges are covered with battens. Building paper may also be omitted if the siding is installed over panel sheathing.

**Applying Panel Siding.** Both plywood and hardboard panels are applied the same way. All edges of panels must be sealed at the time of application with either paint primer or a water-repellent preservative. Allow a 1⁄8″ joint spacing along the panel ends and edges to provide for panel expansion from absorption of moisture. (Some manufacturers suggest only 1⁄16″ allowance.)

Nail the panels 6″ O.C. along

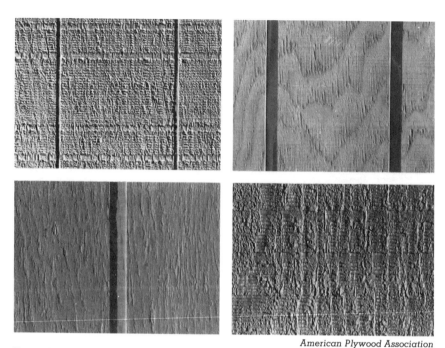

*American Plywood Association*

Figure 57–29. Some examples of surface textures and patterns available in plywood panel siding.

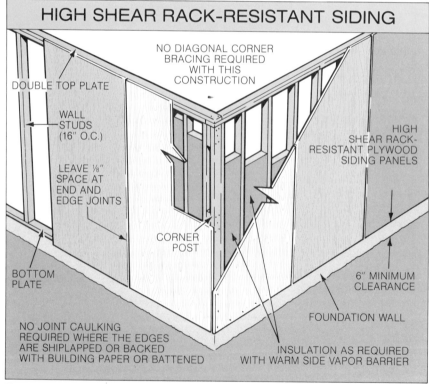

Figure 57–30. APA Sturd-I-Wall plywood siding system. These panels have high shear strength and are rack-resistant. No structural sheathing is required underneath, and diagonal bracing is not necessary in the framed wall.

## RECOMMENDED NAILING SEQUENCE FOR PLYWOOD PANEL SIDING

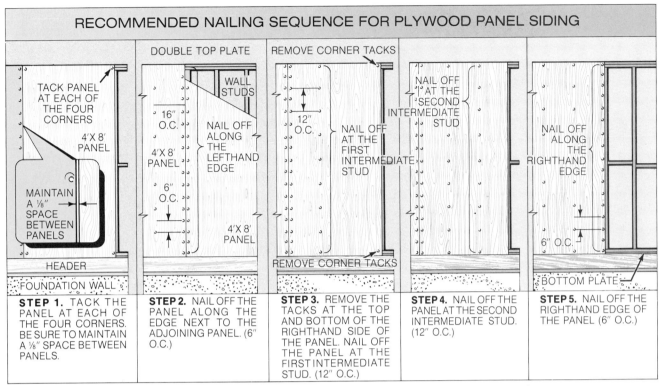

STEP 1. TACK THE PANEL AT EACH OF THE FOUR CORNERS. BE SURE TO MAINTAIN A ⅛" SPACE BETWEEN PANELS.

STEP 2. NAIL OFF THE PANEL ALONG THE EDGE NEXT TO THE ADJOINING PANEL. (6" O.C.)

STEP 3. REMOVE THE TACKS AT THE TOP AND BOTTOM OF THE RIGHTHAND SIDE OF THE PANEL. NAIL OFF THE PANEL AT THE FIRST INTERMEDIATE STUD. (12" O.C.)

STEP 4. NAIL OFF THE PANEL AT THE SECOND INTERMEDIATE STUD. (12" O.C.)

STEP 5. NAIL OFF THE RIGHTHAND EDGE OF THE PANEL (6" O.C.)

Figure 57–31. Nailing sequence for plywood panel siding to ensure a flat, even surface.

the panel edges, ½" from the edge, and 12" O.C. at the intermediate studs. Use 6d non-staining box, casing, or siding nails for panels ½" or less in thickness. For thicker panels use 8d nails. The panels must be nailed properly for a uniform and flat appearance. A recommended nailing sequence is shown in Figure 57–31. A building in the process of being covered with panel siding is shown in Figure 57–32.

Panels may have either

American Plywood Association

Figure 57–32. Panel siding can serve as structural sheathing as well as siding. Note the walls being framed over the garage floor slab.

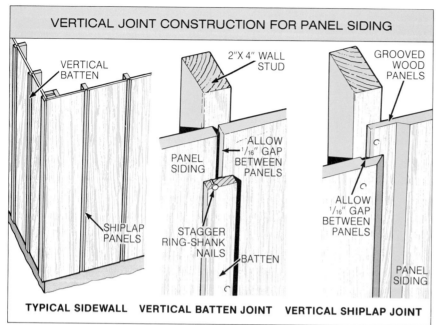

VERTICAL JOINT CONSTRUCTION FOR PANEL SIDING

VERTICAL BATTEN

SHIPLAP PANELS

2"X 4" WALL STUD

PANEL SIDING

ALLOW ¹⁄₁₆" GAP BETWEEN PANELS

STAGGER RING-SHANK NAILS

BATTEN

GROOVED WOOD PANELS

ALLOW ¹⁄₁₆" GAP BETWEEN PANELS

PANEL SIDING

TYPICAL SIDEWALL   VERTICAL BATTEN JOINT   VERTICAL SHIPLAP JOINT

Figure 57–33. Vertical joint details for panel siding. All vertical joints should fall on a stud.

square, shiplap, or tongue-and-groove edges on their long sides. They are usually placed with their long sides in a vertical position, and all vertical joints should fall on a stud.

For panels with square edges, the joint should be covered with a piece of batten. See Figure 57–33. Where horizontal joints occur between the tops of the lower panels and the bottoms of the panels above, blocks for nailing should be placed between the studs. To prevent water leakage, the panels should be lapped or flashing should be applied. See Figure 57–34. Corner joints are finished off using the same methods described for board siding. The procedure for installing panel siding is shown in Figure 57–35.

## Wood Shingles or Shakes

Wood shingles or shakes produce an attractive and durable covering for exterior walls. See Figure 57–36. The same kinds of shingles and shakes described

for roof covering in a previous unit are used for exterior walls. The traditional method of application is to place one shingle (or shake) at a time. Panel systems are also available in which several shingles are glued to each panel backing.

### Applying Individual Shingles.
A layer of asphalt-saturated felt (tar paper) is usually placed under the shingles (or shakes). Shingles can be nailed directly onto structural sheathing. Where nonstructural sheathing is used, horizontal nailing strips are required.

Two methods of individual shingle (or shake) application are *single coursing* and *double coursing.* In single coursing the wall consists of a single layer of tapered shingles, except for the starting course, which is double. See Figure 57–37. Shingles should be applied with about ¹⁄₈" to ¼" space between their vertical edges to allow for expansion from water absorption.

In double coursing two layers of shingles are applied to the entire wall. See Figure 57–38. The first layer (undercourse) is applied directly to the sheathing or nailing strips. These shingles can be of a lower grade. For the second layer, first grade shingles are applied with the butt ends extending ¼" to ½" below the undercourse.

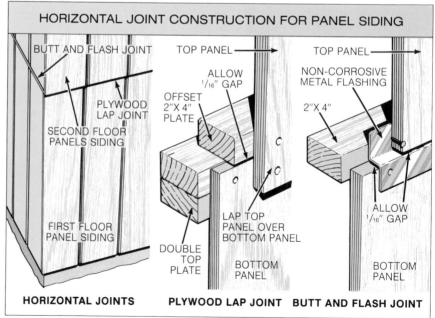

HORIZONTAL JOINT CONSTRUCTION FOR PANEL SIDING

BUTT AND FLASH JOINT

PLYWOOD LAP JOINT

SECOND FLOOR PANELS SIDING

FIRST FLOOR PANEL SIDING

TOP PANEL

ALLOW ¹⁄₁₆" GAP

OFFSET 2"X 4" PLATE

LAP TOP PANEL OVER BOTTOM PANEL

DOUBLE TOP PLATE

BOTTOM PANEL

TOP PANEL

NON-CORROSIVE METAL FLASHING

2"X 4"

ALLOW ¹⁄₁₆" GAP

BOTTOM PANEL

HORIZONTAL JOINTS   PLYWOOD LAP JOINT   BUTT AND FLASH JOINT

Figure 57–34. Horizontal joint details for panel siding. To prevent water leakage, panels should be lapped or flashing should be applied.

478

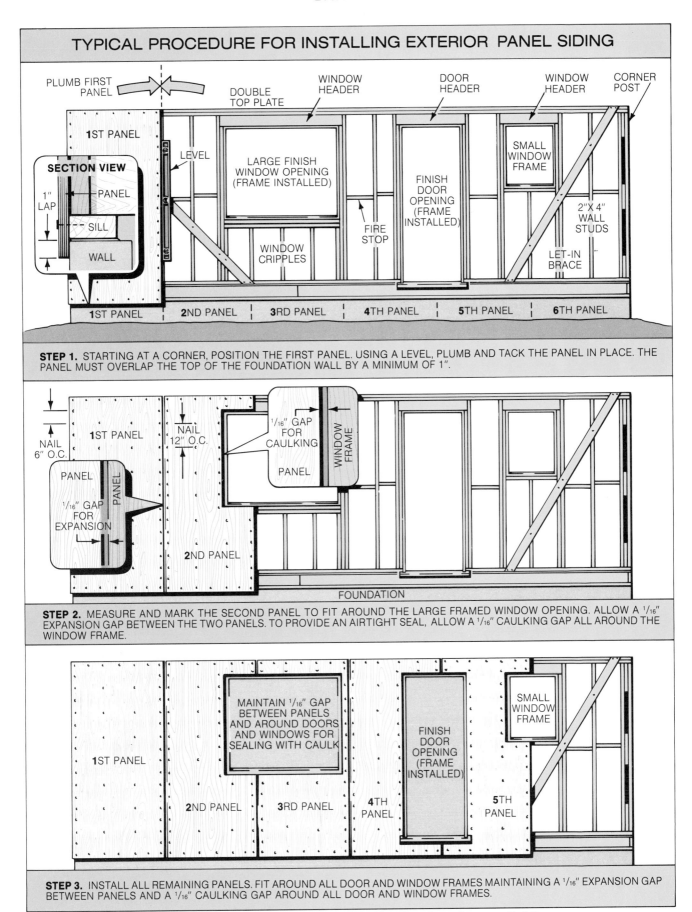

# TYPICAL PROCEDURE FOR INSTALLING EXTERIOR PANEL SIDING

**STEP 1.** STARTING AT A CORNER, POSITION THE FIRST PANEL. USING A LEVEL, PLUMB AND TACK THE PANEL IN PLACE. THE PANEL MUST OVERLAP THE TOP OF THE FOUNDATION WALL BY A MINIMUM OF 1".

**STEP 2.** MEASURE AND MARK THE SECOND PANEL TO FIT AROUND THE LARGE FRAMED WINDOW OPENING. ALLOW A 1/16" EXPANSION GAP BETWEEN THE TWO PANELS. TO PROVIDE AN AIRTIGHT SEAL, ALLOW A 1/16" CAULKING GAP ALL AROUND THE WINDOW FRAME.

**STEP 3.** INSTALL ALL REMAINING PANELS. FIT AROUND ALL DOOR AND WINDOW FRAMES MAINTAINING A 1/16" EXPANSION GAP BETWEEN PANELS AND A 1/16" CAULKING GAP AROUND ALL DOOR AND WINDOW FRAMES.

Figure 57–35. Applying panel siding. Both plywood and hardboard panels are applied the same way.

*Red Cedar Shingle & Handsplit Shake Bureau*
Figure 57–36. Cedar shingles give an attractive rustic appearance to the sidewalls of this house.

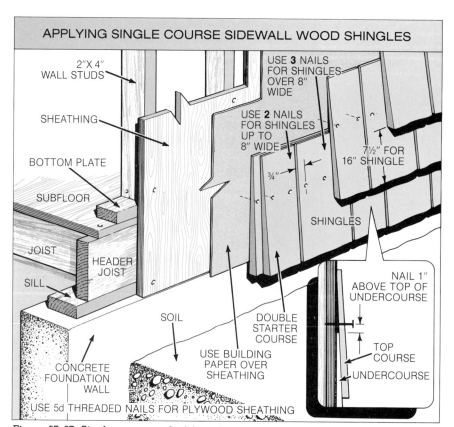

Figure 57–37. Single-course method for sidewall shingles. Note the double starting course.

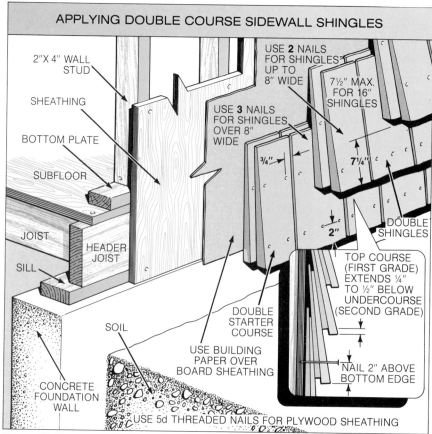

Figure 57–38. Double-course method for sidewall shingles. Shingles in the first course can be of a lower grade than those in the second course.

Recommended exposure for shingles depends on the length of the shingles, and whether single coursing or double coursing is used. The greater the exposure of the shingle, the more it will tend to curl up. See Figure 57–39.

Shingles or shakes should be applied with noncorrosive rust-resistant nails or staples long enough to penentrate into the sheathing or nailing strips. In single coursing, 3d or 4d zinc-coated shingle nails are commonly used.

Nails are placed ¾" from the edge of the shingle and 1" above the butt line of the following course. In this way the nail will be covered by the next course. In double coursing the heads of the nails driven into the outer layer of shingles are exposed. A 5d nail with a small, flat head is placed 1" above the butt line of the next higher course.

Corners for shingled or shaked walls may be woven or mitered

| SHINGLE AND SHAKE EXPOSURES | | | MAXIMUM EXPOSURE | |
|---|---|---|---|---|
| | | | DOUBLE COURSING | |
| MATERIAL | LENGTH | SINGLE COURSING | NO. 1 GRADE | NO. 2 GRADE |
| SHINGLES | 16″ | 7½″ | 12″ | 10″ |
| | 18″ | 8½″ | 14″ | 11″ |
| | 24″ | 11½″ | 16″ | 14″ |
| SHAKES (HAND SPLIT AND RESAWN) | 18″ | 8½″ | 14″ | -------- |
| | 24″ | 11½″ | 20″ | -------- |
| | 32″ | 15″ | -------- | -------- |

Figure 57–39. Recommended exposure distance for wood shingles and shakes applied to sidewalls.

as shown in Figure 57–40. Metal corners are also available. Boards are also used to finish off the corners of shingled or shaked sidewalls. See Figure 57–41.

**Applying Panels of Shingles.** Shingle (or shake) panels consist of several individual shingles (or shakes) that have been bonded to a wood panel backing. They are available in various designs.

Panels with a single row of shingles are available in 4′ and 8′ widths. See Figure 57–42.

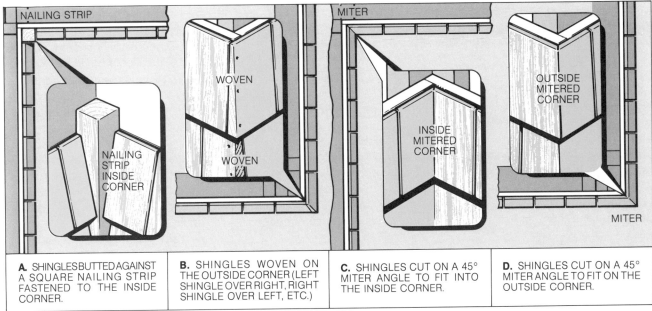

| A. SHINGLES BUTTED AGAINST A SQUARE NAILING STRIP FASTENED TO THE INSIDE CORNER. | B. SHINGLES WOVEN ON THE OUTSIDE CORNER (LEFT SHINGLE OVER RIGHT, RIGHT SHINGLE OVER LEFT, ETC.) | C. SHINGLES CUT ON A 45° MITER ANGLE TO FIT INTO THE INSIDE CORNER. | D. SHINGLES CUT ON A 45° MITER ANGLE TO FIT ON THE OUTSIDE CORNER. |
|---|---|---|---|

Figure 57–40. Corners of sidewall shingles and shakes may be woven or mitered.

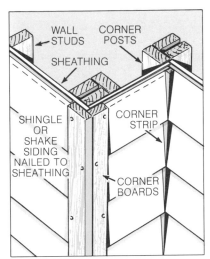

Figure 57–41. Boards are sometimes used to finish off the corners of shingled or shaked sidewalls.

*Shakertown Corporation*

Figure 57–42. Single-course shingle panels are applied the same way individual shingles are applied.

*Shakertown Corporation*

Figure 57–43. Double-course shingle panels are applied the way lap siding is applied, two laps at a time. Since the rabbeted bottom edge aligns the panel, it is not necessary to snap lines after the first course has been applied.

The procedure for laying out and placing these panels is similar to that for individual shingles. Panels are also available with two rows of shingles or shakes in an 8′ width. See Figure 57–43. The bottoms of these panels are rabbeted, making them self-aligning. As a result, snapping or stretching lines on the wall is eliminated. For corners, a special type of single-ply corner piece is used with shingle panels. See Figure 57–44.

## NON-WOOD SIDING

The types of non-wood siding installed by carpenters include products made of aluminum, vinyl, and asphalt.

## Aluminum Siding

Advantages claimed for aluminum siding are low maintenance, durability, light weight, and resistance to corrosion. Most aluminum siding is designed to give the appearance of wood siding. The pre-finished, baked-on surfaces are either smooth or made to look like woodgrain.

Aluminum siding is produced with or without an insulating backboard. Different styles are available for horizontal and vertical application. Aluminum panels, 12″ x 48″, are also available that give the appearance of cedar shakes.

Aluminum siding can be applied over sheathing or directly to the stud framing. Sheathing paper should be used under the same conditions that make sheathing paper necessary for wood siding. Aluminum siding is very popular for residential re-siding projects, in which it is attached to furring strips nailed to the old wall.

All aluminum siding manufacturers offer accessory items with their siding systems. These include starter strips, corner pieces, flashing, stiffeners, and trim pieces.

Some building codes require that aluminum siding be grounded against the possibility of electric shock. Grounding can be done by connecting a No. 8 copper wire to a cold water pipe or steel rod embedded in the ground.

Manufacturer's instructions should be carefully followed for applying aluminum siding. An electric handsaw, tin snips, or a utility knife can be used to cut the panels. See Figure 57–45. An electric handsaw is the quickest way to make precise cuts. A work table should be set up with a jig that will keep the saw base clear of the siding to avoid scratching its surface. A 10-point aluminum cutting blade should be used.

If the siding is cut with tin snips, the duckbill type of tin snips is best for straight cuts. If a utility knife is used, the panel should be heavily scored with the knife, then bent back and forth until it snaps cleanly along the scored line. Aluminum siding comes with factory-slotted holes for nails. Aluminum nails are driven every 16″ to 24″ through these holes. See Figure 57–46. For plain-shank nails driven into wood studs, penetration should be at least ¾″. Screw-shank nails should be used to fasten the siding to plywood. Nails should not be driven in tightly. The siding must be able to expand or contract with temperature changes.

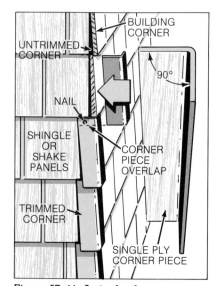

Figure 57–44. A single-ply corner piece is used for corners when panels of shingles or shakes are applied.

Aluminum Association, Inc.

Figure 57–45. A utility knife, tin snips, or an electric handsaw can be used to cut aluminum siding.

The first course of the siding should be securely locked into a starter strip. Joints between panels should be staggered. Corner caps are used to finish off the outside corners of the walls. See Figure 57–47. They must be fitted and installed as each course of siding is placed. The bottom flanges of the corner cap are slipped under the butt of each panel. When the butts of the corner cap and panel are flush, a nail is driven through a prepunched hole at the top of the corner cap.

## Solid Vinyl Siding

Solid vinyl siding is a pre-finished plastic product designed to look like board siding. Vinyl siding is durable, resists dents and abrasions, and does not scratch easily. Many color choices are avail-

483

Aluminum Association, Inc.

Figure 57-46. Aluminum nails are driven every 16″ to 24″ through factory-slotted holes in aluminum siding.

able, including imitation woodgrain patterns.

All vinyl siding systems include accessory pieces. The procedure for application is similar to that for aluminum siding.

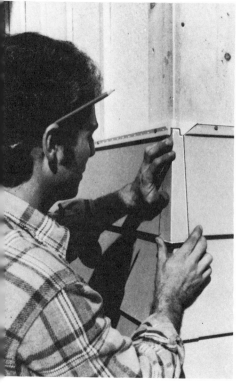

Aluminum Association, Inc.

Figure 57-47. Corner caps for aluminum siding must be fitted and installed as each course of siding is placed.

## WATER TABLE

On some framed buildings a *water table* is placed below the bottom course of the board, panel, or shingle siding. See Figure 57-48. It may also be used to finish off the bottom of a wall that has a stucco finish. The water table diverts water drainage from the face of the foundation wall. It consists of a board placed beneath a regular drip cap. The board may be 3/4″ or 1 1/4″ thick and 6″ or 8″ wide. The *quirk,* a narrow groove beneath the drip cap, helps prevent water from flowing back to the joint between the board and drip cap.

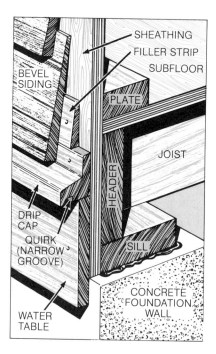

Figure 57-48. A water table may be placed below the bottom course of siding. In this example, a board 3/4″ or 1 1/4″ thick and 6″ or 8″ wide is placed beneath a regular drip cap. The quirk helps prevent water from flowing back to the joint between the board and drip cap.

## DECKS

Decks provide outdoor living space for recreation, entertaining, and dining. They may have rails, stairways, built-in benches, or flower containers. See Figure 57-49 and 57-50.

The location of decks should not prevent access to underground electrical and telephone lines and plumbing and gas pipes. Always check with the appropriate utility company for location and depth of utilities before digging.

The slope of the lot and the style of the house are important considerations when planning and building a deck. Decks may be built for all styles of houses and may range from a few inches to several feet above grade level. See Figure 57-51.

### Wood

Pressure-treated wood is an ideal construction material for decks because it is highly resistant to rot, decay, and termites, while being less expensive than redwood or cypress. Most pressure-treated wood is treated with chromated copper arsenate (CCA) and is guaranteed by the manufacturer for long-term resistance against wood-destroying organisms.

A popular wood for deck construction is Wolmanized® pressure-treated lumber. The chromated copper arsenate is fixed in the wood and resists leaching and evaporating. Wolmanized® pressure-treated lumber is commercially available for a large variety of applications. For example, greater concentrations of CCA are required for below-ground posts and immersion in salt water.

Treated lumber is available in the same sizes as untreated lumber. For example, nominal thicknesses are 1″, 1 1/4″, 2″, 4″, 6″, etc. Nominal widths are 4″, 6″, 8″, 10″, etc. The 1″, 1 1/4″ (5/4″), and 2″ nominal thicknesses are used for decking.

### Fasteners and Connectors

All fasteners and connectors used for deck construction must be

*California Redwood Association*

Figure 57-49. Low deck with stairway and bench with no back.

*California Redwood Association*

Figure 57-50. Deck with a built-in bench forming the rails.

hot-dipped zinc-coated or rust proof. Ordinary fasteners and connectors are weakened by corrosion and will stain the wood. See Figure 57-52.

## Constructing a Deck

The ledger board/header determines the length and height of the deck. Lag screws, expansion bolts, or carriage bolts may be used to secure the ledger board/header against the house.

The vertical placement of the ledger board/header determines the height of the deck floor. Place the ledger board/header $1^3/8''$ below the top of the finished deck floor for a $5/4''$ floor and $1^5/8''$ below the top of the finished deck for a $1^1/2''$ floor. See Figure 57-53.

Mark off the deck area using strings and batterboards. Square the strings by the 3-4-5 method. Prepare the site by removing approximately $2''-3''$ of sod. A sheet of black polyethylene is often placed over the ground to prevent vegetation from growing up through the deck.

Locate and dig holes for posts. Holes should be at least $24''$ deep. The actual depth of the holes is determined by the frost line. Always dig the holes deeper than the frost line. Posts may be set in gravel or concrete or set on concrete piers. See Figure 57-54. Set the posts plumb and in alignment with one another.

Secure beams to the posts and attach joists. Joist hangers may be used or the joists may be screwed or nailed to the headers. Place joists on $16''$ or $24''$ centers. See Figure 57-55.

Decking should be placed with the growth rings facing down. Provide $1/8''$ space between decking boards to allow for expansion. A 10d nail may be used for a spacer.

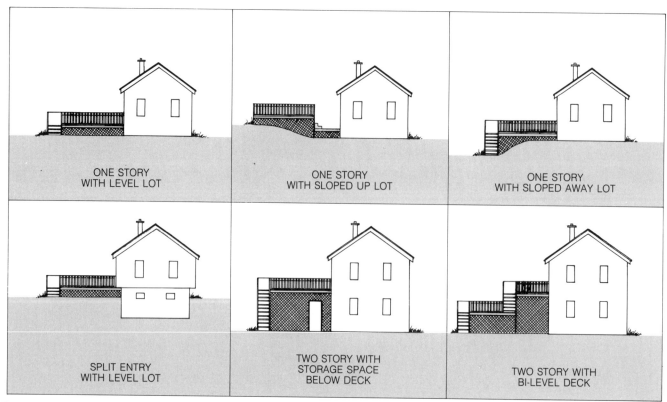

Figure 57-51. Decks must fit the slope of the lot.

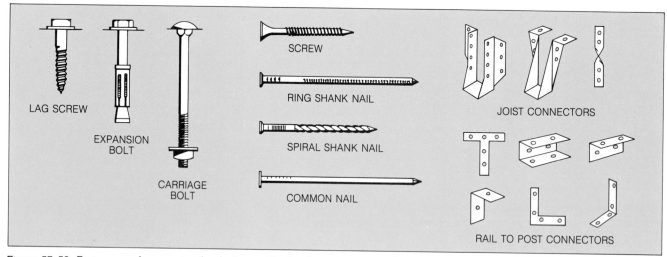

Figure 57-52. Fasteners and connectors for decks must be hot-dipped zinc-coated or rust proof.

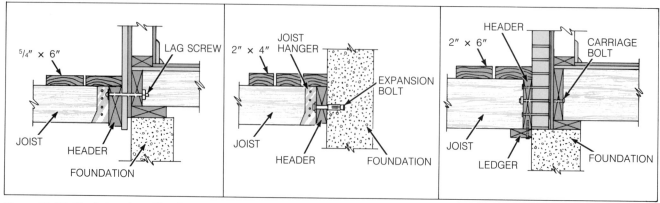

Figure 57-53. The ledger board/header must be securely fastened to the house.

Decking may be fastened with spiral shank nails, ring shank nails, or coated screws. Follow the pressure-treated wood manufacturer's suggestions for nail spacing and size of nails.

Decking may be placed in a variety of patterns. For example, a herringbone pattern provides a contemporary appearance, while decking perpendicular or parallel to the house provides a more traditional appearance. See Figure 57-56. Attach rails and stairs to complete the deck.

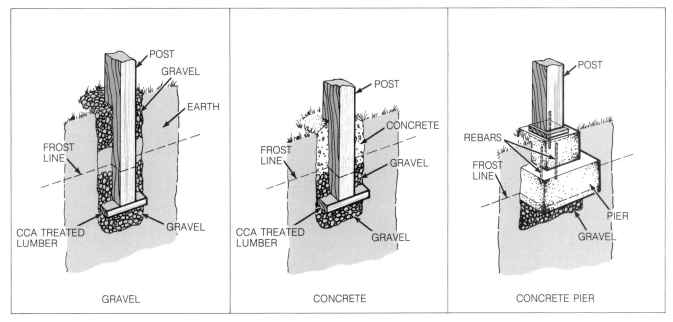

Figure 57-54. Posts should be set below the frost line.

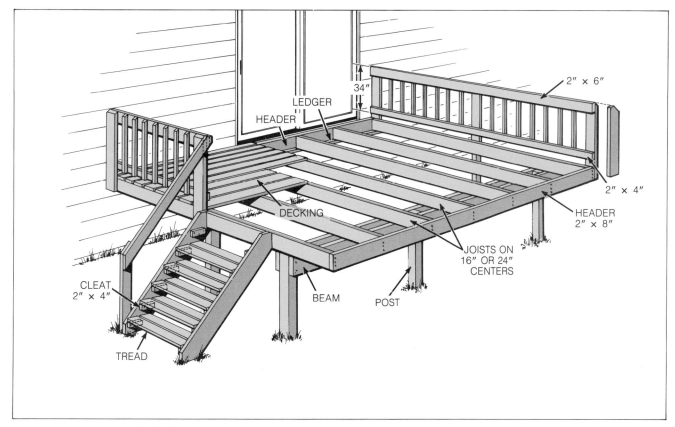

Figure 57-55. The parts of a deck.

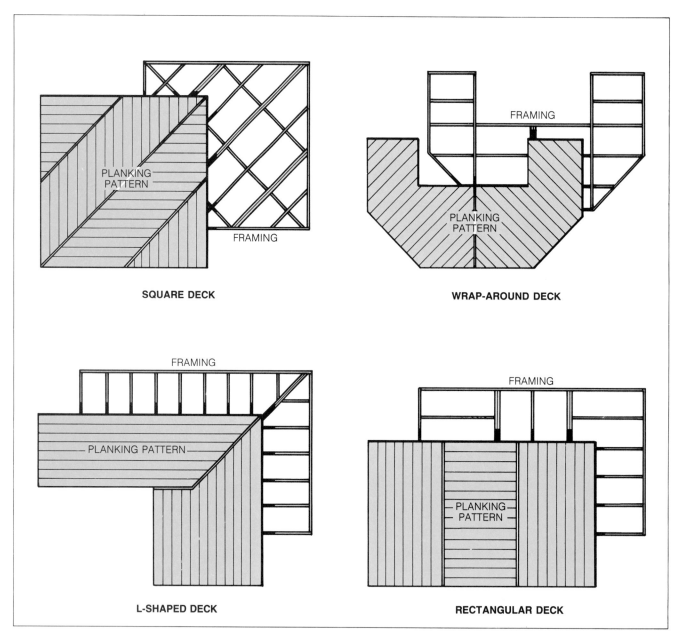

**SQUARE DECK**

**WRAP-AROUND DECK**

**L-SHAPED DECK**

**RECTANGULAR DECK**

Figure 57–57. Joists and blocking must be framed to accommodate the deck pattern.

## SECTION 13

# Interior Finish

Interior finish work is the final stage in the construction of a building. It should not begin until the building is completely enclosed and all windows and exterior doors have been installed. Interior finish includes all the surface materials placed on the walls, floors, and ceilings. Operations such as drywall application, paneling, interior door hanging, door and window trimming, and installing cabinets and finish hardware are involved.

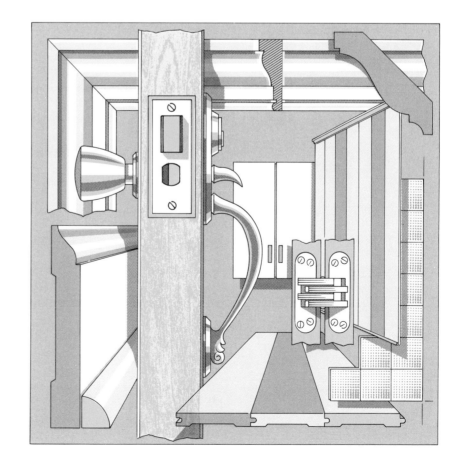

# UNIT 58

# Interior Wall and Ceiling Finish

*Gypsum Association*

Figure 58–1. Plasterers applying a finish coat of plaster to the lath nailed to the walls. Today gypsum board has generally replaced plaster and lath.

*Dow Chemical Company*

Figure 58–2. Gypsum board is fastened directly to wood or metal framing members and provides a smooth surface for paint or wallpaper.

Interior wall and ceiling finish includes the methods and materials used to finish off the wall and ceiling surfaces. Gypsum board, plaster, wood paneling, and plastic laminate are commonly used materials.

## GYPSUM BOARD WALL COVERING

In the past, a plaster finish was usually used over a base *(lath)* if the wall and ceiling surfaces were to be painted or covered with wallpaper. See Figure 58–1. Today *gypsum board* (often called *drywall* or *wallboard*) has generally replaced plaster and lath for this purpose. See Figure 58–2. More than 80% of new houses are constructed with gypsum board wall finish. It can be fastened to wood, metal, or even directly to concrete or masonry.

Gypsum is a mineral-derived product. It is crushed, heated, combined with other chemicals, and then sandwiched between two sheets of specially treated paper to create gypsum board. When properly applied, gypsum board presents a smooth, durable surface suitable for a paint or wallpaper finish. It has good fire resistance and provides

some degree of sound insulation. Pre-decorated gypsum board panels are also available.

Standard gypsum boards are 4′ wide and 8′, 10′, 12′ or 14′ long. Thicknesses range from ¼″ to 1″. For new wall finish, boards ½″ or ⅝″ thick are usually used.

Gypsum boards have differently shaped edges along their longer side depending on their purpose. See Figure 58–3. Boards designed to receive a painted finish usually have a

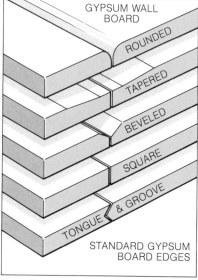

Figure 58–3. Gypsum boards have differently shaped edges depending on their purpose.

490

rounded or tapered edge. This provides a recessed joint between the long edges of the boards. The recessed joint receives the paper tape and joint compound that will be used to cover the joints between the boards. Boards to be covered with other materials, such as plastic or wood paneling, have tongue-and-groove edges. Predecorated boards have beveled edges.

## Applying Gypsum Board

On large construction jobs where large quantities of gypsum board are used, workers specializing in gypsum board application do the installation rather than regular carpenters. Where smaller amounts of gypsum board are required, however, carpenters often perform the installation.

After a gypsum board panel is measured and marked, it is usually cut by scoring the face with a sharp knife, then snapping off the waste piece. See Figure 58–4. Jagged edges should be smoothed with a rasp or knife. Sometimes cuts can be made more conveniently with a saw.

The *single-ply* system of application is most often used for gypsum board (wallboard) in residential and other light construction. See Figure 58–5. A single layer is considered adequate for fire resistance and sound control. The boards are applied with the long edge in a vertical or horizontal position.

The *double-ply* system provides greater fire resistance and sound control than the single-ply system. The first layer of wallboard is nailed to the studs. The second layer is then applied with an adhesive. See Figure 58–6.

Panels are applied to the ceiling first and then to the walls. The wall panels should butt

Figure 58–4. Gypsum board is cut by scoring it on the face side then snapping the waste piece off.

tightly against the ceiling panels. A *foot lift* device can be used to raise the wall panels up against the ceiling panels to ensure a tight fit. See Figure 58–7.

In residential construction the rough ceiling height is usually 8'-1". This height allows enough clearance between the ceiling and the floor for a 1/2" or 5/8" panel nailed to the ceiling and for 8' long panels nailed to the walls. Wall panels may be placed with the 8' length running vertically or horizontally, whichever way results in fewer joints between panels. Also, the joints should be staggered if possible.

**Fastening Methods.** Nails, screws, or adhesives are used to fasten gypsum board to interior walls. Staples are sometimes used for the first layer of gypsum board in double-ply applications.

*Nails.* Nails used to fasten gypsum board to wood framing members must be long enough

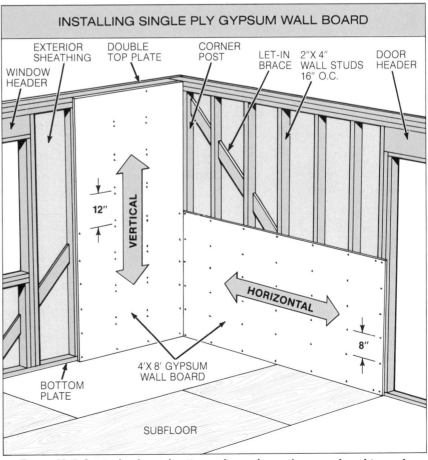

INSTALLING SINGLE PLY GYPSUM WALL BOARD

EXTERIOR SHEATHING — WINDOW HEADER — DOUBLE TOP PLATE — CORNER POST — LET-IN BRACE — 2"X 4" WALL STUDS 16" O.C. — DOOR HEADER — 12" — VERTICAL — HORIZONTAL — 8" — 4'X 8' GYPSUM WALL BOARD — BOTTOM PLATE — SUBFLOOR

Figure 58–5. In single-ply application, only one layer of gypsum board is used.

## INSTALLING DOUBLE PLY GYPSUM WALL BOARD

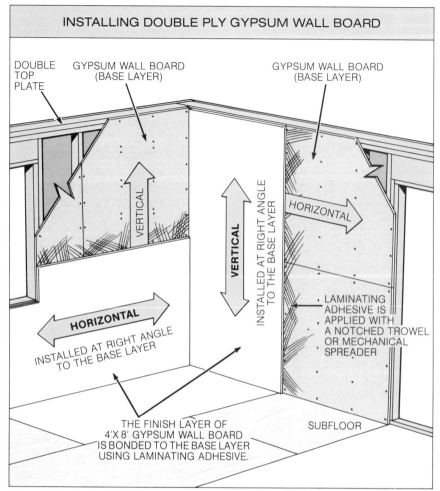

DOUBLE TOP PLATE

GYPSUM WALL BOARD (BASE LAYER)

GYPSUM WALL BOARD (BASE LAYER)

VERTICAL

VERTICAL

INSTALLED AT RIGHT ANGLE TO THE BASE LAYER

HORIZONTAL

HORIZONTAL

INSTALLED AT RIGHT ANGLE TO THE BASE LAYER

LAMINATING ADHESIVE IS APPLIED WITH A NOTCHED TROWEL OR MECHANICAL SPREADER

THE FINISH LAYER OF 4' X 8' GYPSUM WALL BOARD IS BONDED TO THE BASE LAYER USING LAMINATING ADHESIVE.

SUBFLOOR

Figure 58–6. In double-ply application, an adhesive is used to bond the second layer of gypsum board to the first layer. The second layer runs at a right angle to the first layer.

*National Gypsum Company*
Figure 58–7. A metal foot lift is used to raise a gypsum board panel up tightly against the ceiling. The space below the panel will later be covered by base molding.

to go through the gypsum board and penetrate ¾″ to ⅞″ into the wood. The heads should be at

least ¼″ in diameter, flat or concave, and thin at the rim. See Figure 58–8.

The gypsum board should be pressed tightly against the wood while the nail is driven in. The nail should be driven in far enough to cause a dimple on the surface of the board but not far enough for the head to cut into the face paper. See Figure 58–9. The dimple is later filled with putty, hiding the head of the nail.

Improperly driven nails can result in loose gypsum boards. As a result, cracks may appear later in the finished wall. Nails that do not catch the wood properly can

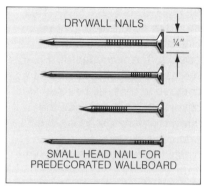

DRYWALL NAILS

¼″

SMALL HEAD NAIL FOR PREDECORATED WALLBOARD

Figure 58–8. Nails for fastening gypsum board to wood framing members. The small-headed nails are used with predecorated panels.

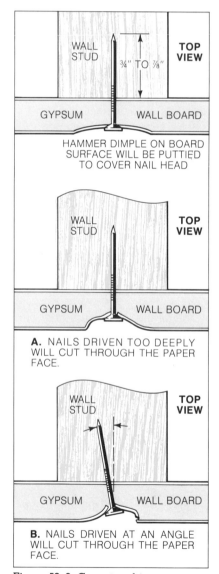

WALL STUD     ¾″ TO ⅞″     TOP VIEW

GYPSUM     WALL BOARD

HAMMER DIMPLE ON BOARD SURFACE WILL BE PUTTIED TO COVER NAIL HEAD

WALL STUD     TOP VIEW

GYPSUM     WALL BOARD

**A.** NAILS DRIVEN TOO DEEPLY WILL CUT THROUGH THE PAPER FACE.

WALL STUD     TOP VIEW

GYPSUM     WALL BOARD

**B.** NAILS DRIVEN AT AN ANGLE WILL CUT THROUGH THE PAPER FACE.

Figure 58–9. Correct and incorrect procedures for nailing gypsum board to a wood framing member. Nails driven too deeply, or at an angle, will cut through the face paper.

work loose and eventually pop out. See Figure 58–10.

A *single-nailing* (Figure 58–11) or *double-nailing* (Figure 58–12) method may be used. With either method, first drive the nails in the center of the board and then outward toward the edges. This prevents a sag or bulge at the center of the gypsum board panel.

*Screws.* For many years metal stud walls covered with gypsum board have been used as interior partitions in office buildings. See Figure 58–13. Today, the use of metal framing systems is spreading into residential and other light construction. Gypsum board is fastened to metal (and also sometimes to wood) framing members with self-drilling, self-tapping Phillips screws. See Figure 58–14. Adhesives are sometimes used in addition to the screws. The screws can be applied with drywall electric screw-drivers designed for this purpose. A self-feeding model is also available. See Figure 58–15.

*Adhesives.* Adhesives can be used to bond gypsum board directly to wood, metal, masonry, and other wall materials. Manufacturer's recommendations must be followed to choose the correct adhesive for a particular job.

A *stud adhesive* can be used together with screws or nails to fasten single-ply gypsum to wood or steel. The adhesive is in canisters that fit into a caulking gun. A ¼″ to ⅜″ bead is applied to each stud.

*Laminating adhesives* are used for multi-ply application. See Figure 58–16. Some laminating adhesives come in a dry powdered form that must be mixed with water. Contact-type laminating adhesives are also available.

Both laminating and stud adhesives are used to fasten gypsum board directly to concrete or masonry. Temporary bracing is often required until the bond develops strength.

**Finishing off.** After all ceiling and wall panels are nailed in place, the joints between panels are finished off by painters or by workers called *tapers* who specialize in finishing off gypsum board. The joints are covered with a bedding compound (Figure 58–17) and a strip of reinforcing tape (Figure 58–18), followed by two additional layers of

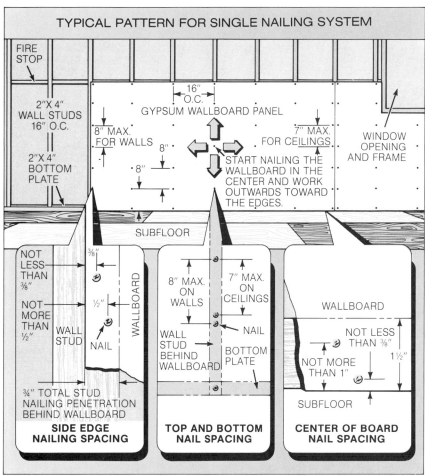

Figure 58–10. Loose nails are caused by a poor fit of the gypsum board to the framing surface or by nails missing the framing members.

Figure 58–11. Single-nailing method of fastening gypsum board.

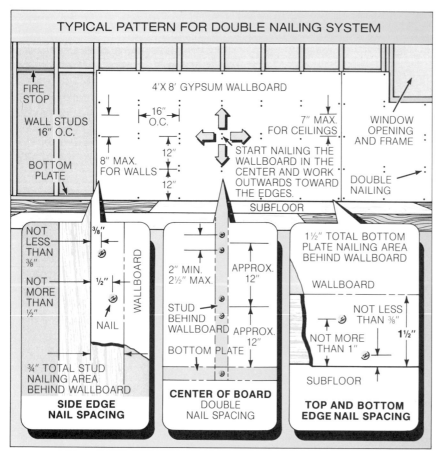

TYPICAL PATTERN FOR DOUBLE NAILING SYSTEM

FIRE STOP

WALL STUDS 16" O.C.

BOTTOM PLATE

4'X 8' GYPSUM WALLBOARD

16" O.C.

8" MAX. FOR WALLS

12"

12"

7" MAX. FOR CEILINGS

START NAILING THE WALLBOARD IN THE CENTER AND WORK OUTWARDS TOWARD THE EDGES.

WINDOW OPENING AND FRAME

DOUBLE NAILING

SUBFLOOR

NOT LESS THAN 3/8"

NOT MORE THAN 1/2"

3/8"

1/2"

NAIL

WALLBOARD

3/4" TOTAL STUD NAILING AREA BEHIND WALLBOARD

**SIDE EDGE NAIL SPACING**

2" MIN. 2 1/2" MAX.

STUD BEHIND WALLBOARD

BOTTOM PLATE

APPROX. 12"

APPROX. 12"

**CENTER OF BOARD** DOUBLE NAIL SPACING

1 1/2" TOTAL BOTTOM PLATE NAILING AREA BEHIND WALLBOARD

WALLBOARD

NOT LESS THAN 3/8"

NOT MORE THAN 1"

1 1/2"

SUBFLOOR

**TOP AND BOTTOM EDGE NAIL SPACING**

Figure 58–12. Double-nailing method of fastening gypsum board.

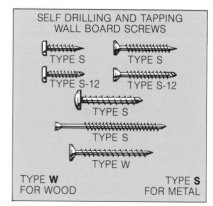

SELF DRILLING AND TAPPING WALL BOARD SCREWS

TYPE S          TYPE S

TYPE S-12     TYPE S-12

TYPE S

TYPE S

TYPE W

TYPE **W** FOR WOOD          TYPE **S** FOR METAL

Figure 58–14. Screws for fastening gypsum board to wood or metal.

*Duo-Fast Corporation*

Figure 58–15. Carpenter using a self-feeding screwdriver to fasten gypsum board to studs. This model drives No. 6 bugle-head screws 1", 1 1/4", and 1 5/8" long. The coil from which the screws are fed holds 150 screws.

*Gypsum Association*

Figure 58–13. Metal stud interior walls ready to be covered with gypsum board (shown stacked on the floor).

topping compound (Figure 58–19). Each layer of compound must dry completely and may require light sanding before the next layer is applied. The third (finish) coat of compound is feathered out on each side of the joint. The nail dimples are also filled in with compound and lightly sanded.

Bedding compound is also applied to finish off the inside cor-

*National Gypsum Company*
Figure 58-16. A laminating adhesive may be used to bond the second layer of gypsum board in double-ply application.

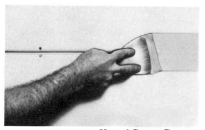

*United States Gypsum*
Figure 58-17. Bedding compound is applied to gypsum board joints.

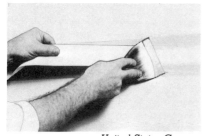

*United States Gypsum*
Figure 58-18. Reinforcement tape is pressed into the bedding compound.

*United States Gypsum*
Figure 58-19. A topping coat is applied after the bedding coat has dried.

ners. Reinforcing tape is folded and pressed into the compound. Drywall metal corner beads are nailed to reinforce outside corners and then covered with joint compound. When all the joints, corners, and nail dimples have been treated and sanded, the wall is ready for a paint finish.

## Pre-decorated Gypsum Board

Pre-decorated gypsum board has a decorated surface that does not require any other kind of finish. A popular product of this type is a gypsum panel covered with vinyl plastic. A wide variety of patterns, colors, and textures is offered. Pre-decorated panels can be applied directly to wall studs or can be applied as the outside layer in a multi-ply drywall system. They are fastened with adhesives or special nails with colored heads that match the wallboard.

Panels are usually applied with the long edge in a vertical position. The joints may be exposed or covered with the types of molding shown in Figure 58-20.

## WALL PANELING

Paneling is one of the most attractive ways to finish off an interior wall. Wood panel systems include full-size plywood or hardboard panels, as well as different kinds of solid board applications. Panels surfaced with plastic laminate are also available.

## Plywood Paneling

Most plywood wall panels have a hardwood veneer such as oak, ash, beech, walnut, birch, pecan, or mahogany. However, softwood-veneer panels such as redwood, cedar, fir, and southern

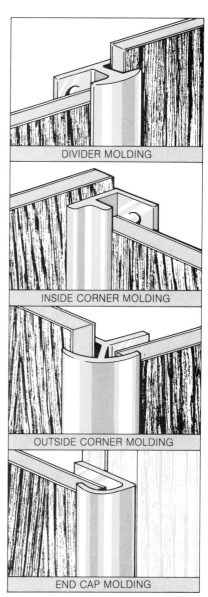

DIVIDER MOLDING

INSIDE CORNER MOLDING

OUTSIDE CORNER MOLDING

END CAP MOLDING

Figure 58-20. Molding used to cover joints and corners of pre-decorated wallboard.

pine are also used. The surface may be plain or textured. See Figure 58-21.

Plywood panels are usually ¼″ or ⅜″ thick. Thicker panels, such as ¾″, are also available. Standard sheet sizes are 4′ × 8′ and 4′ × 10′. Longer lengths are available by special order.

In the past, plywood panels were finished off by the painters after they had been fastened to the walls. Today, most of these panels are pre-finished. They are stained, sealed, and varnished at

*Champion Paneling—Champion International Corporation*

Figure 58–21. Grooved hardwood-veneer (top) and softwood-veneer (bottom) plywood wall panels create pleasant shadow lines.

after the panels have been nailed in place.

## Hardboard Paneling

Hardboard panels are manufactured with many different patterns, including grain finishes that appear almost identical to natural wood grain. Panels range from ⅛″ to ¼″ in thickness. The ⅛″ thickness should be applied only against a solid surface such as plywood, gypsum board, or plaster. Solid backing or

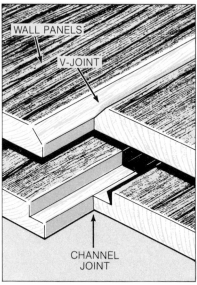

Figure 58–22. Typical edge joints for wall paneling.

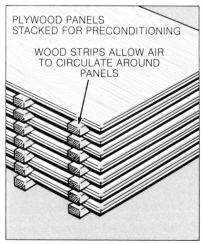

Figure 58–23. Plywood panels are stacked with wood strips between them for preconditioning to room temperature.

the fabricating plant. Therefore, extra care must be taken not to scratch or damage these panels during installation.

Panels may have a rabbeted panel edge or a slightly beveled edge. When placed together, rabbeted edges form a channel joint. Beveled edges form a V-joint. See Figure 58–22. Both create an attractive shadow line between the panels.

**Preconditioning.** Plywood panels should be preconditioned to room temperature for several days before they are applied to the wall. A good procedure is to stack the panels with strips between them. See Figure 58–23. This allows air to reach the faces and backs of the sheets, conditioning them to room temperature and humidity. Preconditioning eliminates significant shrinkage

furring strips spaced 12″ O.C. are recommended for ¼″ thick panels. Standard size sheets are 4′ × 8′.

**Preconditioning.** Hardboard is a wood-based product. Therefore, it will expand and contract with changes in the moisture content of the surrounding air. Panels should be separated and stacked on their edges to allow a free flow of air on both sides of the panels for at least 48 hours. See Figure 58–24.

## Applying Wall Paneling

The procedure for applying plywood panels is about the same as for hardboard panels. The same methods and tools are used to cut and fit both materials. Both materials can be fastened directly to wood studs or *furring strips* (usually 1″ × 2″ wood strips), using nails or adhesives. A popular system that has the advantage of increasing thermal and sound insulation is to first apply gypsum board as a base, then glue the panels to the gypsum board.

All studs must be straight and plumb before panels are applied to them. Crooked framing material produces an uneven paneled wall surface. Correct any unevenness before applying panels. Badly warped or twisted studs should be replaced.

Panels can be fastened directly to masonry walls with adhesives. This is not advisable, however, if the wall is uneven or subject to constantly damp conditions. In this case furring strips should be nailed or glued to the masonry and the panels fastened to the furring strips.

In remodeling work, panels are often placed directly over the plastered or gypsum-covered surface of the old walls. Frequently these surfaces are very uneven, and furring strips should be used. A procedure for installing furring strips is shown in Figure 58–25. Shim shingles (shingles tapered from ½″ thick to about ⅛″ thick) are used to plumb and straighten the strips where necessary.

Before installing panels, determine their arrangement on the walls. Figure 58–26 shows two possible ways for placing panels on the same wall. Whatever arrangement is used, the sheets should be set up around the walls so that panels with better matching grain patterns are next to each other. When the final panel arrangement has been determined, number the backs of the sheets so that they can later be installed in that order.

The first panel must be set in plumb. Subsequent panels do not have to be checked for plumbness if the first panel is plumb. Each panel must be permanently fastened into position before the next one is placed.

Holes for electrical outlets, switches, or other openings must be laid out on the panel, marked, and cut before the panel is set in place. See Figure 58–27 for the layout procedure. See Figure 58–28 for marking and cutting procedures. The item requiring an opening in the panel is used as a pattern for marking the opening. The outline is made slightly larger than the item. Starter holes are drilled in the inside corners of the outline. Then an electric jig saw is used to cut along the marked lines. If a jig saw is not available, a keyhole saw can be used, although it is not as efficient. A typical method for applying wall panels is shown in Figure 58–29.

*Masonite Corporation*

Figure 58–24. Hardboard panels should be separated and stacked on their edges for at least 48 hours before they are applied to a wall.

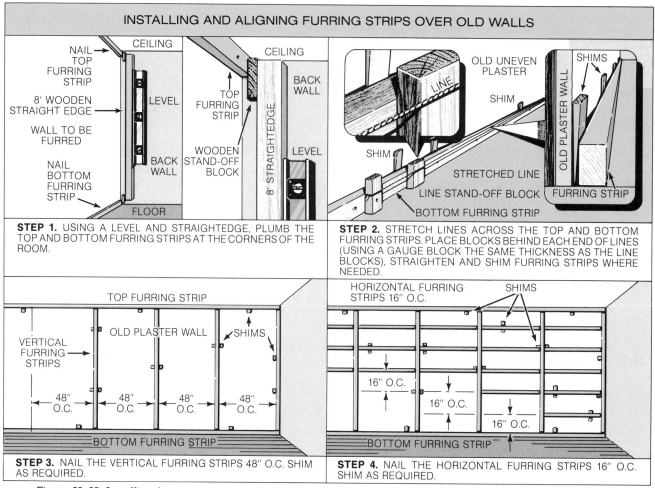

## INSTALLING AND ALIGNING FURRING STRIPS OVER OLD WALLS

**STEP 1.** USING A LEVEL AND STRAIGHTEDGE, PLUMB THE TOP AND BOTTOM FURRING STRIPS AT THE CORNERS OF THE ROOM.

**STEP 2.** STRETCH LINES ACROSS THE TOP AND BOTTOM FURRING STRIPS. PLACE BLOCKS BEHIND EACH END OF LINES (USING A GAUGE BLOCK THE SAME THICKNESS AS THE LINE BLOCKS), STRAIGHTEN AND SHIM FURRING STRIPS WHERE NEEDED.

**STEP 3.** NAIL THE VERTICAL FURRING STRIPS 48" O.C. SHIM AS REQUIRED.

**STEP 4.** NAIL THE HORIZONTAL FURRING STRIPS 16" O.C. SHIM AS REQUIRED.

Figure 58–25. Installing furring strips over old walls. Furring strips are necessary when the wall surfaces are uneven.

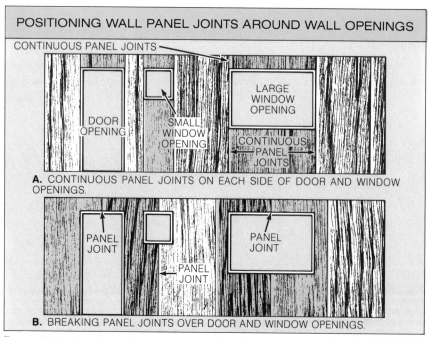

## POSITIONING WALL PANEL JOINTS AROUND WALL OPENINGS

**A.** CONTINUOUS PANEL JOINTS ON EACH SIDE OF DOOR AND WINDOW OPENINGS.

**B.** BREAKING PANEL JOINTS OVER DOOR AND WINDOW OPENINGS.

Figure 58–26. Panels can be positioned so that (A) a continuous joint occurs on each side of the door, or (B) joints break over the openings.

**Joints and Corners.** New panels have a straight "factory" edge. Therefore, joints between panels usually do not require any additional planing in order to fit properly against each other. Any slight imperfection will be hidden by the shadow line.

The bottom of the panel is covered by base molding, so it does not have to be *scribed* (carefully fitted by using a scriber). (Scribers are discussed in Section 3.) If ceiling molding is used, the top of the sheet does not have to be scribed either. See Figure 58–30. In much contemporary construction, the ceiling molding is eliminated, so the top of the panel must be scribed to the ceiling. Refer again to Figure 58–29.

## LAYING OUT SWITCH AND OUTLET BOXES ON WALL PANELS

POSITION OF SECOND WALL PANEL — DOUBLE TOP PLATE

FIRST WALL PANEL (INSTALLED)

POSITION OF WALL STUDS BEHIND PANEL

**SECOND WALL PANEL**

MEASURE OVER FROM PANEL EDGE

8'-0"

MEASURE OVER FOR SWITCH BOX ①

WALL STUDS

② SWITCH BOX CUT-OUT

MEASURE UP FOR SWITCH BOX ③

② ④ MEASURE UP FROM FLOOR

① OUTLET BOX

③

④

OUTLET BOX    BOTTOM PLATE

4'-0" STUD WALL

**STEP 1.** MEASURE OVER FROM THE PREVIOUS PANEL OR WALL TO ONE SIDE OF THE SWITCH OR OUTLET BOX.

**STEP 2.** TRANSFER THIS MEASUREMENT TO THE SECOND PANEL AND ALSO MARK WIDTH OF BOX.

**STEP 3.** MEASURE UP FROM THE FLOOR TO THE BOTTOM OF THE SWITCH OR OUTLET BOX AND ALSO MARK VERTICAL DIMENSION OF BOX.

**STEP 4.** TRANSFER THIS MEASUREMENT TO THE SECOND PANEL AND ALSO MARK VERTICAL DIMENSION OF BOX. MARK THE OUTLINE OF THE BOXES FOR THE HOLE OPENING CUT-OUTS. CUT HOLES SLIGHTLY LARGER THAN BOXES.

Figure 58–27. Laying out hole openings on panels for switch and outlet boxes. Openings must be cut before the panel is set in place.

Inside and outside corners do not have to be fitted carefully if they are to be covered by moldings. See Figure 58–31. If no molding is used on an inside corner, a scriber must be used as shown in Figure 58–32. If no molding is used on an outside corner, the corner may be mitered. See Figure 58–33. However, these corners are difficult to miter properly and can be easily damaged after mitering. A better method is the flush type shown in Figure 58–34.

**Fastening Methods.** Nails or adhesives may be used to fasten wall paneling. Finish nails for paneling should be long enough to penetrate ¾" into the stud or furring strip. When using grooved panels, drive nails into the grooves located in the body and along the edge of each panel. If panel edges are chamfered, place the nail along the chamfer. See Figure 58–35. All nails except those that are colored to match the

*Masonite Corporation*
Figure 58–28. Marking and cutting openings on a panel.

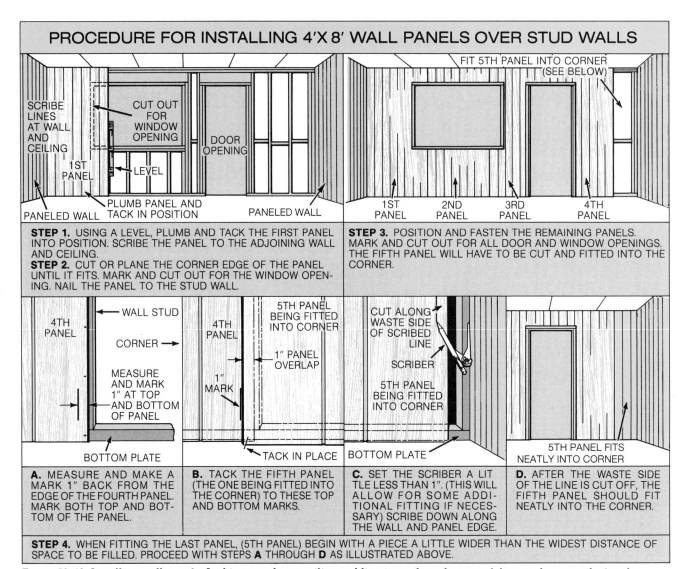

## PROCEDURE FOR INSTALLING 4' X 8' WALL PANELS OVER STUD WALLS

**STEP 1.** USING A LEVEL, PLUMB AND TACK THE FIRST PANEL INTO POSITION. SCRIBE THE PANEL TO THE ADJOINING WALL AND CEILING.

**STEP 2.** CUT OR PLANE THE CORNER EDGE OF THE PANEL UNTIL IT FITS. MARK AND CUT OUT FOR THE WINDOW OPENING. NAIL THE PANEL TO THE STUD WALL.

**STEP 3.** POSITION AND FASTEN THE REMAINING PANELS. MARK AND CUT OUT FOR ALL DOOR AND WINDOW OPENINGS. THE FIFTH PANEL WILL HAVE TO BE CUT AND FITTED INTO THE CORNER.

**A.** MEASURE AND MAKE A MARK 1" BACK FROM THE EDGE OF THE FOURTH PANEL. MARK BOTH TOP AND BOTTOM OF THE PANEL.

**B.** TACK THE FIFTH PANEL (THE ONE BEING FITTED INTO THE CORNER) TO THESE TOP AND BOTTOM MARKS.

**C.** SET THE SCRIBER A LITTLE LESS THAN 1". (THIS WILL ALLOW FOR SOME ADDITIONAL FITTING IF NECESSARY) SCRIBE DOWN ALONG THE WALL AND PANEL EDGE.

**D.** AFTER THE WASTE SIDE OF THE LINE IS CUT OFF, THE FIFTH PANEL SHOULD FIT NEATLY INTO THE CORNER.

**STEP 4.** WHEN FITTING THE LAST PANEL, (5TH PANEL) BEGIN WITH A PIECE A LITTLE WIDER THAN THE WIDEST DISTANCE OF SPACE TO BE FILLED. PROCEED WITH STEPS **A** THROUGH **D** AS ILLUSTRATED ABOVE.

Figure 58–29. Installing wall panels. In this example, no ceiling molding is used, so the tops of the panels are scribed to the ceiling.

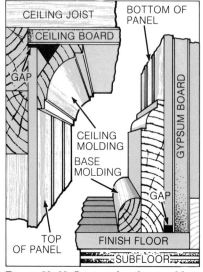

Figure 58–30. Base and ceiling molding eliminate the need to scribe the top and bottom of the panel.

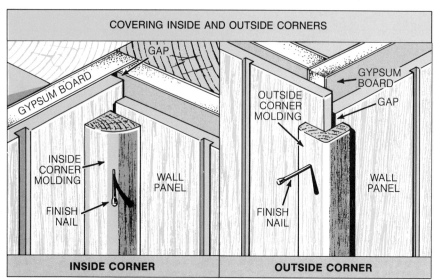

Figure 58–31. Molding is often used to cover the inside and outside corners of interior paneling.

500

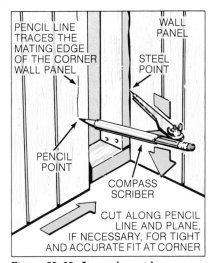

Figure 58–32. A panel must be carefully scribed to the other wall when no molding covers the inside corner.

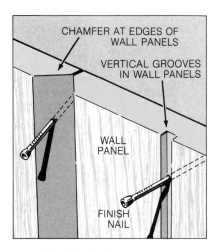

Masonite Corporation

Figure 58–36. Adhesive is applied for the side and bottom edges of a panel in a ⅛″ continuous strip behind the panel joints and at the top and bottom plates.

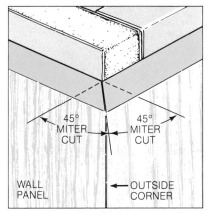

Figure 58–33. If molding is not used on an outside corner, the panels may be mitered.

Masonite Corporation

Figure 58–35. Nails are less noticeable when driven along the bevel or into the grooves of a panel.

paneling must be set below the surface with a nail set. They are later concealed with putty that matches the finish of the panel.

For plywood panels, space the nails 6″ O.C. along the panel edges and every 12″ O.C. at the intermediate studs or furring strips. For hardboard panels, space the nails 4″ O.C. along the edges and 8″ O.C. at the intermediate points.

Most paneling, especially hardboard paneling, is fastened with adhesive rather than nails. (A few nails may be required to hold the panels in place until the adhesive sets up.) Adhesive is applied for the side and bottom edges of the panel in an ⅛″ continuous strip behind the panel joints and at the top and bottom plates. See Figure 58–36. It is applied in 3″ long beads, 6″ apart, at the intermediate studs or furring strips behind the panel. See Figure 58–37. The panel is pressed into place with firm, uniform pressure so that the adhesive will spread evenly. See Figure 58–38. This procedure is made easier by tacking the panel at the top. Next, the panel is grasped at the bottom along both edges and slowly pulled away

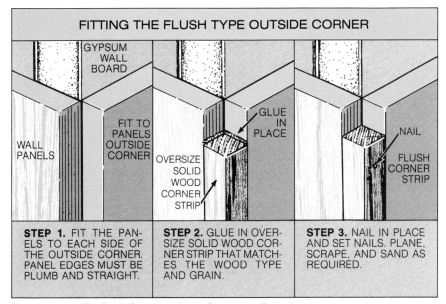

### FITTING THE FLUSH TYPE OUTSIDE CORNER

**STEP 1.** FIT THE PANELS TO EACH SIDE OF THE OUTSIDE CORNER. PANEL EDGES MUST BE PLUMB AND STRAIGHT.

**STEP 2.** GLUE IN OVERSIZE SOLID WOOD CORNER STRIP THAT MATCHES THE WOOD TYPE AND GRAIN.

**STEP 3.** NAIL IN PLACE AND SET NAILS. PLANE, SCRAPE, AND SAND AS REQUIRED.

Figure 58–34. Flush application of panels on outside corners is easier than mitering. Also, flush panels are less susceptible to later damage.

*Masonite Corporation*

Figure 58–37. Adhesive is applied in 3″ long beads, 6″ apart, at the intermediate studs or at the furring strips behind the panel.

from the stud. See Figure 58–39. After two minutes, it is pressed back again.

## Solid Board Paneling

Solid board paneling consists of solid wood boards, usually ¾″ thick and 4″ to 12″ wide. Softwood species such as redwood, fir, pine, hemlock, spruce, and cedar are used. Finishes are smooth, textured, or rough (resawn). Usually the boards are placed horizontally (Figure 58–40) or vertically, although diagonal designs are also used sometimes. Four common types of solid board panels are:

1. Board-on-board
2. Board-and-batten
3. Tongue-and-groove
4. Channel-rustic

The board-on-board and board-and-batten systems must be applied vertically. The tongue-and-groove and channel-rustic systems can be applied vertically or horizontally. Nails are driven into each stud or every row of blocking or furring strips. Face nails should be set below the surface and puttied.

In the board-on-board system, one 8d nail is driven at the cen-

*Masonite Corporation*

Figure 58–38. The panel is pressed into place after the adhesive has been applied.

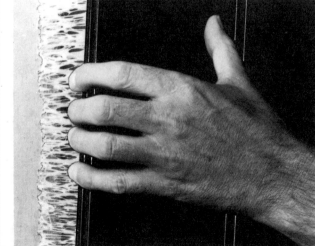

*Masonite Corporation*

Figure 58–39. After the panel has been pressed against the adhesive, it is grasped at the bottom along both edges and slowly pulled away from the stud. After two minutes the panel is pressed back again.

ter of each underboard. See Figure 58–41. The top boards should overlap the underboards a minimum of 1″. The top board is fastened with two 10d nails. Make sure the nails clear the underboard to allow for expansion and contraction.

In the board-and-batten system, one 8d nail is driven at the centers of each underboard. See Figure 58–42. The underboards are spaced approximately ½″ apart. One 10d nail is driven at

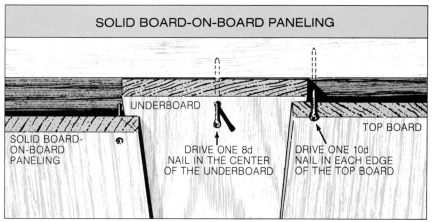

Figure 58–41. Board-on-board wall paneling. Drive an 8d nail at the center of the underboard. The top boards should overlap the underboard a minimum of 1″. Fasten the top board with two 10d nails. Make sure the nails clear the underboard to allow for expansion and contraction.

*California Redwood Association*
Figure 58–40. Redwood solid board paneling applied horizontally.

the center of the batt strip.

In the tongue-and-groove system, boards 4″ to 6″ wide are blind-nailed with 6d finish nails driven at a 45° angle. See Figure 58–43. This eliminates the need to countersink and putty the face nails.

In the channel-rustic system, boards up to 6″ require only one face nail. Boards 8″ or wider also require a face nail at the center of each board. See Figure 58–44.

**Vertical Application.** When boards are fastened vertically to a stud wall, nailing blocks must be placed between the studs. See Figure 58–45. Masonry walls require horizontal furring strips.

Before solid board paneling is

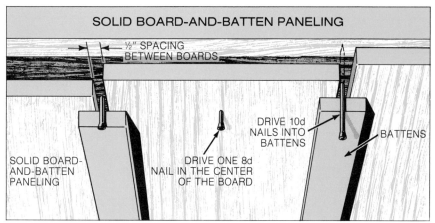

Figure 58–42. Board-and-batten wall paneling. Drive an 8d nail at the center of the underboard. The underboards are spaced approximately ½″ apart. Drive a 10d nail through the center of the batt strip.

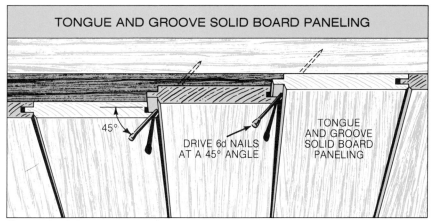

Figure 58–43. Tongue-and-groove wall paneling. Widths of 4″ to 6″ can be blind-nailed with 6d finish nails driven at a 45° angle. This method eliminates the need to countersink and putty face nails.

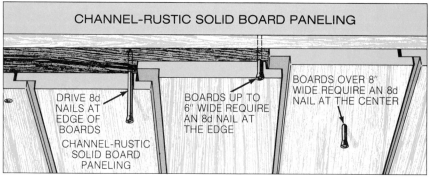

**CHANNEL-RUSTIC SOLID BOARD PANELING**

DRIVE 8d NAILS AT EDGE OF BOARDS

CHANNEL-RUSTIC SOLID BOARD PANELING

BOARDS UP TO 6" WIDE REQUIRE AN 8d NAIL AT THE EDGE

BOARDS OVER 8" WIDE REQUIRE AN 8d NAIL AT THE CENTER

Figure 58–44. Channel-rustic wall paneling. For widths up to 6", one face nail as shown is considered adequate. Widths over 8" require a face nail at the center of each board.

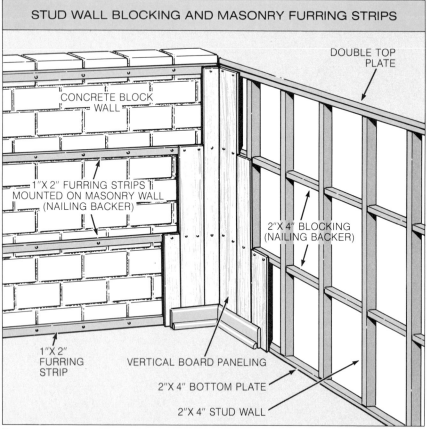

**STUD WALL BLOCKING AND MASONRY FURRING STRIPS**

CONCRETE BLOCK WALL

DOUBLE TOP PLATE

1"X 2" FURRING STRIPS MOUNTED ON MASONRY WALL (NAILING BACKER)

2"X 4" BLOCKING (NAILING BACKER)

1"X 2" FURRING STRIP

VERTICAL BOARD PANELING

2"X 4" BOTTOM PLATE

2"X 4" STUD WALL

Figure 58–45. Blocking must be placed in a stud wall to provide a nailing base for vertically placed board paneling.

*California Redwood Association*

*California Redwood Association*

Figure 58–46. These boards have tongue-and-groove edges. Before application, they are arranged against the wall for the best match of woodgrain patterns.

*California Redwood Association*

Figure 58–47. Holes are cut out for electrical outlets and vents before board panels are applied.

placed, the boards should first be arranged along the wall so that they can be matched as well as possible. See Figure 58–46. Sections are cut out in the panels for electrical outlets and vents before the boards are placed. See Figure 58–47.

The inside corner board is placed first, in plumb position, and scribed if necessary. It is fastened securely before the next board is placed. If boards are to be glued, the adhesive is applied to the backs of the prefitted boards. See Figure 58–48. Press the board against the wall so that the adhesive spreads evenly. Remove the board, wait a few minutes, then tap the board back into place. When tapping a board panel into place,

*California Redwood Association*

Figure 58–48. Adhesive is applied to the backs of pre-fitted boards.

*California Redwood Association*

Figure 58–49. When tapping a board panel into place, use a piece of material with a grooved edge against the board.

*California Redwood Association*

Figure 58–50. Ceiling molding may finish off the top of a board-paneled wall.

use a piece of material with a grooved edge against the board. See Figure 58–49. The grooved edge will protect the board from damage. Facenail the board at the top and bottom with 8d finish nails.

The bottoms of the boards may be covered by, or may rest on top of, a base molding. The tops of the boards may also be covered by a ceiling molding. See Figure 58–50.

**Horizontal Application.** The procedure for placing interior board panels horizontally is similar to that for placing exterior board panels (siding) horizontally. The first row must be perfectly level. All butt joints between the boards should be staggered and fall over a stud.

## PLASTIC LAMINATE WALL COVERING

Originally, plastic laminate was used almost exclusively to surface kitchen and bathroom countertops. Today it is also used to cover shelves, doors, cabinets, and entire wall surfaces.

Plastic laminate is very hard and smooth. It is composed of three or four layers of plastic material bonded under high heat and pressure. Many different patterns are available, including imitation woodgrain. See Figure 58–51.

Plastic laminate comes in sheets 1/32″ or 1/16″ thick. (The 1/32″ thick sheets are used for vertical application only.) The sheets can be fitted and applied directly to a plaster or gypsum wall surface with contact cement. More often, however, plastic laminate is pre-mounted on 3/8″ to 1/2″ plywood or particleboard panels, 16″ to 48″ wide and 8′ to 10′ long. The panels can be

mounted directly to studs or furring strips. See Figure 58–52. Panels are also available with tongue-and-groove edges, which

*Formica Corporation*

Figure 58–51. Plastic laminate wall panels are available with an imitation woodgrain finish.

*Formica Corporation*

Figure 58–52. Furring strips, 1″ x 4″ except for 1″ x 6″ at the base, are fastened to the wall as backing for plastic laminate panels. Note the shingles used to align and straighten the furring strips.

Formica Corporation

Figure 58–53. Partitions constructed of plastic laminate panels are set in metal tracks that are secured to the floor and ceiling. The door hangs in a metal frame fastened to the panels. The panel at the right has been cut out to give a better view of the core. In this example, the core is 1⁹⁄₁₆″ thick and consists of paper honeycomb material. The panel is reinforced by a solid wood or metal frame around the edges and stiles running horizontally every 2′ to 3′.

12″X 12″ CEILING TILE WITH TONGUE AND GROOVE EDGES

NARROW TONGUE

12″

TYPICAL CEILING TILE

12″

WIDE STAPLING FLANGE

STAPLE DRIVEN INTO FURRING STRIPS

WIDE STAPLING FLANGE

GROOVE (FOR TONGUE OF MATING TILE)

CEILING TILE

SECTION VIEW

FURRING STRIP — STAPLE

CEILING TILE

CEILING LIGHT FIXTURE

Figure 58–54. A typical 12″ x 12″ ceiling tile with tongue-and-groove edges. Staples are driven through wide flanges into the furring strips.

allow blind-nailing of the tongue edge.

Most plastic laminate panel systems require a combination of nails, adhesives, and molding to fasten the panels into place. Panels sometimes used for office partitions are covered with plastic laminate on both sides and are set in tracks secured to the floor and ceiling. See Figure 58–53.

## CEILING TILE

Although most ceilings in new residential buildings are finished off with gypsum board, acoustical and decorative ceiling tiles are becoming steadily more popular. Ceiling tiles are particularly practical in remodeling work, since they can be directly applied to the old plaster or drywall ceiling. They may also be used to form a suspended ceiling. In commercial construction, tile ceiling systems have been widely used for many years.

Applying ceiling tile is considered part of the carpentry trade, and a high degree of specialization has developed in this field. Carpenters' apprenticeship programs in many parts of the country supervise special trainee programs for this type of work.

Ceiling tiles are fabricated from fiberboard, mineral fiber, glass fiber, or asbestos, among other materials. Various colors and designs are available. The acoustical types of tiles absorb sound from the room directly below the ceiling and are an effective sound control agent.

### Applying Ceiling Tiles for a Nonsuspended Ceiling

For direct application to furring strips (or to an old plaster or drywall ceiling), in an average size room, 12″ × 12″ tiles, ½″ or ¾″ thick, with tongue-and-groove edges, are usually the most practical. See Figure 58–54.

(Larger tiles are used for suspended ceilings, discussed later in this unit.)

Centerlines, at right angles to each other, must be snapped on the ceiling in order to begin the first row of tiles straight and square to each other.

Most room dimensions are in feet and inches. Therefore, the border tiles against the wall will not be a full 12″. For a balanced appearance, the border tiles at both ends of the room should be the same size. Also, they should be at least 6″ wide. To assure this result, add the width of one tile (12″) to the leftover inches and divide by 2. The answer will be the correct size of the border tiles. Figure 58–55 shows how to establish lines before setting the ceiling tiles in rows of even-foot dimensions. Figure 58–56 shows the procedure for odd-foot dimensions. Figure 58–57 shows the procedure for foot-and-inch dimensions.

A border piece can be cut by scoring it with a sharp knife and then snapping the tile. A handsaw or power saw can also be used. The tile may be safely undercut a small amount, since molding usually covers the joint between the wall and the tile. The section of tile cut off will contain one of the tongued edges. Curved cuts, such as required around light fixtures, can be made with a compass saw or saber saw.

If the old plaster or drywall ceiling is still firm and in good condition, the tile can be directly applied to the ceiling. Manufacturers have developed special adhesives for this purpose. Again, the usual procedure is to start from lines established at the center of the room and work in the direction of the walls. A walnut-sized pat of adhesive is placed about 2″ in from each corner of the tile. It is then firmly pressed to the ceiling and

against the adjoining tiles.

In new construction furring strips are nailed directly to the ceiling joists. In remodeling work furring strips are often used when the old plaster or drywall

ceiling is wavy, cracked, or flaking. The furring strips are nailed directly over the old ceiling with nails that are long enough to penetrate into the joists. Shims are used to straighten out the

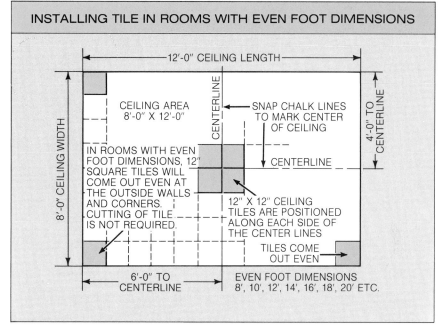

Figure 58–55. In a room with even-foot dimensions, the center tiles are positioned at each side of the center lines.

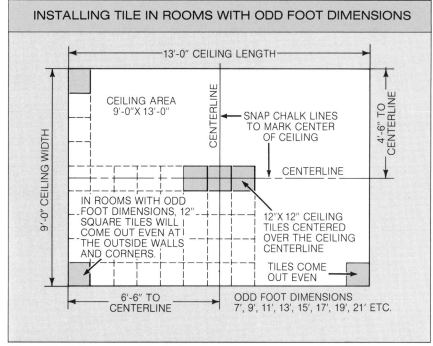

Figure 58–56. In a room with odd-foot dimensions, the center tiles are positioned so that the center lines fall in the middle of the tiles.

low spots. Furring strips are placed 12″ O.C. according to the layout established for the tile lines.

Ceiling tiles are usually fastened to furring strips with staples driven through the wide flange at the grooved edge. Refer again to Figure 58–54. A recommended procedure for this type of installation is to start with a corner tile and then place tiles in two directions as shown in Figure 58–58.

## Applying Ceiling Tiles for a Suspended Ceiling

For a suspended ceiling, a light metal grid is hung by wire from the original ceiling. Tiles that are usually 2′ × 2′ or 2′ × 4′ are then placed in the frames of the metal grid. Suspended ceilings are often used in offices and public buildings and in residential structures with high original ceilings.

One advantage of a suspended ceiling is that it reduces the sound traveling from the floor above and increases the insulating capability of the ceiling. It also allows the use of recessed lighting that fits into the grid system. Pipes, wires, and air conditioning ducts can be conveniently run above the suspended ceiling.

The first step in installing a suspended ceiling is to snap lines on the wall to establish the correct height for the ceiling. Then molding is nailed to the wall. See Figure 58–59. The main runners of the metal grid are suspended from the ceiling with hanger wires. See Figure 58–60. Cross ties are placed between the main runners. Tabs at the ends of the cross ties engage in slots located in the runners. See Figure 58–61. Finally, the tiles are placed in the grid

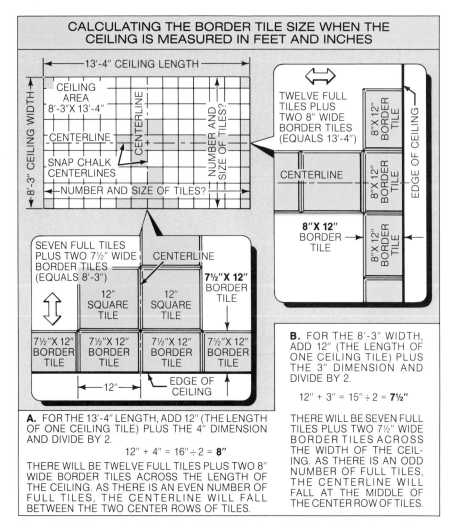

A. FOR THE 13'-4" LENGTH, ADD 12″ (THE LENGTH OF ONE CEILING TILE) PLUS THE 4″ DIMENSION AND DIVIDE BY 2.

$$12″ + 4″ = 16″ \div 2 = \mathbf{8″}$$

THERE WILL BE TWELVE FULL TILES PLUS TWO 8″ WIDE BORDER TILES ACROSS THE LENGTH OF THE CEILING. AS THERE IS AN EVEN NUMBER OF FULL TILES, THE CENTERLINE WILL FALL BETWEEN THE TWO CENTER ROWS OF TILES.

B. FOR THE 8'-3" WIDTH, ADD 12″ (THE LENGTH OF ONE CEILING TILE) PLUS THE 3″ DIMENSION AND DIVIDE BY 2.

$$12″ + 3″ = 15″ \div 2 = \mathbf{7½″}$$

THERE WILL BE SEVEN FULL TILES PLUS TWO 7½″ WIDE BORDER TILES ACROSS THE WIDTH OF THE CEILING. AS THERE IS AN ODD NUMBER OF FULL TILES, THE CENTERLINE WILL FALL AT THE MIDDLE OF THE CENTER ROW OF TILES.

Figure 58–57. Examples of how to calculate the size of the border tiles for a ceiling with foot-and-inch dimensions.

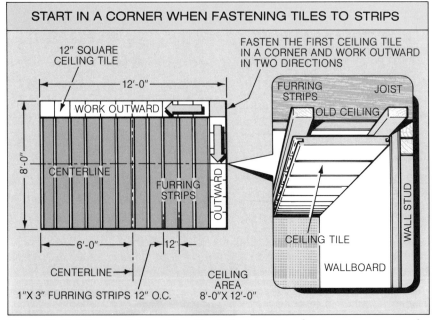

Figure 58–58. When applying ceiling tiles to furring strips, begin in one corner and place the border tiles in two directions.

*Armstrong World Industries*

Figure 58–59. Molding is nailed to the wall for a suspended ceiling system after lines are snapped to establish the correct height.

*Armstrong World Industries*

Figure 58–62. Acoustical tile panels are placed in grid flanges.

*Armstrong World Industries*

Figure 58–63. These workers are completing the grid system in an office building.

*Armstrong World Industries*

Figure 58–60. The main runners of the metal grid for a suspended ceiling are suspended from the ceiling with hanger wires.

*Armstrong World Industries*

Figure 58–64. A completed suspended ceiling in an office. Note the recessed lighting system.

*Armstrong World Industries*

Figure 58–61. Cross ties are placed between the main runners of the grid.

flanges. See Figure 58–62. Figure 58–63 shows construction of a grid system in progress. Figure 58–64 shows a completed suspended ceiling in an office building.

# UNIT 59

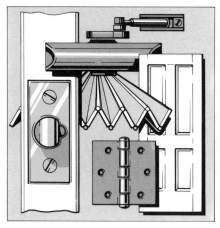

# Interior Doors and Hardware

Interior doors are usually installed after the walls have received their finish covering. The basic construction of an interior door differs little from that of an exterior door. However, waterproof adhesive, required for exterior doors, is not always used for interior doors. Also, interior doors are usually 1⅜" thick, while exterior doors are usually 1¾" thick. (Some interior doors, however, are as thick as exterior doors.)

Stock sizes of interior doors range from 1'-6" to 3'-0" in width. The usual door height is 6'-8" in residential construction and 7'-0" in commercial construction.

An interior door unit includes a *jamb, stops,* and a *casing.* See Figure 59–1. The jamb is the finish frame in which the door hangs. Stops hold the closed door in its proper position. The casing is the trim placed around the door jamb.

## INTERIOR DOOR STYLES

There are many styles of interior doors, but all can be categorized as either *flush* or *panel.* In modern construction, flush doors, which have a smooth surface, are often used. However, traditional panel styles are still much in demand. See Figure 59–2. Plywood veneer or hardboard are the most common finishes for all types of interior doors.

### Flush Doors

A flush door consists of a frame covered with plywood or hardboard face panels *(skins)* about ⅛" in thickness. The two main types of flush doors are *hollow-core* and *solid-core.*

**Hollow-core Doors.** Hollow-core doors are less expensive than solid-core doors and are used more often. One type of hollow-core door has a mesh (cellular) core that provides backing for the face panels. See Figure 59–3. The mesh may not consist of wood or insulation board material. Hollow-core doors are not recommended for exterior doors.

**Solid-core Doors.** Solid-core doors are heavier than the hollow-core doors. They provide

*National Woodwork Manufacturers Association*
Figure 59–2. Traditional panel doors are still popular.

Figure 59–1. An interior door unit includes a jamb, stops, and a casing.

[Figure 59-1 labels: TOP CASING, KERFS, HEAD JAMB, SIDE CASING, SIDE JAMB, DOOR, DOOR STOP]

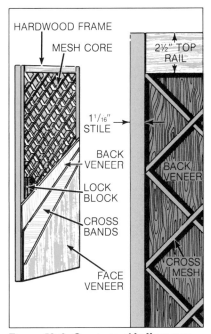

Figure 59-3. One type of hollow-core door has a hardwood frame and a mesh core. It is covered by face panels of either plywood veneer or hardboard. A lock block is located where the hole will be drilled for the door lock.

better sound insulation and have less tendency to warp. The core material is usually particleboard or staggered wood blocks. See Figure 59-4. One type of solid-core door has a special fire-resistant mineral core and is known as a *fire door.* See Figure 59-5.

## Panel Doors

Panel doors are also known as *stile-and-rail doors,* because stiles (vertical members) and rails (cross members) are doweled and glued together to make up the frame. See Figure 59-6. Panels fit into the grooved edges of this frame. Panel doors are manufactured in many designs. See Figure 59-7.

Material for stiles and rails is usually 1⅜" or 1¾" thick. It is available in solid pieces or in pieces made up of laminated lay-

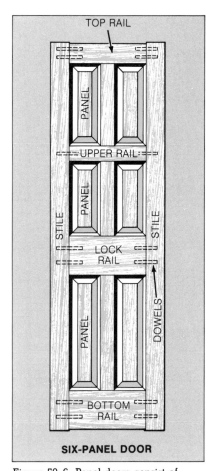

Figure 59-6. Panel doors consist of rails, stiles, and panels.

ers and covered with a veneer face. See Figure 59-8. Panels may be plain or *raised.* Raised panels are usually ¾" thick plywood tapered at the ends to fit into the grooves of the frame. Plain panels are usually ¼" thick plywood. Refer again to Figure 59-8.

## Flush Doors Made to Resemble Panel Doors

An effect similar to a panel door can be obtained by nailing molding to the surface of a flush door. Panel designs may also be routed out on solid-core flush doors. See Figure 59-9.

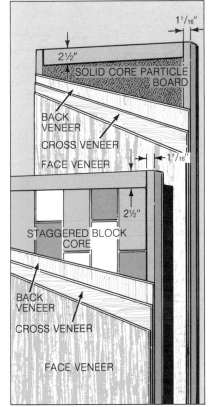

Figure 59-4. Solid-core doors may have a particleboard or staggered-block core.

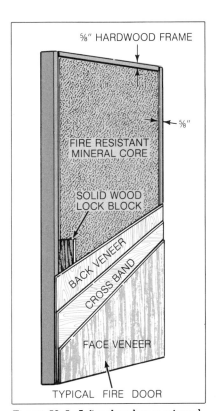

Figure 59-5. A fire door has a mineral core.

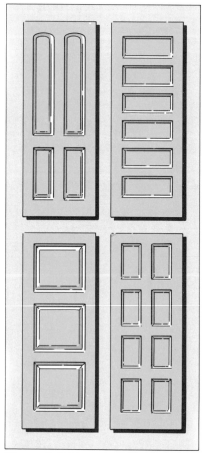

Figure 59–7. Examples of panel door designs.

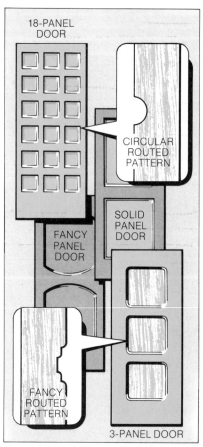

Figure 59–9. Solid-core flush doors may be routed out to give the effect of panel doors.

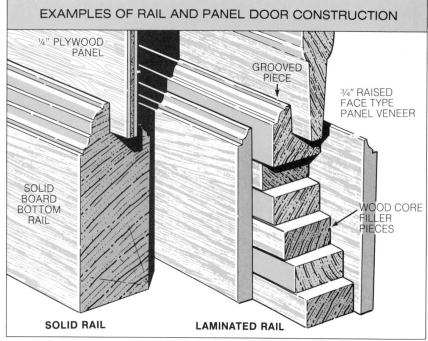

Figure 59–8. Stiles and rails may be solid pieces or may consist of laminated layers. Panels may be plain or raised. Raised panels are ¾" plywood tapered at the ends to fit into the grooves of the frame. Plain panels are usually ¼" thick plywood.

## FINISHING THE DOOR OPENING

An interior door opening may be finished several different ways. The traditional method is to set the jamb first, nail on the casing and stops, and then hang the door. This method is still frequently used, particularly in remodeling and commercial work. In residential and other light construction, various types of *prehung* door systems are usually used. Finish carpenters must be skilled in all the different methods of door installation.

### Blueprint Data on Interior Doors

Blueprints often include a door schedule giving the widths, heights, and thicknesses of all the doors to be hung in the building. Some blueprints indicate door sizes on the floor plan. The direction in which a door swings is always shown on the floor plan. As a rule, section drawings are also provided, giving all the necessary information about the jamb, casing, and stop material to be used. (Refer to Section 6, which includes a door schedule and an interior door frame section drawing.)

### Jamb

A jamb assembly includes a head jamb and two side jambs. See Figure 59–10. The two side jambs are dadoed to receive the head piece. Grooves *(kerfs)* on the back of the jamb stock help prevent later *cupping* of the material. It is good practice to slightly bevel the back edge of the jamb. This ensures a tight fit when the casing material is nailed against the jamb.

Two typical jamb designs are shown in Figure 59–11. One is a rabbeted jamb, and the other is a loose-stop arrangement. The rabbeted design does not require a stop to be attached.

The total width of the jamb material includes the width of the wall stud, the finish wall material, plus ⅛" to allow for small variations in the wall. A standard width available for 2 × 4 stud walls covered with ½" gypsum board is 4⅝".

**Assembling the Jamb.** Usually side pieces for the jamb are delivered from the supplier in a size that is a few inches longer than that needed for the door opening. To figure the exact length needed, the carpenter on the job adds the necessary clearance under the door to the height of the door. See Figure 59–12. If a thin finish floor material, such as ⅛" vinyl tile, is to be laid, ½" clearance between the door and subfloor is adequate. If the floor is to be finished off with thicker finish material, such as carpet or hardwood flooring, greater clearance is necessary. Adequate clearance eliminates the task of

cutting the door bottom after installation.

The width of the finish opening should be ⅛" greater than the door width. This reduces the amount of planing required to fit the door. The total length of the jamb head piece includes the door width, the depth of the two dado cuts in the side pieces, plus ⅛". Refer again to Figure 59–12.

When the head piece and the two side pieces have been cut to their proper lengths, fasten them together with 8d box or casing nails.

**Installing the Jamb.** Tools needed for installing the door jamb are a hand level, a straight-edge with a block nailed at each end, and a steel square. Shim shingles are used to plumb and straighten the side pieces. For nailing the jamb into the trimmers of the rough opening, 8d finish nails are recommended.

When first placing the shim shingles to plumb and align the side pieces of the jamb, some carpenters drive a single nail at the center of the stock above or below (but not through) the shin-

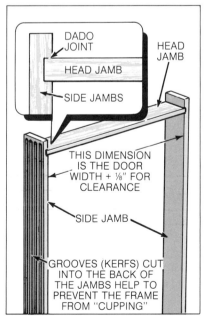

Figure 59–10. A typical jamb assembly includes a head jamb and two side jambs.

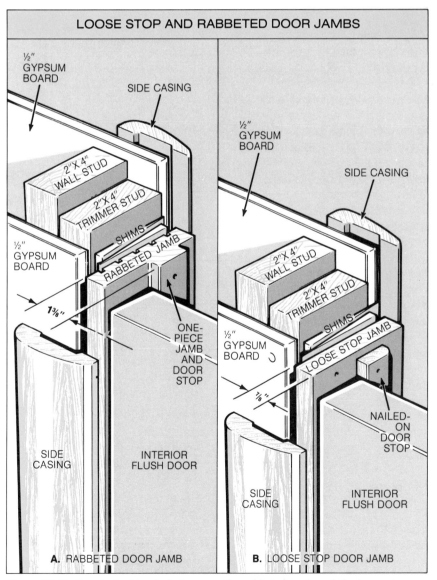

Figure 59–11. Two typical jamb designs are the rabbeted jamb and the loose-stop jamb.

gles. This practice allows the shingles to be readjusted, if necessary, without splitting them.

After all the shingles have been placed and the frame is perfectly plumb and straight, nails are driven near the two edges of the jamb and through the shingles. The shingles are then cut flush with the jamb. The saw should be slightly tilted so as not to scar the edge of the jamb. A complete procedure for installing door jambs is shown in Figure 59–13.

## Stops

Jambs that are not rabbeted to receive the door require door stops. To lay out the position of the stops, measure back the amount of the thickness of the door plus an additional 1/16". Mark the jamb at intervals. Cut and fasten the top piece and then the two side pieces. Tack, rather than nail, the stops in place, as they may have to be shifted after the door is hung. When the stop on the hinge side is permanently nailed, it should be 1/16" away from the door. Many carpenters prefer to wait until the door lock has been fitted before permanently nailing the stop on the lock side. The stop is then adjusted to the door, allowing for a slight amount of rattle when the door is in a closed position.

Butt joints are most often used with door stops, but miter joints may also be used. See Figure 59–14.

## FITTING AND INSTALLING THE DOOR

Power tools are normally furnished by the building contractors. If they are not available, hand tools may be used to fit and install doors, although they

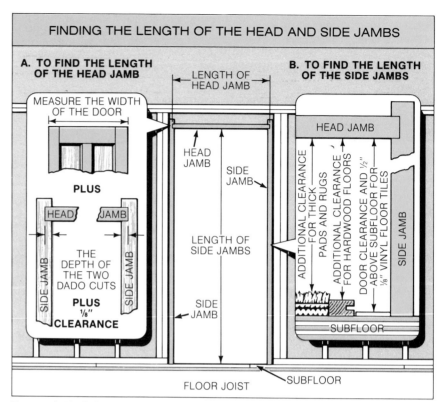

Figure 59–12. Calculating the lengths of the head (A) and side (B) jamb pieces.

are not as efficient as power tools.

Whether power tools or hand tools are used, the door must be held in a stable position while it is being worked on. A *door holder* (or *door jack*) performs this function. Manufactured devices are available, or a job-built door holder can be made with a wedge and a notched piece of 2 × 4. See Figure 59–15. Other tools required to fit and install a door are a jack plane or fore plane, butt gauge or butt marker, a wide chisel, a hammer, and a spiral-ratchet screwdriver.

### Fitting the Door

Place the door in the frame. The door will easily slip into place if the jamb has been set 1/8" wider than the door width. If the door is too wide, plane it to fit.

With the door in the opening, check the sides and top. Mark

where the door will have to be fitted. See Figure 59–16. Remove the door, place it in the holder, and plane where needed. Do not plane off too much at one time. Frequently check the door in the opening to assure proper fit and clearance. Bevel the door 1/8" on the lock side to prevent the inside edge from scraping against the jamb as the door is closed. See Figure 59–17.

### Hanging the Door

The size hinges *(butts)* required to hang a door depends upon the thickness, width, and total weight of the door. See Figure 59–18. Most doors are hung with *loose-pin* hinges. The pin that fits into the barrel of this type of hinge is removable. As a result, the door can be removed from the frame without the hinges being unscrewed. See Figure 59–19.

# INSTALLING AN INTERIOR DOOR JAMB

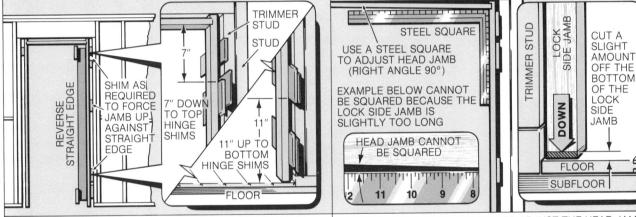

**STEP 1.** SET THE DOOR JAMB IN THE ROUGH OPENING. INSERT AND ADJUST SHIMS AT EACH SIDE OF THE HEAD JAMB SO THAT THERE IS EQUAL CLEARANCE ON BOTH SIDES OF THE DOOR JAMB (CENTERED IN THE ROUGH OPENING).

**STEP 2.** USING A LEVEL AND STRAIGHTEDGE WITH STAND-OFF BLOCKS, PLUMB THE HINGE SIDE JAMB BY ADJUSTING A PAIR OF SHIMS AT THE BOTTOM OF THE JAMB. DRIVE A FINISH NAIL ABOVE THE SHIMS.

**STEP 3.** REVERSE THE STRAIGHTEDGE AND PLACE IT AGAINST THE HINGE SIDE JAMB. DRIVE IN SHIMS AS REQUIRED TO FORCE THE JAMB UP AGAINST THE STRAIGHT-EDGE. BE SURE THERE ARE SHIMS 7" DOWN FROM THE HEAD JAMB AND 11" UP FROM THE FLOOR.

**STEP 4.** USE A STEEL SQUARE TO ADJUST THE HEAD JAMB SO THAT IT IS SQUARE WITH THE SIDE HINGE JAMB. IT MAY BE NECESSARY TO CUT A SLIGHT AMOUNT OFF THE BOTTOM OF THE LOCK SIDE JAMB TO ALLOW IT TO DROP DOWN OR IT MAY BE NECESSARY TO LIFT THE JAMB SLIGHTLY OFF THE FLOOR.

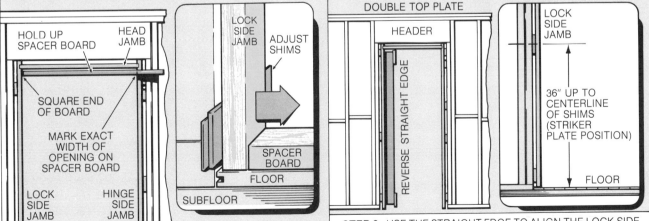

**STEP 5.** SQUARE ONE END OF A PIECE OF MATERIAL THAT IS AS WIDE AS THE JAMB STOCK. HOLD IT AT THE TOP OF THE JAMB AND MARK THE EXACT WIDTH OF THE OPENING. SQUARE AND CUT ON THIS MARK. PLACE THE SPACER BOARD BETWEEN THE BOTTOMS OF THE TWO SIDE JAMBS.

**STEP 6.** USE THE STRAIGHT EDGE TO ALIGN THE LOCK SIDE JAMB, DRIVING IN SHIMS WHERE NEEDED TO FORCE THE JAMB UP AGAINST THE STRAIGHT EDGE. BE SURE THERE IS A SET OF SHIMS POSITIONED 36" UP FROM THE FLOOR WHERE THE LOCK STRIKE PLATE WILL BE FASTENED. THIS COMPLETES THE PLUMBING AND ALIGNING OF THE ENTIRE JAMB.

Figure 59–13. Installing a door jamb. Tools needed are a hand level, a straightedge with a block nailed at each end, and a steel square.

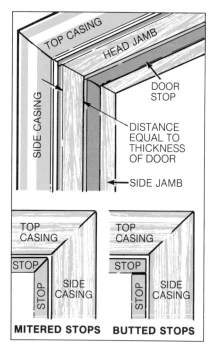

Figure 59–14. Typical door stops. When using a butt joint, always place the top piece first.

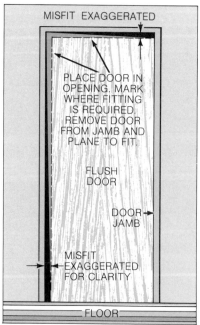

Figure 59–16. With the door in the opening, check the sides and top to see where the door may need to be fitted.

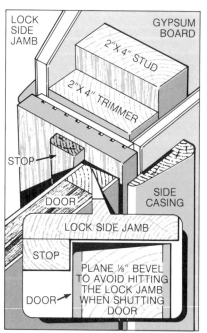

Figure 59–17. The inside edge of the lock side of the door should be beveled ⅛″ to avoid hitting the jamb when the door is closed.

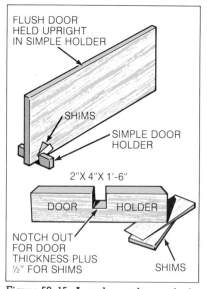

Figure 59–15. A wedge and a notched piece of 2″ x 4″ material can serve as a door holder.

| DOOR HINGE SIZES | | |
|---|---|---|
| FRAME THICKNESS OF DOOR | WIDTH OF DOOR | SIZE OF HINGE |
| 1⅛″ TO 1⅜″ | UP TO 32″ | 3½ |
| 1⅛″ TO 1⅜″ | 32″ TO 37″ | 4 |
| 1⅜″ TO 1⅞″ | UP TO 32″ | 4½ |
| 1⅜″ TO 1⅞″ | 32″ TO 37″ | 5 |
| 1⅜″ TO 1⅞″ | 37″ TO 43″ | 5 EXTRA HEAVY |
| OVER    1⅞″ | UP TO 43″ | 5 EXTRA HEAVY |
| 1⅞″ | OVER 43″ | 6 EXTRA HEAVY |

The size of a loose-pin butt mortise hinge is designated by the length (longer dimension) of the leaf in inches.

Figure 59–18. Hinge sizes for doors. The hinge size is identified by its length dimension. Hollow core doors use two hinges. All exterior doors must be solid core and have three hinges.

**Location of Hinges.** After the door has been fitted, the location of the hinges can be laid out. Traditional practice places the upper hinge 7″ from the top of the door and the lower hinge 11″ from the bottom. Another accepted method is to place the hinges the same distance (8″ to 10″) from the top of the door as from the bottom of the door. Heavier doors, and doors over 6′–8″ in height, require a third hinge centered between the top and bottom hinges. The hinge

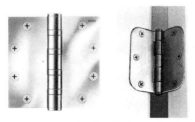

Stanley Hardware division of
The Stanley Works
Figure 59–19. Two types of loose-pin hinges. The hinge with rounded corners is used for mortises made by an electric router.

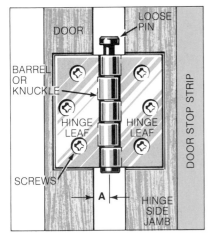

Figure 59–20. Hinge leaves must project from the door and jamb (distance A) by at least one-half of the casing thickness.

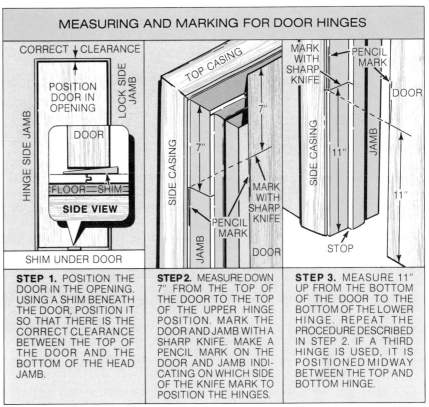

## MEASURING AND MARKING FOR DOOR HINGES

**STEP 1.** POSITION THE DOOR IN THE OPENING. USING A SHIM BENEATH THE DOOR, POSITION IT SO THAT THERE IS THE CORRECT CLEARANCE BETWEEN THE TOP OF THE DOOR AND THE BOTTOM OF THE HEAD JAMB.

**STEP 2.** MEASURE DOWN 7" FROM THE TOP OF THE DOOR TO THE TOP OF THE UPPER HINGE POSITION. MARK THE DOOR AND JAMB WITH A SHARP KNIFE. MAKE A PENCIL MARK ON THE DOOR AND JAMB INDICATING ON WHICH SIDE OF THE KNIFE MARK TO POSITION THE HINGES.

**STEP 3.** MEASURE 11" UP FROM THE BOTTOM OF THE DOOR TO THE BOTTOM OF THE LOWER HINGE. REPEAT THE PROCEDURE DESCRIBED IN STEP 2. IF A THIRD HINGE IS USED, IT IS POSITIONED MIDWAY BETWEEN THE TOP AND BOTTOM HINGE.

Figure 59–21. Laying out door hinges. In this example, 7" and 11" hinge measurements are used.

leaves must project by at least one-half the thickness of the casing material in the direction that the door swings. See Figure 59–20.

A procedure for marking hinge locations on the door and jamb is shown in Figure 59–21. In this example, 7" and 11" hinge measurements are used. The door must be placed in the opening at the exact position it will hang. It is a good idea at this time to lightly pencil an arrow indicating the top of the door. The arrow will prevent a mistake in positioning the door when it is taken out of the opening and placed in the door holder.

**Mortising Gains and Installing Hinges.** The door and jamb must be mortised (notched out) for the hinge to be flush with the surface

of the wood. One method of layout is to place the hinge in position and mark around it with a sharp knife. See Figure 59–22.

Two tools that make hinge layout and mortising easier are a *butt gauge* and a *butt marker.* Both tools quickly mark the door and jamb with the outline of the hinge and its depth of gain. The butt gauge has two cutters. One is adjusted to mark the depth. See Figure 59–23. The butt marker automatically marks the hinge outline and depth of gain when it is hit with a hammer. See Figure 59–24.

The gain should be notched out with a butt chisel. Figure 59–25 shows a procedure for mortising gains and installing hinges. After the gain is mortised, the hinge leaf is screwed in place. See Figure 59–26. After one leaf of each hinge has been attached to the door, the jamb is

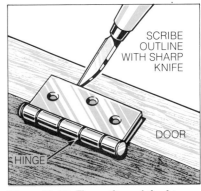

Figure 59–22. The outline of the hinge can be marked with a sharp knife.

mortised and the matching leaves attached. When the door is hung, the leaves of the hinges are lined up. Then the leaves are pushed together and the loose pins replaced. See Figure 59–27.

**Adjusting Hinge Clearance.** A properly fitted door has equal

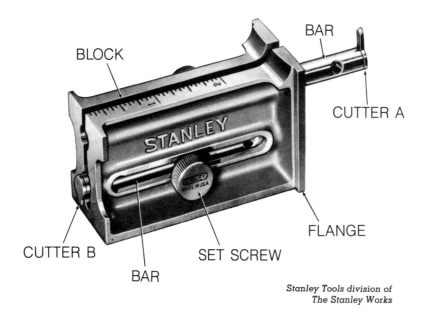

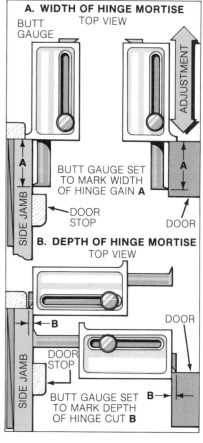

Figure 59–23. Hinge layout is facilitated with a butt gauge. One of the cutters (A) is adjusted to mark the width of the hinge gain. The other cutter (B) is adjusted to mark the depth.

*Stanley Tools division of The Stanley Works*

clearance along the top and the two sides. A little under ⅛″ is recommended for painted doors and ¹⁄₁₆″ for stained doors. Too little clearance makes the doors bind. Too much clearance looks sloppy.

Clearance should be checked after the door is hung. Too little or too much clearance can be easily corrected without removing the door. Loosen the screws of the hinge leaf on the jamb and place a thin cardboard strip toward the front or back of the leaf.

Then retighten the screw. The leaf will tilt in or out as shown in Figure 59–28.

*Stanley Tools division of The Stanley Works*

Figure 59–24. This type of butt marker is used for 3″, 3½″, and 4″ hinge sizes. The hinge outline and depth of gain is marked into the door and jamb when the butt marker is hit with a hammer.

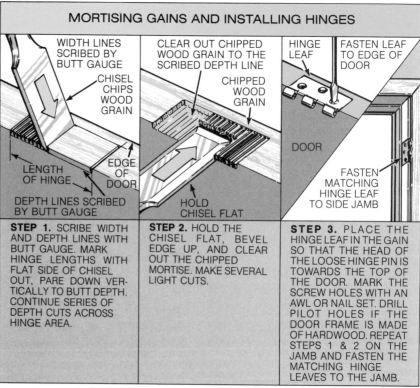

## MORTISING GAINS AND INSTALLING HINGES

**STEP 1.** SCRIBE WIDTH AND DEPTH LINES WITH BUTT GAUGE. MARK HINGE LENGTHS WITH FLAT SIDE OF CHISEL OUT, PARE DOWN VERTICALLY TO BUTT DEPTH. CONTINUE SERIES OF DEPTH CUTS ACROSS HINGE AREA.

**STEP 2.** HOLD THE CHISEL FLAT, BEVEL EDGE UP, AND CLEAR OUT THE CHIPPED MORTISE. MAKE SEVERAL LIGHT CUTS.

**STEP 3.** PLACE THE HINGE LEAF IN THE GAIN SO THAT THE HEAD OF THE LOOSE HINGE PIN IS TOWARDS THE TOP OF THE DOOR. MARK THE SCREW HOLES WITH AN AWL OR NAIL SET. DRILL PILOT HOLES IF THE DOOR FRAME IS MADE OF HARDWOOD. REPEAT STEPS 1 & 2 ON THE JAMB AND FASTEN THE MATCHING HINGE LEAVES TO THE JAMB.

Figure 59–25. Mortising the gain and installing the hinge. A butt chisel is used for mortising.

Figure 59-26. Hinges are fastened to the door with a spiral-ratchet screwdriver.

*Stanley Tools division of The Stanley Works*

*Rockwell International, Power Tool Division*

Figure 59-29. An electric power plane is used to fit the door.

hinges with an electric router that is guided by a hinge-mortising template. Template kits are available from a number of manufacturers. They are adjusted for hinge size and for the spacing between the hinges. The template is usually placed on the door first. See Figure 59-30. Gains are then cut with a router that has been set to the correct

## Advantage of Power Tools in Fitting and Hanging Doors

Doors can be fitted and hung much more quickly and easily with power tools than with hand tools. An electric power plane is used to fit the door. See Figure 59-29. Gains are mortised for

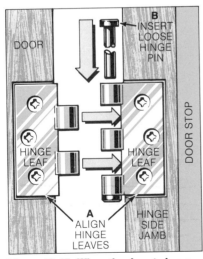

Figure 59-27. When the door is hung, line up the leaves of the door and jamb. Push the leaves together and replace the pin.

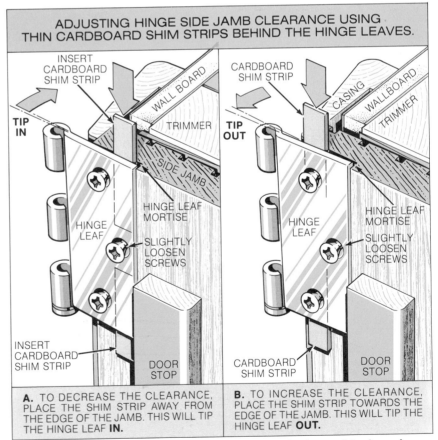

ADJUSTING HINGE SIDE JAMB CLEARANCE USING THIN CARDBOARD SHIM STRIPS BEHIND THE HINGE LEAVES.

**A.** TO DECREASE THE CLEARANCE, PLACE THE SHIM STRIP AWAY FROM THE EDGE OF THE JAMB. THIS WILL TIP THE HINGE LEAF **IN.**

**B.** TO INCREASE THE CLEARANCE, PLACE THE SHIM STRIP TOWARDS THE EDGE OF THE JAMB. THIS WILL TIP THE HINGE LEAF **OUT.**

Figure 59-28. The clearance between the hinge side of the door and jamb can be corrected by placing a thin cardboard strip behind the leaf fastened to the jamb.

depth for the hinges. Hinge screws are driven in by an electric screwdriver.

## Installing Pre-hung Door Units

In a pre-hung door unit the door is already hinged in the frame. In addition, usually the hole is pre-drilled for the lock, and the door stops are tacked to the jamb. The unit may or may not be adjustable.

A section view of a nonadjustable pre-hung door unit looks the same as that of a conventional jamb assembly. (Figure 59–31 is a pictorial drawing based on a section view of a nonadjustable unit.) The casing material, mitered and cut to length, is delivered with the rest of the unit. It is nailed around the jamb after the jamb is installed.

Adjustable *(split-jamb)* units can be adapted to different wall widths. They are delivered with the casing already nailed to the jamb section. Pictorial drawings based on section views of three kinds of adjustable units are shown in Figures 59–32, 59–33, and 59–34.

Pre-hung units can be installed much more rapidly than doors hung in the traditional method. Tools required are a hammer, nail set, straightedge, and hand level. Cedar shim shingles are used to plumb and straighten the jamb. A procedure for installing a nonadjustable unit is shown in Figure 59–35. A procedure for installing an adjustable (split-jamb) unit is shown in Figure 59–36.

## SPECIALTY DOORS

Double doors, sliding doors, and folding doors may be used for closets or when necessitated by

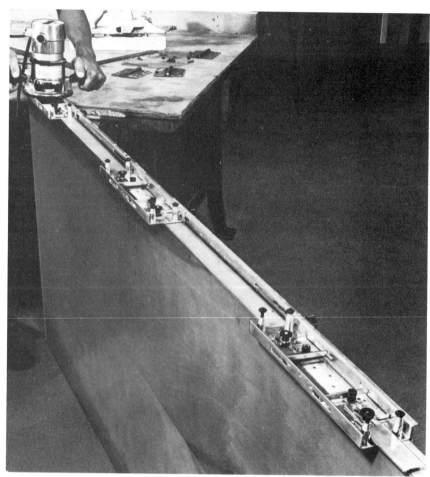

*Porter-Cable Corporation*

Figure 59–30. A hinge mortising template and a portable electric router are used to mortise out the door edge for the door hinges.

certain wall conditions. These doors often require specialized hardware, and information on their installation is normally supplied by the manufacturer.

## Double Doors

Two swinging doors hung in a single frame are often found in office and public buildings. Often one door is *inactive*. It is held in place by *flush bolts*, a type of hardware that is mortised into the upper and lower part of the door edge. (Flush bolts are discussed later in this unit.) The inactive door can be opened when necessary by retracting the flush bolts. Usually, however, only the active door is used for passage

in and out of the room. The active door generally contains a lock with a knob or handle.

The procedure for installing double doors is basically the same as for installing a single door. The inactive door is fitted and set in place first. The active door is fitted to the inactive door and hung in the frame.

## Pocket Sliding Doors

When a swing-type door is not practical, a pocket sliding door, also called a recessed door, may be used. Hangers with small wheels are fastened to the top of the door, enabling it to slide back and forth on a track mounted in the upper part of the frame. See

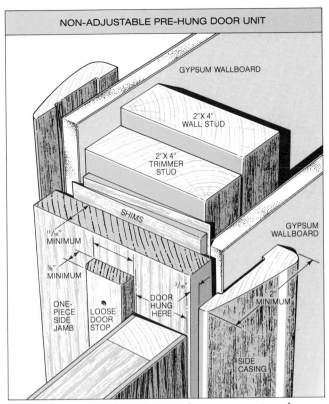

Figure 59–31. Pictorial drawing based on a section view of a rabbeted non-adjustable pre-hung unit.

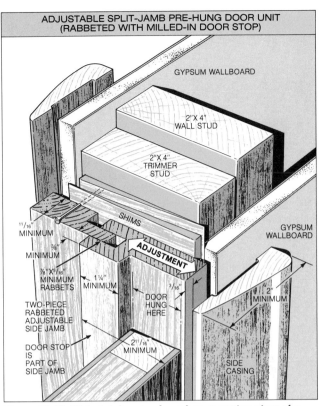

Figure 59–32. Pictorial drawing based on a section view of a rabbeted adjustable pre-hung unit. A separate door stop is not needed with this unit.

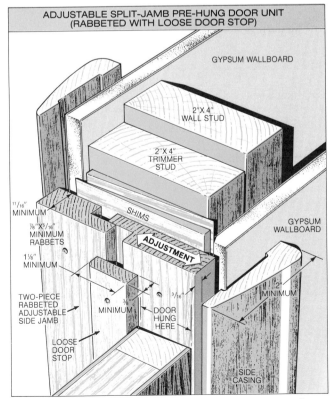

Figure 59–33. Pictorial drawing based on a section view of an adjustable pre-hung unit. The loose stop will be nailed on afterwards.

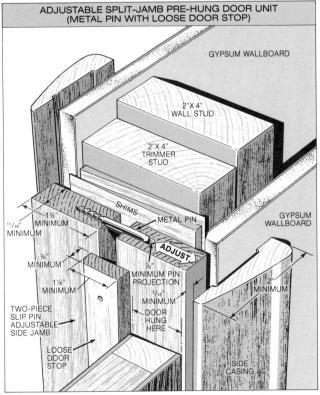

Figure 59–34. Pictorial drawing based on a section view of an adjustable pre-hung door unit with a metal pin.

Figure 59–37. When the door is opened, it slides into a pocket frame in the wall.

The Three-bedroom House

Plan in Section 6 shows a pocket sliding door in the wall between the family room and the kitchen, and between the kitchen and the

utility room. A "recessed door detail" is provided in the plan. This plan is a good example of the circumstances under which

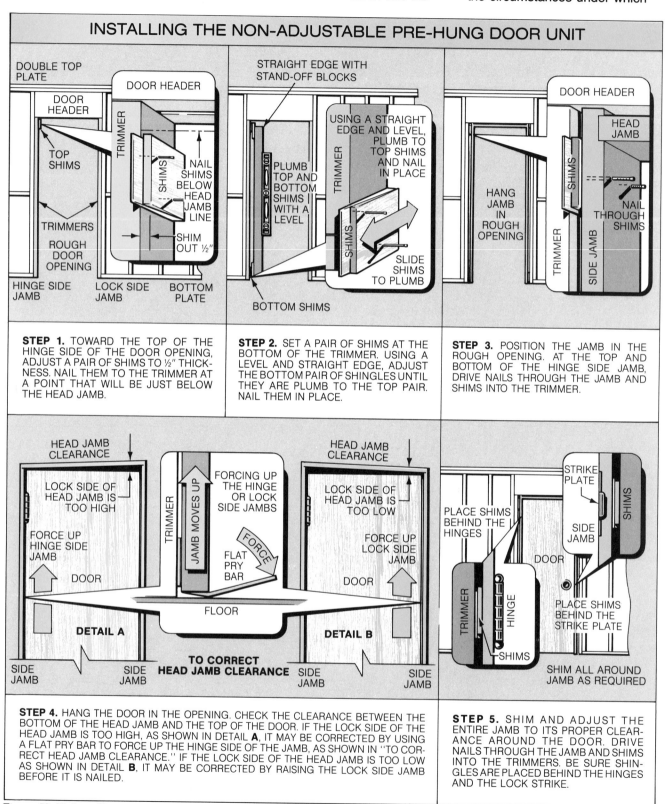

## INSTALLING THE NON-ADJUSTABLE PRE-HUNG DOOR UNIT

**STEP 1.** TOWARD THE TOP OF THE HINGE SIDE OF THE DOOR OPENING, ADJUST A PAIR OF SHIMS TO ½" THICKNESS. NAIL THEM TO THE TRIMMER AT A POINT THAT WILL BE JUST BELOW THE HEAD JAMB.

**STEP 2.** SET A PAIR OF SHIMS AT THE BOTTOM OF THE TRIMMER. USING A LEVEL AND STRAIGHT EDGE, ADJUST THE BOTTOM PAIR OF SHINGLES UNTIL THEY ARE PLUMB TO THE TOP PAIR. NAIL THEM IN PLACE.

**STEP 3.** POSITION THE JAMB IN THE ROUGH OPENING. AT THE TOP AND BOTTOM OF THE HINGE SIDE JAMB, DRIVE NAILS THROUGH THE JAMB AND SHIMS INTO THE TRIMMER.

**STEP 4.** HANG THE DOOR IN THE OPENING. CHECK THE CLEARANCE BETWEEN THE BOTTOM OF THE HEAD JAMB AND THE TOP OF THE DOOR. IF THE LOCK SIDE OF THE HEAD JAMB IS TOO HIGH, AS SHOWN IN DETAIL **A**, IT MAY BE CORRECTED BY USING A FLAT PRY BAR TO FORCE UP THE HINGE SIDE OF THE JAMB, AS SHOWN IN "TO CORRECT HEAD JAMB CLEARANCE." IF THE LOCK SIDE OF THE HEAD JAMB IS TOO LOW AS SHOWN IN DETAIL **B**, IT MAY BE CORRECTED BY RAISING THE LOCK SIDE JAMB BEFORE IT IS NAILED.

**STEP 5.** SHIM AND ADJUST THE ENTIRE JAMB TO ITS PROPER CLEARANCE AROUND THE DOOR. DRIVE NAILS THROUGH THE JAMB AND SHIMS INTO THE TRIMMERS. BE SURE SHINGLES ARE PLACED BEHIND THE HINGES AND THE LOCK STRIKE.

Figure 59–35. Installing a nonadjustable pre-hung door unit. Pre-hung units can be installed much more rapidly than doors hung in the traditional manner.

522

# INSTALLING A TYPICAL SPLIT-JAMB PRE-HUNG DOOR UNIT

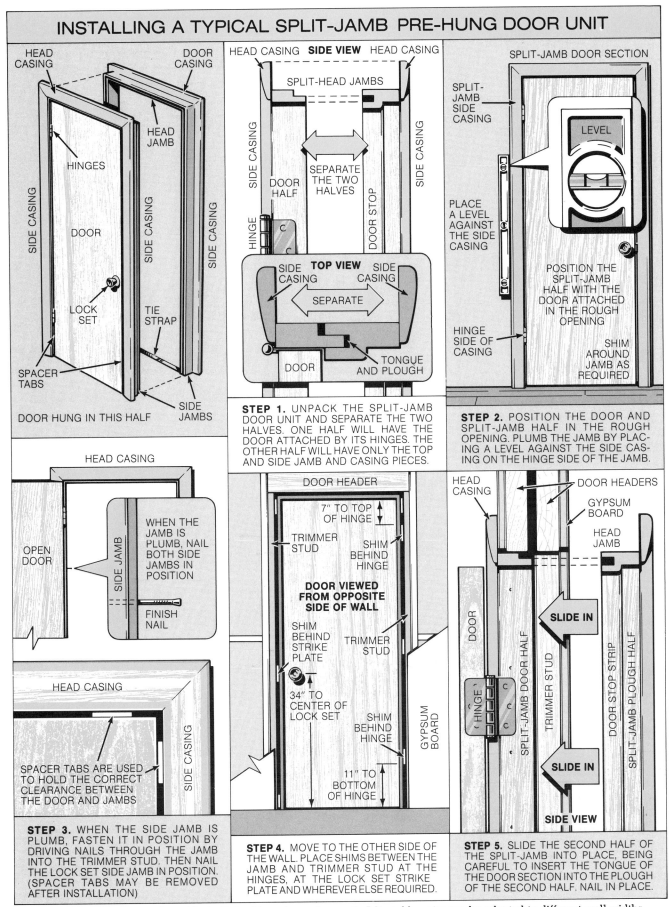

**STEP 1.** UNPACK THE SPLIT-JAMB DOOR UNIT AND SEPARATE THE TWO HALVES. ONE HALF WILL HAVE THE DOOR ATTACHED BY ITS HINGES. THE OTHER HALF WILL HAVE ONLY THE TOP AND SIDE JAMB AND CASING PIECES.

**STEP 2.** POSITION THE DOOR AND SPLIT-JAMB HALF IN THE ROUGH OPENING. PLUMB THE JAMB BY PLACING A LEVEL AGAINST THE SIDE CASING ON THE HINGE SIDE OF THE JAMB.

**STEP 3.** WHEN THE SIDE JAMB IS PLUMB, FASTEN IT IN POSITION BY DRIVING NAILS THROUGH THE JAMB INTO THE TRIMMER STUD. THEN NAIL THE LOCK SET SIDE JAMB IN POSITION. (SPACER TABS MAY BE REMOVED AFTER INSTALLATION)

**STEP 4.** MOVE TO THE OTHER SIDE OF THE WALL. PLACE SHIMS BETWEEN THE JAMB AND TRIMMER STUD AT THE HINGES, AT THE LOCK SET STRIKE PLATE AND WHEREVER ELSE REQUIRED.

**STEP 5.** SLIDE THE SECOND HALF OF THE SPLIT-JAMB INTO PLACE, BEING CAREFUL TO INSERT THE TONGUE OF THE DOOR SECTION INTO THE PLOUGH OF THE SECOND HALF. NAIL IN PLACE.

Figure 59–36. Installing an adjustable (split-jamb) pre-hung door unit. Adjustable units can be adapted to different wall widths.

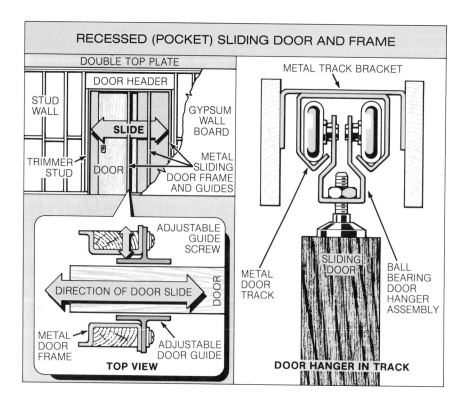

*Stanley Hardware division of The Stanley Works*

Figure 59–37. Pocket sliding door and frame. The frame is set in place at the time the wall is framed.

this type of door is used. A regular swinging door in these particular walls would eliminate needed working space along the walls.

Pocket sliding doors are usually produced in standard sizes. Units include the frame, hangers, track, and other required hardware. The frame for the door is set in place at the time that the wall is framed.

## Bypass Sliding Doors

Where two or more doors are used for a wide closet opening, bypass sliding doors are convenient. The doors are suspended from tracks with roller-type hangers. Floor guides hold the bottoms of the doors in position. See Figure 59–38.

## Folding Doors

Folding doors, particularly *bi-fold* doors, are frequently used for closet doors. They come in flush, panel, or louver designs. See Figure 59–39. Thicknesses most often used are 1 1/8" and 1 3/8". A plastic type of bi-fold door is also available.

In bi-fold installation, the door next to the jamb is secured at the top and bottom with a pivot pin and bracket. The second door is hinged to the first door. The top of the second door has a roller guide that slides along a track mounted to the head jamb. See Figure 59–40.

*Multi-folding* doors work in a similar manner to bi-fold doors. They are used for closets, room entrances, and as room dividers. See Figure 59–41. The floor plan in Section 6 shows that all the bedroom wardrobe closets have wood folding doors. A folding door section view is provided.

Multi-folding doors are supported by wheeled hangers that run along an overhead track. The

door panels usually have a wood core and are surfaced with wood veneer, vinyl, or plastic laminate. Common thickness of door panels are 3/8" and 9/16". Common widths range from 3 5/8" to 5 5/8". Spring hinges run horizontally through the panels and connect with a vertical wood molding that runs the entire length of the door.

## METAL DOOR UNITS

Metal door units have been used in commercial construction for many years. The metal frames can be fastened to concrete masonry, as well as to metal and wood stud partitions. Only recently, however, have metal frames been used in a significant number of residential and other light construction buildings. The metal frames usually used in light construction are designed to be fastened to walls that have a

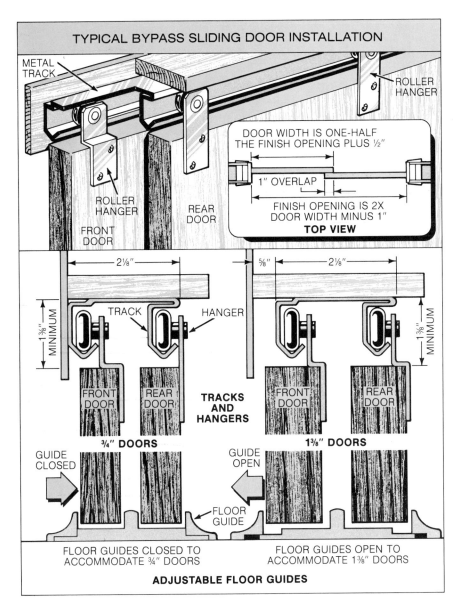

**TYPICAL BYPASS SLIDING DOOR INSTALLATION**

METAL TRACK

ROLLER HANGER

ROLLER HANGER

REAR DOOR

FRONT DOOR

DOOR WIDTH IS ONE-HALF THE FINISH OPENING PLUS ½"

1" OVERLAP

FINISH OPENING IS 2X DOOR WIDTH MINUS 1"

**TOP VIEW**

2⅛"

⅝"    2⅛"

1⅜" MINIMUM

TRACK    HANGER

1⅜" MINIMUM

FRONT DOOR    REAR DOOR

**TRACKS AND HANGERS**

FRONT DOOR    REAR DOOR

**¾" DOORS**    **1⅜" DOORS**

GUIDE CLOSED

GUIDE OPEN

FLOOR GUIDE

FLOOR GUIDES CLOSED TO ACCOMMODATE ¾" DOORS

FLOOR GUIDES OPEN TO ACCOMMODATE 1⅜" DOORS

**ADJUSTABLE FLOOR GUIDES**

*Maywood Industries*

*Maywood Industries*

Figure 59–39. Flush design and combination louver-and-panel design bi-fold doors.

gypsum board (drywall) finish. They do not have to be trimmed afterwards, as the casing is molded into the frame. Either wood or metal doors can be hung in the frame.

A typical metal frame for drywall is made of 16-ga. hot-dipped galvanized steel. See

ROLLER TYPE OF HANGERS    TRACK

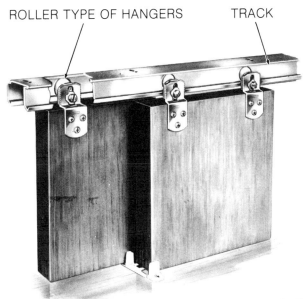

*Stanley Hardware division of The Stanley Works*

Figure 59–38. Bypass sliding doors are suspended from tracks with roller hangers.

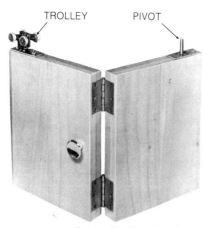

TROLLEY      PIVOT

*Stanley Hardware division of
The Stanley Works*

Figure 59–40. This sales sample model shows the working parts of a bi-fold door. The door next to the jamb swings on pivots (only the top pivot is shown). The second door is hinged to the first door and is supported by a trolley that runs along a track fastened to the top of the frame.

Figure 59–42. The bottom of the frame is secured by screws through the casing. Most metal frames for drywall have adjustable anchors so that the frame can be plumbed and straightened as it is being screwed into the metal or wood door openings.

These frames are available in most standard door sizes. Each frame is delivered in a three-piece package and is installed on the job according to instructions provided by the manufacturer. See Figure 59–43.

## Steel Doors

Steel doors are used most often in commercial construction. They are available in most standard sizes and in 1⅜″ and 1¾″ thicknesses. Steel doors have prepared hinge mortises and lock holes. Pieces of reinforcing steel are placed behind the sections where hinges, locks, door closers, and other type of hardware may be placed. See Figure 59–44.

## DOOR LOCKS

Locks are installed in doors after the doors have been hung. A *lock set* consists of the lock's working components and its *trim pieces* (visible parts of the lock). Trim pieces include the knob or handle, escutcheons, rose, and cylinder. Escutcheons are the decorative metal plates placed against the door face behind the knob and lock. The rose fits against the door and, in some types of lock sets, holds the lock in place. The cylinder is the part of the lock that contains the keyhole and tumbler mechanism.

In new construction a common procedure is to install the lock set in the unfinished door unit and then remove it while the door is painted or stained. After the door is finished, the lock set is reinstalled. Another procedure is to make the first and final installation on the finished door.

The *door hand* is the direction in which a door swings. See Figure 59–45. A *left-hand door* hinges on left and opens inward. A *left-hand reverse door* hinges on left and opens outward. A *right-hand door* hinges on right and opens inward. A *right-hand reverse door* hinges on right and opens outward.

Many lock sets are reversible, meaning that they can be adjusted to operate in doors that swing in either direction. Other lock sets, as well as other kinds

*Pella—Rolscreen Company*      *Pella—Rolscreen Company*      *Pella—Rolscreen Company*

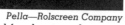
Figure 59–41. Multi-folding doors are used for closets and room entrances, and as room dividers.

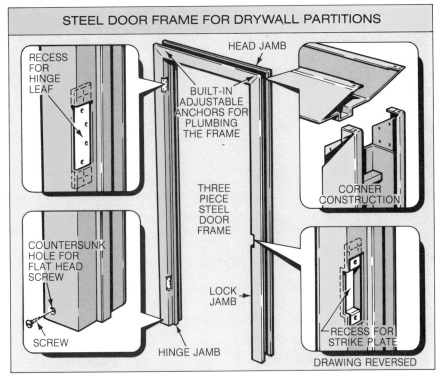

STEEL DOOR FRAME FOR DRYWALL PARTITIONS

RECESS FOR HINGE LEAF

HEAD JAMB

BUILT-IN ADJUSTABLE ANCHORS FOR PLUMBING THE FRAME

THREE PIECE STEEL DOOR FRAME

CORNER CONSTRUCTION

COUNTERSUNK HOLE FOR FLAT HEAD SCREW

SCREW

LOCK JAMB

HINGE JAMB

RECESS FOR STRIKE PLATE

DRAWING REVERSED

Figure 59–42. Typical steel door frame for gypsum board (drywall) partitions. The bottom is secured by screws through the casing. The rest of the jamb is plumbed, straightened, and fastened with adjustable anchors. Note the recessed areas for the hinges and the lock strike plate.

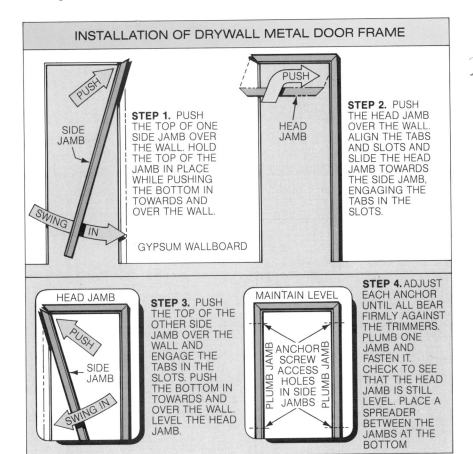

INSTALLATION OF DRYWALL METAL DOOR FRAME

PUSH

SIDE JAMB

STEP 1. PUSH THE TOP OF ONE SIDE JAMB OVER THE WALL. HOLD THE TOP OF THE JAMB IN PLACE WHILE PUSHING THE BOTTOM IN TOWARDS AND OVER THE WALL.

SWING IN

GYPSUM WALLBOARD

PUSH

HEAD JAMB

STEP 2. PUSH THE HEAD JAMB OVER THE WALL. ALIGN THE TABS AND SLOTS AND SLIDE THE HEAD JAMB TOWARDS THE SIDE JAMB, ENGAGING THE TABS IN THE SLOTS.

HEAD JAMB

PUSH

SIDE JAMB

SWING IN

STEP 3. PUSH THE TOP OF THE OTHER SIDE JAMB OVER THE WALL AND ENGAGE THE TABS IN THE SLOTS. PUSH THE BOTTOM IN TOWARDS AND OVER THE WALL. LEVEL THE HEAD JAMB.

MAINTAIN LEVEL

PLUMB JAMB   ANCHOR SCREW ACCESS HOLES IN SIDE JAMBS   PLUMB JAMB

STEP 4. ADJUST EACH ANCHOR UNTIL ALL BEAR FIRMLY AGAINST THE TRIMMERS. PLUMB ONE JAMB AND FASTEN IT. CHECK TO SEE THAT THE HEAD JAMB IS STILL LEVEL. PLACE A SPREADER BETWEEN THE JAMBS AT THE BOTTOM

Figure 59–43. Installing a typical metal door frame in a gypsum board wall.

of door hardware, are designed to work only on doors that swing right or only on doors that swing left. It is necessary to identify the hand of the door before ordering this type of hardware.

To determine the hand of the door, always consider that you are standing on the *outside (exterior)* of the door. The outside of a door would be a hallway or passageway or the street side of an entrance door. In doors that open between rooms, the keyed side of the lock is considered the outside of the door.

The main types of locks are *cylindrical, tubular, mortised,* and *dead-bolt.* For greater security, more than one of these types may be used on a door. Each type is available in a variety of designs and finishes.

## Cylindrical and Tubular Locks

Both cylindrical and tubular locks are *bored* locks, meaning that holes must be drilled in the door to install the locks. See Figure 59–46. Cylindrical and tubular locks are also referred to as *key-in-knob* locks because in both types of locks the cylinder is located in the knob section of the lock. The cylinder contains the tumbler mechanism in which the key fits. If the keys are lost or stolen, the cylinder can be replaced, and the lock will operate with a new set of keys.

Another type of bored lock is the *grip-handle* lock. See Figure 59–47. In this lock the latch is retracted by a thumbpiece on the outside of the door or by an inside knob. It is not considered a key-in-knob lock, because the cylinder is located above the knob.

The cylindrical lock is considered a better quality lock than the tubular lock. It is designed for long life and trouble-free opera-

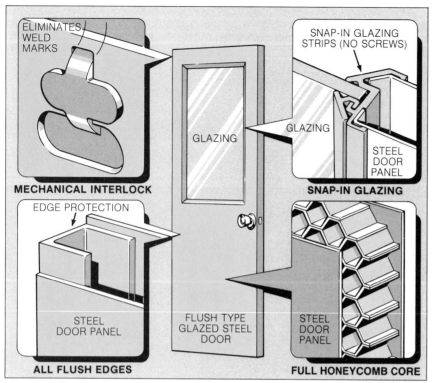

MECHANICAL INTERLOCK

ELIMINATES WELD MARKS

EDGE PROTECTION

STEEL DOOR PANEL

ALL FLUSH EDGES

GLAZING

GLAZING

FLUSH TYPE GLAZED STEEL DOOR

SNAP-IN GLAZING STRIPS (NO SCREWS)

STEEL DOOR PANEL

SNAP-IN GLAZING

STEEL DOOR PANEL

FULL HONEYCOMB CORE

Figure 59–44. A flush steel door with a full honeycomb core. The face material is made of 20-ga. steel sheets. Pieces of 12-ga. steel reinforce sections where the lock and hinges are to be installed. The holes are drilled and tapped to receive flat-head machine screws.

Figure 59–46. Cylindrical lock set. The cylinder is located in the knob section of the lock. It contains the tumbler mechanism in which the key fits.

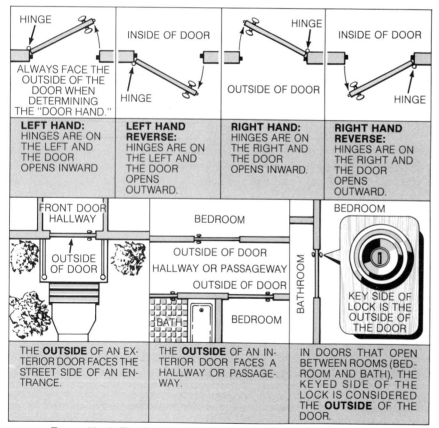

HINGE

ALWAYS FACE THE OUTSIDE OF THE DOOR WHEN DETERMINING THE "DOOR HAND."

INSIDE OF DOOR

HINGE

HINGE

OUTSIDE OF DOOR

INSIDE OF DOOR

HINGE

**LEFT HAND:** HINGES ARE ON THE LEFT AND THE DOOR OPENS INWARD

**LEFT HAND REVERSE:** HINGES ARE ON THE LEFT AND THE DOOR OPENS OUTWARD.

**RIGHT HAND:** HINGES ARE ON THE RIGHT AND THE DOOR OPENS INWARD.

**RIGHT HAND REVERSE:** HINGES ARE ON THE RIGHT AND THE DOOR OPENS OUTWARD.

FRONT DOOR HALLWAY

OUTSIDE OF DOOR

BEDROOM

OUTSIDE OF DOOR

HALLWAY OR PASSAGEWAY

OUTSIDE OF DOOR

BATH

BEDROOM

BEDROOM

BATHROOM

KEY SIDE OF LOCK IS THE OUTSIDE OF THE DOOR

THE **OUTSIDE** OF AN EXTERIOR DOOR FACES THE STREET SIDE OF AN ENTRANCE.

THE **OUTSIDE** OF AN INTERIOR DOOR FACES A HALLWAY OR PASSAGEWAY.

IN DOORS THAT OPEN BETWEEN ROOMS (BEDROOM AND BATH), THE KEYED SIDE OF THE LOCK IS CONSIDERED THE **OUTSIDE** OF THE DOOR.

Figure 59–45. The door hand is the direction in which a door swings.

Figure 59–47. Grip-handle lock. Note that the cylinder is separate from the knob.

tion. The tubular lock is less expensive, more simply constructed, and does not have the strength or smooth working action of the cylindrical lock.

**Operation.** In both cylindrical and tubular locks, the latch unit engages with a latch-detractor device in the main body of the lock. When the door knob or key is turned, the latch bolt is pulled back into the door. When the knob is released, the latch bolt springs back to its original position.

When the door is in a closed position, the latch bolt is engaged by the *strike,* which is a metal plate mortised and screwed into the door jamb. See Figure 59–48.

Two types of latch units are the *spring latch* and the *dead-locking latch* (or *dead-bolt latch*). See Figure 59–49. The spring latch can easily be forced open by a piece of plastic. Therefore, it should not be used on entrance-door locks. The dead-locking latch has a small plunger (guard bolt). When the door is closed, and the latch bolt has slipped into the strike, the plunger remains in a retracted position. The plunger locks the latch in place so that it cannot be moved.

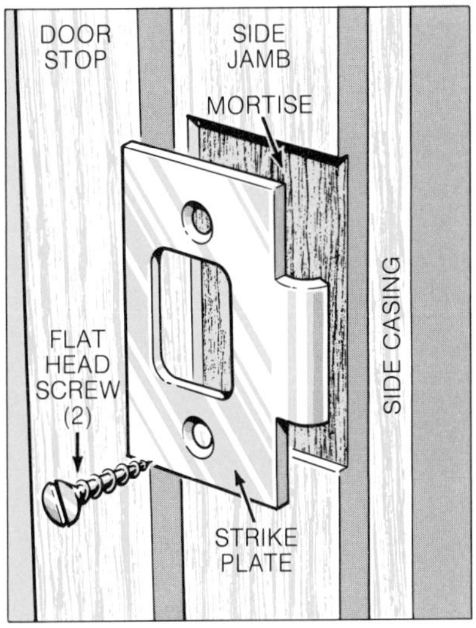

Figure 59–48. The strike plate is mortised into the jamb.

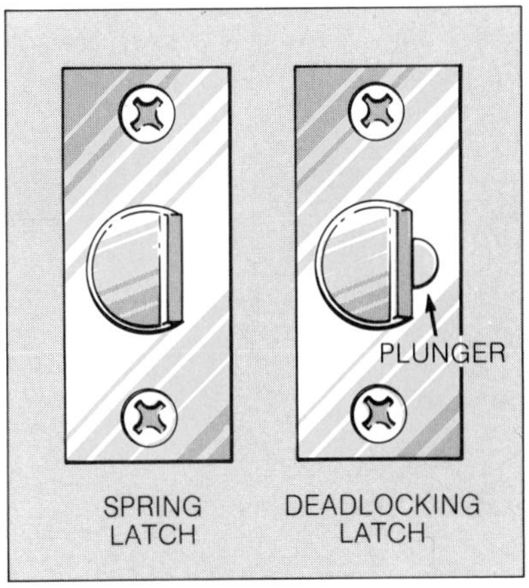

Figure 59–49. A dead-locking (or dead-bolt) latch provides greater security than a spring latch.

**Keyed and Non-keyed Types.** In a typical cylindrical or tubular lock used on entrance doors, the key cylinder is on the exterior side of the door. A *pushbutton* or *turnbutton* is located in the inside knob for the purpose of locking the door from that side. These devices disengage and unlock the door when the inside knob is turned, or when the key is inserted and turned. Non-keyed locks may have pushbutton or turnbutton devices on one side, or none on either side.

**Installation.** In residential construction the height of the lock from the floor to its center is usually 36″. In public and office buildings the height is usually 38″. New lock sets are supplied with manufacturer's instructions for installation and a template for marking holes to be drilled. The usual procedure (for both cylindrical and tubular locks) is to drill the large hole first. A hand brace and expansive bit can be used for this purpose. There are also special lock set bits that fit into a brace or electric drill to bore standard size cylindrical holes. A regular auger bit is used for the smaller latch unit hole.

Sometimes a carpenter must install a lock set without a template or instructions from the manufacturer. A procedure for laying out and boring the holes and for marking the mortise for the latch unit face plate is shown in Figure 59–50.

**Setting the Strike.** After the lock is installed, the position of the strike is located and the door jamb is mortised for it. If this is done accurately, the door will be exactly flush with the edge of the jamb when the latch engages the strike plate. A procedure for establishing the position of the strike plate is shown in Figure 59–51.

**Installation Kits.** Most manufacturers offer installation kits that include a boring jig that clamps to the door, drill bits for boring the holes, and special marking chisels for the latch and strike plates. The boring jig can be adapted for a number of different backsets, door thicknesses, and latch unit holes of different diameters. It facilitates rapid and accurate drilling of the lock holes. The marking chisels score the door and jamb for the face plate of the latch unit and strike plate. The strike locator is used to mark the correct position of the strike plate on the jamb. With the help of an electric drill, installation kits permit rapid and neat lock installation.

## Mortised Locks

The mortised lock is a sturdy, long-lasting type of lock that has been used for many years. See Figure 59–52. It is more costly and difficult to install than cylindrical or tubular locks and thus is not used as often. However, together with heavier entrance doors, mortised locks provide greater security than cylindrical or tubular locks.

Mortised lock sets are supplied with a template and instructions from the manufacturers. In remodeling work, however, an old

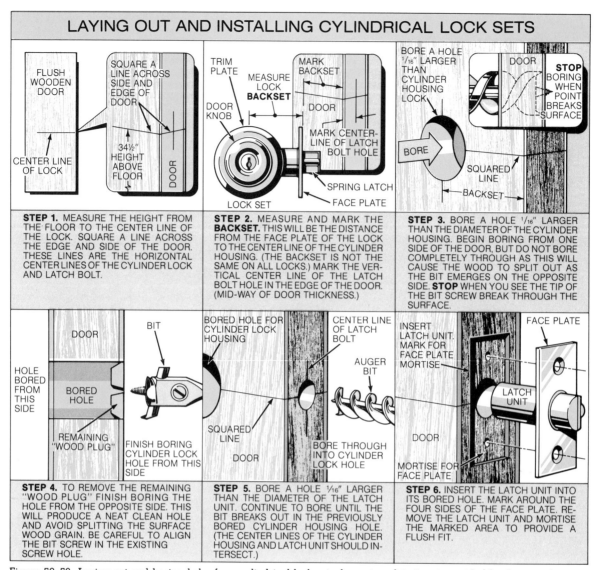

# LAYING OUT AND INSTALLING CYLINDRICAL LOCK SETS

**STEP 1.** MEASURE THE HEIGHT FROM THE FLOOR TO THE CENTER LINE OF THE LOCK. SQUARE A LINE ACROSS THE EDGE AND SIDE OF THE DOOR. THESE LINES ARE THE HORIZONTAL CENTER LINES OF THE CYLINDER LOCK AND LATCH BOLT.

**STEP 2.** MEASURE AND MARK THE **BACKSET.** THIS WILL BE THE DISTANCE FROM THE FACE PLATE OF THE LOCK TO THE CENTER LINE OF THE CYLINDER HOUSING. (THE BACKSET IS NOT THE SAME ON ALL LOCKS.) MARK THE VERTICAL CENTER LINE OF THE LATCH BOLT HOLE IN THE EDGE OF THE DOOR. (MID-WAY OF DOOR THICKNESS.)

**STEP 3.** BORE A HOLE ¹/₁₆" LARGER THAN THE DIAMETER OF THE CYLINDER HOUSING. BEGIN BORING FROM ONE SIDE OF THE DOOR, BUT DO NOT BORE COMPLETELY THROUGH AS THIS WILL CAUSE THE WOOD TO SPLIT OUT AS THE BIT EMERGES ON THE OPPOSITE SIDE. **STOP** WHEN YOU SEE THE TIP OF THE BIT SCREW BREAK THROUGH THE SURFACE.

**STEP 4.** TO REMOVE THE REMAINING "WOOD PLUG" FINISH BORING THE HOLE FROM THE OPPOSITE SIDE. THIS WILL PRODUCE A NEAT CLEAN HOLE AND AVOID SPLITTING THE SURFACE WOOD GRAIN. BE CAREFUL TO ALIGN THE BIT SCREW IN THE EXISTING SCREW HOLE.

**STEP 5.** BORE A HOLE ¹/₁₆" LARGER THAN THE DIAMETER OF THE LATCH UNIT. CONTINUE TO BORE UNTIL THE BIT BREAKS OUT IN THE PREVIOUSLY BORED CYLINDER HOUSING HOLE. (THE CENTER LINES OF THE CYLINDER HOUSING AND LATCH UNIT SHOULD INTERSECT.)

**STEP 6.** INSERT THE LATCH UNIT INTO ITS BORED HOLE. MARK AROUND THE FOUR SIDES OF THE FACE PLATE. REMOVE THE LATCH UNIT AND MORTISE THE MARKED AREA TO PROVIDE A FLUSH FIT.

Figure 59-50. Laying out and boring holes for a cylindrical lock set when a template is not provided by the manufacturer.

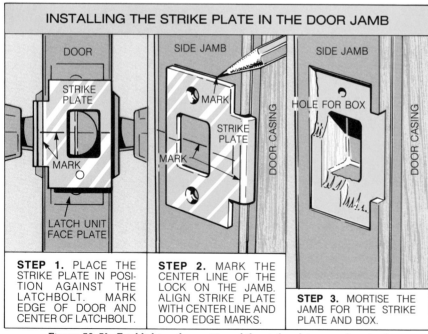

# INSTALLING THE STRIKE PLATE IN THE DOOR JAMB

**STEP 1.** PLACE THE STRIKE PLATE IN POSITION AGAINST THE LATCHBOLT. MARK EDGE OF DOOR AND CENTER OF LATCHBOLT.

**STEP 2.** MARK THE CENTER LINE OF THE LOCK ON THE JAMB. ALIGN STRIKE PLATE WITH CENTER LINE AND DOOR EDGE MARKS.

**STEP 3.** MORTISE THE JAMB FOR THE STRIKE PLATE AND BOX.

Figure 59-51. Establishing the position of the strike plate on a door jamb.

mortised lock is often removed from one door and placed in another. For this reason, a carpenter must know how to install a mortised lock without a template. Figure 59–53 shows a procedure for installing a mortised lock.

## Dead Bolts

A common way to increase the security of an exterior door is to install a dead bolt in addition to a cylindrical lock. A dead bolt consists of a solid metal bar that must be thrown (locked) and retracted (unlocked) with a knob or key. Dead bolts are usually keyed on the outside of the door and have a knob or handle on the inside. See Figure 59–54.

*Schlage Lock Company*
Figure 59–52. A modern heavy-duty mortised lock with a lever-type handle. Knobs and grip-type handles are also used with mortised locks.

They are also available with a double cylinder keyed on both sides.

## Combination Locks

High-security locks are available that combine the mechanisms of a cylindrical lock and a dead bolt. See Figure 59–55. The lock mechanism is protected by an armored plate on both sides of the door. Additional features make forceful entry very difficult.

## OTHER DOOR HARDWARE

In addition to locks, many other pieces of hardware exist for the operation and protection of doors.

## Special Hinges

*Ball-bearing hinges* are used for heavy doors that are subjected to a great deal of use, such as in schools and public buildings. Ball bearings in this type of hinge help prevent wear at the knuckle joints.

*Self-closing spring hinges* are also often used on doors in pub-

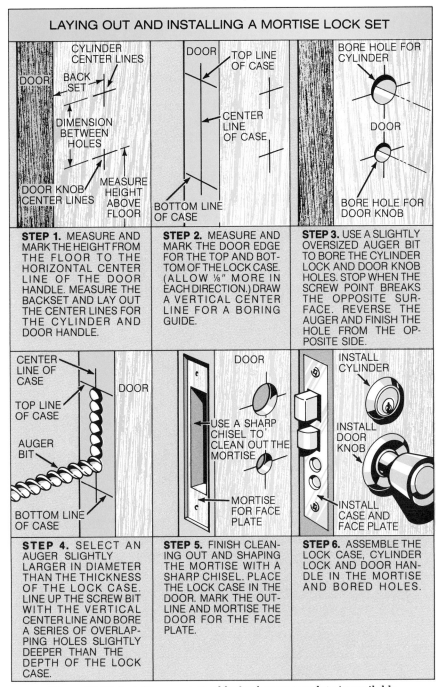

### LAYING OUT AND INSTALLING A MORTISE LOCK SET

**STEP 1.** MEASURE AND MARK THE HEIGHT FROM THE FLOOR TO THE HORIZONTAL CENTER LINE OF THE DOOR HANDLE. MEASURE THE BACKSET AND LAY OUT THE CENTER LINES FOR THE CYLINDER AND DOOR HANDLE.

**STEP 2.** MEASURE AND MARK THE DOOR EDGE FOR THE TOP AND BOTTOM OF THE LOCK CASE. (ALLOW ⅛" MORE IN EACH DIRECTION.) DRAW A VERTICAL CENTER LINE FOR A BORING GUIDE.

**STEP 3.** USE A SLIGHTLY OVERSIZED AUGER BIT TO BORE THE CYLINDER LOCK AND DOOR KNOB HOLES. STOP WHEN THE SCREW POINT BREAKS THE OPPOSITE SURFACE. REVERSE THE AUGER AND FINISH THE HOLE FROM THE OPPOSITE SIDE.

**STEP 4.** SELECT AN AUGER SLIGHTLY LARGER IN DIAMETER THAN THE THICKNESS OF THE LOCK CASE. LINE UP THE SCREW BIT WITH THE VERTICAL CENTER LINE AND BORE A SERIES OF OVERLAPPING HOLES SLIGHTLY DEEPER THAN THE DEPTH OF THE LOCK CASE.

**STEP 5.** FINISH CLEANING OUT AND SHAPING THE MORTISE WITH A SHARP CHISEL. PLACE THE LOCK CASE IN THE DOOR. MARK THE OUTLINE AND MORTISE THE DOOR FOR THE FACE PLATE.

**STEP 6.** ASSEMBLE THE LOCK CASE, CYLINDER LOCK AND DOOR HANDLE IN THE MORTISE AND BORED HOLES.

Figure 59–53. Installing a mortised lock when no template is available.

lic buildings. A door with these hinges will automatically return to a closed position after it is opened.

A coil spring mechanism in the barrel of the hinge can be adjusted for faster or slower closing action. An Allen wrench, provided with the hinge, is inserted into a hex opening at the top of the hinge and the coil is tight-

*Schlage Lock Company*
Figure 59–54. Dead bolts greatly increase the security of an entrance door.

Schlage Lock Company

Figure 59–55. This high-security lock combines the features of a dead bolt with those of a cylindrical lock.

ened for the desired adjustment. A pin, also provided with the hinge, holds the spring in the adjusted position. See Figure 59–56.

A *double-acting floor hinge* allows the door to be opened in two directions. The bottom of the door is notched to receive the hinge, and it is fastened to the floor with a floor plate. The top of the door is held in place by a pivot piece attached to the door, which fits into a pivot socket mortised into the jamb. See Figure 59–57.

*Concealed hinges* are mortised into the door and jamb. No

Stanley Hardware division of The Stanley Works

Figure 59–56. A self-closing spring hinge is often installed on doors in public buildings.

parts of these hinges are exposed when the door is closed. For this reason, concealed hinges are tamperproof and are ideal for security purposes. See Figure 59–58.

A *flush bolt* holds the inactive door of a double door in place. The handle mechanism is mortised into the edge of the door. A hole is drilled for the bolt. The handle operates the mechanism that moves the bolt up and down.

Stanley Hardware division of The Stanley Works

## TOP PIVOT

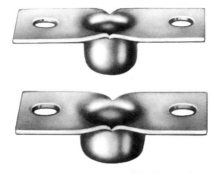

## TOP-PIVOT SOCKET

Stanley Hardware division of The Stanley Works

Figure 59–57. A double-acting floor hinge allows the door to be opened in two directions. The bottom of the door is notched to receive the hinge, and it is fastened to the floor with a floor plate. The top of the door is held in place by a pivot piece attached to the door, which fits into a pivot socket mortised into the jamb.

## Door Closers

Door closers are frequently used in schools, hospitals, and other types of public buildings. Many styles are available. One type is operated by a hydraulic spring action that automatically closes the door. See Figure 59–59. It can be adjusted for faster or slower movement.

## Exit Devices

Exit devices (also called *fire-exit bolts*) are often mounted on doors of public buildings. Slight pressure on the touch bar of an exit device will retract the latch bolts at the top and bottom of the bolts. Two types are shown in Figure 59–60.

## Door Holders and Door Stops

Door *holders* keep the door in an open position. A common type is attached to the door with screws and has a lever that drops to the floor to wedge the door open. See Figure 59–61. Another type of holder is mounted on the door and engages with a strike attached to the wall or floor. A

Stanley Hardware division of The Stanley Works

Figure 59–58. Concealed hinges are mortised into the door and jamb. No parts of the hinge can be seen when the door is closed. For this reason it is also tamper-proof and is ideal for security purposes.

plunger type of holder can be set into action or relieved by foot pressure. A spring in the plunger holds the rubber tip firmly against the floor. See Figure 59–62.

Door *stops* prevent the door from banging against the wall when fully opened. They are

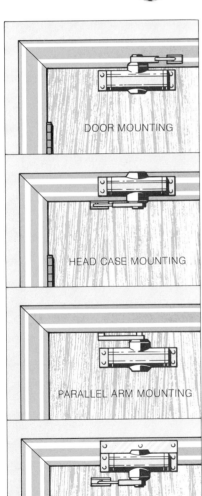

DOOR MOUNTING

HEAD CASE MOUNTING

PARALLEL ARM MOUNTING

DROP PLATE MOUNTING

*Emhart Hardware Group, Russwin Division*
Figure 59–59. This type of door closer is operated by hydraulic spring action. It can be mounted various ways.

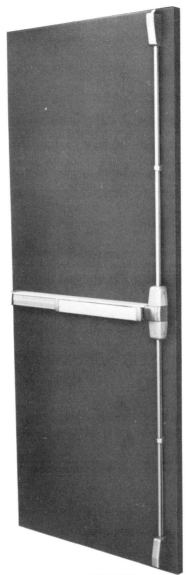

TOUCH BAR RETRACTS LATCH BOLTS AT TOP AND BOTTOM ENDS OF RODS.

TOUCH BAR RETRACTS LATCH BOLT AT SIDE OF DOOR.

*Von Duprin Exit Devices*
Figure 59–60. Exit devices of this type (often called fire-exit bolts) are usually mounted on doors of public buildings. Slight pressure on the touch bar will retract the latch bolts.

screwed into or otherwise mounted on the wall, wall baseboard, door, or floor. See Figure 59–63. A combination door stop and holder is also available. See Figure 59–64.

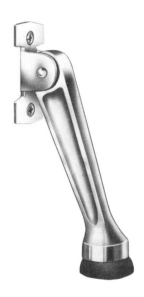

*Stanley Hardware division of The Stanley Works*
Figure 59–61. This type of door holder is screwed to the door. When the lever is dropped down to the floor, it wedges the door open.

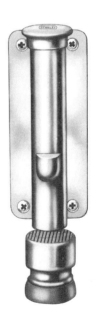

*Stanley Hardware division of The Stanley Works*
Figure 59–62. The plunger device on this door holder is set or released by foot pressure. A spring in the plunger holds the rubber tip firmly against the floor.

Figure 59-63. The door stop on the left is screwed into the floor. The one on the right is screwed into the wall, baseboard, or into the door.

Figure 59–64. This combination door stop and holder has two parts. One screws into the door and one is mounted on the wall.

## Push Plates, Kick Plates, and Door Pulls

While door stops are designed primarily to protect the wall, push plates, kick plates, and door pulls

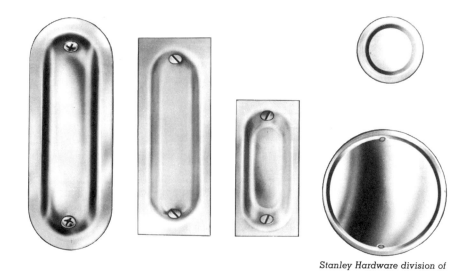

Figure 59–65. Flush pulls are mortised into sliding doors to provide a fingerhold for opening and closing the door.

are designed to protect the door. They also are an attractive addition to the door's appearance. These devices are frequently used in schools and other types of public buildings where the doors are in constant use.

Kick plates look like push plates but are mounted at the bottom of the door. Door pulls may be flush (Figure 59–65) or may have a handle that projects. Flush door pulls are mortised into sliding doors to provide a fingerhold for opening and closing the door.

# UNIT 60

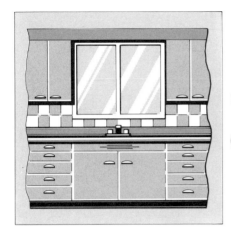

# Cabinet and Countertop Installation

Cabinets and countertops are usually constructed in a cabinet shop or factory, delivered to the job site, and installed by carpenters. In residential construction most of the cabinets and countertops used in the building are located in the kitchen (Figure 60–1) and bathroom. Elsewhere in the building, built-in bookcases, storage units, and wardrobes may be required, and these units are installed in similar fashion to kitchen or bathroom cabinets. In commercial construction such as hospitals, banks, and stores, many more cabinets and countertops are required than in residential construction.

## CONSTRUCTING CABINETS

Although carpenters are required to install, not build cabinets, they should have some knowledge of how cabinets are built. Cabinet parts are fastened together with screws or nails. Nails are usually spaced 6″ apart. They are set below the surface, and the holes are filled with putty. Glue is used at all joints. Clamps are used to produce better fitting glued joints.

A better quality cabinet is rab-beted where the top, bottom, back, and side pieces come together. See Figure 60–2. However, butt joints are also used. See Figure 60–3. If panels are less than ¾″ thick, a reinforcing block should be used with the butt joint. Fixed shelves are da-doed into the sides. See Figure 60–4.

All cabinets have a *face frame* that fits into the front of the cabinet and a *web frame* that fits on the top. See Figure 60–5. The rails and stiles of the face frame are joined by mortise-and-tenon joints or dowel joints. See Figure 60–6.

*Riviera Kitchens, an Evans Products Company*
Figure 60–1. Most cabinets and countertops used in residential construction are lo-cated in the kitchen.

Cabinet backs are best attached by rabbeting the side pieces of the cabinet. See Figure 60–7. A deeper rabbet makes it possible to scribe the side pieces

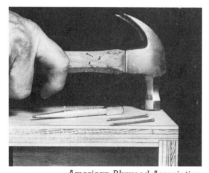

American Plywood Association

Figure 60–2. Rabbet joints make an attractive and sturdy fit for cabinet corners.

American Plywood Association

Figure 60–3. Butt joints are the simplest method of fastening cabinet corners. If plywood corners are thinner than ¾" (as shown at right), a reinforcing block should be used.

American Plywood Association

Figure 60–4. Dado joints provide good support for cabinet shelves.

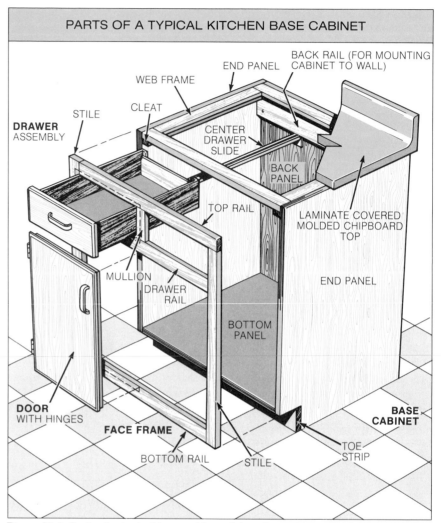

PARTS OF A TYPICAL KITCHEN BASE CABINET

BACK RAIL (FOR MOUNTING CABINET TO WALL)
END PANEL
WEB FRAME
CLEAT
STILE
DRAWER ASSEMBLY
CENTER DRAWER SLIDE
BACK PANEL
TOP RAIL
LAMINATE COVERED MOLDED CHIPBOARD TOP
MULLION
DRAWER RAIL
END PANEL
BOTTOM PANEL
DOOR WITH HINGES
FACE FRAME
BOTTOM RAIL
STILE
BASE CABINET
TOE STRIP

Figure 60–5. Parts of a typical base cabinet. A face frame fits into the front of the cabinet. A web frame fits into the top of the cabinet.

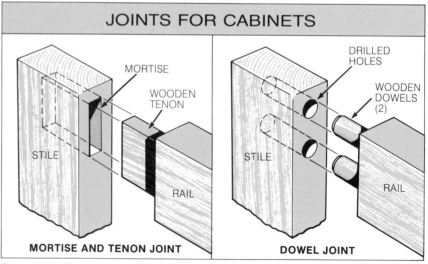

JOINTS FOR CABINETS

MORTISE
WOODEN TENON
STILE
RAIL
MORTISE AND TENON JOINT

DRILLED HOLES
WOODEN DOWELS (2)
STILE
RAIL
DOWEL JOINT

Figure 60–6. The rails and stiles of the face frame of a cabinet may be joined by mortise-and-tenon joints or by dowel joints.

against an uneven or out-of-plumb wall.

There are many methods of drawer construction. Three common methods are (1) multiple dovetail, (2) lock-shouldered, and (3) square-shouldered. See Figure 60–8.

Several types of drawer guides are available. See Figure 60–9. All can be easily installed prior to cabinet installation. Manufacturer's instructions provided with the guides give allowance dimensions for drawer openings and depth and installation procedures.

Drawer guides are designed for loads for various weights. Generally, side-mounted drawer guides with ball-bearing rollers carry more weight than bottom-mounted, single-rail drawer guides.

Wood drawer guides are seldom used in kitchen cabinets because of the construction time required. Also, drawers with wood drawer guides may stick in areas with high humidity.

## INSTALLING KITCHEN CABINETS

A kitchen usually requires more cabinets than all the other rooms of a house combined. Therefore, cabinet installation in general is best explained from the viewpoint of installing kitchen cabinets.

A typical kitchen has a series of *base* cabinets and a series of *wall* cabinets running along one or several walls of the room. Refer again to Figure 60–1. Basic principles of installation are:

1. Set cabinets plumb and level, using shim shingles where necessary. See Figure 60–10.

2. Temporarily tack a 1″ board *(cleat)* to the wall along the bottom line of the wall cabinets. The cleat provides support for the

*American Plywood Association*

Figure 60–7. Cabinet backs are best attached by rabbeting the side pieces. At left, the back piece is flush with the edge of the sides. At right, the rabbet has been made deeper so that the side piece projects about ¼″ past the back piece. This deeper rabbet makes it possible to scribe the side pieces against an uneven or out-of-plumb wall.

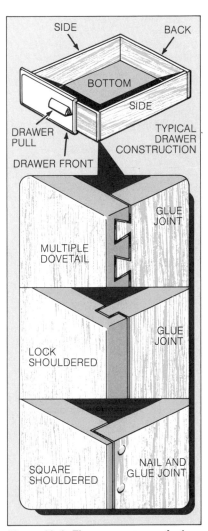

Figure 60–8. Three common methods of drawer construction.

cabinets while they are being fastened to the wall. See Figure 60–11.

3. Make up one or two *stiff legs,* which are boards slightly longer than the distance from the floor to the bottom of the cabinet. The stiff legs prevent the cabinet from tipping forward while it is being fastened to the wall. Refer again to Figure 60–11.

4. Use wood screws long enough to penetrate into the studs. Better quality cabinets have *mounting rails,* also called *hanging rails,* at the top and bottom. See Figure 60–12.

5. Some carpenters prefer to fasten a set of wall cabinets together on the floor, then hang the set as one unit on the wall. Others install one cabinet at a time on the wall. Whichever system is used, wood screws are driven through the stiles at the front frames of the cabinets to fasten the cabinets to each other. C-clamps are used to temporarily hold the frames tightly together. See Figure 60–13.

A typical kitchen cabinet detail drawing, taken from the Three-bedroom House Plan in Section 6, is shown in Figure 60–14. The drawing gives the widths and heights of the cabinets and the height at which the wall cabinets are to be placed.

The floor plan from Section 6 shows that the cabinets go against the north kitchen wall. This wall includes a window and space for a refrigerator.

It is good practice to check the combined widths of the cabinets against the total distance of the wall on which they are to be mounted. This can be easily done by cutting a rod the exact distance of the wall and marking on the rod all the cabinet widths and the width of any windows or any space provided for other purposes (such as an appliance). Figure 60–15 shows a procedure for installing kitchen cabinets.

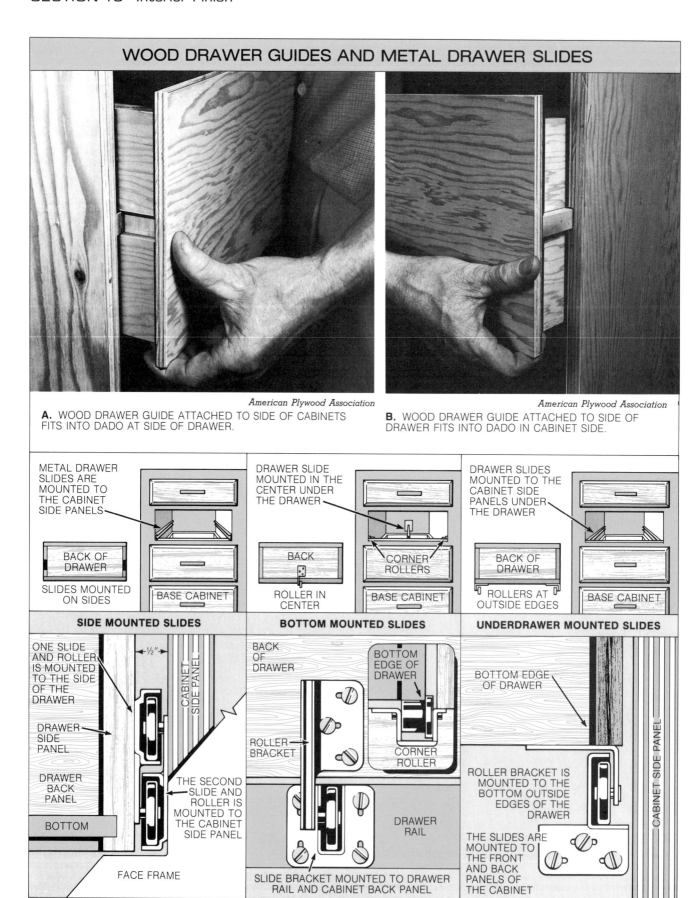

## WOOD DRAWER GUIDES AND METAL DRAWER SLIDES

*American Plywood Association*

**A.** WOOD DRAWER GUIDE ATTACHED TO SIDE OF CABINETS FITS INTO DADO AT SIDE OF DRAWER.

*American Plywood Association*

**B.** WOOD DRAWER GUIDE ATTACHED TO SIDE OF DRAWER FITS INTO DADO IN CABINET SIDE.

METAL DRAWER SLIDES ARE MOUNTED TO THE CABINET SIDE PANELS

BACK OF DRAWER

SLIDES MOUNTED ON SIDES

BASE CABINET

DRAWER SLIDE MOUNTED IN THE CENTER UNDER THE DRAWER

BACK

CORNER ROLLERS

ROLLER IN CENTER

BASE CABINET

DRAWER SLIDES MOUNTED TO THE CABINET SIDE PANELS UNDER THE DRAWER

BACK OF DRAWER

ROLLERS AT OUTSIDE EDGES

BASE CABINET

### SIDE MOUNTED SLIDES

ONE SLIDE AND ROLLER IS MOUNTED TO THE SIDE OF THE DRAWER

½"

CABINET SIDE PANEL

DRAWER SIDE PANEL

DRAWER BACK PANEL

BOTTOM

THE SECOND SLIDE AND ROLLER IS MOUNTED TO THE CABINET SIDE PANEL

FACE FRAME

### BOTTOM MOUNTED SLIDES

BACK OF DRAWER

BOTTOM EDGE OF DRAWER

ROLLER BRACKET

CORNER ROLLER

DRAWER RAIL

SLIDE BRACKET MOUNTED TO DRAWER RAIL AND CABINET BACK PANEL

### UNDERDRAWER MOUNTED SLIDES

BOTTOM EDGE OF DRAWER

ROLLER BRACKET IS MOUNTED TO THE BOTTOM OUTSIDE EDGES OF THE DRAWER

CABINET SIDE PANEL

THE SLIDES ARE MOUNTED TO THE FRONT AND BACK PANELS OF THE CABINET

Figure 60–9. Drawer guides used in cabinets. Photograph at left shows a dado cut in the drawer and a guide strip fastened to the side of the cabinet. Photograph at right shows a dado cut in the side of the cabinet and a guide strip fastened to the drawer.

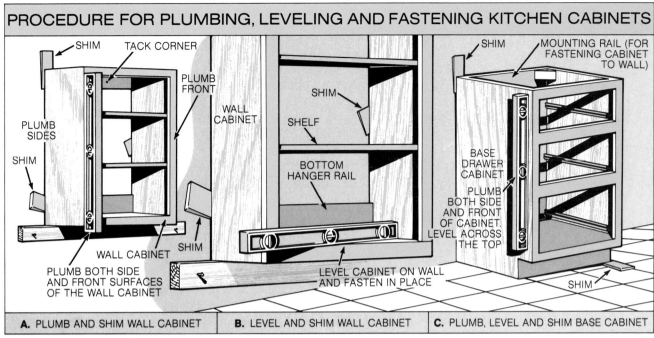

PROCEDURE FOR PLUMBING, LEVELING AND FASTENING KITCHEN CABINETS

A. PLUMB AND SHIM WALL CABINET    B. LEVEL AND SHIM WALL CABINET    C. PLUMB, LEVEL AND SHIM BASE CABINET

Figure 60-10. Shim shingles are used where necessary to plumb and level cabinets.

## Cabinet Doors

Cabinet doors may be hung in the cabinet shop or by carpenters on the job. If cabinets are racked out of square during transportation or installation, pre-hung doors may be thrown out of alignment. For this reason, many carpenters prefer to hang cabinet doors on the job.

The types of doors commonly used on cabinets are *flush, lipped, overlay,* and *sliding* doors. See Figure 60-16. Various types of hinges and other hardware are available for each type of door.

**Flush Doors.** Flush doors are similar to the other swinging

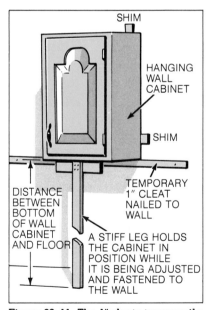

Figure 60-11. The 1" cleats temporarily tacked to the wall, and the stiff leg, hold the cabinet in place while it is adjusted and fastened to the wall.

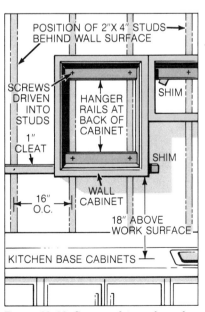

Figure 60-12. Screws, driven through hanger rails (mounting rails), fasten the cabinets to the walls.

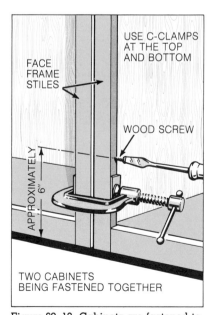

Figure 60-13. Cabinets are fastened to each other with wood screws driven through the face frame stiles. C-clamps are used to temporarily hold them tightly together.

## DRAWING PROVIDES INFORMATION FOR LAYOUT AND PLACEMENT

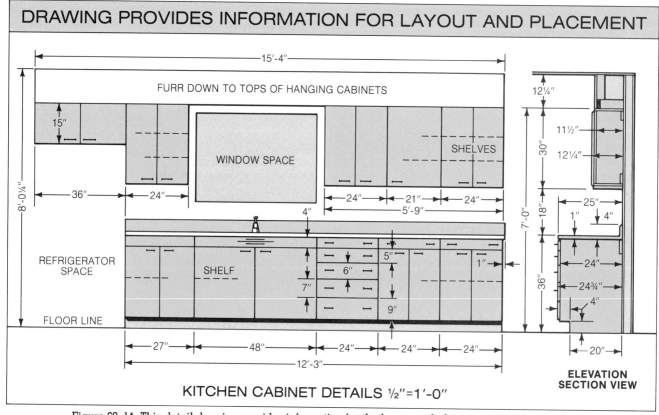

Figure 60–14. This detail drawing provides information for the layout and placement of kitchen cabinets.

doors and are the most difficult to hang. They must be fitted in the cabinet opening with $\frac{1}{16}''$ clearance around all four edges.

An ornamental *surface* hinge, a *concealed* hinge, or a *semi-concealed,* loose-pin hinge may be used on a flush cabinet door. A semi-concealed, loose-pin hinge looks like a regular loose-pin hinge when the door is closed, but its design allows greater holding power. See Figure 60–17.

**Lipped Doors.** Lipped doors are simpler to install than flush doors, since the lip feature allows for a certain amount of adjustment. A semi-concealed hinge is often used with lipped doors. See Figure 60–18. The hinge may be self-closing or may require a catch to hold the door in a closed position.

**Overlay Doors.** Overlay doors are designed to cover the edges of the face frame. Semi-concealed pivot hinges are often used with this type of door. See Figure 60–19. A small-angle cut is required where they are placed at the top and bottom edges of the door. In addition, a special type of pivot hinge can be placed at the center of a larger door.

**Sliding Doors.** Several types of sliding doors are used on cabinets. One type of sliding door is rabbeted to fit into grooves at the top and bottom of the cabinet. See Figure 60–20. A thinner type of door slides in a plastic track set into the bottom of the cabinet. See Figure 60–21. The top groove is always made to allow the door to be removed by lifting it up and pulling the bottom

out. In another type of sliding door, a T-shaped device guides the door, which is grooved on its bottom edge, along a track.

**Door Pulls and Knobs.** Door pulls and knobs for cabinets are available in many designs and finishes. See Figure 60–22. Holes must be accurately laid out for the machine screws (or bolts) used to fasten the knobs and pulls. To prevent the wood of a cabinet from splitting when a drill bit goes through it, clamp a block to the back of the piece being drilled. See Figure 60–23.

**Door Catches.** Door catches are necessary to hold the door in place when it is shut, unless self-closing hinges are used on the door. Door catches operate by means of rubber rollers, friction, or magnets. See Figure 60–24.

# LAYING OUT AND INSTALLING KITCHEN WALL AND BASE CABINETS

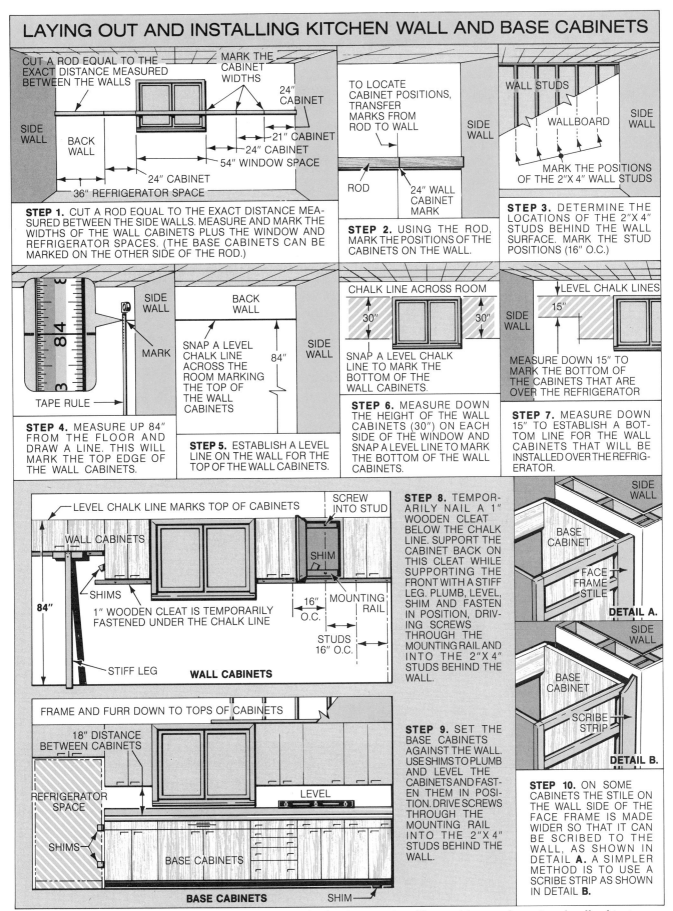

**STEP 1.** CUT A ROD EQUAL TO THE EXACT DISTANCE MEASURED BETWEEN THE SIDE WALLS. MEASURE AND MARK THE WIDTHS OF THE WALL CABINETS PLUS THE WINDOW AND REFRIGERATOR SPACES. (THE BASE CABINETS CAN BE MARKED ON THE OTHER SIDE OF THE ROD.)

**STEP 2.** USING THE ROD, MARK THE POSITIONS OF THE CABINETS ON THE WALL.

**STEP 3.** DETERMINE THE LOCATIONS OF THE 2"X 4" STUDS BEHIND THE WALL SURFACE. MARK THE STUD POSITIONS (16" O.C.)

**STEP 4.** MEASURE UP 84" FROM THE FLOOR AND DRAW A LINE. THIS WILL MARK THE TOP EDGE OF THE WALL CABINETS.

**STEP 5.** ESTABLISH A LEVEL LINE ON THE WALL FOR THE TOP OF THE WALL CABINETS.

**STEP 6.** MEASURE DOWN THE HEIGHT OF THE WALL CABINETS (30") ON EACH SIDE OF THE WINDOW AND SNAP A LEVEL LINE TO MARK THE BOTTOM OF THE WALL CABINETS.

**STEP 7.** MEASURE DOWN 15" TO ESTABLISH A BOTTOM LINE FOR THE WALL CABINETS THAT WILL BE INSTALLED OVER THE REFRIGERATOR.

**STEP 8.** TEMPORARILY NAIL A 1" WOODEN CLEAT BELOW THE CHALK LINE. SUPPORT THE CABINET BACK ON THIS CLEAT WHILE SUPPORTING THE FRONT WITH A STIFF LEG. PLUMB, LEVEL, SHIM AND FASTEN IN POSITION, DRIVING SCREWS THROUGH THE MOUNTING RAIL AND INTO THE 2"X 4" STUDS BEHIND THE WALL.

**STEP 9.** SET THE BASE CABINETS AGAINST THE WALL. USE SHIMS TO PLUMB AND LEVEL THE CABINETS AND FASTEN THEM IN POSITION. DRIVE SCREWS THROUGH THE MOUNTING RAIL INTO THE 2"X 4" STUDS BEHIND THE WALL.

**STEP 10.** ON SOME CABINETS THE STILE ON THE WALL SIDE OF THE FACE FRAME IS MADE WIDER SO THAT IT CAN BE SCRIBED TO THE WALL, AS SHOWN IN DETAIL **A**. A SIMPLER METHOD IS TO USE A SCRIBE STRIP AS SHOWN IN DETAIL **B**.

Figure 60–15. Installing kitchen cabinets. Typically, a kitchen has a series of base cabinets and a series of wall cabinets.

## INSTALLING OTHER CABINETS AND SIMILAR UNITS

Wardrobe closets, built-in bookcases, and storage cabinets are installed about the same way kitchen cabinets are installed. These permanently attached units often serve as room dividers as well. For example, a long base storage cabinet may divide

Figure 60–18. A semi-concealed hinge is often used for lipped cabinet doors.

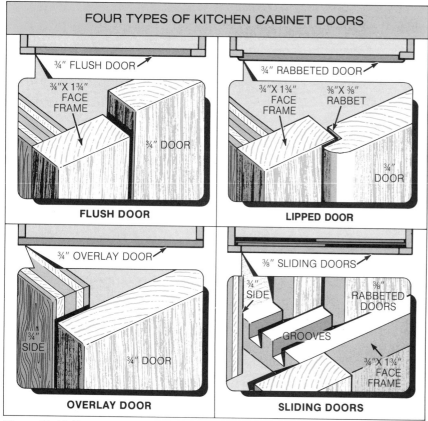

FOUR TYPES OF KITCHEN CABINET DOORS

¾" FLUSH DOOR

¾"X 1¾" FACE FRAME

¾" DOOR

**FLUSH DOOR**

¾" RABBETED DOOR

¾"X 1¾" FACE FRAME

⅜"X ⅜" RABBET

¾" DOOR

**LIPPED DOOR**

¾" OVERLAY DOOR

¾" SIDE

¾" DOOR

**OVERLAY DOOR**

⅜" SLIDING DOORS

¾" SIDE

⅜" RABBETED DOORS

GROOVES

¾"X 1¾" FACE FRAME

**SLIDING DOORS**

Figure 60–16. Four types of doors often used for base and wall cabinets. Various types of hardware are available for each type of door.

Figure 60–19. A semi-concealed pivot hinge is often used on overlay cabinet doors.

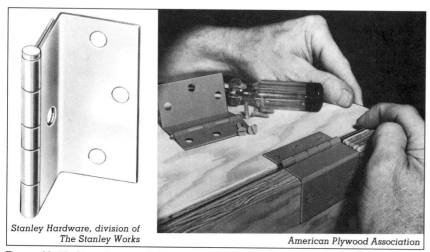

Figure 60–17. A semi-concealed, loose-pin hinge gives the appearance of a regular loose-pin hinge when the door is closed. The offset allows the door side of the hinge to be screwed into the flat plywood grain, which gives greater holding power.

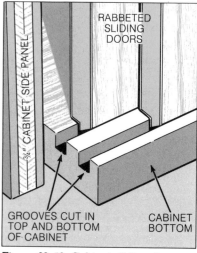

RABBETED SLIDING DOORS

¾" CABINET SIDE PANEL

GROOVES CUT IN TOP AND BOTTOM OF CABINET

CABINET BOTTOM

Figure 60–20. Cabinet sliding doors at least ¾" thick may be rabbeted to slide in grooves at the top and bottom of the cabinet.

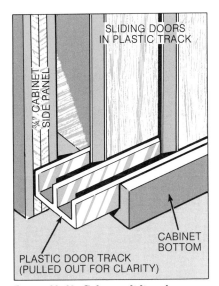

Figure 60–21. Cabinet sliding doors thinner than ¾″ thick may be inserted in plastic tracks.

the work area of the kitchen from the dining room.

Sometimes these built-in units cannot be securely fastened to a wall the way kitchen cabinets can. One method of securing such a unit is to nail 2 × 4s to the floor inside the unit, then drive screws through the toekicks (toeboards) into the 2 × 4s. See Figure 60–25.

## INSTALLING COUNTERTOPS

Countertops for kitchen and bathroom cabinets are set in place after the base cabinets have been secured. Plastic laminate is used most frequently as the surfacing material. It is hard, durable, highly resistant to heat, and cannot be easily stained or damaged.

As a rule, countertops are constructed in a shop and delivered to the job. They usually come with a *backsplash,* which is a piece attached to the back edges of the countertop. The backsplash rests against the wall when the countertop is installed. See Figure 60–26.

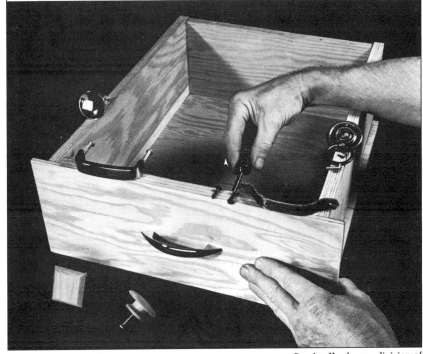

Stanley Hardware, division of the Stanley Works

Figure 60–22. Door pulls and knobs for cabinets are also used on cabinet drawers.

### Applying Plastic Laminate

Plastic laminate for countertops is normally about ¹⁄₁₆″ thick. It

can be cut with a table saw, radial-arm saw, electric handsaw, or saber saw. See Figure 60–27. A fine-toothed blade must be used. If no power tool is avail-

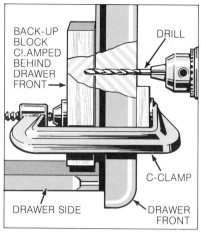

Figure 60–23. A block clamped to the back of the cabinet piece being drilled will prevent the wood from splitting.

able, a finish handsaw or hacksaw can be used.

Contact cement is used to bond laminate to its base core, which is usually ¾″ plywood or particleboard. Cement is applied to the back of the laminate and to the surface of the wood. An animal-hair paintbrush or short-nap roller can be used. Although manufacturers' instructions vary slightly, 15 to 20 minutes is usually an adequate drying period under normal conditions. To test for dryness, lightly press a piece

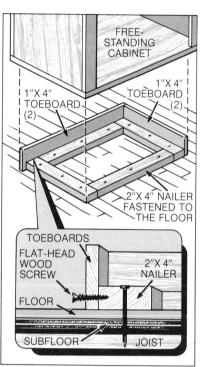

Figure 60–25. A free-standing cabinet can be secured by driving screws through the toekicks (toeboards) into 2 x 4s fastened to the floor.

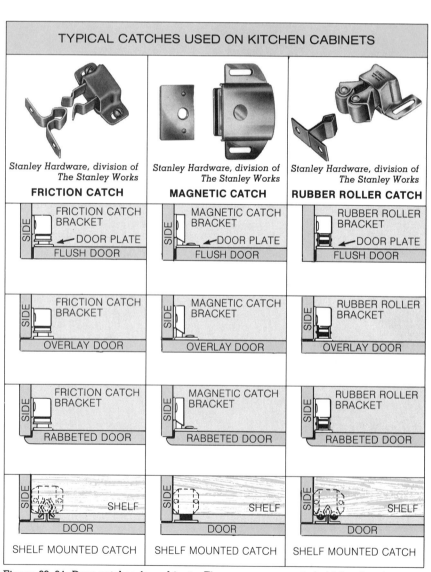

Figure 60–24. Door catches for cabinets. They are necessary to hold the door in place when it is shut, unless self-closing hinges are used on the door.

of kraft paper against the adhesive. If no glue adheres to the paper, the surfaces are ready for bonding.

Plastic laminate must be positioned very carefully when being bonded to wood. Once the glued surface of the laminate contacts the glued surface of the wood, it can no longer be moved. To ensure correct positioning before bonding occurs, place sticks on the wood at intervals, then place the laminate over the sticks. When the laminate is correctly positioned, pull the sticks out one at a time. The laminate can be either rolled or tapped down. Edge strips are applied first, then the top sheet.

Plastic laminate is always cut slightly larger than the base core. After it is bonded to the core, the laminate is trimmed with a router equipped with a special bit. A laminate trimmer (special-purpose router) may be used to both cut and bevel the edges.

Figure 60–26. Plastic laminate countertop with backsplash. The hole for the sink is cut out when the countertop is installed.

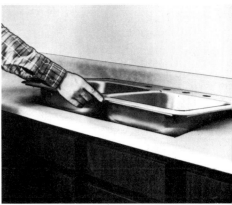

*Formica Corporation*

Figure 60–28. Setting the sink unit in the counter cutout. Waterproof caulking should be placed around the edge of the opening before the sink unit is placed. It will then be secured with hold-down lugs at the bottom surface of the counter.

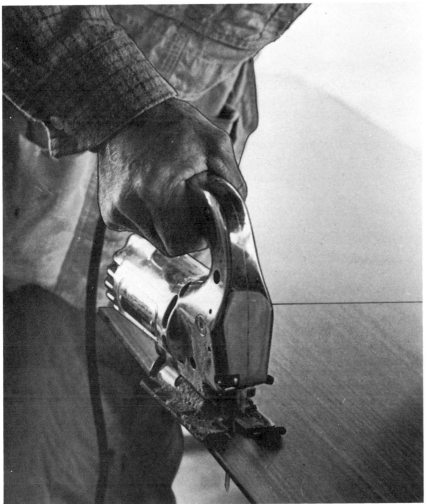

*Formica Corporation*

Figure 60–27. Plastic laminate may be cut with a saber saw. A fine-toothed blade must be used.

## Cutting Opening for Sink

Carpenters on the job often cut the opening for the sink. The opening can be pre-cut at the factory, but it is better practice to delay cutting until the sink unit has been delivered to the job site. Then the exact size of the opening required can be marked on the countertop.

After the opening is cut out, waterproof caulking is placed around the edge of the opening and the sink unit is set in place. The sink is then secured with hold-down lugs at the bottom surface of the counter. See Figure 60–28.

## Attaching Backsplash

The pieces that make up the backsplash are often fastened to the countertop when it is constructed in the shop. Some carpenters, however, prefer to fasten the backsplash to the countertop on the job so that they can make adjustments if the walls are out of square.

If the wall behind the backsplash is very irregular, the laminate strip at the top edge of the

545

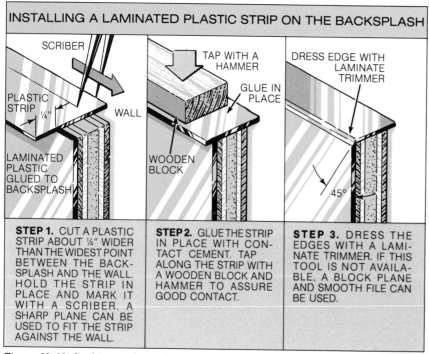

Figure 60–29. Scribing and attaching the top edge of the backsplash to the counter-top.

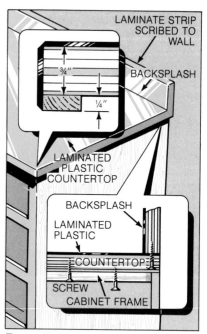

Figure 60–30. Countertops are attached by driving screws up from 1″ thick pieces attached to the tops of the base cabinets.

backsplash may require scribing. For this reason, this strip is sometimes omitted. Figure 60–29 shows a procedure for scribing and attaching this strip after the backsplash is secured to the countertop.

## Attaching Countertops

A base cabinet usually has a narrow 1″ frame at the top or a triangular block at each corner. To attach the countertop, drill holes through these pieces and drive wood screws up into the countertop. See Figure 60–30. Be certain that the screws are the right length. The countertop will be permanently damaged if the screws are too long and penetrate the surface.

# UNIT 61

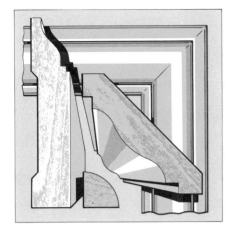

## Interior Trim

Interior trim is the molding material that completes the finish around the doors, windows, cabinets, and tops and bottoms of walls. The molding should harmonize with the building's overall design. For modern buildings, less molding, in simpler patterns, is appropriate. For buildings of traditional design, a greater amount of molding, in more elaborate patterns, is desirable.

Wood is the major material used for interior trim in residential construction. In commercial construction, plastic or metal trim is often used. Wood molding is manufactured from both softwood and hardwood lumber. A strip of lumber is resawn into a

*blank,* which is a piece that will produce the desired pattern of molding with the least possible amount of waste. See Figure 61–1. The blank is shaped into molding on a machine called a *molder* (or *sticker*). The molder is equipped with cutter heads and knives that rotate at high speed to create the molding pattern.

Examples of where molding is placed are shown in Figure 61–2. The *casing* goes around the doors and windows. The interior of the sill of a traditional window is finished off with a *stool* and *apron*. Double windows often have a *mullion* between the two sections of the window.

*Base* molding is placed at the bottoms of the walls. *Ceiling* molding may be placed at the tops of the walls, although it is often omitted. If paneling is used on the walls, *corner* molding is frequently used to cover the inside and outside corners.

Molding is usually fastened with finish nails (staples are sometimes used) that must be set below the surface. The nail holes are later filled with putty that matches the paint or stain finish. In the case of pre-finished molding, nails of a matching color are used, and they are not set below the surface.

## TRIMMING DOOR OPENINGS

Door casing is available in many designs. See Figure 61–3. The material is usually backed out. Backing out the material produces a tight fit even if there is some unevenness between the jamb and the wall.

When nailed to the jamb, the casing should be held back ¼″ from the edge. This ¼″ space is called a *reveal.* It creates a better appearance and puts the cas-

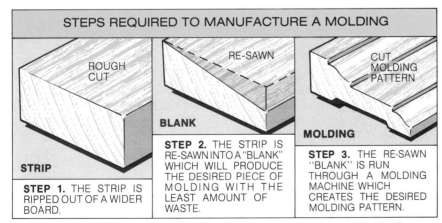

Figure 61–1. Manufacture of wood molding. Both softwood and hardwood lumber are used.

## VARIOUS MOLDINGS USED FOR INTERIOR TRIM

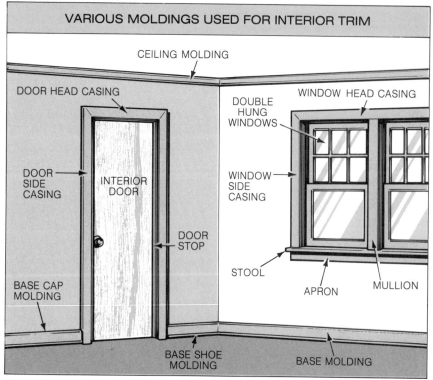

CEILING MOLDING

DOOR HEAD CASING

DOUBLE HUNG WINDOWS

WINDOW HEAD CASING

DOOR SIDE CASING

INTERIOR DOOR

WINDOW SIDE CASING

DOOR STOP

BASE CAP MOLDING

STOOL

APRON

MULLION

BASE SHOE MOLDING

BASE MOLDING

Figure 61–2. Places where molding is used for interior trim. It also may be used on inside and outside wall corners if wall paneling is used.

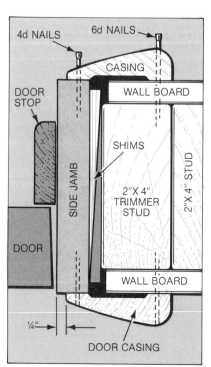

4d NAILS

6d NAILS

CASING

DOOR STOP

WALL BOARD

SIDE JAMB

SHIMS

2″X 4″ STUD

2″X 4″ TRIMMER STUD

DOOR

WALL BOARD

¼″

DOOR CASING

Figure 61–4. Section view of casing nailed in place.

## EXAMPLES OF CONTEMPORARY AND TRADITIONAL CASINGS

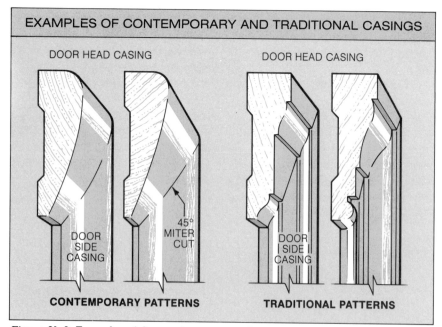

DOOR HEAD CASING

DOOR HEAD CASING

45° MITER CUT

DOOR SIDE CASING

DOOR SIDE CASING

**CONTEMPORARY PATTERNS**

**TRADITIONAL PATTERNS**

Figure 61–3. Examples of door casing patterns. Door casing is usually backed out to produce a tight fit.

Figure 61–5. A 45° miter cut is usually required at the joints between the top piece and the side pieces of casing.

ing out of the way of the door hinges. When tapered casing is used, the narrow edge is nailed to the jamb with 4d or 6d nails and the other edge to the wall studs with 8d nails. See Figure 61–4.

When fastening the side pieces of casing, begin nailing from the top and work down. If the piece must be straightened, nail into the trimmer first, then drive the adjoining nail into the jamb. Space the nails 16″ O.C.

A 45° miter cut is usually required at the joints between the top piece and the side pieces. See Figure 61–5. This cut can be laid out with a combination square or, more quickly, with a *miter box*. Manufactured steel miter boxes are often furnished by the carpenter's employer. (Refer to Section 3.) An even more efficient tool is the *motorized* miter box. (Refer to Section 4.)

If a steel miter box is not available, a wood miter box can easily be constructed on the job. Figure 61–6 shows a procedure for constructing a wood miter box. Use 1″ thick hardwood, if available, for the sides, and 2″ × 4″ or 2″ × 6″ material for the bottom. Cut slots at 45° angles toward the two ends of the miter box. Cut a 90° slot at the center.

If a miter joint between the head and side pieces of the door casing does not fit properly, use a block plane to get a better fit. When nailing the casing to the jamb, also drive a nail down from the top piece into the side pieces at the miter joint. The nail will help prevent the joint from opening up later on. Applying glue at the joint is also helpful. A procedure for trimming the door opening is shown in Figure 61–7.

## TRIMMING WINDOW OPENINGS

Window casing should be the same material and design as door casing. Most windows set in wood frames are trimmed in either *traditional* or *contemporary (picture-frame)* style. See Figure 61–8. Many aluminum window units have no casing. See Figure 61–9.

The procedure for placing casing around a picture-frame window is similar to that for a door. The only difference is that the window requires a fourth piece fitted at the bottom of the frame which is not required in trimming a door.

Traditional windows require a rabbeted stool before the casing can be nailed. Some stool designs are shown in Figure 61–10. A procedure for trimming a tradi-

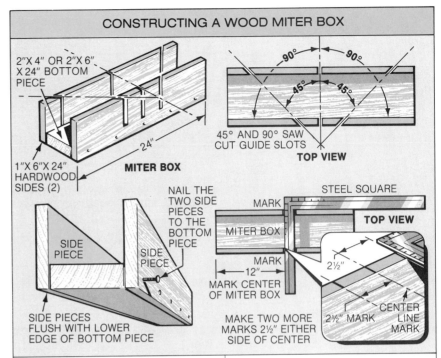

### CONSTRUCTING A WOOD MITER BOX

2″X 4″ OR 2″X 6″ X 24″ BOTTOM PIECE

24″

1″X 6″X 24″ HARDWOOD SIDES (2)

**MITER BOX**

90° 90° 45° 45°

45° AND 90° SAW CUT GUIDE SLOTS

**TOP VIEW**

NAIL THE TWO SIDE PIECES TO THE BOTTOM PIECE

SIDE PIECE

SIDE PIECE

SIDE PIECES FLUSH WITH LOWER EDGE OF BOTTOM PIECE

STEEL SQUARE

MARK

MITER BOX

**TOP VIEW**

MARK

12″

MARK CENTER OF MITER BOX

2½″

2½″ MARK

CENTER LINE MARK

MAKE TWO MORE MARKS 2½″ EITHER SIDE OF CENTER

**STEP 1.** NAIL THE TWO SIDE PIECES TO THE BOTTOM PIECE. USE PERFECTLY STRAIGHT PIECES, AND BE SURE THAT THE BOTTOMS OF THE SIDE PIECES ARE FLUSH WITH THE LOWER EDGE OF THE BOTTOM PIECE.

**STEP 2.** POSITION THE STEEL SQUARE AS ILLUSTRATED AND MARK A 90° ANGLE AT THE CENTER OF THE MITER BOX (12″ FROM EITHER END). MEASURE AND MARK TWO ADDITIONAL LINES 2½″ ON EITHER SIDE OF THE CENTER LINE MARK.

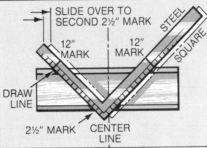

SLIDE OVER TO SECOND 2½″ MARK

STEEL SQUARE

12″ MARK

12″ MARK

DRAW LINE

2½″ MARK

CENTER LINE

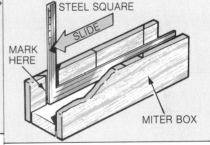

STEEL SQUARE

SLIDE

MARK HERE

MITER BOX

**STEP 3.** TO MARK THE 45° ANGLE CUT, LINE UP THE SAME NUMBER ON THE TONGUE AND BLADE OF THE STEEL SQUARE ALONG ONE EDGE OF THE SIDE PIECE AND THE 2½″ MARK ON THE OTHER SIDE PIECE. (IN THIS EXAMPLE THE NUMBERS 12 AND 12 ARE USED.)

**STEP 4.** REST THE STEEL SQUARE ON THE BOTTOM OF THE MITER BOX. SQUARE LINES DOWN FROM THE 45° AND 90° ANGLE MARKS AT THE INSIDE OF THE MITER BOX.

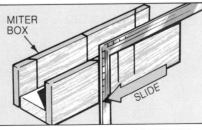

MITER BOX

SLIDE

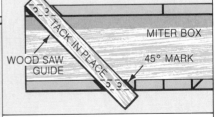

TACK IN PLACE

MITER BOX

WOOD SAW GUIDE

45° MARK

**STEP 5.** REST THE STEEL SQUARE ON THE TOP EDGE OF THE SIDE PIECE. SQUARE LINES DOWN AT THE OUTSIDE OF THE MITER BOX.

**STEP 6.** BEFORE SAWING A SLOT, TEMPORARILY TACK A STRAIGHT PIECE OF 1″ THICK MATERIAL AGAINST THE LINE TO BE CUT. THIS WILL HELP GUIDE THE SAW WHEN BEGINNING THE CUT.

Figure 61–6. Job-built wood miter box. Use 1″ thick hardwood, if available, for the sides, and a 2″ x 4″ or 2″ x 6″ piece of lumber for the bottom. Slots at 45° angles are cut toward the two ends of the miter box. A 90° slot is cut at the center.

tional window is shown in Figure 61–11. The stool has a 45° miter joint at each end to make a 90°

return. The return covers the exposed grain on the end of the stool.

## BASE MOLDING

Base molding *(baseboard)* is held tightly to the floor and fastened to the wall with 6d or 8d

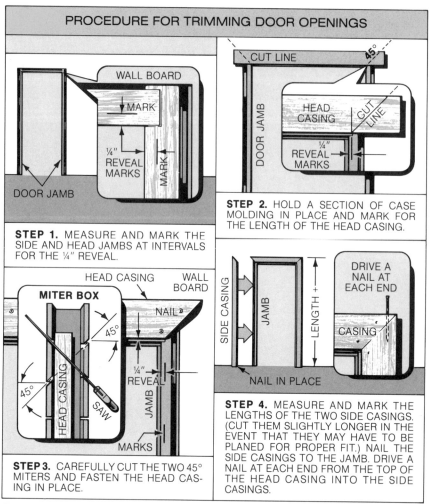

### PROCEDURE FOR TRIMMING DOOR OPENINGS

**STEP 1.** MEASURE AND MARK THE SIDE AND HEAD JAMBS AT INTERVALS FOR THE ¼″ REVEAL.

**STEP 2.** HOLD A SECTION OF CASE MOLDING IN PLACE AND MARK FOR THE LENGTH OF THE HEAD CASING.

**STEP 3.** CAREFULLY CUT THE TWO 45° MITERS AND FASTEN THE HEAD CASING IN PLACE.

**STEP 4.** MEASURE AND MARK THE LENGTHS OF THE TWO SIDE CASINGS. (CUT THEM SLIGHTLY LONGER IN THE EVENT THAT THEY MAY HAVE TO BE PLANED FOR PROPER FIT.) NAIL THE SIDE CASINGS TO THE JAMB. DRIVE A NAIL AT EACH END FROM THE TOP OF THE HEAD CASING INTO THE SIDE CASINGS.

Figure 61–7. Trimming a door opening. The first step is to mark the jamb at intervals for a ¼″ reveal.

### CONTEMPORARY AND TRADITIONAL WINDOW TRIM

Figure 61–8. Interior window trim for windows in wood frames may be either traditional or contemporary (picture-frame) in design. Window trim should match door trim in the building.

Figure 61–9. Aluminum window set in a metal frame. Only a sill piece and apron are used to trim the opening.

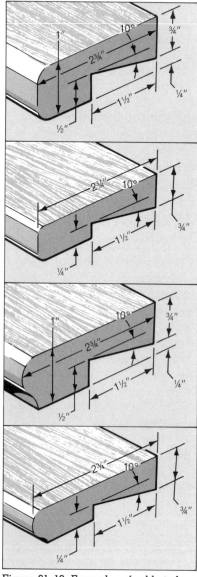

Figure 61–10. Examples of rabbeted stool patterns. Traditional window trim requires a rabbeted stool.

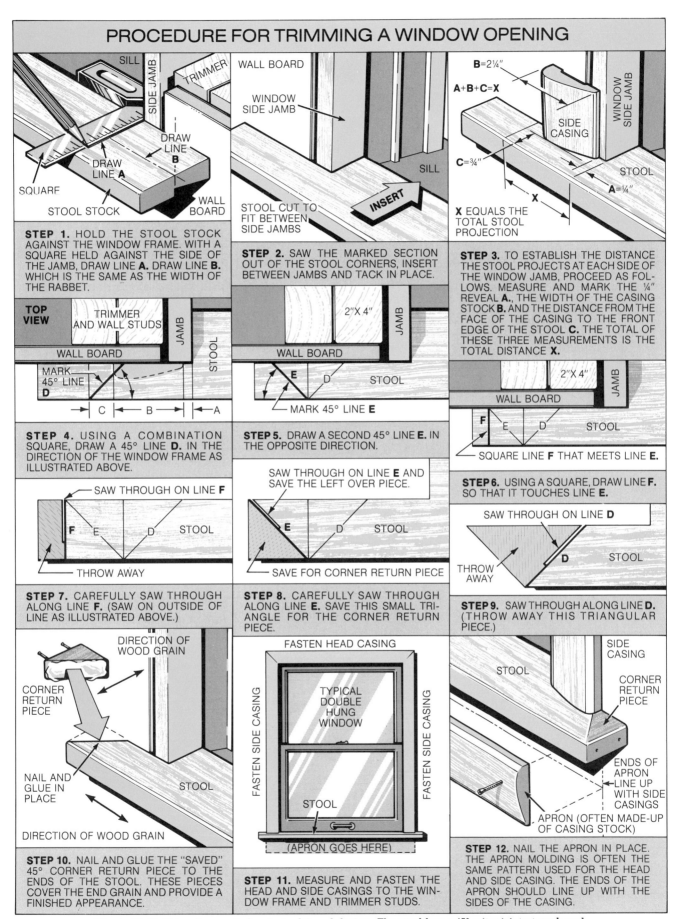

# PROCEDURE FOR TRIMMING A WINDOW OPENING

**STEP 1.** HOLD THE STOOL STOCK AGAINST THE WINDOW FRAME. WITH A SQUARE HELD AGAINST THE SIDE OF THE JAMB, DRAW LINE **A**. DRAW LINE **B**. WHICH IS THE SAME AS THE WIDTH OF THE RABBET.

**STEP 2.** SAW THE MARKED SECTION OUT OF THE STOOL CORNERS, INSERT BETWEEN JAMBS AND TACK IN PLACE.

**STEP 3.** TO ESTABLISH THE DISTANCE THE STOOL PROJECTS AT EACH SIDE OF THE WINDOW JAMB, PROCEED AS FOLLOWS. MEASURE AND MARK THE ¼" REVEAL **A**., THE WIDTH OF THE CASING STOCK **B**. AND THE DISTANCE FROM THE FACE OF THE CASING TO THE FRONT EDGE OF THE STOOL **C**. THE TOTAL OF THESE THREE MEASUREMENTS IS THE TOTAL DISTANCE **X**.

**STEP 4.** USING A COMBINATION SQUARE, DRAW A 45° LINE **D**. IN THE DIRECTION OF THE WINDOW FRAME AS ILLUSTRATED ABOVE.

**STEP 5.** DRAW A SECOND 45° LINE **E**. IN THE OPPOSITE DIRECTION.

**STEP 6.** USING A SQUARE, DRAW LINE **F**. SO THAT IT TOUCHES LINE **E**.

**STEP 7.** CAREFULLY SAW THROUGH ALONG LINE **F**. (SAW ON OUTSIDE OF LINE AS ILLUSTRATED ABOVE.)

**STEP 8.** CAREFULLY SAW THROUGH ALONG LINE **E**. SAVE THIS SMALL TRIANGLE FOR THE CORNER RETURN PIECE.

**STEP 9.** SAW THROUGH ALONG LINE **D**. (THROW AWAY THIS TRIANGULAR PIECE.)

**STEP 10.** NAIL AND GLUE THE "SAVED" 45° CORNER RETURN PIECE TO THE ENDS OF THE STOOL. THESE PIECES COVER THE END GRAIN AND PROVIDE A FINISHED APPEARANCE.

**STEP 11.** MEASURE AND FASTEN THE HEAD AND SIDE CASINGS TO THE WINDOW FRAME AND TRIMMER STUDS.

**STEP 12.** NAIL THE APRON IN PLACE. THE APRON MOLDING IS OFTEN THE SAME PATTERN USED FOR THE HEAD AND SIDE CASING. THE ENDS OF THE APRON SHOULD LINE UP WITH THE SIDES OF THE CASING.

Figure 61–11. Trimming a window in traditional design. The stool has a 45° miter joint at each end.

551

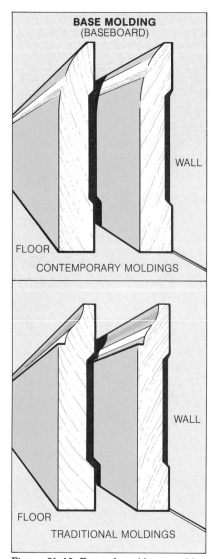

**BASE MOLDING**
(BASEBOARD)

WALL

FLOOR

CONTEMPORARY MOLDINGS

WALL

FLOOR

TRADITIONAL MOLDINGS

Figure 61–12. Examples of base molding patterns.

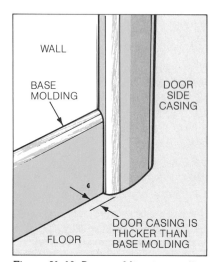

WALL

BASE MOLDING

DOOR SIDE CASING

DOOR CASING IS THICKER THAN BASE MOLDING

FLOOR

Figure 61–13. Base molding is usually thinner than the door casing it butts against.

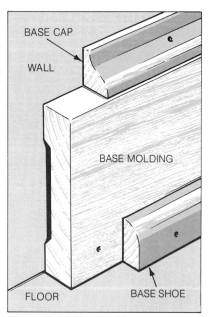

BASE CAP

WALL

BASE MOLDING

FLOOR

BASE SHOE

Figure 61–14. Base molding is sometimes applied with a base shoe and base cap.

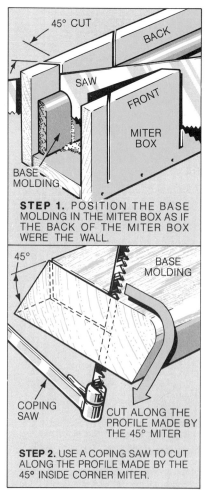

45° CUT

BACK

SAW

FRONT

MITER BOX

BASE MOLDING

**STEP 1.** POSITION THE BASE MOLDING IN THE MITER BOX AS IF THE BACK OF THE MITER BOX WERE THE WALL.

45°

BASE MOLDING

COPING SAW

CUT ALONG THE PROFILE MADE BY THE 45° MITER

**STEP 2.** USE A COPING SAW TO CUT ALONG THE PROFILE MADE BY THE 45° INSIDE CORNER MITER.

Figure 61–15. Coping the base molding for an inside corner fit.

nails driven into the bottom plate and stud. Some baseboard patterns are shown in Figure 61–12.

The base is installed after the door casing has been nailed in place. Base material is usually thinner than the outside edge of the door casing it butts against. See Figure 61–13. A *base shoe* is sometimes nailed into the floor at the bottom of the base molding. See Figure 61–14. It helps conceal any unevenness between the base and the floor. A decorative *base cap* may also be used if molding is traditional in style. Refer again to Figure 61–14.

Miter joints are required at the outside corners of base molding but are not suitable for the inside corners. At the inside corners, miter joints tend to open when being nailed. Later shrinkage of the wood also may cause inside corner miter joints to open. For this reason, a *coped* joint is recommended for inside corners. A coped piece is cut to the shape of the piece it fits against. Figure 61–15 shows a procedure for coping.

On long walls where several pieces of baseboard are required, the joints between the pieces should fall over a stud. The best type of fit is a *scarf joint,* which is made by overlapping two 45° angles. See Figure 61–16. Figure 61–17 shows a procedure for placing baseboard in a room with inside and outside corners.

## CEILING MOLDING

Ceiling molding is available in three basic shapes: *rectangular* with beveled edges, *cove,* or *crown.* See Figure 61–18.

The procedure for placing ceiling molding is similar to that for base molding. Inside corners are usually coped and outside cor-

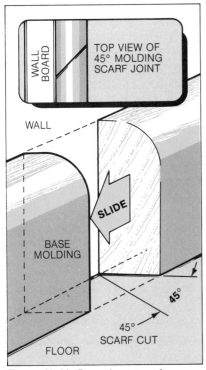

Figure 61-16. A scarf joint produces the best fit between two pieces of base molding joined along a wall.

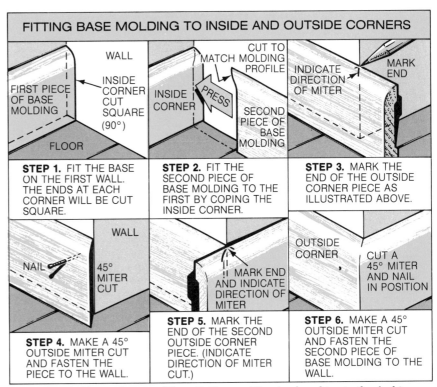

## FITTING BASE MOLDING TO INSIDE AND OUTSIDE CORNERS

**STEP 1.** FIT THE BASE ON THE FIRST WALL. THE ENDS AT EACH CORNER WILL BE CUT SQUARE.

**STEP 2.** FIT THE SECOND PIECE OF BASE MOLDING TO THE FIRST BY COPING THE INSIDE CORNER.

**STEP 3.** MARK THE END OF THE OUTSIDE CORNER PIECE AS ILLUSTRATED ABOVE.

**STEP 4.** MAKE A 45° OUTSIDE MITER CUT AND FASTEN THE PIECE TO THE WALL.

**STEP 5.** MARK THE END OF THE SECOND OUTSIDE CORNER PIECE. (INDICATE DIRECTION OF MITER CUT.)

**STEP 6.** MAKE A 45° OUTSIDE MITER CUT AND FASTEN THE SECOND PIECE OF BASE MOLDING TO THE WALL.

Figure 61-17. Placing base molding. Ceiling molding is placed in similar fashion.

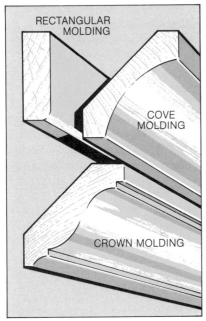

Figure 61-18. Examples of ceiling molding.

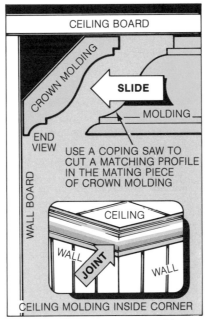

Figure 61-19. Placing crown molding at the ceiling.

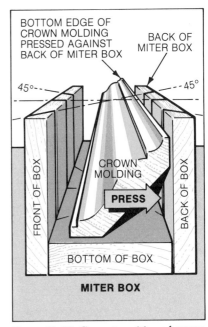

Figure 61-20. Correct position of crown molding in a miter box when cutting a 45° angle.

ners are mitered. A scarf joint is recommended for joints between pieces on longer walls. The pieces are attached to the wall with 6d or 8d nails driven into the wall studs. See Figure 61-19.

The lower side of crown or cove molding fits against the wall. It has a wider surface than the top side, which rests against the ceiling. When placing the molding in the miter box to cut a

45° angle, position it as if the back of the miter box were the wall and as if the bottom of the miter box were the ceiling. See Figure 61-20.

# UNIT 62

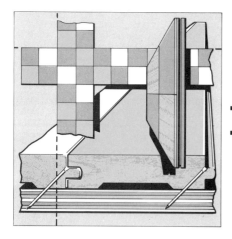

# Finish Flooring

Finish flooring is applied when the construction of a building is almost completed. Many different kinds of material are used. Wood is widely used. So are various resilient tile products such as vinyl, asphalt, rubber, and linoleum. Some resilient products are laid in wheet form, such as linoleum, vinyl, and rubber. Ceramic tiles (especially in bathrooms), brick, slate, and flagstone may be used to create special effects in different sections of the building. Wall-to-wall carpeting applied directly on top of the subfloor is also considered a finish floor material.

In some parts of the country, carpeting and resilient tile are placed by carpenters, but usually this work is done by specialists. Hardwood flooring is always placed by carpenters. In some areas of the United States, placing hardwood flooring has become a specialty within the carpentry trade. The Carpenters' Union has even established separate locals for carpenters who specialize in this work.

## WOOD FINISH FLOORING

Softwood finish flooring is manufactured mostly from southern pine, Douglas fir, and western hemlock. Hardwood finish, which is more durable, is manufactured from oak, beech, birch, and maple. Wood finish flooring comes in strips, planks, or blocks, each of which produces a different appearance.

## Strip Flooring

Finish strips are available in widths of 1½″ to 3½″ and in random lengths up to 16′. Standard thicknesses are ⁵⁄₁₆″, ⅜″, ½″, and ²⁵⁄₃₂″. The thinner pieces have square edges and require top-nailing. Top-nailing requires that the nails be set and the holes later puttied.

The most durable and attractive strip floor is provided by strips ²⁵⁄₃₂″ thick and 2¼″ wide with tongue-and-groove edges. See Figure 62–1. These pieces are also *end-matched,* meaning there is a tongue-and-groove fit where the pieces butt against each other. With the exception of the wall edge of the starter piece, tongue-and-groove strips are blind-nailed. See Figure 62–2. Blind-nailing produces a more attractive appearance than top-nailing. Better quality strip flooring is also *hollow-backed,* which allows the pieces to lie flat and fit snugly against an irregular surface.

Manufacturers recommend that strip flooring be delivered to the job several days before installation. It should be spread loosely around the floor area so that its moisture content can adapt to the humidity of the air in the building.

The subfloor must be cleaned and checked for protruding nails. Next, it is covered with a layer of

*Bruce Hardwood Floors, a division of Triangle Pacific Corporation*
Figure 62–1. Typical strip finish flooring.

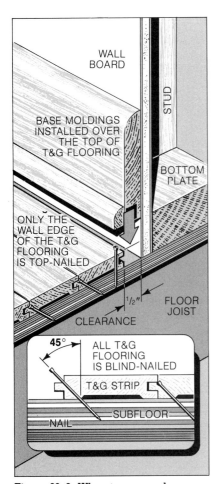

Figure 62–2. When tongue-and-groove strip flooring is used, only the wall edges of the first and last strip must be top-nailed. The rest of the strips are blind-nailed.

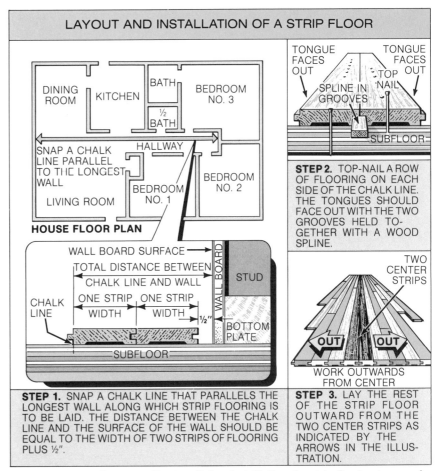

**LAYOUT AND INSTALLATION OF A STRIP FLOOR**

HOUSE FLOOR PLAN

DINING ROOM — KITCHEN — BATH — BEDROOM NO. 3 — ½ BATH — SNAP A CHALK LINE PARALLEL TO THE LONGEST WALL — HALLWAY — LIVING ROOM — BEDROOM NO. 1 — BEDROOM NO. 2

WALL BOARD SURFACE — TOTAL DISTANCE BETWEEN CHALK LINE AND WALL — ONE STRIP WIDTH — ONE STRIP WIDTH — CHALK LINE — WALL BOARD — STUD — ½" — BOTTOM PLATE — SUBFLOOR

**STEP 1.** SNAP A CHALK LINE THAT PARALLELS THE LONGEST WALL ALONG WHICH STRIP FLOORING IS TO BE LAID. THE DISTANCE BETWEEN THE CHALK LINE AND THE SURFACE OF THE WALL SHOULD BE EQUAL TO THE WIDTH OF TWO STRIPS OF FLOORING PLUS ½".

TONGUE FACES OUT — TONGUE FACES OUT — SPLINE IN GROOVES — TOP NAIL — SUBFLOOR

**STEP 2.** TOP-NAIL A ROW OF FLOORING ON EACH SIDE OF THE CHALK LINE. THE TONGUES SHOULD FACE OUT WITH THE TWO GROOVES HELD TOGETHER WITH A WOOD SPLINE.

TWO CENTER STRIPS — OUT — OUT — WORK OUTWARDS FROM CENTER

**STEP 3.** LAY THE REST OF THE STRIP FLOOR OUTWARD FROM THE TWO CENTER STRIPS AS INDICATED BY THE ARROWS IN THE ILLUSTRATION.

Figure 62–3. Installing strip flooring. Before installation, the strip flooring is spread loosely around the floor area so that its moisture content can adapt to the surrounding air.

15-pound waterproof felt. The rows of felt should overlap by 4".

To determine where to begin nailing the strips, consider the floor layout of the building and the number of rooms to receive strip flooring. The strips are laid at right angles to the joists. Snap joist lines to ensure that nails will be driven over joists wherever possible.

Strips nailed along the wall are placed ½" away from the wall to allow for expansion. The ½" gap will be covered by the baseboard. Joints between boards are staggered and are no closer than 6" to joints in adjacent rows.

Drive nails at a 45° to 50° angle. Use a nail set to avoid damaging the edge of the flooring material. The nails should be long enough to go through the subfloor and penetrate into the

joists. Use a short piece of flooring material as a driving block to fit the strips tightly against each other. Proper nailing will reduce future squeaks.

Figure 62–3 shows a procedure for installing hardwood strip flooring. In this example, strip flooring is placed in all rooms of the building except the kitchen and bathroom.

When two floorlayers work together, one often cuts and fits the pieces while the other nails them in place. An efficient method for a floorlayer working alone is to cut and fit six or eight rows of boards at a time and then nail them in place.

**Installation over Concrete.** To apply strip flooring to concrete, nailing strips called *screeds* are used. The screeds are pieces of

1" × 4" or 2" × 4" material, usually treated beforehand with a wood preservative. They are fastened to the concrete floor at 16" O.C. intervals with concrete nails or mastic adhesive. In some cases the screeds may have to be shimmed because of irregularities in the floor.

Before the screeds are placed, waterproof mastic must be spread over the entire concrete floor. In addition, a polyethylene membrane film is strongly recommended. Figure 62–4 shows two procedures for preparing concrete to receive strip flooring. One procedure uses nails to fasten the screeds, and the other uses mastic.

## Plank Flooring

Plank flooring is normally used in buildings of traditional design.

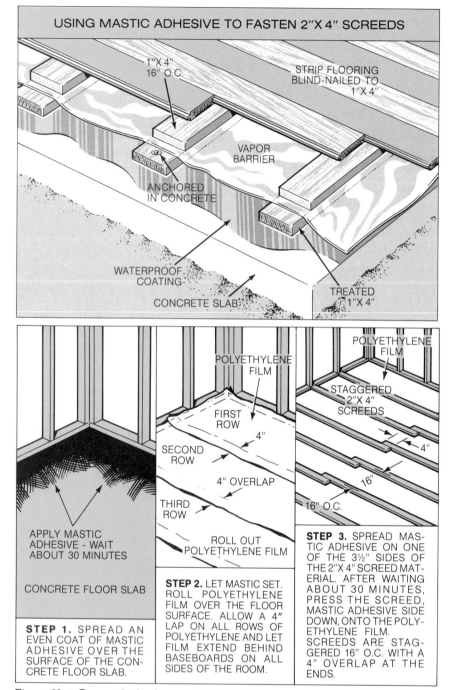

USING MASTIC ADHESIVE TO FASTEN 2"X 4" SCREEDS

1"X 4"
16" O.C.

STRIP FLOORING
BLIND-NAILED TO
1"X 4"

VAPOR
BARRIER

ANCHORED
IN CONCRETE

WATERPROOF
COATING

CONCRETE SLAB

TREATED
1"X 4"

APPLY MASTIC
ADHESIVE - WAIT
ABOUT 30 MINUTES

CONCRETE FLOOR SLAB

**STEP 1.** SPREAD AN
EVEN COAT OF MASTIC
ADHESIVE OVER THE
SURFACE OF THE CON-
CRETE FLOOR SLAB.

POLYETHYLENE
FILM

FIRST
ROW

SECOND
ROW

4"

4" OVERLAP

THIRD
ROW

ROLL OUT
POLYETHYLENE FILM

**STEP 2.** LET MASTIC SET.
ROLL POLYETHYLENE
FILM OVER THE FLOOR
SURFACE. ALLOW A 4"
LAP ON ALL ROWS OF
POLYETHYLENE AND LET
FILM EXTEND BEHIND
BASEBOARDS ON ALL
SIDES OF THE ROOM.

POLYETHYLENE
FILM

STAGGERED
2"X 4"
SCREEDS

4"

16"

16" O.C.

**STEP 3.** SPREAD MAS-
TIC ADHESIVE ON ONE
OF THE 3½" SIDES OF
THE 2"X 4" SCREED MAT-
ERIAL. AFTER WAITING
ABOUT 30 MINUTES,
PRESS THE SCREED,
MASTIC ADHESIVE SIDE
DOWN, ONTO THE POLY-
ETHYLENE FILM.
SCREEDS ARE STAG-
GERED 16" O.C. WITH A
4" OVERLAP AT THE
ENDS.

Figure 62–4. Two methods of preparing concrete to receive strip flooring. One method uses nails, and the other uses mastic.

See Figure 62–5. It was first used in houses built during the colonial period of United States history. Plank pieces are available in widths from 3" to 8" and in thicknesses of ⁵⁄₁₆" or ²⁵⁄₃₂". Planks ²⁵⁄₃₂" thick have tongue-and-groove edges and ends.

Plank flooring is installed in a similar manner to strip flooring.

However, screws may be placed at the ends of the plank pieces. The screws are countersunk and covered with wood plugs. Often plank flooring has simulated plugs for a colonial appearance.

**Installation over Concrete.**
Plank flooring is installed over concrete the same way strip

flooring is installed, unless special *glue-down* planks are obtained. These planks adhere directly to the concrete, making screeds unnecessary. A 150-pound roller is used to apply glue-down planks.

## Block Flooring

Three basic types of block flooring are solid-unit, laminated, and slat. See Figure 62–6. All are fastened to the floor (wood subfloor or concrete slab) with a mastic adhesive. Most block flooring has tongue-and-groove edges to assure alignment between the squares.

Before the adhesive is spread, work lines must be snapped at right angles to each other and at the center of the room. The method for laying out these lines is the same as that used to establish work lines for resilient tile

WOOD
PLUGS

COUNTERBORED
HOLES

PLANK
FLOORING

SUBFLOOR

COUNTERBORED FLATHEAD
SCREWS COVERED WITH
WOOD PLUGS

T&G PLANK FLOORING

⁵⁄₁₆" TO ²⁵⁄₃₂"
THICK

3" TO 8"
WIDE PLANKS

Figure 62–5. In plank flooring, butt joints are used when planks are less than ²⁵⁄₃₂" thick. Tongue-and-groove joints are used for ²⁵⁄₃₂" thick planks.

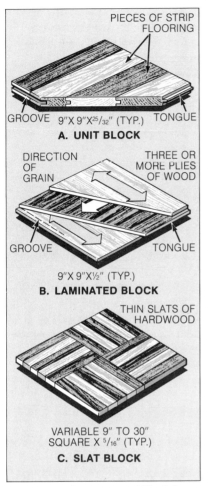

Figure 62–6. Basic types of block floor-ing. All are fastened to the floor with mastic.

*Bruce Hardwood Floors, a division of Triangle Pacific Corporation*

Figure 62–7. A parquet floor is made of slat blocks in which wood strips are ar-ranged in various patterns.

flooring, which is discussed next in this unit.

**Solid-unit Blocks.** Strips ²⁵⁄₃₂″ thick make up most solid-unit blocks. The strips are held to-gether with a wood or metal spline embedded in the lower edge of the material.

**Laminated Blocks.** Laminated blocks are made of plywood, of-ten with an oak face veneer. Be-cause of its cross-laminated con-struction, which reduces swelling or shrinkage, laminated block flooring is recommended for damp locations such as concrete slabs resting directly on the ground.

**Slat Blocks.** Slat blocks are also called *mosaic* or *parquet* floor-ing. The squares are made of wood strips arranged in various patterns. See Figure 62–7. In some slat blocks the strips are held together with face paper that is pulled off after the squares have been fastened to the floor. In other types the strips are held together by mechanical attachments or a textile web backing.

## RESILIENT FINISH FLOORING

Resilient, non-wood products are classified according to their basic ingredients as either vi-nyl, rubber, or cork flooring. They are produced in *tile*

and *sheet* form. Tile sizes range from 9″ × 9″ to 36″ × 36″. Sheets generally come in 6′, 9′, or 12′ widths. A wide variety of patterns and colors is available.

The word *resilient* describes the material's ability to yield when pressure is applied and then return to its original condi-tion. Resilient materials hold up very well against weight and in-dentation caused by falling ob-jects. They are also effective in reducing sound produced by foot traffic and other types of impact.

Resilient flooring is usually laid by specialists outside the car-pentry trade. In some parts of the country, however, it is laid by carpenters.

If the subfloor is uneven, ¼″

557

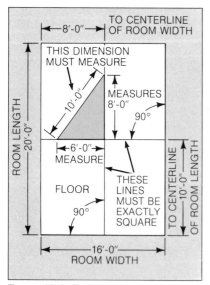

Figure 62–8. To lay out resilient floor tiles, snap lines at the centers of the width and length of the room. These lines must be exactly square to each other.

Armstrong World Industries

Figure 62–10. Finishing off a wall edge with a cove base molding.

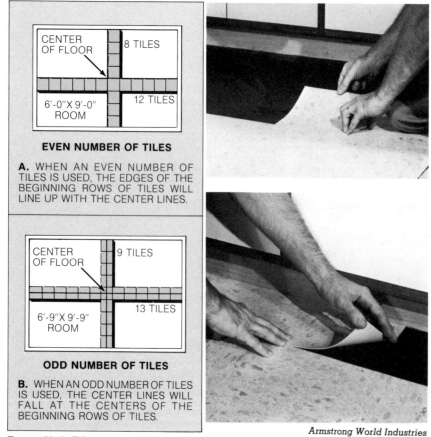

**EVEN NUMBER OF TILES**

**A.** WHEN AN EVEN NUMBER OF TILES IS USED, THE EDGES OF THE BEGINNING ROWS OF TILES WILL LINE UP WITH THE CENTER LINES.

**ODD NUMBER OF TILES**

**B.** WHEN AN ODD NUMBER OF TILES IS USED, THE CENTER LINES WILL FALL AT THE CENTERS OF THE BEGINNING ROWS OF TILES.

Armstrong World Industries

Figure 62–9. Tiles are applied beginning at the center of the room and progressing toward the walls.

underlayment panels (of plywood, hardboard, particleboard, etc.) must be nailed or stapled to the subfloor before the resilient flooring is installed.

The procedure for laying out resilient floor tiles is similar to that used for ceiling tiles. Center lines are snapped on the floor as shown in Figure 62–8. The tiles are placed starting from the center of the room and working toward the wall. See Figure 62–9. This method ensures that when border tiles are less than full size, they will be equal in width to the border tiles at the opposite end of the wall.

Most resilient products can be cut with a pair of scissors or sharp knife. They are fastened with a mastic adhesive. Manufacturers usually recommend a specific adhesive.

When all the tiles have been placed, the wall edges are finished off with a base molding that is usually the same material as the tile. See Figure 62–10.

# SECTION 14

# Stairway Construction

Most buildings with a basement or more than one story have a stairway of some type. Stairway construction is considered one of the more highly skilled areas of carpentry work. Much of this construction is done by contractors who prefabricate the stairways in a shop and directly employ the carpenters who install them.

# UNIT 63

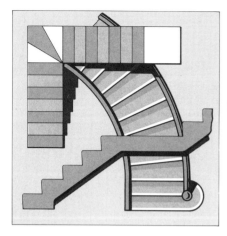

# Types of Stairways

*Main* or *primary* stairways serve the inhabited areas of a building. *Service* stairways serve the non-inhabited areas such as a basement or attic. *Interior* stairways are located inside the building. See Figure 63–1. *Exterior* stairways are located outside the building and lead to entrances or porches. See Figure 63–2.

The blueprints of a building give information on laying out and constructing stairways. Working drawings include plan and elevation views of the stairways as well as detail drawings. The Three-bedroom House Plan in Section 6 includes some typical stairway drawings.

*California Redwood Association*
Figure 63–2. Exterior stairway leading to a porch.

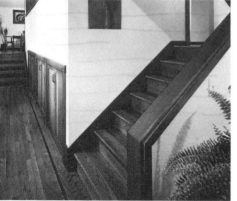

*California Redwood Association*
Figure 63–1. Interior stairways leading to upper level hallways.

## BASIC STAIRWAY ARRANGEMENTS

The architectural design of a room helps determine the stairway arrangement. Figure 63–3 shows the basic arrangements possible. The simplest stairway to construct is a *straight-flight* (or *straight-run*) stairway with no landing. This type of stairway runs in a direct line from one floor to another.

In a straight-flight stairway with a landing, one section of the stairway runs from the floor to the landing, and the other section runs from the landing to the next floor. A straight-flight stairway with landing is often used in public buildings where there is a long distance (15 or more steps) between stories. Landings are placed at the halfway point to

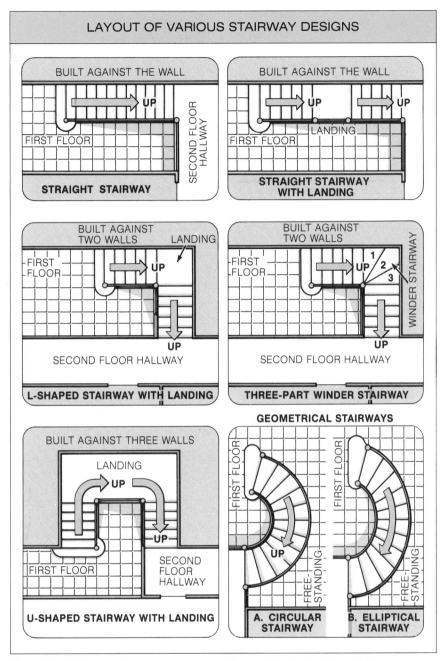

**LAYOUT OF VARIOUS STAIRWAY DESIGNS**

BUILT AGAINST THE WALL

UP

FIRST FLOOR

SECOND FLOOR HALLWAY

**STRAIGHT STAIRWAY**

BUILT AGAINST THE WALL

UP        UP

FIRST FLOOR        LANDING

**STRAIGHT STAIRWAY
WITH LANDING**

BUILT AGAINST
TWO WALLS        LANDING

FIRST
FLOOR        UP

UP

SECOND FLOOR HALLWAY

**L-SHAPED STAIRWAY WITH LANDING**

BUILT AGAINST
TWO WALLS        1
2        WINDER STAIRWAY
3

FIRST
FLOOR        UP

UP

SECOND FLOOR HALLWAY

**THREE-PART WINDER STAIRWAY**

BUILT AGAINST THREE WALLS

LANDING
UP

UP

FIRST FLOOR        SECOND
FLOOR
HALLWAY

**U-SHAPED STAIRWAY WITH LANDING**

**GEOMETRICAL STAIRWAYS**

FIRST FLOOR

UP        FREE-STANDING

**A. CIRCULAR
STAIRWAY**

FIRST FLOOR

FREE-STANDING

**B. ELLIPTICAL
STAIRWAY**

Figure 63–3. Basic stairway arrangements. The architectural design of the room helps determine which stairway arrangement should be used.

*California Redwood Association*
Figure 63-4. L-Shaped stairway with winders instead of a landing.

Figure 63-5. Geometrical winding stairway during construction.

break the climb.

A straight-flight stairway with landing may also be arranged in the shape of an L. An L-shaped stairway runs along two walls rather than one. Another type of straight-flight stairway with landing is the U-shaped stairway. It turns two corners instead of one.

Instead of a landing, a stairway may have a winding section at the turn formed with wedge-shaped steps called *winders.* Stairways with winders are usually L-shaped. See Figure 63–4. Other types of winding stairways are called *geometrical* stairways. See Figure 63–5. They may be circular or elliptical. They are very complicated to build and are usually prefabricated in a shop.

Stairways may be enclosed by walls or railings. See Figure 63–6. If a stairway is enclosed by walls on both sides, railings are attached to the walls. In another type of stairway, a short wall may form the railing for one or both sides of the stairway. Still another type has a *baluster* railing on one or both sides. Balusters (or *bannisters*) are upright pieces that run between the handrail and the treads. The balusters and handrail form a *balustrade.*

## COMPONENTS OF A STAIRWAY

All types of stairways have *treads,* the portion of the stairway that people place their feet on, and *stringers,* the portion that supports the treads. Some stairways also have landings, railings, and balusters. The term *stairway* (or *staircase*) includes all of these components.

Figure 63–7 illustrates the parts of an interior staircase with a balustrade. An explanation of these parts follows. Numbers in this explanation correspond to the numbers on the drawing.

1. *Gooseneck:* A curved or bent section on the handrail.

2. *Handrail:* A rail that is grasped by a person's hand for support in using a stairway.

3. *Landing:* A platform that breaks the stair flight from one floor to another.

4. *Starting newel post:* The main post supporting the handrail of a stair at the bottom of the stairway.

5. *Closed stringer:* A finish board nailed against the wall side of the stairway. The top of this board is parallel to the slope of the stairs.

6. *Tread:* The horizontal surface of a step. The tread is the portion of a step that a person places his or her foot on.

7. *Riser:* The piece forming the vertical face of the step.

8. *Nosing:* The projection of

the tread beyond the face of the riser. This is usually 1″ to 1¼″.

9. *Baluster:* The upright pieces that run between the handrail and the treads. The balusters plus the handrail make up the *balustrade.*

10. *Landing newel post:* The main post supporting the handrail at the landing.

11. *Nosing return:* The projection over the face of the stringer at the end of the tread. This is also called the *end* nosing.

12. *Cove molding:* The trim material sometimes used to

cover the joint between the tread and the stringer and the joint between the tread and the top of the riser.

13. *Open stringer:* The piece that has been cut out to support the open side of the stairway.

## Stringers, Treads, and Risers

The *stringers* (the *carriage*) are the most important parts of the stairway. They provide the main support for the stairs. Finish *treads* and *risers* are nailed to

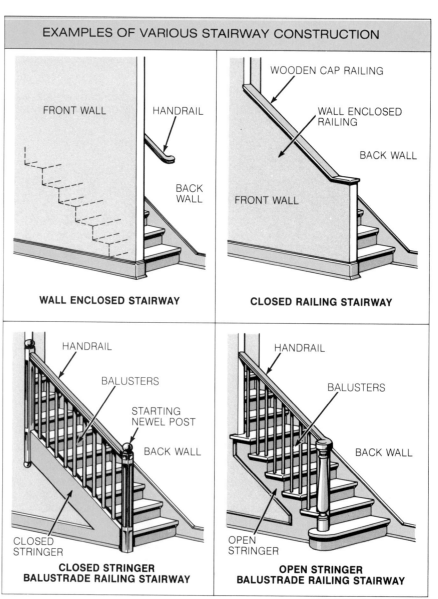

**EXAMPLES OF VARIOUS STAIRWAY CONSTRUCTION**

FRONT WALL    HANDRAIL    BACK WALL

**WALL ENCLOSED STAIRWAY**

WOODEN CAP RAILING    WALL ENCLOSED RAILING    BACK WALL    FRONT WALL

**CLOSED RAILING STAIRWAY**

HANDRAIL    BALUSTERS    STARTING NEWEL POST    BACK WALL    CLOSED STRINGER

**CLOSED STRINGER BALUSTRADE RAILING STAIRWAY**

HANDRAIL    BALUSTERS    BACK WALL    OPEN STRINGER

**OPEN STRINGER BALUSTRADE RAILING STAIRWAY**

Figure 63–6. Stairways may be enclosed by walls or railings.

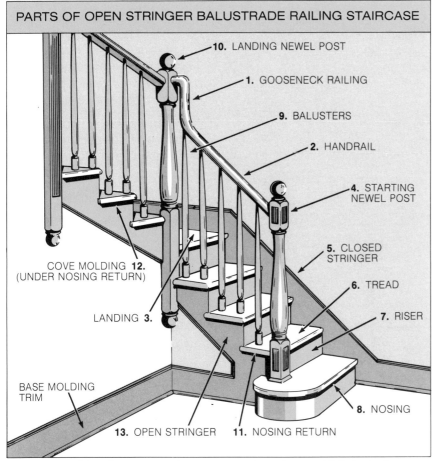

PARTS OF OPEN STRINGER BALUSTRADE RAILING STAIRCASE

**10.** LANDING NEWEL POST

**1.** GOOSENECK RAILING

**9.** BALUSTERS

**2.** HANDRAIL

**4.** STARTING NEWEL POST

**5.** CLOSED STRINGER

**6.** TREAD

**7.** RISER

COVE MOLDING **12.** (UNDER NOSING RETURN)

LANDING **3.**

BASE MOLDING TRIM

**8.** NOSING

**13.** OPEN STRINGER   **11.** NOSING RETURN

Figure 63–7. Components of an interior main stairway with balustrade.

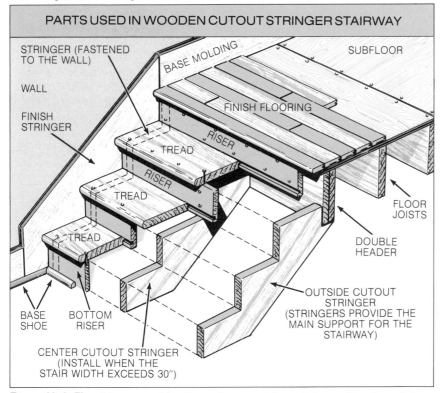

PARTS USED IN WOODEN CUTOUT STRINGER STAIRWAY

STRINGER (FASTENED TO THE WALL)

WALL

FINISH STRINGER

BASE MOLDING

SUBFLOOR

FINISH FLOORING

RISER

TREAD

RISER

TREAD

TREAD

BASE SHOE

BOTTOM RISER

CENTER CUTOUT STRINGER (INSTALL WHEN THE STAIR WIDTH EXCEEDS 30″)

OUTSIDE CUTOUT STRINGER (STRINGERS PROVIDE THE MAIN SUPPORT FOR THE STAIRWAY)

DOUBLE HEADER

FLOOR JOISTS

Figure 63–8. The stringer provides the main support for a stairway. Treads and risers are nailed to the stringer.

the stringers. See Figure 63–8.

On particularly wide stairways, one or more additional stringers may be added in the span between the two outside stringers. The thickness of the tread material is a factor influencing the allowable distance between stringers. A common code ruling calls for stringers every 30″ when $1\frac{1}{16}$″ thick tread material is used, and every 36″ when $1\frac{1}{2}$″ thick tread material is used.

Stringers may be *cleated, dadoed, cut-out,* or *housed.* See Figure 63–9. A simple cleated stringer is a board with cleats nailed to it to support the treads. Another simple design is the dadoed stringer, in which treads fit into dadoes cut into the stringer. As a rule, cleated and dadoed stringers are used for stairways to basements and other noninhabited areas of a house. However, attractive and modern dadoed stringer designs are sometimes used for main stairways as well. See Figure 63–10.

Cut-out stringers are widely used for interior and exterior stairways. Usually 2″ × 10″ or 2″ × 12″ planks are cut out to the width of the tread (minus nosing) and the height of the riser. A minimum of $3\frac{1}{2}$″ measured at a right angle from the edge of the plank must be left after the cuts have been made.

Housed stringers are also widely used for interior and exterior stairways. Prefabricated, shop-constructed stairways usually have housed stringers. These stringers are routed out to receive the treads and risers, which are wedged and glued into place. The housed method of stair construction is efficient and produces an attractive stairway.

## Stringer Layout

Stringer layout includes marking off treads and risers. All risers

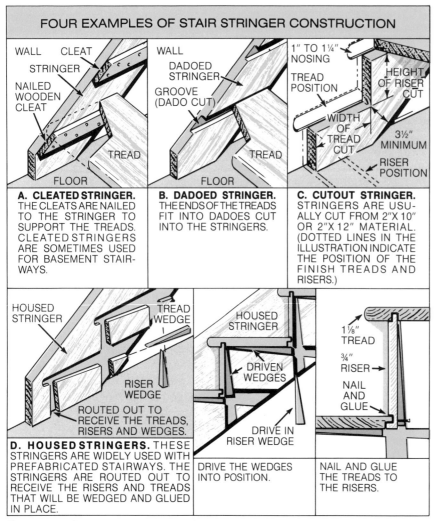

## FOUR EXAMPLES OF STAIR STRINGER CONSTRUCTION

WALL   CLEAT
STRINGER
NAILED
WOODEN
CLEAT
TREAD
FLOOR

**A. CLEATED STRINGER.** THE CLEATS ARE NAILED TO THE STRINGER TO SUPPORT THE TREADS. CLEATED STRINGERS ARE SOMETIMES USED FOR BASEMENT STAIRWAYS.

WALL
DADOED
STRINGER
GROOVE
(DADO CUT)
TREAD
FLOOR

**B. DADOED STRINGER.** THE ENDS OF THE TREADS FIT INTO DADOES CUT INTO THE STRINGERS.

1" TO 1¼"
NOSING
TREAD
POSITION
HEIGHT OF RISER CUT
WIDTH OF TREAD CUT
3½" MINIMUM
RISER POSITION

**C. CUTOUT STRINGER.** STRINGERS ARE USUALLY CUT FROM 2"X 10" OR 2"X 12" MATERIAL. (DOTTED LINES IN THE ILLUSTRATION INDICATE THE POSITION OF THE FINISH TREADS AND RISERS.)

HOUSED STRINGER
TREAD WEDGE
RISER WEDGE
ROUTED OUT TO RECEIVE THE TREADS, RISERS AND WEDGES.

**D. HOUSED STRINGERS.** THESE STRINGERS ARE WIDELY USED WITH PREFABRICATED STAIRWAYS. THE STRINGERS ARE ROUTED OUT TO RECEIVE THE RISERS AND TREADS THAT WILL BE WEDGED AND GLUED IN PLACE.

HOUSED STRINGER
DRIVEN WEDGES
DRIVE IN RISER WEDGE

DRIVE THE WEDGES INTO POSITION.

1⅛" TREAD
¾" RISER
NAIL AND GLUE

NAIL AND GLUE THE TREADS TO THE RISERS.

Figure 63–9. Four types of stringers. Cleated or dadoed stringers are often used for stairways to basements and other noninhabited areas of a house. Cut-out or housed stringers are usually used for other stairways.

must be the same height, and all treads must be the same width.

The height of the riser is found by dividing the *total rise* of the stairway by the number of risers. See Figure 63–11. The riser height is also referred to as the *unit rise*. The total rise is the vertical distance of a stairway from one floor to the floor above.

The width of the treads can be found by dividing the *total run* of a stairway by the number of treads. The total run is the horizontal length measured from the foot of the stairway to a point directly below where the stairway ends at a deck or landing above.

Stringers should be laid out so that a set of stairs will rise at a

safe and comfortable angle. Figure 63–12 shows *preferred* versus *critical* angles. Preferred angles are those that afford the most comfortable walking angles. Critical angles are the minimum and maximum angles for walking safety. The preferred angle for a stairway is 30° to 35°. Critical angles are 20° to 30° and 35° to 50°. For a stairway to rise at a certain angle, the proper combination of tread widths and riser heights must be used.

Recommended riser height is 7" to 7½". Sometimes it is not possible to divide the total rise of a stairway into equal riser heights that will fall between these measurements, so a

higher or lower riser must be used. Many codes set the maximum riser height at 8¼" and the minimum at 6". Recommended tread width is 10" to 12". Minimum width is 9".

**Calculating Tread and Riser Combinations.** A widely accepted formula for determining tread and riser combinations states that the *width of the tread plus the height of the riser shall equal no less than 17" and no more than 18"*. Following are a few examples of stair and riser sizes based on the formula:

| Riser | Tread | Total |
|-------|-------|-------|
| 7" | 11" | 17" |
| 7¼ | 10 | 17¼ |
| 7⅜ | 10½ | 17⅞ |
| 7½ | 10½ | 18 |

For a preferred-angle stairway, a shorter riser must be combined with a wider tread, and a higher riser must be combined with a narrower tread.

Exact tread and riser sizes are based on the total rise and run of a stairway. Figure 63–13 shows the procedure for calculating tread and riser sizes for a stairway with a total rise of 8'-11". The total rise is found by adding the rough floor-to-ceiling height (8'-1"), the actual width of the nominal 2" × 10" headers and joists (9¼") and the thickness of the subfloor (¾").

**Story Pole.** Riser height must be calculated with precision, particularly for long stairways that require many risers. For example, suppose riser height is calculated to be 1/16" less than required for a stringer with 15 risers. On one riser, the error is barely noticeable. However, the top step of the stringer is 15/16" shy of its correct distance from the landing. To avoid this type of error when

*California Redwood Association*

Figure 63-10. Stairway with dadoed stringers.

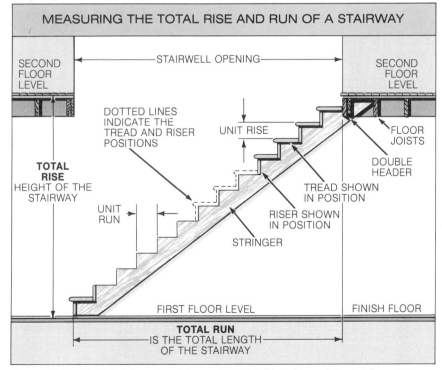

Figure 63-11. The total rise of a stairway is its total height, and the total run is its length. The unit rise is the height of each riser, and the unit run is the width of each tread.

calculating riser heights for long stairways, many carpenters check their calculated measurements by marking off a story pole with a pair of dividers. In this way an adjustment can be made if necessary. Figure 63–13 shows the procedure for laying out a story pole.

**Marking Treads and Risers.** A framing square is used to mark treads and risers on a stringer. The square is positioned so that the tread measurement on the blade and the riser measurement on the tongue are held to one edge of the stringer. Square gauges (also called *stair gauges*) should be used during this operation. Figure 63–14 shows a procedure for marking treads and risers on a cut-out stair stringer with seven treads (each 10½″) and eight risers (each 7¼″).

**Dropping the Stringer.** A certain amount may have to be deducted from the bottom step so that all the finished riser heights will be the same after the tread material is nailed to the stringer. This calculation is called *dropping the stringer.* See Figure 63–15.

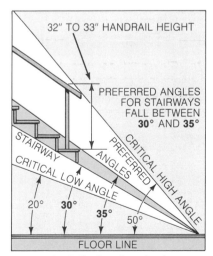

Figure 63-12. Preferred angles for stairs fall between 30° and 35°.

565

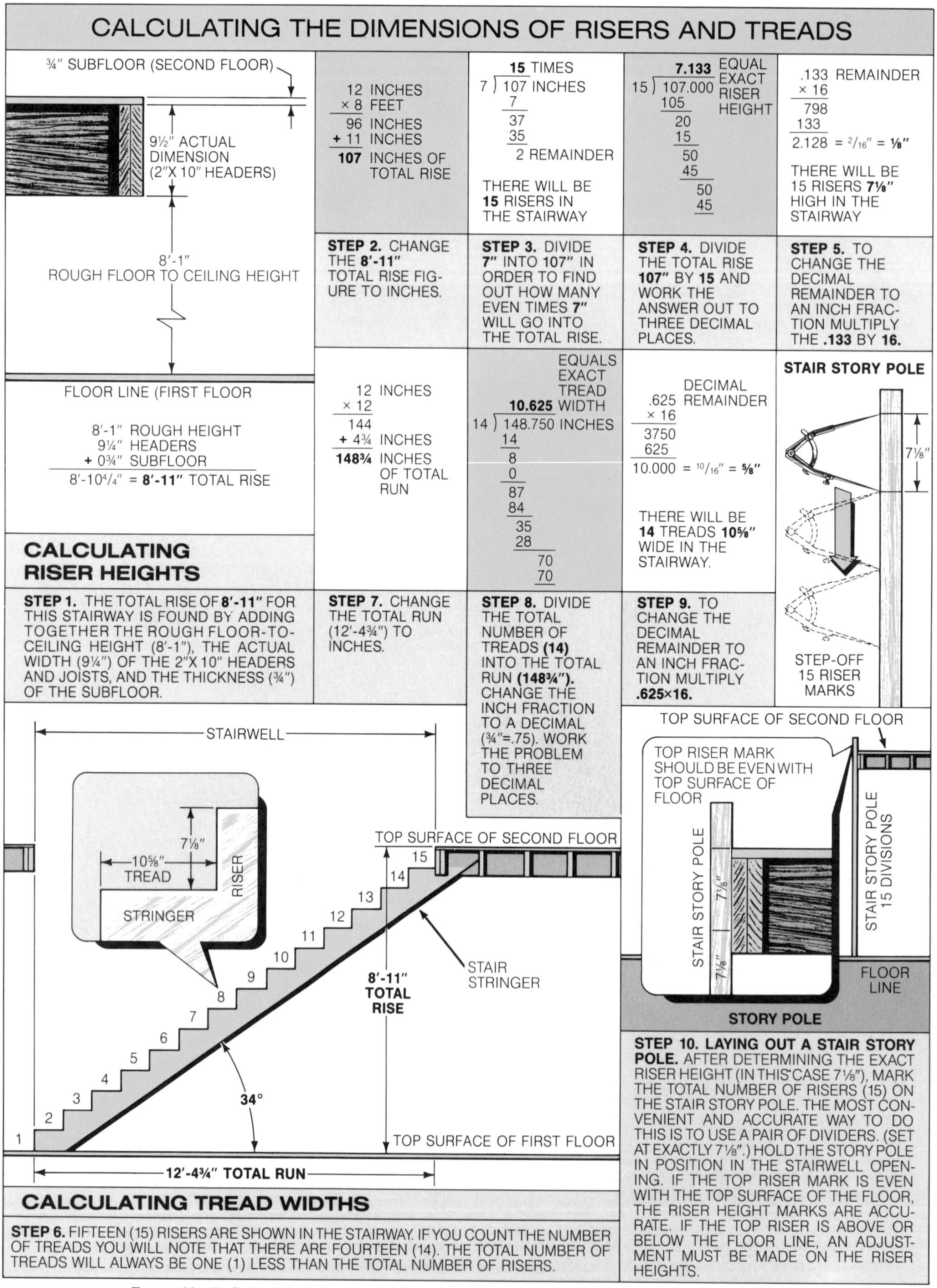
CALCULATING THE DIMENSIONS OF RISERS AND TREADS
¾" SUBFLOOR (SECOND FLOOR)
9½" ACTUAL DIMENSION (2"X 10" HEADERS)
8'-1" ROUGH FLOOR TO CEILING HEIGHT
FLOOR LINE (FIRST FLOOR
8'-1" ROUGH HEIGHT
9¼" HEADERS
+ 0¾" SUBFLOOR
8'-10⁴/₄" = 8'-11" TOTAL RISE
CALCULATING RISER HEIGHTS
STEP 1. THE TOTAL RISE OF 8'-11" FOR THIS STAIRWAY IS FOUND BY ADDING TOGETHER THE ROUGH FLOOR-TO-CEILING HEIGHT (8'-1"), THE ACTUAL WIDTH (9¼") OF THE 2"X 10" HEADERS AND JOISTS, AND THE THICKNESS (¾") OF THE SUBFLOOR.
12 INCHES
× 8 FEET
96 INCHES
+ 11 INCHES
107 INCHES OF TOTAL RISE
STEP 2. CHANGE THE 8'-11" TOTAL RISE FIGURE TO INCHES.
15 TIMES
7 ) 107 INCHES
7
37
35
2 REMAINDER
THERE WILL BE 15 RISERS IN THE STAIRWAY
STEP 3. DIVIDE 7" INTO 107" IN ORDER TO FIND OUT HOW MANY EVEN TIMES 7" WILL GO INTO THE TOTAL RISE.
7.133 EQUAL EXACT RISER HEIGHT
15 ) 107.000
105
20
15
50
45
50
45
STEP 4. DIVIDE THE TOTAL RISE 107" BY 15 AND WORK THE ANSWER OUT TO THREE DECIMAL PLACES.
.133 REMAINDER
× 16
798
133
2.128 = ²/₁₆" = ⅛"
THERE WILL BE 15 RISERS 7⅛" HIGH IN THE STAIRWAY
STEP 5. TO CHANGE THE DECIMAL REMAINDER TO AN INCH FRACTION MULTIPLY THE .133 BY 16.
12 INCHES
× 12
144
+ 4¾ INCHES
148¾ INCHES OF TOTAL RUN
STEP 7. CHANGE THE TOTAL RUN (12'-4¾") TO INCHES.
EQUALS EXACT TREAD 10.625 WIDTH
14 ) 148.750 INCHES
14
8
0
87
84
35
28
70
70
STEP 8. DIVIDE THE TOTAL NUMBER OF TREADS (14) INTO THE TOTAL RUN (148¾"). CHANGE THE INCH FRACTION TO A DECIMAL (¾"=.75). WORK THE PROBLEM TO THREE DECIMAL PLACES.
DECIMAL .625 REMAINDER
× 16
3750
625
10.000 = ¹⁰/₁₆" = ⅝"
THERE WILL BE 14 TREADS 10⅝" WIDE IN THE STAIRWAY.
STEP 9. TO CHANGE THE DECIMAL REMAINDER TO AN INCH FRACTION MULTIPLY .625×16.
STAIR STORY POLE
7⅛"
STEP-OFF 15 RISER MARKS
STAIRWELL
7⅛"
10⅝" TREAD
RISER
STRINGER
TOP SURFACE OF SECOND FLOOR
1 2 3 4 5 6 7 8 9 10 11 12 13 14 15
8'-11" TOTAL RISE
STAIR STRINGER
34°
TOP SURFACE OF FIRST FLOOR
12'-4¾" TOTAL RUN
CALCULATING TREAD WIDTHS
STEP 6. FIFTEEN (15) RISERS ARE SHOWN IN THE STAIRWAY. IF YOU COUNT THE NUMBER OF TREADS YOU WILL NOTE THAT THERE ARE FOURTEEN (14). THE TOTAL NUMBER OF TREADS WILL ALWAYS BE ONE (1) LESS THAN THE TOTAL NUMBER OF RISERS.
TOP SURFACE OF SECOND FLOOR
TOP RISER MARK SHOULD BE EVEN WITH TOP SURFACE OF FLOOR
STAIR STORY POLE
7⅛"
7⅛"
STAIR STORY POLE 15 DIVISIONS
FLOOR LINE
STORY POLE
STEP 10. LAYING OUT A STAIR STORY POLE. AFTER DETERMINING THE EXACT RISER HEIGHT (IN THIS CASE 7⅛"), MARK THE TOTAL NUMBER OF RISERS (15) ON THE STAIR STORY POLE. THE MOST CONVENIENT AND ACCURATE WAY TO DO THIS IS TO USE A PAIR OF DIVIDERS. (SET AT EXACTLY 7⅛".) HOLD THE STORY POLE IN POSITION IN THE STAIRWELL OPENING. IF THE TOP RISER MARK IS EVEN WITH THE TOP SURFACE OF THE FLOOR, THE RISER HEIGHT MARKS ARE ACCURATE. IF THE TOP RISER IS ABOVE OR BELOW THE FLOOR LINE, AN ADJUSTMENT MUST BE MADE ON THE RISER HEIGHTS.

Figure 63–13. Calculating the sizes of treads and risers for a stairway with a total rise of 8'-11".

## LAYING OUT THE TREADS AND RISERS ON A STAIR STRINGER

**STEP 1.** FASTEN THE SQUARE GAUGES AT 7¼" ON THE TONGUE OF THE SQUARE AND 10½" ON THE BLADE OF THE SQUARE.

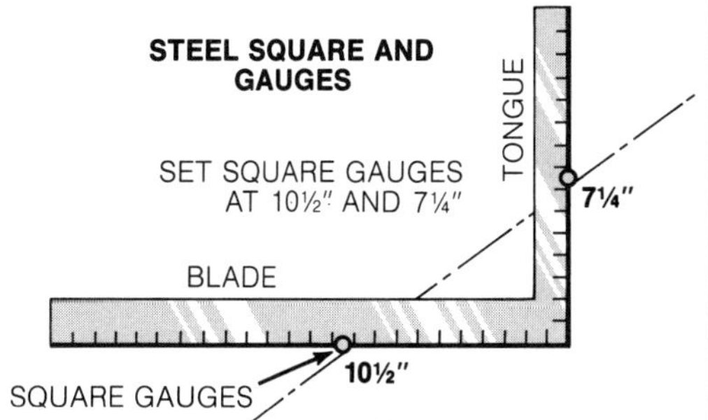

**STEEL SQUARE AND GAUGES**

SET SQUARE GAUGES AT 10½" AND 7¼"

TONGUE

7¼"

BLADE

SQUARE GAUGES

10½"

**STEP 2.** PLACE THE SQUARE ON THE STRINGER WITH THE GAUGES PRESSED FIRMLY AGAINST THE TOP EDGE. MARK THE FIRST RISER.

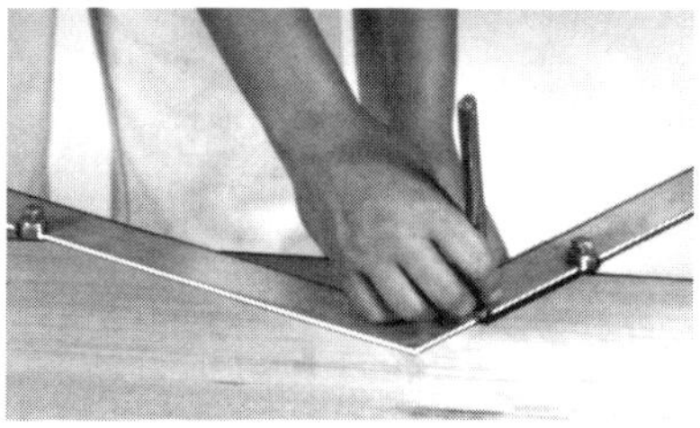

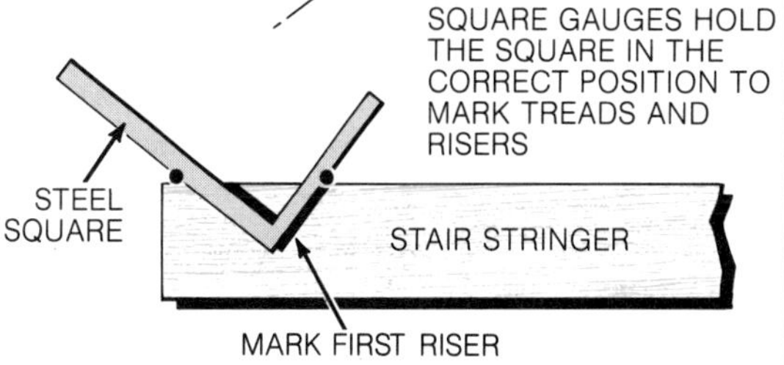

SQUARE GAUGES HOLD THE SQUARE IN THE CORRECT POSITION TO MARK TREADS AND RISERS

STEEL SQUARE

STAIR STRINGER

MARK FIRST RISER

**STEP 3.** SLIDE THE SQUARE TO THE RIGHT SO THAT THE 10½" MEASUREMENT ON THE BLADE LINES UP WITH THE LAST RISER LINE. MARK THE NEXT TREAD AND RISER.

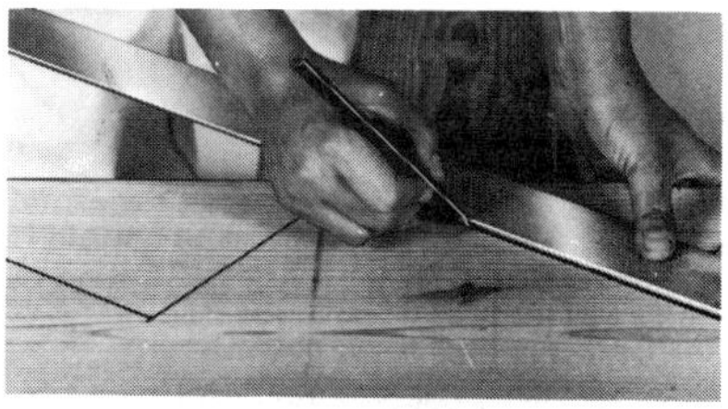

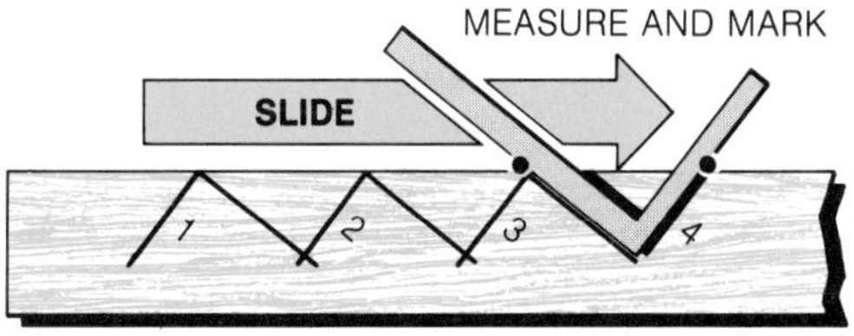

MEASURE AND MARK

**SLIDE**

1  2  3  4

**STEP 4.** REPEAT THIS PROCESS UNTIL 8 RISERS HAVE BEEN LAID OUT ON THE STRINGER. THE LAST (TOP) TREAD MARK IS PART OF THE LANDING.

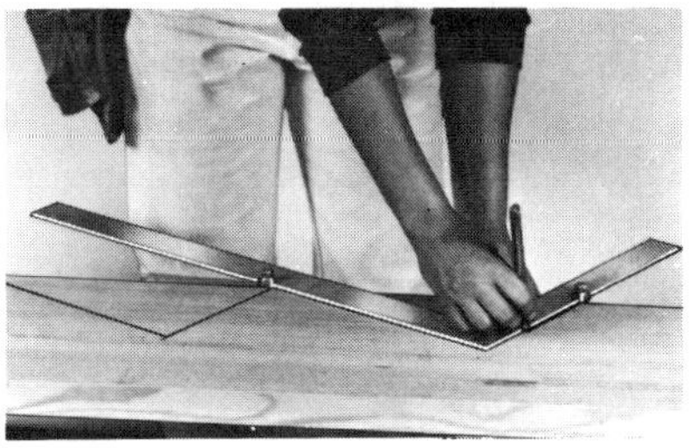

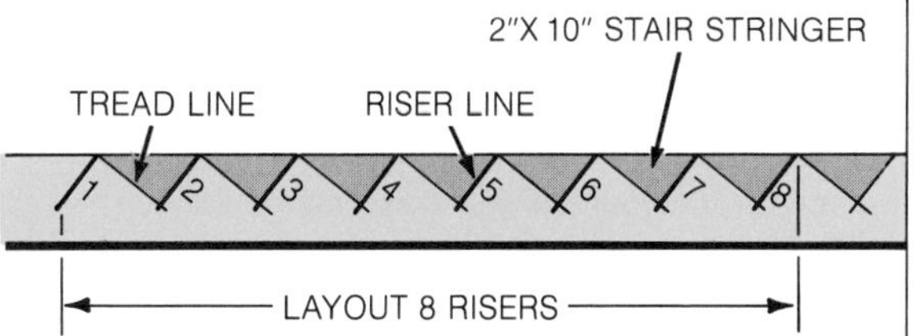

2"X 10" STAIR STRINGER

TREAD LINE    RISER LINE

1  2  3  4  5  6  7  8

LAYOUT 8 RISERS

**STEP 5.** REVERSE THE SQUARE AND MARK THE BOTTOM OF THE FIRST STEP. MOVE THE SQUARE TO THE TOP END OF THE STRINGER. ALIGN THE TONGUE WITH THE LAST RISER LINE AND MARK THE TOP CUT.

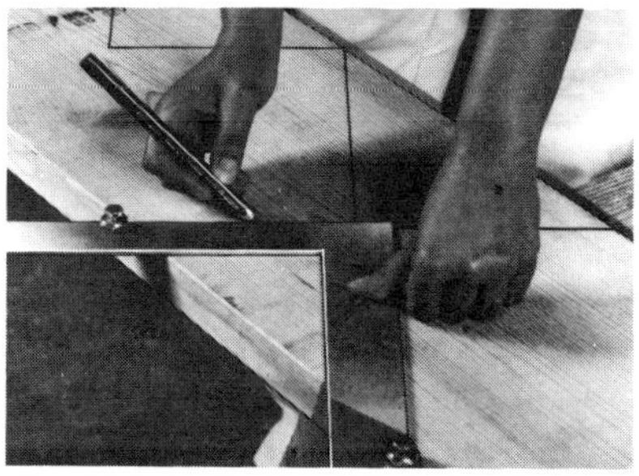

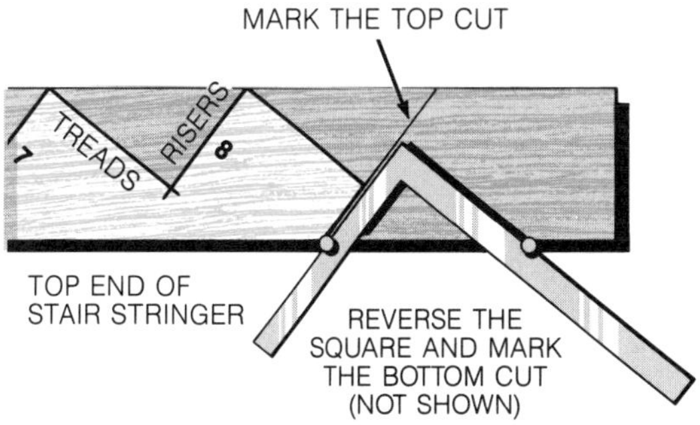

MARK THE TOP CUT

TREADS  RISERS  8

7

TOP END OF STAIR STRINGER

REVERSE THE SQUARE AND MARK THE BOTTOM CUT (NOT SHOWN)

Figure 63–14. Marking treads and risers on a stair stringer. The risers are 7¼" and the treads are 10½". The stairway has seven treads and eight risers.

**Attaching the Stringer.** After the stringers are cut to the correct size, they must be fastened in place. Stringers carry the main load of the staircase and must be securely nailed for strength. Figure 63–16 shows some common methods used for fastening stair stringers. Treads may be fastened with their full or partial width against the stairwell *header.* The header

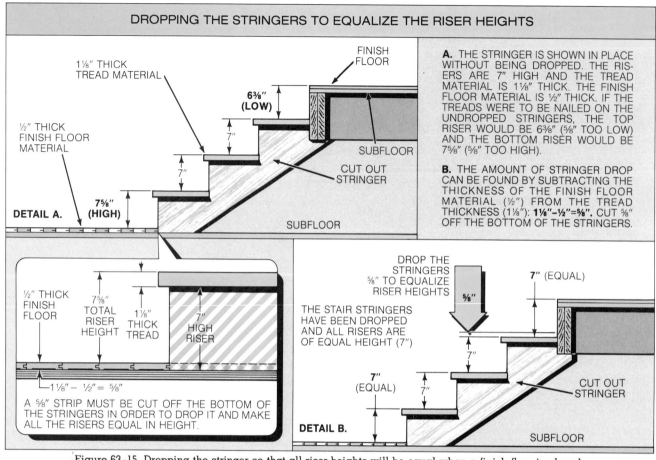

## DROPPING THE STRINGERS TO EQUALIZE THE RISER HEIGHTS

1⅛" THICK TREAD MATERIAL

6⅜" (LOW)

FINISH FLOOR

7"

½" THICK FINISH FLOOR MATERIAL

7"

SUBFLOOR

CUT OUT STRINGER

DETAIL A.

7⅝" (HIGH)

SUBFLOOR

**A.** THE STRINGER IS SHOWN IN PLACE WITHOUT BEING DROPPED. THE RISERS ARE 7" HIGH AND THE TREAD MATERIAL IS 1⅛" THICK. THE FINISH FLOOR MATERIAL IS ½" THICK. IF THE TREADS WERE TO BE NAILED ON THE UNDROPPED STRINGERS, THE TOP RISER WOULD BE 6⅜" (⅝" TOO LOW) AND THE BOTTOM RISER WOULD BE 7⅝" (⅝" TOO HIGH).

**B.** THE AMOUNT OF STRINGER DROP CAN BE FOUND BY SUBTRACTING THE THICKNESS OF THE FINISH FLOOR MATERIAL (½") FROM THE TREAD THICKNESS (1⅛"): **1⅛"–½"=⅝".** CUT ⅝" OFF THE BOTTOM OF THE STRINGERS.

½" THICK FINISH FLOOR

7⅝" TOTAL RISER HEIGHT

1⅛" THICK TREAD

7" HIGH RISER

└ 1⅛" – ½" = ⅝" ┘

A ⅝" STRIP MUST BE CUT OFF THE BOTTOM OF THE STRINGERS IN ORDER TO DROP IT AND MAKE ALL THE RISERS EQUAL IN HEIGHT.

DROP THE STRINGERS ⅝" TO EQUALIZE RISER HEIGHTS

⅝"

7" (EQUAL)

THE STAIR STRINGERS HAVE BEEN DROPPED AND ALL RISERS ARE OF EQUAL HEIGHT (7")

7"

7" (EQUAL)

7"

CUT OUT STRINGER

DETAIL B.

SUBFLOOR

Figure 63–15. Dropping the stringer so that all riser heights will be equal when a finish floor is placed.

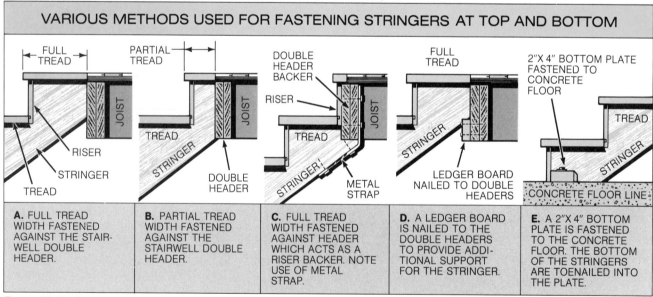

## VARIOUS METHODS USED FOR FASTENING STRINGERS AT TOP AND BOTTOM

FULL TREAD

PARTIAL TREAD

DOUBLE HEADER BACKER

RISER

FULL TREAD

2"X 4" BOTTOM PLATE FASTENED TO CONCRETE FLOOR

JOIST

JOIST

JOIST

TREAD

RISER

STRINGER

TREAD

TREAD

STRINGER

STRINGER

TREAD

STRINGER

LEDGER BOARD NAILED TO DOUBLE HEADERS

STRINGER

TREAD

RISER

STRINGER

TREAD

DOUBLE HEADER

METAL STRAP

CONCRETE FLOOR LINE

**A.** FULL TREAD WIDTH FASTENED AGAINST THE STAIRWELL DOUBLE HEADER.

**B.** PARTIAL TREAD WIDTH FASTENED AGAINST THE STAIRWELL DOUBLE HEADER.

**C.** FULL TREAD WIDTH FASTENED AGAINST HEADER WHICH ACTS AS A RISER BACKER. NOTE USE OF METAL STRAP.

**D.** A LEDGER BOARD IS NAILED TO THE DOUBLE HEADERS TO PROVIDE ADDITIONAL SUPPORT FOR THE STRINGER.

**E.** A 2"X 4" BOTTOM PLATE IS FASTENED TO THE CONCRETE FLOOR. THE BOTTOM OF THE STRINGERS ARE TOENAILED INTO THE PLATE.

Figure 63–16. Some methods of fastening stair stringers. Fastening must be secure since stringers carry the main load of the stairway.

may act as a backing for the top riser. For stairs resting on concrete floors, a wood plate is fastened to the concrete, and the bottom of the stringer is toenailed to the plate. Sometimes a *ledger board* is nailed to the header to provide additional support for the stringer.

# UNIT 64

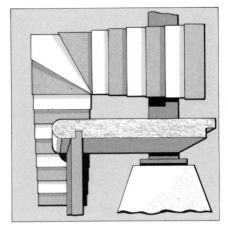

# Stairway Construction

Safety is the most important concern in the design of any stairway. A high percentage of home accidents occur on stairways. Most building codes include detailed requirements for stairway construction.

## CONSTRUCTING INTERIOR STAIRWAYS

General code requirements that apply to interior stairs concern the width of the stairway, the height of handrails, the size and locations of landings, and the amount of head room over the stairs. See Figure 64–1.

### Width of Stairway

The type of building and the greatest number of people that are expected to occupy the building at any given time determine the required width of a stairway. The Uniform Building Code states:

"Stairways serving an occupant load of more than 50 shall be not less than 44 inches wide. Stairways serving an occupant load of 50 or less may be 36 inches wide. Private stairways serving an occupant load of less than 10 may be 30 inches wide."

Based on these code requirements, most residential stairways

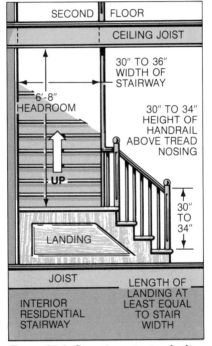

Figure 64–1. Some important code dimensions for residential stairways.

are 30" to 36" wide. Stairways in public buildings are 44" or more wide.

### Handrails

The open sides of any stairway must have handrails. An enclosed residential stairway must have a handrail on at least one side when there are more than four risers to the stairway. Handrails are usually required on both

sides of stairways in public buildings. A center handrail must be placed in public stairways that are more than 88" wide.

A handrail may project as much as 3½" into the established width of a stairway. The space between the handrail and the wall must be at least 1½". The handrail must have a cross-section width of no less than 1¼" but no more than 2". The Uniform Building Code recommends that its height be no less than 30" but no more than 34" above the tread nosing.

### Landings

When landings are used with a stairway, they should be at least as long as the stairway is wide, but they must not exceed 4' in length if there is no change in the direction of the stairway. Landings must be used to break any stairway that rises 12' or more. If possible, the landings should be placed halfway between the top and bottom of the stairway.

### Head Room

Head room is the minimum vertical clearance required from any tread on the stairway to any part of the ceiling structure above the stairway. If there is a stairwell

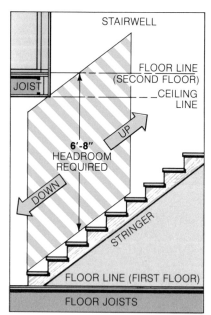

Figure 64–2. When there is a stairwell opening above the stairway, the head room is the minimum vertical distance between the tread closest to the ceiling at the beginning of the stairwell opening.

opening above the stairway, the required distance is measured from the tread nosing to a point at the beginning of the stairwell opening. See Figure 64–2. If there are parallel flights of stairways, the minimum head room must be maintained between the two flights. Local building codes specify head room dimensions. In most areas 6'-8" is considered sufficient head room for main stairs and 6'-4" for basement and service stairs.

## Stairwell Opening

A stairwell opening must be framed in the floor to which a stairway runs. In new construction, the width and length of this opening is shown in the blueprints. (A procedure for framing a stairwell opening is shown in Section 9.) In remodeling work, a stairwell opening may have to be cut into an existing floor. The width of the opening should be the same as the rough width of the stairs. The length of the opening must be calculated to

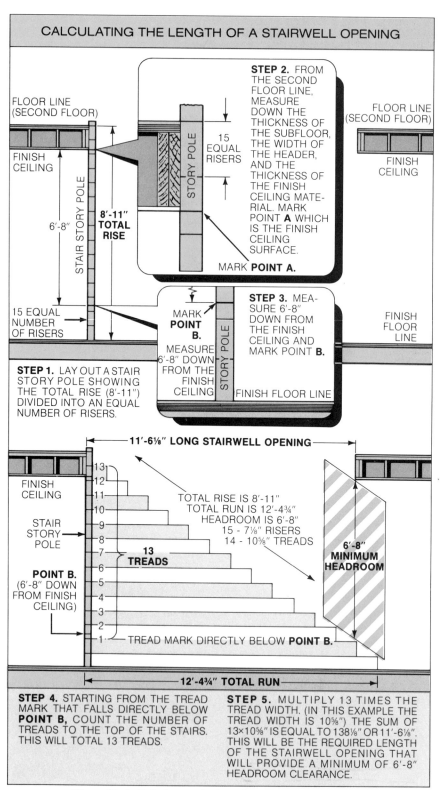

CALCULATING THE LENGTH OF A STAIRWELL OPENING

FLOOR LINE (SECOND FLOOR)
FINISH CEILING
6'-8"
8'-11" TOTAL RISE
STAIR STORY POLE
15 EQUAL NUMBER OF RISERS

**STEP 2.** FROM THE SECOND FLOOR LINE, MEASURE DOWN THE THICKNESS OF THE SUBFLOOR, THE WIDTH OF THE HEADER, AND THE THICKNESS OF THE FINISH CEILING MATERIAL. MARK POINT **A** WHICH IS THE FINISH CEILING SURFACE.
15 EQUAL RISERS
STORY POLE
MARK **POINT A.**

FLOOR LINE (SECOND FLOOR)
FINISH CEILING

**STEP 3.** MEASURE 6'-8" DOWN FROM THE FINISH CEILING AND MARK POINT **B.**
FINISH FLOOR LINE

MARK **POINT B.**
MEASURE 6'-8" DOWN FROM THE FINISH CEILING
STORY POLE
FINISH FLOOR LINE

**STEP 1.** LAY OUT A STAIR STORY POLE SHOWING THE TOTAL RISE (8'-11") DIVIDED INTO AN EQUAL NUMBER OF RISERS.

11'-6⅛" LONG STAIRWELL OPENING

FINISH CEILING
STAIR STORY POLE
POINT B. (6'-8" DOWN FROM FINISH CEILING)
13 12 11 10 9 8 7 6 5 4 3 2 1
**13 TREADS**
TREAD MARK DIRECTLY BELOW **POINT B.**

TOTAL RISE IS 8'-11"
TOTAL RUN IS 12'-4¾"
HEADROOM IS 6'-8"
15 - 7⅛" RISERS
14 - 10⅝" TREADS

6'-8" MINIMUM HEADROOM

12'-4¾" TOTAL RUN

**STEP 4.** STARTING FROM THE TREAD MARK THAT FALLS DIRECTLY BELOW **POINT B,** COUNT THE NUMBER OF TREADS TO THE TOP OF THE STAIRS. THIS WILL TOTAL 13 TREADS.

**STEP 5.** MULTIPLY 13 TIMES THE TREAD WIDTH. (IN THIS EXAMPLE THE TREAD WIDTH IS 10⅝") THE SUM OF 13×10⅝" IS EQUAL TO 138⅛" OR 11'-6⅛". THIS WILL BE THE REQUIRED LENGTH OF THE STAIRWELL OPENING THAT WILL PROVIDE A MINIMUM OF 6'-8" HEADROOM CLEARANCE.

Figure 64–3. Calculating the length of a stairwell opening.

ensure the required amount of head room between the lower steps of the stairway and the end of the opening above. See Figure 64–3.

## Straight-flight Stairways

A straight-flight stairway is the simplest type to build. Figure 64–4 shows one procedure for its construction. For a straight-flight stairway with a landing, the first step is to frame the landing. Figure 64–5 shows a procedure for constructing an L-shaped stairway with a landing. The stairway in this example is open on one side. The first step in constructing an L-shaped stairway with a landing is to frame the landing.

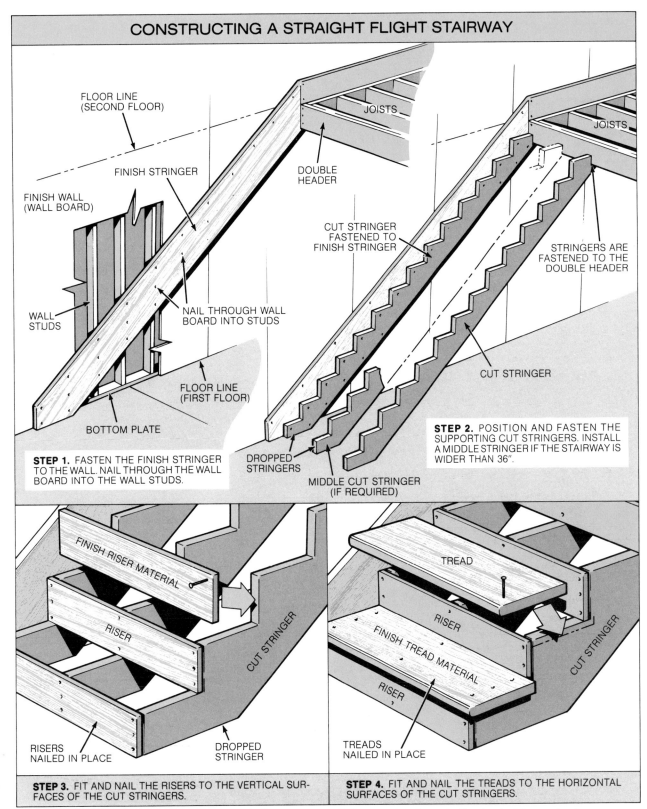

CONSTRUCTING A STRAIGHT FLIGHT STAIRWAY

FLOOR LINE (SECOND FLOOR)

FINISH STRINGER

FINISH WALL (WALL BOARD)

WALL STUDS

NAIL THROUGH WALL BOARD INTO STUDS

FLOOR LINE (FIRST FLOOR)

BOTTOM PLATE

JOISTS

DOUBLE HEADER

CUT STRINGER FASTENED TO FINISH STRINGER

STRINGERS ARE FASTENED TO THE DOUBLE HEADER

CUT STRINGER

JOISTS

**STEP 1.** FASTEN THE FINISH STRINGER TO THE WALL. NAIL THROUGH THE WALL BOARD INTO THE WALL STUDS.

DROPPED STRINGERS

MIDDLE CUT STRINGER (IF REQUIRED)

**STEP 2.** POSITION AND FASTEN THE SUPPORTING CUT STRINGERS. INSTALL A MIDDLE STRINGER IF THE STAIRWAY IS WIDER THAN 36″.

FINISH RISER MATERIAL

RISER

CUT STRINGER

RISERS NAILED IN PLACE

DROPPED STRINGER

TREAD

RISER

FINISH TREAD MATERIAL

RISER

CUT STRINGER

TREADS NAILED IN PLACE

**STEP 3.** FIT AND NAIL THE RISERS TO THE VERTICAL SURFACES OF THE CUT STRINGERS.

**STEP 4.** FIT AND NAIL THE TREADS TO THE HORIZONTAL SURFACES OF THE CUT STRINGERS.

Figure 64–4. Constructing a straight-flight stairway. This type of stairway is the simplest to construct.

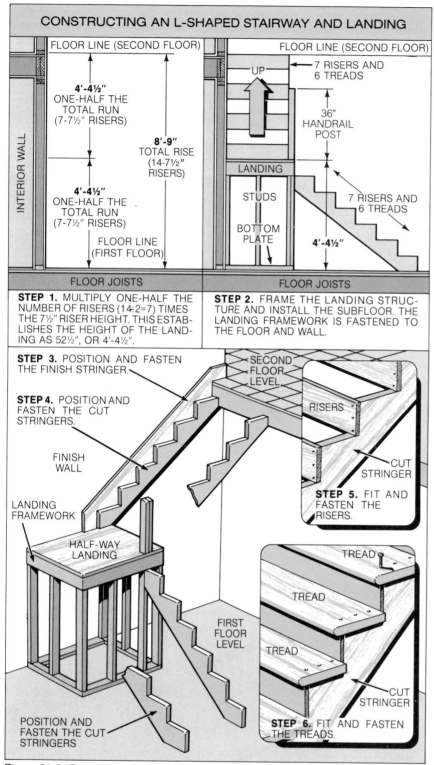

**CONSTRUCTING AN L-SHAPED STAIRWAY AND LANDING**

FLOOR LINE (SECOND FLOOR)

4'-4½"
ONE-HALF THE
TOTAL RUN
(7-7½" RISERS)

8'-9"
TOTAL RISE
(14-7½"
RISERS)

4'-4½"
ONE-HALF THE
TOTAL RUN
(7-7½" RISERS)

FLOOR LINE
(FIRST FLOOR)

INTERIOR WALL

FLOOR JOISTS

**STEP 1.** MULTIPLY ONE-HALF THE NUMBER OF RISERS (14÷2=7) TIMES THE 7½" RISER HEIGHT. THIS ESTABLISHES THE HEIGHT OF THE LANDING AS 52½", OR 4'-4½".

FLOOR LINE (SECOND FLOOR)

UP

7 RISERS AND
6 TREADS

36"
HANDRAIL
POST

LANDING

STUDS

BOTTOM
PLATE

7 RISERS AND
6 TREADS

4'-4½"

FLOOR JOISTS

**STEP 2.** FRAME THE LANDING STRUCTURE AND INSTALL THE SUBFLOOR. THE LANDING FRAMEWORK IS FASTENED TO THE FLOOR AND WALL.

**STEP 3.** POSITION AND FASTEN THE FINISH STRINGER.

**STEP 4.** POSITION AND FASTEN THE CUT STRINGERS.

FINISH WALL

LANDING FRAMEWORK

HALF-WAY LANDING

POSITION AND FASTEN THE CUT STRINGERS

SECOND FLOOR LEVEL

RISERS

CUT STRINGER

**STEP 5.** FIT AND FASTEN THE RISERS.

FIRST FLOOR LEVEL

TREAD

TREAD

TREAD

CUT STRINGER

**STEP 6.** FIT AND FASTEN THE TREADS.

Figure 64–5. Constructing an L-shaped stairway with a landing.

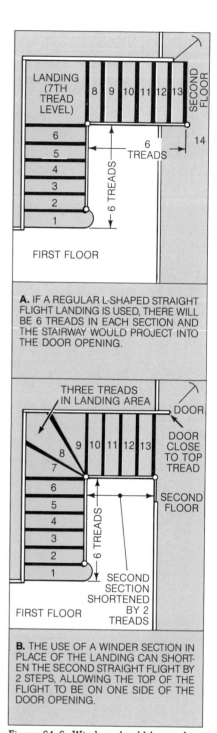

LANDING (7TH TREAD LEVEL)  8  9  10  11  12  13  SECOND FLOOR

6 TREADS  14

6  5  4  3  2  1

6 TREADS

FIRST FLOOR

**A.** IF A REGULAR L-SHAPED STRAIGHT FLIGHT LANDING IS USED, THERE WILL BE 6 TREADS IN EACH SECTION AND THE STAIRWAY WOULD PROJECT INTO THE DOOR OPENING.

THREE TREADS IN LANDING AREA

DOOR

DOOR CLOSE TO TOP TREAD

8  7  9  10  11  12  13

SECOND FLOOR

6  5  4  3  2  1

6 TREADS

SECOND SECTION SHORTENED BY 2 TREADS

FIRST FLOOR

**B.** THE USE OF A WINDER SECTION IN PLACE OF THE LANDING CAN SHORTEN THE SECOND STRAIGHT FLIGHT BY 2 STEPS, ALLOWING THE TOP OF THE FLIGHT TO BE ON ONE SIDE OF THE DOOR OPENING.

Figure 64–6. Winders should be used only when space does not allow a straight-flight stairway.

## Stairways with Winders

Most stairways with winders are L-shaped. Instead of a platform separating the flights (as in an L-shaped or straight-flight stairway with a landing), a series of winding steps makes the turn. The winder section of most L-shaped stairways consists of three (sometimes four) treads. See Figure 64–6.

Stairways with winders are not considered as safe as straight-flight stairways. The width of a winder tread varies from one end to the other. Stairways with winders are usually installed only where

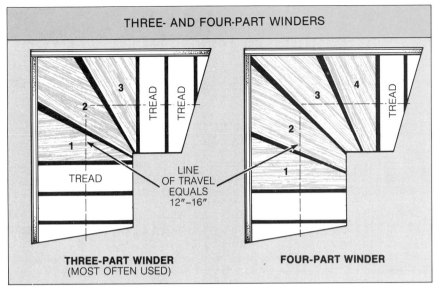

THREE- AND FOUR-PART WINDERS

**THREE-PART WINDER**
(MOST OFTEN USED)

**FOUR-PART WINDER**

LINE OF TRAVEL EQUALS 12"–16"

TREAD

Figure 64–7. The line of travel for a stairway with winders may vary from 12" to 16".

space does not permit a straight-flight-with-landing stairway.

An important dimension that must be established in designing a stairway with winders is the *line of travel.* See Figure 64–7. The line of travel is the distance from the turn at the narrow ends of the winder steps to the place where a person is likely to walk when using the stairway. Acceptable distances for the line of travel vary from 12" to 16". Whenever possible, the width of the winder tread at the line of travel should be the same as the width of the regular treads along the straight-flight sections of the stairway. This width should be as close to 10" as possible. The narrow ends of the winder treads should be as wide as possible.

Some building codes call for a 6" minimum width at the narrow ends of winder treads when three winders are used at the turn. In this case, it is not possible to maintain a 10" tread width with a 12" to 16" line of travel; therefore, the winder treads at the line of travel must be wider than the tread width of the straight-flight section of the stairway.

If a stairway with winders is

open on one side, the narrow ends must be mortised into a post. If the stairway is enclosed by walls on both sides, supporting stringers are required at the narrow as well as at the wide parts of the winder section. The stringers may be the cut-out or housed type. Winder stairways that are prefabricated in a shop are usually built with housed stringers.

Before the stairway is installed, the winder section should be laid out to full scale in the floor area where the stair is to be installed. Only from a full-scale layout can dimensions be obtained for the cuts of the stringer.

General steps in the layout and construction of a stairway with a three-part winder section are shown in Figure 64–8.

## INSTALLING PREFABRICATED STAIRWAYS

Stairways with housed stringers (Figure 64–9) are usually prefabricated in a shop and installed by carpenters who work directly for the stairway contractor. The

stringers are routed out with an electric router, using a special type of template.

The entire staircase is delivered to the job in sections. These sections include the stringers, finish treads and risers, and railing parts. The stringers of the stairway are installed first. Next, the pre-cut treads and risers are set into the grooves of the stringers and secured with glue and wedges. See Figure 64–10. Glue blocks are used to help fasten the bottoms of the risers to the edges of the treads. (Because fewer nails are used with a housed-stringer stairway, it will squeak less than other stairways. Stair squeaks are usually caused by loose nails.) If one side of the stairway is to be open, a *mitered* stringer is used in which the corner joints of the risers and stringers are mitered. See Figure 64–11.

Geometrical stairways are so complex in construction that they are usually prefabricated in a shop and installed as a complete unit.

## CONSTRUCTING EXTERIOR STAIRWAYS

Various types of exterior stairways are used for access to front and back entrances, decks, and porches. An exterior stairway may be all wood (Figure 64–12), or its treads and risers may be finished off with a tile material (Figure 64–13). Concrete stairs are also widely used. Their construction is discussed in Section 8 and Section 16.

The basic layout methods for constructing wood exterior stairs are similar to those for interior stairways. One difference is that the riser height for exterior stairs should be between 6" and 7" rather than the 7" to 7½" recommended for interior stairs.

The steps of an all-wood stair-

# LAYOUT AND CONSTRUCTION OF A THREE-PART WINDER STAIRWAY

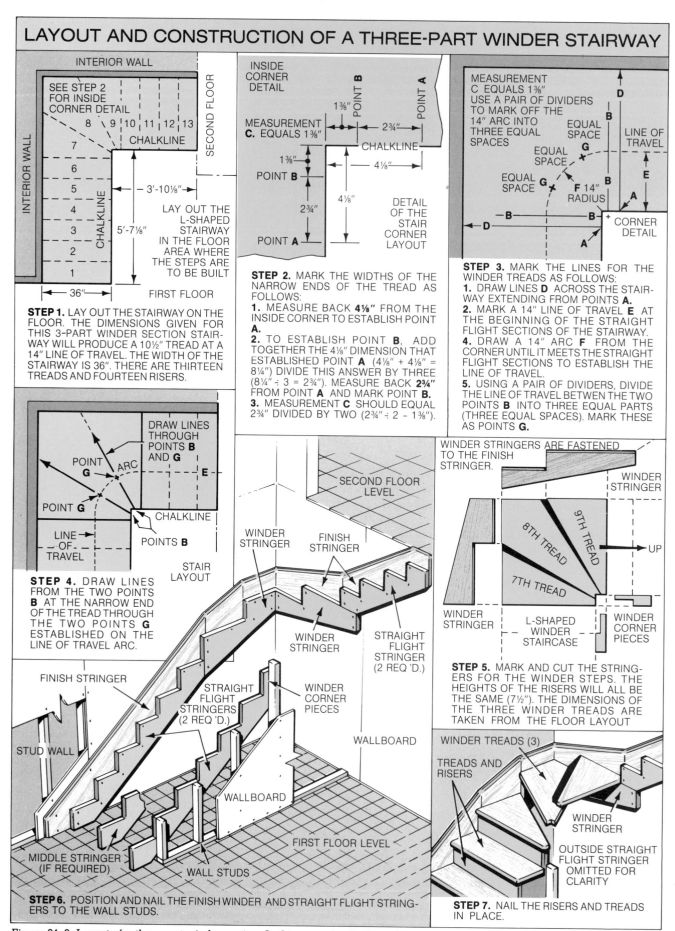

**STEP 1.** LAY OUT THE STAIRWAY ON THE FLOOR. THE DIMENSIONS GIVEN FOR THIS 3-PART WINDER SECTION STAIRWAY WILL PRODUCE A 10½" TREAD AT A 14" LINE OF TRAVEL. THE WIDTH OF THE STAIRWAY IS 36". THERE ARE THIRTEEN TREADS AND FOURTEEN RISERS.

**STEP 2.** MARK THE WIDTHS OF THE NARROW ENDS OF THE TREAD AS FOLLOWS:
1. MEASURE BACK 4⅛" FROM THE INSIDE CORNER TO ESTABLISH POINT **A**.
2. TO ESTABLISH POINT **B**, ADD TOGETHER THE 4⅛" DIMENSION THAT ESTABLISHED POINT **A** (4⅛" + 4⅛" = 8¼") DIVIDE THIS ANSWER BY THREE (8¼" ÷ 3 = 2¾"). MEASURE BACK **2¾"** FROM POINT **A** AND MARK POINT **B**.
3. MEASUREMENT **C** SHOULD EQUAL 2¾" DIVIDED BY TWO (2¾" ÷ 2 – 1⅜").

**STEP 3.** MARK THE LINES FOR THE WINDER TREADS AS FOLLOWS:
1. DRAW LINES **D** ACROSS THE STAIRWAY EXTENDING FROM POINTS **A**.
2. MARK A 14" LINE OF TRAVEL **E** AT THE BEGINNING OF THE STRAIGHT FLIGHT SECTIONS OF THE STAIRWAY.
4. DRAW A 14" ARC **F** FROM THE CORNER UNTIL IT MEETS THE STRAIGHT FLIGHT SECTIONS TO ESTABLISH THE LINE OF TRAVEL.
5. USING A PAIR OF DIVIDERS, DIVIDE THE LINE OF TRAVEL BETWEEN THE TWO POINTS **B** INTO THREE EQUAL PARTS (THREE EQUAL SPACES). MARK THESE AS POINTS **G**.

**STEP 4.** DRAW LINES FROM THE TWO POINTS **B** AT THE NARROW END OF THE TREAD THROUGH THE TWO POINTS **G** ESTABLISHED ON THE LINE OF TRAVEL ARC.

**STEP 5.** MARK AND CUT THE STRINGERS FOR THE WINDER STEPS. THE HEIGHTS OF THE RISERS WILL ALL BE THE SAME (7½"). THE DIMENSIONS OF THE THREE WINDER TREADS ARE TAKEN FROM THE FLOOR LAYOUT

**STEP 6.** POSITION AND NAIL THE FINISH WINDER AND STRAIGHT FLIGHT STRINGERS TO THE WALL STUDS.

**STEP 7.** NAIL THE RISERS AND TREADS IN PLACE.

Figure 64–8. Layout of a three-part winder section. In this example, the line of travel is 14". The tread width along the winder and straight-flight line of travel will work out to 10". The total width of the stairway is 36".

574

Figure 64-9. A finished housed stringer. Usually stairways with housed stringers are prefabricated in a shop.

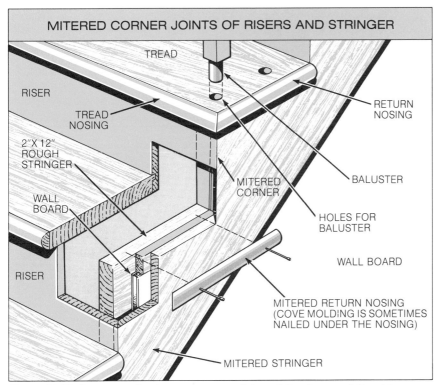

MITERED CORNER JOINTS OF RISERS AND STRINGER

TREAD

RISER

TREAD NOSING

2"X 12" ROUGH STRINGER

WALL BOARD

RISER

RETURN NOSING

MITERED CORNER

BALUSTER

HOLES FOR BALUSTER

WALL BOARD

MITERED RETURN NOSING (COVE MOLDING IS SOMETIMES NAILED UNDER THE NOSING)

MITERED STRINGER

Figure 64-11. In a mitered stringer, the corner joints of the risers and the stringer are mitered.

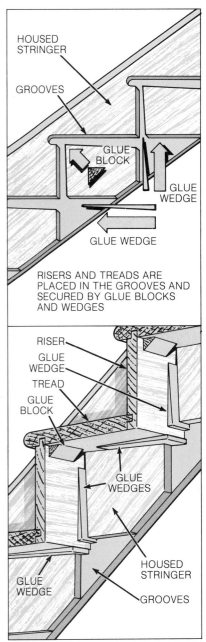

HOUSED STRINGER

GROOVES

GLUE BLOCK

GLUE WEDGE

GLUE WEDGE

RISERS AND TREADS ARE PLACED IN THE GROOVES AND SECURED BY GLUE BLOCKS AND WEDGES

RISER

GLUE WEDGE

TREAD

GLUE BLOCK

GLUE WEDGES

HOUSED STRINGER

GLUE WEDGE

GROOVES

Figure 64-10. Assembling a stairway with housed stringers. The risers and treads are placed in the grooves and secured with glue and wedges. Glue blocks are used to help fasten the bottoms of the risers to the edges of the treads.

way may be finished off with an enclosed tread-and-riser finish, or the risers may be left open. See Figure 64-14. If the stairway is closed off by tread and riser materials, the stringer is cut out so that there will be a ⅛" slope on the tread to allow for water drainage.

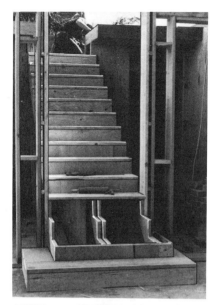

Figure 64–13. Exterior stairway during construction (top) and after it is finished off with a tile material (bottom).

*California Redwood Association*

Figure 64–12. The bottom of this all-wood exterior stairway rests on a concrete slab. Note the concrete piers placed under poles supporting the landing.

A concrete bottom step is strongly recommended for wood stairways if the bottom of the stringer does not rest on a concrete slab. If wood posts are used as part of the stairway structure, they should be supported by concrete piers that extend down into firm soil. Refer again to Figure 64–12.

One procedure for constructing an exterior stairway is shown in Figure 64–15. In this example, the stairway leads to a landing at the rear of the building.

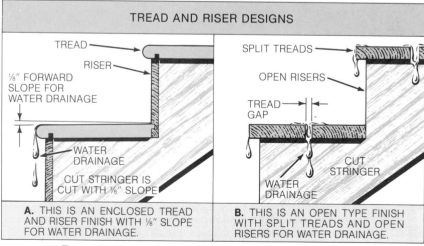

TREAD AND RISER DESIGNS

TREAD

RISER

⅛" FORWARD SLOPE FOR WATER DRAINAGE

WATER DRAINAGE

CUT STRINGER IS CUT WITH ⅛" SLOPE

SPLIT TREADS

OPEN RISERS

TREAD GAP

CUT STRINGER

WATER DRAINAGE

**A.** THIS IS AN ENCLOSED TREAD AND RISER FINISH WITH ⅛" SLOPE FOR WATER DRAINAGE.

**B.** THIS IS AN OPEN TYPE FINISH WITH SPLIT TREADS AND OPEN RISERS FOR WATER DRAINAGE.

Figure 64–14. Tread and riser designs for exterior stairways.

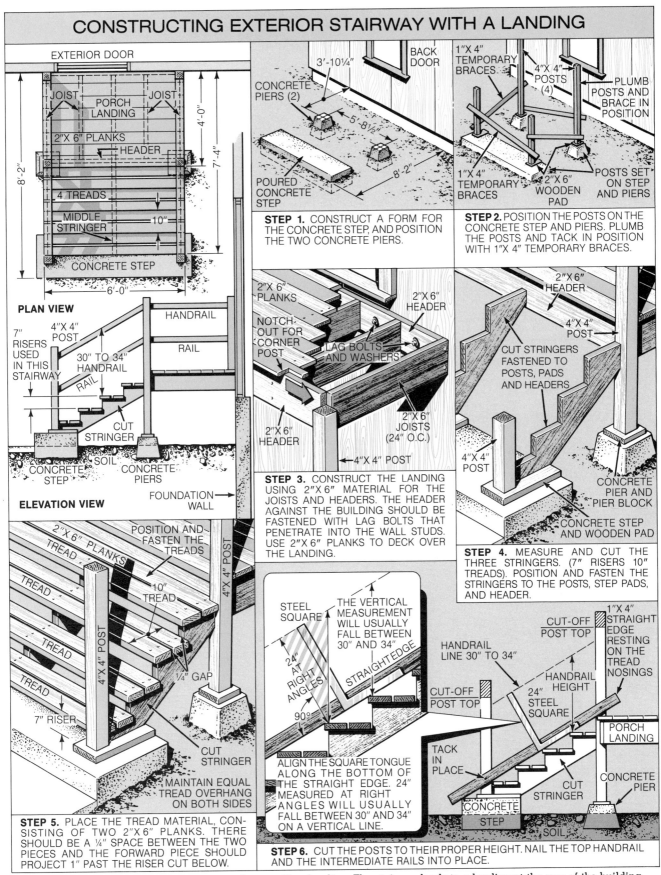

# CONSTRUCTING EXTERIOR STAIRWAY WITH A LANDING

**PLAN VIEW**

EXTERIOR DOOR
JOIST
PORCH LANDING
JOIST
2"X 6" PLANKS
HEADER
4 TREADS
MIDDLE STRINGER
CONCRETE STEP
8'-2"
4'-0"
7'-4"
10"
6'-0"

**ELEVATION VIEW**

HANDRAIL
RAIL
7" RISERS USED IN THIS STAIRWAY
4"X 4" POST
30" TO 34" HANDRAIL
RAIL
CUT STRINGER
CONCRETE STEP
SOIL
CONCRETE PIERS
FOUNDATION WALL

**STEP 1.** CONSTRUCT A FORM FOR THE CONCRETE STEP, AND POSITION THE TWO CONCRETE PIERS.

BACK DOOR
3'-10¼"
CONCRETE PIERS (2)
5'-8½"
8'-2"
POURED CONCRETE STEP

**STEP 2.** POSITION THE POSTS ON THE CONCRETE STEP AND PIERS. PLUMB THE POSTS AND TACK IN POSITION WITH 1"X 4" TEMPORARY BRACES.

1"X 4" TEMPORARY BRACES
4"X 4" POSTS (4)
PLUMB POSTS AND BRACE IN POSITION
1"X 4" TEMPORARY BRACES
2"X 6" WOODEN PAD
POSTS SET ON STEP AND PIERS

**STEP 3.** CONSTRUCT THE LANDING USING 2"X 6" MATERIAL FOR THE JOISTS AND HEADERS. THE HEADER AGAINST THE BUILDING SHOULD BE FASTENED WITH LAG BOLTS THAT PENETRATE INTO THE WALL STUDS. USE 2"X 6" PLANKS TO DECK OVER THE LANDING.

2"X 6" PLANKS
NOTCH-OUT FOR CORNER POST
LAG BOLTS AND WASHERS
2"X 6" HEADER
2"X 6" HEADER
2"X 6" JOISTS (24" O.C.)
4"X 4" POST

**STEP 4.** MEASURE AND CUT THE THREE STRINGERS. (7" RISERS 10" TREADS). POSITION AND FASTEN THE STRINGERS TO THE POSTS, STEP PADS, AND HEADER.

2"X 6" HEADER
4"X 4" POST
CUT STRINGERS FASTENED TO POSTS, PADS AND HEADERS
4"X 4" POST
CONCRETE PIER AND PIER BLOCK
CONCRETE STEP AND WOODEN PAD

**STEP 5.** PLACE THE TREAD MATERIAL, CONSISTING OF TWO 2"X 6" PLANKS. THERE SHOULD BE A ¼" SPACE BETWEEN THE TWO PIECES AND THE FORWARD PIECE SHOULD PROJECT 1" PAST THE RISER CUT BELOW.

2"X 6" PLANKS
POSITION AND FASTEN THE TREADS
TREAD
TREAD
10" TREAD
TREAD
TREAD
4"X 4" POST
4"X 4" POST
¼" GAP
7" RISER
CUT STRINGER
MAINTAIN EQUAL TREAD OVERHANG ON BOTH SIDES

ALIGN THE SQUARE TONGUE ALONG THE BOTTOM OF THE STRAIGHT EDGE. 24" MEASURED AT RIGHT ANGLES WILL USUALLY FALL BETWEEN 30" AND 34" ON A VERTICAL LINE.

STEEL SQUARE
THE VERTICAL MEASUREMENT WILL USUALLY FALL BETWEEN 30" AND 34"
24" AT RIGHT ANGLES
STRAIGHTEDGE
90°

**STEP 6.** CUT THE POSTS TO THEIR PROPER HEIGHT. NAIL THE TOP HANDRAIL AND THE INTERMEDIATE RAILS INTO PLACE.

CUT-OFF POST TOP
1"X 4" STRAIGHT EDGE RESTING ON THE TREAD NOSINGS
HANDRAIL LINE 30" TO 34"
HANDRAIL HEIGHT
CUT-OFF POST TOP
24" STEEL SQUARE
TACK IN PLACE
CUT STRINGER
PORCH LANDING
CONCRETE PIER
CONCRETE STEP
SOIL

Figure 64–15. Constructing an exterior stairway with a landing. The stairway leads to a landing at the rear of the building.

# SECTION 15

# Post-and-Beam Construction

Post-and-beam construction systems used today have developed from some of the oldest building methods known in the United States. The basic structure of many early colonial homes was a post-and-beam framework held together with wood dowels. Later a post-and-beam method known as *heavy timber construction* was used to erect factories, warehouses, mills, bridges, railroad trestles, and other large structures.

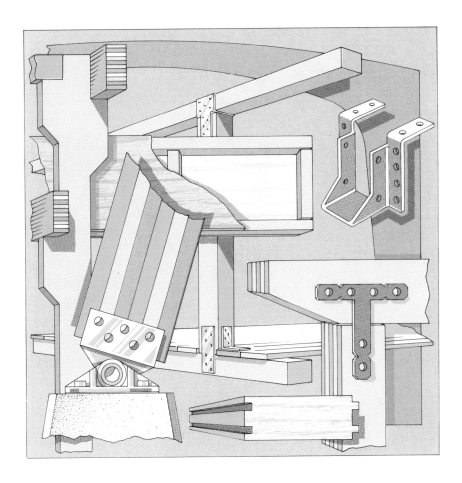

# UNIT 65

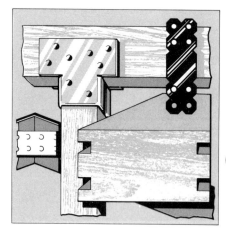

# Residential Post-and-Beam Construction

In residential post-and-beam construction the basic framework of the building consists of vertical posts and horizontal or sloping beams. The posts in the outside walls of the house support the beams that are part of the roof and ceiling. Posts may also be used beneath beams that support the floor. Figure 65–1 compares the framework of a conventional, platform-framed house

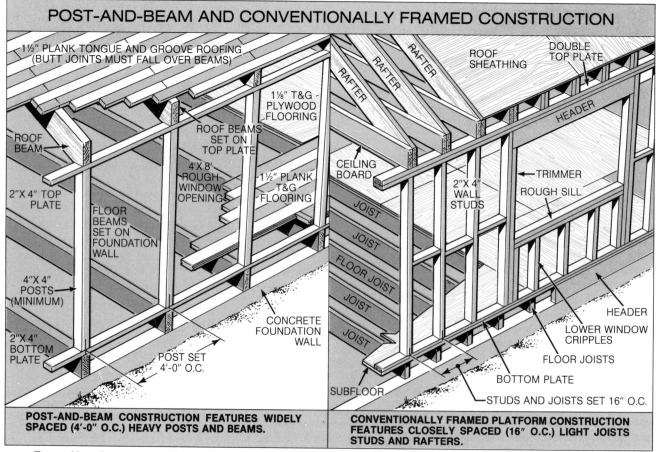

**POST-AND-BEAM AND CONVENTIONALLY FRAMED CONSTRUCTION**

1½" PLANK TONGUE AND GROOVE ROOFING
(BUTT JOINTS MUST FALL OVER BEAMS)

RAFTER

RAFTER

RAFTER

RAFTER

ROOF SHEATHING

DOUBLE TOP PLATE

1⅛" T&G PLYWOOD FLOORING

ROOF BEAMS SET ON TOP PLATE

HEADER

ROOF BEAM

CEILING BOARD

4'X 8' ROUGH WINDOW OPENING

1½" PLANK T&G FLOORING

TRIMMER

2"X 4" WALL STUDS

ROUGH SILL

2"X 4" TOP PLATE

JOIST

FLOOR BEAMS SET ON FOUNDATION WALL

JOIST

FLOOR JOIST

4"X 4" POSTS (MINIMUM)

JOIST

HEADER

JOIST

LOWER WINDOW CRIPPLES

CONCRETE FOUNDATION WALL

FLOOR JOISTS

2"X 4" BOTTOM PLATE

POST SET 4'-0" O.C.

BOTTOM PLATE

SUBFLOOR

STUDS AND JOISTS SET 16" O.C.

**POST-AND-BEAM CONSTRUCTION FEATURES WIDELY SPACED (4'-0" O.C.) HEAVY POSTS AND BEAMS.**

**CONVENTIONALLY FRAMED PLATFORM CONSTRUCTION FEATURES CLOSELY SPACED (16" O.C.) LIGHT JOISTS STUDS AND RAFTERS.**

Figure 65–1. A comparison of the wall and roof sections of a post-and-beam house and a conventionally framed house.

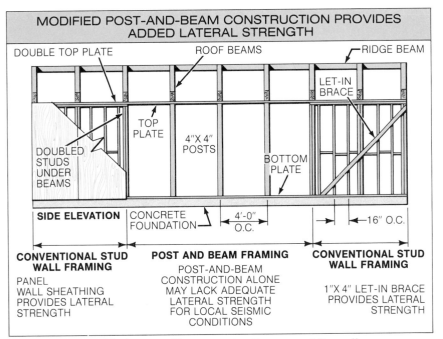

### MODIFIED POST-AND-BEAM CONSTRUCTION PROVIDES ADDED LATERAL STRENGTH

DOUBLE TOP PLATE ROOF BEAMS RIDGE BEAM

LET-IN BRACE

TOP PLATE

4"X 4" POSTS

DOUBLED STUDS UNDER BEAMS

BOTTOM PLATE

SIDE ELEVATION CONCRETE FOUNDATION

4'-0" O.C.

16" O.C.

CONVENTIONAL STUD WALL FRAMING
PANEL WALL SHEATHING PROVIDES LATERAL STRENGTH

POST AND BEAM FRAMING
POST-AND-BEAM CONSTRUCTION ALONE MAY LACK ADEQUATE LATERAL STRENGTH FOR LOCAL SEISMIC CONDITIONS

CONVENTIONAL STUD WALL FRAMING
1"X 4" LET-IN BRACE PROVIDES LATERAL STRENGTH

Figure 65–2. In modified post-and-beam construction, some of the wall areas are framed with studs. Bracing or plywood can be applied to give greater lateral strength to the building.

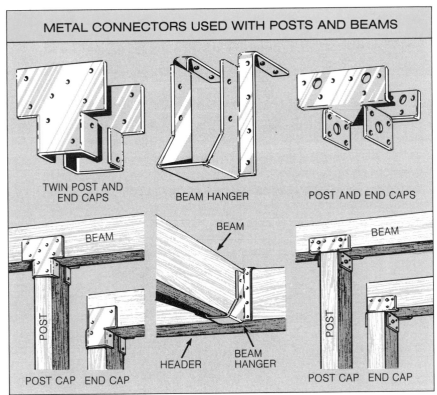

### METAL CONNECTORS USED WITH POSTS AND BEAMS

TWIN POST AND END CAPS

BEAM HANGER

POST AND END CAPS

BEAM

POST

POST CAP END CAP

BEAM

HEADER BEAM HANGER

BEAM

POST

POST CAP END CAP

Figure 65–3. Metal fasteners add strength to post-and-beam construction.

to-ceiling windows can be set between the wall posts, providing a great amount of interior natural light. The high ceilings produce a pleasant, open feeling.

In the past, tongue-and-groove planks were always used for the floor and roof covering of post-and-beam houses. For this reason, post-and-beam construction is still sometimes referred to as *plank-and-beam* construction.

One drawback of the post-and-beam design is that it does not lend itself to the effective bracing methods possible in platform framing. To answer this problem, the building can be planned so that some of the corner sections and other parts of the exterior walls are framed as stud walls. These stud walls permit the use of let-in bracing or plywood sheathing to give greater lateral strength to the building. This method of incorporating some stud walls in the post-and-beam design is known as *modified post-and-beam construction*. See Figure 65–2.

Another way to add strength to a post-and-beam building is to use metal fasteners (connectors) wherever the key structural members are fastened together. The metal fasteners are attached with nails and with machine or lag bolts. See Figures 65–3 and 65–4.

## POST-AND-BEAM FLOORS

Certain types of post-and-beam floors are less expensive to construct than standard floors. Instead of floor joists, solid or laminated beams are placed over posts that are supported by concrete piers. The beams are usually spaced 4' O.C. They are tied to the posts with metal connectors. The outside ends of the beams rest on top of the founda-

with that of a post-and-beam house.

The exposed wood timbers in

post-and-beam construction give an attractive, natural look to the inside of a structure. Large floor-

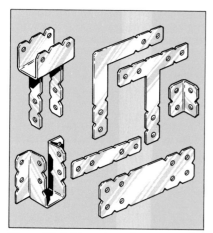

Figure 65–4. Examples of ornamental metal connectors used in post-and-beam construction.

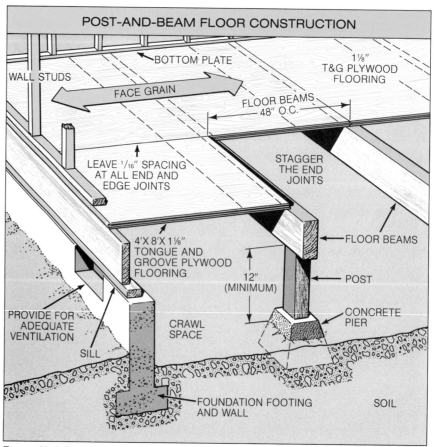

Figure 65–5. In a post-and-beam floor system, floor beams are used instead of joists. In this example, the floor beams are spaced 4' O.C. They are supported by posts resting on concrete piers. The deck material is 1⅛" thick plywood.

tion walls. Wood planks or plywood flooring materials are nailed to the tops of the beams. See Figure 65–5.

Post-and-beam systems are often used beneath the first floor of a platform-framed building. In some sections of the United States posts and beams over piers serve as the outside foundation support for the building.

The size of the beams used for a floor depends on the weight being carried and on the span between supports. Beams with 4" × 8" nominal dimensions are often used when the spacing between beams is 4' O.C. Some beams consist of laminated layers, some are a solid piece, and one type (a *box* beam) is hollow inside. See Figure 65–6. These beam types are discussed further in the following unit.

## Covering a Post-and-Beam Floor

At one time 2" thick tongue-and-groove planks were always used for the covering of a post-and-beam floor. Today 1⅛" tongue-and-groove plywood panels are also used. Plank flooring is placed so that joints between butt ends are staggered. Each

plank should span at least two openings between floor beams. See Figure 65–7.

Plank or plywood flooring nailed over widely spaced beams is not capable of carrying heavy loads. If heavy loads are expected (such as a bearing partition, a bathtub, or a refrigerator),

additional framing should be constructed beneath the floor to transmit the load to the beams.

## POST-AND-BEAM WALLS

The walls of a post-and-beam building are constructed of posts

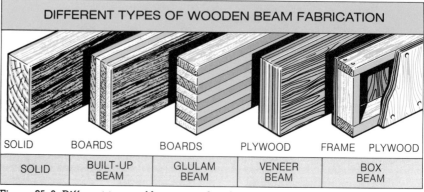

Figure 65–6. Different types of beams used with post-and-beam floor systems.

582

spaced 4', 6', or 8' apart. The post material may be solid or consist of laminated layers. It should be no less than 4" × 4" nominal size. The size of the post needed is determined by the load it must support and by its length. The farther upward a post extends without lateral bracing, the greater will be its tendency to buckle. This tendency to buckle can be compensated for by using a wider post. See Figure 65–8.

One method of constructing the walls is to nail the posts to a bottom (sole) and top plate. See Figure 65–9. Another method is to anchor the bottom of the posts directly to the floor. Posts that butt directly to beams should be reinforced with a strap or bracket. See Figure 65–10. The spaces between the posts are often used for wide floor-to-ceiling windows. However, some of these spaces may be framed and the outside surfaces finished off with rustic stone, brick, or other exterior finish material.

## POST-AND-BEAM ROOFS

The main structural members of a post-and-beam roof are the exposed ridge beam, the roof beams, and the planks or panels used for the roof covering. See Figures 65–11 and 65–12. This type of combined roof-ceiling design, also called a *cathedral* roof, is frequently used over buildings that have regular stud-framed walls. See Figure 65–13.

## Types of Post-and-Beam Roofs

The two basic post-and-beam roof designs are *longitudinal* and *transverse*. See Figure 65–14. Longitudinal roof beams run the

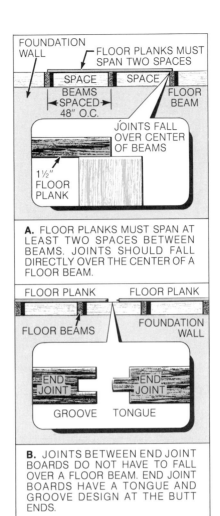

**A.** FLOOR PLANKS MUST SPAN AT LEAST TWO SPACES BETWEEN BEAMS. JOINTS SHOULD FALL DIRECTLY OVER THE CENTER OF A FLOOR BEAM.

**B.** JOINTS BETWEEN END JOINT BOARDS DO NOT HAVE TO FALL OVER A FLOOR BEAM. END JOINT BOARDS HAVE A TONGUE AND GROOVE DESIGN AT THE BUTT ENDS.

Figure 65–7. Placing plank flooring. Joints between butt ends are staggered.

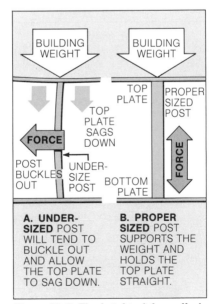

**A. UNDER-SIZED** POST WILL TEND TO BUCKLE OUT AND ALLOW THE TOP PLATE TO SAG DOWN.

**B. PROPER SIZED** POST SUPPORTS THE WEIGHT AND HOLDS THE TOP PLATE STRAIGHT.

Figure 65–8. The height of the wall of a post-and-beam building and the weight it must support determine the size of the post that should be used. An undersized post may buckle.

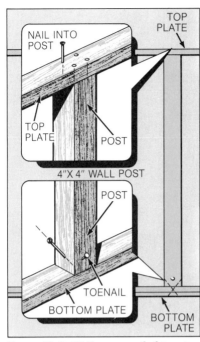

Figure 65–9. Wall posts nailed to a top and bottom plate.

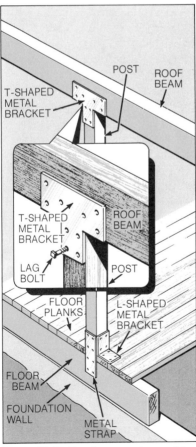

Figure 65–10. Brackets may be used to fasten the posts directly to the floor and to the beam at the top of the posts.

*California Redwood Association*

Figure 65–11. Interior view of a post-and-beam house showing the exposed roof beams and plank decking material. Note the great amount of window area at left. A sliding door leads to a deck.

full length of the building and are supported by posts at each end. In larger buildings it may be necessary to provide intermediate support between the ends. In one longitudinal method the roof planks are nailed directly to the beams as shown in Figure 65–14. In another method the rafters are nailed on top of the beams and the decking is nailed to the top of the rafters. See Figure 65–15.

Transverse roof beams run from the outside walls to a ridge beam. One end of each beam rests over a post in an outside wall. The other end butts against or rests on top of the ridge beam, as shown in Figure 65–16. The procedures for calculating the length of a transverse beam and marking the an-

*California Redwood Association*

Figure 65–12. Queen-post trusses are shown in the exposed ceiling. This is another variation of post-and-beam roof design.

Figure 65–13. Interior view of the ridge beam and roof beams.

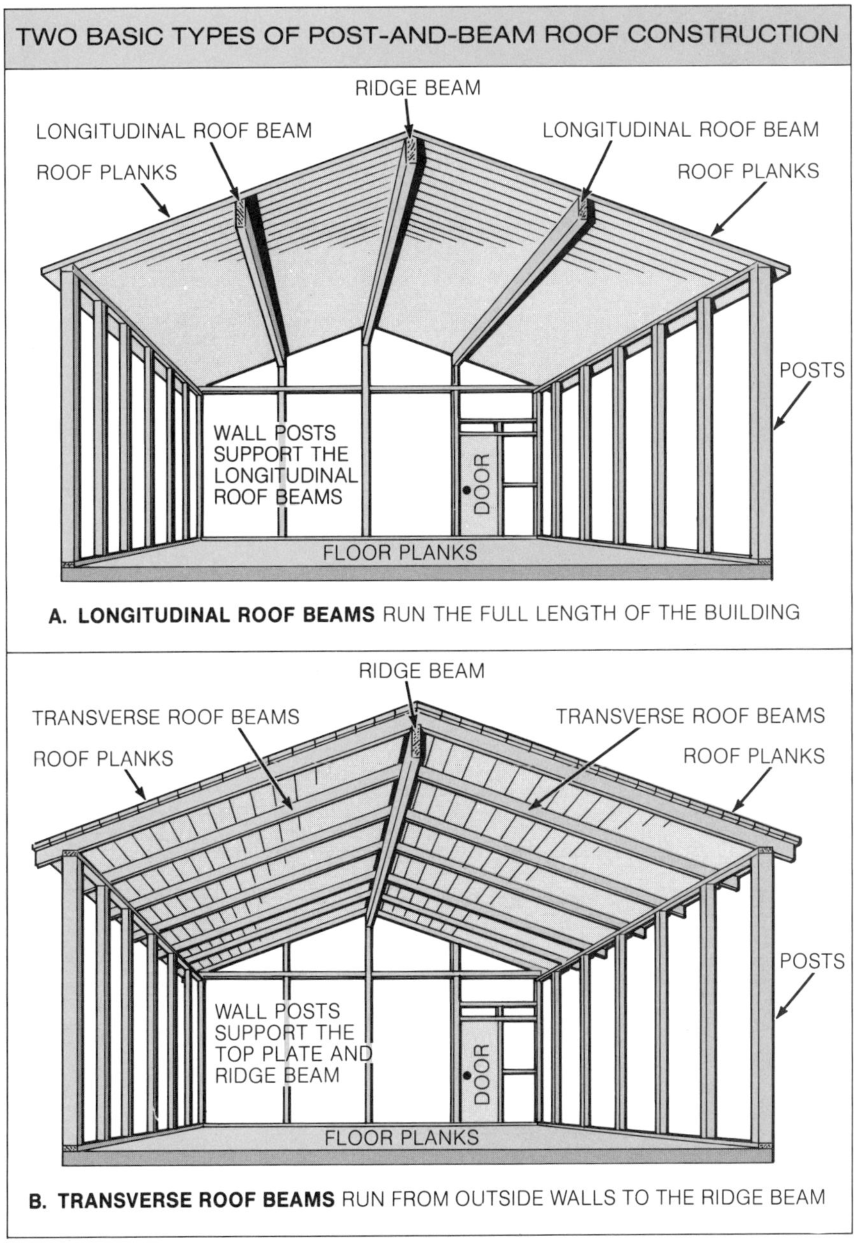

## TWO BASIC TYPES OF POST-AND-BEAM ROOF CONSTRUCTION

RIDGE BEAM

LONGITUDINAL ROOF BEAM                    LONGITUDINAL ROOF BEAM

ROOF PLANKS                                        ROOF PLANKS

                                                              POSTS

WALL POSTS
SUPPORT THE
LONGITUDINAL
ROOF BEAMS

DOOR

FLOOR PLANKS

**A. LONGITUDINAL ROOF BEAMS** RUN THE FULL LENGTH OF THE BUILDING

RIDGE BEAM

TRANSVERSE ROOF BEAMS                    TRANSVERSE ROOF BEAMS

ROOF PLANKS                                        ROOF PLANKS

                                                              POSTS

WALL POSTS
SUPPORT THE
TOP PLATE AND
RIDGE BEAM

DOOR

FLOOR PLANKS

**B. TRANSVERSE ROOF BEAMS** RUN FROM OUTSIDE WALLS TO THE RIDGE BEAM

Figure 65–14. Two basic post-and-beam roof designs are the longitudinal and transverse designs.

gle cuts are basically the same as for common rafters. Reviews of these layout methods as they apply to transverse beams are shown in Figures 65–17 and 65–18.

## Roof Beams

A good appearance grade of timber should be used for all exposed beams. Solid beams are usually practical for the sizes required in residential construction. Where larger sizes are required, because of greater weight and stress factors, it is advisable to use material consisting of laminated layers for the beams.

## Covering a Post-and-Beam Roof

Tongue-and-groove planks are the traditional material for roof

*California Redwood Association*
Figure 65–15. Building with longitudinal roof beams supporting the rafters. Tongue-and-groove planks are nailed to the tops of the rafters.

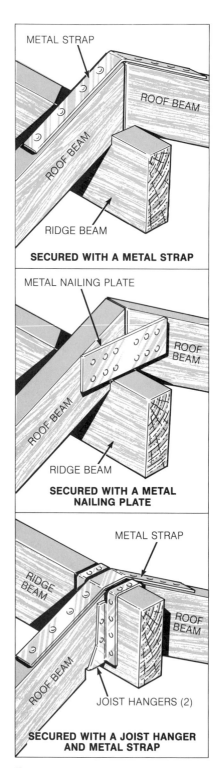

**SECURED WITH A METAL STRAP**

**SECURED WITH A METAL NAILING PLATE**

**SECURED WITH A JOIST HANGER AND METAL STRAP**

Figure 65–16. Methods of securing the ridge ends of transverse beams.

decking, and they are still widely used today. The planks come in thicknesses of 1½″, 2½″, and 3½″. See Figure 65–19. A more

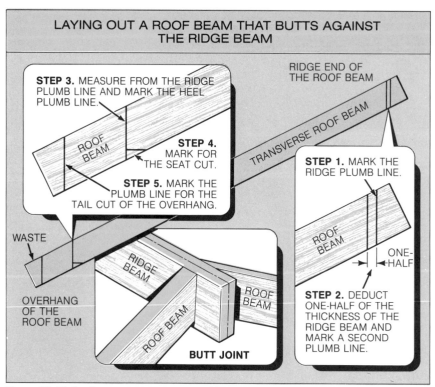

Figure 65–17. Laying out a transverse roof beam that butts against the ridge beam.

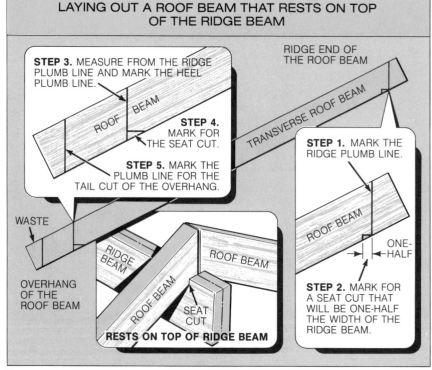

Figure 65–18. Laying out the ridge end of a transverse beam that rests on top of the ridge beam.

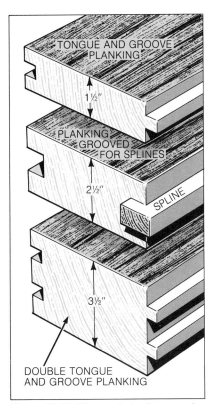

Figure 65–19. Types of planking used for plank-and-beam roofs.

economical material for covering a post-and-beam roof is 1⅛" thick tongue-and-groove plywood panels.

## Insulating a Post-and-Beam Roof

If the underside of the roof decking is to be exposed in order to give a natural look, insulation material must be placed on top of the deck. A standard procedure is to first place a vapor barrier and then a rigid type of insulation material. See Figure 65–20. After the insulation is placed, the finish roof covering can be applied.

## SEQUENCE OF POST-AND-BEAM CONSTRUCTION

Figure 65–21 shows a procedure for constructing a post-and-beam framework. However, the proce-

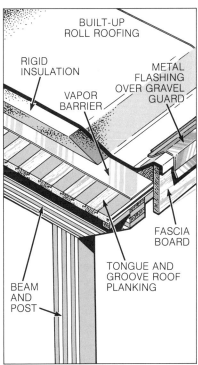

Figure 65–20. Typical insulation over a plank-and-beam roof consists of a vapor barrier and rigid insulation placed under the roofing material.

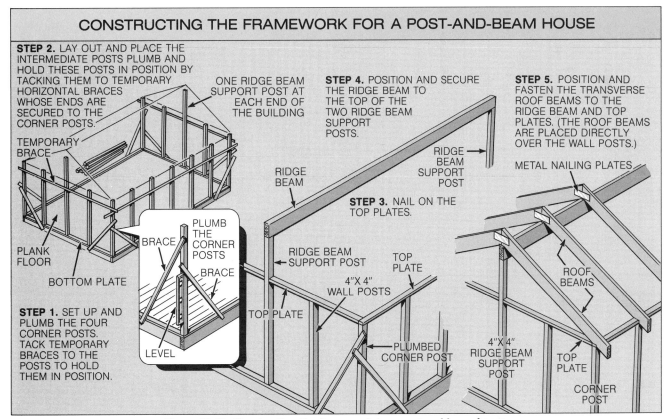

Figure 65–21. Constructing the framework for a post-and-beam house.

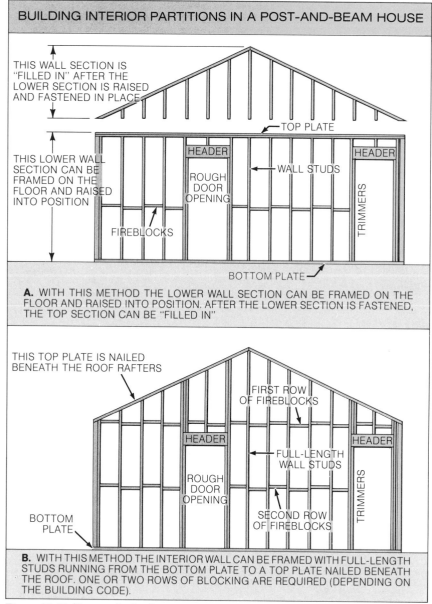

**BUILDING INTERIOR PARTITIONS IN A POST-AND-BEAM HOUSE**

THIS WALL SECTION IS "FILLED IN" AFTER THE LOWER SECTION IS RAISED AND FASTENED IN PLACE

THIS LOWER WALL SECTION CAN BE FRAMED ON THE FLOOR AND RAISED INTO POSITION

TOP PLATE

HEADER

HEADER

ROUGH DOOR OPENING

WALL STUDS

TRIMMERS

FIREBLOCKS

BOTTOM PLATE

**A.** WITH THIS METHOD THE LOWER WALL SECTION CAN BE FRAMED ON THE FLOOR AND RAISED INTO POSITION. AFTER THE LOWER SECTION IS FASTENED, THE TOP SECTION CAN BE "FILLED IN"

THIS TOP PLATE IS NAILED BENEATH THE ROOF RAFTERS

FIRST ROW OF FIREBLOCKS

HEADER

HEADER

ROUGH DOOR OPENING

FULL-LENGTH WALL STUDS

TRIMMERS

SECOND ROW OF FIREBLOCKS

BOTTOM PLATE

**B.** WITH THIS METHOD THE INTERIOR WALL CAN BE FRAMED WITH FULL-LENGTH STUDS RUNNING FROM THE BOTTOM PLATE TO A TOP PLATE NAILED BENEATH THE ROOF. ONE OR TWO ROWS OF BLOCKING ARE REQUIRED (DEPENDING ON THE BUILDING CODE).

Figure 65–22. Two methods of constructing interior partitions in an exposed-roof, post-and-beam house.

dure may vary depending on the design of a particular structure.

Interior partitions are usually erected after the outside wall and roof construction has been finished. The tops of partitions placed across the width of the house must be sloped to the shape of the roof. Figure 65–22 shows two methods for constructing interior partitions. In one method the lower wall section is framed on the floor and raised into place. Then the top section is filled in. In another method the interior wall is framed with full-length studs running from the bottom plate to a top plate nailed below the roof. One or two rows of blocking are required in this method.

A post-and-beam house generally requires less material than a platform-framed house of similar size. However, the savings in material can be offset by the disadvantage of slower building methods involved in post-and-beam construction. Walls cannot always be framed on the floor and raised in place. Posts must be positioned and plumbed individually and beams raised into place. Since so much of the carpentry work is exposed, extra care must be taken to avoid hammer marks and ill-fitting joints.

# UNIT 66

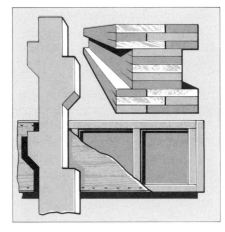

# Heavy Timber Construction and Pole Construction

## HEAVY TIMBER CONSTRUCTION

In heavy timber and pole construction post-and-beam principles are applied to large buildings and structures such as small bridges and piers. This type of construction is one of the oldest methods used in the United States. Today it is becoming increasingly popular for schools, office buildings, houses of worship, auditoriums, gymnasiums, arenas, and supermarkets. See Figure 66–1. In addition, it continues to be widely used in building small bridges, trestles, and waterfront docks and piers. See Figure 66–2.

## Glued, Laminated Timbers

Improved technology has produced glued, laminated (glulam) timbers that are probably the main reason for the increasing use of heavy timber construction. See Figure 66–3. In the laminated process, several pieces of lumber are joined together to make a heavier framing member. For example, three 2 × 6's (actual size 1¹/₂″ × 5¹/₂″) laminated together would produce a 4¹/₂″ × 5¹/₂″ piece. After surfacing the sides, the dimensions of the timber would be reduced to 4¹/₂″ × 5¹/₈″. The layers of wood are held together with a very strong bonding adhesive. The layers are joined edge to edge or face to face in a variety of arrangements. See Figure 66–4.

The thicknesses, widths, and lengths of solid pieces of lumber are limited to the diameter and

*Standard Structures, Inc.*

Figure 66–1. Heavy timber construction methods can be used to erect large buildings.

589

*American Institute of Timber Construction*

Figure 66–2. Heavy timber construction is still widely used for wood bridges.

*Weyerhaeuser Company*

Figure 66–3. The exposed ceiling beams in this building are made of glued, laminated lumber.

height of the treetrunk from which they are manufactured. Laminated timbers, however, can be made in shapes and dimensions that are not possible with a solid wood member. In addition, laminated timbers have higher fire-resistance than solid lumber. Although, being wood, they will burn, tests have shown that they retain their strength longer than an unprotected metal beam. See Figure 66–5.

**Fabrication.** Laminated timbers are usually made from softwood lumber. Some of the species commonly used are Douglas fir, southern pine, and California redwood. The lumber is kiln-dried before being glued together. Its moisture content must not exceed 16%.

The individual layers are not more than 2″ thick. The ends are fastened together by a *scarf joint* or a *finger joint*. A stepped scarf joint is made by notching and lapping the ends of the layers. See Figure 66–6. Each scarf joint should be at least 24″ away from the joint in an adjacent layer. Finger joints are made by interlocking multiple notches in the ends of the layers. See Figure 66–7.

Long layers that have been joined together are run through planers to guarantee a smooth, uniform thickness. Adhesive is then applied, and the layers of wood are clamped to each other. The completed member is run through a planer for a final surfacing (Figure 66–8) and is then prepared for shipment.

**Grades.** A uniform grading system has been established by the American Institute of Timber Construction (AITC). Since all the material used for laminated timbers must conform to high-strength structural standards, the grading system refers only to the

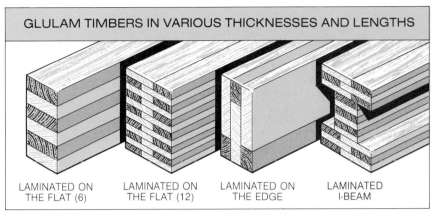

GLULAM TIMBERS IN VARIOUS THICKNESSES AND LENGTHS

LAMINATED ON
THE FLAT (6)

LAMINATED ON
THE FLAT (12)

LAMINATED ON
THE EDGE

LAMINATED
I-BEAM

Figure 66–4. Variations of glued, laminated (glulam) timbers.

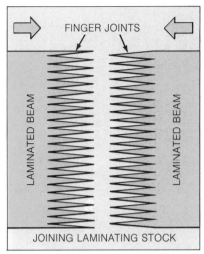

FINGER JOINTS

LAMINATED BEAM

LAMINATED BEAM

JOINING LAMINATING STOCK

Figure 66–7. A finger joint is made by interlocking fingers in the ends of lumber bonded with an adhesive.

*American Institute of Timber Construction*
Figure 66–5. As a result of fire, the steel beams of this building twisted and bent. In contrast, the charred wood beam retained much of its strength.

surface appearance of the member. *Industrial* grades are used in warehouses and factories. *Architectural* and *premium* appearance grades are used where an attractive surface is important.

## Heavy Timber Construction Methods

Heavy timbers may be used for the basic framework of an entire building (Figure 66–9) or may be combined with stud-framed or concrete walls. Various types of heavy timber roofs are frequently combined with walls made by other construction methods. See Figure 66–10. Large truss systems are often placed over these walls.

A popular system in commer-

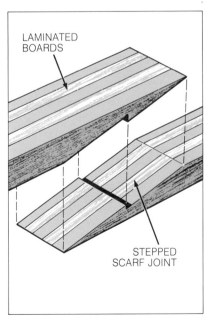

LAMINATED BOARDS

STEPPED SCARF JOINT

Figure 66–6. A stepped scarf joint is made by notching and lapping the ends of the layers of glulam boards.

*American Institute of Timber Construction*
Figure 66–8. Glulam timbers are run through a planer for a final surfacing.

American Institute of Timber Construction

Figure 66–9. Curved glulam timbers provide the basic framework for this building.

cial construction places a *beam-and-purlin* roof on top of wood-framed or metal-framed walls. The heavy timbers are lifted into place by crane. The purlins fit between the beams and rest on heavy metal hangers fastened to the beams. Wood planking or plywood is usually applied as decking material. See Figure 66–11.

Metal connectors are used extensively in heavy timber construction. Through-bolts and nuts are recommended wherever possible. See Figure 66–12. Usually glued, laminated members are prefabricated at a plant according to blueprint and job specifications. They are delivered to the site and put in place with the aid of mechanical cranes or lifts. See Figures 66–13 through 66–15.

Western Wood Products Association

Figure 66–10. Erecting a commercial building that will combine a heavy timber post, beam, and girder system with stud-framed wall construction.

## Veneered Beams, Box Beams, and Stressed-skin Panels

Several types of engineered beam or panel units have been developed in recent years. Some have proven practical for residential and commercial buildings. *Veneered beams, box beams,* and *stressed-skin panels* are used as structural parts in heavy timber construction or are combined with other framing methods.

**Veneered Beams.** Veneered beams are made up of wood veneers dried to a 6% moisture

content and positioned with all the grains running in the same direction. See Figure 66–16. They are coated with a waterproof adhesive and fed into a combination press-and-curing oven. Veneered beams can be manufactured in lengths up to 80′ and widths up to 24″. Thicknesses vary from ¾″ to 2½″ at ¼″ intervals. See Figure 66–17. Veneered beams conform to the structural requirements of major building codes. They offer the advantage of being virtually free of warp, crook, or check.

**Box Beams.** Box beams resemble a hollow box. Naturally, they are lighter to handle than solid

beams, and are sometimes used in place of solid beams in regular post-and-beam construction. They are also used as headers to span wide openings for double garage doors. Flat roofs can be constructed with box instead of solid beams.

Box beams are made up of plywood sides, a top and bottom chord, and vertical stiffeners. They are easy to build on or off the building site. See Figure 66–18. Figure 66–19 shows a procedure for constructing a box beam.

**Stressed-skin Panels.** Stressed-skin panels are designed primarily for floor and roof systems. Under certain conditions they are also used to form wall sections. They are made of plywood panels (skins) nailed and glued with special rigid structural adhesive on one or both sides of nominal 2″ x 4″ pieces. See Figures 66–20 and 66–21. Stressed-skin panels must be fabricated in a shop under careful quality control.

## POLE CONSTRUCTION

Pole construction is a timber building method of ancient origin. Evidence exists that it was first used by prehistoric lake dwellers. Since that time, in many cultures, people have sunk poles into the earth and built structures off the ground to protect themselves from other unfriendly humans, wild animals, and storm and flood conditions.

In the past the biggest problem in pole construction was the fairly rapid decay of the wood poles embedded in the ground, even when decay-resistant species of wood such as redwood, cypress, and cedar were used. Pole construction became practical as a modern building method because of effective methods developed

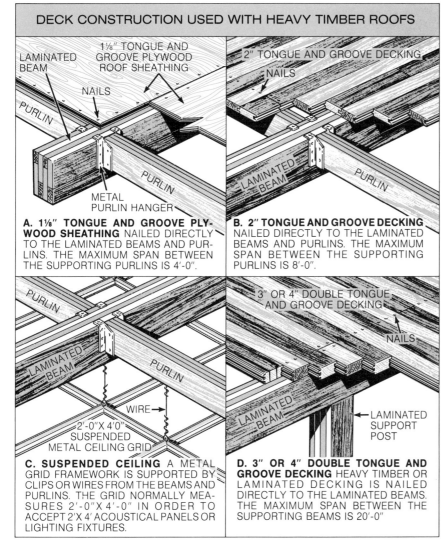

**DECK CONSTRUCTION USED WITH HEAVY TIMBER ROOFS**

LAMINATED BEAM
1⅛″ TONGUE AND GROOVE PLYWOOD ROOF SHEATHING
NAILS
PURLIN
PURLIN
METAL PURLIN HANGER

**A. 1⅛″ TONGUE AND GROOVE PLYWOOD SHEATHING** NAILED DIRECTLY TO THE LAMINATED BEAMS AND PURLINS. THE MAXIMUM SPAN BETWEEN THE SUPPORTING PURLINS IS 4′-0″.

2″ TONGUE AND GROOVE DECKING
NAILS
LAMINATED BEAM
PURLIN

**B. 2″ TONGUE AND GROOVE DECKING** NAILED DIRECTLY TO THE LAMINATED BEAMS AND PURLINS. THE MAXIMUM SPAN BETWEEN THE SUPPORTING PURLINS IS 8′-0″.

PURLIN
LAMINATED BEAM
PURLIN
WIRE
2′-0″X 4′0″ SUSPENDED METAL CEILING GRID

**C. SUSPENDED CEILING** A METAL GRID FRAMEWORK IS SUPPORTED BY CLIPS OR WIRES FROM THE BEAMS AND PURLINS. THE GRID NORMALLY MEASURES 2′-0″X 4′-0″ IN ORDER TO ACCEPT 2′X 4′ ACOUSTICAL PANELS OR LIGHTING FIXTURES.

3″ OR 4″ DOUBLE TONGUE AND GROOVE DECKING
NAILS
LAMINATED BEAM
LAMINATED SUPPORT POST

**D. 3″ OR 4″ DOUBLE TONGUE AND GROOVE DECKING** HEAVY TIMBER OR LAMINATED DECKING IS NAILED DIRECTLY TO THE LAMINATED BEAMS. THE MAXIMUM SPAN BETWEEN THE SUPPORTING BEAMS IS 20′-0″

Figure 66–11. Decking systems used with heavy timber roofs.

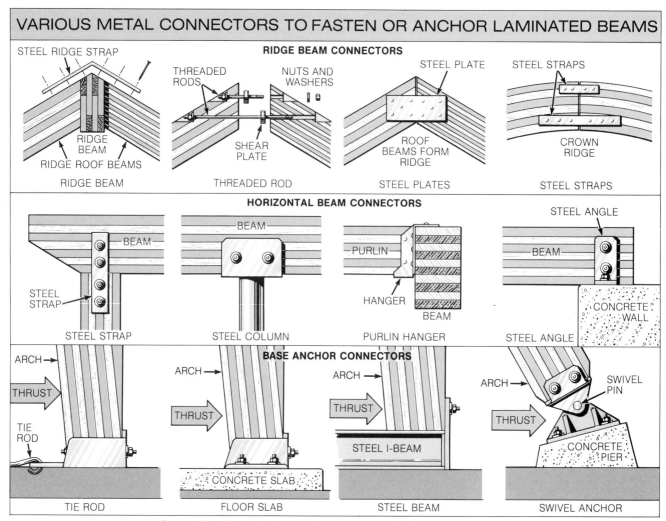

**VARIOUS METAL CONNECTORS TO FASTEN OR ANCHOR LAMINATED BEAMS**

**RIDGE BEAM CONNECTORS**

STEEL RIDGE STRAP

THREADED RODS

NUTS AND WASHERS

STEEL PLATE

STEEL STRAPS

RIDGE BEAM

RIDGE ROOF BEAMS

SHEAR PLATE

ROOF BEAMS FORM RIDGE

CROWN RIDGE

RIDGE BEAM          THREADED ROD          STEEL PLATES          STEEL STRAPS

**HORIZONTAL BEAM CONNECTORS**

BEAM

STEEL STRAP

BEAM

PURLIN

HANGER

BEAM

STEEL ANGLE

BEAM

CONCRETE WALL

STEEL STRAP          STEEL COLUMN          PURLIN HANGER          STEEL ANGLE

**BASE ANCHOR CONNECTORS**

ARCH

THRUST

TIE ROD

ARCH

THRUST

CONCRETE SLAB

ARCH

THRUST

STEEL I-BEAM

ARCH

THRUST

SWIVEL PIN

CONCRETE PIER

TIE ROD          FLOOR SLAB          STEEL BEAM          SWIVEL ANCHOR

Figure 66–12. Details of connector systems used with glulam timbers.

Figure 66–13. Nailing metal joist hangers to a glulam beam before it is lifted into place.

*Western Wood Products Association*

Figure 66–14. Glulam girders and beams will support the roof of this building. Note the steel connectors tieing the beams to the girders.

Figure 66–15. Glulam beam resting on a header in the outside wall. Note the metal connector fastening the timbers together.

*Trus Joist Corporation, Micro-Lam Division*

Figure 66–17. Veneered beams in a basement supporting floor joists. The floor joists shown here are also made of laminated veneered lumber pressed between top and bottom flanges.

*Trus Joist Corporation, Micro-Lam Division*

Figure 66–16. Veneered beams are made up of wood veneers bonded together with a waterproof adhesive.

*American Plywood Association*

Figure 66–18. Nailing together box beams on the job.

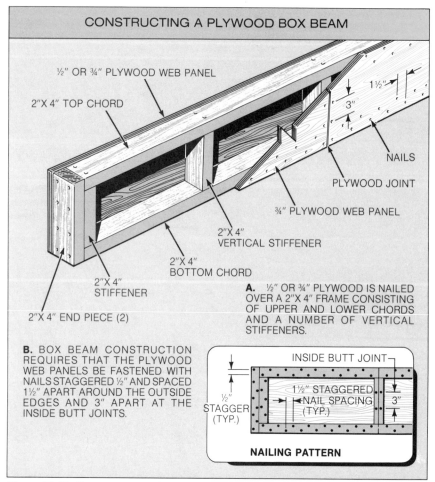

CONSTRUCTING A PLYWOOD BOX BEAM

½" OR ¾" PLYWOOD WEB PANEL

2"X 4" TOP CHORD

1½"

3"

NAILS

PLYWOOD JOINT

¾" PLYWOOD WEB PANEL

2"X 4" VERTICAL STIFFENER

2"X 4" BOTTOM CHORD

2"X 4" STIFFENER

2"X 4" END PIECE (2)

**A.** ½" OR ¾" PLYWOOD IS NAILED OVER A 2"X 4" FRAME CONSISTING OF UPPER AND LOWER CHORDS AND A NUMBER OF VERTICAL STIFFENERS.

**B.** BOX BEAM CONSTRUCTION REQUIRES THAT THE PLYWOOD WEB PANELS BE FASTENED WITH NAILS STAGGERED ½" AND SPACED 1½" APART AROUND THE OUTSIDE EDGES AND 3" APART AT THE INSIDE BUTT JOINTS.

INSIDE BUTT JOINT

1½" STAGGERED NAIL SPACING (TYP.)

3"

½" STAGGER (TYP.)

**NAILING PATTERN**

Figure 66–19. Constructing a plywood box beam. It consists of plywood sides, top and bottom chords, and vertical stiffeners.

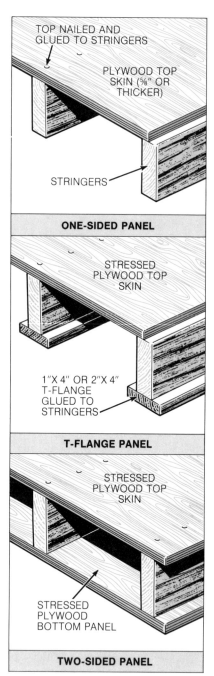

TOP NAILED AND GLUED TO STRINGERS

PLYWOOD TOP SKIN (⅝" OR THICKER)

STRINGERS

**ONE-SIDED PANEL**

STRESSED PLYWOOD TOP SKIN

1"X 4" OR 2"X 4" T-FLANGE GLUED TO STRINGERS

**T-FLANGE PANEL**

STRESSED PLYWOOD TOP SKIN

STRESSED PLYWOOD BOTTOM PANEL

**TWO-SIDED PANEL**

Figure 66–20. Stressed-skin panel designs. These panels are prefabricated in a shop.

Figure 66–21. Stressed-skin panels being placed on a building of heavy timber construction featuring an A-frame design.

for pressure-treating lumber with chemical preservatives to resist decay. Pressure-treated poles became available after World War II, resulting in a large increase of pole-constructed farm buildings and beach and country cottages. The use of chemically-

596

Figure 66–22. Example of pole construction on a steep hillside lot. The house is supported by pressure-treated poles that extend deep into the ground.

treated poles further increased when pole-constructed houses were built on the West Coast in the 1950s on steep slopes in areas subject to floods and earthquakes. See Figures 66–22 and 66–23. West Coast houses convinced many architects and builders that pole construction is a safe and economical method.

## Pole Construction Methods

The poles used in pole construction range in diameter from 6″ to 12″ and are spaced 6′ to 12′ apart. Figure 66–24 shows how the poles are laid out for a building. Batter boards and building lines must be set up. Pole loca-

tions are measured from the building lines. Holes are dug with a power auger or by hand, and the poles are dropped into place. Poles are then plumbed and held in position with temporary braces until the holes can be backfilled and the soil tamped down. See Figure 66–25.

In some cases soil conditions

597

Koppers Company, Inc.

Figure 66–23. Pressure-treated poles act as the foundation of this house. The poles around the porch area extend upward to help support the roof section over the porch area.

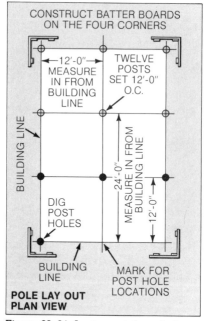

Figure 66–24. Laying out the pole locations for a pole-constructed building.

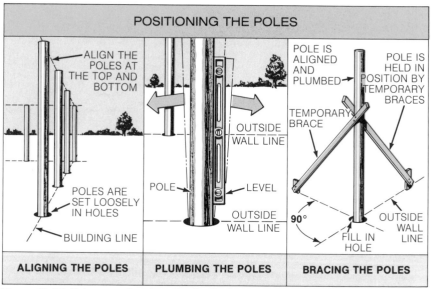

Figure 66–25. Poles must be aligned, plumbed, and temporarily braced before the holes can be back-filled.

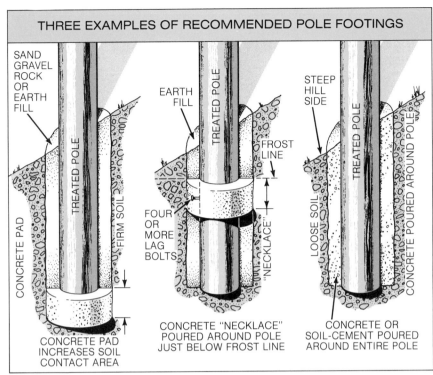

THREE EXAMPLES OF RECOMMENDED POLE FOOTINGS

SAND GRAVEL ROCK OR EARTH FILL

CONCRETE PAD

TREATED POLE

FIRM SOIL

CONCRETE PAD INCREASES SOIL CONTACT AREA

EARTH FILL

TREATED POLE

FROST LINE

FOUR OR MORE LAG BOLTS

"NECKLACE"

CONCRETE "NECKLACE" POURED AROUND POLE JUST BELOW FROST LINE

STEEP HILL SIDE

TREATED POLE

LOOSE SOIL

CONCRETE POURED AROUND POLE

CONCRETE OR SOIL-CEMENT POURED AROUND ENTIRE POLE

Figure 66–26. Examples of recommended pole footings. These may be required on steep slopes or where unstable soil conditions exist.

may require that a concrete *pad* be placed under the pole to increase contact area with the soil. A concrete *necklace* around the pole may be used for the same purpose. On very steep slopes, soil conditions may require concrete poured around the entire embedded section of the pole. See Figure 66–26.

In pole platform construction the poles act as a foundation. They extend only to the first floor of the building. See Figure 66–27. Header joists are bolted to the poles, the rest of the floor unit is framed, and the subfloor is placed. After the subfloor is placed, conventional platform framing methods can be used to complete the building.

In pole-framed buildings all the poles around the outside of the building, and some of the interior poles, extend to the roof. The floor, wall, ceiling, and roof framing members are tied into the poles. Headers bolted to the poles with lag bolts provide the main support for the floor and ceiling joists, as well as for the roof rafters. See Figure 66–28.

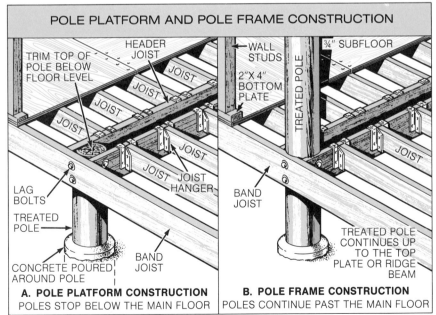

POLE PLATFORM AND POLE FRAME CONSTRUCTION

HEADER JOIST

TRIM TOP OF POLE BELOW FLOOR LEVEL

JOIST

JOIST

JOIST

JOIST

JOIST

JOIST HANGER

LAG BOLTS

TREATED POLE

CONCRETE POURED AROUND POLE

BAND JOIST

**A. POLE PLATFORM CONSTRUCTION**
POLES STOP BELOW THE MAIN FLOOR

WALL STUDS

¾" SUBFLOOR

2"X 4" BOTTOM PLATE

TREATED POLE

BAND JOIST

JOIST

TREATED POLE CONTINUES UP TO THE TOP PLATE OR RIDGE BEAM

**B. POLE FRAME CONSTRUCTION**
POLES CONTINUE PAST THE MAIN FLOOR

Figure 66–27. In a pole-platform building, poles do not extend past the main floor. In a pole-frame building, the poles continue past the main floor. Headers are bolted to poles with lag bolts. Metal joist hangers may be utilized.

*Koppers Company, Inc.*

Figure 66–28. The floor beams of a pole-constructed house are bolted to pressure-treated poles that extend into the ground.

Metal fasteners are recommended wherever framing members are fastened to the headers.

Metal fasteners are available in a variety of configurations to meet varied job requirements. These fasteners may be either bolted or nailed, depending upon the size of the fastener and the size of the framing member to which the fastener is attached.

# SECTION 16

# Concrete Heavy Construction

Concrete heavy construction is used to erect large office buildings, factories, apartment buildings, schools, hospitals, churches, warehouses, and many other kinds of commercial structures. In addition, many heavy construction procedures are used in building dams, bridges, and overpasses.

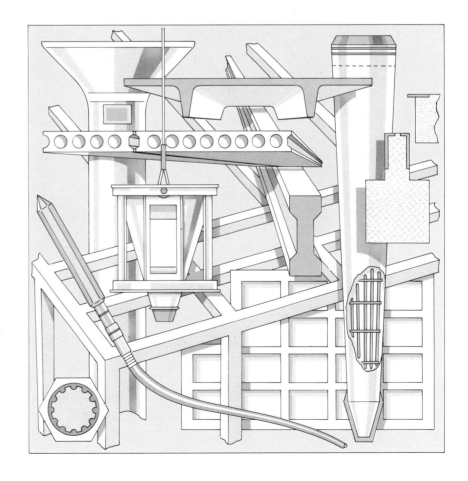

# UNIT 67

# Foundation Design for Heavy Construction

The major considerations in designing a foundation for a large concrete building are the size and weight of the building and the soil conditions on the job site. (Soil types are discussed in Section 8.) Some types of soil can support smaller and lighter concrete buildings with little excavation. Larger and heavier buildings on the same type of soil may require very deep excavations before the foundation footings can be placed. See Figures 67–1 and 67–2. (General

Figure 67–1. Deep excavation for a high-rise building. Note the earth ramp and roadway that allow vehicles and machinery to move in and out of the excavation.

*Spencer, White, & Prentis, Inc.*

Figure 67-2. Several pieces of heavy equipment are constantly operating during the excavation stage of a large construction project. A transit-mix concrete truck is delivering concrete to be placed in the forms at left. A power shovel is digging out a section at the bank. A truck-mounted telescope crane is lifting and placing materials for the form construction at left. A mobile crane with an auger attachment is boring holes in the ground for drilled caissons that are part of the foundation system.

safety considerations and shoring methods around deep excavations are discussed in Section 5. Concrete foundations for light construction are discussed in Section 8.)

## PILES

Many tall concrete buildings are designed with deep basements and heavy column loads. However, because of general soil conditions and the closeness of existing surrounding buildings, it may not be possible to excavate the entire site area to the depth required to reach bearing soil. Under these conditions *piles* are often used.

Piles are long structural members that penetrate deep into the soil. They are placed beneath grade beams that support bearing walls. See Figure 67–3. Piles may also be grouped and joined together with a concrete cap. See Figure 67–4. This is necessary when columns provide the main structural support for the building above and the column load exceeds the capacity of an individual pile.

Piles are placed into the ground with special pile-driving equipment. The equipment has a drop, mechanical, or vibratory hammer that directly drives a complete pile or pile casing into the ground. Some of the power sources used for pile-driving hammers are compressed air,

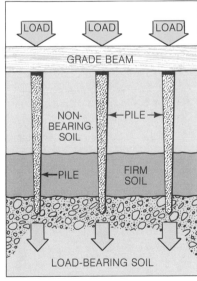

Figure 67-3. Piles are column-like structural members that carry building loads through non-bearing soil to lower levels of bearing soil. Piles receive their loads from columns or grade beams.

steam, diesel pistons, and hydraulic fluid under pressure. Although there are machines specifically designed for pile driving, mobile cranes with additional equipment can also be used for this purpose.

## Types of Piles

The three major types of piles are *bearing, friction,* and *sheet piles.* They are made of wood, steel, or concrete. A composite type has a wood lower section and a concrete upper section.

**Bearing Piles.** Bearing piles are the most common type used.

They must penetrate completely through unstable soil layers to the firm, bearing soil below.

**Friction Piles.** Friction piles need not penetrate to bearing soil. They need only reach a point where soil resistance is sufficient to support a portion of the load. A major share of the load is supported by the surrounding soil pressing against the pile. A comparison between bearing and friction piles is shown in Figure 67–5.

**Sheet Piles.** Sheet piles are not intended to carry vertical loads. Their primary purpose is to resist horizontal pressure. They are often used to hold back earth around the perimeters of deep excavations. (Refer to Section 5.)

**Wood Piles.** Wood piles were used by the Romans over 2,000 years ago and are still used today. Species such as Douglas fir, southern pine, red pine, western hemlock, and larch are commonly used for wood piles. The wood is pressure-treated for protection against decay and deterioration.

Wood piles are light, relatively inexpensive, and readily available in many areas. They have an indefinite life expectancy under water, so are particularly suitable for wharves, docks, and other structures built over water. They are not often used as part of the foundation system for concrete buildings constructed on dry land.

**Concrete Piles.** Concrete piles may be *precast* or *cast-in-place.* Precast piles are usually fabricated in a factory. See Figures 67–6 and 67–7. They are then delivered to the job site, where

they are driven into the ground with a pile-driving rig. Some precast shapes are shown in Figure 67–8.

Cast-in-place concrete piles are *shell* or *shell-less* piles. The shell type is made by first driving a steel shell called a *casing* into the ground. Casings vary from 20" to 5' in diameter. As the casing is being driven into the ground, the soil is removed within the casing with a special grab mechanism. Concrete is then placed in the casing, which acts as a form and prevents mud

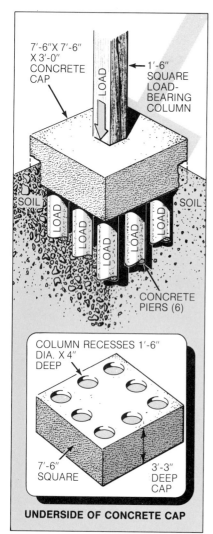

Figure 67–4. Concrete piles grouped under a concrete cap. A load-bearing column rests on top of the cap.

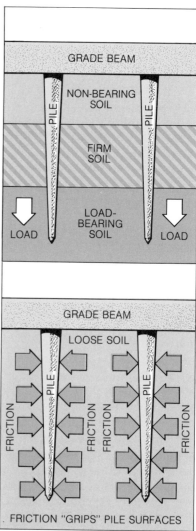

Figure 67–5. Bearing piles transmit their loads directly to firm, bearing soil. Friction piles rely largely on pressure and friction from surrounding soil to help support the load.

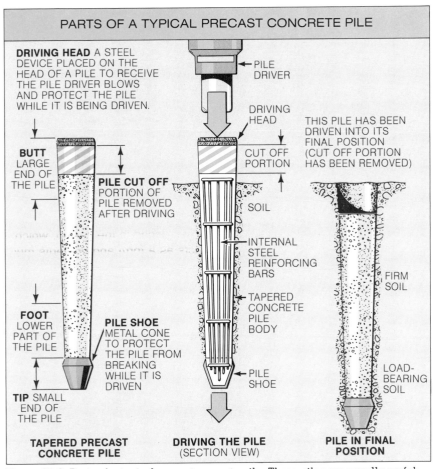

## PARTS OF A TYPICAL PRECAST CONCRETE PILE

**DRIVING HEAD** A STEEL DEVICE PLACED ON THE HEAD OF A PILE TO RECEIVE THE PILE DRIVER BLOWS AND PROTECT THE PILE WHILE IT IS BEING DRIVEN.

PILE DRIVER

DRIVING HEAD

THIS PILE HAS BEEN DRIVEN INTO ITS FINAL POSITION (CUT OFF PORTION HAS BEEN REMOVED)

**BUTT** LARGE END OF THE PILE

**PILE CUT OFF** PORTION OF PILE REMOVED AFTER DRIVING

CUT OFF PORTION

SOIL

INTERNAL STEEL REINFORCING BARS

TAPERED CONCRETE PILE BODY

**FOOT** LOWER PART OF THE PILE

**PILE SHOE** METAL CONE TO PROTECT THE PILE FROM BREAKING WHILE IT IS DRIVEN

PILE SHOE

FIRM SOIL

LOAD-BEARING SOIL

**TIP** SMALL END OF THE PILE

**TAPERED PRECAST CONCRETE PILE**

**DRIVING THE PILE** (SECTION VIEW)

**PILE IN FINAL POSITION**

Figure 67–6. Parts of a typical precast concrete pile. These piles are usually prefabricated in a factory.

*Kaiser Cement Corporation*
Figure 67–7. These precast concrete piles are ready for delivery to the construction site.

and water from mixing with the concrete. After the concrete hardens, the casing (shell) is left in place. Examples of shell concrete piles are shown in Figure 67–9.

For shell-less piles, the casing is withdrawn and the hole is filled with concrete. Two methods can be used. The casing can be completely withdrawn before the concrete is placed, or the concrete can be poured as the casing is gradually being withdrawn. In either case, shell-less piles can be used only with firm, cohesive soil.

**Steel piles.** Steel piles may be *H-shaped* or *tubular*. The tubular piles, also referred to as *pipe piles*, are normally filled with concrete after they have been placed. See Figures 67–10 and

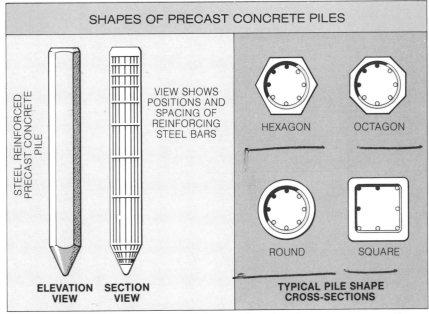

## SHAPES OF PRECAST CONCRETE PILES

STEEL REINFORCED PRECAST CONCRETE PILE

VIEW SHOWS POSITIONS AND SPACING OF REINFORCING STEEL BARS

HEXAGON

OCTAGON

ROUND

SQUARE

**ELEVATION VIEW**   **SECTION VIEW**

**TYPICAL PILE SHAPE CROSS-SECTIONS**

Figure 67–8. Typical shapes and designs of steel-reinforced precast concrete piles.

67–11. When driven into the ground, the lower ends of tubular piles are closed off with a steel

*boot* to prevent earth from filling the inside of the pile. Tubular piles are sometimes placed in

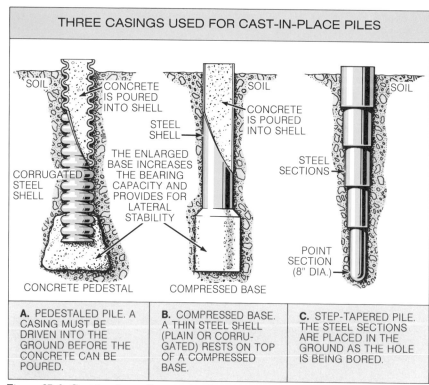

**THREE CASINGS USED FOR CAST-IN-PLACE PILES**

SOIL — CONCRETE IS POURED INTO SHELL

CORRUGATED STEEL SHELL

CONCRETE PEDESTAL

SOIL — CONCRETE IS POURED INTO SHELL

STEEL SHELL

THE ENLARGED BASE INCREASES THE BEARING CAPACITY AND PROVIDES FOR LATERAL STABILITY

COMPRESSED BASE

SOIL

STEEL SECTIONS

POINT SECTION (8″ DIA.)

**A.** PEDESTALED PILE. A CASING MUST BE DRIVEN INTO THE GROUND BEFORE THE CONCRETE CAN BE POURED.

**B.** COMPRESSED BASE. A THIN STEEL SHELL (PLAIN OR CORRUGATED) RESTS ON TOP OF A COMPRESSED BASE.

**C.** STEP-TAPERED PILE. THE STEEL SECTIONS ARE PLACED IN THE GROUND AS THE HOLE IS BEING BORED.

Figure 67–9. Cast-in-place piles require that a casing be driven into the ground before the concrete can be placed.

prepared holes drilled by an earth auger.

## CAISSONS

For higher buildings with increased column loads it is sometimes necessary to go so deep to reach good bearing soil that pile systems are inadequate. Under these conditions *caissons* are used. A caisson is a cylindrical or box-like casing that is placed in the ground. It is then filled with concrete and serves the same function as a cast-in-place pile. The main difference between piles and caissons is their size. Caissons can be constructed to much wider diameters and they extend deeper than piles.

BOX LEADS FRAME

MOBILE CRANE

COMPRESSED AIR LINE

HAMMER

SLEEVE

*Spencer, White, & Prentis, Inc.*

Figure 67–10. Driving tubular (pipe) piles into the ground with a mobile crane outfitted with fixed box leads. The box leads apparatus is attached at its top to the crane boom and is held in position at the bottom by a hydraulically operated beam. The air-driven hammer that drives the pile rides up and down within the leads. The piles shown above are being driven in sections until the desired depth is reached. The sections are joined together by a prefabricated sleeve welded into place.

## Bored Caissons

Excavation for bored caissons is made by machines capable of drilling large-diameter holes in the ground. The casing, which is usually cylindrical in shape and made of metal, is then placed in the hole. See Figures 67–12 and 67–13. Drilling machinery and methods have been devised that make it possible to bore holes 10′ in diameter to a depth of over 150′. The casing for caissons extending to greater depths is made up of sections that are added as the drilling proceeds.

## Belled Caissons

The belled design provides a greater bearing area at the bottom of the caisson. See Figure 67–14. After the caisson hole has been drilled to the desired depth, a belling tool is attached to the drilling head in order to dig out the shape of the bell. It is often necessary to clear out the bell hole with hand tools, in which case a worker must be lowered into the excavation.

## SPREAD FOUNDATIONS

If building loads and soil conditions do not require a system of piles or caissons, some type of spread foundation design is usually adequate for concrete buildings. The spread foundation usually consists of a foundation wall resting on top of a wider base

GROUPED PILES

HYDRAULIC BEAM SECURING BOTTOM OF BOX LEADS

*Spencer, White, & Prentis, Inc.*

Figure 67–11. Piles are being driven in groups on this construction site. Each group will be covered and joined together with a concrete cap. Note the section being welded into place where piles are driven at center.

*Calweld, Inc.*

Figure 67–12. An auger 5' in diameter is being used to bore holes for the caissons on this construction site. A crane attachment with a 70' long, double-telescoping Kelly bar drives the auger. Note the top of the caisson casing extending above the ground to the left of the auger.

Figure 67–13. This reamer-and-bucket device is drilling a caisson hole 10'-6" in diameter

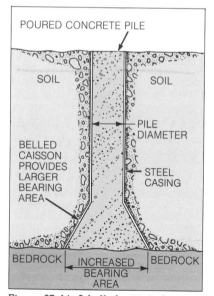

Figure 67–14. A belled caisson has a greater bearing area at its bottom.

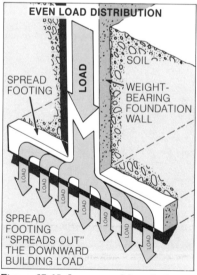

Figure 67–15. In a spread foundation, the footing below the foundation wall distributes the building load over a wider area.

(footing). This design is basically the same as the inverted T-type of design often used for residential and other light construction foundations. The main difference is the size of the footings. They are wider and thicker for heavy construction than for light construction. See Figure 67–15.

## Matt Foundations

Matt foundations are used in soils of low bearing strength. This type of spread foundation consists of a solid slab of heavily reinforced concrete placed beneath the entire building area. In some cases the slab may be 3' to 8' thick.

# UNIT 68

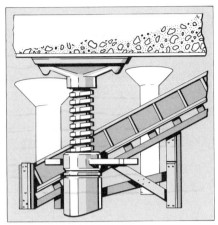

# Forms for Concrete Heavy Construction

The basic structural parts of a concrete building are the walls, floors, beams, and columns. Forms must be constructed to the shapes of these parts (unless prefabricated systems are used, as discussed in Unit 70). Concrete is then placed in the forms. See Figure 68–1.

Forms must be rigid enough to resist bulging, twisting, or moving during the concrete pour. They must be tightly constructed to prevent concrete from leaking through the joints. Leaks will cause ridges on the surface of the hardened concrete. All debris such as wood scraps, sawdust, and nails should be removed from the inside of the form before the concrete is placed. This is done by cutting clean-out holes at the bottom of the forms and using compressed air hoses to blow out the debris. The clean-out holes are then blocked off.

Forms must also be designed for easy removal after the concrete has hardened. Removal must be accomplished without damaging the concrete.

Wood is still the most widely used material for building forms. However, the use of patented forming systems of other materials such as metal or plastic is increasing.

Steel reinforcing bars are used to strengthen the concrete and also to tie together sections of a concrete building. See Unit 37 of Section 8 for more information on reinforced concrete.

## PLYWOOD FORMS

Most wood forms today are constructed of plywood backed by various combinations of studs, walers, strongbacks, and some type of bracing. Plywood presents a large, smooth surface that can withstand rough treatment without splitting. Another advantage of plywood is its ability to bend when curved surfaces are required in the form design. See Figure 68–2. Textured plywood can also be used for a textured finish on the concrete. See Figure 68–3.

Most exterior grade plywood can be used for form construction. However, the plywood industry produces a product called *Plyform* specifically for concrete formwork. Plyform is an exterior type of plywood fabricated from wood species and veneer grades

*The Burke Company*

Figure 68–1. Formwork in progress during construction of a reinforced concrete building. A completed form section is shown at upper left. At right is a section of concrete wall that has already hardened (the forms have been removed). At bottom a worker is placing steel reinforcing bars before the wall forms are doubled up.

that assure high performance.

Plyform panels can be reused about 10 times if properly cared for. Oiling the face of the panels reduces moisture penetration.

American Plywood Association

Figure 68-2. An advantage of plywood in form construction is that it can be bent when a curved surface is required, as in this ramp being constructed as the base for a curved stairway.

## FOUNDATION AND BASEMENT FORMS

The procedure for laying out and constructing footing forms for large concrete buildings is basically the same as that described in Section 8 for residential and other light construction work. The major difference is the greater width and depth required for footings for large concrete buildings.

In constructing basement walls, the first step is to place all the outside panel sections. If possible, holes for patented ties should be drilled in the panels before they are set up. See Figures 68-5 through 68-7.

After all the panels for a wall have been positioned and the walers and strongbacks (if required) have been attached, the

The Burke Company

Figure 68-3. Special textured plywood was used to create this finish on the surface of the concrete.

The oil also acts as a release agent when the panels are later stripped from the hardened concrete. Various other types of release agents are available, including waxes, oil emulsions, cream emulsions, and water-soluble emulsions. Panels should always be treated with one of the products before they are reused. See Figure 68-4.

Symons Corporation

Figure 68-4. Spraying a release agent on panels to be used for concrete forms.

610

Figure 68–5. Carpenters placing ties into holes drilled in the form panel before raising the panel into place.

## Doors and Windows

After the outside form panels for the basement walls are set, provisions are made for any door or window openings required. Refer to Section 8 for a description of this procedure. Traditional rough window and door wood bucks for openings in concrete walls are shown in Figure 68–10. A rectangle at the bottom provides an

Figure 68–6. Setting a panel for the outside form wall.

wall is aligned and braced. Braces are usually placed 6' to 8' apart. A piece of 2" × 4" material nailed to the wall and a wood stake driven into the ground may serve as a brace, or a patented brace may be used. One type of patented wood brace features a turnbuckle tightening attachment that is fastened to the ground with steel stakes. See Figure 68–8. A brace plate is used to fasten the brace to a metal stiffback on the wall. Another type of brace is made completely of steel. See Figure 68–9.

Figure 68–7. Panels in place for the outside form walls.

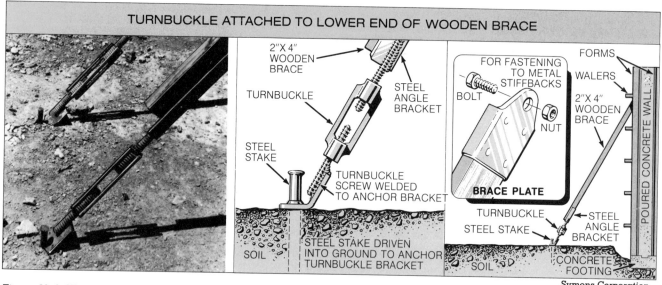

## TURNBUCKLE ATTACHED TO LOWER END OF WOODEN BRACE

2"X 4" WOODEN BRACE

TURNBUCKLE

STEEL STAKE

STEEL ANGLE BRACKET

TURNBUCKLE SCREW WELDED TO ANCHOR BRACKET

SOIL

STEEL STAKE DRIVEN INTO GROUND TO ANCHOR TURNBUCKLE BRACKET

FOR FASTENING TO METAL STIFFBACKS

BOLT

NUT

**BRACE PLATE**

FORMS

WALERS

2"X 4" WOODEN BRACE

POURED CONCRETE WALL

TURNBUCKLE

STEEL STAKE

STEEL ANGLE BRACKET

SOIL

CONCRETE FOOTING

*Symons Corporation*

Figure 68–8. Wood braces featuring a turnbuckle tightening attachment at their lower ends. The turnbuckle is fastened to the ground with steel stakes. Where the brace fastens to a metal stiffback on the wall, a brace plate can be used as shown at right.

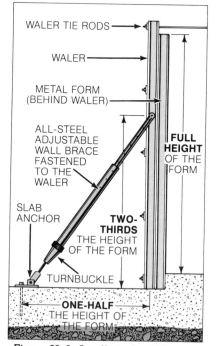

WALER TIE RODS

WALER

METAL FORM (BEHIND WALER)

ALL-STEEL ADJUSTABLE WALL BRACE FASTENED TO THE WALER

SLAB ANCHOR

**TWO-THIRDS** THE HEIGHT OF THE FORM

TURNBUCKLE

**FULL HEIGHT** OF THE FORM

**ONE-HALF** THE HEIGHT OF THE FORM

Figure 68–9. An all-steel wall brace.

opening to observe the flow and consolidation of concrete beneath the buck. When the concrete reaches the bottom level of the buck, the piece that has been cut out is replaced and cleated down. The entire buck is removed when the forms are stripped from the hardened concrete walls. The wedge-shaped key strip attached to the buck remains in the concrete and serves

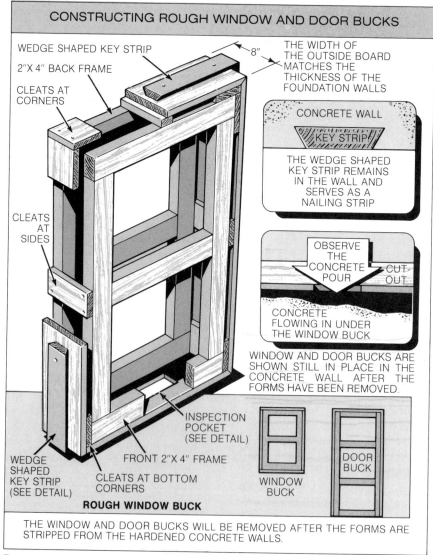

## CONSTRUCTING ROUGH WINDOW AND DOOR BUCKS

WEDGE SHAPED KEY STRIP

2"X 4" BACK FRAME

CLEATS AT CORNERS

CLEATS AT SIDES

8"

THE WIDTH OF THE OUTSIDE BOARD MATCHES THE THICKNESS OF THE FOUNDATION WALLS

CONCRETE WALL

KEY STRIP

THE WEDGE SHAPED KEY STRIP REMAINS IN THE WALL AND SERVES AS A NAILING STRIP

OBSERVE THE CONCRETE POUR

CUT OUT

CONCRETE FLOWING IN UNDER THE WINDOW BUCK

WINDOW AND DOOR BUCKS ARE SHOWN STILL IN PLACE IN THE CONCRETE WALL AFTER THE FORMS HAVE BEEN REMOVED.

INSPECTION POCKET (SEE DETAIL)

FRONT 2"X 4" FRAME

WEDGE SHAPED KEY STRIP (SEE DETAIL)

CLEATS AT BOTTOM CORNERS

**ROUGH WINDOW BUCK**

WINDOW BUCK

DOOR BUCK

THE WINDOW AND DOOR BUCKS WILL BE REMOVED AFTER THE FORMS ARE STRIPPED FROM THE HARDENED CONCRETE WALLS.

Figure 68–10. Door and window bucks use similar construction techniques. Door bucks do not require a bottom piece with an inspection pocket.

as a nailing strip for the finish window frame.

## Reinforcing Steel

Reinforcing steel bars (rebars) are placed by steelworkers after the outside form panels for the walls are set. See Figure 68–11. With some forming systems the steel rebars can be placed first and the forms built afterwards. See Figure 68–12. See Unit 37 in Section 8 for more information on reinforcing steel.

## Construction Joints

Tall buildings are poured one story at a time; therefore, a *horizontal* construction joint occurs at each floor level. These wall sections are structurally tied together by continuous reinforcing steel bars.

It is often necessary to place concrete for long high walls in sections, creating *vertical* construction joints. In this event, carpenters must place a *bulkhead* inside the form. See Figure 68–13. The bulkhead is made of 1″ × 8″ pieces notched around the rebars.

*HCB Contractors*

Figure 68–12. Rebars (at rear) are placed before the wall forms are constructed with this forming system.

## Electrical and Plumbing Work

Provisions for water pipes, electrical conduit, outlet boxes, and sleeves for holes in the wall must also be placed after the outside form panels are set.

## Doubling up the Walls

When all the preceding operations are completed, carpenters can proceed to set the inside walls in place. This procedure is called *doubling up* the walls. The inside walers are attached. Clamps, wedges, or other devices are placed at both ends of the ties, spacing and fastening together the opposite form walls. See Figures 68–14 and 68–15.

## PATENTED TIES AND WALER SYSTEMS

Form ties combine with walers to secure and space the opposing form walls. A great variety of pa-

tented form tie systems are available. The two basic designs used most often in heavy construction form work are the *continuous*

Figure 68–11. Rebars placed after the outside form walls are set. Note the rebars for columns at front and rear.

Figure 68–13. Building a bulkhead for a vertical construction joint. The bulkhead is made of 1″ × 8″ pieces notched around the rebars.

613

Figure 68-14. Carpenters doubling up the form walls. As the panel is tilted into place, the ties are fed into holes drilled in the inside panel.

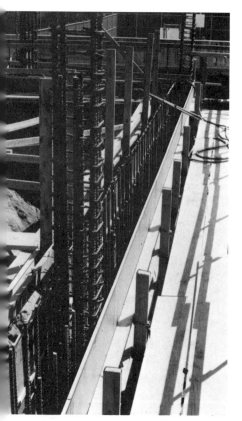

Figure 68-15. Top view of completed form walls with rebars projecting from the top. Note the steel placed for a column at front.

*single member* and *internal disconnecting* types.

Snap ties are the most commonly used continuous single member ties. They are used with double or single waler systems and are available for wall thicknesses ranging from 6″ to 26″. A snap tie is a metal rod extending from the outside surfaces of the opposing walers. It is tightened by driving slotted metal wedges behind the buttons at the ends of the ties. Small plastic cones or metal washers act as spacers between the walls. A breakback consisting of a grooved section next to the tapered edge of the cone allows the tie to be broken off (snapped) after the concrete has hardened and the forms have been removed. A variation of the snap tie design is the penta-tie system, which features a nut button that enables the tie to be snapped by twisting the button with a socket wrench while the forms are still in place. See Figure 68-16.

Internal disconnecting ties are used for heavier and thicker walls where greater pressure is exerted during concrete placement. They consist of two external rods or bolts that screw into an internal threaded device. The internal threaded device is available in graduated lengths to accommodate wall sizes ranging from 8″ to 36″.

Large hex nut washers are

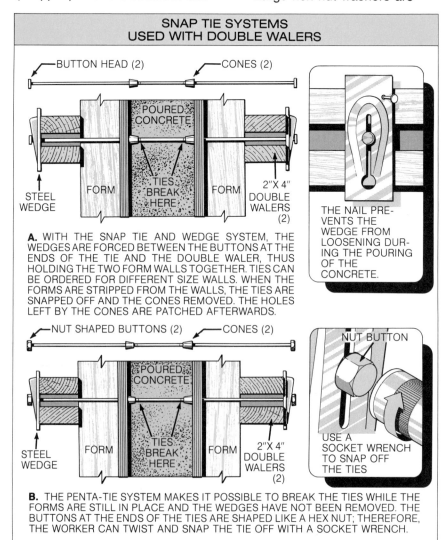

## SNAP TIE SYSTEMS USED WITH DOUBLE WALERS

BUTTON HEAD (2) — CONES (2)

POURED CONCRETE

STEEL WEDGE — FORM — TIES BREAK HERE — FORM — 2″X 4″ DOUBLE WALERS (2)

THE NAIL PREVENTS THE WEDGE FROM LOOSENING DURING THE POURING OF THE CONCRETE.

**A.** WITH THE SNAP TIE AND WEDGE SYSTEM, THE WEDGES ARE FORCED BETWEEN THE BUTTONS AT THE ENDS OF THE TIE AND THE DOUBLE WALER, THUS HOLDING THE TWO FORM WALLS TOGETHER. TIES CAN BE ORDERED FOR DIFFERENT SIZE WALLS. WHEN THE FORMS ARE STRIPPED FROM THE WALLS, THE TIES ARE SNAPPED OFF AND THE CONES REMOVED. THE HOLES LEFT BY THE CONES ARE PATCHED AFTERWARDS.

NUT SHAPED BUTTONS (2) — CONES (2)

POURED CONCRETE

STEEL WEDGE — FORM — TIES BREAK HERE — FORM — 2″X 4″ DOUBLE WALERS (2)

NUT BUTTON

USE A SOCKET WRENCH TO SNAP OFF THE TIES

**B.** THE PENTA-TIE SYSTEM MAKES IT POSSIBLE TO BREAK THE TIES WHILE THE FORMS ARE STILL IN PLACE AND THE WEDGES HAVE NOT BEEN REMOVED. THE BUTTONS AT THE ENDS OF THE TIES ARE SHAPED LIKE A HEX NUT; THEREFORE, THE WORKER CAN TWIST AND SNAP THE TIE OFF WITH A SOCKET WRENCH.

Figure 68-16. Two tie systems used with double walers.

tightened against the walers to secure the walls. Metal spreaders are placed between the walls to maintain the proper width. The outer sections of the tie can be screwed out after the concrete has hardened and the internal device remains in the concrete wall. Two internal disconnecting designs widely used are the waler rod and coil tie systems. See Figure 68–17.

The more traditional heavy construction panel forming method uses vertical 2″ × 4″ studs spaced 12″ to 16″ O.C. The studs are backed up by walers and strongbacks. Snap ties, waler rods, and coil ties are commonly used with this system. See Figure 68–18.

Another widely used system features a single waler system that eliminates the need for studs under the walers. After a series of panels is set in place, snap brackets and walers are attached. Strongbacks and braces are then placed at intervals. See Figures 68–19 and 68–20.

Pilasters are sometimes used with waler systems. Pilasters add strength to the walls and may also support the end of a beam. See Figure 68–21.

## COLUMN AND BEAM FORMS

Columns and beams are a major part of the basic structure of a concrete building. The columns support the beams, which in turn tie into and support the floor and roof systems. Beams also help tie together the walls of the building.

## Columns

Column forms are constructed of plywood, prefabricated steel-framed plywood, tubular fiber, fiberglass, and metal. Column forms are subject to much

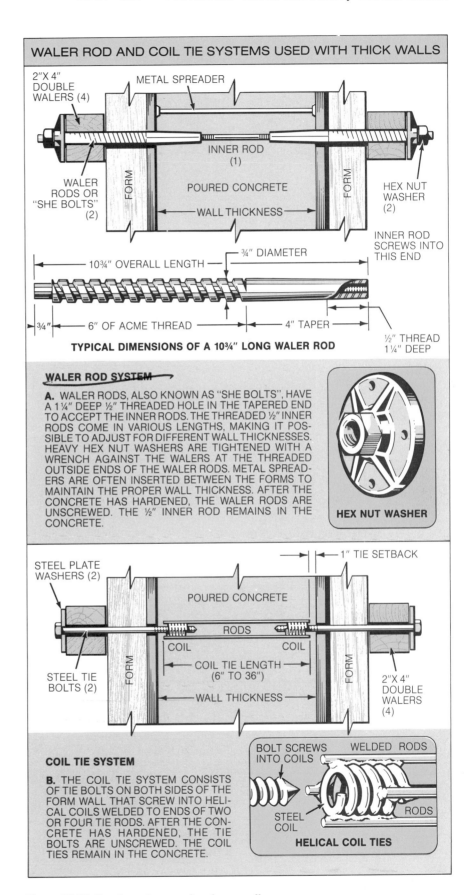

Figure 68-17. Two tie systems used on heavy walls.

Figure 68-18. Concrete building under construction using a double-waler snap-tie-and-wedge system. Note the

Figure 68-19. Section of form wall using a single-waler system.    *The Burke Company*

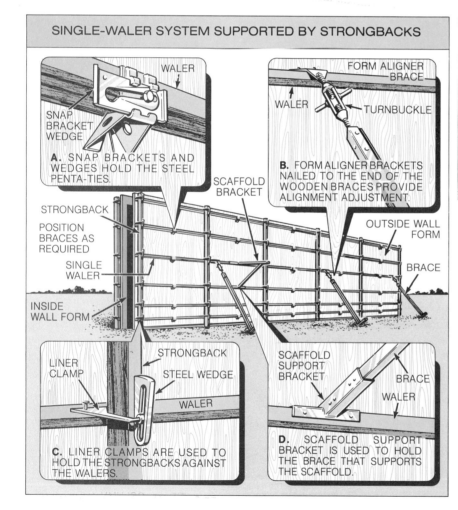

## SINGLE-WALER SYSTEM SUPPORTED BY STRONGBACKS

WALER

SNAP BRACKET WEDGE

**A.** SNAP BRACKETS AND WEDGES HOLD THE STEEL PENTA-TIES.

FORM ALIGNER BRACE

WALER    TURNBUCKLE

**B.** FORM ALIGNER BRACKETS NAILED TO THE END OF THE WOODEN BRACES PROVIDE ALIGNMENT ADJUSTMENT.

SCAFFOLD BRACKET

STRONGBACK

POSITION BRACES AS REQUIRED

SINGLE WALER

INSIDE WALL FORM

OUTSIDE WALL FORM

BRACE

LINER CLAMP

STRONGBACK

STEEL WEDGE

WALER

**C.** LINER CLAMPS ARE USED TO HOLD THE STRONGBACKS AGAINST THE WALERS.

SCAFFOLD SUPPORT BRACKET

BRACE

WALER

**D.** SCAFFOLD SUPPORT BRACKET IS USED TO HOLD THE BRACE THAT SUPPORTS THE SCAFFOLD.

Figure 68-20. Studs are not required in a single-waler system supported by strongbacks.

*The Burke Company*

Figure 68-21. Pilaster being formed along with a form wall constructed with a single-waler system.

greater pressure than wall forms when concrete is being placed. Tight joints and strong tie supports around the form are necessary.

Wooden forms are used for rectangular columns. The sides are constructed of plywood and backed with 2″ × 4″ stiffeners. The stiffeners may be omitted for smaller columns that require a lighter form assembly. After the sides are assembled, they are placed in a template fastened to the column footing. See Figure 68–22.

Hinged metal scissor clamps are often used to tie together the form sides. The scissor clamps are tightened and held in place by wedges driven into slots located in the clamp. See Figures 68–23, 68–24, and 68–25. After the concrete has hardened sufficiently and the forms have been removed, it is common practice to temporarily protect the corners with wooden strips. See Figure 68–26.

Tubular fiber forms are frequently used to construct round concrete columns. They are positioned and held in place with wood or metal braces. When stripped, they are cut with a knife or saw and carefully pried from the concrete. See Figures 68–27 and 68–28.

Single-piece molded fiberglass forms are also used for the construction of round columns. The form can be pulled apart and placed in position around previously installed rebars. The edges are then secured with bolts at closure flanges reinforced with a predrilled steel bar. The form is then secured with braces tied to a steel bracing collar. Forms are stripped after the concrete has set by removing the flange bolts and then pulling apart the flanges. See Figure 68–29. Forms made entirely of steel are used for the construction of rectangular and round columns.

## Beams and Girders

Beams and girders are tied to and rest on top of the columns. Two pouring methods can be used. In one method concrete for the beams and girders is poured separately after the concrete for

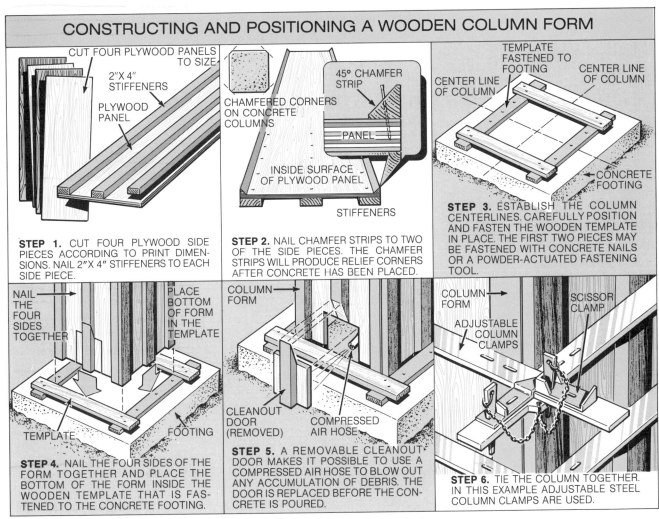

CONSTRUCTING AND POSITIONING A WOODEN COLUMN FORM

CUT FOUR PLYWOOD PANELS TO SIZE
2″X 4″ STIFFENERS
PLYWOOD PANEL

**STEP 1.** CUT FOUR PLYWOOD SIDE PIECES ACCORDING TO PRINT DIMENSIONS. NAIL 2″X 4″ STIFFENERS TO EACH SIDE PIECE.

CHAMFERED CORNERS ON CONCRETE COLUMNS
45° CHAMFER STRIP
PANEL
INSIDE SURFACE OF PLYWOOD PANEL
STIFFENERS

**STEP 2.** NAIL CHAMFER STRIPS TO TWO OF THE SIDE PIECES. THE CHAMFER STRIPS WILL PRODUCE RELIEF CORNERS AFTER CONCRETE HAS BEEN PLACED.

TEMPLATE FASTENED TO FOOTING
CENTER LINE OF COLUMN
CENTER LINE OF COLUMN
CONCRETE FOOTING

**STEP 3.** ESTABLISH THE COLUMN CENTERLINES. CAREFULLY POSITION AND FASTEN THE WOODEN TEMPLATE IN PLACE. THE FIRST TWO PIECES MAY BE FASTENED WITH CONCRETE NAILS OR A POWDER-ACTUATED FASTENING TOOL.

NAIL THE FOUR SIDES TOGETHER
PLACE BOTTOM OF FORM IN THE TEMPLATE
TEMPLATE
FOOTING

**STEP 4.** NAIL THE FOUR SIDES OF THE FORM TOGETHER AND PLACE THE BOTTOM OF THE FORM INSIDE THE WOODEN TEMPLATE THAT IS FASTENED TO THE CONCRETE FOOTING.

COLUMN FORM
CLEANOUT DOOR (REMOVED)
COMPRESSED AIR HOSE

**STEP 5.** A REMOVABLE CLEANOUT DOOR MAKES IT POSSIBLE TO USE A COMPRESSED AIR HOSE TO BLOW OUT ANY ACCUMULATION OF DEBRIS. THE DOOR IS REPLACED BEFORE THE CONCRETE IS POURED.

COLUMN FORM
SCISSOR CLAMP
ADJUSTABLE COLUMN CLAMPS

**STEP 6.** TIE THE COLUMN TOGETHER. IN THIS EXAMPLE ADJUSTABLE STEEL COLUMN CLAMPS ARE USED.

Figure 68–22. Assembling and placing a wood column form. Tight joints and strong tie supports around the form are necessary.

Figure 68-23. Tying up a column form with hinged steel scissor-clamps. Note the form brace in the foreground.

Figure 68-24. Scissor-clamps are tightened and held in place by wedges driven into slots in the clamp.

Figure 68-25. Completed wood column form ready for the placement of concrete.

Figure 68-26. Concrete columns after the wood forms have been stripped. The carpenter is clamping wood pieces at the corners to prevent damage while surrounding work is going on. Note the protruding rebars for the second floor columns.

*The Burke Company*

Figure 68-27. These tubular fiber forms for round columns have been set in place and braced. Rebars will be set inside the tubes, and then the concrete will be placed.

*The Burke Company*

Figure 68-28. Carpenter stripping a fiber form from a concrete column.

the columns has hardened. In the other methods, the concrete is poured *monolithically*, meaning it is poured for columns, beams, and girders at the same time.

Although the words *beam* and *girder* are often used interchangeably, they have distinct meanings. A beam is a horizontal member that supports a

bending load over an opening, as from column to column. A girder is a heavy beam that supports other beams and girders. See Figure 68–30.

Figure 68-29. Column spring forms are stripped by removing the flange bolts and pulling apart the flanges.

Beam and girder forms consist of a bottom piece (soffit) and sides. See Figure 68-31. The entire unit is supported by shores placed at intervals. Wood T-shores can be used with wedges underneath, although patented metal types are also available. Beams and girders are heavily reinforced with steel bars that tie into the rebars of the column below and the floor above. An example of beam-and-column concrete construction is shown in Figure 68-32.

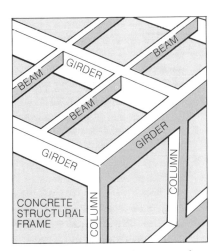

Figure 68-30. Example of a structural frame for a concrete building tying together columns, beams, and girders.

## FLOOR AND ROOF FORMS

Several different basic designs are used for concrete floors and roofs. All require formwork consisting of a deck supported by joists and shores. The formwork must be strong enough to support the weight of the form material, the concrete, and the load imposed by workers and equipment.

## Beam-and-Slab Floors

Beam-and-slab systems are suitable for floors that will bear heavy loads. The floor slab rests on top of closely spaced beams that tie into girders supported by columns. A lighter design developed more recently features concrete joists tied into girders. See Figure 68-33.

## Flat-slab Floor

Beams and girders are not used with a flat-slab system. The slab receives its main support from the columns and the thickened sections over the columns called *drop panels.* See Figure 68-34. Another variation of this system uses column *capitals,* sometimes call *drop heads,* over the column. See Figure 68-35.

## Concrete Joist Systems

Concrete joist systems consist of concrete joists placed monolithically with a floor slab, beams, girders, and columns. This allows for a thinner and lighter floor slab with a high bearing capacity. The two basic designs are the *one-way* and *two-way* (waffle) joist systems. Both systems are formed with reusable metal or fiberglass pans.

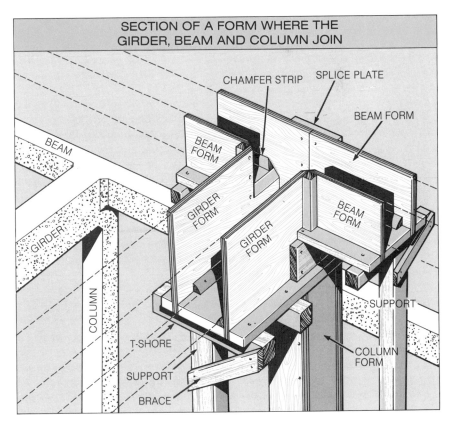

SECTION OF A FORM WHERE THE GIRDER, BEAM AND COLUMN JOIN

Figure 68-31. Section of a form where a girder and beams join on top of a column.

Figure 68–32. This monolithic concrete building relies on girders, beams, and columns for its structural support. A series of closely spaced column forms are shown at right. Girder forms and more widely spaced column forms are being constructed at front.

Figure 68–33. This concrete floor system features a slab resting on concrete joists that tie into girders supported by columns.

Figure 68–34. Flat-slab floor supported by drop panels over the columns.

Figure 68–35. Flat-slab floor supported by drop panels and capitals over the columns.

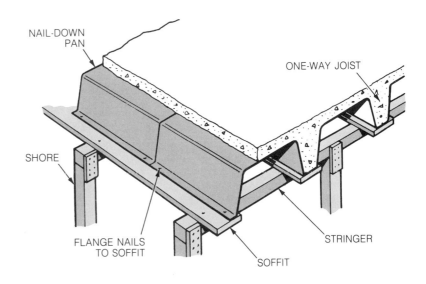

Figure 68–36. Long pans used for one-way joist systems rest on 2″ thick soffits supported by shores and stringers.

Long pans are used in the forming of one-way joist systems. They are nailed to the tops or sides of 2″ thick soffits that are supported by shores and stringers. See Figure 68–36.

Long pans commonly frame into girder forms.

Dome pans are used in the forming of two-way joist systems and are supported in the same manner as long pans. Because two-way joists do not include beams or girders, a solid area equal to the thickness of the slab and joist is formed around the supporting columns. See Figure 68–37.

Figure 68–37. Fiberglass dome forms are in place for a waffle-slab floor system.

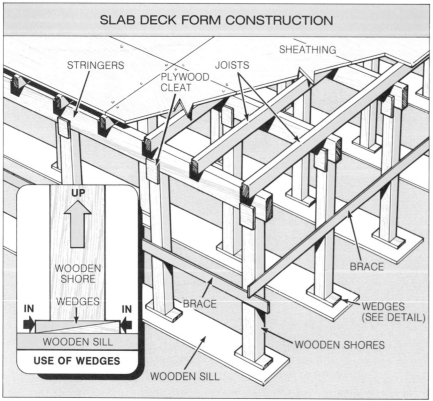

SLAB DECK FORM CONSTRUCTION

STRINGERS

PLYWOOD CLEAT

JOISTS

SHEATHING

UP

WOODEN SHORE

WEDGES

IN            IN

WOODEN SILL

USE OF WEDGES

BRACE

WOODEN SILL

BRACE

WEDGES (SEE DETAIL)

WOODEN SHORES

Figure 68–38. Example of slab-deck form construction supported by wood shores and stringers.

## Slab Decks and Shoring

When constructing formwork for a slab deck, first place the shores and temporarily brace them. Shores are cut short to allow for wedges over a wood sill. The wedges are used to drive shores up tightly and to line up the floor above. Plywood cleats secure the stringers to the shores. Stringers can then be set on top of the shores. Joists are laid across the stringers. The plywood deck is then nailed down on the joists. See Figure 68–38.

In an aluminum beam-and-stringer system used to support wood panel forms, the stringers are supported by metal shores that are brought up tightly with an adjusting screw at the top. See Figure 68–39. Aluminum beams come in lengths of 8′ to 30′ in 2′ increments. Their light weight of 4 pounds per foot

Symons Corporation

Figure 68–39. Lightweight aluminum beam-and-stringer system used to support wood panel forms.

Figure 68–40. Completed beam-and-slab deck form. The next step will be to place the rebars over the deck. Note the steel extending up for the columns.

makes them easy to handle.

When the deck formwork has been completed, reinforcing steel is placed over the deck. The floor reinforcing bars are tied to the rebars in the walls, beams, and columns. See Figures 68–40 through 68–42.

## SEQUENCE OF FORM CONSTRUCTION

The general sequence for heavy construction formwork is shown in Figure 68–43. Steps for forming the first floor of the building are shown.

When the reinforcing steel bars are placed in the walls, they will extend several feet above the floor line of the second floor. In the case of the first floor columns, the steel may project several feet or even to the top height of the second floor columns. See Figure 68–44.

When forms are constructed for the outside walls of the second story, the bottoms of the outside panels must be secured to the top of the first story wall. J-bolts must be provided in the concrete for the panels to be secured. See Figure 68–45. The J-bolts extend through walers at the bottom of the panels. Large nuts are tightened over the walers.

On many concrete construction jobs large panel sections known

*Portland Cement Association*

Figure 68–41. Rebars are placed over the deck of the floor form. Note the bulkhead toward the center of the deck. Concrete will first be placed at the left side of the bulkhead.

*Portland Cement Association*

Figure 68–42. Placing concrete for a section of the floor slab. After the concrete has set up in this section, the bulkhead will be removed, and the concrete for the remaining sections will be placed. This method provides for a construction or expansion joint where the two sections meet.

# SEQUENCE IN THE CONSTRUCTION OF HEAVY CONCRETE FORMS

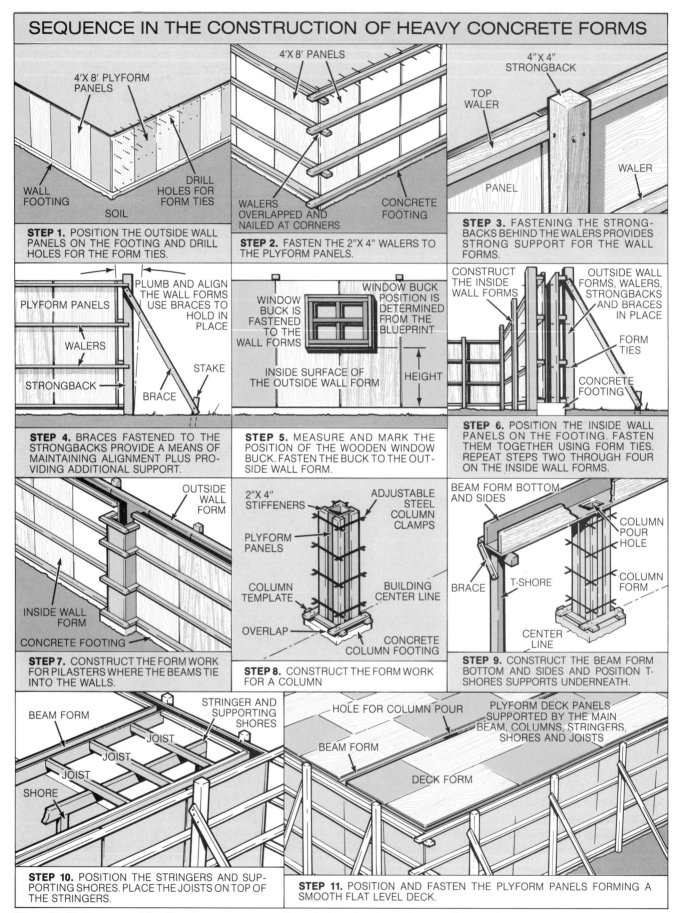

**STEP 1.** POSITION THE OUTSIDE WALL PANELS ON THE FOOTING AND DRILL HOLES FOR THE FORM TIES.

**STEP 2.** FASTEN THE 2"X 4" WALERS TO THE PLYFORM PANELS.

**STEP 3.** FASTENING THE STRONG-BACKS BEHIND THE WALERS PROVIDES STRONG SUPPORT FOR THE WALL FORMS.

**STEP 4.** BRACES FASTENED TO THE STRONGBACKS PROVIDE A MEANS OF MAINTAINING ALIGNMENT PLUS PRO-VIDING ADDITIONAL SUPPORT.

**STEP 5.** MEASURE AND MARK THE POSITION OF THE WOODEN WINDOW BUCK. FASTEN THE BUCK TO THE OUT-SIDE WALL FORM.

**STEP 6.** POSITION THE INSIDE WALL PANELS ON THE FOOTING. FASTEN THEM TOGETHER USING FORM TIES. REPEAT STEPS TWO THROUGH FOUR ON THE INSIDE WALL FORMS.

**STEP 7.** CONSTRUCT THE FORM WORK FOR PILASTERS WHERE THE BEAMS TIE INTO THE WALLS.

**STEP 8.** CONSTRUCT THE FORM WORK FOR A COLUMN

**STEP 9.** CONSTRUCT THE BEAM FORM BOTTOM AND SIDES AND POSITION T-SHORES SUPPORTS UNDERNEATH.

**STEP 10.** POSITION THE STRINGERS AND SUP-PORTING SHORES. PLACE THE JOISTS ON TOP OF THE STRINGERS.

**STEP 11.** POSITION AND FASTEN THE PLYFORM PANELS FORMING A SMOOTH FLAT LEVEL DECK.

Figure 68–43. Sequence of first-floor heavy construction forms built on foundation footings. Note reinforcing steel not shown.

*HCB Contractors*

Figure 68–44. The rebars in these rectangular columns are set so they will extend past the floor and beams that are supported by the columns. Note the completed column at left rear. The steel extending from this column will be tied to additional steel placed inside the form for the column of the floor above.

as *gang forms* (discussed later in this unit) cover a wide wall area. The entire section must be lifted by crane. It is set into position so that the bottom of the gang form can be bolted to the concrete wall. See Figures 68–46 and 68–47.

## STAIRWAY FORMS

The layout for a concrete stairway is similar to that for a con-

ventional wood stairway. (Refer to Section 14.) For both concrete and wood stairways, the same procedure is used to calculate the tread and riser dimensions based on the total rise and run of the stairway. Construction of a form for a concrete stairway, however, can be more complicated than construction of a wood stairway. Figure 68–48 shows one type of concrete form for an open stairway. Figure 68–

49 shows a procedure for laying out and constructing the form in Figure 68–48. Figure 68–50 shows a procedure for constructing forms for stairways placed against one wall or between two walls.

## PREFABRICATED PANEL FORMING SYSTEMS

With few exceptions, concrete formwork today uses some type

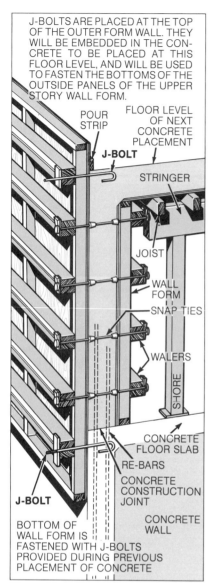

J-BOLTS ARE PLACED AT THE TOP OF THE OUTER FORM WALL. THEY WILL BE EMBEDDED IN THE CONCRETE TO BE PLACED AT THIS FLOOR LEVEL, AND WILL BE USED TO FASTEN THE BOTTOMS OF THE OUTSIDE PANELS OF THE UPPER STORY WALL FORM.

POUR STRIP

FLOOR LEVEL OF NEXT CONCRETE PLACEMENT

**J-BOLT**

STRINGER

JOIST

WALL FORM

SNAP TIES

WALERS

SHORE

CONCRETE FLOOR SLAB

RE-BARS

CONCRETE CONSTRUCTION JOINT

CONCRETE WALL

**J-BOLT**

BOTTOM OF WALL FORM IS FASTENED WITH J-BOLTS PROVIDED DURING PREVIOUS PLACEMENT OF CONCRETE

Figure 68–45. Fastening an outside wall panel over a floor already placed.

of prefabricated panel forming system. These systems range from individual panel units that can be placed by hand to much larger units that must be raised in place by crane. The systems are constructed of wood, metal, or plastic, or sometimes a combination of these materials.

## Gang Forms

On high structures where concrete must be placed in two or more stages, *gang forms* are often used. Gang forms are large

*The Burke Company*

Figure 68–46. Setting a large gang form section into place by crane.

Figure 68-47. These large panels were lifted by crane after the concrete wall has set. Note the lines showing the impression of the construction joint at each floor level.

Figure 68-48. This side view of an open stairway form shows a side panel backed by a top and bottom plate with studs between the plates. Joists, stringers, and shoring support the bottom of the form.

panels made up by fastening together a series of smaller panels. Walers, strongbacks, lifting brackets, and scaffold brackets are then fastened to the panels. The entire unit is lifted in place by crane. Figures 68–51 through 68–53 show carpenters assembling a gang form. Figures 68–54 and 68–55 show all-wood gang forms being placed in position.

## Flying Forms

*Flying forms* are total floor units that usually consist of a wood deck with a metal support system. See Figure 68–56. After a freshly placed concrete floor slab has hardened sufficiently, the

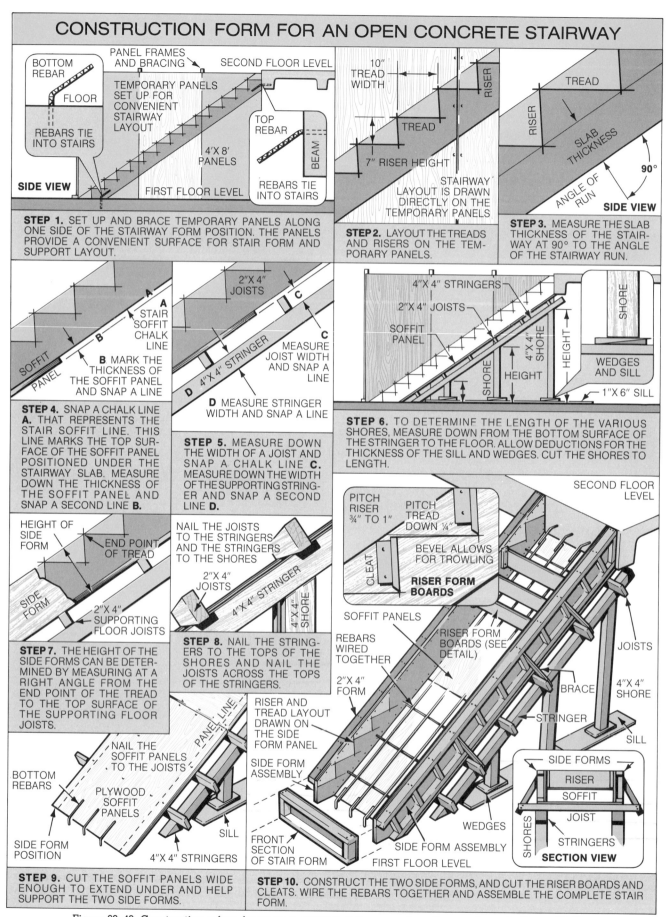

# CONSTRUCTION FORM FOR AN OPEN CONCRETE STAIRWAY

**SIDE VIEW**

BOTTOM REBAR
FLOOR
REBARS TIE INTO STAIRS

PANEL FRAMES AND BRACING
TEMPORARY PANELS SET UP FOR CONVENIENT STAIRWAY LAYOUT
SECOND FLOOR LEVEL
TOP REBAR
BEAM
REBARS TIE INTO STAIRS
4'X 8' PANELS
FIRST FLOOR LEVEL

**STEP 1.** SET UP AND BRACE TEMPORARY PANELS ALONG ONE SIDE OF THE STAIRWAY FORM POSITION. THE PANELS PROVIDE A CONVENIENT SURFACE FOR STAIR FORM AND SUPPORT LAYOUT.

10" TREAD WIDTH
RISER
TREAD
7" RISER HEIGHT
STAIRWAY LAYOUT IS DRAWN DIRECTLY ON THE TEMPORARY PANELS

**STEP 2.** LAYOUT THE TREADS AND RISERS ON THE TEMPORARY PANELS.

TREAD
RISER
SLAB THICKNESS
ANGLE OF RUN
90°
**SIDE VIEW**

**STEP 3.** MEASURE THE SLAB THICKNESS OF THE STAIRWAY AT 90° TO THE ANGLE OF THE STAIRWAY RUN.

A
STAIR SOFFIT CHALK LINE
B MARK THE THICKNESS OF THE SOFFIT PANEL AND SNAP A LINE
SOFFIT PANEL

**STEP 4.** SNAP A CHALK LINE A. THAT REPRESENTS THE STAIR SOFFIT LINE. THIS LINE MARKS THE TOP SURFACE OF THE SOFFIT PANEL POSITIONED UNDER THE STAIRWAY SLAB. MEASURE DOWN THE THICKNESS OF THE SOFFIT PANEL AND SNAP A SECOND LINE B.

2"X 4" JOISTS
C
MEASURE JOIST WIDTH AND SNAP A LINE
4"X 4" STRINGER
D MEASURE STRINGER WIDTH AND SNAP A LINE

**STEP 5.** MEASURE DOWN THE WIDTH OF A JOIST AND SNAP A CHALK LINE C. MEASURE DOWN THE WIDTH OF THE SUPPORTING STRINGER AND SNAP A SECOND LINE D.

4"X 4" STRINGERS
2"X 4" JOISTS
SOFFIT PANEL
4"X 4" SHORE
SHORE
HEIGHT
HEIGHT
SHORE
HEIGHT
WEDGES AND SILL
1"X 6" SILL

**STEP 6.** TO DETERMINE THE LENGTH OF THE VARIOUS SHORES, MEASURE DOWN FROM THE BOTTOM SURFACE OF THE STRINGER TO THE FLOOR. ALLOW DEDUCTIONS FOR THE THICKNESS OF THE SILL AND WEDGES. CUT THE SHORES TO LENGTH.

HEIGHT OF SIDE FORM
END POINT OF TREAD
SIDE FORM
2"X 4" SUPPORTING FLOOR JOISTS

**STEP 7.** THE HEIGHT OF THE SIDE FORMS CAN BE DETERMINED BY MEASURING AT A RIGHT ANGLE FROM THE END POINT OF THE TREAD TO THE TOP SURFACE OF THE SUPPORTING FLOOR JOISTS.

NAIL THE JOISTS TO THE STRINGERS AND THE STRINGERS TO THE SHORES
2"X 4" JOISTS
4"X 4" STRINGER
4"X 4" SHORE

**STEP 8.** NAIL THE STRINGERS TO THE TOPS OF THE SHORES AND NAIL THE JOISTS ACROSS THE TOPS OF THE STRINGERS.

PITCH RISER ¾" TO 1"
PITCH TREAD DOWN ¼"
CLEAT
BEVEL ALLOWS FOR TROWLING
**RISER FORM BOARDS**

SECOND FLOOR LEVEL
SOFFIT PANELS
REBARS WIRED TOGETHER
2"X 4" FORM
RISER FORM BOARDS (SEE DETAIL)
JOISTS
4"X 4" SHORE
BRACE
STRINGER
SILL

PANEL LINE
NAIL THE SOFFIT PANELS TO THE JOISTS
BOTTOM REBARS
PLYWOOD SOFFIT PANELS
SIDE FORM POSITION
4"X 4" STRINGERS
SILL

RISER AND TREAD LAYOUT DRAWN ON THE SIDE FORM PANEL
SIDE FORM ASSEMBLY
FRONT SECTION OF STAIR FORM
WEDGES
SIDE FORM ASSEMBLY
FIRST FLOOR LEVEL

SIDE FORMS
RISER
SOFFIT
SHORES
JOIST
STRINGERS
**SECTION VIEW**

**STEP 9.** CUT THE SOFFIT PANELS WIDE ENOUGH TO EXTEND UNDER AND HELP SUPPORT THE TWO SIDE FORMS.

**STEP 10.** CONSTRUCT THE TWO SIDE FORMS, AND CUT THE RISER BOARDS AND CLEATS. WIRE THE REBARS TOGETHER AND ASSEMBLE THE COMPLETE STAIR FORM.

Figure 68–49. Constructing a form for an open concrete stairway such as the one shown in Figure 68–48.

# CONSTRUCTING THE FORM FOR A CLOSED CONCRETE STAIRWAY

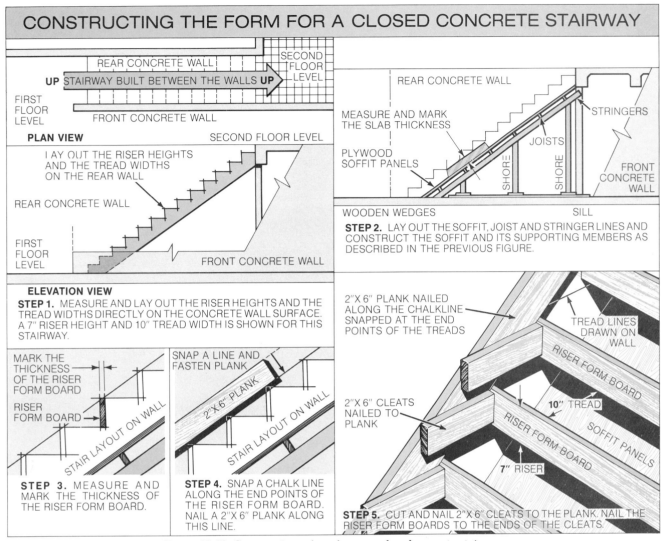

**PLAN VIEW**

REAR CONCRETE WALL

UP STAIRWAY BUILT BETWEEN THE WALLS UP

SECOND FLOOR LEVEL

FIRST FLOOR LEVEL

FRONT CONCRETE WALL

SECOND FLOOR LEVEL

LAY OUT THE RISER HEIGHTS AND THE TREAD WIDTHS ON THE REAR WALL

REAR CONCRETE WALL

FIRST FLOOR LEVEL

FRONT CONCRETE WALL

**ELEVATION VIEW**

**STEP 1.** MEASURE AND LAY OUT THE RISER HEIGHTS AND THE TREAD WIDTHS DIRECTLY ON THE CONCRETE WALL SURFACE. A 7" RISER HEIGHT AND 10" TREAD WIDTH IS SHOWN FOR THIS STAIRWAY.

MARK THE THICKNESS OF THE RISER FORM BOARD

RISER FORM BOARD

STAIR LAYOUT ON WALL

**STEP 3.** MEASURE AND MARK THE THICKNESS OF THE RISER FORM BOARD.

SNAP A LINE AND FASTEN PLANK

2" X 6" PLANK

STAIR LAYOUT ON WALL

**STEP 4.** SNAP A CHALK LINE ALONG THE END POINTS OF THE RISER FORM BOARD. NAIL A 2" X 6" PLANK ALONG THIS LINE.

REAR CONCRETE WALL

MEASURE AND MARK THE SLAB THICKNESS

PLYWOOD SOFFIT PANELS

STRINGERS

JOISTS

SHORE   SHORE

FRONT CONCRETE WALL

WOODEN WEDGES        SILL

**STEP 2.** LAY OUT THE SOFFIT, JOIST AND STRINGER LINES AND CONSTRUCT THE SOFFIT AND ITS SUPPORTING MEMBERS AS DESCRIBED IN THE PREVIOUS FIGURE.

2" X 6" PLANK NAILED ALONG THE CHALKLINE SNAPPED AT THE END POINTS OF THE TREADS

TREAD LINES DRAWN ON WALL

RISER FORM BOARD

2" X 6" CLEATS NAILED TO PLANK

10" TREAD

RISER FORM BOARD

SOFFIT PANELS

7" RISER

**STEP 5.** CUT AND NAIL 2" X 6" CLEATS TO THE PLANK. NAIL THE RISER FORM BOARDS TO THE ENDS OF THE CLEATS.

Figure 68–50. Constructing a form for an enclosed concrete stairway.

Figure 68–51. Sections of this gang form are tied together with wedge bolts.

*Symons Corporation*

Figure 68–52. Walers are bolted to the panel side rails of a gang form.

*Symons Corporation*

Figure 68–53. Strongbacks are bolted to the walers, completing assembly of the gang form.

Figure 68–54. All-wood gang form unit set in position.

The Burke Company

*The Burke Company*
Figure 68–55. Series of all-wood gang forms set in position and tied together.

*Portland Cement Association*
Figure 68–57. Flying form being set into place by crane. At right is the corner of a flying form that has been secured next to a column.

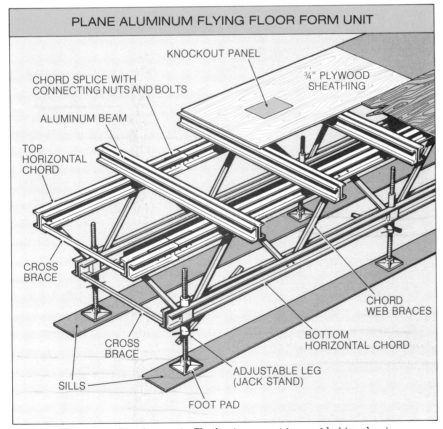

Figure 68–56. Flying floor form unit. The basic support is provided by aluminum trusses placed on either side, and aluminum beams placed across the trusses. A plywood deck is fastened on top of the beams. Jacks are used to raise the unit into position.

whole flying form unit can be removed and transported to the next floor level. See Figure 68–57. The structural design of many high-rise buildings is a continuous system of columns supporting floor slabs with the same floor design at each level. The use of flying forms greatly increases production on high-rise buildings.

## Slip Forms

Slip forms were originally developed for the construction of curved concrete structures such as silos and towers. Slip forming methods have expanded to the construction of rectangular buildings, caissons,

631

*Portland Cement Association*
Figure 68–58. Building under construction using slip forms. Note the concrete bucket being raised by crane at left.

building cores, underground shafts, shear wall buildings, communication towers, and a variety of other structures. See Figure 68–58. Slip forming can save significant labor, time, and material cost on construction projects that lend themselves to this method.

Most slip forms consist of 4′ high inner and outer walls of $3/8″$ to $3/4″$ plywood skins backed by a stud and waler system. The walls are held together with steel cross beams and yoke legs. The yoke legs can be adjusted to the wall width and are fastened to the cross beams. Hydraulic jacks are mounted on the cross beams. The forms are raised by electrically- or pneumatically-powered hydraulic jacks that climb the jackrods extending into the form. New lengths can be fastened to the threaded ends of the jackrods when required. See Figure 68–59.

Slip forms can climb while the concrete is being placed. Speeds may range from 2″ to 70″ per hour. The climbing speed depends on the type of structure,

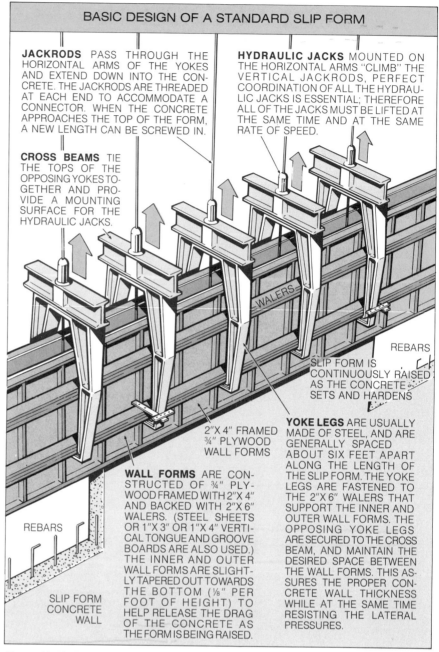

**BASIC DESIGN OF A STANDARD SLIP FORM**

**JACKRODS** PASS THROUGH THE HORIZONTAL ARMS OF THE YOKES AND EXTEND DOWN INTO THE CONCRETE. THE JACKRODS ARE THREADED AT EACH END TO ACCOMMODATE A CONNECTOR. WHEN THE CONCRETE APPROACHES THE TOP OF THE FORM, A NEW LENGTH CAN BE SCREWED IN.

**CROSS BEAMS** TIE THE TOPS OF THE OPPOSING YOKES TOGETHER AND PROVIDE A MOUNTING SURFACE FOR THE HYDRAULIC JACKS.

**HYDRAULIC JACKS** MOUNTED ON THE HORIZONTAL ARMS "CLIMB" THE VERTICAL JACKRODS, PERFECT COORDINATION OF ALL THE HYDRAULIC JACKS IS ESSENTIAL; THEREFORE ALL OF THE JACKS MUST BE LIFTED AT THE SAME TIME AND AT THE SAME RATE OF SPEED.

WALERS

REBARS

SLIP FORM IS CONTINUOUSLY RAISED AS THE CONCRETE SETS AND HARDENS

2″X 4″ FRAMED $3/4″$ PLYWOOD WALL FORMS

REBARS

**WALL FORMS** ARE CONSTRUCTED OF $3/4″$ PLYWOOD FRAMED WITH 2″X 4″ AND BACKED WITH 2″X 6″ WALERS. (STEEL SHEETS OR 1″X 3″ OR 1″X 4″ VERTICAL TONGUE AND GROOVE BOARDS ARE ALSO USED.) THE INNER AND OUTER WALL FORMS ARE SLIGHTLY TAPERED OUT TOWARDS THE BOTTOM ($1/8″$ PER FOOT OF HEIGHT) TO HELP RELEASE THE DRAG OF THE CONCRETE AS THE FORM IS BEING RAISED.

**YOKE LEGS** ARE USUALLY MADE OF STEEL, AND ARE GENERALLY SPACED ABOUT SIX FEET APART ALONG THE LENGTH OF THE SLIP FORM. THE YOKE LEGS ARE FASTENED TO THE 2″X 6″ WALERS THAT SUPPORT THE INNER AND OUTER WALL FORMS. THE OPPOSING YOKE LEGS ARE SECURED TO THE CROSS BEAM, AND MAINTAIN THE DESIRED SPACE BETWEEN THE WALL FORMS. THIS ASSURES THE PROPER CONCRETE WALL THICKNESS WHILE AT THE SAME TIME RESISTING THE LATERAL PRESSURES.

SLIP FORM CONCRETE WALL

Figure 68–59. Basic design of a standard slip form. Additional features, such as scaffolding, are custom made for the structure being erected.

the rate of concrete placement, and how quickly the reinforcing steel and built-ins can be placed. *Built-ins* consist primarily of door and window bucks and beam pockets. Provisions must also be made for the placement of brackets, anchors, electrical conduits, plumbing, blockouts,

etc. Slip forms do not operate 24 hours a day on most types of slip form construction. Therefore, construction joints occur on the surface of the structure. A complete slip form unit must be custom-made to the shape and requirement of the structure being erected.

# UNIT 69

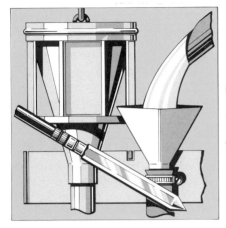

# Concrete Placement for Heavy Construction

Concrete for heavy construction is prepared to specification at a *batch plant* and delivered to the job site by a *transit-mix* truck. See Figure 69–1. For projects in isolated areas, stationary mixers are sometimes used to mix the concrete on the job site. See Figure 69–2. The composition of concrete and various formulas for the concrete mixture (called

the *mix*) are discussed in Unit 37 of Section 8. Precautions for pouring concrete in very hot or cold weather are also discussed in Unit 37.

## POURING OPERATIONS

For the basement and lower floors of a large concrete building, concrete is usually poured

(placed) by a pumping system. See Figures 69–3 and 69–4. As the building gains in height, either *climber* or *tower cranes* together with buckets are usually used for pouring operations. In these crane-and-bucket operations, the concrete is dispatched from the transit-mix truck into large buckets. Usually the buckets have a capacity of ¾ cubic yard or 1 cubic yard, although buckets as large as 1¾ cubic yards are sometimes used.

*Portland Cement Association*

Figure 69–1. Ready-mixed concrete is prepared in a batch plant and deposited in a transit-mix truck for delivery to the job site.

*Portland Cement Association*

Figure 69–2. Concrete being mixed on the job site in a large stationary mixer.

Figure 69–3. The stabilizers are positioned before the boom is extended to pump the concrete.

which the sand-cement ingredients of the concrete separate from the gravel. Improper pouring procedures can also cause segregation. If the concrete has been properly mixed before being discharged, the problem of segregation can usually be avoided by following these pouring procedures:

1. The drop of the concrete should always be vertical (straight down), and not angled.

2. Drop chutes should be used to prevent concrete from striking the reinforcement steel or the side of the form above the level of placement. See Figures 69–8 and 69–9.

3. The free-fall distance of concrete should be from 4' to 6'.

All the concrete for a form should not be deposited in one corner. Instead it should be discharged from different positions until an even layer (*lift*) has been

Figure 69–4. Concrete is pumped through hoses like the one shown here. The hose is supported by a boom and cables.

The bucket is raised by cable and positioned, if possible, directly over the wall form or floor slab where the concrete is to be placed. The concrete is then released by opening the gate at the bottom of the bucket. See Figures 69–5 and 69–6.

In cases where the concrete cannot be discharged directly in place from the bucket, motorized concrete buggies called *power buggies* are often used. The concrete is deposited in the buggy and is then moved to its final location. See Figure 69–7.

Concrete is a mixture of a small amount of cement added to a much larger quantity of sand and gravel. When water is added to this mixture, the cement acts as the agent that bonds together the sand and gravel. The cement, sand, and gravel ingredients must be thoroughly mixed together for the concrete to *set up* (dry). Incomplete mixing can cause *segregation,* a condition in

*The Burke Company*

Figure 69–5. Placing concrete by bucket into a column form. The bucket being used on this job is equipped with a side dump chute that makes placement more convenient. The worker at right is pulling down on the handle that opens the gate at the bottom of the bucket to release the concrete. The worker at left is using a vibrator to help consolidate the concrete as it is placed in the form.

*Portland Cement Association*

Figure 69–6. Placing concrete by bucket for a floor slab.

*Portland Cement Association*

Figure 69–7. Power buggies are used to transfer concrete over short distances and deposit it in place.

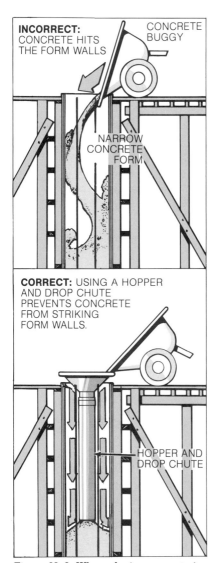

**INCORRECT:** CONCRETE HITS THE FORM WALLS

CONCRETE BUGGY

NARROW CONCRETE FORM

**CORRECT:** USING A HOPPER AND DROP CHUTE PREVENTS CONCRETE FROM STRIKING FORM WALLS.

HOPPER AND DROP CHUTE

Figure 69–8. When placing concrete in a narrow form, use a hopper and drop chute to prevent concrete from striking against the steel and the form walls.

placed in the form. See Figure 69–10. The procedure should then be repeated until enough lifts have been placed to fill the form.

Thinner lifts are recommended in forms containing heavier or more closely spaced rebars. Thinner lifts are also recommended when the concrete is poured from a higher than usual distance.

While concrete is being deposited, it must be compacted and worked so that each lift is *consolidated* with the lift below. *Vibration* is an effective method for

635

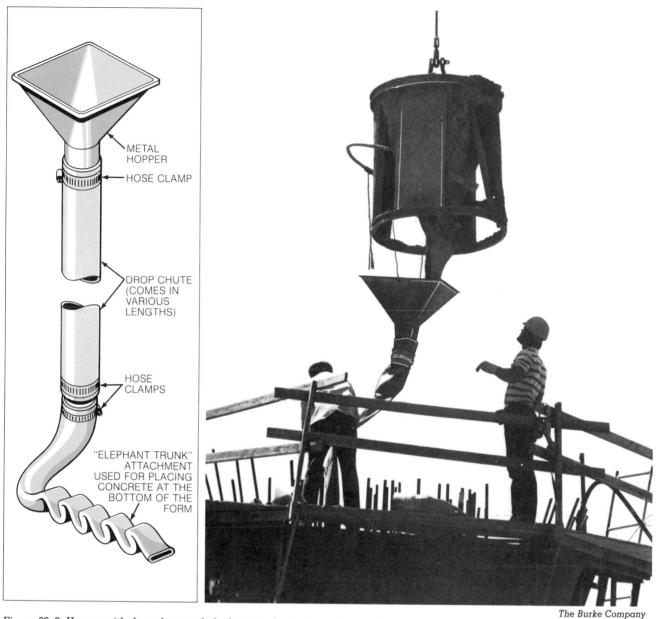

The Burke Company

Figure 69–9. Hopper with drop chute and elephant-trunk attachment for placing concrete at the bottom of a deep form such as a high column form.

consolidating concrete. The type of mechanical vibrator used most often on heavy construction work is the *immersion vibrator,* also called an *internal vibrator.* It is powered by electricity, gas, or compressed air. Refer again to Figure 69–5.

A systematic vibration of each new lift of concrete is necessary. The vibrator should penetrate a few inches into the previous lift at regular intervals. Hit-or-miss penetration with the vibrator at

varying angles and without sufficient depth will result in improper consolidation between the two layers. To aid consolidation in concrete placed in a sloping layer, start vibration at the bottom of the slope so that compaction is increased by the weight of the newly added concrete. See Figure 69–11.

Typical surface defects resulting from improper consolidation are honeycombing, air pockets, and sand or gravel streaks.

These defects can also be caused by faulty form construction or improper concrete mix.

## Concrete Admixtures

Admixtures are ingredients that may be added to the basic materials (cement, aggregates, and water) of the concrete mix immediately before or during mixing. Admixtures are designed to increase the effectiveness of the

636

concrete under certain conditions.

**Air-entraining Admixtures.** Air-entraining admixtures produce microscopic air bubbles in the concrete. The air bubbles improve the workability of the concrete as well as its resistance to freezing and thawing. Air-entraining admixtures are of particular benefit in concrete being placed during cold weather. (The effects of hot and cold weather on concrete are discussed later in this unit.)

**Water-reducing Admixtures.** Admixtures that reduce the amount of water required to produce a workable mix increase the strength of the concrete. Many water-reducing admixtures also retard the setting time of the concrete, which allows a slower curing process.

**Accelerating Admixtures.** Concrete gains strength very slowly at low temperatures. Accelerating admixtures improve the strength of concrete at an early stage. Therefore, this type of admixture is of particular advantage in concrete construction during cold weather.

**Pozzolans.** Pozzolans used as admixtures in concrete are permicite, volcanic ashes, tuffs, diatomaceous earth, fly ash, calcined clays, and shale. By themselves, pozzolans have little or no cementing qualities. However, in the presence of moisture and under ordinary temperature conditions, they react with other cement materials to form compounds that have cementing properties. As a result, pozzolans can sometimes be used as a partial replacement for cement. Also, they are sometimes used to help reduce internal temperatures when concrete is being

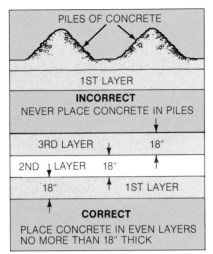

Figure 69-10. Methods of depositing concrete.

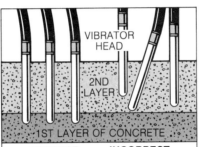

A SYSTEMATIC VIBRATION OF EACH NEW LAYER OF CONCRETE (LIFT) IS NECESSARY. THE VIBRATOR SHOULD PENETRATE A FEW INCHES INTO THE PREVIOUSLY PLACED CONCRETE AT REGULAR INTERVALS AND FOR PERIODS OF 5 TO 15 SECONDS TO CAUSE PROPER CONSOLIDATION. HIT-OR-MISS PENETRATION WITH A VIBRATOR AT ALL ANGLES AND WITHOUT SUFFICIENT DEPTH WILL NOT RESULT IN PROPER CONSOLIDATION BETWEEN THE TWO LAYERS OF CONCRETE.

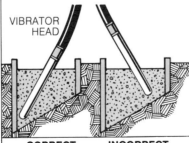

WHEN PLACING CONCRETE IN A SLOPING LAYER, START AT THE BOTTOM OF THE SLOPE SO THAT COMPACTION IS INCREASED BY THE WEIGHT OF THE NEWLY ADDED CONCRETE.

Figure 69-11. The metal head of an internal vibrator is placed into the concrete for periods of 5 to 15 seconds.

placed in massive structures such as dams.

Some pozzolans are used to reduce concrete expansion caused by alkali-reactive aggregates. Other pozzolans improve the sulfate resistance of concrete. However, pozzolans also may create a disadvantage. As a rule, concrete with pozzolans requires more mixing water to achieve the same *slump* (consistency) as concrete without pozzolans. Also, when hardened, concrete with pozzolans has a tendency to crack because of greater contraction while drying.

**Other Admixtures.** Other admixtures are available to improve the workability of the concrete, to help waterproof it, and to increase bonding strength when new concrete is placed against old concrete. Grouting agents are available to improve the quality of grout used for a variety of filling, patching, and repair purposes. Gas-forming agents are sometimes added to concrete and grout to cause a slight expansion before hardening.

## JOB-SITE TESTING OF CONCRETE

Various testing procedures are involved in the manufacture, mixing, and placement of concrete. Most are conducted in a laboratory. However, two important tests are often carried out at the time the concrete is placed: the *slump test* and the *compression strength test*. As a rule, carpenters are not involved in these tests, but they should understand why the tests are conducted.

### Slump Test

The slump test measures the consistency of the concrete. Consistency directly affects flow-

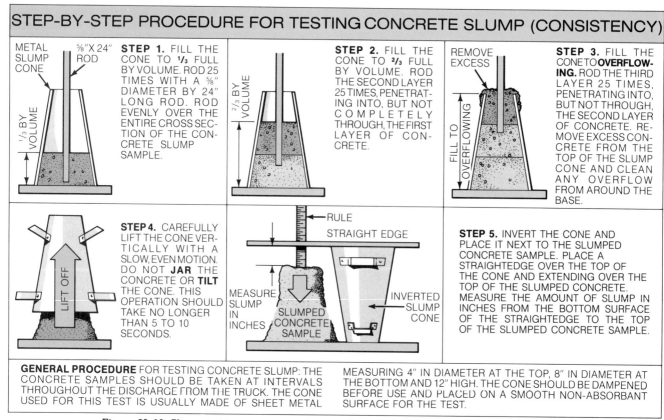

STEP-BY-STEP PROCEDURE FOR TESTING CONCRETE SLUMP (CONSISTENCY)

**STEP 1.** FILL THE CONE TO 1/3 FULL BY VOLUME. ROD 25 TIMES WITH A 5/8" DIAMETER BY 24" LONG ROD. ROD EVENLY OVER THE ENTIRE CROSS SECTION OF THE CONCRETE SLUMP SAMPLE.

**STEP 2.** FILL THE CONE TO 2/3 FULL BY VOLUME. ROD THE SECOND LAYER 25 TIMES, PENETRATING INTO, BUT NOT COMPLETELY THROUGH, THE FIRST LAYER OF CONCRETE.

**STEP 3.** FILL THE CONE TO **OVERFLOWING.** ROD THE THIRD LAYER 25 TIMES, PENETRATING INTO, BUT NOT THROUGH, THE SECOND LAYER OF CONCRETE. REMOVE EXCESS CONCRETE FROM THE TOP OF THE SLUMP CONE AND CLEAN ANY OVERFLOW FROM AROUND THE BASE.

**STEP 4.** CAREFULLY LIFT THE CONE VERTICALLY WITH A SLOW, EVEN MOTION. DO NOT **JAR** THE CONCRETE OR **TILT** THE CONE. THIS OPERATION SHOULD TAKE NO LONGER THAN 5 TO 10 SECONDS.

**STEP 5.** INVERT THE CONE AND PLACE IT NEXT TO THE SLUMPED CONCRETE SAMPLE. PLACE A STRAIGHTEDGE OVER THE TOP OF THE CONE AND EXTENDING OVER THE TOP OF THE SLUMPED CONCRETE. MEASURE THE AMOUNT OF SLUMP IN INCHES FROM THE BOTTOM SURFACE OF THE STRAIGHTEDGE TO THE TOP OF THE SLUMPED CONCRETE SAMPLE.

**GENERAL PROCEDURE** FOR TESTING CONCRETE SLUMP: THE CONCRETE SAMPLES SHOULD BE TAKEN AT INTERVALS THROUGHOUT THE DISCHARGE FROM THE TRUCK. THE CONE USED FOR THIS TEST IS USUALLY MADE OF SHEET METAL MEASURING 4" IN DIAMETER AT THE TOP, 8" IN DIAMETER AT THE BOTTOM AND 12" HIGH. THE CONE SHOULD BE DAMPENED BEFORE USE AND PLACED ON A SMOOTH NON-ABSORBANT SURFACE FOR THE TEST.

Figure 69–12. Slump test to measure consistency of concrete. Consistency affects flowability.

ability during placement. Samples are taken at intervals during discharge of the concrete into the forms. A *slump cone* made of sheet metal, 4″ in diameter at the top, 8″ diameter at the bottom, and 12″ high, is used. It is dampened before use and placed on a smooth, nonabsorbent surface for the test. Concrete placed in the cone is *rodded*. A 24″ long rod, 5⁄8″ in diameter, is moved up and down in the concrete. Rodding differs from mixing in that it is an up-and-down motion rather than a circular one. See Figure 69–12. Some allowable slump ranges for various types of construction are shown in the table in Figure 69–13.

## Compression Strength Test

A concrete mix varies in strength according to the proportions of its ingredients. These proportions are determined for each job according to certain factors such as the shape and size of structural members to be built (walls, slabs, beams, columns, etc.), their required strength, and the environmental conditions they will face.

Engineers involved in designing the structure or concrete field specialists calculate the *compression strength* required for its members. Compression strength is the pounds per square inch (psi) of force the concrete can withstand 28 days after it has been placed. Twenty-eight days is the average period of time required for concrete to gain its full strength.

The *water-cement ratio* largely determines compression strength. This ratio is calculated by dividing the weight of the water used in one cubic yard of

| CONCRETE CONSTRUCTION | SLUMP | |
|---|---|---|
| | MAXIMUM* | MINIMUM |
| REINFORCED FOUNDATION WALLS AND FOOTINGS | 3″ | 1″ |
| PLAIN FOOTINGS, CAISSONS, AND SUBSTRUCTURE WALLS | 3″ | 1″ |
| BEAMS AND REINFORCED WALLS | 4″ | 1″ |
| BUILDING COLUMNS | 4″ | 1″ |
| PAVEMENTS AND SLABS | 3″ | 1″ |
| MASS CONCRETE | 2″ | 1″ |

*MAY BE INCREASED 1″ FOR CONSOLIDATION BY HAND METHODS SUCH AS RODDING AND SPADING. 1″ = 25 MM

*Portland Cement Association*
Figure 69–13. Allowable slump ranges for different types of construction.

| COMPRESSIVE STRENGTH AT 28 DAYS, PSI* | WATER-CEMENT RATIO, BY WEIGHT | |
| --- | --- | --- |
| | NON-AIR-ENTRAINED CONCRETE | AIR-ENTRAINED CONCRETE |
| 6000 | 0.41 | — |
| 5000 | 0.48 | 0.40 |
| 4000 | 0.57 | 0.48 |
| 3000 | 0.68 | 0.59 |
| 2000 | 0.82 | 0.74 |

*POUNDS PER SQUARE INCH

*American Concrete Institute*

Figure 69–14. Relationship between water-cement ratio and compressive strength of concrete. Note that as compressive strength increases, water-cement ratio decreases. Note also that air-entrained concrete requires a lower water-cement ratio than non-air-entrained concrete.

concrete by the weight of the cement used in one cubic yard. Thus, if the weight of the water used in one cubic yard is 10 pounds, and the weight of the cement is 20 pounds, the water-cement ratio is .50 (10 ÷ 20 = .50). Figure 69–14 shows a table of several water-cement ratios and their relation to compression strength. This table is from a design manual published by the American Concrete Institute (ACI), which recommends design standards for the concrete industry.

For the compression strength test, at least three samples are taken and placed into cylinders while the concrete is being discharged. The cylinders are covered with a lid and carefully stored on the job site in a place where they are protected from jarring. After 24 hours the cylinders are taken to a laboratory, where the concrete sample is removed from the cylinder and is cured for 28 days. (Curing is discussed later in this unit.)

At the end of the 28-day curing period, the specimen is capped with a thin layer of capping compound. After the cap has hardened, the sample is placed in a compression testing machine. Pressure is exerted until the sample is broken. A dial on the machine registers the amount of load required to break the sample. See Figure 69–15.

## EFFECT OF WEATHER ON CONCRETE

Extreme hot or cold weather can create special problems for pouring concrete if proper precautions are not taken.

### Effect of Hot Weather

Some of the difficulties caused by hot weather are increased water demand, early slump loss, and increased rate of setting. All these factors can result in strength loss and the possibility of cracking. The ideal concrete temperature at the time of placement is 50°F to 70°F. However, many specifications require only that temperatures be less than 85°F or 90°F.

The best way to maintain low concrete temperature is to keep the concrete materials as cool as possible before mixing. Water, before being placed in the mix, can be cooled by refrigeration, liquid nitrogen, or crushed ice. Aggregates can be stockpiled in a shady place before used and kept moist by sprinkling.

Equipment such as mixers, chutes, hoppers, and pump lines should be shaded or covered with wet burlap prior to placing the concrete. The forms, the reinforcing steel, and the subgrade should also be fogged with cool water before the pour begins. (Water-reducing admixtures, discussed earlier in this unit, may also be used.)

### Effect of Cold Weather

Hydration takes place at a slower rate when temperatures are low. When the temperature drops to zero or below, practically no chemical action occurs. To ensure proper hydration in concrete placed during freezing temperatures, the concrete mixture is usually heated at the time it is placed. (Air-entraining or accelerating admixtures, discussed earlier in this unit, may also be used.) The forms are then covered with tarpaulins or plastic film material. If necessary, the enclosed area can be further warmed with gasoline-powered

*Portland Cement Association*

Figure 69–15. Concrete specimen placed in a compression testing machine. Pressure will be placed on the sample until it breaks. The machine will then register the amount of load that was required to break the sample.

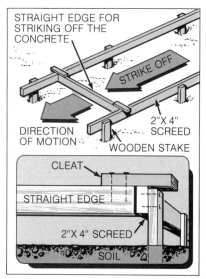

Figure 69-16. Screeds are placed at intervals with their bottoms set to the desired level of the concrete slab. The straightedge resting on the screeds is used to level off (strike off) the concrete. The screeds are removed from the poured section as soon as the area has been struck off.

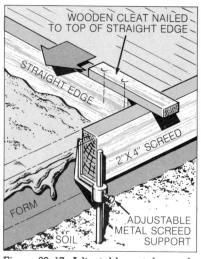

Figure 69-17. Adjustable metal screed support used for slab-at-grade concrete placement.

heaters.

Concrete should not be placed on frozen soil, so precautions must be taken to prevent frost from setting into the ground where the foundation walls are to be erected. After the forms have been built, the insides of the forms and the reinforcing steel must be kept free of ice. De-icing

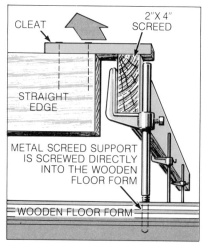

Figure 69-18. Adjustable metal screed supports used over floor forms. The bottoms of the supports screw into the wood deck.

chemicals are often used for this purpose.

## CONCRETE FLOOR SLABS

A concrete floor slab may be laid directly on the ground (slab-at-grade floor) or it may be supported by walls, beams, and columns.

### Slab-at-grade Floor

Slab-at-grade concrete floors are placed at the basement or first floor level and receive direct support from the ground. They may be placed at the initial stages of the foundation work, in which case the outside of the foundation wall also acts as the perimeter form for the floor slab.

Ground preparation is of extreme importance to a slab-at-grade floor, as the ground below the slab (the subgrade) must be able to uniformly support the slab and any other weight carried by the slab. The subgrade must be properly graded, compacted (pressed down), and moistened. Vegetation and any other foreign matter must be removed. Soft or

muddy spots must be dug out and replaced with well-compacted fill.

Before a slab-at-grade floor is placed, a layer of granular rock (gravel) is laid down. Then a vapor barrier in the form of a waterproof membrane (polyethylene, butyl rubber, asphalt-impregnated sheets, etc.) is placed on top of the gravel.

The perimeter form of the floor slab also establishes the outside finish levels of the slab. To maintain the proper floor levels throughout, a system of screeds and screed supports is set up. The screed system consists of a straightedge that moves along guide strips held by some type of support. See Figure 69-16. Screeds are placed at intervals with their bottoms set to the desired level of the concrete slab. The straightedge resting on the screeds levels off (strikes off) the concrete. The screeds are removed from the poured section as soon as the area has been struck off. See Figures 69-17 and 69-18. Mechanical screeds that help consolidate as well as level off the concrete are also available.

As in all concrete construction, the concrete for a slab-at-grade floor should be placed as close as possible to its final location. This practice reduces unnecessary labor. It also helps to prevent segregation (discussed previously in this unit), which can occur with concrete movement after placement. Concrete deposited for floor slabs must also be consolidated (discussed previously in this unit). Vibrators can be used for consolidation. See Figure 69-19.

As the concrete is being placed, cement finishers strike off the concrete with screeds. When the screeds are removed, the concrete is finished off. A hand tamping tool is used to help compact the concrete. The tamp-

*Portland Cement Association*
Figure 69-19. Vibrating concrete placed for a floor slab.

*Portland Cement Association*
Figure 69-21. Using a saw to cut a control joint in a concrete slab.

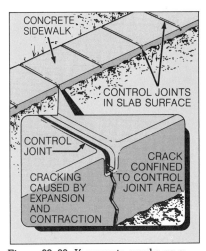

Figure 69-22. If concrete cracks as a result of expansion and contraction, the crack will be confined to the area of the control joint.

*Portland Cement Association*
Figure 69-20. Cement masons at work finishing off a concrete slab. Some workers are hand-troweling the surface of the concrete. At center a power trowel is being used.

ing tools consists of mesh stretched over a frame, and the operation is commonly referred to as *jitterbugging*. A *bullfloat* tool is used to eliminate high and low areas in the concrete. See Figure 69-20. Power trowels and then hand trowels produce the final finished surface of the concrete.

## Control Joints

Like all other construction materials, concrete expands and contracts with moisture and temperature changes. To prevent cracking from this expansion and contraction, one or more *control joints* are provided in a concrete slab.

A control joint is made by inserting a wood, metal, or tar paper strip into the top of the slab at the time that the concrete is being placed. Another method is to cut into the slab with a saw fitted with a special masonry blade after the concrete has sufficiently hardened. See Figure 69-21. If cracking later occurs, it will be confined to the area of the control joint. See Figure 69-22.

641

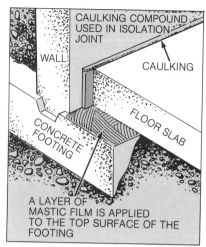

Figure 69–23. Example of a caulked expansion joint between a floor slab and the wall.

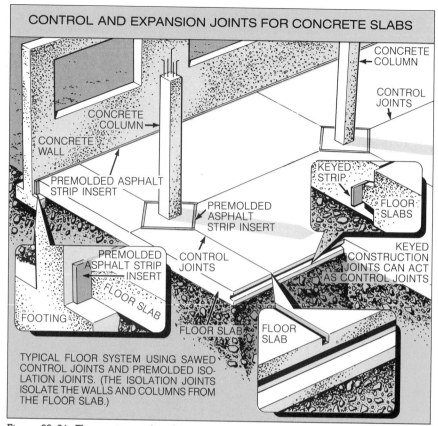

Figure 69–24. Floor system with isolation joints between the slab and the walls and around columns. Control joints are placed at intervals in the slab.

## Isolation Joints

*Isolation joints* are required when a concrete slab is laid for a basement floor after the foundation walls, piers, and columns have been erected. An isolation joint extends through the entire thickness of the slab. It provides space for the slab to expand as a result of temperature changes without exerting pressure that can damage the walls and other parts of the structure in contact with the floor slab.

An isolation joint is commonly formed using an asphalt-impregnated strip around the slab perimeters before the concrete is placed. The strip remains in place after the concrete has set. See Figures 69–23 and 69–24.

## Structurally Supported Slab Floor

The main difference between a slab-at-grade floor and a structurally supported slab floor is that in the structurally supported floor concrete is placed on top of the deck forms and is monolithically joined with the walls, columns, and beams. Pouring, screeding, and finishing off operations are the same as described for slab-at-grade floors.

## CURING CONCRETE AND REMOVING FORMS

*Curing* is the process of keeping concrete moist long enough to allow proper *hydration* to occur. See Unit 37 of Section 8 for a discussion of hydration and further information on curing.

The first three days after discharge are the most critical to the quality of concrete. During this early period, it is most vulnerable to damage. At seven days after discharge, it reaches about 70% of its strength and at fourteen days about 85%. Under normal conditions maximum strength is reached at twenty-eight days.

Curing floor slabs is more difficult than curing walls. To ensure that the concrete slab retains moisture for a minimum of three days, several different methods may be used. See Figure 69–25.

The slab may be misted or flooded with water by placing a pipe with a series of spray nozzles down the center of the slab. It may be covered with waterproof paper or plastic sheeting (polyethylene film) to help it retain moisture and also to protect it from damage from frost, direct sun, traffic, and debris. Chemical sealing compounds for curing can be applied by hand sprays or, on large projects, by automatic self-powered sprays.

Form removal *(stripping)* must be done carefully to avoid injuring workers or damaging the concrete. In general, stripping of walls, columns, and beam sides is recommended no sooner than three days after the concrete has been placed, stripping of floor

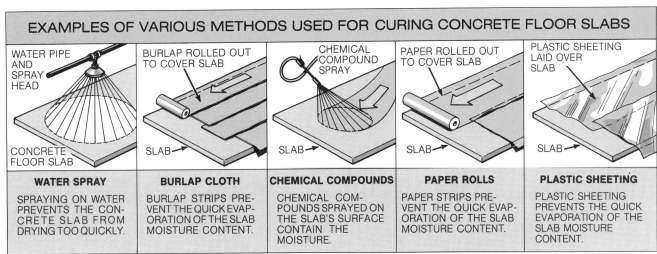

## EXAMPLES OF VARIOUS METHODS USED FOR CURING CONCRETE FLOOR SLABS

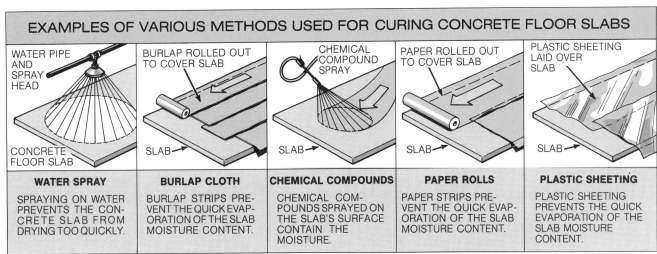

| WATER SPRAY | BURLAP CLOTH | CHEMICAL COMPOUNDS | PAPER ROLLS | PLASTIC SHEETING |
|---|---|---|---|---|
| SPRAYING ON WATER PREVENTS THE CONCRETE SLAB FROM DRYING TOO QUICKLY. | BURLAP STRIPS PREVENT THE QUICK EVAPORATION OF THE SLAB MOISTURE CONTENT. | CHEMICAL COMPOUNDS SPRAYED ON THE SLAB'S SURFACE CONTAIN THE MOISTURE. | PAPER STRIPS PREVENT THE QUICK EVAPORATION OF THE SLAB MOISTURE CONTENT. | PLASTIC SHEETING PREVENTS THE QUICK EVAPORATION OF THE SLAB MOISTURE CONTENT. |

Figure 69–25. Various methods used for curing floor slabs. Floor slabs are more difficult to cure than walls.

slabs no sooner than seven days, and stripping of arch centers and beam bottoms no sooner than 14 days. On large construction projects recommended times for stripping removal may be included in the specifications by the architect or engineer.

# UNIT 70

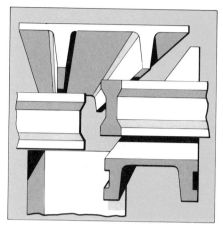

# Prefabricated Concrete Systems

Many concrete structures today are partially or completely constructed with prefabricated *(precast)* concrete systems instead of the *cast-in-place* systems discussed so far in this section. Some prefabricated concrete components, such as wall and floor sections and beams and columns, are made in *casting beds* (forms) at a prefabricating plant. They are then transported by truck to the construction area. Other prefabricated components are *sitecast* in forms that are constructed on the job. These *on-site* precast systems (sitecast systems) are usually limited to the production of wall and floor panels. Figure 70–1 shows several types of precast structural members. Figure 70–2 shows two types of precast beams: L-shaped and T-shaped.

Some buildings are constructed with precast beam-and-column framework that may or may not include steel beams. See Figures 70–3 through 70–5. One common type of construction is a completely steel-framed building with precast concrete wall panels *(curtains)*. See Figures 70–6 and 70–7. Precast components are also used for sections of buildings that have an unusual or complex architectural design. See Figure 70–8.

Precast members must be raised in place by crane. See Figures 70–9 and 70–10. Bolts, clips, or other types of inserts must be provided for the crane hookup. Hollow-core units, which combine strength with light weight, are quicker and easier to install. See Figure 70–11.

After the precast members are set in place (Figure 70–12), various methods are used to connect them. Most of these methods involve either bolting or welding procedures. For exam-

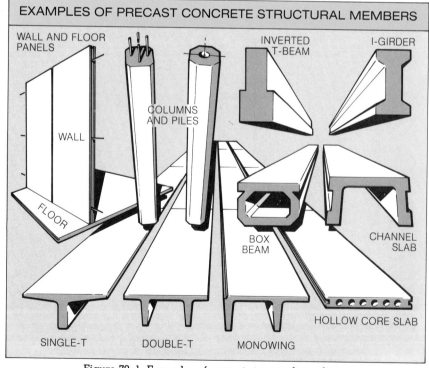

Figure 70–1. Examples of precast structural members.

644

*Prestressed Concrete Institute*

Figure 70–2. Completed precast L-shaped concrete beams are shown at front. A single T-shaped beam is being moved at rear.

*Economy Forms Corporation*

Figure 70–4. Precast beam-and-column framework.

*Portland Cement Association*

Figure 70–3. The basic structure of this building consists of precast walls, floor slabs, beams, and columns.

*Precast Concrete Institute*

Figure 70–5. Close-up view of joints between beams and columns being set in place. Steel beams are also a part of the framework of this building.

Figure 70–6. Steel-framed building with precast concrete wall panels. The wall panels have brackets that are bolted or welded to the steel frame. Note the wall panel being lifted from the flat-bed truck at right.

Figure 70–7. High-rise steel-framed building under construction. Concrete curtain walls have already been placed in the lower portion of the building.

Figure 70–8. The precast concrete sections for this building of unusual architectural design were formed in a casting yard.

*The Burke Company*

Figure 70–9. Lifting a precast concrete wall section from a stack on the job site. The panels are being moved by crane and will be fastened to the steel framework.

*Bethlehem Steel Corporation*

Figure 70–11. These hollow-core slab sections are supported by the girders of a steel-framed building. After the slabs are fastened to the girders, the joints between the slabs will be filled with grout.

*Prestressed Concrete Institute*

Figure 70–10. Single T-beam being lifted by crane after delivery to the job site. Special inserts at the top of the beam have been provided for the crane hookup.

ple, the bottom of a column may have a metal anchor that fastens to bolts which were set in the foundation footing at the time of the concrete pour. Other examples are shown in Figures 70–13 and 70–14.

*Kaiser Cement Corporation*

Figure 70–12. Single T-beams placed over columns. Various methods using bolts or welds can be used to fasten them.

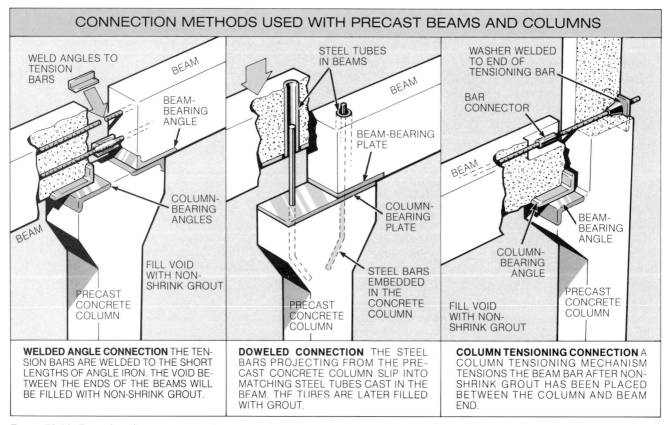

**CONNECTION METHODS USED WITH PRECAST BEAMS AND COLUMNS**

WELD ANGLES TO TENSION BARS

BEAM

BEAM-BEARING ANGLE

COLUMN-BEARING ANGLES

BEAM

PRECAST CONCRETE COLUMN

FILL VOID WITH NON-SHRINK GROUT

STEEL TUBES IN BEAMS

BEAM

BEAM-BEARING PLATE

COLUMN-BEARING PLATE

STEEL BARS EMBEDDED IN THE CONCRETE COLUMN

PRECAST CONCRETE COLUMN

WASHER WELDED TO END OF TENSIONING BAR

BAR CONNECTOR

BEAM

BEAM-BEARING ANGLE

COLUMN-BEARING ANGLE

PRECAST CONCRETE COLUMN

FILL VOID WITH NON-SHRINK GROUT

**WELDED ANGLE CONNECTION** THE TENSION BARS ARE WELDED TO THE SHORT LENGTHS OF ANGLE IRON. THE VOID BETWEEN THE ENDS OF THE BEAMS WILL BE FILLED WITH NON-SHRINK GROUT.

**DOWELED CONNECTION** THE STEEL BARS PROJECTING FROM THE PRECAST CONCRETE COLUMN SLIP INTO MATCHING STEEL TUBES CAST IN THE BEAM. THE TUBES ARE LATER FILLED WITH GROUT.

**COLUMN TENSIONING CONNECTION** A COLUMN TENSIONING MECHANISM TENSIONS THE BEAM BAR AFTER NON-SHRINK GROUT HAS BEEN PLACED BETWEEN THE COLUMN AND BEAM END.

Figure 70–13. Examples of connecting methods used to fasten precast beams to columns.

## PRESTRESSED CONCRETE

Prestressed concrete is precast concrete placed in a high state of compression with high tensile steel cables. Additional reinforcement with rebars is not required. Prestressed members have much greater resistance to lateral pressures than conventional reinforced members. As a result, lighter structural members can be used. Two methods used to prestress concrete are *pretensioning* and *post-tensioning.* Pretensioning is commonly done in fabricating plants. Post-tensioning is more common to on-site precasting.

In pretensioning, cables are placed in forms (casting beds). Powerful jacks stretch the cables until they are under the prescribed tension. See Figure 70–15. The concrete is placed in the form and bonds to the cables as it

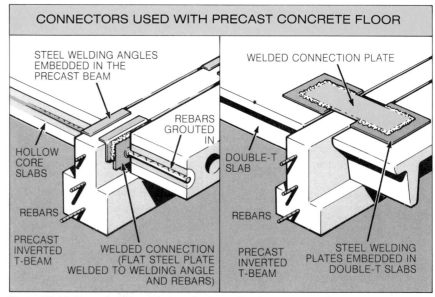

**CONNECTORS USED WITH PRECAST CONCRETE FLOOR**

STEEL WELDING ANGLES EMBEDDED IN THE PRECAST BEAM

HOLLOW CORE SLABS

REBARS

PRECAST INVERTED T-BEAM

REBARS GROUTED IN

WELDED CONNECTION (FLAT STEEL PLATE WELDED TO WELDING ANGLE AND REBARS)

WELDED CONNECTION PLATE

DOUBLE-T SLAB

REBARS

PRECAST INVERTED T-BEAM

STEEL WELDING PLATES EMBEDDED IN DOUBLE-T SLABS

Figure 70–14. Examples of welding methods used to fasten precast beams to precast floors.

hardens. When the concrete has set to its specified strength, the tensioned cables are released. The backward pull of the cables places the concrete under compression. See Figure 70–16.

In post-tensioning, concrete is placed around unstressed cables enclosed in flexible metal or plastic ducts. The cables are stressed and anchored at both ends after concrete has set.

*Portland Cement Association*

Figure 70-15. A hydraulic tension jack is used to stretch the cables extending from a precast structural member.

## PRETENSIONING A CONCRETE STRUCTURAL MEMBER

HYDRAULIC JACK          CONCRETE BEAM FORM                                    JACK

(STRETCHING CABLE)

**TENSION FORCE**        CONCRETE IS                          **TENSION FORCE**
                         POURED INTO THE                      TENSILE CABLE IS
                         CASTING BED AFTER THE                BEING STRETCHED
                         CABLE IS STRETCHED

CONCRETE BEAM
REMOVED FROM CASTING BED                          INTERNAL TENSIONED CABLE

**COMPRESSION FORCE**

BECAUSE THEY ARE NOW BONDED TO THE HARDENED CONCRETE, THE CABLES CREATE A COMPRESSIVE FORCE BY PULLING FROM BOTH ENDS OF THE MEMBER TOWARD THE CENTER. THIS COMPRESSIVE FORCE ADDS CONSIDERABLE STRENGTH TO THE MEMBER. AS A RESULT, PRESTRESSED CONCRETE BEAMS AND FLOOR SLAB SECTIONS CAN SUPPORT HEAVIER LOADS. PRESTRESSED COLUMNS AND WALL PANELS HAVE GREATER RESISTANCE TO LATERAL PRESSURES.

Figure 70-16. Pretensioning adds strength to precast concrete members.

## TILT-UP CONSTRUCTION

Tilt-up construction is an economical, on-site, precast construction method used most often in the erection of one- or two-story buildings with a slab-at-grade foundation system. The walls are usually constructed in casting panels on the floor slab and then raised in place by crane. Walls may be cast outside the building on a temporary concrete slab, wood platform, or well-compacted fill. Tilt-up form work consists mainly of 2″ thick edge forms fastened to the slab. A traditional method of securing the edge forms consists of flat planks bolted or pinned down behind the edge form. Short 1″ × 4″ braces secure the top of the edge forms to the flat planks. More current methods include triangular plywood brackets nailed to short 2″ × 4″ pads secured to the slab or triangular metal brackets fastened directly to the slab. See Figure 70-17. The basic steps in tilt-up construction are:

1. Casting panels for walls are formed on the floor slab. Two coats of *bond-breaker liquid* are

*The Burke Company*

Figure 70-17. The form for this tilt-up wall section consists of 2″ × 12″ planks set on edge and resting against the concrete slab. Other planks, lying flat, have been fastened to the slab and hold the bottom of the form planks in position. Short pieces nailed to the top of the form and the flat planks complete the bracing of the form. A window buck, braced and in position, can be seen. Rebars are also in place, and the carpenter at left is setting a coil-lifting insert. These inserts will later receive the bolts that secure the lift plates to the wall.

*The Burke Company*

Figure 70–18. Preparing to lift a tilt-up wall section. Double-angle lift plates have been secured to the wall for the crane hookup.

5. Concrete is placed in the casting panels. The concrete is vibrated for proper consolidation and worked to the desired finish.

6. After the concrete has hardened and reached the proper strength, the precast wall panels are raised by crane and maneuvered over the wall footings. See Figures 70–18 and 70–19. Once they have been set in position on top of the footings, the wall panels must be held in position with temporary braces. See Figure 70–20.

applied to the floor slab to prevent the precast wall panels from sticking to it.

2. Vertical and horizontal steel reinforcement is place. Horizontal bars should extend 12" or more beyond the edges of the panels if columns are going to be cast in place between the wall sections after they have been raised.

3. Openings for doors and windows are provided by steel-framed or wood bucks.

4. Electrical conduits and outlet boxes are set in place. Inserts for the crane hookup are also provided.

*The Burke Company*

Figure 70–20. This tilt-up wall section is in position and is being braced. In this particular construction, the wall does not rest on top of the slab but instead is placed on an independent footing to one side of the slab. The bent steel bars toward the bottom of the wall will be straightened and tied to the bars extending from the slab. The space between the edge of the slab and wall will then be filled with concrete. Note the swivel lift plates in the wall section still resting on the slab at front. They are secured by bolts to the coil inserts that were placed in the wall.

*The Burke Company*

Figure 70–19. Maneuvering a tilt-up wall section toward its final position.

650

## Structural Connections

Tilt-up wall panels are secured to the floor slab and foundation after they have been raised. One method is to rest the wall panels on grout pads placed at intervals on top of the foundation walls. The grout pads are placed to level the wall. The space between the wall bottom and foundation is then filled with grout. Rebars protruding from the wall are tied to rebars extending from the concrete floor slab that has been placed to within a few feet of the foundation wall. The area between the foundation wall and floor slab is then filled with concrete. A variation of this design is to set the bottom of the wall in a slot provided at the top of the foundation. Older methods include slipping the wall panel over dowels extending vertically from the foundation, or securing the wall bottom with welding plates.

Columns or chord bars are used to tie together the vertical edges of the wall panels. Cast-in-place columns that require a form to be built between the wall panels were common in the early tilt-up construction. Contemporary designs use independent precast columns or columns cast with the walls. See Figure 70–21. Independent precast columns are formed with oversized recesses that accommodate the wall panel edges. Welding plates are used to secure the walls to the columns. When columns are cast as part of the wall panel, one half of a column is formed at both ends of each panel. The half columns at the panel ends are then joined with welding plates after the panels have been raised. Columns are not required when using chord bars. The chord bars extend horizontally through the panel sections

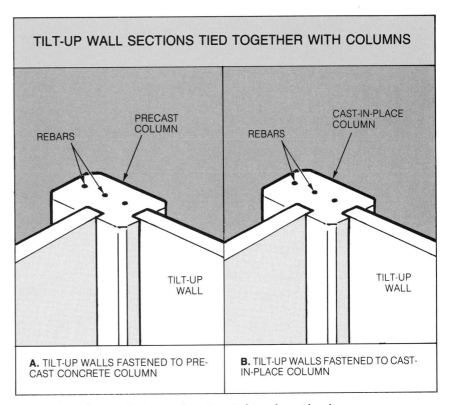

**TILT-UP WALL SECTIONS TIED TOGETHER WITH COLUMNS**

REBARS  PRECAST COLUMN  TILT-UP WALL

REBARS  CAST-IN-PLACE COLUMN  TILT-UP WALL

**A.** TILT-UP WALLS FASTENED TO PRE-CAST CONCRETE COLUMN

**B.** TILT-UP WALLS FASTENED TO CAST-IN-PLACE COLUMN

Figure 70–21. Structural tilt-up wall sections tied together with columns.

and small pockets around the ends of the bars are blocked out when forming the panel sections. The bars are spliced and welded together after the walls have been raised and the pockets are filled with concrete.

The second floors of two-story tilt-up buildings are usually cast-in-place, requiring the construction of a suspended floor slab form. Other methods include wood or metal floor joists, or trusses supported by steel brackets and angles secured to the walls. Most roofs of tilt-up structures built today are framed with glulam timbers and sheathed with plywood.

## LIFT-SLAB CONSTRUCTION

Lift-slab construction combines precast concrete or steel columns with floor slabs cast on the job site. See Figure 70–22. After the foundation work has been completed and the ground floor

slab has been placed, columns extending the entire height of the building are set up. On high-rise structures the columns cannot be set up for the entire height at one time, so they are erected in sections as the floors are being raised.

*Portland Cement Association*

Figure 70–22. Lift-slab building under construction. The floor slabs for the building were stack-cast on the ground floor and lifted into place with special lifting jacks.

The floor slabs for the building are *stack-cast* (cast directly on top of each other). Resin-type compounds are used as bond breakers between the slabs. Slab thicknesses vary from 7″ to 10″, depending upon span and load conditions. Maximum slab strength can be attained by post-tensioning the slabs after they have hardened.

Hydralic jacks are placed at the top of each column. Lifting rods that go from the jacks are connected to lifting collars that have been placed in the slabs around the columns. The lifting jacks are connected to a central, electrically controlled console that simultaneously operates all the jacks within a ¼″ tolerance. Normal lifting rates are 7′ to 10′ per hour.

On smaller buildings the roof slab is usually lifted to its final position first. The lifting sequence that follows varies according to the building's height. For column stability, a number of stacked layers may be raised to a higher floor level.

To secure the floor slab, usually steel shear bars are inserted through the columns and under the slab after the slab has been elevated to its final position. These shear bars are designed to provide a permanent attachment between the slab and column. When steel columns are used, the floor slab may be secured by welding a steel collar under the slab to the steel column. After the slabs have been permanently attached to the column, any gaps between the slab and the column are filled with grout.

Exterior walls in lift-slab construction are curtain walls of various materials, usually metal. Figure 70–23 reviews the general procedure for constructing a lift-slab building.

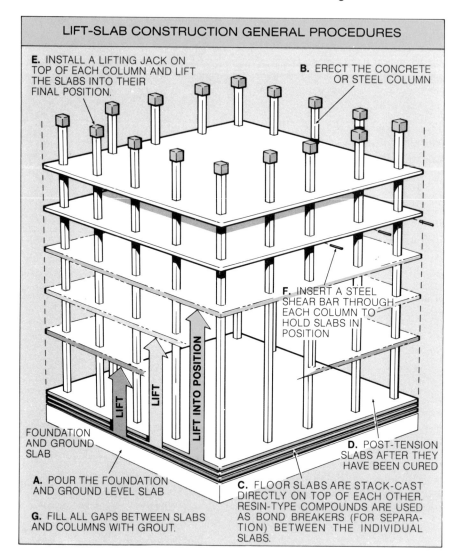

Figure 70-23. General procedure for a lift-slab construction.

# APPENDIX A
# WWPA WESTERN LUMBER

## DIMENSION LUMBER GRADES

| Product | Grades | WWPA Western Lumber Grading Rules Section Reference | Uses |
|---|---|---|---|
| Structural Light Framing (SLF)<br><br>2″ to 4″ thick<br>2″ to 4″ wide | SELECT STRUCTURAL<br>NO. 1<br>NO. 2<br>NO. 3 | (42.10)<br>(42.11)<br>(42.12)<br>(42.13) | Structural applications where highest design values are needed in light framing sizes. |
| Light Framing (LF)<br><br>2″ to 4″ thick<br>2″ to 4″ wide | CONSTRUCTION<br>STANDARD<br>UTILITY | (40.11)<br>(40.12)<br>(40.13) | Where high-strength values are not required, such as wall framing, plates, sills, cripples, blocking, etc. |
| Stud<br><br>2″ to 4″ thick<br>2″ and wider | STUD | (41.13) | An optional all-purpose grade designed primarily for stud uses, including bearing walls. |
| Structural Joists and Planks (SJ&P) | SELECT STRUCTURAL<br>NO. 1<br>NO. 2<br>NO. 3 | (62.10)<br>(62.11)<br>(62.12)<br>(62.13) | Intended to fit engineering applications for lumber 5″ and wider, such as joists, rafters, headers, beams, trusses, and general framing. |

## STRUCTURAL DECKING GRADES

| Product | Grades | WWPA Western Lumber Grading Rules Section Reference | Uses |
|---|---|---|---|
| Structural Decking<br><br>2″ to 4″ thick<br>4″ to 12″ wide | SELECTED DECKING | (55.11) | Used where the appearance of the best face is of primary importance. |
| | COMMERCIAL DECKING | (55.12) | Customarily used when appearance is not of primary importance. |

## TIMBER GRADES

| Product | Grades | WWPA Western Lumber Grading Rules Section Reference | Uses |
|---|---|---|---|
| Beams and Stringers<br><br>5″ and thicker, width more than 2″ greater than thickness | DENSE SELECT STRUCTURAL*<br>DENSE NO. 1*<br>DENSE NO. 2*<br>SELECT STRUCTURAL<br>NO. 1<br>NO. 2 | (53.00 & 170.00)<br>(53.00 & 170.00)<br>(53.00 & 170.00)<br>(70.10)<br>(70.11)<br>(70.12) | Grades are designed for beam and stringer type uses when sizes larger than 4″ nominal thickness are required. |
| Post and Timbers<br><br>5″ × 5″ and larger, width not more than 2″ greater than thickness | DENSE SELECT STRUCTURAL*<br>DENSE NO. 1*<br>DENSE NO. 2*<br>SELECT STRUCTURAL<br>NO. 1<br>NO. 2 | (53.00 & 170.00)<br>(53.00 & 170.00)<br>(53.00 & 170.00)<br>(80.10)<br>(80.11)<br>(80.12) | Grades are designed for vertically loaded applications where sizes larger than 4″ nominal thickness are required. |

*Douglas Fir or Douglas Fir-Larch only

# APPENDIX B
## SOUTHERN PINE GRADE DESCRIPTIONS

| Product | Grade | Grade Characteristics and Typical Uses |
|---|---|---|
| **Dimension Lumber: 2″ to 4″ thick, 2″ and wider** *See table 1 for design values* | | |
| | * Dense Select Structural<br>Select Structural<br>* NonDense Select Structural | High quality, relatively free of characteristics which impair strength or stiffness. Recommended for uses where high strength, stiffness and good appearance are desired. |
| | * No. 1 Dense<br>No. 1<br>* No. 1 NonDense | Recommended for general utility and construction where high strength, stiffness and good appearance are desired. |
| | * No. 2 Dense<br>No. 2<br>* No. 2 NonDense | Recommended for most general construction uses where moderately high design values are required. Allows well-spaced knots of any quality. |
| | No. 3 | Assigned design values meet a wide range of design requirements. Recommended for general construction purposes where appearance is not a controlling factor. Many pieces included in this grade would qualify as No. 2 except for a single limiting characteristic. |
| | Stud | Suitable for stud uses including use in load-bearing walls. Composite of No. 3 strength and No. 1 nailing edge characteristics. |
| | * Construction (2″ to 4″ wide only) | Recommended for general framing purposes. Good appearance, but graded primarily for strength and serviceability. |
| | * Standard (2″ to 4″ wide only) | Recommended for same purposes as Construction grade. Characteristics are limited to provide good strength and excellent serviceability. |
| | * Utility (2″ to 4″ wide only) | Recommended where a combination of economical construction and good strength is desired. Used for such purposes as studding, blocking, plates, bracing and rafters. |
| | *Design values are not assigned*<br>Economy | Usable lengths suitable for bracing, blocking, bulk heading and other utility purposes where strength and appearance are not controlling factors. |
| **\*Timbers: 5″ × 5″ and larger** *See table 2 for design values* | | |
| | Dense Select Structural<br>Select Structural | Recommended where high strength, stiffness and good appearance are desired. |
| | No. 1 Dense<br>No. 1<br>No. 2 Dense<br>No. 2 | No. 1 and No. 2 are similar in appearance to corresponding grades of 2″ thick Dimension Lumber. Recommended for general construction uses. |
| | *Design values are not assigned*<br>No. 3 | Non-stress rated, but economical for general construction purposes such as blocking, fillers, etc. |
| **\*Structural Lumber: 2″ and thicker, 2″ and wider** *See SPIB Special Product Rules for design values* | | |
| | Dense Structural 86<br>Dense Structural 72<br>Dense Structural 65 | Premier structural grades. Provides good appearance with some of the highest design values available in any softwood species. |
| **Radius Edge Decking 1-¼″ thick, 4″ to 6″ wide** *See SPIB Special Product Rules for recommended spans* | | |
| | Premium | High-quality product, recommended where smallest knots are desired and appearance is of utmost importance. Excellent for painting or staining. |
| | Standard | Slightly less restrictive than premium grade. A very good product to use where appearance is not the major factor. Excellent for painting or staining. |

*Most mills do not manufacture all products and make all grade separations. Those products and grades not manufactured by most mills are noted.

# APPENDIX B (continued)
## SOUTHERN PINE GRADE DESCRIPTIONS

| Product | Grade | Grade Characteristics and Typical Uses |
|---|---|---|
| **Finish: ⅜″ to 4″ thick, 2″ and wider** | | |
| *Design values are not assigned* | | |
| | * B&B | Highest recognized grade of finish. Generally clear, although a limited number of pin knots are permitted. Finest quality for natural or stain finish. |
| | C | Excellent for painting or natural finish where requirements are less exacting. Reasonably clear, but permits limited number of surface checks and small tight knots. |
| | C&Btr | Combination for B&B and C grades; satisfies requirements for high-quality finish. |
| | D | Economical, serviceable grade for natural or painted finish. |
| **Flooring, Drop Siding, Paneling, Ceiling and Partition, OG Batts, Bevel Siding, Miscellaneous Millwork** | | |
| *Design values are not assigned* | | |
| | * B&B, C C&Btr, D | See Finish grades for face side; reverse side wane limitations are lower. |
| | No. 1 | No. 1 Flooring and Paneling not provided under SPIB Grading Rules as a separate grade, but if specified, will be designated and graded as D; No. 1 Drop Siding is graded as No. 1 Boards. |
| | No. 2 | Graded as No. 2 Boards. High utility value where appearance is not a factor. |
| | No. 3 | More manufacturing imperfections allowed than in No. 2, but suitable for economical use. |
| **\*Shop and Moulding** | | |
| *Design values are not assigned* | | |
| | No. 1 | Recommended for moulding and millwork applications. Currently graded according to rules developed by the Western Wood Products Association. |
| | No. 2 | |
| | No. 3 | |
| **\*Marine Grades: 1″ to 20″ thick, 2″ to 20″ wide** | | |
| *See tables 1 and 2 for design values* | | |
| | Any grade of Dimension Lumber or Timbers | All four longitudinal faces must be free of pith and/or heartwood. Application of the product requires pressure treatment by an approved treating process and preservative for marine usage. |
| **\*Decking, Heavy Roofing and Heavy Shiplap: 2″ to 4″ thick, 2″ and wider** | | |
| *See SPIB Grading Rules for design values* | | |
| | Dense Standard Decking | High-quality product, suitable for plank floor where face serves as finish floor. Has a better appearance than No. 1 Dense Dimension Lumber because of additional restrictions on pitch, knots, pith and wane. |
| | Dense Select Decking Select Decking | An excellent decking grade that can be used face side down for roof decking or face side up for floor decking. |
| | Dense Commercial Decking Commercial Decking | An economical roof decking which conforms to No. 2 Dimension Lumber characteristics. |
| **Boards: 1″ to 1-½″ thick, 2″ and wider** | | |
| *See SPIB Grading Rules for design values* | | |
| | Industrial 55 | Graded as per No. 1 Dimension. |
| | Industrial 45 | Graded as per No. 2 Dimension. |
| | Industrial 26 | Graded as per No. 3 Dimension. |
| *Design values are not assigned* | | |
| | No. 1 | High quality with good appearance characteristics. Generally sound and tight-knotted. Largest hole permitted is ¹⁄₁₆″. Superior product suitable for a wide range of uses including shelving, crating, and form lumber. |
| | No. 2 | Good-quality sheathing, fencing, shelving and other general purpose uses. |

\*Most mills do not manufacture all products and make all grade separations. Those products and grades not manufactured by most mills are noted.

| Product | Grade | Grade Characteristics and Typical Uses |
|---|---|---|
| | No. 3 | Good, serviceable sheathing; usable for many economical applications without waste. |
| | No. 4 | Admits pieces below a No. 3 grade which can be used without waste, or which contain less than 25% waste by cutting. |

### *Industrial Lumber: 4″ and less in thickness, 12″ and less in width
*See SPIB Special Product Rules for design values*

| Product | Grade | Grade Characteristics and Typical Uses |
|---|---|---|
| | Industrial 86 | Appearance is same as B&B Finish. Larger sizes conform to Dense Structural 86 Structural Lumber except for dense grain requirement. |
| | Industrial 72 | Appearance is same as C Finish. Larger sizes conform to Dense Structural 72 Structural Lumber except for dense grain requirement. |
| | Industrial 65 | Appearance is same as D Finish. Larger sizes conform to Dense Structural 65 Structural Lumber except for dense grain requirement. |

### *Mechanically Graded Lumber – Machine Stress Rated (MSR) Lumber: 2″ and less in thickness, 2″ and wider
*See table 3 for design values*

| Product | Grade | Grade Characteristics and Typical Uses |
|---|---|---|
| | 1650f – 1.5E thru 3000f – 2.4E | Machine Stress Rated (MSR) lumber is lumber that has been evaluated by mechanical stress rating equipment. MSR lumber is distinguished from visually stress graded lumber in that each piece is non-destructively tested. MSR lumber is also required to meet certain visual grading requirements. |

### *Mechanically Graded Lumber – Machine Evaluated Lumber (MEL): 2″ and less in thickness, 2″ and wider
*See SPIB Grading Rules for design values*

| Product | Grade | Grade Characteristics and Typical Uses |
|---|---|---|
| | M – 5 thru M – 28 | Well-manufactured material evaluated by calibrated mechanical grading equipment which measures certain properties and sorts the lumber into various strength classifications. Machine Evaluated Lumber is also required to meet certain visual requirements. |

### *Scaffold Plank: 2″ and 3″ thick, 8″ and wider
*See table 4 for design values*

| Product | Grade | Grade Characteristics and Typical Uses |
|---|---|---|
| | Dense Industrial 72 Dense Industrial 65 | All Scaffold Plank design values are calculated using ASTM Standards D245 and D2555. These values are modified using procedures shown in "Calculating Apparent Reliability of Wood Scaffold Planks," as published by the Journal on Structural Safety, 2 (1984) 47-57. |
| | MSR: 2400f – 2.0E MSR: 2200f – 1.8E | Dressed to standard dry size prior to machine stress rating, and visually graded to assure that characteristics affecting strength are no more serious than the limiting characteristics for each grade. MSR Scaffold Plank is available 2″ thick only. |

### *Stadium Grade: 2″ thick, 4″ to 12″ wide
*See table 1 for design values*

| Product | Grade | Grade Characteristics and Typical Uses |
|---|---|---|
| | No. 1 Dense No. 1 | For outdoor seating. Free of pitch pockets, pitch streaks and medium pitch on one wide face, but otherwise conforms to No. 1 Dense or No. 1 Dimension Lumber. |

### *Prime Dimension: 2″ to 4″ thick, 2″ to 12″ wide
*See table 1 for design values*

| Product | Grade | Grade Characteristics and Typical Uses |
|---|---|---|
| | No. 1 Prime | Grade based on No. 1 Dimension Lumber characteristics except that holes, skip and wane are closely limited to provide a high-quality product. |
| | No. 2 Prime | Grade based on No. 2 Dimension Lumber characteristics except that holes, skip and wane are closely limited to provide a high-quality product. |

*Most mills do not manufacture all products and make all grade separations. Those products and grades not manufactured by most mills are noted.

# APPENDIX C
## GUIDE TO APA PERFORMANCE RATED PANELS[(a)(b)]

**APA RATED SHEATHING**
Typical Trademark

Specially designed for subflooring and wall and roof sheathing. Also good for a broad range of other construction and industrial applications. Can be manufactured as plywood, as a composite, or as OSB. EXPOSURE DURABILITY CLASSIFICATIONS: Exterior, Exposure 1, Exposure 2. COMMON THICKNESSES: 5/16, 3/8, 7/16, 15/32, 1/2, 19/32, 5/8, 23/32, 3/4.

**APA STRUCTURAL I RATED SHEATHING[(c)]**
Typical Trademark

Unsanded grade for use where shear and cross-panel strength properties are of maximum importance, such as panelized roofs and diaphragms. Can be manufactured as plywood, as a composite, or as OSB. EXPOSURE DURABILITY CLASSIFICATIONS: Exterior, Exposure 1. COMMON THICKNESSES: 5/16, 3/8, 7/16, 15/32, 1/2, 19/32, 5/8, 23/32, 3/4.

**APA RATED STURD-I-FLOOR**
Typical Trademark

Specially designed as combination subfloor-underlayment. Provides smooth surface for application of carpet and pad and possesses high concentrated and impact load resistance. Can be manufactured as plywood, as a composite, or as OSB. Available square edge or tongue-and-groove. EXPOSURE DURABILITY CLASSIFICATIONS: Exterior, Exposure 1, Exposure 2. COMMON THICKNESSES: 19/32, 5/8, 23/32, 3/4, 1, 1-1/8.

**APA RATED SIDING**
Typical Trademark

For exterior siding, fencing, etc. Can be manufactured as plywood, as a composite or as an overlaid OSB. Both panel and lap siding available. Special surface treatment such as V-groove, channel groove, deep groove (such as APA Texture 1-11), brushed, rough sawn and overlaid (MDO) with smooth- or texture-embossed face. Span Rating (stud spacing for siding qualified for APA Sturd-I-Wall applications) and face grade classification (for veneer-faced siding) indicated in trademark. EXPOSURE DURABILITY CLASSIFICATION: Exterior. COMMON THICKNESSES: 11/32, 3/8, 7/16, 15/32, 1/2, 19/32, 5/8.

(a) Specific grades, thicknesses and exposure durability classifications may be in limited supply in some areas. Check with your supplier before specifying.

(b) Specify Performance Rated Panels by thickness and Span Rating. Span Ratings are based on panel strength and stiffness. Since these properties are a function of panel composition and configuration as well as thickness, the same Span Rating may appear on panels of different thickness. Conversely, panels of the same thickness may be marked with different Span Ratings.

(c) All plies in Structural I plywood panels are special improved grades and panels marked PS 1 are limited to Group 1 species. Other panels marked Structural I Rated quality through special performance testing. Structural II plywood panels are also provided for, but rarely manufactured. Application recommendations for Structural II plywood are identical to those for APA RATED SHEATHING plywood.

# APPENDIX D
## GUIDE TO APA SANDED & TOUCH-SANDED PLYWOOD PANELS[(a)(b)(c)]

**APA A-A**
Typical Trademark

Use where appearance of both sides is important for interior applications such as built-ins, cabinets, furniture, partitions; and exterior applications such as fences, signs, boats, shipping containers, tanks, ducts, etc. Smooth surfaces suitable for painting. EXPOSURE DURABILITY CLASSIFICATIONS: Interior, Exposure 1, Exterior. COMMON THICKNESSES: 1/4, 11/32, 3/8, 15/32, 1/2, 19/32, 5/8, 23/32, 3/4.

A-A • G-1 • EXPOSURE 1-APA • 000 • PS1-95

**APA A-B**
Typical Trademark

For use where appearance of one side is less important but where two solid surfaces are necessary. EXPOSURE DURABILITY CLASSIFICATIONS: Interior, Exposure 1, Exterior. COMMON THICKNESSES: 1/4, 11/32, 3/8, 15/32, 1/2, 19/32, 5/8, 23/32, 3/4.

A-B • G-1 • EXPOSURE 1-APA • 000 • PS1-95

**APA A-C**
Typical Trademark

For use where appearance of only one side is important in exterior or interior applications, such as soffits, fences, farm buildings, etc.[(f)] EXPOSURE DURABILITY CLASSIFICATION: Exterior. COMMON THICKNESSES: 1/4, 11/32, 3/8, 15/32, 1/2, 19/32, 5/8, 23/32, 3/4.

*American Plywood Association*

# APPENDIX D (continued)
## GUIDE TO APA SANDED & TOUCH-SANDED PLYWOOD PANELS[(a)(b)(c)]

APA A-D
Typical Trademark

For use where appearance of only one side is important in interior applications, such as paneling, built-ins, shelving, partitions, flow racks, etc.[(f)] EXPOSURE DURABILITY CLASSIFICATIONS: Interior, Exposure 1. COMMON THICKNESSES: 1/4, 11/32, 3/8, 15/32, 1/2, 19/32, 5/8, 23/32, 3/4.

APA B-B
Typical Trademark

B-B • G-2 • EXPOSURE 1-APA • 000 • PS1-95

Utility panels with two solid sides. EXPOSURE DURABILITY CLASSIFICATIONS: Interior, Exposure 1, Exterior. COMMON THICKNESSES: 1/4, 11/32, 3/8, 15/32, 1/2, 19/32, 5/8, 23/32, 3/4.

APA B-C
Typical Trademark

Utility panel for farm service and work buildings, boxcar and truck linings, containers, tanks, agricultural equipment, as a base for exterior coatings and other exterior uses or applications subject to high or continuous moisture.[(f)] EXPOSURE DURABILITY CLASSIFICATION: Exterior. COMMON THICKNESSES: 1/4, 11/32, 3/8, 15/32, 1/2, 19/32, 5/8, 23/32, 3/4.

APA B-D
Typical Trademark

Utility panel for backing, sides of built-ins, industry shelving, slip sheets, separator boards, bins and other interior or protected applications.[(f)] EXPOSURE DURABILITY CLASSIFICATIONS: Interior, Exposure 1. COMMON THICKNESSES: 1/4, 11/32, 3/8, 15/32, 1/2, 19/32, 5/8, 23/32, 3/4.

APA UNDERLAYMENT
Typical Trademark

For application over structural subfloor. Provides smooth surface for application of carpet and pad and possesses high concentrated and impact load resistance. For areas to be covered with resilient flooring, specify panels with "sanded face."[(e)] EXPOSURE DURABILITY CLASSIFICATIONS: Interior, Exposure 1. COMMON THICKNESSES[(d)]: 1/4, 11/32, 3/8, 15/32, 1/2, 19/32, 5/8, 23/32, 3/4.

APA C-C PLUGGED[(g)]
Typical Trademark

For use as an underlayment over structural subfloor, refrigerated or controlled atmosphere storage rooms, pallet fruit bins, tanks, boxcar and truck floors and linings, open soffits, and other similar applications where continuous or severe moisture may be present. Provides smooth surface for application of carpet and pad and possesses high concentrated and impact load resistance. For areas to be covered with resilient flooring, specify panels with "sanded face."[(e)] EXPOSURE DURABILITY CLASSIFICATION: Exterior. COMMON THICKNESSES[(d)] : 11/32, 3/8, 15/32, 1/2, 19/32, 5/8, 23/32, 3/4.

APA C-D PLUGGED
Typical Trademark

For open soffits, built-ins, cable reels, separator boards and other interior or protected applications. Not a substitute for Underlayment or APA Rated Sturd-I-Floor as it lacks their puncture resistance. EXPOSURE DURABILITY CLASSIFICATIONS: Interior, Exposure 1. COMMON THICKNESSES: 3/8, 15/32, 1/2, 19/32, 5/8, 23/32, 3/4.

(a) Specific plywood grades, thicknesses and exposure durability classifications may be in limited supply in some areas. Check with your supplier before specifying.

(b) Sanded Exterior plywood panels, C-C Plugged, C-D Plugged and Underlayment grades can also be manufactured in Structural I (all plies limited to Group 1 species).

(c) Some manufacturers also produce plywood panels with premium N-grade veneer on one or both faces. Available only by special order. Check with the manufacturer.

(d) Some panels 1/2 inch and thicker are Span Rated and do not contain species group number in trademark.

(e) Also available in Underlayment A-C or Underlayment B-C grades, marked either "touch sanded" or "sanded face."

(f) For nonstructural floor underlayment, or other applications requiring improved inner ply construction, specify panels marked either "plugged inner plies"
(may also be designated plugged crossbands under face or plugged crossbands or core); or "meets underlayment requirements."

(g) Also may be designated APA Underlayment C-C Plugged.

*American Plywood Association*

# APPENDIX E
## GUIDE TO APA SPECIALTY PLYWOOD PANELS[(a)]

**APA DECORATIVE**
Typical Trademark

Rough-sawn, brushed, grooved, or striated faces. For paneling, interior accent walls, built-ins, counter facing, exhibit displays. Can also be made by some manufacturers in Exterior for exterior siding, gable ends, fences and other exterior applications. Use recommendations for Exterior panels vary with the particular product. Check with the manufacturer. EXPOSURE DURABILITY CLASSIFICATIONS: Interior, Exposure 1, Exterior. COMMON THICKNESSES: 5/16, 3/8, 1/2, 5/8.

**APA HIGH DENSITY OVERLAY (HDO)[(b)]**
Typical Trademark

HDO • A-A • G-1 • EXT-APA • 000 • PS 1-95

Has a hard semi-opaque resin-fiber overlay on both faces. Abrasion resistant. For concrete forms, cabinets, countertops, signs, tanks. Also available with skid-resistant screen-grid surface. EXPOSURE DURABILITY CLASSIFICATION: Exterior. COMMON THICKNESSES: 3/8, 1/2, 5/8, 3/4.

**APA MEDIUM DENSITY OVERLAY (MDO)[(b)]**
Typical Trademark

Smooth, opaque, resin-fiber overlay on one or both faces. Ideal base for paint, both indoors and outdoors. For exterior siding, paneling, shelving, exhibit displays, cabinets, signs. EXPOSURE DURABILITY CLASSIFICATION: Exterior. COMMON THICKNESSES: 11/32, 3/8, 15/32, 1/2, 19/32, 5/8, 23/32, 3/4.

**APA MARINE**
Typical Trademark

MARINE • A-A • EXT-APA • PS 1-95

Ideal for boat hulls. Made only with Douglas-fir or western larch. Subject to special limitations on core gaps and face repairs. Also available with HDO or MDO faces. EXPOSURE DURABILITY CLASSIFICATION: Exterior. COMMON THICKNESSES: 1/4, 3/8, 1/2, 5/8, 3/4.

**APA B-B PLYFORM CLASS I[(b)]**
Typical Trademark

Concrete form grades with high reuse factor. Sanded both faces and mill-oiled unless otherwise specified. Special restrictions on species. Also available in HDO for very smooth concrete finish, and with special overlays. EXPOSURE DURABILITY CLASSIFICATION: Exterior. COMMON THICKNESSES: 19/32, 5/8, 23/32, 3/4.

**APA PLYRON**
Typical Trademark

PLYRON • EXPOSURE 1-APA • 000

Hardboard face on both sides. Faces tempered, untempered, smooth or screened. For countertops, shelving, cabinet doors, flooring. EXPOSURE DURABILITY CLASSIFICATIONS: Interior, Exposure 1, Exterior. COMMON THICKNESSES: 1/2, 5/8, 3/4.

(a) Specific plywood grades, thicknesses and exposure durability classifications may be in limited supply in some areas. Check with your supplier before specifying.

(b) Can also be manufactured in Structural I (all plies limited to Group 1 species).

*American Plywood Association*

# Glossary and Trade Tips

## A

**Abrasive paper.** Paper or cloth with an abrasive material glued to one side. Used primarily for smoothing wood surfaces.

**Accelerator (concrete).** Admixture for concrete that speeds the curing process and increases the rate of strength gain.

**Acoustical tile (ceiling).** Ceiling tile with small holes and fissures that act as sound traps to reduce the reflection of sound.

**Acoustics (room).** Pertains to sounds created within a room space.

**Active door.** Door (of a double door unit) used for normal traffic.

**Actual length (rafters).** Length of a main rafter or valley jack rafter after it has been shortened because of ridge board and/or rafter thickness.

**Actual lumber size.** Thickness and width of a piece of lumber after shrinkage resulting from drying and after surfacing at a planing mill.

**Aggregates.** Sand and gravel in a concrete mixture.

**Airborne sound transmission.** Pertains to sounds conducted through the air as pressure waves that strike against a wall, floor, or ceiling. Examples are speech and music.

**Air compressor.** Machine used to compress air for pneumatic (air-powered) equipment.

**Air-dried lumber.** Lumber that has been dried to acceptable moisture content by being stacked for a period of time in the open air.

**Air duct.** Pipe, usually rectangular and made of sheet metal, used to conduct hot or cold air in heating or cooling systems.

**Air-entraining admixture (concrete).** Admixture that produces microscopic air bubbles in concrete to improve its workability and its resistance to freezing and thawing.

**Air pocket (concrete).** Void in hardened concrete caused by poor consolidation when the concrete was placed.

**Allowable span.** Distance allowed between supporting points for various sizes of girders, joists, and roof rafters.

**Alteration work.** Change or addition to an existing building. Also called *remodeling.*

**Aluminum siding.** Pre-finished aluminum siding pieces that look like wood board siding.

**Anchor bolt.** Bolt used to secure sill plates, columns, and girders to concrete or other masonry. It is hooked or has a welded plate at the non-threaded end embedded in the concrete.

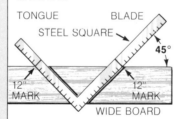

## TRADE TIP

### LAYING OUT A 45° ANGLE

A METHOD FOR USING THE STEEL SQUARE TO MARK A 45° ANGLE ON A WIDER PIECE OF MATERIAL.

TONGUE    BLADE
STEEL SQUARE
45°
12" MARK    12" MARK
WIDE BOARD

ALIGN THE SAME NUMBER MARK ON THE BLADE AND TONGUE OF THE SQUARE WITH THE EDGE OF THE MATERIAL. IN THIS EXAMPLE THE **12"** NUMBER MARK IS USED.

**Anchor clip (sill).** Straplike device embedded into the top of a foundation wall and used to fasten sill plates.

**Angle iron.** Structural steel bent to form a 90° angle. Used for support or fastening purposes.

**Annual ring.** Layer of wood seen in cross section of a treetrunk indicating a year's growth.

**Annular nails.** Nails with circular ridges on the shank to give greater holding power.

**Apprenticeship.** On-the-job training program with related instruction leading to journeyman status in a trade.

**Apron (window).** Trim piece placed below the stool of a finished window frame.

**Arbor.** Shaft on which a cutting tool is mounted, such as the blade of a circular saw.

**Arc.** Part of the circumference of a circle.

**Arc welding.** Electrical welding procedure in which metal is melted by the heat of an electric arc. Molten metal from the tip of an electrode provides filler at the joint.

**Arch.** Curved structure designed to support the weight above an opening.

**Architect.** Person qualified and licensed to design and oversee construction of a building.

**Archway.** Passage area under an arch.

**Areaway.** Small sunken area allowing light or air into a basement window that is partially below the grade level around the building.

**Asbestos cement.** See *mineral fiber.*

**Asphalt.** Petroleum product obtained from crude oil. It is waterproof and is the base for many products used for roof, wall, and floor covering.

**Asphalt shingle.** Shingle made of felt saturated with asphalt and surfaced with mineral granules. Usually used as a finish roofing material.

**Asphalt tile.** Resilient floor tile. Its basic ingredients are asphaltic and/or resin binders combined with asbestos fibers and limestone fillers.

**Associated General Contractors.** Industry organization of general contractors who perform heavy construction work.

**Astragal.** Piece of molding attached to the edge of the inactive door of a pair of double doors. It serves as a stop for the active door.

**Atrium.** See *greenhouse*.

**Attic.** Space between a roof and a ceiling.

**Awning window.** Window that is hinged at the top and swings out at the bottom.

# B

**Backfill.** Soil or gravel used to fill the space between the completed foundation wall and the excavated areas on one or both sides of the wall.

**Backing (ceiling).** Pieces nailed over the top wall plates to provide a nailing surface for the edges of ceiling finish materials.

**Backing (hip rafters).** Beveling each side of the top edge of a hip rafter so that the ends of the roof sheathing will not be pushed up and out of line with the rest of the roof.

**Backing (wall).** Pieces nailed between studs to provide a surface for fastening plumbing fixtures. Also, pieces sometimes nailed behind studs to provide a surface for corner nailing.

**Back priming.** Painting or treating with a sealer the backs of wood siding materials

before they are nailed to walls.

**Backsplash.** Pieces that extend up from a countertop and are fastened to the wall.

**Balcony.** Deck projecting above ground from the wall of a building.

**Ball-bearing butt hinge.** Butt hinge with ball bearings at the knuckle joints to prevent wear. Used most often on heavy doors in public buildings.

**Balloon framing.** Framing method in which studs extend from the sill plate to the roof. Second floor joists are spiked into the studs, but they receive their main support from a ribbon notched into and nailed to the studs.

**Balusters.** Upright pieces that run between the handrails and the treads. Also called *bannisters*.

**Balustrade railing.** Railing with newel posts at both ends and balusters in between.

**Band joist.** See *header joist*.

**Bannisters.** See *balusters*.

**Barge rafter.** See *fascia rafter*.

**Bargeboard.** See *fascia rafter*.

**Bark.** Surface covering of a treetrunk. Consists of an outer layer of dead, dry tissue and a thin inner layer of living tissue.

**Barrel (hinge).** Rounded hollow section located between the leaves of a hinge. It holds the pin. Also called the *knuckle*.

**Barricade.** Structure set up around a construction job to prevent unauthorized persons from entering working areas. Covered barricades also protect the public from falling objects.

**Baseboard.** Molding placed at the base of a wall and fitted to the floor. Also called *base molding*.

**Base cabinet.** Cabinet placed against a wall and resting on the floor.

**Base cap.** Small decorative molding sometimes nailed at the top of a baseboard.

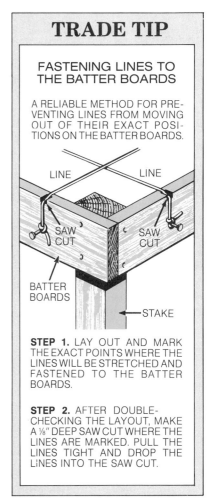

**TRADE TIP**

**FASTENING LINES TO THE BATTER BOARDS**

A RELIABLE METHOD FOR PREVENTING LINES FROM MOVING OUT OF THEIR EXACT POSITIONS ON THE BATTER BOARDS.

LINE    LINE

SAW CUT    SAW CUT

BATTER BOARDS

STAKE

**STEP 1.** LAY OUT AND MARK THE EXACT POINTS WHERE THE LINES WILL BE STRETCHED AND FASTENED TO THE BATTER BOARDS.

**STEP 2.** AFTER DOUBLE-CHECKING THE LAYOUT, MAKE A 1/8" DEEP SAW CUT WHERE THE LINES ARE MARKED. PULL THE LINES TIGHT AND DROP THE LINES INTO THE SAW CUT.

**Basement.** A building's lowest story, which is partly or completely below the ground.

**Base molding.** See *baseboard*.

**Base shoe.** Strip of molding sometimes nailed at the joint between a baseboard and the floor.

**Batch plant.** Plant where ready-mixed concrete is mixed to specification, then discharged into transit-mix trucks for delivery to the job site.

**Batt insulation.** Blanket-like sections of mineral fiber material placed between studs.

**Batten.** Thin, narrow strip of lumber usually used to cover the joint between wider boards.

**Batter board.** Level board nailed to stakes driven into the ground. String or wire is attached to batter boards to identify property lines, building lines, and pier locations.

**Battered foundation.** Foundation with inside sloped walls to provide a wide base.

**Bay window.** Window projecting from a wall. It creates a recessed area that is square, rectangular, or polygonal in shape.

**Beam ceiling.** Ceiling in which the beams are exposed to view.

**Beam hanger.** Metal strap used to support beams that butt against a girder.

**Beam pocket (concrete).** See *girder pocket.*

**Bearing pile.** Pile that penetrates through layers of unstable soil until it reaches firm, bearing soil.

**Bed molding.** Molding placed over a frieze board and up against the cornice soffit. May be placed between the rafters when the frieze board is notched into the rafters of an open cornice.

**Bedrock.** Solid layer of rock beneath the earthen materials on a job site.

**Belled caisson.** Concrete caisson flared at its bottom to provide a greater bearing area.

**Berm.** Earth slope built against the exterior wall of a building to provide insulation.

**Bevel.** Angled surface across the edge of a piece of material.

**Bevel siding.** Tapered siding board that laps over the board below.

**Bib.** Threaded faucet for a hose attachment.

**Bid.** Contractor's offer to a customer to do construction work for a specified price.

**Bi-fold doors.** Pair of folding doors in which only one door is hung from the jamb. The second door is hinged to the first door and is further supported by a roller guide that slides along a track mounted to the head jamb.

**Bind (saw).** Interference of cutting action as a result of wood fibers pressing against the blade of the saw.

**Bird's mouth.** Cut-out formed by the seat and heel plumb lines of a rafter where the lower end of the roof rafter fits over the top plates.

**Bit.** Drilling device with a screw point that is held in the jaws of a brace or drill.

**Bituminous.** Pertains to materials that contain asphalt and tar and are used for waterproofing.

**Blanket insulation.** Insulation material usually backed by treated paper that serves as a vapor barrier. It comes in rolls and is fastened between the studs and ceiling joists.

**Blind nailing.** Driving a nail in such a way that its head will be concealed. For example, a toenail driven at the inside corner of the tongue of tongue-and-groove flooring would be blind-nailed.

**Blinds.** See *shutter.*

**Blind stop.** Rectangular piece of molding placed toward the exterior of the head and side pieces of a window frame. Its main purpose is to serve as a stop for screens or storm windows.

**Blind valley construction.** Method of building an intersecting roof. The main roof is sheathed and the intersecting roof section is constructed on top of the sheathing. Valley rafters are not required.

**Block flooring.** Small squares made of narrow wood strips held together by splines or backing material. Usually installed with a mastic adhesive.

**Blockout (form).** Piece placed inside a form to provide a small recess or ledge at the top of the concrete wall to later support the end of a beam or other structural member.

**Blue stain.** Nonharmful blue-gray discoloration of lumber caused by a fungus growth on unseasoned wood.

**Board.** Lumber no more than 1″ thick and 4″ to 12″ wide.

**Board-and-batten siding.** Siding design that features wide vertical boards. The joints between the boards are covered with a narrow piece called the batten.

**Board foot.** Unit of cost for lumber, based on the volume of a piece 12″ square and 1″ thick.

**Bottom plate.** See *sole plate.*

**Bow (warp).** Distortion of a piece of lumber resulting in a curve from end to end across the flat plane of the wide side.

**Bow windows.** Series of windows set in a curved bow projecting from the building.

**Box beam.** Hollow beam made of plywood sides nailed to a framework of 2″ × 4″ chords and stiffeners.

**Box nail.** Flat-head nail with a thinner shank and head than a common nail.

**Bracket.** Projecting L-shaped support for a shelf or any other kind of weight.

**Brad.** Small finish nail ranging from ½″ to 2″ in length.

**Breezeway.** Roofed passageway, open at ends or sides, extending from a house to a garage.

**Bridging (joist).** Solid wood blocking or other types of wood or metal pieces nailed between joists to stiffen and hold them in position.

**British thermal unit (Btu).** Measure of heat transfer based on the amount of heat required to raise the temperature of one pound of water to one degree Fahrenheit.

**Buck.** Frame placed inside a concrete form to provide an opening for a door or window.

**Bucket (concrete).** Large

container into which concrete is discharged from a mixer or transit-mix truck. The bucket is then raised by crane and moved to where the concrete is to be placed.

**Buggy (concrete).** Two-wheeled conveyance pushed by a single person and used on the job to carry fresh concrete to the point where it is to be placed.

**Building code.** Set of regulations that establish the required standards for the materials and methods of construction in a city, county, or state. Building codes are enforceable by law.

**Building line.** Line set up on batter boards to represent the outside face of the exterior wall of a building.

**Building paper.** See *sheathing paper.*

**Building permit.** Permit legally required by local or state building authorities for any kind of new construction work.

**Built-in.** Pertains to cabinets, cupboards, and other furniture that is built into the wall structure.

**Built-up Girder.** Girder made of two or more planks nailed or bolted together. Also called *built-up beam.*

**Built-up header.** Header over a rough opening made up of two or more pieces spiked together.

**Built-up roofing.** Roofing consisting of three to five layers of felt mopped down with hot tar or asphalt. Usually used on flat deck roofs.

**Bulkhead (form).** Vertical piece set inside a form to stop the concrete pour at a certain place, thus providing a vertical construction joint. Also called *shut off.*

**Butt hinge.** See *door hinge.*

**Butt joint.** Joint in which one piece butts squarely against another.

---

## TRADE TIP

### MARKING SIDE PIECES OF DOOR CASING

A FAST AND ACCURATE METHOD FOR CUTTING PIECES OF SIDE CASING WHEN A NUMBER OF DOOR OPENINGS ARE THE SAME HEIGHT.

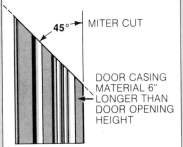

**STEP 1.** CUT A 45° MITER ANGLE AT ONE END OF THE CASING MATERIAL. THE LENGTH FROM THE LONG POINT OF THE MITER CUT TO THE OTHER END OF THE CASING MATERIAL SHOULD BE ABOUT 6" LONGER THAN THE DOOR OPENING HEIGHT.

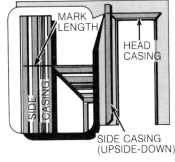

**STEP 2.** PLACE THE SIDE PIECE OF CASING IN AN UPSIDE-DOWN POSITION NEXT TO THE DOOR FRAME. (THE HEAD CASING MUST ALREADY BE NAILED IN PLACE). MARK THE SIDE PIECE OF CASING EVEN WITH THE UPPER EDGE OF THE HEAD CASING. SQUARE A LINE AND CUT IT TO LENGTH.

---

**Butterfly roof.** Roof with two sides sloping toward the central part of the building.

**Buttress.** Projecting masonry structure built against a wall to give additional lateral strength and support.

**Bypass sliding doors.** Doors suspended from tracks with roller hangers, often used with closets. Two or three bypass sliding doors may be used to cover an opening, with the tracks designed to let the doors slide past each other.

---

## C

**Cabinetmaker.** Person who works in a cabinet shop and is skilled in the layout and construction of wood cabinets.

**Caisson (concrete).** Large cylindrical or box-like casing placed in the ground and filled with concrete. Caissons have wide diameters and usually extend deeper than piles.

**Camber.** Slight arch in a beam or any other kind of horizontal timber. Prevents the member from eventually bending from its own weight in addition to the load it must carry.

**Cambium.** Layer of tissue equal to a one-cell thickness and located between the bark and sapwood of a tree.

**Cant strip.** Triangular piece placed where a flat roof deck meets a wall. Used to prevent a sharp angle for the roofing material.

**Cantilevered joists.** Joists projecting from the wall below in order to support a floor or balcony which extends past the wall below.

**Cap plate.** See *double plate.*

**Capillary action.** Natural process in which water rises from the water table to the surface of the ground. Occurs in most types of soil.

**Capital (column).** Flared section at the top of a concrete column that helps support the floor slab above. Also called *drop head.*

**Carbide tip.** Tungsten carbide metal tip that is braised to the end of each tooth of a circular saw blade and to the cutting edges of masonry drills to prolong sharpness.

**Carpenters' union.** United

## TRADE TIP

### FINDING CENTER POINT

A SIMPLE AND ACCURATE METHOD FOR ESTABLISHING CENTER LINES BETWEEN WALLS AND OTHER POINTS.

**STEP 1.** MEASURE FROM WALL **A** AND MARK POINT **X** ON THE FLOOR AT AN EVEN FOOT NUMBER (IN THIS EXAMPLE 5'-0") CLOSE TO THE CENTER OF THE ROOM.

**STEP 2.** MEASURE FROM WALL **B** THE SAME DISTANCE AS MEASURED FROM WALL (5'-0"). MARK POINT **Y** ON THE FLOOR.

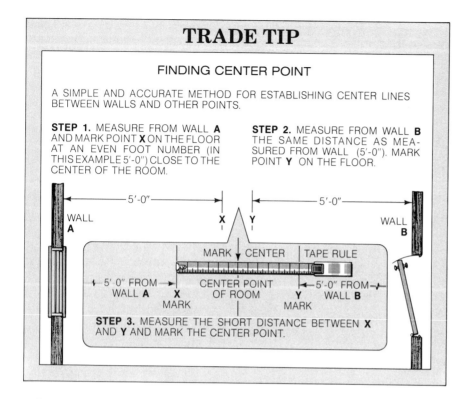

**STEP 3.** MEASURE THE SHORT DISTANCE BETWEEN **X** AND **Y** AND MARK THE CENTER POINT.

Brotherhood of Carpenters and Joiners of America, AFL-CIO. Labor organization representing union carpenters in the United States mainland, Puerto Rico, Panama, Canada, Alaska, and Hawaii.

**Carport.** Roofed car shelter, not enclosed at the sides.

**Carriage bolt.** Used only to fasten wood pieces. A square section below the oval head embeds itself into the wood as the nut is drawn up.

**Carriage (stair).** See *stringer.*

**Casement window.** Window hinged on one side. It has the same swing action as a door.

**Casing (molding).** Molding that covers the space between the finished door and window frames and the wall.

**Casing nail.** Nail with a flared head. Used for outside finish work and finish flooring.

**Casing (for piles).** Steel shell driven into the ground or placed in a pre-bored hole. Acts as a form to receive the concrete for a cast-in-place concrete pile.

**Cast-in-place piles.** Piles

formed by placing concrete into steel casings that have been driven into the ground.

**Cathead.** Flanged nut that threads onto a waler rod and is tightened against the walers.

**Caulking compounds.** Plastic materials applied at joints and cracks to prevent air and water leakage.

**Cedar shingles.** Shingles cut from western red cedar logs and used as a finish roofing and siding material.

**Ceiling grid.** Light metal framework supporting the tiles for a suspended ceiling.

**Ceiling molding.** Molding placed at the top of a wall close to or against the ceiling.

**Ceiling tile.** Rectangular or square fibrous pieces used to finish off ceilings.

**Cellulose.** Principal substance in the walls of wood cells.

**Cement.** Ingredient that binds together the sand and gravel in a concrete mixture after water is added.

**Central heating.** System in which heat comes from a single source and is distributed

by ducts or pipes to all parts of a building.

**Ceramic tile.** Thin tile made of fired clay used as a finish material on floors, walls, and countertops.

**Chamfer.** Cut made from the face to the side of the board.

**Chamfer strip.** In construction, a piece placed in the corners of beam and column forms to produce a beveled edge.

**Channel molding.** Metal channels placed under the lower edges of mineral fiber shingles.

**Channel rustic siding.** Rabbeted type of board siding that creates a grooved channel effect between the boards.

**Channel slab.** Precast,

## TRADE TIP

### MARKING THE CIRCUMFERENCE OF A CIRCLE

A METHOD FOR LAYING OUT LARGE CIRCULAR OPENINGS THAT MAY HAVE TO BE CUT IN FLOORS, WALLS OR ROOFS.

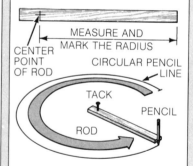

**STEP 1.** MEASURE THE RADIUS OF THE CIRCLE FROM ONE END OF A WOOD ROD AND MAKE A MARK. MEASURE THE WIDTH OF THE ROD AND MARK THE CENTER POINT.

**STEP 2.** TACK THE ROD AT THE CENTER POINT WHERE THE CIRCLE IS TO BE LAID OUT. HOLD A PENCIL AT THE OTHER END AS YOU MOVE THE ROD IN A COMPLETE CIRCLE AROUND THE TACKED END.

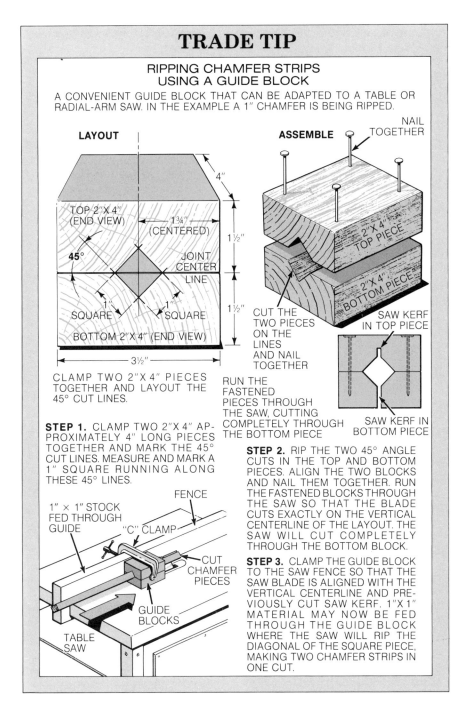

# TRADE TIP

## RIPPING CHAMFER STRIPS USING A GUIDE BLOCK

A CONVENIENT GUIDE BLOCK THAT CAN BE ADAPTED TO A TABLE OR RADIAL-ARM SAW. IN THE EXAMPLE A 1″ CHAMFER IS BEING RIPPED.

**LAYOUT**

TOP 2″X 4″ (END VIEW)
1¾″ (CENTERED)
45°
JOINT CENTER LINE
1″ SQUARE
1″ SQUARE
BOTTOM 2″X 4″ (END VIEW)
4″
1½″
1½″
3½″

CLAMP TWO 2″X 4″ PIECES TOGETHER AND LAYOUT THE 45° CUT LINES.

**STEP 1.** CLAMP TWO 2″X 4″ APPROXIMATELY 4″ LONG PIECES TOGETHER AND MARK THE 45° CUT LINES. MEASURE AND MARK A 1″ SQUARE RUNNING ALONG THESE 45° LINES.

**ASSEMBLE**

NAIL TOGETHER
2″X 4″ TOP PIECE
2″X 4″ BOTTOM PIECE
SAW KERF IN TOP PIECE
SAW KERF IN BOTTOM PIECE

CUT THE TWO PIECES ON THE LINES AND NAIL TOGETHER

RUN THE FASTENED PIECES THROUGH THE SAW, CUTTING COMPLETELY THROUGH THE BOTTOM PIECE

**STEP 2.** RIP THE TWO 45° ANGLE CUTS IN THE TOP AND BOTTOM PIECES. ALIGN THE TWO BLOCKS AND NAIL THEM TOGETHER. RUN THE FASTENED BLOCKS THROUGH THE SAW SO THAT THE BLADE CUTS EXACTLY ON THE VERTICAL CENTERLINE OF THE LAYOUT. THE SAW WILL CUT COMPLETELY THROUGH THE BOTTOM BLOCK.

FENCE
1″ × 1″ STOCK FED THROUGH GUIDE
"C" CLAMP
CUT CHAMFER PIECES
GUIDE BLOCKS
TABLE SAW

**STEP 3.** CLAMP THE GUIDE BLOCK TO THE SAW FENCE SO THAT THE SAW BLADE IS ALIGNED WITH THE VERTICAL CENTERLINE AND PREVIOUSLY CUT SAW KERF. 1″X 1″ MATERIAL MAY NOW BE FED THROUGH THE GUIDE BLOCK WHERE THE SAW WILL RIP THE DIAGONAL OF THE SQUARE PIECE, MAKING TWO CHAMFER STRIPS IN ONE CUT.

---

prestressed concrete floor slab shaped like an inverted U. The two sections extending down from each side of the slab give beam support to the slab.

**Check (lumber).** Lumber defect caused by a lengthwise separation of the wood across the annual growth rings during seasoning.

**Check rails.** Bottom rail of the top sash of a window and top rail of the bottom sash. These rails are next to each other when both window sections are in a closed position.

**Chemical compounds (curing).** Chemicals sprayed on a freshly placed concrete slab to retain moisture in the slab.

**Chipboard.** See *particleboard.*

**Chords (truss).** Top and bottom members of a roof or floor truss.

**Circular stairs.** Type of winding stairway in which all the steps radiate from a common center.

**Circumference.** Perimeter of a circle.

**Clay.** Fine-grained natural earth material that is plastic when wet and compact and brittle when dry.

**Clean-out hole (formwork).** Small opening cut out at the base of a wall or column form to allow debris to be blown out. The hole will be filled in before concrete is placed.

**Cleat.** Narrow wood piece nailed across another board (or several boards) to strengthen it, or to provide support for a shelf or some other object.

**Cleated stringer.** Stringer made of cleats nailed to a plank. The cleats support the ends of the treads.

**Climbing tower crane.** Tower crane located within the building structure that can be raised with hydraulic jacks as the building gains in height.

**Clinching nails.** Bending over the points of nails that have been driven through boards.

**Closed panel construction.** Prefabricated panelized construction method. The interior and exterior wall sections are finished off and plumbing and wiring are installed at the factory.

**Closed railing.** Stairway railing in which the section below the rail is framed in and covered with drywall, plaster, or paneling.

**Closed stringer.** Finish board nailed against the wall side of a stairway. Its top edge is parallel to the slope of the stairs.

**Coil tie (formwork).** Device used to hold opposite form walls in position. It consists of bolts that extend through the walers on both sides and screw into a coil in the inside of the wall.

**Cold joint.** Joint formed when concrete is placed on top of or next to a previous placement of concrete that has already hardened.

**Collar beam.** See *collar tie.*

**Collar tie.** Horizontal member connecting two opposite rafters. It is usually placed at every second or third pair of rafters. Also called *collar beam.*

**Column clamp (formwork).** Adjustable metal device used to tie together column forms.

**Combination blade (circular).** Multipurpose blade used with power saws. Suitable for both cross-cutting and ripping.

**Common length difference (jack rafters).** The amount that jack rafters should increase or decrease in length when the same center-to-center measurements are maintained.

**Common nail.** Flat-head nail used most often in rough work.

**Common rafter.** Rafter that extends from the top wall plates to the ridge of a gable roof.

**Compass direction.** Direction based on the compass points north, south, east, and west.

**Component (building).** Construction unit usually preassembled before being set in place.

**Composite board.** Panel consisting of a reconstituted wood core with a face and back of softwood veneer. (*Comply* is a trade name.)

**Compression.** Stress caused by a squeezing together or crushing force.

**Compression strength.** In construction, the greatest amount of compressive force that a material can withstand before it fractures.

**Concrete.** Construction material made of cement, sand, and gravel. Used for foundations, entire buildings, flatwork and

many other types of structures.

**Concrete blocks.** Precast blocks, solid or hollow, used in the construction of walls.

**Concrete mix.** Proportions of cement, sand, and gravel in a mixture of concrete.

**Concrete pour.** Trade term for placing concrete.

**Condensation.** Formation of drops of water on the insides of exterior walls and windows. This occurs when the warm moisture-laden air within the building comes in contact with the cooler surfaces of the exterior walls.

**Condominium.** Apartment building in which each apartment is owned individually.

**Conductance (C) (heat).** Degree to which a nonhomogeneous material transfers heat.

**Conduction (heat).** Movement of heat through a solid or liquid substance.

**Conductivity (k) (heat).** Degree to which a material transfers heat.

**Conifers.** Species of softwood trees that bear cones and

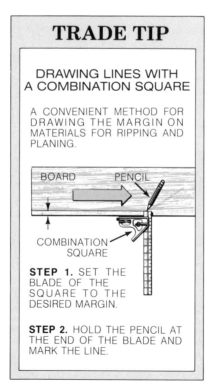

## TRADE TIP

### DRAWING LINES WITH A COMBINATION SQUARE

A CONVENIENT METHOD FOR DRAWING THE MARGIN ON MATERIALS FOR RIPPING AND PLANING.

BOARD     PENCIL

COMBINATION SQUARE

**STEP 1.** SET THE BLADE OF THE SQUARE TO THE DESIRED MARGIN.

**STEP 2.** HOLD THE PENCIL AT THE END OF THE BLADE AND MARK THE LINE.

have thin, needle-shaped leaves.

**Consistency (concrete).** Wetness and flowability of concrete at the time it is placed.

**Consolidating concrete.** Working freshly placed concrete so that each layer is compacted with the layer below and voids caused by water or air pockets are eliminated.

**Construction joint.** Joint that occurs between separate pours of reinforced concrete.

**Contact cement.** Rubber- or butane-based liquid adhesive that adheres instantly on contact.

**Contour lines.** Lines drawn on a survey plan and some plot plans that pass through points having the same elevation on a lot.

**Control joint.** Groove cut into the surface of concrete flatwork. It helps to control cracking due to shrinkage of the concrete as it hardens.

**Convection (heat).** Movement of heat from one area to another by air or water.

**Coped joint.** Joint made by cutting the end of a piece of molding to the shape of the piece it will fit against. A coping saw is commonly used for this purpose.

**Core (plywood).** Innermost layer of plywood. It may consist of hardwood or softwood sawed lumber, veneer, or reconstituted wood.

**Corner bead.** Metal or plastic strip placed to reinforce the outside corner of a wall with a drywall or plastered finish.

**Corner post.** Combination of studs or studs and blocks placed at the outside and inside corners of framed buildings.

**Cornice (roof).** Finish applied to the area under the eaves where the roof and sidewalls meet.

**Counterflashing.** Flashing usually used where chimneys meet the roofline. It covers the shingle flashing and prevents infiltration of moisture.

**Countersink.** To make a depression in wood or metal where a screw or bolt is to be placed so that the head of the screw or bolt will be flush with or below the surface of the material.

**Course.** Layer or row of exterior finish material such as shingles, board siding, or brick.

**Court.** Open space partly or entirely surrounded by buildings.

**Cove base.** Resilient material placed at the bottom of a wall next to a floor finished off with resilient tiles.

**Cove molding.** Molding with a concave curve on its exposed side.

**Crane.** Machine for lifting and moving heavy objects by means of cables extending from a movable boom.

**Crank operator.** Device often used to open casement windows and hold them open.

**Crawl space.** Narrow space between a floor unit and the ground.

**Cricket.** See *saddle*.

**Cripple studs.** Studs placed between headers and top plates or between rough sills and bottom plates.

**Crook.** Distortion (warpage) resulting in a curve from end to end along the narrow edge of a piece of lumber.

**Crossband.** In plywood, the veneer layers placed between the core and the face plies.

**Cross bridging.** Pair of narrow wood or metal pieces set diagonally between floor joists. They help to stiffen and hold the joists in position.

**Crosscut blade (circular).** Blade with teeth that form a knife-like edge with a face

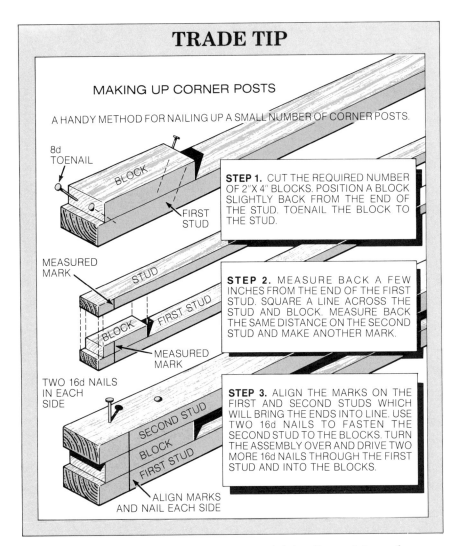

**TRADE TIP**

MAKING UP CORNER POSTS

A HANDY METHOD FOR NAILING UP A SMALL NUMBER OF CORNER POSTS.

8d TOENAIL

BLOCK

FIRST STUD

**STEP 1.** CUT THE REQUIRED NUMBER OF 2″X 4″ BLOCKS. POSITION A BLOCK SLIGHTLY BACK FROM THE END OF THE STUD. TOENAIL THE BLOCK TO THE STUD.

MEASURED MARK

STUD

BLOCK

FIRST STUD

MEASURED MARK

**STEP 2.** MEASURE BACK A FEW INCHES FROM THE END OF THE FIRST STUD. SQUARE A LINE ACROSS THE STUD AND BLOCK. MEASURE BACK THE SAME DISTANCE ON THE SECOND STUD AND MAKE ANOTHER MARK.

TWO 16d NAILS IN EACH SIDE

SECOND STUD

BLOCK

FIRST STUD

**STEP 3.** ALIGN THE MARKS ON THE FIRST AND SECOND STUDS WHICH WILL BRING THE ENDS INTO LINE. USE TWO 16d NAILS TO FASTEN THE SECOND STUD TO THE BLOCKS. TURN THE ASSEMBLY OVER AND DRIVE TWO MORE 16d NAILS THROUGH THE FIRST STUD AND INTO THE BLOCKS.

ALIGN MARKS AND NAIL EACH SIDE

bevel of 10° to 15°. It is used with power saws to cut across the grain of materials.

**Crosscutting.** Cutting with a saw across the grain of lumber.

**Cross slope.** Slope across the width of a sidewalk or patio to allow for water drainage.

**Crown (joist).** Bow along the edge of a joist. Joists should be placed so that the crown is facing up.

**Crown molding.** Ornate molding characterized by a convex curve. Used in traditional roof cornice finish and at the top of interior walls.

**Cup.** Distortion (warpage) resulting in a curve across the grain from edge to edge of a piece of lumber.

**Curing concrete.** Process of retaining the moisture of freshly placed concrete to ensure proper hydration.

**Curtain wall.** Light, non-bearing exterior wall set into and attached to the steel or concrete structural members of a building.

**Custom-built house.** House designed by an architect for a specific customer.

**Cut-in brace.** Older type of diagonal brace with pieces cut in between the studs of a wood-framed wall.

**Cut-out stringer.** Stringer that has been cut to receive all the treads and risers of a stairway.

**Cutting plane line.** Line (identified by letters) that cuts through a part of a structure on an elevation or plan view drawing. It refers to a separate section view or detail drawing given for that area. Also called *section line.*

**Cylinder knob.** Doorknob that contains the lock cylinder.

**Cylinder (lock).** Part of a lock that contains the keyhole and tumbler mechanism.

**Cylindrical lock.** Bored lock with a cylinder that contains the tumbler mechanism into which the key fits. The cylinder is located in the knob section of the lock.

# D

**Dado.** Groove, rectangular in shape, cut across the grain of a board.

**Dadoed stringer.** Stringer with dadoes cut to receive and support the ends of the treads of a stairway.

**Dado joint.** Joint in which a tongue cut into the end of one member fits into a dado grooved in the other member.

**Datum point.** Point of elevation reference, established by local authorities, from which other elevations in the area are measured.

**Dead bolt.** Locking device consisting of a solid metal bar that can be thrown or retracted with a knob or key. Considered a good security provision for any door.

**Dead load.** Weight of the permanent structure of a building, including all materials that make up the units for walls, floors, ceilings, roofs, etc.

**Decay.** Disintegration and breakdown of wood caused by a wood-destroying fungus. Also called *dry rot.*

**Decibel.** Unit used to measure sound intensity.

**Deciduous.** Tree species that shed their leaves during the fall and remain leafless until the following spring.

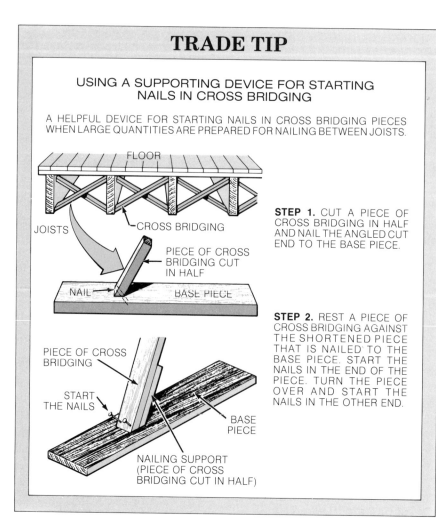

**TRADE TIP**

USING A SUPPORTING DEVICE FOR STARTING NAILS IN CROSS BRIDGING

A HELPFUL DEVICE FOR STARTING NAILS IN CROSS BRIDGING PIECES WHEN LARGE QUANTITIES ARE PREPARED FOR NAILING BETWEEN JOISTS.

FLOOR

JOISTS — CROSS BRIDGING

PIECE OF CROSS BRIDGING CUT IN HALF

NAIL — BASE PIECE

STEP 1. CUT A PIECE OF CROSS BRIDGING IN HALF AND NAIL THE ANGLED CUT END TO THE BASE PIECE.

PIECE OF CROSS BRIDGING

START THE NAILS

BASE PIECE

NAILING SUPPORT (PIECE OF CROSS BRIDGING CUT IN HALF)

STEP 2. REST A PIECE OF CROSS BRIDGING AGAINST THE SHORTENED PIECE THAT IS NAILED TO THE BASE PIECE. START THE NAILS IN THE END OF THE PIECE. TURN THE PIECE OVER AND START THE NAILS IN THE OTHER END.

**Deflection.** Amount of bending that occurs in a wood member subjected to a load or stress.

**Degree (angles).** Unit by which angles are measured.

**Derrick.** Hoisting equipment that consists of a vertical mast and swinging boom to which cables are attached.

**Detail drawing.** In blueprints, an enlarged picture of a part of the structure that cannot be adequately explained in more general plan or elevation view drawings.

**Dew point.** Temperature at which condensation occurs in a given area.

**Diagonal brace.** Brace placed in a framed wall to increase lateral strength.

**Dimple.** Slight depression made by a hammer head when it is used to drive nails into gypsum

board.

**Dolly varden siding.** Beveled board siding with a rabbeted bottom edge.

**Dome form.** Dome-shaped sections used for constructing waffle slab forms. Also called *pans.*

**Door check.** See *door closer.*

**Door closer.** Device mounted on a door and its jamb to return the door to a closed position. It controls the speed and also prevents the door from slamming. Also called *door check.*

**Door hand.** Direction in which a door swings.

**Door header.** Wood member placed across the top of a rough door or window opening in a framed wall. It supports the weight from structures above the opening. Also called

## DIVIDING MATERIAL INTO TWO EQUAL PARTS

A METHOD FOR MARKING A PIECE OF MATERIAL INTO TWO EQUAL PIECES. IN THIS EXAMPLE A BOARD 3½" WIDE IS USED.

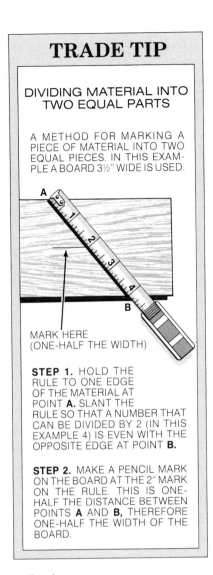

MARK HERE (ONE-HALF THE WIDTH)

**STEP 1.** HOLD THE RULE TO ONE EDGE OF THE MATERIAL AT POINT **A**. SLANT THE RULE SO THAT A NUMBER THAT CAN BE DIVIDED BY 2 (IN THIS EXAMPLE 4) IS EVEN WITH THE OPPOSITE EDGE AT POINT **B**.

**STEP 2.** MAKE A PENCIL MARK ON THE BOARD AT THE 2" MARK ON THE RULE. THIS IS ONE-HALF THE DISTANCE BETWEEN POINTS **A** AND **B**, THEREFORE ONE-HALF THE WIDTH OF THE BOARD.

*lintel.*

**Door holder.** Device fastened near the bottom of a door to hold the door in an open position.

**Door pull.** Metal handle, not part of a lock, used to pull open a door.

**Door sill.** See *threshold.*

**Door stop (jamb).** Narrow piece nailed to a non-rabbeted door jamb to stop and hold the door in position when the door is closed.

**Dormer.** Shed or gable framework projecting from the side of a roof to add light, ventilation, and space to an attic area.

**Double-acting floor hinge.** Self-closing hinge which also allows a door to swing in both directions. The hinge mechanism fits into a notch at the bottom of the door.

**Double coursing (sidewall shingles).** Shingles applied in two layers to a sidewall.

**Double-headed nail.** See duplex nail.

**Double-hung window.** A window unit with upper and lower sash sections that move up and down.

**Double-ply application (drywall).** Double layer of gypsum board applied to a wall.

**Doubler plate.** Horizontal plate nailed to the upper surface of the top plate of a wood-framed wall. It strengthens the load-bearing capacity of the upper section of the wall. Also called *cap* plate. Together, the doubler plate plus the top plate

## TRADE TIP

### DIVIDING MATERIAL INTO THREE OR FOUR EQUAL PARTS

A METHOD FOR RIPPING A PIECE OF MATERIAL INTO THREE OR FOUR EQUAL PIECES. IN THIS EXAMPLE A 5½" WIDE BOARD IS USED.

**STEP 1.** TO DIVIDE THE BOARD INTO THREE EQUAL PARTS, SLANT THE RULE SO THAT A NUMBER THAT CAN BE DIVIDED BY 3 IS EVEN WITH THE TWO EDGES OF THE BOARD. IN THIS EXAMPLE 6 IS USED.

**STEP 1.**

MARK
MARK
MARK

**STEP 2.**

**STEP 2.** TO DIVIDE THE BOARD INTO FOUR EQUAL PARTS, NUMBER THAT CAN BE DIVIDED BY 4 IS EVEN WITH THE TWO EDGES OF THE BOARD. IN THIS EXAMPLE 8 IS USED.

## TRADE TIP

### WEDGING DOOR BOTTOM

AN EFFECTIVE METHOD FOR HOLDING THE DOOR IN A FIXED POSITION WHILE BORING THE LOCK HOLES.

UPWARD FORCE HOLDS THE DOOR IN A FIXED POSITION

EDGE OF DOOR

PLACE A SMALL WOODEN BLOCK UNDER THE FIRST PIECE

WOODEN PIECE

FLOOR

PLACE A WOODEN PIECE UNDER THE BOTTOM OF THE DOOR. PLACE A SMALL BLOCK SNUGLY UNDER THE FIRST PIECE, FORCING THE DOOR UP.

are called a *double top plate.*

**Doubling up walls.** Trade term for placing the second (opposite) wall of a concrete form.

**Dovetail joint.** Wood joint similar to a mortise-and-tenon joint except that its interlocking parts are narrower at the heel.

**Dowel.** Wood or metal pin used to strengthen a joint between two pieces.

**Doweled joint.** Joint in which glued dowels extend into both pieces.

**Downspout.** Pipe carrying water from a roof gutter to the ground.

**Drain tile.** Pipe made of clay, plastic, or concrete sections. Placed alongside the foundation footing for the purpose of moving away water collecting around the foundation.

**Dressed lumber.** See *surfaced lumber.*

**Drill.** Device with cutting edges to make holes in wood, metal,

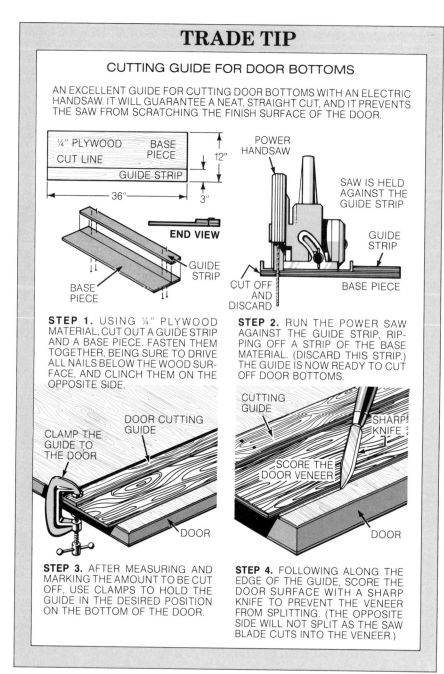

## TRADE TIP

### CUTTING GUIDE FOR DOOR BOTTOMS

AN EXCELLENT GUIDE FOR CUTTING DOOR BOTTOMS WITH AN ELECTRIC HANDSAW. IT WILL GUARANTEE A NEAT, STRAIGHT CUT, AND IT PREVENTS THE SAW FROM SCRATCHING THE FINISH SURFACE OF THE DOOR.

¼" PLYWOOD — BASE PIECE
CUT LINE
GUIDE STRIP
36"
12"
3"

END VIEW

GUIDE STRIP
BASE PIECE

POWER HANDSAW
SAW IS HELD AGAINST THE GUIDE STRIP
GUIDE STRIP
CUT OFF AND DISCARD
BASE PIECE

**STEP 1.** USING ¼" PLYWOOD MATERIAL, CUT OUT A GUIDE STRIP AND A BASE PIECE. FASTEN THEM TOGETHER, BEING SURE TO DRIVE ALL NAILS BELOW THE WOOD SURFACE, AND CLINCH THEM ON THE OPPOSITE SIDE.

**STEP 2.** RUN THE POWER SAW AGAINST THE GUIDE STRIP, RIPPING OFF A STRIP OF THE BASE MATERIAL. (DISCARD THIS STRIP.) THE GUIDE IS NOW READY TO CUT OFF DOOR BOTTOMS.

CLAMP THE GUIDE TO THE DOOR
DOOR CUTTING GUIDE
DOOR

CUTTING GUIDE
SHARP KNIFE
SCORE THE DOOR VENEER
DOOR

**STEP 3.** AFTER MEASURING AND MARKING THE AMOUNT TO BE CUT OFF, USE CLAMPS TO HOLD THE GUIDE IN THE DESIRED POSITION ON THE BOTTOM OF THE DOOR.

**STEP 4.** FOLLOWING ALONG THE EDGE OF THE GUIDE, SCORE THE DOOR SURFACE WITH A SHARP KNIFE TO PREVENT THE VENEER FROM SPLITTING. (THE OPPOSITE SIDE WILL NOT SPLIT AS THE SAW BLADE CUTS INTO THE VENEER.)

or plastic materials. The hand or power-operated tool for holding the device is also called a drill.

**D-ring.** Metal ring attached to a safety belt to which a lifeline is attached.

**Drip cap.** Piece of molding placed at the top of an exterior wood frame to direct water away from the door and window openings.

**Drip edge.** Metal piece bent over the exposed edges of roof sheathing. Placed after felt underlayment has been put down.

**Drive pin.** Nail made of hardened steel that can be driven into concrete. It is usually shot into the concrete with a powder-actuated fastening tool.

**Drop chute.** Long chute usually attached to the bottom of a hopper. Used to prevent concrete from striking against the reinforcing steel and form walls when it is being placed in higher walls.

**Drop head.** See *capital.*

**Drop panel.** Thickened section over concrete columns that help support the floor slab above.

**Dropping hip rafters.** Deducting an amount from the seat cut so that the rafter drops. As a result, the ends of the roof sheathing can rest on the corners of the dropped rafter and be in line with the rest of the roof.

**Dropping stringer.** Deducting an amount from the height of the first riser so that the first and last finished risers will equal the riser heights of the rest of the stairway.

**Drop siding.** Board siding design that comes in tongue-and-groove or shiplap patterns.

**Dry rot.** See *decay.*

**Drywall.** See *gypsum board.*

**Drywall frame.** Metal door jamb designed for walls finished off with gypsum board.

**Duplex nail.** Double-headed nail designed to be pulled out easily. Used in temporary construction. Also called *double-headed nail.*

## E

**Easement.** Legal right-of-way provision on another person's property.

**Eave.** Part of a roof that projects past the side walls of a building.

**Eave trough.** See *gutter.*

**Edge-grained lumber.** Lumber produced by quartering a softwood log lengthwise, then cutting boards out of each section.

**Electrode.** Thin metal rod used in arc welding to provide filler at the welded joint between metals.

**Elephant trunk.** Flexible hose-like device sometimes attached to the bottom of a drop chute. Used when concrete is being dropped from great heights.

**Elevations (heights).** Heights established for different levels of a building.

**Elevation view.** Blueprint drawing giving a view from the side of a structure.

**End-matched lumber.** Boards or panels with tongue-and-groove edges at their end sides.

**English measurement.** Measurement based on yards, feet, inches, and inch-fractions. Also called *customary measurement*.

**Equilibrium moisture content.** Point at which the moisture content in wood is at the same level as the moisture in the surrounding air.

**Escutcheon plate.** Decorative metal plate placed against the door face behind the knob and lock.

**Evergreens.** Softwood species of trees that retain their leaves during all seasons of the year.

**Excavation.** Cavity dug in the ground.

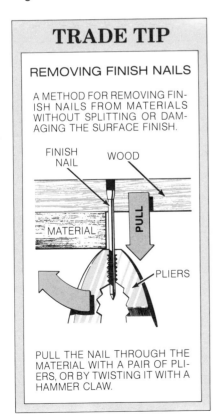

**TRADE TIP**

REMOVING FINISH NAILS

A METHOD FOR REMOVING FINISH NAILS FROM MATERIALS WITHOUT SPLITTING OR DAMAGING THE SURFACE FINISH.

FINISH NAIL

WOOD

PULL

MATERIAL

PLIERS

PULL THE NAIL THROUGH THE MATERIAL WITH A PAIR OF PLIERS, OR BY TWISTING IT WITH A HAMMER CLAW.

**Existing grade.** See *natural grade*.

**Exit device.** Mechanism with a horizontal bar that operates one or more door latches. When pushed in, the bar retracts the latches, enabling the door to be opened. Mainly used in theaters, schools, and public buildings. Sometimes called *panic-exit door bolt*.

**Expansion anchor.** Device placed in concrete and other types of solid masonry walls for fastening purposes. As a screw or bolt is driven into the anchor, it expands and presses against the concrete.

**Expansion joint.** Joint extending through the entire thickness of a concrete slab wherever the slab butts against walls, piers, or columns. Expansion joints allow for expansion of a slab due to temperature changes.

**Exposure (shingle).** Distance between exposed edges of overlapping shingles.

**Exterior finish.** In carpentry, materials such as roof shingles, wall siding, and window and exterior door frames.

# F

**Facade.** Exterior of the front of a building.

**Face frame (cabinet).** Frame attached to the front of a cabinet. It holds the doors and contains the drawer openings. Also called *front frame*.

**Face-nailing.** Nailing on the surface of the lumber.

**Face shield.** Transparent plastic device that covers and protects the entire face from the hazard of flying particles.

**Face side.** Side of a board or other building material that has the best appearance.

**Factory edge.** Straight edge of a panel after it has been manufactured.

**Factory and shop lumber.**

Lumber that will be further manufactured into millwork such as door jambs, window sashes and frames, and molding.

**Fascia board.** Horizontal finish piece nailed to the tail end of the roof rafters.

**Fascia rafter.** Rafter placed toward the outside face of the framework for a gable end overhang. Also called *barge rafter* or *bargeboard*.

**Fiberglass.** Insulating material made of spun glass fibers.

**Fiber saturation point.** Point at which water has evaporated from the cell cavities of wood but the cell walls still retain water.

**Fiber (wood).** Narrow tapered wood cell closed at both ends. Also called *tracheid*.

**Filler rod.** In concrete formwork, a threaded rod made in various lengths that joins

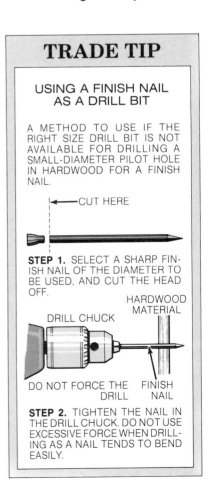

**TRADE TIP**

USING A FINISH NAIL AS A DRILL BIT

A METHOD TO USE IF THE RIGHT SIZE DRILL BIT IS NOT AVAILABLE FOR DRILLING A SMALL-DIAMETER PILOT HOLE IN HARDWOOD FOR A FINISH NAIL.

CUT HERE

**STEP 1.** SELECT A SHARP FINISH NAIL OF THE DIAMETER TO BE USED, AND CUT THE HEAD OFF.

HARDWOOD MATERIAL

DRILL CHUCK

FINISH NAIL

DO NOT FORCE THE DRILL

**STEP 2.** TIGHTEN THE NAIL IN THE DRILL CHUCK. DO NOT USE EXCESSIVE FORCE WHEN DRILLING AS A NAIL TENDS TO BEND EASILY.

together the two outer sections of a waler rod assembly.

**Finger joint.** Interlocking joint used to splice the ends of lumber pieces together.

**Finish floor elevation.** Height of the first floor after finish materials have been applied, in relation to the bench mark established on the construction site.

**Finish flooring.** Material used for the exposed, finished surface of a floor. Some examples are hardwood flooring, tile, and rugs.

**Finish grades.** Various levels of the lot surface after grading work has been completed.

**Finish nail.** Nail with a small barrel-shaped head that allows it to be set below the surface. Used for finish work.

**Finish stringer.** Finish board placed against a wall and behind a cut-out stringer.

**Fink truss.** Roof truss sloping in two directions with web members that form a W shape. Also called *W-truss*.

**Fire blocks.** Horizontal pieces placed between the studs to slow down the passage of flames in case the structure catches on fire. Also called *fire stops*.

**Fire cut.** Angled cut at the end of a joist where it is set into a brick wall. Prevents the joist from damaging the wall in the event it collapses due to fire.

**Fire-retardant lumber.** Lumber treated with a fire-retardant chemical.

**Fire stops.** See *fire blocks*.

**Fire wall.** Wall constructed of fire-resistant materials and subdividing a building to help retard the spread of fire.

**Fixed-sash window.** Window that cannot be opened.

**Fixture.** In construction, electrical or plumbing device attached to a wall, floor, or ceiling. Examples of fixtures are lights, toilets, and sink

basins.

**Flagstone.** Flat, fine-grained, evenly split rocks. When set in mortar, flagstone can be used for the finished surface of outside walks and terraces and for sections of an interior floor.

**Flakeboard.** See *particleboard*.

**Flange (steel beam).** Projecting rim at the top and bottom webs of a steel beam.

**Flashing.** Strips of metal, plastic, or asphalt-saturated felt placed at roof areas vulnerable to water leakage and around window and door openings.

**Flat-grained lumber.** Lumber produced by cutting a softwood log so that the annual growth rings are at an angle of 45° or less to the wide surface of the boards.

**Flat slab floor.** Concrete slab supported mainly by drop panels over columns.

**Flatwork.** Work connected with concrete slabs used for walks, driveways, patios, and floors.

**Flexible insulation.** Insulating materials that come in a blanket or batt form and are placed between studs and joists.

**Flight (stair).** Unbroken and continuous series of steps from one floor to another or from a floor to a landing.

**Floor beams.** Used in a post-and-beam floor system. Beams are spaced apart 4' on centers or more. Flooring usually consists of 1½" tongue-and-groove planks or 1⅛" panels.

**Floor guide (sliding door).** Metal or plastic device that holds the bottom of a sliding door in correct position as it is moved.

**Floor plans.** Drawings in a set of blueprints that give a plan view of each floor of the building.

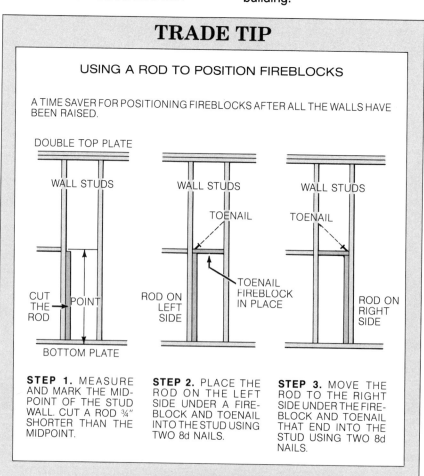

# TRADE TIP

### USING A ROD TO POSITION FIREBLOCKS

A TIME SAVER FOR POSITIONING FIREBLOCKS AFTER ALL THE WALLS HAVE BEEN RAISED.

**STEP 1.** MEASURE AND MARK THE MIDPOINT OF THE STUD WALL. CUT A ROD ¾" SHORTER THAN THE MIDPOINT.

**STEP 2.** PLACE THE ROD ON THE LEFT SIDE UNDER A FIREBLOCK AND TOENAIL INTO THE STUD USING TWO 8d NAILS.

**STEP 3.** MOVE THE ROD TO THE RIGHT SIDE UNDER THE FIREBLOCK AND TOENAIL THAT END INTO THE STUD USING TWO 8d NAILS.

## TRADE TIP

### PLACING FLOOR PANELS

A METHOD FOR ASSURING THAT THE FIRST ROW OF FLOOR PANELS WILL BE SET IN A STRAIGHT LINE EVEN IF THE OUTSIDE WALL IS SLIGHTLY IRREGULAR.

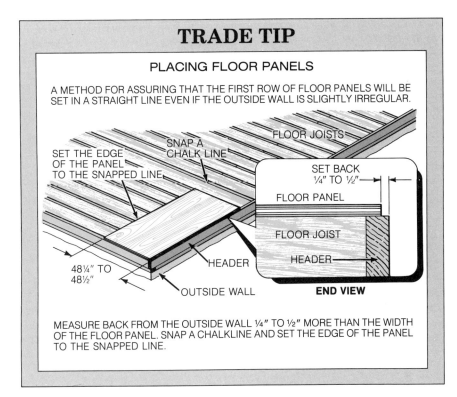

MEASURE BACK FROM THE OUTSIDE WALL ¼″ TO ½″ MORE THAN THE WIDTH OF THE FLOOR PANEL. SNAP A CHALKLINE AND SET THE EDGE OF THE PANEL TO THE SNAPPED LINE.

**Floor truss.** Prefabricated structural unit made of top and bottom chords tied together with web members. Used for the same purpose as floor joists or beams.

**Flue.** Passage through a chimney for gas, fumes, or smoke to rise.

**Flush bolt.** Sliding bolt mechanism mortised into a door at its top and bottom edge. Usually used to hold in a fixed position the inactive door of a pair of double doors.

**Flush door.** Door with a flat surface made of a frame covered with plywood or hardboard face panels.

**Flush pull.** Recessed device mortised into sliding doors to provide a finger hold for moving the doors.

**Flying form.** Complete form unit consisting of a wood deck and metal supports. The entire assembly is raised and set into place by crane.

**Foam insulation.** Plastic chemical foam poured or blown into wall cavities for insulating purposes.

**Footing.** Base of a foundation system. It bears directly on the soil.

**Form (concrete).** Braced structure built to the shape of the structural member into which concrete is placed.

**Formwork.** Construction of concrete forms.

**Foundation.** The part of a building that rests on and extends into the ground. It provides support for the structural loads above.

**Foundation plan.** Drawing in a set of blueprints that gives a plan view as well as section views of the foundation of a building.

**Frame construction.** Framed wall, floor, ceiling, and roof units.

**Framing anchors.** Metal devices used to strengthen the ties between structural members of a wood-framed building.

**Friction catch.** Catch used to hold cabinet doors or other types of light doors in position. A metal tongue mounted inside the door engages a jaw device mounted inside the cabinet.

**Friction pile.** Pile that relies on surrounding soil pressure for a major part of its bearing capacity.

**Frieze board.** Finish piece nailed below or notched between the rafters of an open cornice.

**Front frame (cabinet).** See *face frame*.

**Front setback.** Distance from the property line to the front of a building.

**Frost line.** Depth to which soil freezes in a particular area.

**Furring strips.** Narrow wood strips nailed to a wall or ceiling

## TRADE TIP

### DIVIDING IMPROPER FRACTIONS

A SHORT CUT FOR DIVIDING IMPROPER FRACTIONS INTO TWO EQUAL PARTS. IN THIS EXAMPLE 5⅝″ IS USED.

**STEP 1.** DETERMINE THE HIGHEST WHOLE NUMBER THAT WILL GO INTO 5. THE ANSWER IS 2 (IGNORE THE REMAINDER).

$$2\overline{)5} \quad \genfrac{}{}{0pt}{}{2}{} $$

$$\begin{array}{r} 2 \\ 2\overline{)5} \\ 4 \\ \hline 1 \end{array}$$ (IGNORE THIS REMAINDER)

**STEP 2.** ADD THE NUMERATOR (5) AND THE DENOMINATOR (8), WHICH EQUALS **13**.

$$\begin{array}{r} 5 \text{ NUMERATOR} \\ +8 \text{ DENOMINATOR} \\ \hline 13 \end{array}$$

**STEP 3.** DOUBLE THE DENOMINATOR (8), WHICH EQUALS **16**.

$$\begin{array}{r} 8 \text{ DENOMINATOR} \\ +8 \text{ DENOMINATOR} \\ \hline 16 \end{array}$$

**STEP 4.** THE FINAL ANSWER IS 2¹³/₁₆″.

$$5\tfrac{5}{8}'' \div 2 = 2\tfrac{13}{16}''$$

surface as a nailing base for finish materials.

# G

**Gable end.** Triangular upper section of the end wall of a building with a gable roof.

**Gable roof.** Roof that has a ridge at the center and slopes in two directions.

**Gable stud.** Stud that extends from the top plate of an end wall to the bottom of a common rafter at the gable end of a gable roof.

**Gain.** Recess cut into a piece of wood for door hinges and other hardware. Also a notch cut into lumber, such as a notch cut into a stud for a ribbon board.

**Galvanizing.** Coating iron or steel with zinc to prevent rusting.

**Gambrel roof.** Roof similar to a gable roof but with a break in each slope at an intermediate point between the ridge and

the two exterior side walls of the building.

**Gang form.** Large form panel made of smaller panel sections.

**Gauge.** Standard of measure for metal thickness and the diameter of wire. Also spelled *gage.*

**General contractor.** Licensed individual or firm that can enter into legal contracts to do construction work, and is in charge of the overall organization and supervision of a construction project. Also called *building contractor.*

**Geometrical stairway.** Winding stairway of circular or elliptical design.

**Girder.** Large horizontal wood, steel, or concrete member used to help support a floor, ceiling, roof, or other load.

**Girder pocket (concrete).** Opening prepared in a concrete wall to receive the end of a wood girder so that the top of the girder will be at the same level as the top of

the wall. Also called *beam pocket.*

**Glazing.** Installing glass in a window sash or door and applying putty to hold the panes in position.

**Glue block.** Small piece of wood glued at the joint of two pieces of wood that meet at an angle.

**Glue bond.** Bond between pieces of wood in products such as veneered panels and glued, laminated (glulam) timbers.

**Glue-nailing.** Fastening system that combines the use of an adhesive with nailing.

**Glued, laminated timber.** Heavy timber made up of planks joined together by a very strong bonding adhesive. Often referred to in the abbreviated form of *glulam.*

**Glulam** See *glued, laminated timber.*

**Gooseneck.** Curved or bent section on a handrail.

**Grade (lumber).** The identification of the quality of a piece of lumber. The grade determines the lumber's price and what it can be used for.

**Grade beam.** Reinforced concrete foundation wall placed at ground level. It rests on and is tied to deeply penetrating piers or piles that provide the main structural support for the beam and building loads above.

**Grading (job site).** Removing or adding soil to the surface of the lot so that there is enough slope for surface water to flow away from the building.

**Grain (wood).** Direction, size, and arrangement of the wood fibers in a piece of lumber.

**Gravel.** Crushed rock. Particles range in size from ¼″ to 1½″ in diameter.

**Gravel streak (concrete).** Surface defect in hardened concrete caused by poor consolidation when the concrete was placed.

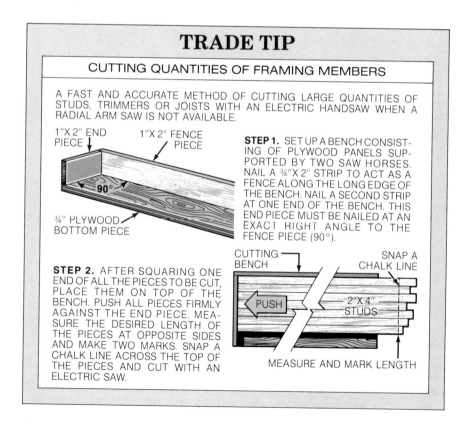

# TRADE TIP

## CUTTING QUANTITIES OF FRAMING MEMBERS

A FAST AND ACCURATE METHOD OF CUTTING LARGE QUANTITIES OF STUDS, TRIMMERS OR JOISTS WITH AN ELECTRIC HANDSAW WHEN A RADIAL ARM SAW IS NOT AVAILABLE.

1″X 2″ END PIECE
1″X 2″ FENCE PIECE
90°
¾″ PLYWOOD BOTTOM PIECE

**STEP 1.** SET UP A BENCH CONSISTING OF PLYWOOD PANELS SUPPORTED BY TWO SAW HORSES. NAIL A ¾″X 2″ STRIP TO ACT AS A FENCE ALONG THE LONG EDGE OF THE BENCH. NAIL A SECOND STRIP AT ONE END OF THE BENCH. THIS END PIECE MUST BE NAILED AT AN EXACT RIGHT ANGLE TO THE FENCE PIECE (90°).

**STEP 2.** AFTER SQUARING ONE END OF ALL THE PIECES TO BE CUT, PLACE THEM ON TOP OF THE BENCH. PUSH ALL PIECES FIRMLY AGAINST THE END PIECE. MEASURE THE DESIRED LENGTH OF THE PIECES AT OPPOSITE SIDES AND MAKE TWO MARKS. SNAP A CHALK LINE ACROSS THE TOP OF THE PIECES AND CUT WITH AN ELECTRIC SAW.

CUTTING BENCH
SNAP A CHALK LINE
PUSH
2″X 4″ STUDS
MEASURE AND MARK LENGTH

**Green concrete.** Trade term for freshly placed concrete.

**Greenhouse (solar).** Glassed-in area for collecting solar heat. Also called *atrium, sunspace, solar room,* and *solarium.*

**Green lumber.** Trade term for unseasoned lumber that has not been air- or kiln-dried.

**Grip-handle lock.** Lockset with a grip handle on one or both sides. The latch is retracted with a thumbpiece.

**Ground (electrical).** Safety feature to prevent shock due to a fault in an electrical system. It consists of an added ground wire running from a plug or equipment to the ground.

**Ground (plaster).** Narrow piece of wood nailed to a framed wall to serve as a thickness gauge for plaster. Also provides backing and a nailing surface for wall molding.

**Ground clamp.** In arc welding, a device for securing to the workpiece the end of the ground cable running from the welding machine.

**Grout.** Thin mixture of cement, sand, and water used for patching and leveling.

**Guardrail.** Temporary railing placed around floor openings, across exterior door and window openings, and on scaffolding during construction.

**Guide strip (drawer).** Narrow strip used to hold a cabinet drawer in position as it slides in and out. If the strip is nailed to the drawer, a dado is cut in the side of the cabinet. If the strip is nailed to the cabinet side, a dado is cut in the drawer.

**Gusset plates (truss).** Plywood plates glued and nailed at the joints between truss members.

**Gutter.** Wood or metal trough attached to the eaves to receive water runoff from the roof. Also called *eaves trough.*

**Gypsum board.** Panels made of

### TRADE TIP

**MARKING HOLE OPENINGS ON GYPSUM BOARD**

A QUICK METHOD FOR MARKING HOLES TO BE CUT OUT FOR ELECTRICAL BOXES. IT ELIMINATES THE NEED FOR TAKING MEASUREMENTS.

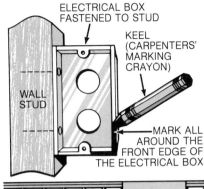

**STEP 1.** MARK ALL AROUND THE FRONT EDGE OF THE ELECTRICAL BOX USING A CARPENTER'S KEEL. (MAKE SURE THERE IS ENOUGH CRAYON MARKING TO TRANSFER THE IMPRINT TO THE BACK SURFACE OF THE BOARD).

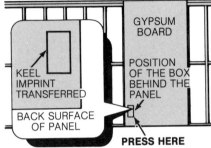

**STEP 2.** PLACE THE GYPSUM PANEL INTO ITS PROPER POSITION AND PRESS IT AGAINST THE ELECTRICAL BOX. THE KEEL IMPRINT WILL TRANSFER FROM THE FRONT EDGE OF THE ELECTRICAL BOX TO THE BACK SURFACE OF THE GYPSUM PANEL.

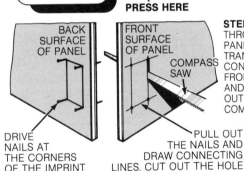

**STEP 3.** DRIVE FOUR NAILS THROUGH THE BACK OF THE PANEL AT THE CORNERS OF THE TRANSFERRED IMPRINT. DRAW CONNECTING LINES ON THE FRONT SURFACE OF THE PANEL AND REMOVE THE NAILS. CUT OUT THE HOLE WITH A KNIFE OR COMPASS SAW.

a gypsum rock base sandwiched between specially treated paper. They have largely replaced plaster as a finish interior wall covering. Also called *wallboard* and *drywall.*

## H

**Handrail.** Rail that is grasped by the hand to provide support when a person uses a stairway.

**Hanger wire.** Wire used to hang grids or other objects from a ceiling.

**Hanging rail (cabinet).** See *mounting rail.*

**Hardboard.** Nonstructural reconstituted wood panel product used for interior paneling, cabinets, underlayment, and exterior siding.

**Hard hat.** Hat made of plastic or metal that protects a worker from injury caused by falling objects.

**Hardwood lumber.** Lumber that comes from deciduous broad-leaved tree species.

**Header joists.** Continuous

pieces of lumber that are the same size as the floor joists and are nailed into the ends of the floor joists to prevent them from rolling or tipping. Also called *rim joists* and *band joists*.

**Header joists for floor openings.** Double joists nailed between the trimmer joists of floor or ceiling openings.

**Headlap (shingle).** Distance in inches from the lower edge of an overlapping shingle to the upper edge of the shingle in the second course below.

**Head room.** Minimum vertical clearance required from any tread on a stairway to any part of the ceiling structure above the stairway.

**Heartwood.** Part of the treetrunk that is between the pith and the sapwood and no longer contributes to the growth process of the tree.

**Heat flow.** Movement of heat toward colder air.

**Heavy construction.** Concrete and heavy timber construction methods used to erect structures such as office and apartment buildings, factories, bridges, freeways, and dams.

**Heavy timber construction.** System using heavy posts and girders for the basic structural members of buildings, bridges, trestles, and waterfront docks, and piers.

**Heel plumb line (rafter).** Mark indicating the rafter plumb cut at the building line.

**Hinges (door).** Metal devices with movable joints attached to the door and jamb. They secure the door to the jamb and allow the door to swing back and forth. Also called *butt hinges*.

**Hip jack rafter.** Jack rafter that extends from the top wall plate to a hip rafter.

**Hip rafter.** Rafter that runs at a 45° angle from the corner of a building to the ridge of a hip

roof.

**Hip roof.** Roof that slopes in four directions from a central ridge.

**Hip-valley jack rafter.** Short rafter that extends from a hip rafter to a valley rafter.

**Hollow-backed.** Depression between the two edges of the back of a piece of wood molding or flooring material. The depression allows for a flatter, tighter fit against an irregular surface.

**Hollow-core door.** Lightweight, less expensive type of flush door. With the exception of the outside frame, the space between the face panels is filled with a mesh or cellular material.

**Hollow-core slabs.** Lightweight precast, prestressed concrete slab section used for floors,

ceilings, and walls. Cores in the slab provide for air distribution and raceways for power and communication wiring.

**Honeycomb (concrete).** Surface defect in hardened concrete caused by poor consolidation when the concrete was placed.

**Honeycomb (lumber).** White pits and specks on the surface of lumber. Similar to defect called *white speck*, but pits are deeper or larger. It is caused by a tree fungus.

**Hopper (concrete).** Funnel-shaped box used when placing concrete into a form.

**Hopper window.** Window that hinges at the bottom and swings out at the top.

**Horizontal sliding windows.** Windows that open and shut

## TRADE TIP

### TRANSFERRING HINGE LAYOUT

A METHOD FOR TRANSFERRING HINGE LAYOUT FROM THE DOOR TO THE SIDE JAMB. IT CAN BE PARTICULARLY HELPFUL WHEN FITTING VERY HEAVY DOORS.

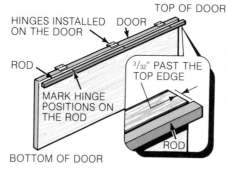

**STEP 1.** PLACE A ROD ALONGSIDE THE DOOR, HOLDING IT 3/32″ PAST THE TOP EDGE OF THE DOOR. TRANSFER MARKS FROM THE TOPS OF THE HINGES TO THE ROD.

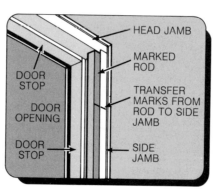

**STEP 2.** PUSH THE ROD UP AGAINST THE HEAD JAMB. HOLDING IT FIRMLY IN POSITION, TRANSFER THE HINGE MARKS ON THE ROD TO THE SIDE JAMB. THE ADDITIONAL 3/32″ ALLOWED FOR ON THE ROD WILL INSURE AMPLE CLEARANCE BETWEEN THE TOP OF THE DOOR AND THE BOTTOM SURFACE OF THE HEAD JAMB.

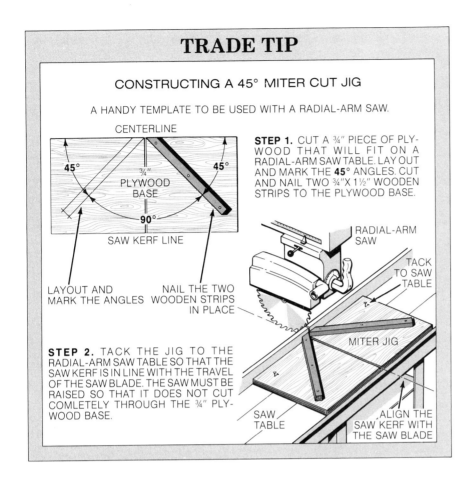

## TRADE TIP

### CONSTRUCTING A 45° MITER CUT JIG

A HANDY TEMPLATE TO BE USED WITH A RADIAL-ARM SAW.

CENTERLINE

45°      ¾"      45°
PLYWOOD
BASE

90°

SAW KERF LINE

LAYOUT AND
MARK THE ANGLES

NAIL THE TWO
WOODEN STRIPS
IN PLACE

**STEP 1.** CUT A ¾" PIECE OF PLYWOOD THAT WILL FIT ON A RADIAL-ARM SAW TABLE. LAY OUT AND MARK THE **45°** ANGLES. CUT AND NAIL TWO ¾" X 1½" WOODEN STRIPS TO THE PLYWOOD BASE.

RADIAL-ARM
SAW

TACK
TO SAW
TABLE

MITER JIG

**STEP 2.** TACK THE JIG TO THE RADIAL-ARM SAW TABLE SO THAT THE SAW KERF IS IN LINE WITH THE TRAVEL OF THE SAW BLADE. THE SAW MUST BE RAISED SO THAT IT DOES NOT CUT COMLETELY THROUGH THE ¾" PLYWOOD BASE.

SAW
TABLE

ALIGN THE
SAW KERF WITH
THE SAW BLADE

by sliding horizontally in upper and lower tracks.

**Housed stringer.** Stringer that has been routed out to receive treads and risers that are glued and wedged into place.

**Housing tract.** Residential development where many houses are constructed for sale on a common tract of land.

**Hub.** Stake placed in the corner of a lot when the lot is being surveyed and its exact boundaries established.

**Humidity.** Amount of dampness in the air.

**Hydration.** Chemical reaction that takes place when water is combined with cement, sand, and gravel in a concrete mix. Hydration causes the concrete to harden.

## I

**Immersion vibrator.** Tool used to consolidate freshly poured concrete. It consists of an electrically or pneumatically activated metal vibrating head that is dipped into the concrete.

**Impact sound transmission.** Sound produced when part of a building structure is set into vibration by a direct impact. For example, dropping an object on the floor produces impact sound transmission.

**Impermeable.** Pertains to materials that are completely vapor-resistant, allowing no water to pass through.

**Inactive door.** The door of a double door unit that is held in place by flush bolts and will not be used for normal traffic.

**Infiltration.** Heat loss as a result of leakage through cracks around door and window frames.

**Inserts (lifting).** Bolt-like devices set into precast concrete wall sections to secure the lifting plates needed for crane hookups.

**Inspector (building).** Local official authorized to inspect and approve construction work.

**Interior finish.** In carpentry, the application of finish wall covering, molding, cabinets, and interior door jambs. Also included are the hanging of doors and installation of finish hardware.

**Invisible hinge.** Hinge mortised into the door and jamb with no parts visible when the door is in a closed position.

**Isolation joint.** Joint placed around the perimeters of concrete work where it butts up to pre-existing hardened concrete. The joint commonly is filled with an asphalt-impregnated strip.

## J

**Jack rafters.** Short rafters that extend from a main rafter to the ridge or top plate, or from one main rafter to another.

**Jalousie window.** Window with a series of small glass slats that are opened and closed by a crank operator.

**Jamb.** Finish frame of a door opening.

**Joint.** The place where two pieces of material meet or are joined together.

**Joint Apprenticeship and Training Committee (JATC).** Labor-management organization responsible for the supervision and organization of the apprentice programs in a trade.

**Jointing lumber.** Process of making the edges of boards straight and true before they are fitted together. Also, process of planing door edges when fitting and hanging doors.

**Jointing saws.** Process of filing

saw teeth to make them even when refitting saws.

**Joist.** Horizontal plank placed on edge to which subfloor and ceiling materials are nailed.

**Joist hanger.** Metal strip used to support the end of a joist that is to be flush with and nailed against another joist or girder.

**Joist tie.** Wood or metal piece notched into the tops of joists where they butt together over a wall or girder.

**Journeyman.** Worker who has completed an apprenticeship training course and/or passed certification requirements for working in the trade.

## K

**Kerf.** Groove or notch made by a saw.

**Keyway.** Groove formed in concrete at the top surface of a spread footing. It helps to secure the bottom of the foundation wall to be placed on top of the footing.

**Kick plate.** Metal or plastic plate mounted at the bottom of a door face to prevent damage from foot pressure against the door. Usually found on doors in public buildings.

**Kiln-dried lumber.** Lumber dried in a temperature-controlled building called a kiln.

**Kingpost truss.** Simple roof truss sloping in two directions with a vertical post extending from the ridge to the bottom chord.

**Knob (cabinet).** Projecting metal, plastic, or wood handle fastened to the face of a cabinet door or drawer for opening and closing purposes.

**Knocked-down unit.** Unassembled structural unit. Components are prefabricated in a shop but are assembled on the job site.

**Knot.** Lumber defect caused by a section of broken limb

remaining embedded in the treetrunk.

**Knuckle (hinge).** See *barrel.*

## L

**Ladder jack.** Metal device that hooks to the ladder rungs and provides support for planks that may support a single worker doing light work.

**Lagging (excavations).** Heavy wood planks placed between soldier piles to form a shoring system around excavations.

**Lally column.** Steel pipe column used to support wood or steel girders.

**Lanyard.** Rope lifeline. One end is attached to the D ring of a safety belt and the other end is secured to the structure.

**Lap.** Amount that materials extend over or past each other.

**Lap joint.** Joint made by overlapping two pieces of material.

**Latch bolt (lock).** Retractable device that is part of a lockset. It engages the strike plate on the door jamb and holds the door in a closed position.

**Latch retractor.** The part of a lock mechanism that engages the latch unit and retracts the bolt.

**Latch unit (lock).** The part of a lockset that contains a retractable latch bolt.

**Lateral force.** In construction, a sideward pressure against a wall or other structural member.

**Leaf (hinge).** The flat pieces of a door hinge that screw into the jamb and door.

**Lean-to roof.** See *shed roof.*

**Ledger.** Lumber nailed between the uprights of a wooden scaffold that directly support the scaffold planks.

**Ledger strip.** Strip of lumber nailed along the bottom edge of a girder to give support to joists butting against the

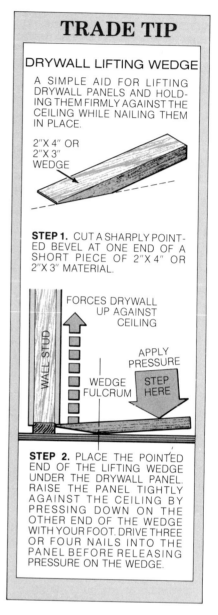

## TRADE TIP

**DRYWALL LIFTING WEDGE**

A SIMPLE AID FOR LIFTING DRYWALL PANELS AND HOLDING THEM FIRMLY AGAINST THE CEILING WHILE NAILING THEM IN PLACE.

2"X 4" OR 2"X 3" WEDGE

**STEP 1.** CUT A SHARPLY POINTED BEVEL AT ONE END OF A SHORT PIECE OF 2"X 4" OR 2"X 3" MATERIAL.

FORCES DRYWALL UP AGAINST CEILING

WALL STUD

APPLY PRESSURE

WEDGE FULCRUM

STEP HERE

**STEP 2.** PLACE THE POINTED END OF THE LIFTING WEDGE UNDER THE DRYWALL PANEL. RAISE THE PANEL TIGHTLY AGAINST THE CEILING BY PRESSING DOWN ON THE OTHER END OF THE WEDGE WITH YOUR FOOT. DRIVE THREE OR FOUR NAILS INTO THE PANEL BEFORE RELEASING PRESSURE ON THE WEDGE.

girder.

**Let-in brace.** Diagonal brace notched into the studs of a wood-framed wall.

**Level.** (1) Line or plane that would be parallel to still water. (2) Tool used for leveling and plumbing purposes.

**Lifting bracket.** Metal device attached to a gang form to provide for a crane hookup.

**Lift plates.** Devices fastened to inserts embedded in precast concrete wall sections and used for a crane hookup.

**Lift-slab construction.** Construction method in which floor slabs are stack-cast, then lifted into place by hydraulic jacks and anchored to pre-set columns.

**Light (window).** Section of window glass. Also called *pane.*

**Light construction.** Residential buildings and small to medium-sized commercial buildings.

**Lignin.** Second most abundant substance found in wood. It covers the cell walls and cements them together.

**Lineal foot.** Refers to a line one foot in length as differentiated from a square foot or cubic foot.

**Line of sight.** An imaginary straight line extending from the telescope of a builder's level or transit-level to the object being sighted.

**Line of travel.** Area that a person is likely to walk on when using a winding stairway.

**Linoleum.** Composite material used as a finish floor covering. It comes in tile or sheet form.

**Lintel.** Wood, stone or steel member placed across the top of a rough door or window opening. It supports the weight from above.

**Lipped door (cabinet).** Cabinet door with a rabbeted edge.

**Live load.** All moving and changing loads that may be placed on different sections of a building. Live load factors may be people, furnishings, snow, and wind.

**Lock block.** Solid piece of wood placed between the face panels of a hollow-core door where the holes will be drilled for the door lock.

**Locking latch.** Device that locks horizontal sliding windows.

**Lockset.** Entire lock unit, including the locks, strike plate, and all the trim pieces.

**Longitudinal roof beam.** Roof beam that runs the length of a post-and-beam roof structure.

**Longitudinal section.** Section view in a blueprint drawing

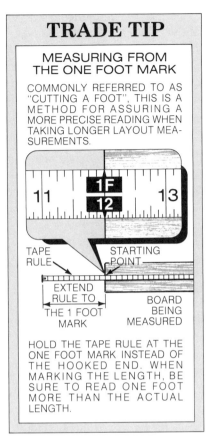

**TRADE TIP**

**MEASURING FROM THE ONE FOOT MARK**

COMMONLY REFERRED TO AS "CUTTING A FOOT", THIS IS A METHOD FOR ASSURING A MORE PRECISE READING WHEN TAKING LONGER LAYOUT MEASUREMENTS.

1F 12

11    13

TAPE RULE    STARTING POINT

EXTEND RULE TO THE 1 FOOT MARK    BOARD BEING MEASURED

HOLD THE TAPE RULE AT THE ONE FOOT MARK INSTEAD OF THE HOOKED END. WHEN MARKING THE LENGTH, BE SURE TO READ ONE FOOT MORE THAN THE ACTUAL LENGTH.

representing a cut along the length of the building.

**Lookout.** Level pieces nailed between the wall and the ends of the rafters to provide a nailing base for the soffit material of a closed cornice.

**Loose fill insulation.** Insulation material that is poured directly from a bag or is blown into place with a pressurized hose.

**Loose-pin hinge.** Hinge used most often for hanging doors. It consists of a pair of leaves, a barrel or knuckle, and a removable pin.

**Lot.** Piece of land or property having established boundaries.

**Lot survey.** Survey of a piece of property, usually carried out by a qualified surveyor or engineer. Its main purpose is to stake out the corners of the property. Also, grades may be checked and recorded at the corners and at various points within the property.

**Louver.** Slotted opening for ventilation.

**Lumber.** Any materials cut from a log (boards, planks, timbers) and used for construction purposes.

# M

**Machine bolt.** Bolt with a square or hexagonal head that is used to fasten together wood or metal pieces.

**Magnetic catch.** Catch used to hold cabinet or other types of lightweight doors in position. It usually consists of a magnet fastened inside the cabinet and a metal plate fastened inside the door.

**Main entrance door.** Exterior door on the front or street side of a building.

**Main stairs.** Stairs that serve the inhabited (living) areas of a house. Also called *primary stairs.*

**Mansard roof.** Roof similar to a hip roof but with a break in the slopes at an intermediate point between the ridge and the four outside walls of a building.

**Manufactured housing.** General term for the prefabricated housing industry.

**Masonry.** Molded or shaped construction materials such as concrete blocks, bricks, stones, and tiles. Also refers sometimes to poured concrete.

**Masonry veneer.** Exterior finish cover of brick or stone, usually applied to a wood stud wall.

**Mastic.** Thick, pasty adhesive used for fastening wallboard, paneling, ceiling and floor tile, and other finish materials.

**Matt foundation.** Foundation system consisting of a thick, heavily reinforced concrete slab placed beneath the entire building area.

**Mechanical core.** Prefabricated modular unit that contains the hooked-up kitchen and bath fixtures as well as other mechanical equipment

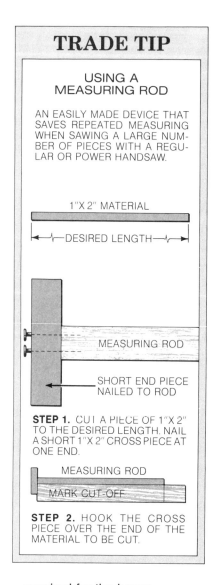

## TRADE TIP

### USING A MEASURING ROD

AN EASILY MADE DEVICE THAT SAVES REPEATED MEASURING WHEN SAWING A LARGE NUMBER OF PIECES WITH A REGULAR OR POWER HANDSAW.

1"X 2" MATERIAL

DESIRED LENGTH

MEASURING ROD

SHORT END PIECE NAILED TO ROD

**STEP 1.** CUT A PIECE OF 1"X 2" TO THE DESIRED LENGTH. NAIL A SHORT 1"X 2" CROSS PIECE AT ONE END.

MEASURING ROD

MARK CUT-OFF

**STEP 2.** HOOK THE CROSS PIECE OVER THE END OF THE MATERIAL TO BE CUT.

required for the house.

**Medullary rays.** Wood rays in a tree that extend radially from the pith toward the outside of the trunk. Their purpose is to store and transport food.

**Mesh core (door).** Crisscross backing for the face panels used in hollow-core doors.

**Metal frame drywall system.** Walls consisting of light-gauge metal tracks and studs, surfaced with gypsum board.

**Mil.** Unit of thickness that is equal to one-thousandth of an inch (.001).

**Millman.** Person working for a millworking firm that manufactures such items as doors and windows, door and window frames, molding, and

other types of interior trim.

**Millwright.** Person who installs machinery and other mechanical equipment in mills and factories.

**Mineral fiber.** Roofing and sidewall material available in sheets or shingles. It is made of 30% asbestos fibers and 70% portland cement. Also called *asbestos-cement.*

**Minute (measurement).** Unit measurement of an angle equal to ¹/₆₀ of a degree.

**Mitered stringer.** Stringer used on open sides of a stairway where the joints between the finish riser material and the stringer are mitered.

**Modified post-and-beam construction.** System combining a post-and-beam design with some stud wall construction to allow for some diagonal bracing or panel sheathing.

**Modular home.** In manufactured housing, a building entirely

constructed of prefabricated modular units. Also called *sectional home.*

**Modular unit.** In manufactured housing, a completed three-dimensional section of a house.

**Moisture content.** Amount of moisture present in wood at a given time. It is usually expressed as a percentage of the dry weight of the wood.

**Moisture meter.** Instrument used in the field to give an instant reading of moisture content in wood.

**Molder.** Machine that produces different patterns of molding from softwood and hardwood lumber. Also called *sticker.*

**Monolithic concrete.** Concrete placed in forms in a continuous pour without construction joints.

**Mortise.** Cavity cut into lumber to receive a similarly shaped tenon projecting from another piece of wood. Also, cavity for

## TRADE TIP

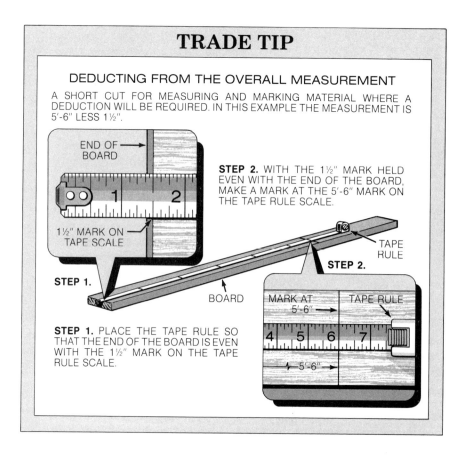

### DEDUCTING FROM THE OVERALL MEASUREMENT

A SHORT CUT FOR MEASURING AND MARKING MATERIAL WHERE A DEDUCTION WILL BE REQUIRED. IN THIS EXAMPLE THE MEASUREMENT IS 5'-6" LESS 1½".

END OF BOARD

1½" MARK ON TAPE SCALE

**STEP 1.**

**STEP 1.** PLACE THE TAPE RULE SO THAT THE END OF THE BOARD IS EVEN WITH THE 1½" MARK ON THE TAPE RULE SCALE.

**STEP 2.** WITH THE 1½" MARK HELD EVEN WITH THE END OF THE BOARD, MAKE A MARK AT THE 5'-6" MARK ON THE TAPE RULE SCALE.

TAPE RULE

**STEP 2.**

BOARD

MARK AT 5'-6"

TAPE RULE

5'-6"

receiving hardware, such as a mortised lock.

**Mortise-and-tenon joint.** Joint between two members where a tenon cut into the end of one piece fits into a cavity mortised in the other piece.

**Mortised lock.** Older type of lock that fits into a mortised cavity in a door.

**Mounting plate (lock).** Plate placed on one side of a door. Holes are provided for two machine screws that engage and tighten into threaded holes in the cylindrical housing, thus securing the lock in place.

**Mounting rail (cabinet).** Wood strip provided at the top and bottom of the back interior of a wall cabinet. Wood screws that fasten the cabinet to the wall penetrate the mounting rails and are driven into the wall studs. Also called *hanging rail.*

**Mudshore (form).** Timber used to brace and straighten a

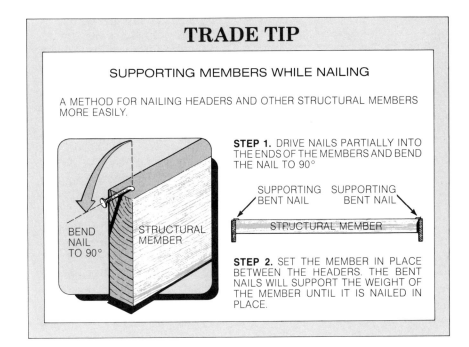

**TRADE TIP**

SUPPORTING MEMBERS WHILE NAILING

A METHOD FOR NAILING HEADERS AND OTHER STRUCTURAL MEMBERS MORE EASILY.

BEND NAIL TO 90°

STRUCTURAL MEMBER

**STEP 1.** DRIVE NAILS PARTIALLY INTO THE ENDS OF THE MEMBERS AND BEND THE NAIL TO 90°

SUPPORTING BENT NAIL    SUPPORTING BENT NAIL

STRUCTURAL MEMBER

**STEP 2.** SET THE MEMBER IN PLACE BETWEEN THE HEADERS. THE BENT NAILS WILL SUPPORT THE WEIGHT OF THE MEMBER UNTIL IT IS NAILED IN PLACE.

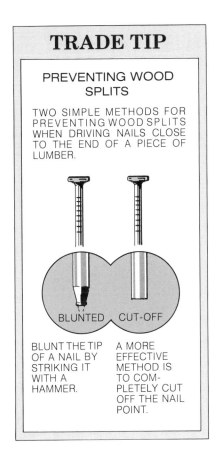

**TRADE TIP**

PREVENTING WOOD SPLITS

TWO SIMPLE METHODS FOR PREVENTING WOOD SPLITS WHEN DRIVING NAILS CLOSE TO THE END OF A PIECE OF LUMBER.

BLUNTED    CUT-OFF

BLUNT THE TIP OF A NAIL BY STRIKING IT WITH A HAMMER.

A MORE EFFECTIVE METHOD IS TO COMPLETELY CUT OFF THE NAIL POINT.

concrete form. One end is nailed to the form and the other end butts against a wood pad placed on the ground.

**Mudsill.** See *sill plate.*

**Mullion.** Vertical bar in a window frame that separates two windows.

**Multi-folding door.** Series of doors hinged to each other and supported by wheeled hangers that run across an overhead track.

**Multiple glazing.** Two or three layers of glass with spaces in between set into a window sash to improve insulation.

**Muntin.** Small pieces that separate window lights enclosed by a window sash.

## N

**Nailing plates (truss).** Metal plates nailed at the joints between truss members.

**National Association of Home Builders.** The major industry association representing general contractors who perform residential construction.

**Natural grade.** The various levels of the lot surface before any finish grading takes place. Also called *existing grade.*

**Newel posts.** Main upright members that support the handrails of a stairway.

**Noise suppressors.** Devices worn in or over the ears to protect workers from ear injuries caused by constant exposure to high noise levels.

**Nominal lumber size.** Commercial size given to a piece of lumber based on its thickness and width at the time it was manufactured in the sawmill, before it was surfaced and seasoned.

**Non-veneered panels.** Panel products manufactured from reconstituted wood.

**Nosing.** Projection of the tread beyond the face of the riser.

**Notch.** Rabbet-like cut made across the edge of a piece of lumber. An example is the notch made in a joist or stud to accommodate a pipe or conduit.

## O

**O.C.** Abbreviation for *on center.*

**Open panel construction.**

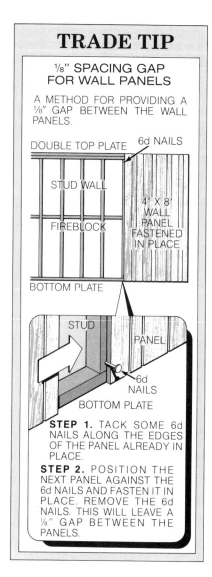

**⅛″ SPACING GAP FOR WALL PANELS**

A METHOD FOR PROVIDING A ⅛″ GAP BETWEEN THE WALL PANELS.

DOUBLE TOP PLATE    6d NAILS

STUD WALL

FIREBLOCK

4' X 8' WALL PANEL FASTENED IN PLACE

BOTTOM PLATE

STUD    PANEL

6d NAILS

BOTTOM PLATE

**STEP 1.** TACK SOME 6d NAILS ALONG THE EDGES OF THE PANEL ALREADY IN PLACE.

**STEP 2.** POSITION THE NEXT PANEL AGAINST THE 6d NAILS AND FASTEN IT IN PLACE. REMOVE THE 6d NAILS. THIS WILL LEAVE A ⅛″ GAP BETWEEN THE PANELS.

Prefabricated panelized construction method. The outsides of the wall sections are finished off, but the insides are left open.

**Open stringer.** Stringer that has been cut out to support the treads on the open side of a stairway.

**Orbital motion.** Circular motion common to some types of electric sanders and jig saws.

**Orientation.** Position of a building on a lot and the direction which the different walls will face.

**Oriented strand board.** Panel product made of layers of wood strands bonded together with a phenolic resin.

**Orthographic drawings.** Two-dimensional drawings used for making up blueprints.

**Oscillating motion.** Back-and-forth movement common to some types of electric finish sanders.

**Overhang (rafter).** Part of the rafter that extends past the building line. Also called *tail* or *tail piece*.

**Overhead garage doors.** Garage doors that swing or roll up when opened.

**Oxyacetylene welding.** Fusion process brought about by a welding flame applied to the edges and surfaces of metal pieces. Filler metal may also be added.

# P

**Pan.** See *dome form*.

**Pane.** See *light*.

**Panel clip.** Device used at the unsupported edges of panel roof sheathing.

**Panel door.** Door made of sunken panels fitted between vertical stiles and horizontal rails. Also called *stile-and-rail door*.

**Paneling (wall).** Panels applied as a finish for interior walls.

**Panelized construction.** Method widely used in manufactured housing. The basic units of this system are framed wall sections built in a factory. They are delivered to the job site where they are to be installed.

**Panic-exit door bolt.** See *exit device*.

**Paraffin.** Wax used by carpenters when driving wood screws. Also used to reduce friction created by moving members such as drawers and sliding cabinet doors.

**Parapet.** Low wall at the edge of a roof.

**Parquet floor.** Finish floor material consisting of squares made of patterns created by inlaid wooden strips.

**Particleboard.** Panel product

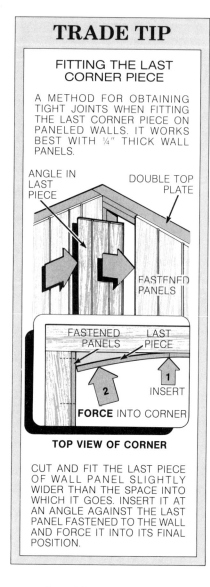

**FITTING THE LAST CORNER PIECE**

A METHOD FOR OBTAINING TIGHT JOINTS WHEN FITTING THE LAST CORNER PIECE ON PANELED WALLS. IT WORKS BEST WITH ¼″ THICK WALL PANELS.

ANGLE IN LAST PIECE

DOUBLE TOP PLATE

FASTENED PANELS

FASTENED PANELS    LAST PIECE

2    1

INSERT

**FORCE** INTO CORNER

**TOP VIEW OF CORNER**

CUT AND FIT THE LAST PIECE OF WALL PANEL SLIGHTLY WIDER THAN THE SPACE INTO WHICH IT GOES. INSERT IT AT AN ANGLE AGAINST THE LAST PANEL FASTENED TO THE WALL AND FORCE IT INTO ITS FINAL POSITION.

produced by combining wood particles such as chips or flakes with a resin binder and hot-pressing them into panels. Also called *flakeboard* or *chipboard*.

**Parting bead.** See *parting strip*.

**Parting strip.** Thin vertical piece of wood that separates the upper and lower sash of a double-hung window. Also called *parting bead*.

**Partition.** Interior wall.

**Patented ties.** Devices used to hold together opposite wall forms during the placement of concrete.

**Patio.** Outdoor paved area adjacent to a house and used for dining and recreation.

**Pattern.** Anything that is used to mark the dimensions of an item on a material in order to produce duplicates. For example, a pattern is used to lay out common rafters.

**Peeler block.** Section of a log from which strips of veneer are peeled off with a special lathe knife.

**Penny.** Measure of nail length. Abbreviated by the letter *d*.

**Perimeter.** Outside boundary of an area.

**Perlite.** Natural volcanic glass material used as a lightweight aggregate in concrete and as loose fill insulation.

**Perm rating.** Measurement of water vapor flow through a material or a combination of materials.

**Phenol formaldehyde.** Synthetic resin commonly used as a binder in reconstituted wood panels.

**Phillips-head screw.** Wood screw with a cross-slot in the head.

**Pictorial drawing.** Three-dimensional view that shows three sides of an object or structure.

**Pier (concrete).** In light construction, a square, round, or battered concrete base set in the soil to directly support posts or columns. Also used to directly support grade beams.

**Pier block.** Wood piece anchored to the top of a pier to provide a nailing surface for the bottom of a post.

**Pilaster.** Column-like projection from a wall. It helps strengthen the wall and may also provide added support for a beam.

**Pile.** Long, slender concrete, steel, or wood structural member that penetrates through unstable soil layers until it rests on firm soil. Piles provide support for grade beams or columns that carry the structural load of a building.

**Pile driver.** Machine used for driving piles.

**Pilot hole.** Hole drilled to facilitate driving a wood screw into wood. The diameter of the hole should be slightly smaller than the screw size.

**Pin (hinge).** Bolt-shaped device placed into the barrel of a hinge. It holds together the two leaves of the hinge.

**Pitch (roof).** Angle that a roof slopes from the outside walls of the building to the ridge of the roof.

**Pitch (wood).** Resinous substance derived from the sap of various coniferous trees.

**Pitch pocket.** Opening that extends parallel to the annual growth rings of a piece of lumber. It contains solid or liquid pitch.

**Pitch streak.** Visible accumulation of pitch in a piece of lumber. Differs from pitch pocket in that the pitch

---

## TRADE TIP

### PARAFFIN WAX IN HAMMER HANDLE

A CONVENIENT SOURCE OF PARAFFIN FOR SCREWS AND NAILS USED IN TRIM OPERATIONS. THIS METHOD CAN ONLY BE USED WITH WOODEN HANDLES.

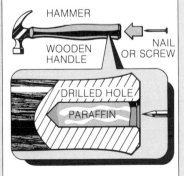

DRILL A HOLE IN THE END OF THE HAMMER HANDLE AND FILL IT WITH PARAFFIN WAX. INSERT THE POINTS OF SCREWS OR NAILS INTO THE PARAFFIN BEFORE DRIVING THEM WITH A SCREWDRIVER OR HAMMER.

---

saturates the section of wood fiber.

**Pith.** Small, soft central core of a treetrunk, branch, or twig.

**Plainsawn lumber.** Hardwood lumber cut out of a log so that the annual growth rings are at an angle of 45° or less to the wide surface of the boards.

**Plank.** Lumber over 1″ thick and 6″ or more in width.

**Plank-and-beam construction.** See *post-and-beam construction*.

**Plank flooring.** Wood finish floor pieces ranging in width from 3″ to 8″ and in thickness from 5/16″ to 25/32″. They give a more traditional effect than strip flooring materials.

**Planter strip.** Area between the street curb and sidewalk where a lawn or other vegetation may be planted.

**Plan view.** Drawing in a set of blueprints that presents a view looking down on an object.

**Plastic laminate.** Product made of three or four layers of plastic material bonded together under high heat and pressure. Used to surface countertops, wall surfaces, shelving, cabinets, etc.

**Plates.** Horizontal pieces at the top and bottom of framed walls to which the upright wall members are fastened.

**Platform framing.** Framing method in which each story of the building is framed as a unit consisting of walls, joists, and subfloor. The subfloor acts as a "platform" for the construction of the story above. Also called *Western framing*.

**Plen-Wood system.** Heating system that delivers hot air from an underfloor plenum area.

**Plot plan.** Plan included in a set of working drawings showing the size of the lot, location of the building on the lot, grades, and all other information needed to perform work required before construction of the foundation begins.

**Plumb.** Exact vertical and perpendicular line. It would be at a 90° angle to a level plane.

**Plyform.** American Plywood Association's trade name for a reusable material for concrete forms.

**Plywood.** Product made of wood layers (veneers) glued and pressed together under high heat and pressure.

**Pneumatic tools.** Tools powered by compressed air.

**Pocket sliding door.** Sliding door suspended from a track by roller type hangers. When opened, it slides into a cavity prepared in the wall. Also called *recessed door.*

**Point of beginning.** See *datum point.*

**Pole construction.** Method in which pressure-treated poles are sunk into the ground to provide the foundation and basic anchor for the framework of a building.

**Polyurethane.** Chemical material used in construction adhesives and foamed-in-place insulation.

**Post.** Upright member used to support beams or girders.

**Post-and-beam construction.** Type of construction in which the basic framework of the building consists of vertical posts and horizontal or sloping beams. Also called *plank-and-beam* construction.

**Post base.** Metal strap or other type of device set into the tops of concrete piers or walls. Its purpose is to provide an anchor for the bottoms of wood posts.

**Post cap.** Metal device used to strengthen the tie between the top of a wood post and the girder above.

**Post-tensioning concrete members.** Procedure of stretching and anchoring the cables used for prestressed

concrete members after the concrete has hardened.

**Pour strip.** Strip of wood tacked inside one of the form walls to indicate the height to which the concrete should be poured.

**Power buggy.** Power-driven vehicle driven by an operator to carry fresh concrete to the place on the job site where it is to be poured.

**Pozzolans.** Admixtures that under certain conditions have cementing properties and can be used as a partial replacement for cement in the concrete mix.

**Precast Concrete.** Concrete structural members that have been cast and cured in a casting yard or factory and

then delivered to the job site.

**Prefabrication.** Construction of building members or units in a yard or factory.

**Pre-finished materials.** Panels, boards, and other types of building materials that have received finish coats of paint, stain, or sealer on their appearance sides at the factory before delivery to the job site.

**Pre-hung door.** Prefabricated door unit. It usually consists of a door already hung in the jamb, with bored lock holes and pre-fitted stops and casing.

**Preservative (wood).** Substance applied to or injected into wood to protect it from fungi

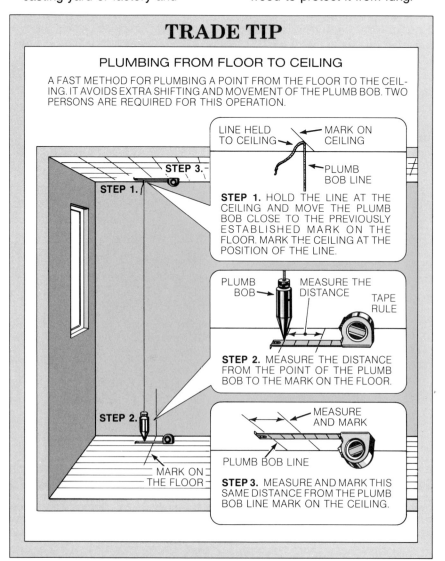

# TRADE TIP

### PLUMBING FROM FLOOR TO CEILING

A FAST METHOD FOR PLUMBING A POINT FROM THE FLOOR TO THE CEILING. IT AVOIDS EXTRA SHIFTING AND MOVEMENT OF THE PLUMB BOB. TWO PERSONS ARE REQUIRED FOR THIS OPERATION.

STEP 3.

STEP 1.

LINE HELD TO CEILING — MARK ON CEILING

PLUMB BOB LINE

**STEP 1.** HOLD THE LINE AT THE CEILING AND MOVE THE PLUMB BOB CLOSE TO THE PREVIOUSLY ESTABLISHED MARK ON THE FLOOR. MARK THE CEILING AT THE POSITION OF THE LINE.

PLUMB BOB — MEASURE THE DISTANCE

TAPE RULE

**STEP 2.** MEASURE THE DISTANCE FROM THE POINT OF THE PLUMB BOB TO THE MARK ON THE FLOOR.

STEP 2.

MARK ON THE FLOOR

MEASURE AND MARK

PLUMB BOB LINE

**STEP 3.** MEASURE AND MARK THIS SAME DISTANCE FROM THE PLUMB BOB LINE MARK ON THE CEILING.

and insects.

**Pressure treatment.** Process of treating lumber with chemical preservatives by applying strong pressure to the pieces placed inside a tank.

**Prestressed concrete.** Precast concrete that is reinforced by high-tensile steel cables placed under great tension.

**Pretensioning (concrete members).** Procedure of stretching the cables used for prestressed concrete members prior to placing the concrete in a casting bed.

**Primary stairs.** See *main stairs.*

**Property lines.** Recorded, legal boundaries of a piece of property.

**Pump jack scaffold.** Type of scaffolding consisting of 4 × 4 uprights and platforms supported by bracket devices that can be raised by a pumping action.

**Purlin (roof).** Horizontal timber held in place by braces. It is

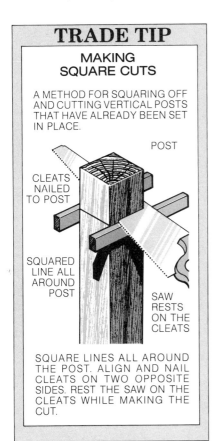

**TRADE TIP**

**MAKING SQUARE CUTS**

A METHOD FOR SQUARING OFF AND CUTTING VERTICAL POSTS THAT HAVE ALREADY BEEN SET IN PLACE.

POST

CLEATS NAILED TO POST

SQUARED LINE ALL AROUND POST

SAW RESTS ON THE CLEATS

SQUARE LINES ALL AROUND THE POST. ALIGN AND NAIL CLEATS ON TWO OPPOSITE SIDES. REST THE SAW ON THE CLEATS WHILE MAKING THE CUT.

placed beneath the roof rafters at an intermediate point between the ridge and outside wall.

**Push bar.** Device used to open a hopper window and hold it in a fixed position.

**Pushbutton (lock).** Device located inside the knob of a cylindrical lock. When pushed in, it locks the door.

**Push plate.** Metal or plastic plate fastened to the lock side face of a door. Mounted at arm level, it protects the door from wear. Used most often on doors in public buildings.

# Q

**Quartersawn lumber.** Lumber produced by quartering a hardwood log lengthwise, then cutting boards out of each section.

# R

**Rabbet.** Groove cut along the edge of a piece of lumber.

**Rabbet joint.** Joint formed at the ends of two pieces that have been rabbeted.

**Radiation (heat).** Transfer of heat through space.

**Rafter tables.** Tables found on the blade of a framing square. Used to compute the lengths of roof rafters.

**Rail (door).** Horizontal piece of a panel door frame.

**Rake end (roof).** Overhang construction at the end of a gable roof.

**Ramp.** Sloping runway from a lower to a higher level for passage of workers and materials on a construction job.

**Ratchet.** Mechanism consisting of a hinged catch that can be adjusted to permit motion in only one direction. Ratchet devices are commonly found in braces and wrenches.

**Ready-mixed concrete.** Concrete mixed at a batch plant and delivered by truck to the job site.

**Reamer and bucket.** Device used for drilling large-diameter holes for caissons. The reamer is situated within the drilling bucket and digs into the ground with a revolving movement.

**Rebars.** See *reinforcing bars.*

**Recessed door.** See *pocket sliding door.*

**Reciprocal motion.** Up-and-down motion such as the action of the projecting blade of a reciprocating saw.

**Reconstituted wood.** Wood products made up of particles, flakes, or strands hot-pressed and bonded together into panel-sized sheets.

**Refitting (saws).** Process of jointing and setting saw teeth before filing.

**Reflective insulation.** Insulation materials that reflect heat rather than absorb it.

**Register.** Grilled frame through which heated or cooled air is released into a room.

**Reinforcing bars.** Deformed steel bars placed in concrete to increase its ability to with-stand weight and pressure. They also help tie together structural concrete members. Also called *rebars.*

**Remodeling.** See *alteration work.*

**Resawn lumber.** Lumber pieces run through a special bevel saw to produce a coarse, textured pattern.

**Residential construction.** Structures in which people live, such as homes, condominiums, and apartment buildings.

**Residue.** Waste portion of a log (chips, bark, trimmings, shavings, sawdust) after the log has been cut into lumber at a sawmill.

**Resilient channels.** Metal

pieces nailed across wood studs or joists to which drywall material is fastened. They help break the sound vibration path through a wall or floor.

**Resilient tile.** Type of floor tile that yields when pressure is applied, then returns to its original position.

**Resin (natural wood.)** Sticky material obtained from the sap of certain trees, especially pine and fir species. Resins are often used in making varnishes and paints.

**Resistance (R).** Ability of a material to hold back heat flow.

**Respiratory protector.** Device worn over the mouth and nose to protect workers from inhaling dangerous dusts or fumes.

**Retaining wall.** Masonry or wood wall constructed to hold back a bank of earth.

**Retarder (concrete).** Admixture added to concrete to delay the stiffening of concrete.

**Ribband.** Piece of lumber, 1″ × 4″, laid flat and nailed to the tops of ceiling joists. It is placed at the center of the spans and helps prevent twisting and bowing of the joists.

**Ribbon.** Narrow board notched into studs to support joists.

**Ridge.** Highest point at the top of a roof where the roof slopes meet.

**Ridge beam.** Beam that is placed between or below the roof beams in post-and-beam construction.

**Ridge board.** Horizontal board or plank placed at the ridge of a roof, to which the upper ends of the rafters are nailed.

**Ridge plumb line.** Marking for the plumb cut made at the ridge end of a roof rafter.

**Rigid insulation.** Insulation materials that come in the form of panels or tile.

**Rim joist.** See *header joist.*

**Rip blade (circular).** Blade with flat-faced teeth shaped like a chisel. It is used with power saws to rip in the direction of the grain.

**Ripping.** Cutting with a saw in the same direction as the woodgrain.

**Riser.** Piece forming the vertical face of a step.

**Roller hanger.** Wheel-like device used to suspend and move sliding doors along tracks attached to the head jamb.

**Roof beam.** Beam that is part of the roof structure of a post-and-beam building and to which the roof decking is nailed.

**Roof truss.** Prefabricated structural roof unit made of top and bottom chords tied together with web members. The top chord acts as a roof rafter and the bottom chord serves as a ceiling joist.

**Rose (lock).** Trim piece that is part of a lockset. It fits against the door and with some types of locks holds the lock in place.

**Rough opening.** Framed opening in a wall into which a finished door or window unit will be placed.

**Rough sill.** Horizontal piece nailed across the bottom of a rough window opening.

**Rubber tile.** Resilient floor tile of synthetic materials or natural rubber combined with mineral fillers.

**Runner track.** U-shaped metal piece attached to the floor. It holds the bottoms of the studs and bottom cripples of a framed metal wall.

**Runway.** Temporary platform-like structure that provides passage for workers and materials on a construction job.

## S

**Saddle.** Structure with a ridge sloping in two directions that is placed between the back side of a chimney and the roof sloping toward it. Its purpose is to divert water away from the chimney. Also called *cricket.*

**Safety belt.** Leather or nylon belt worn when working where falling hazards exist. The belt is equipped with a D-ring to which a line can be attached.

**Safety goggles.** Device made of specially treated glass or plastic worn over the eyes to protect the worker from eye hazards caused by flying particles.

**Safety nets.** Meshed nets sometimes used when safety belts and lifelines are impractical, and placed below work areas that are 25′ or more above the ground, water surface, or a floor below.

**Sandpaper.** Heavy paper with abrasive materials (garnet, emery, flint, etc.) glued to one side. Used with a hand sanding block or electric sander to smooth wood surfaces.

**Sap.** Watery fluid that circulates through a wood plant carrying food and other substances to the tissues.

**Sapwood.** Pale-colored living wood near the outside of a treetrunk.

**Sash.** Wood or metal frame into which glass panes are set.

**Sash lift.** Hooked or bar device fastened to the lower window of a double-hung window unit. It acts as a handhold for lifting and lowering the window.

**Sash lock.** Device attached to a window sash for locking purposes.

**Sash weight.** Weight used to balance the sash so that it will remain in any vertical position when opened. Not used in new construction.

**Sawhorse.** Portable work bench used by carpenters. It consists of a top piece supported by

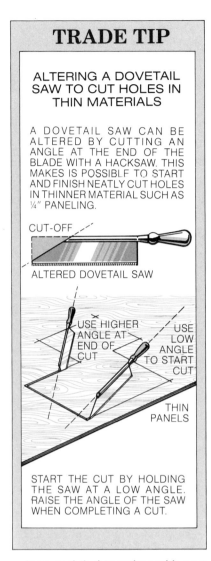

## TRADE TIP

### ALTERING A DOVETAIL SAW TO CUT HOLES IN THIN MATERIALS

A DOVETAIL SAW CAN BE ALTERED BY CUTTING AN ANGLE AT THE END OF THE BLADE WITH A HACKSAW. THIS MAKES IS POSSIBLE TO START AND FINISH NEATLY CUT HOLES IN THINNER MATERIAL SUCH AS ¼" PANELING.

CUT-OFF

ALTERED DOVETAIL SAW

USE HIGHER ANGLE AT END OF CUT

USE LOW ANGLE TO START CUT

THIN PANELS

START THE CUT BY HOLDING THE SAW AT A LOW ANGLE. RAISE THE ANGLE OF THE SAW WHEN COMPLETING A CUT.

legs and tied together with end pieces.

**Sawmill.** Plant where lumber is manufactured from wood logs.

**Scab.** Short piece of wood nailed over a joint between two pieces of lumber to add strength to the joint.

**Scaffolding.** Temporary, braced platforms set up around buildings to enable carpenters to complete work that is out of reach from the floor or ground level.

**Scale drawing.** Drawing in which inches or inch-fractions represent one foot of the actual measurement of a building.

**Scarf joint.** Splice made by notching and lapping the ends of two pieces of timber. Also,

end joint made by overlapping two pieces of molding with 45° angle cuts.

**Scissors truss.** Roof truss design featuring ceilings that slope up from the walls to the center of the span between the walls.

**Screeding concrete.** In concrete flatwork, leveling the newly placed concrete. Also called *striking off.*

**Screeds (concrete).** Temporary wood or metal pieces positioned in an area where a concrete slab is to be placed. The tops of these pieces are set to the finish surface of the concrete. A piece placed across the tops of sidewall forms and used to strike off the concrete is also called a screed.

**Screed support.** Adjustable metal device that supports a screed.

**Screw anchor.** Light-duty anchor for fastening materials to hollow walls.

**Scribe strip.** Thin piece fitted against a wall to cover the space between an end cabinet and the wall.

**Scribing.** Process of fitting the edge of a trim piece against an irregular surface.

**Scuttle.** Small opening with a removable lid located in a ceiling below an attic to allow access into the attic.

**Seasoning lumber.** Process of removing moisture from unseasoned lumber prior to marketing.

**Seat cut (rafter).** Horizontal cut at the lower end of a rafter where the rafter rests on top of the wall plate.

**Second (measurement).** Unit measurement of an angle equal to 1/60 of a minute. (A minute is 1/60 of a degree.)

**Sectional home.** See *modular home.*

**Section line.** See *cutting plane line.*

**Section-view drawing.** Drawing of a vertical or horizontal cut made through a part of a building. It gives information about both the interior and exterior of the structure or object.

**Segregation (concrete).** Separation of sand-cement ingredients from gravel due to improper placement of the concrete.

**Seismic risk zone.** Area of the earth's surface where the conditions for earthquakes exist.

**Self-drilling anchor.** Heavy-duty anchoring device used to fasten material to concrete. It is driven by a rotary hammer. Sharp teeth at the end of the device drill the hole. The anchor is then secured with an expander plug.

**Self-drilling screw.** Screw that, when driven into thinner metals, will drill a hole, cut threads, and fasten in one operation.

**Self-tapping screw.** Screw that, when driven into a prepared hole in thinner metals, will cut threads and fasten in one operation.

**Semi-concealed hinge (cabinet).** Small hinge similar in appearance to a loose-pin butt hinge. An offset on one leaf makes it possible to screw the leaf into the back of a door.

**Serrated forming tools.** The cutting action of these tools is carried out with a steel blade containing hundreds of pre-set razor-sharp teeth that cut as chisels. Holes between the teeth permit the passage of shavings.

**Service door.** Usually a rear-entry exterior door.

**Service stairs.** Stairs that serve noninhabited areas of a house such as a basement or attic.

**Set of saw.** Angle at which saw teeth are alternately bent from side to side. Helps prevent binding while cutting wood.

**Shake (roof and sidewall).** Type of shingle split from red cedar logs. It gives a more rustic appearance than regular shingles.

**Shake (lumber).** Lumber defect caused by a lengthwise separation of the wood, usually between or through the annual growth rings.

**Shank (nail and screw).** Body of a nail or screw. The shank is the part driven into wood.

**Shear wall.** Wall designed to resist lateral forces due to earthquake, wind, or other causes. A shear wall is often created by placing performance-rated panels on one or both sides.

**Sheathing.** Panels or boards placed on the outside of an exterior framed wall or roof to provide greater insulation, strength, and a nailing base for finish materials.

**Sheathing paper.** Layer of water-resistant paper applied over board sheathing or, where no sheathing is used, directly on stud walls. It is placed before the siding is nailed in place. Also called *building paper.*

**She-bolt (formwork).** See *waler rod.*

**Shed roof.** Roof that slopes in only one direction. Also called *lean-to roof.*

**Sheet piling.** Piles that interlock with each other and are not intended to carry vertical loads. Their primary purpose is to resist horizontal pressure, and they are frequently used around excavations.

**Shim shingles.** Narrow strips, usually cut from cedar shingles, used to plumb and straighten door jambs and furring strips.

**Shingle panel.** Panel to which one or two courses of shingles or shakes are bonded.

**Shiplap siding.** Boards rabbeted on both edges so that when laid edge to edge they make a half-lap joint.

**Shore.** Wood timber or metal device placed in a vertical, horizontal, or angled position to provide temporary support.

**Shoring.** System used to prevent the sliding or collapse of the earth banks around an excavation. Also, temporary bracing against a wall or beneath any type of structure that exerts weight from above.

**Shortened valley rafter.** In an intersecting roof with unequal spans, the shorter valley rafter extending from an intersecting corner of the building. The top end butts against the supporting valley rafter.

**Shortening (rafters).** Process of deducting an amount from the ridge plumb line of a common rafter equal to one-half the thickness of the ridge. In the case of hip and valley rafters, or rafter jacks butting against a rafter, the deduction is one-half the 45° thickness.

**Shut-off.** See *bulkhead.*

**Shutter.** Louvered or flush rectangular panels located at each side of a window. They may be hinged in order to close over the windows as added protection against weather. Often they only serve a decorative purpose. Also called *blinds.*

**Side cut (rafters).** Angle cut where a hip or valley rafter comes up against the ridge of a roof. It is also formed where jack rafters are nailed against hip or valley rafters.

**Sidelap.** Area where materials (shingles, underlayment felt, etc.) lap over each other at the side edges.

**Sidewall.** Outside wall of a building.

**Side yard.** Distance from the property line to the side of a building.

**Siding.** All types of exterior board, panel, and shingle wall covering applied by carpenters.

**Siding story pole.** Rod upon which the layout of the siding boards can be determined in relation to the finish window openings.

**Sill anchor.** Anchor bolt positioned in a concrete form at the time the concrete is being placed. Its purpose is to fasten the sill plate to the top of the wall.

**Sill (window).** Slanted piece at the bottom of the finished window frame.

**Sill plate.** Wood plate fastened to the top of a foundation wall. It provides a nailing base for floor joists or studs. Also called *mudsill.*

**Silt.** Earth material consisting of fine mineral particles that are midway in size between sand and clay.

**Single coursing (sidewall shingles).** Applying wood shingles in a single layer.

**Single-ply application (drywall).** Application of a single layer of gypsum board to a wall.

**Site.** In construction work, the location of a construction project.

**Skylight.** Opening in a roof or ceiling, covered with glass or transparent plastic to provide light in the space below. Sometimes also used for ventilation. Some types are called *roof windows.*

**Slab-at-grade.** Foundation system that combines concrete foundation walls with a concrete floor slab that rests directly on a bed of gravel that has been placed over the ground. Also known as slab-on-grade.

**Slat.** Narrow strip of wood or metal.

**Slate.** In construction, a gray-blue rock-based material used as a finish roof covering.

**Sleeper.** Wood strips fastened to a concrete slab to provide a nailing base for wood finish flooring.

**Sleeve (form).** Metal or fiber cylinder set inside a form before the concrete is placed. It provides a hole for the passage of pipes or other objects through the finished concrete wall.

**Slip form.** Forming system that moves continuously upward while the concrete is being placed.

**Slump test.** Test taken at the time the concrete is placed to measure the consistency (slump) of the concrete.

**Snap bracket (formwork).** Wedge used with snap ties in a single-waler form system.

**Snap lock.** Surface-mounted locking device sometimes used with hopper windows or bottom-hinged cabinet doors. It snaps into a locking position when the door or window is shut.

**Snap tie.** Patented tie system with cones acting as spreaders. Grooved breakbacks are provided for breaking off the ends of the ties extending from the hardened concrete wall.

**Soffit.** Underside of a building member such as a cornice, arch, beam, staircase, etc.

**Softwood lumber.** Lumber that comes from coniferous (evergreen) tree species. Most of these trees have needle-shaped leaves.

**Soil auger.** Large device powered by a drilling rig for the purpose of boring pier holes into the ground.

**Solar heating system (active).** Heating system that uses a collector to absorb solar energy. Distribution and storage components are also required.

**Solar house (passive).** Building designed with large south-facing glazed areas to collect and store solar energy.

**Soldier beams.** See *soldier piles*.

**Soldier piles.** Steel members driven into the ground for shoring around excavations. They are driven at intervals and wood planks called *lagging* are placed between the piles. Also called *soldier beams*.

**Sole plate.** Horizontal piece at the base of a framed wall to which the bottoms of the upright members are nailed. Also called *bottom plate*.

**Solid beam.** Beam made of one piece of timber.

**Solid blocking.** Wood pieces, the same thickness and width as the joists, placed between the joists to stiffen and hold the joists in position.

**Solid board paneling.** Solid wood boards, usually ¾″ thick and 8″ to 12″ wide, applied as a finish for interior walls.

**Solid-core door.** Flush doors with a solid core, usually consisting of staggered wood blocks.

**Sound transmission class (STC).** Rating for the sound transmission performance of a wall or ceiling system.

**Spacer strip (siding).** Strip placed underneath the bottom edge of the first row of bevel siding. It should be equal to the thin edge of the siding.

**Specifications.** Written document included with a set of blueprints clarifying the working drawing and supplying additional data.

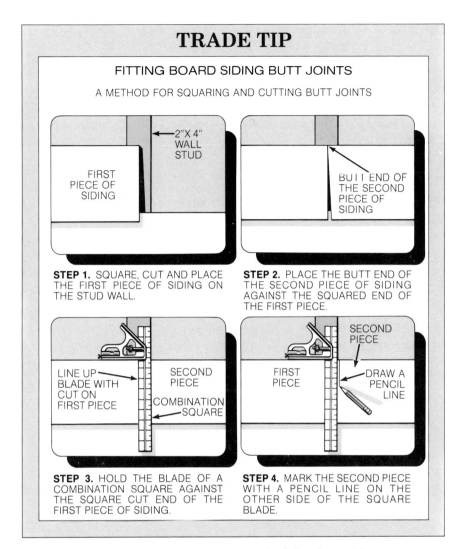

# TRADE TIP

## FITTING BOARD SIDING BUTT JOINTS

A METHOD FOR SQUARING AND CUTTING BUTT JOINTS

2″X 4″ WALL STUD

FIRST PIECE OF SIDING

BUTT END OF THE SECOND PIECE OF SIDING

**STEP 1.** SQUARE, CUT AND PLACE THE FIRST PIECE OF SIDING ON THE STUD WALL.

**STEP 2.** PLACE THE BUTT END OF THE SECOND PIECE OF SIDING AGAINST THE SQUARED END OF THE FIRST PIECE.

LINE UP BLADE WITH CUT ON FIRST PIECE

SECOND PIECE

COMBINATION SQUARE

SECOND PIECE

FIRST PIECE

DRAW A PENCIL LINE

**STEP 3.** HOLD THE BLADE OF A COMBINATION SQUARE AGAINST THE SQUARE CUT END OF THE FIRST PIECE OF SIDING.

**STEP 4.** MARK THE SECOND PIECE WITH A PENCIL LINE ON THE OTHER SIDE OF THE SQUARE BLADE.

**Splash board.** Board placed at the top of a form on the side opposite from where the concrete is being placed. Its purpose is to prevent the concrete from spilling over the opposite side.

**Spline.** Thin strip of wood placed into mortises or grooves that have been cut in boards where they are being joined together.

**Split jamb.** Type of jamb used with adjustable pre-hung door units. The jamb comes in two halves and can be adjusted for the width of the wall.

**Spreader (form).** Wood or metal piece placed between form walls to assure that the finished concrete wall will be the correct thickness.

**Spread footing.** Wide base placed beneath a foundation wall. It spreads the load over a greater ground area.

**Spring balance.** Coiled spring device used to control the up-and-down movement of double-hung window sashes.

**Spring hinge.** Self-closing door hinge operated by a coil mechanism in the barrel. Used most often with doors installed in public buildings.

**Springwood.** Portion of a tree-trunk's annual growth ring that is formed during the early part of the season's growth. It is usually weaker and less dense than summerwood.

**Stack-cast slabs.** Series of concrete slabs cast directly on top of each other. A resin type of compound is used as a bond-breaker to prevent the slabs from adhering to each other.

**Staging.** See *scaffold.*

**Staircase.** Entire assembly of stairs, landings, railings, and balusters.

**Stair flight.** Section of stairs going from one floor or landing to another.

**Stair landing.** Platform between one flight of stairs and another.

**Stair story pole.** Rod upon which the total number of risers for a stairway is marked off. It is checked to see if adjustments are necessary for the exact riser height.

**Stairwell.** Space in which the flights of a stairway are placed.

**Stairwell opening.** Opening in a floor to receive a stairway.

**Staple.** U-shaped metal fastener driven by manually operated or pneumatically powered stapling tools. Used for many fastening operations in construction work.

**Starter course.** First course of sidewall wood shingles. It is usually doubled even if single coursing is used for the rest of the wall.

**Stepped foundation.** A foundation system used on sloped and hillside lots. The walls and footings are shaped like steps.

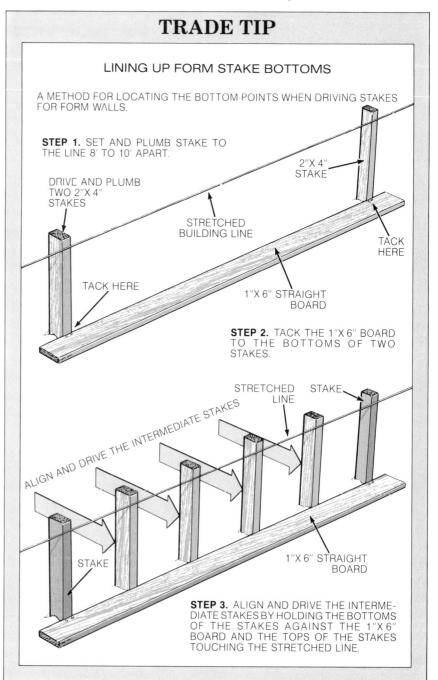

**TRADE TIP**

LINING UP FORM STAKE BOTTOMS

A METHOD FOR LOCATING THE BOTTOM POINTS WHEN DRIVING STAKES FOR FORM WALLS.

**STEP 1.** SET AND PLUMB STAKE TO THE LINE 8' TO 10' APART.

DRIVE AND PLUMB TWO 2"X 4" STAKES

2"X 4" STAKE

STRETCHED BUILDING LINE

TACK HERE

TACK HERE

1"X 6" STRAIGHT BOARD

**STEP 2.** TACK THE 1"X 6" BOARD TO THE BOTTOMS OF TWO STAKES.

STRETCHED LINE

STAKE

ALIGN AND DRIVE THE INTERMEDIATE STAKES

STAKE

1"X 6" STRAIGHT BOARD

**STEP 3.** ALIGN AND DRIVE THE INTERMEDIATE STAKES BY HOLDING THE BOTTOMS OF THE STAKES AGAINST THE 1"X 6" BOARD AND THE TOPS OF THE STAKES TOUCHING THE STRETCHED LINE.

**Sticker (millwork machine).** See *molder.*

**Stickers.** Wood boards placed between layers of lumber in a lumber pile to allow air to circulate.

**Stiffback (formwork).** See *strongback.*

**Stiff leg.** Piece used as a temporary shore to help hold an upper wall cabinet in position while it is being fastened to the wall.

**Stile (door).** Vertical pieces of a panel door frame.

**Stile-and-rail door.** See *panel door.*

**Stock plans.** Existing plans that can be purchased from concerns that produce a variety of working drawings for home construction.

**Stool (window).** Trim piece usually notched over the interior sill edge of a traditionally finished window opening.

**Stop (door).** Device with a soft tip that is fastened near the bottom of a door or into the wall base of the floor. It prevents the door from hitting against the wall when fully opened.

**Storm door.** Additional, exterior door installed during the winter for added protection against cold and wet weather.

**Storm window.** Additional, removable window installed during the winter for added protection against cold and wet weather.

**Story.** Section of a building that runs from one floor to another floor or a roof.

**Stove bolt.** Bolt with a slotted flat or round head. Usually used for lighter work.

**Straightedge.** Long wood or metal piece with straight and parallel edges used to check and set surfaces to a straight line. Also used with a hand level for plumbing.

**Straight-flight stairway.**

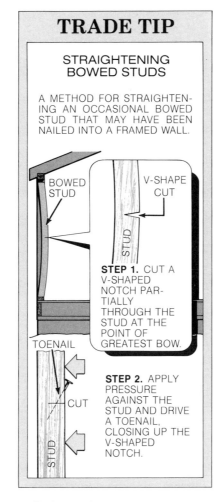

**TRADE TIP**

STRAIGHTENING BOWED STUDS

A METHOD FOR STRAIGHTENING AN OCCASIONAL BOWED STUD THAT MAY HAVE BEEN NAILED INTO A FRAMED WALL.

BOWED STUD

V-SHAPE CUT

STUD

**STEP 1.** CUT A V-SHAPED NOTCH PARTIALLY THROUGH THE STUD AT THE POINT OF GREATEST BOW.

TOENAIL

CUT

STUD

**STEP 2.** APPLY PRESSURE AGAINST THE STUD AND DRIVE A TOENAIL, CLOSING UP THE V-SHAPED NOTCH.

Stairway that runs in a direct line from one floor to another. Also called *straight-run stairway.*

**Straight-run stairway.** See *straight-flight stairway.*

**Strap hinge.** Hinge with long plates that is surface-mounted. Used with heavy doors or a gate.

**Stress.** In construction, an applied force that tends to strain or deform structural materials.

**Stressed-skin panels.** Structural units used in floor, wall, and roof systems. They are made of plywood panels nailed and glued to $2'' \times 4''$ pieces.

**Strike board.** Straightedge used for screeding concrete.

**Strike plate (lock).** Metal piece mortised into a door jamb. It receives the latch bolt when the door is in a closed

position.

**Striking off concrete.** See *screeding concrete.*

**Stringer (formwork).** Timber placed on top of shores to support the joists and panels of a deck form.

**Stringers (stair).** Sloping members that provide the main support for the treads, risers, and other parts of a staircase. Also called *carriage* or *strings.*

**Strings (stair).** See *stringers (stair).*

**Strip flooring.** Wood finish floor pieces ranging in width from $1\frac{1}{2}''$ to $3\frac{1}{2}''$ and in thickness from $\frac{5}{16}''$ to $\frac{25}{32}''$.

**Stripping forms.** Removing forms from a structure after the concrete has adequately cured.

**Strongback (ceiling).** Piece of $2'' \times 4''$ lumber nailed to a wider plank and also fastened down to the tops of ceiling joists at the center of their spans. The strongback keeps the joists in alignment and provides central support.

**Strongback (form).** Piece placed in back of and across walers to reinforce and stiffen a form. Also called *stiffback.*

**Structural insulation.** Panels of insulation materials that meet the structural requirements for wall and roof sheathing.

**Stub joists.** Short joists running at right angles from a regular joist to an outside wall. This may occur under some roof sections when the rafters and ceiling joists do not run in the same direction.

**Stud.** Upright wood or steel member that extends from the bottom to the top plates of a framed wall.

**Stud fastener.** Device usually driven into concrete with a powder-actuated tool. One end is a nail that is embedded in the concrete. The other end is a threaded bolt that receives the nut used to secure the object being fastened.

**Subcontractor.** Licensed individual or firm that can enter into legal contracts to do work such as painting, electrical work, plumbing work, plastering, sheet-metal work, etc.

**Subdivision map.** Map showing the established lots of a section of a city or township. The dimensions of each lot are given and the necessary property setbacks.

**Subfascia.** Horizontal piece the same thickness and width as the roof rafters. It is nailed to the tail ends of the rafter overhangs and is covered with finish fascia material.

**Subfloor.** Consists of structurally rated panels or boards fastened to the tops of the floor joists. It provides a base for the finish floor materials.

**Summerwood.** Portion of a tree trunk's annual growth ring that is formed after the springwood is formed. It is usually stronger and denser than springwood.

**Sump.** Pit located in the basement floor to collect and drain off water. Also called *sump well.*

**Sump pump.** Water pump placed in a sump well to pump out water collecting in the well.

**Supporting valley rafter.** In an intersecting roof with unequal spans, the longer valley rafter extending from one intersecting corner of the building to the main ridge.

**Surfaced lumber.** Lumber that has been smoothed in a planing mill. Also called *dressed lumber.*

**Surface hinge (cabinet).** Hinge with leaves that are mounted to the face frame and the face of the door.

**Survey point.** Small nail driven into the top of a corner stake (hub) to identify the exact corner of the property.

**Suspended ceiling.** Ceiling made of a light metal grid hung by wire from the original ceiling. Ceiling tiles are dropped into the grid frame.

**Swale.** The slopes required on a lot to ensure water drainage away from the building.

# T

**Tail (rafter).** See *overhang.*

**Tail cut (rafter).** Cut at the tail end of a rafter overhang.

**Tail joists.** Members that run from the header to a supporting wall or girder in a floor or ceiling opening.

**Tail piece.** See *overhang (rafter).*

**Tangential cut.** Type of cut produced in plainsawn or flat-grained lumber.

**Taper-ground (saws).** Saw blade that is thinner at the top edge than at the cutting edge. The taper helps prevent binding while sawing wood.

**T-beams (concrete).** Precast, prestressed concrete beam shaped like a T and suitable for spanning longer distances than conventionally shaped precast beams.

**Template (anchor bolt).** Wood piece used to lay out and hold in position anchor bolts that must be accurately set into the tops of concrete piers or walls.

**Tenon.** Projecting part of a piece of lumber cut and shaped to fit into a mortise.

**Tensile strength.** Resistance of a material to forces attempting to tear it apart.

**Tension.** Pulling or stretching force.

**Tensioning jack.** Hydraulic jack used for tensioning cables in prestressed concrete members.

**Termite shield.** Metal shield placed under and extending out from the sill plate to prevent the passage of termites.

**T-foundation.** Foundation system consisting of a rectangular wall resting on a spread footing. The shape of this foundation is an inverted T.

**T-guide (door).** Device used to hold the bottoms of sliding cabinet doors in position. It fastens to the bottom of the cabinet and fits into a groove at the bottom of the door.

**Theoretical length (rafter).** The length of a rafter before it is shortened because of ridge or rafter thicknesses.

**Thermal insulation.** Insulation materials placed in the walls, floors, ceilings, and roof to control heat flow and help maintain a comfortable temperature in the building.

**Thermal storage unit.** Heat storage unit required in an active solar heating system.

**Thermostat.** Electrically operated instrument that can be set to automatically control the room temperatures produced by heating or cooling equipment.

**Threshold.** Wood, metal, or stone piece set between the jamb and the bottom of a door opening. Also called *door sill.*

**Thrombe wall.** Masonry wall used to store heat in an indirect-gain passive solar system.

**Tie-downs.** Metal devices bolted to wood posts in a wall and fastened to bolts anchored in the concrete foundation.

**Tilt-up construction.** Construction method in which the wall sections of a building are cast in place on the job site and lifted into position by crane.

**Timber.** Lumber pieces that are no less than 5″ in their least dimension.

**Toeboard.** Narrow board located at the back of the recessed area under a base cabinet to

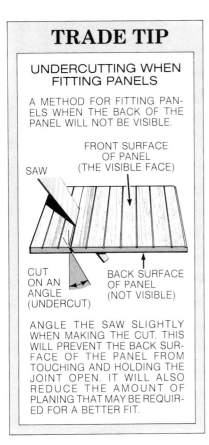

## TRADE TIP

### UNDERCUTTING WHEN FITTING PANELS

A METHOD FOR FITTING PANELS WHEN THE BACK OF THE PANEL WILL NOT BE VISIBLE.

SAW

FRONT SURFACE OF PANEL (THE VISIBLE FACE)

CUT ON AN ANGLE (UNDERCUT)

BACK SURFACE OF PANEL (NOT VISIBLE)

ANGLE THE SAW SLIGHTLY WHEN MAKING THE CUT. THIS WILL PREVENT THE BACK SURFACE OF THE PANEL FROM TOUCHING AND HOLDING THE JOINT OPEN. IT WILL ALSO REDUCE THE AMOUNT OF PLANING THAT MAY BE REQUIRED FOR A BETTER FIT.

provide foot space when a person stands close to the cabinet. Also called *toe strip*.

**Toe holds.** Boards temporarily nailed on top of panels as a safety measure when carpenters are sheathing steeper roofs.

**Toenailing.** Driving nails at a slant into two pieces. For example, studs are toenailed into top and bottom plates.

**Toe strip.** See *toeboard*.

**Toggle bolt.** Anchoring device used to fasten light materials to hollow walls.

**Tongue-and-groove (T & G) lumber.** Boards or planks with a groove in one edge and a tongue on the other edge. The tongue fits into the groove of a matched piece of lumber.

**Toplap.** Area where one shingle overlaps another shingle in the course below the first shingle.

**Top plate.** Horizontal piece to which the top of the vertical members of a framed wall are nailed.

**Total rise (roof).** Ridge height of a roof, measured from the top of the walls.

**Total rise (stairs).** Vertical distance of a stairway from a bottom floor to the floor above.

**Total run (roof).** Distance equal to one-half the total roof span.

**Total run (stairs).** Horizontal length of a stairway measured from the foot of the stairway to a point plumbed down from where the stairway ends at the deck or landing above.

**Total span (roof).** Measurement equal to width of the building.

**Tower crane.** Crane consisting of a high tower and gib. Sections can be added to the tower to achieve greater heights.

**Tracheid.** See *fiber*.

**Track (sliding door).** Metal channel attached to the head jamb. Roller hangers, attached to the doors, move along the tracks.

**Transit-mix truck.** Truck equipped with a large drum concrete mixer for delivery of ready-mixed concrete to the job site.

**Transmittance (U).** A factor that expresses the amount of heat in Btu transferred in one hour through one square foot of the floor, ceiling, wall, or roof area of a building.

**Transom.** Small opening over a door or another window. It usually contains a movable sash or stationary louver.

**Transverse beam.** Roof beam that extends from an outside wall to a ridge beam of a post-and-beam roof structure.

**Transverse section.** Section view in a blueprint drawing representing a cut across the width of a building.

**Tread.** Horizontal walking surface of the step of a stairway.

**Tremie.** Device consisting of a funnel with a tube-like chute attached at the bottom. It is used to place concrete in forms under water.

**Trench.** Ditch dug in the ground down to bearing soil for foundation footings.

**Trenching.** Digging trenches for foundation footings.

**Trestle jack.** Adjustable metal horse used to support low working platforms.

**Trim.** Finish materials such as molding placed around doors and windows and at the top and bottom of the walls.

**Trimmer joists.** Doubled joists placed at floor or ceiling openings to which the header joists are fastened.

**Trimmer studs.** Vertical members of a framed wall nailed to studs and supporting the ends of the headers placed at the top of rough openings.

**Trim ring (lock).** Finish piece placed against a door face or a plate before a cylinder unit is installed.

**Trolley (bi-fold door).** Roller device that fits into an overhead track. It is attached to the top of the second door for support.

**T-shore.** In concrete formwork, a shore with a braced head piece. It is used to support the forms for concrete beams and girders.

**Tubular lock.** Less expensive type of cylindrical lock.

**Turnbuckle.** In construction work a device often used with adjustable braces. It consists of two rods held together with a coupling that can be turned to regulate the tension of the rods.

**Turnbutton.** Device located in the inside knob of a cylindrical lock. When turned in one direction it locks the door.

## U

**Undercutting.** Tilting a saw so that the back edge of the cut is slightly in from the face edge.

**Underlayment (floor).** Thin plywood or non-veneered panels fastened over the subfloor to provide a smooth surface for the placement of the finish floor material.

**Underlayment (roof).** Usually a layer of asphalt-saturated felt paper laid down over the roof sheathing before most types of roof shingles are placed.

**Underpinning.** Wood wall constructed directly over the foundation and below the first floor of the house.

**Unit rise (roof).** Number of inches a common rafter will rise vertically for each foot of run.

**Unit rise (stairs).** Riser height calculated by dividing the total rise by the number of risers in a stairway.

**Unit run (roof).** Unit of the total run, based on 12″.

**Unit run (stairs).** Width of the tread calculated by dividing the total run by the number of treads in the stairway.

**Urea formaldehyde.** Synthetic resin made by combining urea and formaldehyde. It is widely used as a binder in the manufacture of reconstituted wood panels. Also a basic ingredient in a type of foamed-in-place insulation.

**Utilities.** In construction, services provided to the public, requiring electrical and plumbing hookups in a building.

# V

**Valley cripple jack rafter.** Jack rafter sometimes placed between the end sections past the point of intersection of a shortened and supporting valley rafter.

**Valley jack rafter.** Jack rafter extending from a valley rafter to a ridge.

**Valley rafter.** Rafter found on intersecting roofs that runs at an angle from the intersecting corners of a building to the ridge. It is located at the bottom of the valley formed where different sloping roof sections come together.

**Vapor barrier.** Thin moisture-resistant material placed over the ground or in walls to retard the passage of moisture.

**Veneer.** Thin layer of wood.

**Veneered beam.** Beam made of glued veneers.

**Vent.** Opening that provides air ventilation in any space of a building.

**Ventilation.** In construction, the provisions made in a building to permit warm air to escape and to allow the circulation of air in enclosed areas.

**Vermiculite.** Expanded mica material used as a loose fill insulation material. Sometimes used as an aggregate in lightweight concrete and plaster.

**Vernier.** Graduated scale that gives fractions of a degree on leveling instruments.

**Vibration (concrete).** Consolidation of concrete through the use of mechanical vibration equipment. Also see *immersion vibrator.*

**Vinyl-asbestos tile.** Resilient floor tile with basic ingredients of polyvinyl chloride, resin binders mixed with limestone, and asbestos fibers.

**Vinyl-clad window units.** Window units with a frame and sash consisting of a wood core covered with a thin layer of pre-finished rigid vinyl.

**Vinyl siding.** Pre-finished plastic product that gives a similar effect to wood board siding.

**Vinyl tile.** Resilient floor tile with basic ingredients of polyvinyl chloride resin binders and mineral fillers.

**V-joint.** Slightly beveled joint between boards or panels that creates a shadow line.

# W

**Waferboard.** Panel product manufactured by combining wood wafers sliced from logs with an exterior grade of phenolic resin and hot-pressing them into panels.

**Wafers.** Wood flakes used in the manufacture of non-veneered, reconstituted wood panels.

**Waffle slab.** Floor slab characterized by a waffle-like appearance at its underside. Its structural design makes possible the use of lighter concrete.

**Wainscot.** Traditionally, wood panel work covering the lower section of a plastered or gypsum board wall. Wainscoting material may also be plastic laminate, tile, glass, etc.

**Wainscot cap.** Piece of finish molding placed at the top of a wainscot.

**Wales.** See *walers.*

**Walers.** Horizontal pieces placed on the outsides of the form walls to strengthen and stiffen the walls. The form ties are also fastened to the walers. Also called *wales.*

**Waler Rod (formwork).** Device used to secure and hold together opposite form walls. It consists of waler rods (also called *she bolts*) extending past the walers on each side. They are joined inside the wall by a threaded "filler" rod. The whole assembly is tightened by nut washers on opposite sides.

**Wallboard.** See *gypsum board.*

**Wall cabinet.** Cabinet not resting on the floor and fastened to the upper part of a wall.

**Wall framing story pole.** Rod laid out according to information provided by the working drawings for a building. It gives lengths of the studs, trimmers, and top and bottom cripples for the walls.

**Wane.** Lumber defect caused by the presence of bark or a lack of wood on the edge or corner of a piece of lumber.

**Warm side.** Inner surface of the outside shell (walls, roof, etc.) of a building.

**Warp.** Condition in which lumber has bent or twisted out of its original shape. Often caused by uneven shrinkage during seasoning.

**Water-cement (W/C) ratio.** Amount of water used in a concrete mix in relation to the amount of cement. Major factor in the compressive strength of concrete.

**Water wall.** Water containers used to store heat as part of an indirect-gain passive solar system.

**Water table (ground).** Highest point below the surface of the ground that is normally saturated with water in a given area.

**Water table (wood).** Piece placed below the first course of siding to direct water drainage away from the foundation wall.

**W-beam.** Steel beam with flanges at the top and bottom of the web. Also called *wide-flange beam.*

**Weatherstripping.** Metal, felt, or plastic strips used to prevent air or moisture infiltration through the spaces between the outer edges of doors and windows and their finish frames.

**Web (truss).** Truss member that runs between and ties together the top and bottom chords.

**Wedges (form ties).** Devices driven against the walers and at the ends of the form ties. They hold together opposite form walls.

**Wedge Ties.** Patented ties consisting of a strap and

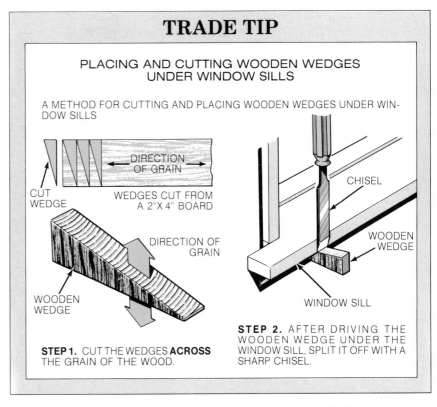

## TRADE TIP

### PLACING AND CUTTING WOODEN WEDGES UNDER WINDOW SILLS

A METHOD FOR CUTTING AND PLACING WOODEN WEDGES UNDER WINDOW SILLS

DIRECTION OF GRAIN

CUT WEDGE

WEDGES CUT FROM A 2"X 4" BOARD

DIRECTION OF GRAIN

WOODEN WEDGE

**STEP 1.** CUT THE WEDGES **ACROSS** THE GRAIN OF THE WOOD.

CHISEL

WOODEN WEDGE

WINDOW SILL

**STEP 2.** AFTER DRIVING THE WOODEN WEDGE UNDER THE WINDOW SILL, SPLIT IT OFF WITH A SHARP CHISEL.

wedges. Used with plank forming systems.

**Weep hole.** Small hole extending through a concrete retaining wall to allow water to drain from behind the wall.

**Welded wire fabric.** See *wire mesh.*

**Western framing.** See *platform framing.*

**White speck.** Small white pits or spots on the surface of the wood caused by a fungus in the living tree.

**Wide-flange beam.** See *W-beam.*

**Winder stairway.** Stairway that has a series of winding steps.

**Wire mesh.** Heavy steel wire welded together in a grid pattern and used to reinforce concrete slabs resting directly on the ground. Also called *welded wire fabric.*

**Wood preservative.** Chemical in liquid form applied to wood for protection against fungi, decay, and insect attack.

**Working drawings.** Set of plans drawn up by an architect. Contains all the dimensions

and structural information needed to complete a construction project. Also called *blueprints.*

**Woven corner (sidewall shingles).** Outside corner finish where alternate corner edges of the shingles are exposed.

**W-truss.** See *fink truss.*

## Y

**Yard lumber.** All lumber sold for structural building purposes.

## Z

**Zoning regulations.** Local regulations that govern the type of buildings and structures that may be erected in different areas of a community. Most zones come under the general categories of residential, commercial, and manufacturing.

# Index

# G

# I

# M

716